Boundary Value Problems for Fractional Differential Equations and Systems

TRENDS IN ABSTRACT AND APPLIED ANALYSIS

ISSN: 2424-8746

Series Editor: John R. Graef
The University of Tennessee at Chattanooga, USA

This series will provide state of the art results and applications on current topics in the broad area of Mathematical Analysis. Of a more focused nature than what is usually found in standard textbooks, these volumes will provide researchers and graduate students a path to the research frontiers in an easily accessible manner. In addition to being useful for individual study, they will also be appropriate for use in graduate and advanced undergraduate courses and research seminars. The volumes in this series will not only be of interest to mathematicians but also to scientists in other areas. For more information, please go to http://www.worldscientific.com/series/taaa

Published

Vol. 9   *Boundary Value Problems for Fractional Differential Equations and Systems*
by Bashir Ahmad, Johnny Henderson & Rodica Luca

Vol. 8   *Ordinary Differential Equations and Boundary Value Problems Volume II: Boundary Value Problems*
by John R. Graef, Johnny Henderson, Lingju Kong & Xueyan Sherry Liu

Vol. 7   *Ordinary Differential Equations and Boundary Value Problems Volume I: Advanced Ordinary Differential Equations*
by John R. Graef, Johnny Henderson, Lingju Kong & Xueyan Sherry Liu

Vol. 6   *The Strong Nonlinear Limit-Point/Limit-Circle Problem*
by Miroslav Bartušek & John R. Graef

Vol. 5   *Higher Order Boundary Value Problems on Unbounded Domains: Types of Solutions, Functional Problems and Applications*
by Feliz Manuel Minhós & Hugo Carrasco

Vol. 4   *Quantum Calculus: New Concepts, Impulsive IVPs and BVPs, Inequalities*
by Bashir Ahmad, Sotiris Ntouyas & Jessada Tariboon

More information on this series can be found at https://www.worldscientific.com/series/taaa

Trends in Abstract
and Applied Analysis
Volume **9**

Boundary Value Problems for Fractional Differential Equations and Systems

Bashir Ahmad

King Abdulaziz University, Saudi Arabia

Johnny Henderson

Baylor University, USA

Rodica Luca

"Gheorghe Asachi" Technical University of Iasi, Romania

W⊖ World Scientific

NEW JERSEY • LONDON • SINGAPORE • BEIJING • SHANGHAI • HONG KONG • TAIPEI • CHENNAI • TOKYO

Published by

World Scientific Publishing Co. Pte. Ltd.
5 Toh Tuck Link, Singapore 596224
USA office: 27 Warren Street, Suite 401-402, Hackensack, NJ 07601
UK office: 57 Shelton Street, Covent Garden, London WC2H 9HE

Library of Congress Cataloging-in-Publication Data
Names: Ahmad, Bashir, author. | Henderson, Johnny, author. | Luca, Rodica, author.
Title: Boundary value problems for fractional differential equations and systems /
  Bashir Ahmad, King Abdulaziz University, Saudi Arabia; Johnny Henderson, Baylor
  University, USA; Rodica Luca, "Gheorghe Asachi" Technical University of Iasi, Romania.
Description: USA : World Scientific, 2021. | Series: Trends in abstract and
  applied analysis, 2424-8746 ; vol 9 | Includes bibliographical references and index.
Identifiers: LCCN 2020041568 (print) | LCCN 2020041569 (ebook) | ISBN
  9789811224454 (hardcover) | ISBN 9789811224461 (ebook) | ISBN
  9789811224478 (ebook other)
Subjects: LCSH: Boundary value problems. | Functional differential equations.
Classification: LCC QA379 .A35 2021 (print) | LCC QA379 (ebook) | DDC 515/.35--dc23
LC record available at https://lccn.loc.gov/2020041568
LC ebook record available at https://lccn.loc.gov/2020041569

British Library Cataloguing-in-Publication Data
A catalogue record for this book is available from the British Library.

For any available supplementary material, please visit
https://www.worldscientific.com/worldscibooks/10.1142/11942#t=suppl

Desk Editors: George Vasu/Lai Fun

Typeset by Stallion Press
Email: enquiries@stallionpress.com

Printed in Singapore

Dedication

Bashir Ahmad dedicates this book to the memory of his parents.

Johnny Henderson dedicates this book to the memory of Kathryn Madora Strunk (February 5, 1991–March 1, 2007).

Rodica Luca dedicates this book to the memory of her parents.

Preface

Fractional-order differential and integral operators, and fractional differential equations have extensive applications in the mathematical modeling of real-world phenomena occurring in scientific and engineering disciplines, such as physics, biophysics, chemistry, biology, medical sciences, ecology, financial economics, control theory, signal and image processing, transport dynamics, aerodynamics, thermodynamics, hydrology, viscoelasticity, electromagnetics, statistical mechanics, astrophysics, cosmology, bioengineering and rheology (see [21, 26, 28, 30, 34, 55, 65, 66, 70, 86, 87, 92, 95, 96, 113]). For some recent developments on the topic, we mention [1, 3, 4, 13, 18–20, 22, 24, 31–33, 36–39, 54, 56–59, 61, 64, 73, 76, 88–91, 94, 97, 100, 101, 106, 111, 115] and the references cited therein. For example, Ding *et al.* [31] and Arafa *et al.* [18] developed a model for the primary infection with HIV which is a virus that targets the white blood cells-CD4$^+$T lymphocytes. This model can be described as a system with three fractional differential equations of different orders $(\alpha, \beta, \gamma > 0)$ in the unknown functions T (the concentration of uninfected CD4$^+$T cells), I (infected CD4$^+$T cells) and V (the free HIV virus particles in the blood). The fractional differential equations are also regarded as a better tool for the description of hereditary properties of various materials and processes than the corresponding integer order differential equations. For some recent results on the existence, nonexistence and multiplicity of solutions or positive solutions for fractional differential equations, inclusions and inequalities, and systems of fractional differential equations supplemented with various boundary conditions we refer the reader to [6, 11, 14–16, 35, 42, 43, 62, 63, 74, 75, 93, 98, 103, 105, 109, 112, 117–119].

vii

In this monograph, we present our results obtained in the last years related to the existence of solutions or positive solutions for various classes of Riemann–Liouville and Caputo fractional differential equations, and systems of fractional differential equations subject to nonlocal boundary conditions. Chapter 1 contains some preliminaries related to the fractional integral and fractional derivatives, and the fixed point theorems that we will use in the subsequent chapters. Chapter 2 deals with the existence of positive solutions or nonnegative solutions for some Riemann–Liouville fractional differential equations with parameters or without parameters, supplemented with multi-point or Riemann–Stieltjes integral boundary conditions, which contain fractional derivatives. The nonlinearities of equations are nonsingular or singular functions, with nonnegative values or real values. Chapter 3 is focused on the existence and multiplicity of positive solutions for systems of two Riemann–Liouville fractional differential equations, subject to uncoupled or coupled multi-point boundary conditions, which contain fractional derivatives, and the nonlinearities of systems are nonsingular or singular functions, with nonnegative values. Chapter 4 is devoted to the existence and nonexistence of positive solutions for systems of two Riemann–Liouville fractional differential equations with p-Laplacian operators and positive parameters, supplemented with uncoupled or coupled multi-point boundary conditions, which contain fractional derivatives, and the nonlinearities are nonnegative functions. In Chapter 5, we investigate the existence and nonexistence of positive solutions for systems of three Riemann–Liouville fractional differential equations with positive parameters, subject to uncoupled multi-point boundary conditions which contain fractional derivatives, and the nonlinearities of systems are nonnegative functions. Chapter 6 is concerned with the existence of solutions for Riemann–Liouville fractional differential equations and systems of Riemann–Liouville fractional differential equations with integral terms, supplemented with nonlocal boundary conditions which contain fractional derivatives and Riemann–Stieltjes integrals. In Chapter 7, we study the existence of solutions for some Caputo fractional differential equations and inclusions, and systems of Caputo fractional differential equations subject to nonlocal boundary conditions which contain Riemann–Stieltjes integrals. In each chapter, various examples are presented, which support the main results. The methods used in the proof of our theorems include results from the fixed point theory and the fixed point index theory.

This book complements the existing literature in fractional differential equations, inclusions and systems. The monograph can serve as a good resource for mathematical and scientific researchers, and for graduate students in mathematics and science interested in the existence of solutions for fractional differential equations and systems.

We would like to express our warm thanks to the editors at World Scientific Publishing: Trends in Abstract and Applied Analysis Series Editor John R. Graef, Executive Editor Rochelle Kronzek, Desk Editor Lai Fun Kwong, and Book Editor George Vasu for their support and work on our book.

Bashir Ahmad
Johnny Henderson
Rodica Luca

Contents

Notations

$\mathbb{N} = \{1, 2, \ldots\}$	the set of natural numbers
$\mathbb{R}$	the set of real numbers
$\mathbb{R}_+ = \{x \in \mathbb{R},\ x \geq 0\}$	the set of nonnegative real numbers
$\mathcal{P}(C)$	the family of all nonempty subsets of C
B_r	the open ball centered at θ of radius r, (θ is the zero element)
$\overline{B}_r$	the closure of B_r
∂B_r	the boundary of B_r
$\overline{\Omega}$	the closure of the set Ω
$\partial\Omega$	the boundary of Ω
$C[0, 1]$	the space of continuous real functions defined on the interval $[0, 1]$
$C(A, B)$	the space of continuous functions defined on the set A with values in the set B
$C^n A$	the space of n-times continuously differentiable real functions defined on A
$AC^n D$	the space of $(n-1)$-times absolutely continuously differentiable real functions defined on D
$L^1(0, 1)$	the space of Lebesgue integrable and measurable real functions (actually, classes of equivalent functions which are equal almost everywhere) on $(0, 1)$

$L^\gamma(0,1),\ \gamma \in (1,\infty)$	the space of measurable real functions f (actually, classes of equivalent functions which are equal almost everywhere) on $(0,1)$, with $|f|^\gamma \in L^1(0,1)$
$L^\infty(0,1)$	the space of measurable almost everywhere bounded real functions (actually, classes of equivalent functions which are equal almost everywhere) on $(0,1)$
$i(f, U, X)$	the fixed point index of f over U with respect to X, (X is a retract of a real Banach space E, and $U \subset X$ is an open set)
$I_{0+}^\alpha f$	the (left-sided) Riemann–Liouville fractional integral of order α of function f
$D_{0+}^\alpha f$	the (left-sided) Riemann–Liouville fractional derivative of order α of function f
$^c D_{0+}^\alpha f$	the (left-sided) Caputo fractional derivative of order α of function f.

Chapter 1

Preliminaries

In this chapter, we present the definitions and some properties of the fractional integral and the Riemann–Liouville and Caputo fractional derivatives, and the fixed point theorems that we will use in the following chapters.

1.1 Fractional Integral and Fractional Derivatives

In this section, we present the definitions of the Riemann–Liouville fractional integral and the Riemann–Liouville and Caputo fractional derivatives and some of their properties.

Definition 1.1.1. The (left-sided) Riemann–Liouville fractional integral of order $\alpha > 0$ of a function $f : (0, \infty) \to \mathbb{R}$ is given by

$$(I_{0+}^{\alpha} f)(t) = \frac{1}{\Gamma(\alpha)} \int_0^t (t - s)^{\alpha - 1} f(s) \, ds, \quad t > 0,$$

provided the right-hand side is pointwise defined on $(0, \infty)$, where $\Gamma(\alpha)$ is the Euler gamma function defined by $\Gamma(\alpha) = \int_0^\infty t^{\alpha - 1} e^{-t} \, dt$, $\alpha > 0$.

Definition 1.1.2. The (left-sided) Riemann–Liouville fractional derivative of order $\alpha \geq 0$ for a function $f : (0, \infty) \to \mathbb{R}$ is given by

$$(D_{0+}^{\alpha} f)(t) = \left(\frac{d}{dt} \right)^n \left(I_{0+}^{n-\alpha} f \right)(t)$$

$$= \frac{1}{\Gamma(n - \alpha)} \left(\frac{d}{dt} \right)^n \int_0^t \frac{f(s)}{(t - s)^{\alpha - n + 1}} \, ds, \quad t > 0,$$

where $n = \lfloor \alpha \rfloor + 1$, provided that the right-hand side is pointwise defined on $(0, \infty)$.

The notation $\lfloor \alpha \rfloor$ stands for the largest integer not greater than α. For $f \in C^m[0, \infty)$, if $\alpha = m \in \mathbb{N}$, then $D^m_{0+} f(t) = f^{(m)}(t)$ for $t > 0$, and if $\alpha = 0$, then $D^0_{0+} f(t) = f(t)$ for $t > 0$.

Definition 1.1.3. For $u \in AC^n[0, \infty)$, the (left-sided) Caputo fractional derivative of order α is defined as

$$^c D^\alpha_{0+} u(t) = \frac{1}{\Gamma(n - \alpha)} \int_0^t (t - s)^{n-\alpha-1} u^{(n)}(s) \, ds,$$

$$t > 0, \quad n - 1 < \alpha < n, \quad n \in \mathbb{N}.$$

For $u \in C^m[0, \infty)$, if $\alpha = m \in \mathbb{N}$ then $^c D^\alpha_{0+} u(t) = u^{(m)}(t)$ for $t > 0$, and if $\alpha = 0$ then $^c D^0_{0+} u(t) = u(t)$ for $t > 0$.

Lemma 1.1.1 ([65]).

(a) If $\alpha > 0$, $\beta > 0$ and $f \in L^p(0,1)$, $(1 \leq p \leq \infty)$, then the relation $(I^\alpha_{0+} I^\beta_{0+} f)(t) = (I^{\alpha+\beta}_{0+} f)(t)$ is satisfied at almost every point $t \in (0, 1)$. If $\alpha + \beta > 1$, then the above relation holds at any point of $[0, 1]$.

(b) If $\alpha > 0$ and $f \in L^p(0,1)$, $(1 \leq p \leq \infty)$, then the relation $(D^\alpha_{0+} I^\alpha_{0+} f)(t) = f(t)$ holds almost everywhere on $(0, 1)$.

(c) If $\alpha > \beta > 0$ and $f \in L^p(0,1)$, $(1 \leq p \leq \infty)$, then the relation $(D^\beta_{0+} I^\alpha_{0+} f)(t) = (I^{\alpha-\beta}_{0+} f)(t)$ holds almost everywhere on $(0, 1)$.

By using the above definitions and a direct computation, we obtain the following lemma.

Lemma 1.1.2.

(a) If $\alpha > 0$ and $\beta > 0$, then $I^\alpha_{0+} t^{\beta-1} = \frac{\Gamma(\beta)}{\Gamma(\alpha+\beta)} t^{\alpha+\beta-1}$.

(b) If $\alpha \in \mathbb{N}, \beta \in \mathbb{R}, \beta \neq 1, 2, \ldots, \alpha$, then
$$D^\alpha_{0+} t^{\beta-1} = (\beta - 1)(\beta - 2) \cdots (\beta - \alpha) t^{\beta-\alpha-1};$$
If $\alpha \in \mathbb{N}$ and $\beta = 1, 2, \ldots, \alpha$, then $D^\alpha_{0+} t^{\beta-1} = 0$;

(c) If $\alpha > 0, \alpha \notin \mathbb{N}, n - 1 < \alpha < n, \beta > 0$ and $\beta \neq \alpha, \alpha - 1, \ldots, \alpha - n + 1$, then
$$D^\alpha_{0+} t^{\beta-1} = \frac{\Gamma(\beta)}{\Gamma(\beta+n-\alpha)} (\beta + n - \alpha - 1)(\beta + n - \alpha - 2) \cdots (\beta - \alpha) t^{\beta-\alpha-1};$$
If $\alpha > 0, \alpha \notin \mathbb{N}, n - 1 < \alpha < n$ and $\beta = \alpha, \alpha - 1, \ldots, \alpha - n + 1$, then $D^\alpha_{0+} t^{\beta-1} = 0$;

(d) If $\alpha \geq 0$ and $\beta > \alpha$, then $D^\alpha_{0+} t^{\beta-1} = \frac{\Gamma(\beta)}{\Gamma(\beta-\alpha)} t^{\beta-\alpha-1}$.

Lemma 1.1.3 ([65]). *Let $\alpha > 0$ and $n = \lfloor \alpha \rfloor + 1$ for $\alpha \notin \mathbb{N}$ and $n = \alpha$ for $\alpha \in \mathbb{N}$; that is, n is the smallest integer greater than or equal to α (namely, $n - 1 < \alpha \le n$). If $u, D_{0+}^{\alpha} u(t) \in C(0,1) \cap L^1(0,1)$, then*

$$I_{0+}^{\alpha} D_{0+}^{\alpha} u(t) = u(t) + \sum_{i=1}^{n} c_i t^{\alpha - i}, \quad 0 < t < 1,$$

where $c_1, \ldots, c_n \in \mathbb{R}$.

Lemma 1.1.4 ([65]). *Let $\alpha > 0$ and $n = \lfloor \alpha \rfloor + 1$ for $\alpha \notin \mathbb{N}$ and $n = \alpha$ for $\alpha \in \mathbb{N}$, (namely, $n - 1 < \alpha \le n$). If $u \in AC^n[0,1]$, then*

$$I_{0+}^{\alpha} {}^{c}D_{0+}^{\alpha} u(t) = u(t) + \sum_{i=0}^{n-1} c_i t^i, \quad 0 < t < 1,$$

where $c_0, \ldots, c_{n-1} \in \mathbb{R}$.

For other properties of the fractional integrals and fractional derivatives, we refer the reader to [65, 96].

1.2   Fixed Point Theorems

In this section, we present the fixed point theorems and the fixed point index theorems that will be used in the following chapters.

The Banach contraction mapping principle

Theorem 1.2.1 (see [29]). *If Y is a nonempty complete metric space with the metric d, and $T : Y \to Y$ is a contraction mapping, then T has a unique fixed point $x^* \in Y, (Tx^* = x^*)$.*

The Guo–Krasnosel'skii fixed point theorem

Let X be a real Banach space with the norm $\| \cdot \|$.

Definition 1.2.1. A nonempty convex closed set $C \subset X$ is a cone if it satisfies the conditions:

(a) if $u \in C$ and $\alpha \ge 0$, then $\alpha u \in C$;
(b) if $u \in C$ and $-u \in C$, then $u = \theta$,

where θ is the zero element of X.

A cone $C \subset X$ defines a partial ordering in X given by $x \leq y$ if and only if $y - x \in C$.

Definition 1.2.2. An operator $A : D(A) \subset X \to X$ is compact if it maps bounded sets into relatively compact sets. An operator $A : D(A) \subset X \to X$ is completely continuous if it is continuous and compact.

Theorem 1.2.2 (Fixed point theorem of cone expansion and compression of norm type — see [41]). *Let X be a real Banach space with the norm $\|\cdot\|$, and let $C \subset X$ be a cone in X. Assume Ω_1 and Ω_2 are bounded open subsets of X with $\theta \in \Omega_1$, $\overline{\Omega}_1 \subset \Omega_2$ and let $\mathcal{A} : C \cap (\overline{\Omega}_2 \setminus \Omega_1) \to C$ be a completely continuous operator such that, either*

(i) $\|\mathcal{A}u\| \leq \|u\|$, $\forall u \in C \cap \partial\Omega_1$, *and* $\|\mathcal{A}u\| \geq \|u\|$, $\forall u \in C \cap \partial\Omega_2$, *or*
(ii) $\|\mathcal{A}u\| \geq \|u\|$, $\forall u \in C \cap \partial\Omega_1$, *and* $\|\mathcal{A}u\| \leq \|u\|$, $\forall u \in C \cap \partial\Omega_2$.

Then $\mathcal{A}$ has at least one fixed point in $C \cap (\overline{\Omega}_2 \setminus \Omega_1)$.

The Krasnosel'skii fixed point theorem for the sum of two operators

Theorem 1.2.3 (see [23, 67, 99]). *Let M be a closed, convex, bounded and nonempty subset of a Banach space X. Let A_1 and A_2 be two operators such that*

(a) $A_1 x + A_2 y \in M$ *for all* $x, y \in M$;
(b) A_1 *is a completely continuous operator;*
(c) A_2 *is a contraction mapping.*

Then there exists $z \in M$ such that $z = A_1 z + A_2 z$.

The Schauder fixed point theorem

Theorem 1.2.4. *Let X be a Banach space and $Y \subset X$ a nonempty, bounded, convex and closed subset. If operator $A : Y \to Y$ is completely continuous, then A has at least one fixed point.*

The Leray–Schauder alternative

Theorem 1.2.5. *Let E be a Banach space and $T : E \to E$ be a completely continuous operator. Let $F = \{x \in E, \ x = \nu T(x) \text{ for some } 0 < \nu < 1\}$. Then either the set F is unbounded or T has at least one fixed point.*

The nonlinear alternative of Leray–Schauder type

Theorem 1.2.6. *Let X be a Banach space and U be an open and bounded subset of X, $\theta \in U$ and $A : \overline{U} \to X$ be a completely continuous operator. Then either A has a fixed point in $\overline{U}$, or there exist $u \in \partial U$ and $\nu \in (0, 1)$ such that $u = \nu Au$.*

Theorems from the fixed point index theory

Let X be a real Banach space with the norm $\| \cdot \|$, $C \subset X$ a cone, "$\leq$" the partial ordering defined by C and θ the zero element of X. For $r > 0$, let $B_r = \{u \in X, \ \|u\| < r\}$ be the open ball of radius r centered at θ, its closure $\overline{B}_r = \{u \in X, \ \|u\| \leq r\}$ and its boundary $\partial B_r = \{u \in X, \ \|u\| = r\}$.

Theorem 1.2.7 (see [17]). *Let $\mathcal{A} : \overline{B}_r \cap C \to C$ be a completely continuous operator. If $\mathcal{A}u \neq \mu u$ for all $u \in \partial B_r \cap C$ and $\mu \geq 1$, then $i(\mathcal{A}, B_r \cap C, C) = 1$.*

Theorem 1.2.8 (see [17]). *Let $\mathcal{A} : \overline{B}_r \cap C \to C$ be a completely continuous operator which has no fixed points on $\partial B_r \cap C$. If $\|\mathcal{A}u\| \leq \|u\|$ for all $u \in \partial B_r \cap C$, then $i(\mathcal{A}, B_r \cap C, C) = 1$.*

Theorem 1.2.9 (see [17]). *Let $\mathcal{A} : \overline{B}_r \cap C \to C$ be a completely continuous operator. If there exists $u_0 \in C \setminus \{\theta\}$ such that $u - \mathcal{A}u \neq \lambda u_0$ for all $\lambda \geq 0$ and $u \in \partial B_r \cap C$, then $i(\mathcal{A}, B_r \cap C, C) = 0$.*

Theorem 1.2.10 ([120]). *Let $A : \overline{B}_r \cap C \to C$ be a completely continuous operator which has no fixed point on $\partial B_r \cap C$. If there exists a linear operator $L : C \to C$ and $u_0 \in C \setminus \{\theta\}$ such that*

$$\text{(i) } u_0 \leq Lu_0, \quad \text{(ii) } Lu \leq Au, \quad \forall u \in \partial B_r \cap C,$$

then $i(A, B_r \cap C, C) = 0$.

Theorem 1.2.11 (see [17, 68]). *Let $\Omega \subset X$ be a bounded open set with $\theta \in \Omega$. Assume that $A : \overline{\Omega} \cap C \to C$ is a completely continuous operator.*

(a) *If $u \not\leq Au$ for all $u \in \partial\Omega \cap C$, then the fixed point index $i(A, \Omega \cap C, C) = 1$.*

(b) *If $Au \not\leq u$ for all $u \in \partial\Omega \cap C$, then the fixed point index $i(A, \Omega \cap C, C) = 0$.*

The Leggett–Williams theorem

Let X be a real Banach space with the norm $\| \cdot \|$, and let P be a cone in X.

Definition 1.2.3. A map ξ is a nonnegative concave functional on the cone P if $\xi : P \to [0, \infty)$ satisfies the condition

$$\xi(tx + (1 - t)y) \geq t\xi(x) + (1 - t)\xi(y)$$

for all $x, y \in P$ and $t \in [0, 1]$.

Let ξ be a nonnegative concave functional on P, and $0 < c < d$. We define the convex sets P_c, $\overline{P}_c$, $P(\xi, c, d)$ and $S(\xi, c, d)$ by

$$P_c = \{u \in P, \; \|u\| < c\}, \quad \overline{P}_c = \{u \in P, \; \|u\| \leq c\},$$

$$P(\xi, c, d) = \{u \in P, \; \xi(u) \geq c, \; \|u\| \leq d\},$$

$$S(\xi, c, d) = \{u \in P, \; \xi(u) > c, \; \|u\| \leq d\}.$$

Theorem 1.2.12 ([72]). *Suppose that* $A : \overline{P}_c \to P$ *is completely continuous, and there exists a concave positive functional* ξ *with* $\xi(u) \leq \|u\|$ *for all* $u \in P$, *and positive constants* $0 < a < b \leq c$ *satisfying the following conditions:*

(1) $S(\xi, a, b) \neq \emptyset$, *and* $\xi(Au) > a$ *if* $u \in P(\xi, a, b)$;
(2) $Au \in \overline{P}_c$ *if* $u \in P(\xi, a, c)$;
(3) $\xi(Au) > a$ *for all* $u \in P(\xi, a, c)$ *with* $\|Au\| > b$.

Then the fixed point index $i(A, S(\xi, a, c), \overline{P}_c) = 1$.

The Krasnosel'skii fixed point index theorem
With the same notations as that used in Theorem 1.2.12, we present the next theorem.

Theorem 1.2.13 (see [41, 118]). *Let* P *be a cone in the real Banach space* X, *and* $A : P \to P$ *be a completely continuous operator. Let* a, b, c *be three positive constants with* $0 < a < b < c$.

(a) *If* $\|Au\| > \|u\|$ *for all* $u \in \partial P_a$, *and* $\|Au\| \leq \|u\|$ *for all* $u \in \partial P_b$, *then*

$$i(A, \overline{P}_b \setminus \overline{P}_a, \overline{P}_b) = 1.$$

(b) *If* $\|Au\| > \|u\|$ *for all* $u \in \partial P_a$, *and* $\|Au\| < \|u\|$ *for all* $u \in \partial P_b$, *then*

$$i(A, P_b \setminus \overline{P}_a, \overline{P}_c) = 1.$$

The nonlinear alternative of Leray–Schauder type for Kakutani maps

Theorem 1.2.14 ([40]). *Let C a closed convex subset of a Banach space E and U be an open subset of C with $\theta \in U$. Suppose that $F : \overline{U} \to \mathcal{P}_{cp,c}(C)$ is an upper semicontinuous compact map. Then either* (i) *F has a fixed point in $\overline{U}$ or* (ii) *there exists $u \in \partial U$ and $\lambda \in (0,1)$ such that $u \in \lambda F(u)$.*

The Covitz–Nadler fixed point theorem

Theorem 1.2.15 ([27]). *Let (X, d) be a complete metric space. If $N : X \to \mathcal{P}_{cl}(X)$ is a contraction, then $\mathrm{Fix}\, N \neq \emptyset$.*

In Theorems 1.2.14 and 1.2.15, $\mathcal{P}_{cp,c}(C) = \{Y \in \mathcal{P}(C), Y$ is compact and convex$\}$, $\mathcal{P}_{cl}(X) = \{Y \in \mathcal{P}(X), Y$ is closed$\}$, and $\mathrm{Fix}\, N$ is the set of the fixed points of operator N.

Chapter 2

Riemann–Liouville Fractional Differential Equations with Nonlocal Boundary Conditions

In this chapter, we investigate the existence of positive solutions or nonnegative solutions for some Riemann-Liouville fractional differential equations with parameters or without parameters, subject to various multi-point or Riemann–Stieltjes integral boundary conditions which contain fractional derivatives. The nonlinearities of equations are nonsingular or singular functions, with nonnegative values or real values.

2.1 Singular Fractional Differential Equations with Parameters and Multi-Point Boundary Conditions

We consider the nonlinear fractional differential equation

$$D_{0+}^\alpha u(t) + \lambda f(t, u(t)) = 0, \quad t \in (0,1), \tag{2.1}$$

with the multi-point fractional boundary conditions

$$u(0) = u'(0) = \cdots = u^{(n-2)}(0) = 0, \quad D_{0+}^p u(1) = \sum_{i=1}^m a_i D_{0+}^q u(\xi_i), \tag{2.2}$$

where λ is a positive parameter, $\alpha \in \mathbb{R}$, $\alpha \in (n-1, n]$, $n \in \mathbb{N}$, $n \geq 3$, $\xi_i \in \mathbb{R}$ for all $i = 1, \ldots, m$, $(m \in \mathbb{N})$, $0 < \xi_1 < \cdots < \xi_m \leq 1$, $p, q \in \mathbb{R}$, $p \in [1, \alpha - 1)$, $q \in [0, p]$, D_{0+}^k denotes the Riemann-Liouville derivative of order k (for $k = \alpha, p, q$), and the nonlinearity f may change sign and may be singular at $t = 0$ and/or $t = 1$.

9

We present intervals for the parameter λ such that problem (2.1), (2.2) has at least one positive solution. Because f is a sign-changing function, our problem (2.1), (2.2) is a semipositone boundary value problem. In the proof of the main existence results, we use the Guo-Krasnosel'skii fixed point theorem (Theorem 1.2.2). By a positive solution of (2.1), (2.2), we mean a function $u \in C[0,1]$ satisfying (2.1) and (2.2), with $u(t) > 0$ for all $t \in (0,1]$. This problem is a generalization of the problem studied in [60], where $p \in \mathbb{N}$ and $q = 1$. Other particular cases of problem (2.1), (2.2) and of problem from [60] were investigated in [105] (where $n = 3$, $p = 0$ and $a_i = 0$ for all $i = 1, \ldots, m$) and in [110] (where $p = 0$ and $a_i = 0$ for all $i = 1, \ldots, m$ and n is an arbitrary natural number, $n \geq 3$).

2.1.1 *Auxiliary results*

In this section, we present some auxiliary results that will be used to prove our main results. We consider the fractional differential equation

$$D_{0+}^{\alpha} u(t) + \widetilde{x}(t) = 0, \quad 0 < t < 1, \tag{2.3}$$

with the boundary conditions (2.2), where $\widetilde{x} \in C(0,1) \cap L^1(0,1)$. We denote by $\Delta = \dfrac{\Gamma(\alpha)}{\Gamma(\alpha-p)} - \dfrac{\Gamma(\alpha)}{\Gamma(\alpha-q)} \sum_{i=1}^{m} a_i \xi_i^{\alpha-q-1}$.

Lemma 2.1.1. *If $\Delta \neq 0$, then the unique solution $u \in C[0,1]$ of problem (2.3), (2.2) is given by*

$$
\begin{aligned}
u(t) = {}& -\frac{1}{\Gamma(\alpha)} \int_0^t (t-s)^{\alpha-1} \widetilde{x}(s)\, ds + \frac{t^{\alpha-1}}{\Delta \Gamma(\alpha-p)} \int_0^1 (1-s)^{\alpha-p-1} \widetilde{x}(s)\, ds \\
& - \frac{t^{\alpha-1}}{\Delta \Gamma(\alpha-q)} \sum_{i=1}^{m} a_i \left(\int_0^{\xi_i} (\xi_i - s)^{\alpha-q-1} \widetilde{x}(s)\, ds \right), \quad t \in [0,1].
\end{aligned}
\tag{2.4}
$$

Proof. By Lemma 1.1.3, we deduce that the solutions $u \in C(0,1) \cap L^1(0,1)$ of the fractional differential equation (2.3) are given by

$$
\begin{aligned}
u(t) &= -I_{0+}^{\alpha} \widetilde{x}(t) + c_1 t^{\alpha-1} + \cdots + c_n t^{\alpha-n} \\
&= -\frac{1}{\Gamma(\alpha)} \int_0^t (t-s)^{\alpha-1} \widetilde{x}(s)\, ds + c_1 t^{\alpha-1} + \cdots + c_n t^{\alpha-n},
\end{aligned}
$$

where $c_1, c_2, \ldots, c_n \in \mathbb{R}$. By using the conditions $u(0) = u'(0) = \cdots = u^{(n-2)}(0) = 0$, we obtain $c_2 = \cdots = c_n = 0$. Then we conclude

$$u(t) = c_1 t^{\alpha-1} - \frac{1}{\Gamma(\alpha)} \int_0^t (t-s)^{\alpha-1} \widetilde{x}(s)\, ds, \quad t \in [0,1]. \tag{2.5}$$

For the obtained function (2.5), we find

$$D_{0+}^p u(t) = c_1 \frac{\Gamma(\alpha)}{\Gamma(\alpha-p)} t^{\alpha-p-1} - I_{0+}^{\alpha-p} \widetilde{x}(t),$$

$$D_{0+}^q u(t) = c_1 \frac{\Gamma(\alpha)}{\Gamma(\alpha-q)} t^{\alpha-q-1} - I_{0+}^{\alpha-q} \widetilde{x}(t).$$

Then the condition $D_{0+}^p u(1) = \sum_{i=1}^m a_i D_{0+}^q u(\xi_i)$ gives us

$$c_1 \frac{\Gamma(\alpha)}{\Gamma(\alpha-p)} - \frac{1}{\Gamma(\alpha-p)} \int_0^1 (1-s)^{\alpha-p-1} \widetilde{x}(s)\, ds$$

$$= \sum_{i=1}^m a_i \left(c_1 \frac{\Gamma(\alpha)}{\Gamma(\alpha-q)} \xi_i^{\alpha-q-1} - \frac{1}{\Gamma(\alpha-q)} \int_0^{\xi_i} (\xi_i-s)^{\alpha-q-1} \widetilde{x}(s)\, ds \right).$$

So, we deduce

$$c_1 = \frac{1}{\Delta\Gamma(\alpha-p)} \int_0^1 (1-s)^{\alpha-p-1} \widetilde{x}(s)\, ds$$

$$- \frac{1}{\Delta\Gamma(\alpha-q)} \sum_{i=1}^m a_i \left(\int_0^{\xi_i} (\xi_i-s)^{\alpha-q-1} \widetilde{x}(s)\, ds \right).$$

Replacing the above constant c_1 in (2.5), we obtain the expression (2.4) for the solution $u \in C[0,1]$ of problem (2.3), (2.2). Conversely, one easily verifies that $u \in C[0,1]$ given by (2.4) satisfies Eq. (2.3) and the boundary conditions (2.2). $\qquad\square$

Lemma 2.1.2. *If $\Delta \neq 0$, then the solution u of problem (2.3), (2.2) given by (2.4) can be written as*

$$u(t) = \int_0^1 G(t,s) \widetilde{x}(s)\, ds, \quad t \in [0,1], \tag{2.6}$$

where the Green function G is

$$G(t,s) = g_1(t,s) + \frac{t^{\alpha-1}}{\Delta} \sum_{i=1}^m a_i g_2(\xi_i, s), \tag{2.7}$$

and

$$g_1(t,s) = \frac{1}{\Gamma(\alpha)} \begin{cases} t^{\alpha-1}(1-s)^{\alpha-p-1} - (t-s)^{\alpha-1}, & 0 \le s \le t \le 1, \\ t^{\alpha-1}(1-s)^{\alpha-p-1}, & 0 \le t \le s \le 1, \end{cases}$$

$$g_2(t,s) = \frac{1}{\Gamma(\alpha-q)} \begin{cases} t^{\alpha-q-1}(1-s)^{\alpha-p-1} - (t-s)^{\alpha-q-1}, & 0 \le s \le t \le 1, \\ t^{\alpha-q-1}(1-s)^{\alpha-p-1}, & 0 \le t \le s \le 1, \end{cases}$$

(2.8)

for all $(t,s) \in [0,1] \times [0,1]$.

Proof. By Lemma 2.1.1 and relation (2.4), we deduce

$$u(t) = \frac{1}{\Gamma(\alpha)} \left\{ \int_0^t [t^{\alpha-1}(1-s)^{\alpha-p-1} - (t-s)^{\alpha-1}]\widetilde{x}(s)\, ds \right.$$

$$\left. + \int_t^1 t^{\alpha-1}(1-s)^{\alpha-p-1}\widetilde{x}(s)\, ds \right\}$$

$$- \frac{1}{\Gamma(\alpha)} \int_0^1 t^{\alpha-1}(1-s)^{\alpha-p-1}\widetilde{x}(s)\, ds$$

$$+ \frac{t^{\alpha-1}}{\Delta\Gamma(\alpha-p)} \int_0^1 (1-s)^{\alpha-p-1}\widetilde{x}(s)\, ds$$

$$- \frac{t^{\alpha-1}}{\Delta\Gamma(\alpha-q)} \sum_{i=1}^m a_i \left(\int_0^{\xi_i} (\xi_i-s)^{\alpha-q-1}\widetilde{x}(s)\, ds \right)$$

$$= \frac{1}{\Gamma(\alpha)} \left\{ \int_0^t [t^{\alpha-1}(1-s)^{\alpha-p-1} - (t-s)^{\alpha-1}]\widetilde{x}(s)\, ds \right.$$

$$\left. + \int_t^1 t^{\alpha-1}(1-s)^{\alpha-p-1}\widetilde{x}(s)\, ds \right\}$$

$$- \frac{1}{\Delta\Gamma(\alpha-p)} \int_0^1 t^{\alpha-1}(1-s)^{\alpha-p-1}\widetilde{x}(s)\, ds$$

$$+ \frac{1}{\Delta\Gamma(\alpha-q)} \left(\sum_{i=1}^m a_i\xi_i^{\alpha-q-1} \right) \int_0^1 t^{\alpha-1}(1-s)^{\alpha-p-1}\widetilde{x}(s)\, ds$$

$$+ \frac{t^{\alpha-1}}{\Delta\Gamma(\alpha-p)} \int_0^1 (1-s)^{\alpha-p-1}\widetilde{x}(s)\, ds$$

$$- \frac{t^{\alpha-1}}{\Delta\Gamma(\alpha-q)} \sum_{i=1}^m a_i \left(\int_0^{\xi_i} (\xi_i-s)^{\alpha-q-1}\widetilde{x}(s)\, ds \right)$$

$$= \frac{1}{\Gamma(\alpha)} \left\{ \int_0^t [t^{\alpha-1}(1-s)^{\alpha-p-1} - (t-s)^{\alpha-1}] \widetilde{x}(s) \, ds \right.$$

$$\left. + \int_t^1 t^{\alpha-1}(1-s)^{\alpha-p-1} \widetilde{x}(s) \, ds \right\}$$

$$+ \frac{t^{\alpha-1}}{\Delta\Gamma(\alpha-q)} \sum_{i=1}^m a_i \left[\int_0^1 \xi_i^{\alpha-q-1}(1-s)^{\alpha-p-1} \widetilde{x}(s) \, ds \right.$$

$$\left. - \int_0^{\xi_i} (\xi_i - s)^{\alpha-q-1} \widetilde{x}(s) \, ds \right]$$

$$= \frac{1}{\Gamma(\alpha)} \left\{ \int_0^t [t^{\alpha-1}(1-s)^{\alpha-p-1} - (t-s)^{\alpha-1}] \widetilde{x}(s) \, ds \right.$$

$$\left. + \int_t^1 t^{\alpha-1}(1-s)^{\alpha-p-1} \widetilde{x}(s) \, ds \right\}$$

$$+ \frac{t^{\alpha-1}}{\Delta\Gamma(\alpha-q)} \sum_{i=1}^m a_i \left\{ \int_0^{\xi_i} [\xi_i^{\alpha-q-1}(1-s)^{\alpha-p-1} - (\xi_i - s)^{\alpha-q-1}] \widetilde{x}(s) \, ds \right.$$

$$\left. + \int_{\xi_i}^1 \xi_i^{\alpha-q-1}(1-s)^{\alpha-p-1} \widetilde{x}(s) \, ds \right\}$$

$$= \int_0^1 g_1(t,s) \widetilde{x}(s) \, ds + \frac{t^{\alpha-1}}{\Delta} \sum_{i=1}^m a_i \int_0^1 g_2(\xi_i, s) \widetilde{x}(s) \, ds$$

$$= \int_0^1 G(t,s) \widetilde{x}(s) \, ds,$$

where G and g_1, g_2 are given in (2.7) and (2.8).

Therefore, we obtain the expression (2.6) for the solution u of problem (2.3), (2.2) given by (2.4). $\qquad\square$

Lemma 2.1.3. *The functions g_1 and g_2 given by (2.8) have the following properties:*

(a) $g_1(t,s) \leq h_1(s)$ *for all t, $s \in [0,1]$, where*

$$h_1(s) = \frac{1}{\Gamma(\alpha)} (1-s)^{\alpha-p-1}(1-(1-s)^p), s \in [0,1];$$

(b) $g_1(t,s) \geq t^{\alpha-1} h_1(s)$ *for all t, $s \in [0,1]$;*

(c) $g_1(t,s) \leq \frac{t^{\alpha-1}}{\Gamma(\alpha)}$, *for all t, $s \in [0,1]$;*

(d) $g_2(t, s) \geq t^{\alpha - q - 1} h_2(s)$ for all $t, s \in [0, 1]$, where

$$h_2(s) = \frac{1}{\Gamma(\alpha - q)} (1 - s)^{\alpha - p - 1} (1 - (1 - s)^{p - q}), s \in [0, 1];$$

(e) $g_2(t, s) \leq \frac{1}{\Gamma(\alpha - q)} t^{\alpha - q - 1}$ for all $t, s \in [0, 1]$;

(f) The functions g_1 and g_2 are continuous on $[0, 1] \times [0, 1]$; $g_1(t, s) \geq 0$, $g_2(t, s) \geq 0$ for all $t, s \in [0, 1]$; $g_1(t, s) > 0$, $g_2(t, s) > 0$ for all $t, s \in (0, 1)$.

Proof. (a) The function g_1 is nondecreasing in the first variable. Indeed, for $s \leq t$, we have

$$\frac{\partial g_1}{\partial t}(t, s) = \frac{1}{\Gamma(\alpha)} [(\alpha - 1) t^{\alpha - 2} (1 - s)^{\alpha - p - 1} - (\alpha - 1)(t - s)^{\alpha - 2}]$$

$$\geq \frac{1}{\Gamma(\alpha - 1)} [t^{\alpha - 2} (1 - s)^{\alpha - 2} - (t - s)^{\alpha - 2}]$$

$$= \frac{1}{\Gamma(\alpha - 1)} [(t - ts)^{\alpha - 2} - (t - s)^{\alpha - 2}] \geq 0.$$

Then, $g_1(t, s) \leq g_1(1, s)$ for all $(t, s) \in [0, 1] \times [0, 1]$ with $s \leq t$.

For $s \geq t$, we obtain

$$\frac{\partial g_1}{\partial t}(t, s) = \frac{1}{\Gamma(\alpha - 1)} t^{\alpha - 2} (1 - s)^{\alpha - p - 1} \geq 0.$$

Hence, $g_1(t, s) \leq g_1(s, s)$ for all $(t, s) \in [0, 1] \times [0, 1]$ with $s \geq t$.

Therefore, we deduce that $g_1(t, s) \leq h_1(s)$ for all $(t, s) \in [0, 1] \times [0, 1]$, where $h_1(s) = g_1(1, s) = \frac{1}{\Gamma(\alpha)} (1 - s)^{\alpha - p - 1} (1 - (1 - s)^p)$ for all $s \in [0, 1]$.

(b) For $s \leq t$, we have

$$g_1(t, s) = \frac{1}{\Gamma(\alpha)} [t^{\alpha - 1} (1 - s)^{\alpha - p - 1} - (t - s)^{\alpha - 1}]$$

$$\geq \frac{1}{\Gamma(\alpha)} [t^{\alpha - 1} (1 - s)^{\alpha - p - 1} - (t - ts)^{\alpha - 1}]$$

$$= \frac{1}{\Gamma(\alpha)} t^{\alpha - 1} [(1 - s)^{\alpha - p - 1} - (1 - s)^{\alpha - 1}]$$

$$= \frac{1}{\Gamma(\alpha)} t^{\alpha - 1} (1 - s)^{\alpha - p - 1} (1 - (1 - s)^p) = t^{\alpha - 1} h_1(s).$$

For $s \geq t$, we obtain

$$g_1(t,s) = \frac{1}{\Gamma(\alpha)}t^{\alpha-1}(1-s)^{\alpha-p-1} \geq \frac{1}{\Gamma(\alpha)}t^{\alpha-1}(1-s)^{\alpha-p-1}(1-(1-s)^p)$$

$$= t^{\alpha-1}h_1(s).$$

Hence, we conclude that $g_1(t,s) \geq t^{\alpha-1}h_1(s)$ for all $(t,s) \in [0,1] \times [0,1]$.

(c) For all $(t,s) \in [0,1] \times [0,1]$ we have

$$g_1(t,s) \leq \frac{1}{\Gamma(\alpha)}t^{\alpha-1}(1-s)^{\alpha-p-1} \leq \frac{t^{\alpha-1}}{\Gamma(\alpha)}.$$

(d) From (2.8), if $s \leq t$, we obtain

$$g_2(t,s) = \frac{1}{\Gamma(\alpha-q)}[t^{\alpha-q-1}(1-s)^{\alpha-p-1} - (t-s)^{\alpha-q-1}]$$

$$\geq \frac{1}{\Gamma(\alpha-q)}[t^{\alpha-q-1}(1-s)^{\alpha-p-1} - (t-ts)^{\alpha-q-1}]$$

$$= \frac{1}{\Gamma(\alpha-q)}t^{\alpha-q-1}[(1-s)^{\alpha-p-1} - (1-s)^{\alpha-q-1}]$$

$$= \frac{1}{\Gamma(\alpha-q)}t^{\alpha-q-1}(1-s)^{\alpha-p-1}(1-(1-s)^{p-q}) = t^{\alpha-q-1}h_2(s),$$

where $h_2(s) = \frac{1}{\Gamma(\alpha-q)}(1-s)^{\alpha-p-1}(1-(1-s)^{p-q})$, $s \in [0,1]$.

If $t \leq s$, we deduce

$$g_2(t,s) = \frac{1}{\Gamma(\alpha-q)}t^{\alpha-q-1}(1-s)^{\alpha-p-1} \geq t^{\alpha-q-1}h_2(s).$$

Hence, we conclude $g_2(t,s) \geq t^{\alpha-q-1}h_2(s)$ for all $t, s \in [0,1]$.

(e) We have

$$g_2(t,s) \leq \frac{1}{\Gamma(\alpha-q)}t^{\alpha-q-1}(1-s)^{\alpha-p-1} \leq \frac{t^{\alpha-q-1}}{\Gamma(\alpha-q)}, \qquad \forall t, s \in [0,1].$$

(f) This property follows from the definitions of g_1 and g_2 and from the properties (b) and (d) above. $\qquad\square$

Lemma 2.1.4. *Assume that $a_i \geq 0$ for all $i = 1, \ldots, m$ and $\Delta > 0$. Then the Green function G given by (2.7) is a continuous function on $[0,1] \times [0,1]$ and satisfies the inequalities:*

(a) $G(t,s) \leq J(s)$ *for all* $t, s \in [0,1]$, *where* $J(s) = h_1(s) + \frac{1}{\Delta}\sum_{i=1}^m a_i g_2(\xi_i, s)$, $s \in [0,1]$;

(b) $G(t,s) \geq t^{\alpha-1}J(s)$ for all t, $s \in [0,1]$;

(c) $G(t,s) \leq \sigma t^{\alpha-1}$, for all t, $s \in [0,1]$, where $\sigma = \frac{1}{\Gamma(\alpha)} + \frac{1}{\Delta\Gamma(\alpha-q)}\sum_{i=1}^{m}a_i\xi_i^{\alpha-q-1}$.

Proof. By definition of the function G, we deduce that G is a continuous function. In addition, by using Lemma 2.1.3, we obtain for all t, $s \in [0,1]$

(a) $G(t,s) \leq h_1(s) + \frac{1}{\Delta}\sum_{i=1}^{m}a_ig_2(\xi_i,s) = J(s)$;

(b) $G(t,s) \geq t^{\alpha-1}h_1(s) + \frac{t^{\alpha-1}}{\Delta}\sum_{i=1}^{m}a_ig_2(\xi_i,s) = t^{\alpha-1}J(s)$;

(c) $G(t,s) \leq \frac{t^{\alpha-1}}{\Gamma(\alpha)} + \frac{t^{\alpha-1}}{\Delta\Gamma(\alpha-q)}\sum_{i=1}^{m}a_i\xi_i^{\alpha-q-1} = \sigma t^{\alpha-1}$.

$\square$

Lemma 2.1.5. *Assume that $a_i \geq 0$ for all $i = 1,\ldots,m$, $\Delta > 0$, $\widetilde{x} \in C(0,1) \cap L^1(0,1)$ and $\widetilde{x}(t) > 0$ for all $t \in (0,1)$. Then the solution u of problem (2.3), (2.2) given by (2.4) satisfies the inequality $u(t) \geq t^{\alpha-1}u(t')$ for all t, $t' \in [0,1]$.*

Proof. By using Lemma 2.1.4, we obtain

$$u(t) = \int_0^1 G(t,s)\widetilde{x}(s)\,ds \geq \int_0^1 t^{\alpha-1}J(s)\widetilde{x}(s)\,ds$$

$$\geq t^{\alpha-1}\int_0^1 G(t',s)\widetilde{x}(s)\,ds = t^{\alpha-1}u(t'),$$

for all t, $t' \in [0,1]$.

$\square$

2.1.2   Existence of positive solutions

In this section, we investigate the existence of positive solutions for our problem (2.1), (2.2). First, we present the assumptions that we shall use in the sequel.

(H1) $\alpha \in \mathbb{R}$, $\alpha \in (n-1,n]$, $n \in \mathbb{N}$, $n \geq 3$, $\xi_i \in \mathbb{R}$ for all $i = 1,\ldots,m$, ($m \in \mathbb{N}$), $0 < \xi_1 < \cdots < \xi_m \leq 1$, p, $q \in \mathbb{R}$, $p \in [1,\alpha-1)$, $q \in [0,p]$, $a_i \geq 0$ for all $i = 1,\ldots,m$, $\lambda > 0$, $\Delta = \frac{\Gamma(\alpha)}{\Gamma(\alpha-p)} - \frac{\Gamma(\alpha)}{\Gamma(\alpha-q)}\sum_{i=1}^{m}a_i\xi_i^{\alpha-q-1} > 0$.

(H2) The function $f \in C((0,1) \times [0,\infty),\mathbb{R})$ may be singular at $t = 0$ and/or $t = 1$, and there exist the functions r, $z \in C((0,1),[0,\infty))$, $g \in C([0,1] \times [0,\infty),[0,\infty))$ such that $-r(t) \leq f(t,x) \leq z(t)g(t,x)$ for all $t \in (0,1)$ and $x \in [0,\infty)$, with $0 < \int_0^1 r(t)\,dt < \infty$, $0 < \int_0^1 z(t)\,dt < \infty$.

(H3) There exists $c \in (0, 1/2)$ such that

$$f_\infty = \lim_{u \to \infty} \min_{t \in [c, 1-c]} \frac{f(t, u)}{u} = \infty.$$

(H4) There exists $c \in (0, 1/2)$ such that $\lim\inf_{u \to \infty} \min_{t \in [c, 1-c]} f(t, u) > L_0$, with

$$L_0 = \left(2\sigma \int_0^1 r(s)\, ds \right) \left(c^{\alpha - 1} \int_c^{1-c} J(s)\, ds \right)^{-1}, \quad \text{and}$$

$$g_\infty = \lim_{u \to \infty} \max_{t \in [0, 1]} \frac{g(t, u)}{u} = 0,$$

where J and σ are given in Lemma 2.1.4.

We consider the fractional differential equation

$$D_{0+}^\alpha x(t) + \lambda(f(t, [x(t) - \lambda w(t)]^*) + r(t)) = 0, \quad 0 < t < 1, \qquad (2.9)$$

with the multi-point fractional boundary conditions

$$x(0) = x'(0) = \cdots = x^{(n-2)}(0) = 0, \quad D_{0+}^p x(1) = \sum_{i=1}^m a_i D_{0+}^q x(\xi_i), \quad (2.10)$$

where $\lambda > 0$ and $\zeta(t)^* = \zeta(t)$ if $\zeta(t) \geq 0$, and $\zeta(t)^* = 0$ if $\zeta(t) < 0$. Here $w(t) = \int_0^1 G(t, s) r(s)\, ds$, $t \in [0, 1]$ is solution of problem

$$D_{0+}^\alpha w(t) + r(t) = 0, \quad 0 < t < 1,$$

$$w(0) = w'(0) = \cdots = w^{(n-2)}(0) = 0, \quad D_{0+}^p w(1) = \sum_{i=1}^m a_i D_{0+}^q w(\xi_i).$$

Under assumptions (H1), (H2), we have $w(t) \geq 0$ for all $t \in [0, 1]$. We shall prove that there exists a solution x of the problem (2.9), (2.10) with $x(t) \geq \lambda w(t)$ on $[0, 1]$ and $x(t) > \lambda w(t)$ on $(0, 1)$. In this case $u = x - \lambda w$ represents a positive solution of the problem (2.1), (2.2). Therefore, in what follows we shall investigate the problem (2.9), (2.10).

By using Lemma 2.1.2, x is a solution of the integral equation

$$x(t) = \lambda \int_0^1 G(t, s)(f(s, [x(s) - \lambda w(s)]^*) + r(s))\, ds, \quad t \in [0, 1],$$

if and only if x is a solution for problem (2.9), (2.10).

We consider the Banach space $X = C[0, 1]$ with the supremum norm $\|u\| = \sup_{t \in [0, 1]} |u(t)|$, and we define the cone

$$P = \{x \in X,\ x(t) \geq t^{\alpha - 1} \|x\|,\ \forall t \in [0, 1]\}.$$

For $\lambda > 0$ we introduce the operator $\mathcal{T} : X \to X$ defined by

$$\mathcal{T}x(t) = \lambda \int_0^1 G(t,s)(f(s,[x(s) - \lambda w(s)]^*) + r(s)) \, ds, \quad t \in [0,1], \ x \in X.$$

It is clear that x is a solution of problem (2.9), (2.10) if and only if x is a fixed point of operator $\mathcal{T}$.

Lemma 2.1.6. *If* (H1) *and* (H2) *hold, then operator* $\mathcal{T} : P \to P$ *is a completely continuous operator.*

Proof. Let $x \in P$ be fixed. By using (H1) and (H2), we deduce that $\mathcal{T}x(t) < \infty$ for all $t \in [0,1]$. Besides, by Lemma 2.1.4, we obtain

$$\mathcal{T}x(t) \leq \lambda \int_0^1 J(s)(f(s,[x(s) - \lambda w(s)]^*) + r(s)) \, ds, \quad \forall t \in [0,1],$$

and

$$\mathcal{T}x(t) \geq \lambda \int_0^1 t^{\alpha-1} J(s)(f(s,[x(s) - \lambda w(s)]^*) + r(s)) ds$$
$$\geq t^{\alpha-1} \mathcal{T}x(t'), \quad \forall t, t' \in [0,1].$$

Therefore, $\mathcal{T}x(t) \geq t^{\alpha-1}\|\mathcal{T}x\|$ for all $t \in [0,1]$. We deduce that $\mathcal{T}x \in P$, and hence $\mathcal{T}(P) \subset P$.

By using standard arguments, we conclude that operator $\mathcal{T} : P \to P$ is a completely continuous operator. $\qquad\square$

Theorem 2.1.1. *Assume that* (H1)–(H3) *hold. Then there exists* $\lambda^* > 0$ *such that, for any* $\lambda \in (0, \lambda^*]$, *the boundary value problem* (2.1), (2.2) *has at least one positive solution.*

Proof. We choose a positive number $R_1 > \sigma \int_0^1 r(s) \, ds > 0$, and we define the set $\Omega_1 = \{x \in X, \ \|x\| < R_1\} (= B_{R_1})$.

We introduce

$$\lambda^* = \min \left\{ 1, R_1 \left(M_1 \int_0^1 J(s)(z(s) + r(s)) \, ds \right)^{-1} \right\},$$

with $M_1 = \max \{ \max_{t \in [0,1], \, u \in [0,R_1]} g(t,u), 1 \}$.

Let $\lambda \in (0, \lambda^*]$. Because $w(t) \leq \sigma t^{\alpha-1} \int_0^1 r(s)\,ds$ for all $t \in [0,1]$, we deduce for any $x \in P \cap \partial\Omega_1$ and $t \in [0,1]$

$$[x(t) - \lambda w(t)]^* \leq x(t) \leq \|x\| \leq R_1,$$

and

$$x(t) - \lambda w(t) \geq t^{\alpha-1}\|x\| - \lambda \sigma t^{\alpha-1} \int_0^1 r(s)\,ds = t^{\alpha-1}\left(R_1 - \lambda \sigma \int_0^1 r(s)\,ds\right)$$

$$\geq t^{\alpha-1}\left(R_1 - \lambda^* \sigma \int_0^1 r(s)\,ds\right)$$

$$\geq t^{\alpha-1}\left(R_1 - \sigma \int_0^1 r(s)\,ds\right) \geq 0.$$

Then for any $x \in P \cap \partial\Omega_1$ and $t \in [0,1]$, we obtain

$$\mathcal{T}x(t) \leq \lambda \int_0^1 J(s)\,(z(s)g(s, [x(s) - \lambda w(s)]^*) + r(s))\,ds$$

$$\leq \lambda^* M_1 \int_0^1 J(s)(z(s) + r(s))\,ds \leq R_1 = \|x\|.$$

Therefore, we conclude

$$\|\mathcal{T}x\| \leq \|x\|, \quad \forall x \in P \cap \partial\Omega_1. \tag{2.11}$$

On the other hand, for c from (H3), we choose a positive constant $L > 0$ such that

$$L \geq 2\left(\lambda c^{2(\alpha-1)} \int_c^{1-c} J(s)\,ds\right)^{-1}.$$

By (H3), we deduce that there exists a constant $M_0 > 0$ such that

$$f(t, u) \geq Lu, \quad \forall t \in [c, 1-c], \quad u \geq M_0. \tag{2.12}$$

Now, we define $R_2 = \max\{2R_1, 2M_0/c^{\alpha-1}\}$ and let $\Omega_2 = \{x \in X, \|x\| < R_2\}(= B_{R_2})$.

Then for any $x \in P \cap \partial \Omega_2$, we obtain

$$x(t) - \lambda w(t) \geq t^{\alpha-1} \|x\| - \lambda \sigma t^{\alpha-1} \int_0^1 r(s) \, ds$$

$$\geq t^{\alpha-1} \left(R_2 - \sigma \int_0^1 r(s) \, ds \right) \geq t^{\alpha-1} \left(R_1 - \sigma \int_0^1 r(s) \, ds \right)$$

$$\geq 0, \quad \forall t \in [0,1].$$

Therefore, we conclude

$$[x(t) - \lambda w(t)]^* = x(t) - \lambda w(t) \geq t^{\alpha-1} \left(R_2 - \sigma \int_0^1 r(s) \, ds \right)$$

$$\geq \frac{c^{\alpha-1} R_2}{2} \geq M_0, \ \forall t \in [c, 1-c]. \tag{2.13}$$

Then for any $x \in P \cap \partial \Omega_2$ and $t \in [c, 1-c]$, by (2.12) and (2.13), we deduce

$$\mathcal{T}x(t) \geq \lambda \int_c^{1-c} G(t,s)(f(x, [x(s) - \lambda w(s)]^*) + r(s)) \, ds$$

$$\geq \lambda \int_c^{1-c} G(t,s) L[x(s) - \lambda w(s)] \, ds \geq \lambda L \int_c^{1-c} t^{\alpha-1} J(s) \frac{c^{\alpha-1} R_2}{2} \, ds$$

$$\geq \frac{\lambda L c^{2(\alpha-1)} R_2}{2} \int_c^{1-c} J(s) \, ds \geq R_2 = \|x\|.$$

Then

$$\|\mathcal{T}x\| \geq \|x\|, \quad \forall x \in P \cap \partial \Omega_2. \tag{2.14}$$

By (2.11), (2.14) and Theorem 1.2.2 (i), we conclude that $\mathcal{T}$ has a fixed point $x_1 \in P \cap (\overline{\Omega}_2 \setminus \Omega_1)$, that is, $R_1 \leq \|x_1\| \leq R_2$. Since $\|x_1\| \geq R_1$, we deduce

$$x_1(t) - \lambda w(t) \geq t^{\alpha-1} \left(\|x_1\| - \sigma \lambda \int_0^1 r(s) \, ds \right)$$

$$\geq t^{\alpha-1} \left(R_1 - \sigma \int_0^1 r(s) \, ds \right) = \Lambda_1 t^{\alpha-1},$$

and so $x_1(t) \geq \lambda w(t) + \Lambda_1 t^{\alpha-1}$ for all $t \in [0,1]$, where $\Lambda_1 = R_1 - \sigma \int_0^1 r(s) \, ds > 0$.

Let $u_1(t) = x_1(t) - \lambda w(t)$ for all $t \in [0,1]$. Then u_1 is a positive solution of problem (2.1), (2.2) with $u_1(t) \geq \Lambda_1 t^{\alpha-1}$ for all $t \in [0,1]$. This completes the proof of Theorem 2.1.1. □

Theorem 2.1.2. *Assume that* (H1), (H2) *and* (H4) *hold. Then there exists* $\lambda_* > 0$ *such that, for any* $\lambda \geq \lambda_*$, *the boundary value problem* (2.1), (2.2) *has at least one positive solution.*

Proof. By (H4) there exists $M_2 > 0$ such that

$$f(t, u) \geq \left(2\sigma \int_0^1 r(s)\,ds\right)\left(c^{\alpha-1}\int_c^{1-c} J(s)\,ds\right)^{-1},$$

$$\forall t \in [c, 1-c], \; u \geq M_2.$$

We define

$$\lambda_* = M_2\left(c^{\alpha-1}\sigma\int_0^1 r(s)\,ds\right)^{-1}.$$

We assume now $\lambda \geq \lambda_*$. Let $R_3 = 2\lambda\sigma\int_0^1 r(s)\,ds$ and $\Omega_3 = \{x \in X, \; \|x\| < R_3\}(= B_{R_3})$. Then for any $x \in P \cap \partial\Omega_3$, we deduce

$$x(t) - \lambda w(t) \geq t^{\alpha-1}\|x\| - \lambda\sigma t^{\alpha-1}\int_0^1 r(s)\,ds = t^{\alpha-1}\left(R_3 - \lambda\sigma\int_0^1 r(s)\,ds\right)$$

$$= t^{\alpha-1}\lambda\sigma\int_0^1 r(s)\,ds \geq t^{\alpha-1}\lambda_*\sigma\int_0^1 r(s)\,ds$$

$$= \frac{M_2 t^{\alpha-1}}{c^{\alpha-1}} \geq 0, \quad \forall t \in [0, 1].$$

Therefore, for any $x \in P \cap \partial\Omega_3$ and $t \in [c, 1-c]$, we have

$$[x(t) - \lambda w(t)]^* = x(t) - \lambda w(t) \geq \frac{M_2 t^{\alpha-1}}{c^{\alpha-1}} \geq M_2.$$

Hence, for any $x \in P \cap \partial\Omega_3$ and $t \in [c, 1-c]$, we conclude

$$\mathcal{T}x(t) \geq \lambda\int_c^{1-c} G(t, s)f(s, [x(s) - \lambda w(s)]^*)\,ds$$

$$\geq \lambda\int_c^{1-c} t^{\alpha-1}J(s)f(s, u(s) - \lambda w(s))\,ds$$

$$\geq \lambda\int_c^{1-c} t^{\alpha-1}J(s)\left(2\sigma\int_0^1 r(s)\,ds\right)\left(c^{\alpha-1}\int_c^{1-c} J(s)\right)^{-1}\,ds$$

$$= \frac{R_3 t^{\alpha-1}}{c^{\alpha-1}} \geq R_3.$$

Therefore, we obtain

$$\|\mathcal{T}x\| \geq \|x\|, \quad \forall x \in P \cap \partial\Omega_3. \tag{2.15}$$

On the other hand, we consider the positive number $\varepsilon = (2\lambda \int_0^1 J(s)z(s)\,ds)^{-1}$. Then by (H4), we deduce that there exists $M_3 > 0$ such that $g(t,u) \leq \varepsilon u$ for all $t \in [0,1]$, $u \geq M_3$. Therefore, we obtain $g(t,u) \leq M_4 + \varepsilon u$ for all $t \in [0,1]$, $u \geq 0$, where $M_4 = \max_{t \in [0,1],\, u \in [0,M_3]} g(t,u)$.

We define now

$$R_4 > \max\left\{ R_3, 2\lambda \max\{M_4, 1\} \int_0^1 J(s)(z(s) + r(s))\,ds \right\}$$

and $\Omega_4 = \{x \in X,\ \|x\| < R_4\}(= B_{R_4})$.

Then for any $x \in P \cap \partial\Omega_4$ we have

$$x(t) - \lambda w(t) \geq t^{\alpha-1}\|x\| - \lambda \sigma t^{\alpha-1} \int_0^1 r(s)\,ds$$

$$= t^{\alpha-1}\left(R_4 - \lambda\sigma \int_0^1 r(s)\,ds \right) \geq t^{\alpha-1}\left(R_3 - \lambda\sigma \int_0^1 r(s)\,ds \right)$$

$$= t^{\alpha-1}\lambda\sigma \int_0^1 r(s)\,ds \geq t^{\alpha-1}\lambda_*\sigma \int_0^1 r(s)\,ds$$

$$= \frac{M_2 t^{\alpha-1}}{c^{\alpha-1}} \geq 0, \quad \forall t \in [0,1].$$

Then for any $x \in P \cap \partial\Omega_4$, we obtain

$$\mathcal{T}x(t) \leq \lambda \int_0^1 J(s)[z(s)g(s,[x(s) - \lambda w(s)]^* + r(s)]\,ds$$

$$\leq \lambda \int_0^1 J(s)[z(s)(M_4 + \varepsilon(x(s) - \lambda w(s))) + r(s)]\,ds$$

$$\leq \lambda \max\{M_4,1\} \int_0^1 J(s)(z(s) + r(s))\,ds + \lambda\varepsilon R_4 \int_0^1 J(s)z(s)\,ds$$

$$\leq \frac{R_4}{2} + \frac{R_4}{2} = R_4 = \|x\|, \quad \forall t \in [0,1].$$

Therefore,

$$\|\mathcal{T}x\| \leq \|x\|, \quad \forall x \in P \cap \partial\Omega_4. \tag{2.16}$$

By (2.15), (2.16) and Theorem 1.2.2 (ii), we conclude that $\mathcal{T}$ has a fixed point $x_1 \in P \cap (\overline{\Omega}_4 \setminus \Omega_3)$, so $R_3 \leq \|x_1\| \leq R_4$.

In addition, we deduce that for all $t \in [0, 1]$

$$x_1(t) - \lambda w(t) \geq x_1(t) - \lambda \sigma t^{\alpha-1} \int_0^1 r(s)\,ds$$

$$\geq t^{\alpha-1}\|x_1\| - \lambda \sigma t^{\alpha-1} \int_0^1 r(s)\,ds$$

$$\geq t^{\alpha-1}R_3 - \lambda \sigma t^{\alpha-1} \int_0^1 r(s)\,ds$$

$$= \lambda \sigma t^{\alpha-1} \int_0^1 r(s)\,ds \geq \lambda_* \sigma t^{\alpha-1} \int_0^1 r(s)\,ds = \frac{M_2 t^{\alpha-1}}{c^{\alpha-1}}.$$

Let $u_1(t) = x_1(t) - \lambda w(t)$ for all $t \in [0, 1]$. Then $u_1(t) \geq \widetilde{\Lambda}_1 t^{\alpha-1}$ for all $t \in [0, 1]$, where $\widetilde{\Lambda}_1 = \frac{M_2}{c^{\alpha-1}}$. Hence, we conclude that u_1 is a positive solution of problem (2.1), (2.2), which completes the proof of Theorem 2.1.2. $\square$

In a similar manner to that used in the proof of Theorem 2.1.2, we deduce the following result.

Theorem 2.1.3. *Assume* (H1), (H2) *and*

$(\widetilde{H4})$ *There exists* $c \in (0, 1/2)$ *such that*

$$\hat{f}_\infty = \lim_{u \to \infty} \min_{t \in [c, 1-c]} f(t, u) = \infty \quad and$$

$$g_\infty = \lim_{u \to \infty} \max_{t \in [0,1]} \frac{g(t, u)}{u} = 0,$$

hold. Then there exists $\widetilde{\lambda}_* > 0$ *such that, for any* $\lambda \geq \widetilde{\lambda}_*$, *the boundary value problem* (2.1), (2.2) *has at least one positive solution.*

2.1.3 Examples

Let $\alpha = 10/3$ $(n = 4)$, $p = 3/2$, $q = 4/3$, $m = 3$, $\xi_1 = 1/4$, $\xi_2 = 1/2$, $\xi_3 = 3/4$, $a_1 = 1$, $a_2 = 1/2$, $a_3 = 1/3$.

We consider the fractional differential equation

$$D_{0+}^{10/3} u(t) + \lambda f(t, u(t)) = 0, \quad t \in (0, 1), \tag{2.17}$$

with the boundary conditions

$$\begin{cases} u(0) = u'(0) = u''(0) = 0, \\ D_{0+}^{3/2} u(1) = D_{0+}^{4/3} u\left(\frac{1}{4}\right) + \frac{1}{2}D_{0+}^{4/3} u\left(\frac{1}{2}\right) + \frac{1}{3}D_{0+}^{4/3} u\left(\frac{3}{4}\right). \end{cases} \tag{2.18}$$

Then we obtain $\Delta = \Gamma(10/3)(1/\Gamma(11/6) - 3/4) \approx 0.86980822$. So assumption (H1) is satisfied. Besides, we deduce

$$g_1(t,s) = \frac{1}{\Gamma(10/3)} \begin{cases} t^{7/3}(1-s)^{5/6} - (t-s)^{7/3}, & 0 \le s \le t \le 1, \\ t^{7/3}(1-s)^{5/6}, & 0 \le t \le s \le 1, \end{cases}$$

$$g_2(t,s) = \begin{cases} t(1-s)^{5/6} - t + s, & 0 \le s \le t \le 1, \\ t(1-s)^{5/6}, & 0 \le t \le s \le 1, \end{cases}$$

$$G(t,s) = g_1(t,s) + \frac{t^{7/3}}{\Delta} \left(g_2\left(\frac{1}{4},s\right) + \frac{1}{2}g_2\left(\frac{1}{2},s\right) + \frac{1}{3}g_2\left(\frac{3}{4},s\right) \right),$$

$$\forall t, s \in [0,1].$$

We also obtain $h_1(s) = \frac{1}{\Gamma(10/3)}(1-s)^{5/6}(1-(1-s)^{3/2})$ for all $s \in [0,1]$, $\sigma = 1/\Gamma(10/3) + 3/(4\Delta) \approx 1.22220971$ and

$$J(s) = \begin{cases} h_1(s) + \frac{1}{\Delta}\left[\frac{3}{4}(1-s)^{5/6} + \frac{22s-9}{12}\right], & 0 \le s < \frac{1}{4}, \\ h_1(s) + \frac{1}{\Delta}\left[\frac{3}{4}(1-s)^{5/6} + \frac{5s-3}{6}\right], & \frac{1}{4} \le s < \frac{1}{2}, \\ h_1(s) + \frac{1}{\Delta}\left[\frac{3}{4}(1-s)^{5/6} + \frac{4s-3}{12}\right], & \frac{1}{2} \le s < \frac{3}{4}, \\ h_1(s) + \frac{3}{4\Delta}(1-s)^{5/6}, & \frac{3}{4} \le s \le 1. \end{cases}$$

Example 1. We consider the function

$$f(t,u) = \frac{u^2 + u + 1}{\sqrt[3]{t(1-t)^2}} + \ln t, \quad t \in (0,1), \quad u \ge 0.$$

We have $r(t) = -\ln t$ and $z(t) = \frac{1}{\sqrt[3]{t(1-t)^2}}$ for all $t \in (0,1)$, $g(t,u) = u^2 + u + 1$ for all $t \in [0,1]$ and $u \ge 0$, $\int_0^1 r(t)\,dt = 1$, $\int_0^1 z(t)\,dt = \Gamma(\frac{2}{3})\Gamma(\frac{1}{3}) \approx 3.63$. Therefore, assumption (H2) is satisfied. In addition, for $c \in (0, 1/2)$ fixed, assumption (H3) is also satisfied ($f_\infty = \infty$).

After some computations, we deduce that $\int_0^1 J(s)(z(s) + r(s))\,ds \approx 1.12036124$. We choose $R_1 = 2$, ($R_1 > \sigma \int_0^1 r(s)\,ds$), and then we obtain $M_1 = 7$ and $\lambda^* \approx 0.255$. By Theorem 2.1.1, we conclude that problem (2.17), (2.18) has at least one positive solution for any $\lambda \in (0, \lambda^*]$.

Example 2. We consider the function

$$f(t,u) = \frac{\sqrt{u+1}}{\sqrt[4]{t^3(1-t)}} - \frac{1}{\sqrt{t}}, \quad t \in (0,1), \quad u \ge 0.$$

Here, we have $r(t) = \frac{1}{\sqrt{t}}$ and $z(t) = \frac{1}{\sqrt[4]{t^3(1-t)}}$ for all $t \in (0,1)$, $g(t,u) = \sqrt{u+1}$ for all $t \in [0,1]$ and $u \geq 0$.. For $c \in (0,1/2)$ fixed, assumptions (H2) and (H4) are satisfied ($\int_0^1 r(t)\,dt = 2$, $\int_0^1 z(t)\,dt \approx 4.44$, $\lim_{u\to\infty}\min_{t\in[c,1-c]} f(t,u) = \infty$ and $g_\infty = 0$).

For $c = 1/4$, we obtain $\int_{1/4}^{3/4} J(s)\,ds \approx 0.23472325$ and $L_0 \approx 529.001$. From the proof of Theorem 2.1.2, we deduce that $M_2 \approx 91277.5$ and $\lambda_* \approx 948407$. Then by Theorem 2.1.2, we conclude that, for any $\lambda \geq \lambda_*$, our problem (2.17), (2.18) has at least one positive solution.

Remark 2.1.1. The results presented in this section, under the assumption $p \in [1, n-2]$ instead of $p \in [1, \alpha - 1)$, were published in [46]. The properties (a)–(c) of Lemma 2.1.3 are from [45].

2.2 A Fractional Differential Equation with Integral Terms and Multi-Point Boundary Conditions

In this section, we consider the nonlinear fractional integro-differential equation

$$D_{0+}^\alpha u(t) + f(t, u(t), Tu(t), Su(t)) = 0, \quad t \in (0,1), \qquad (2.19)$$

with the multi-point boundary conditions

$$u(0) = u'(0) = \cdots = u^{(n-2)}(0) = 0, \quad D_{0+}^p u(1) = \sum_{i=1}^m a_i D_{0+}^q u(\xi_i), \quad (2.20)$$

where $\alpha \in \mathbb{R}$, $\alpha \in (n-1, n]$, $n \in \mathbb{N}$, $n \geq 3$, $a_i, \xi_i \in \mathbb{R}$ for all $i = 1, \ldots, m$, ($m \in \mathbb{N}$), $0 < \xi_1 < \cdots < \xi_m \leq 1$, $p, q \in \mathbb{R}$, $p \in [1, \alpha - 1)$, $q \in [0, p]$, D_{0+}^k denotes the Riemann–Liouville derivative of order k (for $k = \alpha, p, q$), $Tu(t) = \int_0^t K(t,s)u(s)\,ds$, $Su(t) = \int_0^1 H(t,s)u(s)\,ds$ for all $t \in [0,1]$, and f is a nonnegative and nonsingular function.

We study the existence and uniqueness of nonnegative solutions for problem (2.19), (2.20), by using the Banach contraction mapping principle (Theorem 1.2.1) and the Krasnosel'skii fixed point theorem for the sum of two operators (Theorem 1.2.3). By a nonnegative solution of (2.19), (2.20), we mean a function $u \in C([0,1], \mathbb{R}_+)$ satisfying (2.19) and (2.20). In [104], the authors investigated the existence of nonnegative solutions for the Caputo fractional differential equation

$$^cD_{0+}^\alpha u(t) + f(t, u(t), Tu(t), Su(t)) = 0, \quad t \in (0, 1),$$

with the boundary conditions

$$\begin{cases} u(0) = b_0, \ u'(0) = b_1, \ldots, u^{(n-3)}(0) = b_{n-3}, \\ u^{(n-1)}(0) = b_{n-1}, \ u(1) = \mu \int_0^1 u(s)\, ds, \end{cases}$$

where $n - 1 < \alpha \leq n$, $0 \leq \mu < n-1$, $n \geq 3$, $b_i \geq 0$ $(i = 1, 2, \ldots, n-3, n-1)$, $^cD_{0+}^\alpha$ is the Caputo fractional derivative, and the operators T and S are defined as the operators from our problem, given above.

We will use in what follows the notations (Δ, G and σ) and the auxiliary results (Lemmas 2.1.1–2.1.4) from Section 2.1.

2.2.1 *Existence of nonnegative solutions*

In this section, we prove the existence of nonnegative solutions for our problem (2.19), (2.20). First, we present the assumptions that we shall use in the sequel.

(I1) $\alpha \in \mathbb{R}$, $\alpha \in (n - 1, n]$, $n \in \mathbb{N}$, $n \geq 3$, $\xi_i \in \mathbb{R}$ for all $i = 1, \ldots, m$, ($m \in \mathbb{N}$), $0 < \xi_1 < \cdots < \xi_m \leq 1$, $p, q \in \mathbb{R}$, $p \in [1, \alpha - 1)$, $q \in [0, p]$, $a_i \geq 0$ for all $i = 1, \ldots, m$, $\Delta = \frac{\Gamma(\alpha)}{\Gamma(\alpha-p)} - \frac{\Gamma(\alpha)}{\Gamma(\alpha-q)} \sum_{i=1}^m a_i \xi_i^{\alpha-q-1} > 0$.

(I2) $f : [0, 1] \times \mathbb{R}_+^3 \to \mathbb{R}_+$ is measurable with respect to t on $[0, 1]$.

(I3) There exists $\beta \in [1, +\infty]$ and the functions $a, b, c \in L^\beta((0, 1), \mathbb{R}_+)$ such that

$$|f(t, u, v, w) - f(t, \bar{u}, \bar{v}, \bar{w})| \leq a(t)|u - \bar{u}| + b(t)|v - \bar{v}| + c(t)|w - \bar{w}|,$$

a.e. $t \in (0, 1)$ and for all $u, v, w, \bar{u}, \bar{v}, \bar{w} \in \mathbb{R}_+$.

(I4) There exists $\gamma \in [1, +\infty]$ and the function $h \in L^\gamma((0, 1), \mathbb{R}_+)$ such that

$$|f(t, u, v, w)| \leq h(t), \quad \text{a.e. } t \in (0, 1), \ \forall u, v, w \in \mathbb{R}_+.$$

(I5) $K \in C(D, \mathbb{R}_+)$, $D = \{(t, s) \in [0, 1] \times [0, 1], \ t \geq s\}$, and $H \in C([0, 1] \times [0, 1], \mathbb{R}_+)$.

Remark 2.2.1. In the classical case when $\alpha = n \in \mathbb{N}$, $n \geq 3$ and $p, q \in \mathbb{N}$, $p \in [1, \alpha - 1)$, $q \in [0, p]$, the condition from assumption (I1) becomes $\Delta = \frac{(n-1)!}{(n-p-1)!} - \frac{(n-1)!}{(n-q-1)!} \sum_{i=1}^m a_i \xi_i^{n-q-1} > 0$.

We denote by $k^* = \sup_{t \in [0,1]} \int_0^t K(t,s)\,ds$ and $h^* = \sup_{t \in [0,1]} \int_0^1 H(t,s)\,ds$. We also denote by $\|\cdot\|_\gamma$ (for $\gamma \in \mathbb{R}$, $\gamma \geq 1$) and $\|\cdot\|_\infty$ the norms of the Lebesgue spaces $L^\gamma(0,1)$ and $L^\infty(0,1)$, that is $\|g\|_\gamma = (\int_0^1 |g(s)|^\gamma\,ds)^{1/\gamma}$ for $g \in L^\gamma(0,1)$, and $\|g\|_\infty = \operatorname{esssup}_{s \in (0,1)} |g(s)|$ for $g \in L^\infty(0,1)$.

Theorem 2.2.1. *Assume that* (I1)–(I5) *hold. If* $\sigma \omega_0 < 1$, *where* $\omega_0 = \|a + k^*b + h^*c\|_1$, *then problem* (2.19), (2.20) *has a unique nonnegative solution on* $[0,1]$.

Proof. For $u \in C([0,1], \mathbb{R}_+)$ and $t \in [0,1]$, by (I4), we have

(a) for $\gamma = 1$:

$$\int_0^t |(t-s)^{\alpha-1} f(s, u(s), Tu(s), Su(s))|\,ds$$

$$\leq \int_0^t |f(s, u(s), Tu(s), Su(s))|\,ds$$

$$\leq \int_0^1 h(s)\,ds = \|h\|_1;$$

(b) for $\gamma \in (1, \infty)$:

$$\int_0^t |(t-s)^{\alpha-1} f(s, u(s), Tu(s), Su(s))|\,ds$$

$$\leq \int_0^t (t-s)^{\alpha-1} h(s)\,ds$$

$$\leq \left(\int_0^t (t-s)^{\frac{(\alpha-1)\gamma}{\gamma-1}}\,ds \right)^{\frac{\gamma-1}{\gamma}} \left(\int_0^t h^\gamma(s)\,ds \right)^{1/\gamma}$$

$$\leq t^{\frac{\alpha\gamma-1}{\gamma}} \left(\frac{\gamma-1}{\alpha\gamma-1} \right)^{\frac{\gamma-1}{\gamma}} \|h\|_\gamma \leq \left(\frac{\gamma-1}{\alpha\gamma-1} \right)^{\frac{\gamma-1}{\gamma}} \|h\|_\gamma;$$

(c) for $\gamma = \infty$:

$$\int_0^t |(t-s)^{\alpha-1} f(s, u(s), Tu(s), Su(s))|\,ds$$

$$\leq \int_0^t (t-s)^{\alpha-1} h(s)\,ds$$

$$\leq \int_0^t (t-s)^{\alpha-1}\,ds \|h\|_\infty \leq \frac{1}{\alpha} \|h\|_\infty.$$

We deduce that the function $s \to (t-s)^{\alpha-1} f(s, u(s), Tu(s), Su(s))$ is Lebesgue integrable on $[0, t]$ for all $t \in [0, 1]$ and $u \in C([0, 1], \mathbb{R}_+)$. In a similar manner, we show that the function $s \to (1-s)^{\alpha-p-1} f(s, u(s), Tu(s), Su(s))$ is Lebesgue integrable on $[0, 1]$ for all $u \in C([0, 1], \mathbb{R}_+)$, and the function $s \to (\xi_i - s)^{\alpha-q-1} f(s, u(s), Tu(s), Su(s))$ is Lebesgue integrable on $[0, \xi_i]$ for all $u \in C([0, 1], \mathbb{R}_+)$.

We will consider in what follows the integral equation

$$
u(t) = -\frac{1}{\Gamma(\alpha)} \int_0^t (t-s)^{\alpha-1} f(s, u(s), Tu(s), Su(s)) \, ds
$$

$$
+ \frac{t^{\alpha-1}}{\Delta\Gamma(\alpha-p)} \int_0^1 (1-s)^{\alpha-p-1} f(s, u(s), Tu(s), Su(s)) \, ds
$$

$$
- \frac{t^{\alpha-1}}{\Delta\Gamma(\alpha-q)} \sum_{i=1}^m a_i \left(\int_0^{\xi_i} (\xi_i - s)^{\alpha-q-1} f(s, u(s), Tu(s), Su(s)) \, ds \right),
$$

(2.21)

or equivalently

$$
u(t) = \int_0^1 G(t, s) f(s, u(s), Tu(s), Su(s)) \, ds. \tag{2.22}
$$

By Lemmas 2.1.1 and 2.1.2, we know that u is a solution of problem (2.19), (2.20) if and only if u is a solution of equation (2.21) (or equivalently (2.22)).

For any $\gamma \in [1, +\infty]$, we deduce that $h \in L^1(0, 1)$. So, let $r = \sigma\|h\|_1$. We define the operator $\mathcal{A}$ on $\Xi_1 = \{u \in C([0, 1], \mathbb{R}_+), \|u\| \le r\}$, where $\|u\| = \sup_{t \in [0,1]} |u(t)|$, by

$$
\mathcal{A}u(t) = -\frac{1}{\Gamma(\alpha)} \int_0^t (t-s)^{\alpha-1} f(s, u(s), Tu(s), Su(s)) \, ds
$$

$$
+ \frac{t^{\alpha-1}}{\Delta\Gamma(\alpha-p)} \int_0^1 (1-s)^{\alpha-p-1} f(s, u(s), Tu(s), Su(s)) \, ds
$$

$$
- \frac{t^{\alpha-1}}{\Delta\Gamma(\alpha-q)} \sum_{i=1}^m a_i \left(\int_0^{\xi_i} (\xi_i - s)^{\alpha-q-1} f(s, u(s), Tu(s), Su(s)) \, ds \right),
$$

or equivalently

$$
\mathcal{A}u(t) = \int_0^1 G(t, s) f(s, u(s), Tu(s), Su(s)) \, ds.
$$

The function u is a solution of equation (2.21) (or (2.22)) if and only if u is a fixed point of operator $\mathcal{A}$.

We will investigate the existence and uniqueness of fixed points of operator $\mathcal{A}$ by using the Banach contraction mapping principle.

First, we show that for $u \in \Xi_1$, then $\mathcal{A}u \in \Xi_1$. Indeed, we have

$$|\mathcal{A}u(t+\delta) - \mathcal{A}u(t)| = \left| \int_0^1 [G(t+\delta, s) - G(t,s)] f(s, u(s), Tu(s), Su(s)) \, ds \right|$$

$$\leq \int_0^1 |G(t+\delta, s) - G(t,s)| h(s) \, ds$$

$$\leq \begin{cases} \max_{s \in [0,1]} |G(t+\delta, s) - G(t,s)| \cdot \|h\|_1 \to 0 \\ \qquad \text{for } \delta \to 0, \ \text{if } \gamma = 1; \\ \left(\int_0^1 |G(t+\delta, s) - G(t,s)|^{\frac{\gamma}{\gamma-1}} \, ds \right)^{\frac{\gamma-1}{\gamma}} \|h\|_\gamma \to 0 \\ \qquad \text{for } \delta \to 0, \ \text{if } \gamma \in (1, \infty); \\ \int_0^1 |G(t+\delta, s) - G(t,s)| \, ds \cdot \|h\|_\infty \to 0 \\ \qquad \text{for } \delta \to 0, \ \text{if } \gamma = \infty. \end{cases}$$

So, $\mathcal{A}u$ is a continuous function. By (I1), (I2) and Lemma 2.1.4, we obtain $\mathcal{A}u(t) \geq 0$ for all $t \in [0,1]$, so $\mathcal{A}u \in C([0,1], \mathbb{R}_+)$.

Moreover, for any $u \in \Xi_1$ and all $t \in [0,1]$, we deduce

$$(\mathcal{A}u)(t) = \int_0^1 G(t,s) f(s, u(s), Tu(s), Su(s)) \, ds$$

$$\leq \int_0^1 \sigma t^{\alpha-1} f(s, u(s), Tu(s), Su(s)) \, ds$$

$$\leq \sigma t^{\alpha-1} \int_0^1 h(s) \, ds \leq \sigma \|h\|_1 = r,$$

and then $\|\mathcal{A}u\| \leq r$ for all $u \in \Xi_1$, so $\mathcal{A} : \Xi_1 \to \Xi_1$.

Then we show that $\mathcal{A}$ is a contraction mapping on Ξ_1. For $u_1, u_2 \in \Xi_1$, and any $t \in [0,1]$, by using (I3), we obtain

$$|(\mathcal{A}u_1)(t) - (\mathcal{A}u_2)(t)|$$

$$= \left| \int_0^1 G(t,s) f(s, u_1(s), Tu_1(s), Su_1(s)) \, ds \right.$$

$$\left. - \int_0^1 G(t,s) f(s, u_2(s), Tu_2(s), Su_2(s)) \, ds \right|$$

$$\leq \int_0^1 G(t,s)|f(s,u_1(s),Tu_1(s),Su_1(s))$$

$$- f(s,u_2(s),Tu_2(s),Su_2(s))|\,ds$$

$$\leq \int_0^1 G(t,s)\,[a(s)|u_1(s)-u_2(s)|+b(s)|Tu_1(s)$$

$$- Tu_2(s)|+c(s)|Su_1(s)-Su_2(s)|]\,ds$$

$$\leq \|u_1-u_2\|\int_0^1 G(t,s)\,[a(s)+k^*b(s)+h^*c(s)]\,ds$$

$$\leq \|u_1-u_2\|\sigma t^{\alpha-1}\int_0^1 [a(s)+k^*b(s)+h^*c(s)]\,ds$$

$$\leq \sigma\|u_1-u_2\|\cdot\|a+k^*b+h^*c\|_1 = \sigma\omega_0\|u_1-u_2\|,$$

because

$$Tu_1(s)-Tu_2(s)=\int_0^s K(s,\tau)[u_1(\tau)-u_2(\tau)]\,d\tau$$

$$\leq \sup_{\tau\in[0,1]}|u_1(\tau)-u_2(\tau)|\int_0^s K(s,\tau)\,d\tau$$

$$\leq k^*\|u_1-u_2\|,\ \forall s\in[0,1],$$

$$Su_1(s)-Su_2(s)=\int_0^1 H(s,\tau)[u_1(\tau)-u_2(\tau)]\,d\tau$$

$$\leq \sup_{\tau\in[0,1]}|u_1(\tau)-u_2(\tau)|\int_0^1 H(s,\tau)\,d\tau$$

$$\leq h^*\|u_1-u_2\|,\ \forall s\in[0,1],$$

and $a,b,c\in L^\beta(0,1)\subset L^1(0,1)$.
    Then we deduce

$$\|\mathcal{A}u_1-\mathcal{A}u_2\|\leq \sigma\omega_0\|u_1-u_2\|.$$

Because $\sigma\omega_0<1$, we conclude that $\mathcal{A}$ is a contraction mapping. By Theorem 1.2.1, we deduce that $\mathcal{A}$ has a unique fixed point, which is the unique nonnegative solution of problem (2.19), (2.20).   □

Next, we denote by

$$\omega_1 = \frac{1}{\Gamma(\alpha)} \int_0^1 (1-s)^{\alpha-1}(a(s) + k^* b(s) + h^* c(s))\, ds,$$

$$\omega_2 = \begin{cases} \|a + k^* b + h^* c\|_1 \left(\dfrac{1}{\Delta\Gamma(\alpha - p)} + \dfrac{1}{\Delta\Gamma(\alpha - q)} \sum_{i=1}^m a_i \right), & \text{if } \beta = 1, \\[2ex] \|a + k^* b + h^* c\|_\beta \left[\dfrac{1}{\Delta\Gamma(\alpha - p)} \left(\dfrac{\beta - 1}{\alpha\beta - p\beta - 1} \right)^{\frac{\beta-1}{\beta}} \right. \\[2ex] \quad + \dfrac{1}{\Delta\Gamma(\alpha - q)} \left(\sum_{i=1}^m a_i \xi_i^{\frac{\alpha\beta - q\beta - 1}{\beta}} \right) \left(\dfrac{\beta - 1}{\alpha\beta - q\beta - 1} \right)^{\frac{\beta-1}{\beta}} \left. \right], \\[1ex] \quad \text{if } \beta \in (1, \infty), \\[2ex] \|a + k^* b + h^* c\|_\infty \left(\dfrac{1}{\Delta\Gamma(\alpha - p + 1)} + \dfrac{1}{\Delta\Gamma(\alpha - q + 1)} \sum_{i=1}^m a_i \xi_i^{\alpha - q} \right), \\[1ex] \quad \text{if } \beta = \infty. \end{cases}$$

$$(2.23)$$

Theorem 2.2.2. *Assume that assumptions* (I1),

(I2)′ $f : [0,1] \times \mathbb{R}_+^3 \to \mathbb{R}_+$ *is a continuous function,*

and (I3)–(I5) *hold. If* $\max\{\omega_1, \omega_2\} < 1$, *then problem* (2.19), (2.20) *has at least one nonnegative solution on* $[0,1]$.

Proof. We choose $R \geq R_0$, where

$$R_0 = \begin{cases} (1 - \omega_1)^{-1} \left[\dfrac{\psi_0}{\Gamma(\alpha + 1)} + \|h\|_1 \left(\dfrac{1}{\Delta\Gamma(\alpha - p)} + \dfrac{1}{\Delta\Gamma(\alpha - q)} \sum_{i=1}^m a_i \right) \right], \\[1ex] \quad \text{if } \gamma = 1, \\[2ex] (1 - \omega_1)^{-1} \left[\dfrac{\psi_0}{\Gamma(\alpha + 1)} + \|h\|_\gamma \left(\dfrac{1}{\Delta\Gamma(\alpha - p)} \left(\dfrac{\gamma - 1}{\alpha\gamma - p\gamma - 1} \right)^{\frac{\gamma-1}{\gamma}} \right. \right. \\[2ex] \quad \left. \left. + \dfrac{1}{\Delta\Gamma(\alpha - q)} \left(\dfrac{\gamma - 1}{\alpha\gamma - q\gamma - 1} \right)^{\frac{\gamma-1}{\gamma}} \sum_{i=1}^m a_i \xi_i^{\frac{\alpha\gamma - q\gamma - 1}{\gamma}} \right) \right], \\[1ex] \quad \text{if } \gamma \in (1, \infty), \\[2ex] (1 - \omega_1)^{-1} \left[\dfrac{\psi_0}{\Gamma(\alpha + 1)} + \|h\|_\infty \right. \\[2ex] \quad \left. \times \left(\dfrac{1}{\Delta\Gamma(\alpha - p + 1)} + \dfrac{1}{\Delta\Gamma(\alpha - q + 1)} \sum_{i=1}^m a_i \xi_i^{\alpha - q} \right) \right], \\[1ex] \quad \text{if } \gamma = \infty, \end{cases}$$

$\psi_0 = \max\{f(t, 0, 0, 0), \ t \in [0, 1]\}$, and ω_1 is given by (2.23). We consider the set $\Xi_2 = \{u \in C([0, 1], \mathbb{R}_+), \ \|u\| \le R\}$. Then Ξ_2 is a closed, convex, bounded and nonempty subset of $C([0, 1], \mathbb{R}_+)$. We define the operators $\mathcal{B}$ and $\mathcal{C}$ on Ξ_2 by

$$(\mathcal{B}u)(t) = -\frac{1}{\Gamma(\alpha)} \int_0^t (t - s)^{\alpha-1} f(s, u(s), Tu(s), Su(s))\, ds, \quad t \in [0, 1],$$

$$(\mathcal{C}u)(t) = \frac{t^{\alpha-1}}{\Delta\Gamma(\alpha - p)} \int_0^1 (1 - s)^{\alpha-p-1} f(s, u(s), Tu(s), Su(s))\, ds$$

$$-\frac{t^{\alpha-1}}{\Delta\Gamma(\alpha - q)} \sum_{i=1}^m a_i$$

$$\times \left(\int_0^{\xi_i} (\xi_i - s)^{\alpha-q-1} f(s, u(s), Tu(s), Su(s))\, ds \right), \quad t \in [0, 1].$$

By (I1), (I2)' and Lemma 2.1.4, we have $(\mathcal{B}u)(t) + (\mathcal{C}u)(t) \ge 0$ for all $t \in [0, 1]$. For any $u \in \Xi_2$, by using (I3), we deduce

$$|f(t, u(t), Tu(t), Su(t))| \le |f(t, u(t), Tu(t), Su(t)) - f(t, 0, 0, 0)|$$

$$+ |f(t, 0, 0, 0)|$$

$$\le a(t)|u(t)| + b(t)|Tu(t)| + c(t)|Su(t)|$$

$$+ \psi_0, \quad \forall t \in [0, 1].$$

Then for any $u \in \Xi_2$ and all $t \in [0, 1]$, we obtain by using the above inequality

$$|(\mathcal{B}u)(t)| = \left| -\frac{1}{\Gamma(\alpha)} \int_0^t (t - s)^{\alpha-1} f(s, u(s), Tu(s), Su(s))\, ds \right|$$

$$\le \frac{1}{\Gamma(\alpha)} \int_0^t (t - s)^{\alpha-1} (a(s)|u(s)| + b(s)|Tu(s)|$$

$$+ c(s)|Su(s)| + \psi_0)\, ds$$

$$\le \frac{\|u\|}{\Gamma(\alpha)} \int_0^1 (1 - s)^{\alpha-1} (a(s) + k^* b(s) + h^* c(s))\, ds$$

$$+ \frac{1}{\Gamma(\alpha)} \int_0^t (t - s)^{\alpha-1} \psi_0\, ds$$

$$\le \omega_1 R + \frac{t^\alpha \psi_0}{\Gamma(\alpha + 1)} \le \omega_1 R + \frac{\psi_0}{\Gamma(\alpha + 1)},$$

because $(Tu)(t) \le \sup_{t \in [0,1]} u(t) \int_0^t K(t,s)\, ds \le k^* \|u\|$ and $(Su)(t) \le \sup_{t \in [0,1]} u(t) \int_0^1 H(t,s)\, ds \le h^* \|u\|$ for all $t \in [0,1]$.

Then for any $v \in \Xi_2$ and all $t \in [0,1]$, we deduce, by (I4)

(a) for $\gamma = 1$:

$$
\begin{aligned}
|Cv(t)| &\le \frac{t^{\alpha-1}}{\Delta\Gamma(\alpha-p)} \int_0^1 (1-s)^{\alpha-p-1} f(s, v(s), Tv(s), Sv(s))\, ds \\
&\quad + \frac{t^{\alpha-1}}{\Delta\Gamma(\alpha-q)} \sum_{i=1}^m a_i \\
&\quad \times \left(\int_0^{\xi_i} (\xi_i - s)^{\alpha-q-1} f(s, v(s), Tv(s), Sv(s))\, ds \right) \\
&\le \frac{t^{\alpha-1}}{\Delta\Gamma(\alpha-p)} \int_0^1 (1-s)^{\alpha-p-1} h(s)\, ds \\
&\quad + \frac{t^{\alpha-1}}{\Delta\Gamma(\alpha-q)} \sum_{i=1}^m a_i \int_0^{\xi_i} (\xi_i - s)^{\alpha-q-1} h(s)\, ds \\
&\le \frac{1}{\Delta\Gamma(\alpha-p)} \int_0^1 h(s)\, ds + \frac{1}{\Delta\Gamma(\alpha-q)} \sum_{i=1}^m a_i \int_0^1 h(s)\, ds \\
&= \|h\|_1 \left(\frac{1}{\Delta\Gamma(\alpha-p)} + \frac{1}{\Delta\Gamma(\alpha-q)} \sum_{i=1}^m a_i \right);
\end{aligned}
$$

(b) for $\gamma \in (1, \infty)$:

$$
\begin{aligned}
|Cv(t)| &\le \frac{t^{\alpha-1}}{\Delta\Gamma(\alpha-p)} \int_0^1 (1-s)^{\alpha-p-1} h(s)\, ds \\
&\quad + \frac{t^{\alpha-1}}{\Delta\Gamma(\alpha-q)} \sum_{i=1}^m a_i \int_0^{\xi_i} (\xi_i - s)^{\alpha-q-1} h(s)\, ds \\
&\le \frac{1}{\Delta\Gamma(\alpha-p)} \left(\int_0^1 (1-s)^{\frac{\gamma(\alpha-p-1)}{\gamma-1}}\, ds \right)^{\frac{\gamma-1}{\gamma}} \left(\int_0^1 h^\gamma(s)\, ds \right)^{\frac{1}{\gamma}} \\
&\quad + \frac{1}{\Delta\Gamma(\alpha-q)} \sum_{i=1}^m a_i \left(\int_0^{\xi_i} (\xi_i - s)^{\frac{\gamma(\alpha-q-1)}{\gamma-1}}\, ds \right)^{\frac{\gamma-1}{\gamma}} \left(\int_0^1 h^\gamma(s)\, ds \right)^{\frac{1}{\gamma}}
\end{aligned}
$$

$$= \|h\|_\gamma \left(\frac{1}{\Delta\Gamma(\alpha - p)} \left(\frac{\gamma - 1}{\alpha\gamma - p\gamma - 1} \right)^{\frac{\gamma-1}{\gamma}} \right.$$

$$\left. + \frac{1}{\Delta\Gamma(\alpha - q)} \left(\frac{\gamma - 1}{\alpha\gamma - q\gamma - 1} \right)^{\frac{\gamma-1}{\gamma}} \sum_{i=1}^{m} a_i \xi_i^{\frac{\alpha\gamma - q\gamma - 1}{\gamma}} \right);$$

(c) for $\gamma = \infty$:

$$|\mathcal{C}v(t)| \leq \frac{t^{\alpha-1}}{\Delta\Gamma(\alpha - p)} \int_0^1 (1 - s)^{\alpha-p-1} h(s)\, ds$$

$$+ \frac{t^{\alpha-1}}{\Delta\Gamma(\alpha - q)} \sum_{i=1}^{m} a_i \left(\int_0^{\xi_i} (\xi_i - s)^{\alpha-q-1} h(s)\, ds \right)$$

$$\leq \frac{1}{\Delta\Gamma(\alpha - p)} \|h\|_\infty \int_0^1 (1 - s)^{\alpha-p-1}\, ds$$

$$+ \frac{1}{\Delta\Gamma(\alpha - q)} \|h\|_\infty \sum_{i=1}^{m} a_i \int_0^{\xi_i} (\xi_i - s)^{\alpha-q-1}\, ds$$

$$= \|h\|_\infty \left(\frac{1}{\Delta\Gamma(\alpha - p + 1)} + \frac{1}{\Delta\Gamma(\alpha - q + 1)} \sum_{i=1}^{m} a_i \xi_i^{\alpha-q} \right).$$

Then, for $u, v \in \Xi_2$, and $t \in [0, 1]$, we have
(a) for $\gamma = 1$:

$$|\mathcal{B}u(t) + \mathcal{C}v(t)| \leq |\mathcal{B}u(t)| + |\mathcal{C}v(t)| \leq \omega_1 R + \frac{\psi_0}{\Gamma(\alpha + 1)}$$

$$+ \|h\|_1 \left(\frac{1}{\Delta\Gamma(\alpha - p)} + \frac{1}{\Delta\Gamma(\alpha - q)} \sum_{i=1}^{m} a_i \right) \leq R;$$

(b) for $\gamma \in (1, \infty)$:

$$|\mathcal{B}u(t) + \mathcal{C}v(t)| \leq |\mathcal{B}u(t)| + |\mathcal{C}v(t)| \leq \omega_1 R + \frac{\psi_0}{\Gamma(\alpha + 1)}$$

$$+ \|h\|_\gamma \left(\frac{1}{\Delta\Gamma(\alpha - p)} \left(\frac{\gamma - 1}{\alpha\gamma - p\gamma - 1} \right)^{\frac{\gamma-1}{\gamma}} \right.$$

$$\left. + \frac{1}{\Delta\Gamma(\alpha - q)} \left(\frac{\gamma - 1}{\alpha\gamma - q\gamma - 1} \right)^{\frac{\gamma-1}{\gamma}} \sum_{i=1}^{m} a_i \xi_i^{\frac{\alpha\gamma - q\gamma - 1}{\gamma}} \right) \leq R;$$

(c) for $\gamma = \infty$:

$$|\mathcal{B}u(t) + \mathcal{C}v(t)| \leq |\mathcal{B}u(t)| + |\mathcal{C}v(t)| \leq \omega_1 R + \frac{\psi_0}{\Gamma(\alpha+1)}$$

$$+ \|h\|_\infty \left(\frac{1}{\Delta\Gamma(\alpha-p+1)} + \frac{1}{\Delta\Gamma(\alpha-q+1)} \sum_{i=1}^{m} a_i \xi_i^{\alpha-q} \right) \leq R.$$

Therefore, for $v_1, v_2 \in \Xi_2$ and $t \in [0,1]$, by using (I3), we deduce

$$|\mathcal{C}v_1(t) - \mathcal{C}v_2(t)| \leq \frac{t^{\alpha-1}}{\Delta\Gamma(\alpha-p)} \int_0^1 (1-s)^{\alpha-p-1} |f(s, v_1(s), Tv_1(s), Sv_1(s))$$

$$- f(s, v_2(s), Tv_2(s), Sv_2(s))| \, ds$$

$$+ \frac{t^{\alpha-1}}{\Delta\Gamma(\alpha-q)} \sum_{i=1}^{m} a_i \left(\int_0^{\xi_i} (\xi_i - s)^{\alpha-q-1} |f(s, v_1(s), \right.$$

$$\left. Tv_1(s), Sv_1(s)) - f(s, v_2(s), Tv_2(s), Sv_2(s))| \, ds \right)$$

$$\leq \frac{1}{\Delta\Gamma(\alpha-p)} \int_0^1 (1-s)^{\alpha-p-1} \left(a(s)|v_1(s) - v_2(s)| \right.$$

$$+ b(s)|Tv_1(s) - Tv_2(s)| + c(s)|Sv_1(s) - Sv_2(s)| \big) \, ds$$

$$+ \frac{1}{\Delta\Gamma(\alpha-q)} \sum_{i=1}^{m} a_i \left[\int_0^{\xi_i} (\xi_i - s)^{\alpha-q-1} \left(a(s)|v_1(s) \right. \right.$$

$$- v_2(s)| + b(s)|Tv_1(s) - Tv_2(s)|$$

$$+ c(s)|Sv_1(s) - Sv_2(s)| \big) \, ds \Big]$$

$$\leq \frac{\|v_1 - v_2\|}{\Delta\Gamma(\alpha-p)} \int_0^1 (1-s)^{\alpha-p-1} (a(s) + k^* b(s) + h^* c(s)) \, ds$$

$$+ \frac{\|v_1 - v_2\|}{\Delta\Gamma(\alpha-q)} \sum_{i=1}^{m} a_i \left[\int_0^{\xi_i} (\xi_i - s)^{\alpha-q-1} (a(s) \right.$$

$$+ k^* b(s) + h^* c(s)) \, ds \Big].$$

So, we obtain

($\widetilde{a}$) for $\beta = 1$:

$$|\mathcal{C}v_1(t) - \mathcal{C}v_2(t)| \leq \frac{\|v_1 - v_2\|}{\Delta\Gamma(\alpha - p)} \int_0^1 (a(s) + k^*b(s) + h^*c(s))\, ds$$

$$+ \frac{\|v_1 - v_2\|}{\Delta\Gamma(\alpha - q)} \sum_{i=1}^m a_i \int_0^1 (a(s) + k^*b(s) + h^*c(s))\, ds$$

$$= \|v_1 - v_2\| \cdot \|a + k^*b + h^*c\|_1$$

$$\times \left(\frac{1}{\Delta\Gamma(\alpha - p)} + \frac{1}{\Delta\Gamma(\alpha - q)} \sum_{i=1}^m a_i \right) = \omega_2 \|v_1 - v_2\|;$$

($\widetilde{b}$) for $\beta \in (1, \infty)$:

$$|\mathcal{C}v_1(t) - \mathcal{C}v_2(t)| \leq \frac{\|v_1 - v_2\|}{\Delta\Gamma(\alpha - p)} \left(\int_0^1 (1 - s)^{\frac{\beta(\alpha - p - 1)}{\beta - 1}}\, ds \right)^{\frac{\beta - 1}{\beta}}$$

$$\times \left(\int_0^1 (a(s) + k^*b(s) + h^*c(s))^\beta\, ds \right)^{\frac{1}{\beta}}$$

$$+ \frac{\|v_1 - v_2\|}{\Delta\Gamma(\alpha - q)} \sum_{i=1}^m a_i \left(\int_0^{\xi_i} (\xi_i - s)^{\frac{\beta(\alpha - q - 1)}{\beta - 1}}\, ds \right)^{\frac{\beta - 1}{\beta}}$$

$$\times \left(\int_0^1 (a(s) + k^*b(s) + h^*c(s))^\beta\, ds \right)^{\frac{1}{\beta}}$$

$$= \|v_1 - v_2\| \cdot \|a + k^*b + h^*c\|_\beta$$

$$\times \left[\frac{1}{\Delta\Gamma(\alpha - p)} \left(\frac{\beta - 1}{\alpha\beta - p\beta - 1} \right)^{\frac{\beta - 1}{\beta}} \right.$$

$$\left. + \frac{1}{\Delta\Gamma(\alpha - q)} \sum_{i=1}^m a_i \xi_i^{\frac{\alpha\beta - q\beta - 1}{\beta}} \left(\frac{\beta - 1}{\alpha\beta - q\beta - 1} \right)^{\frac{\beta - 1}{\beta}} \right]$$

$$= \omega_2 \|v_1 - v_2\|;$$

($\widetilde{c}$) for $\beta = \infty$:

$$|\mathcal{C}v_1(t) - \mathcal{C}v_2(t)| \leq \frac{\|v_1 - v_2\|}{\Delta\Gamma(\alpha - p)} \|a + k^*b + h^*c\|_\infty \int_0^1 (1 - s)^{\alpha - p - 1}\, ds$$

$$+ \frac{\|v_1 - v_2\|}{\Delta\Gamma(\alpha - q)} \sum_{i=1}^m a_i \|a + k^*b + h^*c\|_\infty \int_0^{\xi_i} (\xi_i - s)^{\alpha - q - 1}\, ds$$

$$= \|v_1 - v_2\| \cdot \|a + k^* b + h^* c\|_\infty$$

$$\times \left(\frac{1}{\Delta \Gamma(\alpha - p + 1)} + \frac{1}{\Delta \Gamma(\alpha - q + 1)} \sum_{i=1}^{m} a_i \xi_i^{\alpha - q} \right)$$

$$= \omega_2 \|v_1 - v_2\|,$$

where ω_2 is given by (2.23).

Because $\omega_2 < 1$, we deduce that $\mathcal{C}$ is a contraction mapping.

By using assumptions (I2)' and (I5), we conclude that $\mathcal{B}$ is a continuous mapping. In addition, $\mathcal{B}$ is uniformly bounded on Ξ_2, because for any $u \in \Xi_2$, we have

$$|\mathcal{B}u(t)| = \left| \frac{1}{\Gamma(\alpha)} \int_0^t (t - s)^{\alpha - 1} f(s, u(s), Tu(s), Su(s)) \, ds \right|$$

$$\leq \frac{1}{\Gamma(\alpha)} \int_0^t (t - s)^{\alpha - 1} h(s) \, ds \leq \frac{1}{\Gamma(\alpha)} \int_0^1 h(s) \, ds$$

$$= \frac{1}{\Gamma(\alpha)} \|h\|_1, \quad \forall t \in [0, 1],$$

so $\|\mathcal{B}u\| \leq \frac{1}{\Gamma(\alpha)} \|h\|_1$ for all $u \in \Xi_2$.

The operator $\mathcal{B}$ is also equicontinuous on Ξ_2. Indeed, let $u \in \Xi_2$, $t_1, t_2 \in [0, 1]$, with $t_1 < t_2$. We obtain

$$|\mathcal{B}u(t_2) - \mathcal{B}u(t_1)| = \left| \frac{1}{\Gamma(\alpha)} \int_0^{t_2} (t_2 - s)^{\alpha - 1} f(s, u(s), Tu(s), Su(s)) \, ds \right.$$

$$\left. - \frac{1}{\Gamma(\alpha)} \int_0^{t_1} (t_1 - s)^{\alpha - 1} f(s, u(s), Tu(s), Su(s)) \, ds \right|$$

$$\leq \frac{1}{\Gamma(\alpha)} \int_0^{t_1} \left[(t_2 - s)^{\alpha - 1} - (t_1 - s)^{\alpha - 1} \right]$$

$$\times f(s, u(s), Tu(s), Su(s)) \, ds$$

$$+ \frac{1}{\Gamma(\alpha)} \int_{t_1}^{t_2} (t_2 - s)^{\alpha - 1} f(s, u(s), Tu(s), Su(s)) \, ds$$

$$\leq \psi_1 \left\{ \frac{1}{\Gamma(\alpha)} \int_0^{t_1} \left[(t_2 - s)^{\alpha - 1} - (t_1 - s)^{\alpha - 1} \right] ds \right.$$

$$\left. + \frac{1}{\Gamma(\alpha)} \int_{t_1}^{t_2} (t_2 - s)^{\alpha - 1} \, ds \right\}$$

$$= \frac{\psi_1}{\Gamma(\alpha + 1)} (t_2^\alpha - t_1^\alpha) \leq \frac{\psi_1 (t_2 - t_1)}{\Gamma(\alpha)},$$

where $\psi_1 = \sup\{f(t, u, v, w), \ t \in [0,1], \ |u| \leq R, \ |v| \leq k^*R, \ |w| \leq h^*R\}$.
So, it is clear that $|\mathcal{B}u(t_2) - \mathcal{B}u(t_1)| \to 0$ as $t_2 - t_1 \to 0$.

By using the Arzela–Ascoli theorem, we deduce that $\mathcal{B}(\Xi_2)$ is relatively compact, and so $\mathcal{B}$ is completely continuous. By Theorem 1.2.3, we conclude that operator $\mathcal{B} + \mathcal{C}$ has at least one fixed point, and then problem (2.19), (2.20) has at least one nonnegative solution. □

2.2.2   An example

Let $\alpha = 10/3$ ($n = 4$), $p = 3/2$, $q = 4/3$, $m = 3$, $\xi_1 = 1/4$, $\xi_2 = 1/2$, $\xi_3 = 3/4$, $a_1 = 1$, $a_2 = 1/2$, $a_3 = 1/3$.

We consider the fractional differential equation

$$D_{0+}^{10/3}u(t) + f(t, u(t), Tu(t), Su(t)) = 0, \quad t \in (0,1), \qquad (2.24)$$

with the boundary conditions

$$\begin{cases} u(0) = u'(0) = u''(0) = 0, \\ D_{0+}^{3/2}u(1) = D_{0+}^{4/3}u\left(\dfrac{1}{4}\right) + \dfrac{1}{2}D_{0+}^{4/3}u\left(\dfrac{1}{2}\right) + \dfrac{1}{3}D_{0+}^{4/3}u\left(\dfrac{3}{4}\right), \end{cases} \qquad (2.25)$$

where $Tu(t) = \int_0^t K(t,s)u(s)\,ds$ and $Su(t) = \int_0^1 H(t,s)u(s)\,ds$ for all $t \in [0,1]$, with $K(t,s) = e^{-t}s$ for all $t, s \in [0,1]$ with $s \leq t$, and $H(t,s) = e^{-2t}(s+1)$ for all $t, s \in [0,1]$.

Then we obtain $\Delta = \Gamma(10/3)(1/\Gamma(11/6) - 3/4) \approx 0.86980822$, and $\sigma = 1/\Gamma(10/3) + 3/(4\Delta) \approx 1.22220971$. So, assumptions (I1) and (I5) are satisfied.

We define the function

$$f(t, u, v, w) = \frac{t}{2} + \frac{e^{-3t}\,u}{(1+k)(1+u)} + \frac{e^{-2t}v}{(1+k^2)(1+v)} + \frac{e^{-t}w}{(1+k^3)(1+w)},$$

for all $t \in [0,1]$ and $u, v, w \in \mathbb{R}_+$, with $k \geq 1$. We deduce that $k^* = \sup_{t \in [0,1]} \int_0^t K(t,s)\,ds = \frac{1}{2e}$ and $h^* = \sup_{t \in [0,1]} \int_0^1 H(t,s)\,ds = \frac{3}{2}$. In addition, we obtain the inequalities

$$|f(t,u,v,w) - f(t,\bar{u},\bar{v},\bar{w})| \leq \frac{e^{-3t}}{1+k}|u - \bar{u}| + \frac{e^{-2t}}{1+k^2}|v - \bar{v}| + \frac{e^{-t}}{1+k^3}|w - \bar{w}|,$$

for all $t \in [0,1]$, $u, v, w, \bar{u}, \bar{v}, \bar{w} \in \mathbb{R}_+$, and

$$|f(t,u,v,w)| \leq \frac{t}{2} + \frac{e^{-3t}}{1+k} + \frac{e^{-2t}}{1+k^2} + \frac{e^{-t}}{1+k^3}, \quad \forall t \in [0,1], \ u, v, w \in \mathbb{R}_+.$$

We define $a(t) = \frac{e^{-3t}}{1+k}$, $b(t) = \frac{e^{-2t}}{1+k^2}$, $c(t) = \frac{e^{-t}}{1+k^3}$, and $h(t) = \frac{t}{2} + \frac{e^{-3t}}{1+k} + \frac{e^{-2t}}{1+k^2} + \frac{e^{-t}}{1+k^3}$, for all $t \in [0,1]$. For β, $\gamma \in [1,\infty]$, we have $a, b, c \in L^\beta(0,1)$ and $h \in L^\gamma(0,1)$. So, assumptions (I2)–(I4) are also satisfied.

Besides, we obtain $\omega_0 = \frac{e^3-1}{3e^3(1+k)} + \frac{e^2-1}{4e^3(1+k^2)} + \frac{3(e-1)}{2e(1+k^3)} \leq \frac{e^3-1}{6e^3} + \frac{e^2-1}{8e^3} + \frac{3(e-1)}{4e} \approx 0.67222$, and so $\omega_0 < 1/\sigma \approx 0.81819$. Therefore, by Theorem 2.2.1, we deduce that problem (2.24), (2.25) has a unique nonnegative and nontrivial solution.

Remark 2.2.2. The results presented in this section were published in [47].

2.3 Semipositone Singular Fractional Boundary Value Problems with Integral Boundary Conditions

In this section, we consider the nonlinear fractional differential equation

$$D_{0+}^\alpha u(t) + f(t, u(t)) = 0, \quad t \in (0,1), \tag{2.26}$$

with the integral boundary conditions

$$u(0) = u'(0) = \cdots = u^{(n-2)}(0) = 0, \quad D_{0+}^p u(1) = \int_0^1 D_{0+}^q u(t) \, dH_0(t), \tag{2.27}$$

where $\alpha \in \mathbb{R}$, $\alpha \in (n-1, n]$, $n \in \mathbb{N}$, $n \geq 3$, $p, q \in \mathbb{R}$, $p \in [1, \alpha - 1)$, $q \in [0, p]$, D_{0+}^k denotes the Riemann–Liouville derivative of order k (for $k = \alpha, p, q$), the integral from the boundary condition (2.27) is the Riemann–Stieltjes integral with H_0 a function of bounded variation, the nonlinearity $f(t, u)$ may change sign and may be singular at the points $t = 0$, $t = 1$ and/or $u = 0$.

We present conditions on the functions f and H_0 such that problem (2.26), (2.27) has at least one positive solution or at least m positive solutions ($m \in \mathbb{N}$, $n \geq 2$). In the proof of our main results, we use an approximation method and the Guo-Krasnosel'skii fixed point theorem (Theorem 1.2.2). By a positive solution of (2.26), (2.27) we mean a function $u \in C[0,1]$ satisfying (2.26) and (2.27) with $u(t) > 0$ for all $t \in (0,1]$. This problem is a generalization of the problem studied in [93], where the function H_0 is a step function, that is, the boundary conditions are of the form

$$u(0) = u'(0) = \cdots = u^{(n-2)}(0) = 0, \quad D_{0+}^p u(1) = \sum_{i=1}^{M} a_i D_{0+}^q u(\xi_i),$$

and $a_i, \xi_i \in \mathbb{R}$ for all $i = 1, \ldots, M$, $(M \in \mathbb{N})$, $0 < \xi_1 < \cdots < \xi_M < 1$.

2.3.1   *Preliminary results*

We consider the fractional differential equation

$$D_{0+}^\alpha u(t) + \widetilde{x}(t) = 0, \quad t \in (0, 1), \tag{2.28}$$

with the integral boundary conditions (2.27), where $\widetilde{x} \in C(0, 1) \cap L^1(0, 1)$. We denote by $\Delta_1 = \frac{\Gamma(\alpha)}{\Gamma(\alpha-p)} - \frac{\Gamma(\alpha)}{\Gamma(\alpha-q)} \int_0^1 s^{\alpha-q-1} dH_0(s)$.

Lemma 2.3.1. *If $\Delta_1 \neq 0$, then the unique solution $u \in C[0, 1]$ of problem* (2.28), (2.27) *is given by*

$$u(t) = -\frac{1}{\Gamma(\alpha)} \int_0^t (t - s)^{\alpha-1} \widetilde{x}(s)\, ds + \frac{t^{\alpha-1}}{\Delta_1 \Gamma(\alpha - p)} \int_0^1 (1 - s)^{\alpha-p-1} \widetilde{x}(s)\, ds$$

$$- \frac{t^{\alpha-1}}{\Delta_1 \Gamma(\alpha - q)} \int_0^1 \left(\int_0^s (s - \tau)^{\alpha-q-1} \widetilde{x}(\tau)\, d\tau \right) dH_0(s), \quad t \in [0, 1]. \tag{2.29}$$

Proof. By Lemma 1.1.3, we deduce that the solutions $u \in C(0, 1) \cap L^1(0, 1)$ of the fractional differential equation (2.28) are given by

$$u(t) = -I_{0+}^\alpha \widetilde{x}(t) + c_1 t^{\alpha-1} + \cdots + c_n t^{\alpha-n}$$

$$= -\frac{1}{\Gamma(\alpha)} \int_0^t (t - s)^{\alpha-1} \widetilde{x}(s)\, ds + c_1 t^{\alpha-1} + \cdots + c_n t^{\alpha-n},$$

where $c_1, c_2, \ldots, c_n \in \mathbb{R}$. By using the conditions $u(0) = u'(0) = \cdots = u^{(n-2)}(0) = 0$, we obtain $c_2 = \cdots = c_n = 0$. Then we conclude

$$u(t) = c_1 t^{\alpha-1} - \frac{1}{\Gamma(\alpha)} \int_0^t (t - s)^{\alpha-1} \widetilde{x}(s)\, ds, \quad t \in [0, 1]. \tag{2.30}$$

For the obtained function (2.30), we find

$$D_{0+}^p u(t) = c_1 \frac{\Gamma(\alpha)}{\Gamma(\alpha - p)} t^{\alpha-p-1} - I_{0+}^{\alpha-p} \widetilde{x}(t),$$

$$D_{0+}^q u(t) = c_1 \frac{\Gamma(\alpha)}{\Gamma(\alpha - q)} t^{\alpha-q-1} - I_{0+}^{\alpha-q} \widetilde{x}(t).$$

Then the condition $D_{0+}^p u(1) = \int_0^1 D_{0+}^q u(t)\, dH_0(t)$ gives us

$$c_1 \frac{\Gamma(\alpha)}{\Gamma(\alpha - p)} - I_{0+}^{\alpha - p}\widetilde{x}(1) = \int_0^1 \left(c_1 \frac{\Gamma(\alpha)}{\Gamma(\alpha - q)} t^{\alpha - q - 1} - I_{0+}^{\alpha - q}\widetilde{x}(t) \right) dH_0(t).$$

So, we find

$$c_1 = \frac{1}{\Delta_1 \Gamma(\alpha - p)} \int_0^1 (1 - s)^{\alpha - p - 1}\widetilde{x}(s)\, ds$$

$$- \frac{1}{\Delta_1 \Gamma(\alpha - q)} \int_0^1 \left(\int_0^s (s - \tau)^{\alpha - q - 1}\widetilde{x}(\tau)\, d\tau \right) dH_0(s).$$

Replacing the above constant c_1 in (2.30), we obtain the expression (2.29) for the solution $u \in C[0, 1]$ of problem (2.28), (2.27). Conversely, one easily verifies that $u \in C[0, 1]$ given by (2.29) satisfies equation (2.28) and the boundary conditions (2.27). $\qquad\square$

Lemma 2.3.2. *If $\Delta_1 \neq 0$, then the solution u of problem (2.28), (2.27) given by (2.29) can be written as*

$$u(t) = \int_0^1 G_1(t, s)\widetilde{x}(s)\, ds, \quad t \in [0, 1], \tag{2.31}$$

where the Green function G_1 is

$$G_1(t, s) = g_1(t, s) + \frac{t^{\alpha - 1}}{\Delta_1} \int_0^1 g_2(\tau, s)\, dH_0(\tau), \tag{2.32}$$

and g_1, g_2 are given by (2.8).

Proof. By Lemma 2.3.1 and relation (2.29), we deduce

$$u(t) = \frac{1}{\Gamma(\alpha)} \left\{ \int_0^t \left[t^{\alpha - 1}(1 - s)^{\alpha - p - 1} - (t - s)^{\alpha - 1} \right] \widetilde{x}(s)\, ds \right.$$

$$\left. + \int_t^1 t^{\alpha - 1}(1 - s)^{\alpha - p - 1}\widetilde{x}(s)\, ds \right\}$$

$$- \frac{1}{\Gamma(\alpha)} \int_0^1 t^{\alpha - 1}(1 - s)^{\alpha - p - 1}\widetilde{x}(s)\, ds$$

$$+ \frac{t^{\alpha - 1}}{\Delta_1 \Gamma(\alpha - p)} \int_0^1 (1 - s)^{\alpha - p - 1}\widetilde{x}(s)\, ds$$

$$- \frac{t^{\alpha - 1}}{\Delta_1 \Gamma(\alpha - q)} \int_0^1 \left(\int_0^s (s - \tau)^{\alpha - q - 1}\widetilde{x}(\tau)\, d\tau \right) dH_0(s)$$

$$= \frac{1}{\Gamma(\alpha)} \left\{ \int_0^t \left[t^{\alpha-1}(1-s)^{\alpha-p-1} - (t-s)^{\alpha-1} \right] \widetilde{x}(s)\, ds \right.$$

$$\left. + \int_t^1 t^{\alpha-1}(1-s)^{\alpha-p-1}\widetilde{x}(s)\, ds \right\} - \frac{1}{\Delta_1 \Gamma(\alpha-p)}$$

$$\times \int_0^1 t^{\alpha-1}(1-s)^{\alpha-p-1}\widetilde{x}(s)\, ds + \frac{1}{\Delta_1 \Gamma(\alpha-q)} \left(\int_0^1 s^{\alpha-q-1}\, dH_0(s) \right)$$

$$\times \int_0^1 t^{\alpha-1}(1-s)^{\alpha-p-1}\widetilde{x}(s)\, ds + \frac{t^{\alpha-1}}{\Delta_1 \Gamma(\alpha-p)} \int_0^1 (1-s)^{\alpha-p-1}\widetilde{x}(s)\, ds$$

$$- \frac{t^{\alpha-1}}{\Delta_1 \Gamma(\alpha-q)} \int_0^1 \left(\int_0^s (s-\tau)^{\alpha-q-1}\widetilde{x}(\tau)\, d\tau \right) dH_0(s)$$

$$= \frac{1}{\Gamma(\alpha)} \left\{ \int_0^t \left[t^{\alpha-1}(1-s)^{\alpha-p-1} - (t-s)^{\alpha-1} \right] \widetilde{x}(s)\, ds \right.$$

$$\left. + \int_t^1 t^{\alpha-1}(1-s)^{\alpha-p-1}\widetilde{x}(s)\, ds \right\}$$

$$+ \frac{t^{\alpha-1}}{\Delta_1 \Gamma(\alpha-q)} \left[\left(\int_0^1 \tau^{\alpha-q-1}\, dH_0(\tau) \right) \int_0^1 (1-s)^{\alpha-p-1}\widetilde{x}(s)\, ds \right.$$

$$\left. - \int_0^1 \left(\int_0^s (s-\tau)^{\alpha-q-1}\widetilde{x}(\tau)\, d\tau \right) dH_0(s) \right]$$

$$= \frac{1}{\Gamma(\alpha)} \left\{ \int_0^t \left[t^{\alpha-1}(1-s)^{\alpha-p-1} - (t-s)^{\alpha-1} \right] \widetilde{x}(s)\, ds \right.$$

$$\left. + \int_t^1 t^{\alpha-1}(1-s)^{\alpha-p-1}\widetilde{x}(s)\, ds \right\}$$

$$+ \frac{t^{\alpha-1}}{\Delta_1 \Gamma(\alpha-q)} \left[\int_0^1 \left(\int_0^1 \tau^{\alpha-q-1}(1-s)^{\alpha-p-1}\, dH_0(\tau) \right) \widetilde{x}(s)\, ds \right.$$

$$\left. - \int_0^1 \left(\int_\tau^1 (s-\tau)^{\alpha-q-1}\, dH_0(s) \right) \widetilde{x}(\tau)\, d\tau \right]$$

$$= \int_0^1 g_1(t,s)\widetilde{x}(s)\, ds + \frac{t^{\alpha-1}}{\Delta_1} \int_0^1 \left(\int_0^1 g_2(\tau,s)\, dH_0(\tau) \right) \widetilde{x}(s)\, ds$$

$$= \int_0^1 G_1(t,s)\widetilde{x}(s)\, ds.$$

Hence, we obtain the expression (2.31) for the solution u of problem (2.28), (2.27) given by (2.29). □

Lemma 2.3.3. *Assume that $H_0 : [0,1] \to \mathbb{R}$ is a nondecreasing function and $\Delta_1 > 0$. Then the function G_1 given by (2.32) is a continuous function on $[0,1] \times [0,1]$ and satisfies the inequalities:*

(a) $G_1(t,s) \le J_1(s)$ *for all* $t, s \in [0,1]$, *where* $J_1(s) = h_1(s) + \frac{1}{\Delta_1}$ $\times \int_0^1 g_2(\tau,s) \, dH_0(\tau)$, $s \in [0,1]$, *and* $h_1(s) = \frac{1}{\Gamma(\alpha)}(1-s)^{\alpha-p-1}(1-(1-s)^p)$, $s \in [0,1]$;

(b) $G_1(t,s) \ge t^{\alpha-1} J_1(s)$ *for all* $t, s \in [0,1]$;

(c) $G_1(t,s) \le \sigma_1 t^{\alpha-1}$, *for all* $t, s \in [0,1]$, *where* $\sigma_1 = \frac{1}{\Gamma(\alpha)} + \frac{1}{\Delta_1 \Gamma(\alpha-q)}$ $\times \int_0^1 \tau^{\alpha-q-1} \, dH_0(\tau)$.

Proof. By the definition of function G_1 we deduce that G_1 is a continuous function. In addition, by using Lemma 2.1.3, we obtain for all $t, s \in [0,1]$

(a) $G_1(t,s) \le h_1(s) + \frac{1}{\Delta_1} \int_0^1 g_2(\tau,s) \, dH_0(\tau) = J_1(s)$;

(b) $G_1(t,s) \ge t^{\alpha-1} h_1(s) + \frac{t^{\alpha-1}}{\Delta_1} \int_0^1 g_2(\tau,s) \, dH_0(\tau) = t^{\alpha-1} J_1(s)$;

(c) $G_1(t,s) \le \frac{t^{\alpha-1}}{\Gamma(\alpha)} + \frac{t^{\alpha-1}}{\Delta_1 \Gamma(\alpha-q)} \int_0^1 \tau^{\alpha-q-1} \, dH_0(\tau) = \sigma_1 t^{\alpha-1}$. $\qquad \square$

Lemma 2.3.4. *Assume that $H_0 : [0,1] \to \mathbb{R}$ is a nondecreasing function, $\Delta_1 > 0$, $\widetilde{x} \in C(0,1) \cap L^1(0,1)$ and $\widetilde{x}(t) \ge 0$ for all $t \in (0,1)$. Then the solution u of problem (2.28), (2.27) given by (2.29) satisfies the inequality $u(t) \ge t^{\alpha-1} u(t')$ for all $t, t' \in [0,1]$.*

Proof. By using Lemma 2.3.3, we obtain for all $t, t' \in [0,1]$

$$u(t) = \int_0^1 G_1(t,s)\widetilde{x}(s) \, ds \ge \int_0^1 t^{\alpha-1} J_1(s)\widetilde{x}(s) \, ds$$

$$\ge t^{\alpha-1} \int_0^1 G_1(t',s)\widetilde{x}(s) \, ds = t^{\alpha-1} u(t').$$

$\qquad \square$

2.3.2 *Existence and multiplicity of positive solutions*

In this section, we prove the existence of at least one or at least m ($m \in \mathbb{N}$, $m \ge 2$) positive solutions for our problem (2.26), (2.27). We introduce the assumptions that we will use in the sequel.

(L1) $\alpha \in \mathbb{R}$, $\alpha \in (n-1, n]$, $n \in \mathbb{N}$, $n \ge 3$, $p, q \in \mathbb{R}$, $p \in [1, \alpha - 1)$, $q \in [0, p]$, $H_0 : [0,1] \to \mathbb{R}$ is a nondecreasing function, and $\Delta_1 = \frac{\Gamma(\alpha)}{\Gamma(\alpha-p)} - \frac{\Gamma(\alpha)}{\Gamma(\alpha-q)} \int_0^1 s^{\alpha-q-1} \, dH_0(s) > 0$.

(L2) The function $f \in C((0,1) \times (0,\infty), \mathbb{R})$ and there exists a function $r \in C((0,1), [0,\infty)) \cap L^1(0,1)$ such that $f(t,x) \geq -r(t)$ for all $t \in (0,1)$ and $x \in (0,\infty)$, and $\|r\|_1 > 0$, ($\|r\|_1 = \int_0^1 |r(t)|\, dt$ is the norm of r in $L^1(0,1)$).

(L3) For any positive numbers $\theta_1, \theta_2, \theta_1 < \theta_2$, there exists a function $h_{\theta_1,\theta_2} \in C((0,1), [0,\infty)) \cap L^1(0,1)$ such that

$$f(t,x) \leq h_{\theta_1,\theta_2}(t), \quad 0 < t < 1, \quad \theta_1 t^{\alpha-1} \leq x \leq \theta_2.$$

(L4) There exists a positive constant $\gamma_1 > \sigma_1 \|r\|_1$ such that $\int_0^1 J_1(s)$ $\varphi(s, \gamma_1)\, ds \leq \gamma_1$, where

$$\varphi(t, \gamma_1) = \max\{f(t,x), \ (\gamma_1 - \sigma_1 \|r\|_1)t^{\alpha-1}$$
$$\leq x \leq \gamma_1\} + r(t), \quad \forall t \in (0,1).$$

(L5) There exists a positive constant $\gamma_2 > \gamma_1$ such that $\int_0^1 J_1(s)$ $\psi(s, \gamma_2)\, ds \geq \gamma_2$, where

$$\psi(t, \gamma_2) = \min\{f(t,x), \ (\gamma_2 - \sigma_1 \|r\|_1)t^{\alpha-1}$$
$$\leq x \leq \gamma_2\} + r(t), \quad \forall t \in (0,1).$$

We approximate the problem (2.26), (2.27) by the approximating fractional differential equation

$$D_{0+}^{\alpha} x(t) + f(t, \chi_n(x(t) - w(t))) + r(t) = 0, \quad 0 < t < 1, \qquad (2.33)$$

with the boundary conditions

$$x(0) = x'(0) = \cdots = x^{(n-2)}(0) = 0, \quad D_{0+}^p x(1) = \int_0^1 D_{0+}^q x(t)\, dH_0(t),$$

$$(2.34)$$

where $\chi_n(x) = \left\{x, \ x \geq \frac{1}{n}; \ \frac{1}{n}, \ x < \frac{1}{n}, \right\}$, $n \in \mathbb{N}$.

Here, $w(t) = \int_0^1 G_1(t,s) r(s)\, ds$, $t \in [0,1]$ is solution of problem

$$\begin{cases} D_{0+}^{\alpha} w(t) + r(t) = 0, \quad 0 < t < 1, \\ w(0) = w'(0) = \cdots = w^{(n-2)}(0) = 0, \quad D_{0+}^p w(1) = \int_0^1 D_{0+}^q w(t)\, dH_0(t). \end{cases}$$

By (L1), (L2), Lemmas 2.3.3 and 2.3.4, we obtain $w(t) \geq 0$ and $w(t) \geq t^{\alpha-1} \max_{s \in [0,1]} w(s)$ for all $t \in [0,1]$, and

$$w(t) = \int_0^1 G_1(t,s) r(s) \, ds \leq \int_0^1 \sigma_1 t^{\alpha-1} r(s) \, ds = \sigma_1 t^{\alpha-1} \|r\|_1, \quad \forall t \in [0,1].$$

By using Lemma 2.3.2, x is solution of the integral equation

$$x(t) = \int_0^1 G_1(t,s)(f(s, \chi_n(x(s) - w(s))) + r(s)) \, ds, \quad t \in [0,1], \quad (2.35)$$

if and only if x is solution for problem (2.33), (2.34).

We consider the Banach space $X = C[0,1]$ with the supremum norm $\| \cdot \|$, and we define the cone

$$P = \{x \in X, \ x(t) \geq t^{\alpha-1} \|x\|, \ \forall t \in [0,1]\}.$$

For $r > 0$, we denote by $B_r = \{x \in X, \ \|x\| < r\}$ and $\overline{B}_r = \{x \in X, \ \|x\| \leq r\}$.

We introduce now the operators $A_n : X \to X$, $n \in \mathbb{N}$ defined by

$$(A_n x)(t) = \int_0^1 G_1(t,s)(f(s, \chi_n(x(s) - w(s))) + r(s)) \, ds,$$

$$\forall t \in [0,1], \quad x \in X.$$

It is clear that x_n is a solution of problem (2.33), (2.34) if and only if x_n is a fixed point of operator A_n.

Lemma 2.3.5. *If* (L1)–(L3) *hold, then for any* θ_1, θ_2 *with* $\sigma \|r\|_1 < \theta_1 < \theta_2$, *the operator* $A_n : P \cap (\overline{B}_{\theta_2} \setminus B_{\theta_1}) \to P$ *for* $n \geq [1/\theta_1] + 1$ *fixed, is completely continuous.*

Proof. Let $n \geq [1/\theta_1] + 1$ be fixed and let $x \in P \cap (\overline{B}_{\theta_2} \setminus B_{\theta_1})$. Then for all $t \in [0,1]$ we obtain

$$x(t) - w(t) \geq t^{\alpha-1} \|x\| - \sigma_1 t^{\alpha-1} \|r\|_1 \geq t^{\alpha-1}(\theta_1 - \sigma_1 \|r\|_1) \geq 0,$$

and

$$t^{\alpha-1}(\theta_1 - \sigma_1 \|r\|_1) \leq \max\left\{x(t) - w(t), \frac{1}{n}\right\} \leq \theta_2,$$

that is

$$t^{\alpha-1}(\theta_1 - \sigma_1 \|r\|_1) \leq \chi_n(x(t) - w(t)) \leq \theta_2.$$

Then by (L1)–(L3) and Lemma 2.3.3, we deduce

$$J_1(s) \leq \frac{1}{\Gamma(\alpha)} + \frac{1}{\Delta_1 \Gamma(\alpha - q)} \int_0^1 \tau^{\alpha - q - 1} \, dH_0(\tau)$$

$$\leq \frac{1}{\Gamma(\alpha)} + \frac{1}{\Delta_1 \Gamma(\alpha - q)}(H_0(1) - H_0(0)) =: M_0, \quad \forall s \in [0, 1],$$

and

$$0 \leq (A_n x)(t) \leq \int_0^1 J_1(s)(f(x, \chi_n(x(s) - w(s))) + r(s)) \, ds$$

$$\leq \int_0^1 J_1(s)(h_{\theta_1 - \sigma_1 \|r\|_1, \theta_2}(s) + r(s)) \, ds$$

$$\leq M_0(\|h_{\theta_1 - \sigma_1 \|r\|_1, \theta_2}\|_1 + \|r\|_1) < \infty, \quad \forall t \in [0, 1]. \quad (2.36)$$

We conclude that $A_n : P \cap (\overline{B}_{\theta_2} \setminus B_{\theta_1}) \to X$ is well defined.
Besides, by Lemma 2.3.3, we obtain for all t, $t' \in [0, 1]$ that

$$(A_n x)(t) \leq \int_0^1 J_1(s)(f(s, \chi_n(x(s) - w(s)) + r(s)) \, ds,$$

$$(A_n x)(t) \geq \int_0^1 t^{\alpha - 1} J_1(s)(f(x, \chi_n(x(s) - w(s)) + r(s)) \, ds \geq t^{\alpha - 1}(A_n x)(t').$$

Therefore, $(A_n x)(t) \geq t^{\alpha - 1}\|A_n x\|$ for all $t \in [0, 1]$. We deduce that $A_n x \in P$, and hence $A_n(P \cap (\overline{B}_{\theta_2} \setminus B_{\theta_1})) \subset P$.

In what follows, we will prove that A_n is a completely continuous operator. By assumptions (L2)–(L3) and the Lebesgue's dominated convergence theorem, we obtain that A_n is continuous in the space X. Besides, A_n is compact, that is, for any bounded set $E \subset P \cap (\overline{B}_{\theta_2} \setminus B_{\theta_1})$, the set $A_n(E)$ is relatively compact. Indeed, let E be a bounded set in $P \cap (\overline{B}_{\theta_2} \setminus B_{\theta_1})$. By (2.36), we deduce that $A_n(E)$ is uniformly bounded. We will prove next that $A_n(E)$ is equicontinuous. For this, we estimate $(A_n x)'$ with $x \in E$, and we obtain

$$|(A_n x)'(t)| = \left| \int_0^1 \frac{\partial G_1}{\partial t}(t, s)(f(s, \chi_n(x(s) - w(s))) + r(s)) \, ds \right|$$

$$\leq \left| \int_0^1 \left[\frac{\partial g_1}{\partial t}(t, s) + \frac{(\alpha - 1)t^{\alpha - 2}}{\Delta_1} \int_0^1 g_2(\tau, s) \, dH_0(\tau) \right] \right.$$

$$\left. \times (f(s, \chi_n(x(s) - w(s))) + r(s)) \, ds \right|$$

$$\leq \int_0^1 (\alpha - 1) \left[t^{\alpha-1}(1-s)^{\alpha-p-1} \right.$$

$$+ (t-s)^{\alpha-2} + \frac{t^{\alpha-2}}{\Delta_1} \int_0^1 g_2(\tau, s)\, dH_0(\tau) \bigg]$$

$$\times (f(s, \chi_n(x(s) - w(s))) + r(s))\, ds$$

$$\leq \int_0^1 (\alpha - 1)[(1-s)^{\alpha-p-1} + (1-s)^{\alpha-2}$$

$$+ \frac{1}{\Delta_1} \int_0^1 g_2(\tau, s)\, dH_0(\tau)](h_{\theta_1 - \sigma_1 \|r\|_1, \theta_2}(s) + r(s))\, ds$$

$$\leq (\alpha - 1) \left[2 + \frac{1}{\Delta_1 \Gamma(\alpha - q)} (H_0(1) - H_0(0)) \right]$$

$$\times (\|h_{\theta_1 - \sigma_1 \|r\|_1, \theta_2}\|_1 + \|r\|_1) =: M_1, \quad \forall t \in (0,1).$$

Then

$$|(A_n x)(t') - (A_n x)(t)| = \left| \int_t^{t'} (A_n x)'(s)\, ds \right|$$

$$\leq M_1 |t' - t|, \quad \forall t, t' \in [0,1], \quad x \in E.$$

Hence, we conclude that $A_n(E)$ is equicontinuous. By using Arzela–Ascoli theorem, we deduce that $A_n(E)$ is relatively compact, and then $A_n : P \cap (\overline{B}_{\theta_2} \setminus B_{\theta_1}) \to P$ is a compact operator, and so it is completely continuous. $\qquad \square$

Lemma 2.3.6. *If* (L1)–(L5) *hold and* $n \geq n_0$ *is fixed, where* $n_0 = [1/\gamma_1] + 1$, *then the operator* $A_n : P \cap (\overline{B}_{\gamma_2} \setminus B_{\gamma_1}) \to P$ *has a fixed point in* $P \cap (\overline{B}_{\gamma_2} \setminus B_{\gamma_1})$.

Proof. Let $x \in P \cap \partial B_{\gamma_1}$. Then we have $x(t) \geq t^{\alpha-1}\|x\| = \gamma_1 t^{\alpha-1}$ for all $t \in [0,1]$, and

$$t^{\alpha-1}(\gamma_1 - \sigma_1 \|r\|_1) \leq \chi_n(x(t) - w(t)) \leq \gamma_1, \quad \forall t \in [0,1].$$

From the definition of $\varphi(t, \gamma_1)$, we deduce

$$f(t, \chi_n(x(t) - w(t))) + r(t) \leq \varphi(t, \gamma_1), \quad \forall t \in (0,1).$$

By using (L4) and Lemma 2.3.3, we obtain

$$(A_n x)(t) = \int_0^1 G_1(t,s)(f(s, \chi_n(x(s) - w(s))) + r(s)) \, ds$$

$$\leq \int_0^1 J_1(s)\varphi(s, \gamma_1) \, ds \leq \gamma_1, \quad \forall t \in [0,1].$$

Then

$$\|A_n x\| \leq \gamma_1 = \|x\|, \quad \forall x \in P \cap \partial B_{\gamma_1}. \tag{2.37}$$

Next, let $x \in P \cap \partial B_{\gamma_2}$. Then we have $x(t) \geq t^{\alpha-1}\|x\| = \gamma_2 t^{\alpha-1}$ for all $t \in [0,1]$, and

$$t^{\alpha-1}(\gamma_2 - \sigma_1\|r\|_1) \leq \chi_n(x(t) - w(t)) \leq \gamma_2.$$

From the definition of $\psi(t, \gamma_2)$, we deduce

$$f(t, \chi_n(x(t) - w(t))) + r(t) \geq \psi(t, \gamma_2), \quad \forall t \in (0,1).$$

By using (L5) and Lemma 2.3.3, we obtain

$$(A_n x)(t) \geq t^{\alpha-1} \int_0^1 J_1(s)(f(s, \chi_n(x(s) - w(s))) + r(s)) \, ds$$

$$\geq t^{\alpha-1} \int_0^1 J_1(s)\psi(s, \gamma_2) \, ds \geq t^{\alpha-1}\gamma_2, \quad \forall t \in [0,1].$$

Then

$$\|A_n x\| \geq \sup_{t \in [0,1]} t^{\alpha-1}\gamma_2 = \gamma_2 = \|x\|, \ \forall x \in P \cap \partial B_{\gamma_2}. \tag{2.38}$$

By (2.37), (2.38), Lemma 2.3.5 and Theorem 1.2.2, we conclude that the operator A_n (with $n \geq n_0$) has a fixed point $x_n \in P \cap (\overline{B}_{\gamma_2} \setminus B_{\gamma_1})$. $\square$

Theorem 2.3.1. *Assume that the assumptions* (L1)–(L5) *hold. Then the problem* (2.26), (2.27) *has at least one positive solution.*

Proof. By Lemma 2.3.6 we know that the operators A_n, $n \geq n_0$ have the fixed points $x_n \in P \cap (\overline{B}_{\gamma_2} \setminus B_{\gamma_1})$, that is, $x_n(t) = (A_n x_n)(t)$ for all $t \in [0,1]$ or

$$x_n(t) = \int_0^1 G_1(t,s)(f(s, \chi_n(x_n(s) - w(s))) + r(s)) \, ds, \quad \forall t \in [0,1].$$

$$\tag{2.39}$$

In addition, the functions x_n, $n \geq n_0$ satisfy the inequalities

$$x_n(t) - w(t) \geq t^{\alpha-1}(\|x_n\| - \sigma_1\|r\|_1) \geq t^{\alpha-1}(\gamma_1 - \sigma_1\|r\|_1), \quad \forall t \in [0,1].$$
(2.40)

By the proof of Lemma 2.3.5 we deduce that the set $\{x_n, \ n \geq n_0\}$ is relatively compact in the space X. Then $(x_n)_{n \geq n_0}$ contains a subsequence $(x_{n_k})_{k \geq n_0}$ which converges in X for $k \to \infty$ to a function $x \in X$.

By taking $k \to \infty$ in relations (2.40) and (2.39) (with $n = n_k$), we obtain

$$x(t) - w(t) \geq t^{\alpha-1}(\gamma_1 - \sigma_1\|r\|_1), \quad \forall t \in [0,1], \qquad (2.41)$$

and

$$x(t) = \int_0^1 G_1(t,s)(f(s,x(s) - w(s)) + r(s))\,ds, \quad \forall t \in [0,1]. \qquad (2.42)$$

By (2.41) and (2.42), we conclude that the function $u(t) = x(t) - w(t)$, $t \in [0,1]$ satisfies the relations $u(t) \geq t^{\alpha-1}(\gamma_1 - \sigma_1\|r\|_1)$ for all $t \in [0,1]$ and

$$u(t) = \int_0^1 G_1(t,s)f(s,u(s))\,ds, \quad t \in [0,1],$$

that is, u is a positive solution of problem (2.26), (2.27).   □

In a similar manner as that used in the proof of Theorem 2.3.1, we obtain the existence of m ($m \in \mathbb{N}$, $m \geq 2$) positive solutions for problem (2.26), (2.27) if we replace the assumptions (L4) and (L5) by the following ones:

(L̃4) There exist $\gamma_i > \sigma_1\|r\|_1$, $i = 1, \ldots, m$ such that $\int_0^1 J_1(s)\varphi(s,\gamma_i)\,ds \leq \gamma_i$, $i = 1, \ldots, m$, where

$$\varphi(t,\gamma_i) = \max\{f(t,x), \ (\gamma_i - \sigma_1\|r\|_1)t^{\alpha-1} \leq x \leq \gamma_i\} + r(t),$$

$$\forall t \in (0,1), \quad i = 1, \ldots, m;$$

(L̃5) There exist Γ_i, $i = 1, \ldots, m$ with $0 < \gamma_1 < \Gamma_1 < \gamma_2 < \Gamma_2 < \cdots < \gamma_m < \Gamma_m$ such that $\int_0^1 J_1(s)\psi(s,\Gamma_i)\,ds \geq \Gamma_i$, $i = 1, \ldots, m$, where

$$\psi(t,\Gamma_i) = \min\{f(t,x), \ (\Gamma_i - \sigma_1\|r\|_1)t^{\alpha-1} \leq x \leq \Gamma_i\} + r(t),$$

$$\forall t \in (0,1), \quad i = 1, \ldots, m.$$

Theorem 2.3.2. *Assume that the assumptions* (L1)–(L3), $(\widetilde{L4})$, $(\widetilde{L5})$ *hold. Then the boundary value problem* (2.26), (2.27) *has at least m positive solutions.*

2.3.3   *An example*

Let $\alpha = 10/3$ ($n = 4$), $p = 5/4$, $q = 1/2$, and $H_0(t) = \{1, \text{ if } t \in [0, 1/4); \ 4/3, \text{ if } t \in [1/4, 1/2); \ 2, \text{ if } t \in [1/2, 3/4]; \ 4t - 1, \text{ if } t \in (3/4, 1]\}$.

We consider the fractional differential equation

$$D_{0+}^{10/3} u(t) + \frac{1}{8\sqrt[4]{t(1-t)^3}}\left(u^3(t) + \frac{1}{\sqrt[5]{u(t)}}\right) - \frac{1}{10\sqrt{t}} = 0, \quad t \in (0,1),$$

$$(2.43)$$

with the boundary conditions

$$
\begin{cases}
u(0) = u'(0) = u''(0) = 0, \\
D_{0+}^{5/4} u(1) = \dfrac{1}{3} D_{0+}^{1/2} u\left(\dfrac{1}{4}\right) + \dfrac{2}{3} D_{0+}^{1/2} u\left(\dfrac{1}{2}\right) + 4 \displaystyle\int_{3/4}^{1} D_{0+}^{1/2} u(t)\, dt.
\end{cases}
\quad (2.44)
$$

Here, $f(t,x) = \frac{1}{8\sqrt[4]{t(1-t)^3}}\left(x^3 + \frac{1}{\sqrt[5]{x}}\right) - \frac{1}{10\sqrt{t}}$ for all $t \in (0,1)$, $x \in (0,\infty)$, and $r(t) = \frac{1}{10\sqrt{t}}$ for all $t \in (0,1)$. We obtain $\Delta_1 \approx 1.06472292 > 0$, $\sigma_1 \approx 0.90470354$, and $\|r\|_1 = 0.2$. Besides, f satisfies the inequality $f(t,x) \le h_{\theta_1,\theta_2}(t)$ for $0 < t < 1$ and $\theta_1 t^{\alpha-1} \le x \le \theta_2$, where

$$h_{\theta_1,\theta_2}(t) = \frac{1}{8\sqrt[4]{t(1-t)^3}}(\theta_2^3 + (\theta_1 t^{7/3})^{-1/5}), \quad \forall t \in (0,1),$$

and $h_{\theta_1,\theta_2} \in L^1(0,1)$.

In addition, we deduce

$$h_1(s) = \frac{1}{\Gamma(10/3)}(1-s)^{13/12}(1 - (1-s)^{5/4}), \quad s \in [0,1],$$

$$g_2(t,s) = \frac{1}{\Gamma(17/6)}\begin{cases} t^{11/6}(1-s)^{13/12} - (t-s)^{11/6}, & 0 \le s \le t \le 1, \\ t^{11/6}(1-s)^{13/12}, & 0 \le t \le s \le 1, \end{cases}$$

$$J_1(s) = h_1(s) + \frac{1}{\Delta_1} \int_0^1 g_2(\tau, s) \, dH_0(\tau),$$

$$= h_1(s) + \frac{1}{\Delta_1} \left[\frac{1}{3} g_2\left(\frac{1}{4}, s\right) + \frac{2}{3} g_2\left(\frac{1}{2}, s\right) + 4 \int_{3/4}^1 g_2(\tau, s) \, d\tau \right]$$

$$= \begin{cases} \frac{1}{\Gamma(10/3)}(1-s)^{13/12}(1-(1-s)^{5/4}) \\ \quad + \frac{1}{\Delta_1 \Gamma(17/6)} \left\{ \frac{1}{3} \left[\left(\frac{1}{4}\right)^{11/6}(1-s)^{13/12} - \left(\frac{1}{4}-s\right)^{11/6} \right] \right. \\ \quad + \frac{2}{3} \left[\left(\frac{1}{2}\right)^{11/6}(1-s)^{13/12} - \left(\frac{1}{2}-s\right)^{11/6} \right] \\ \quad + 4 \left[\frac{6}{17}(1-s)^{13/12} - \frac{6}{17}\left(\frac{3}{4}\right)^{17/6}(1-s)^{13/12} \right. \\ \quad \left. \left. - \frac{6}{17}(1-s)^{17/6} + \frac{6}{17}\left(\frac{3}{4}-s\right)^{17/6} \right] \right\}, \ 0 \le s < \frac{1}{4}, \\[4pt] \frac{1}{\Gamma(10/3)}(1-s)^{13/12}(1-(1-s)^{5/4}) \\ \quad + \frac{1}{\Delta_1 \Gamma(17/6)} \left\{ \frac{1}{3}\left(\frac{1}{4}\right)^{11/6}(1-s)^{13/12} \right. \\ \quad + \frac{2}{3} \left[\left(\frac{1}{2}\right)^{11/6}(1-s)^{13/12} - \left(\frac{1}{2}-s\right)^{11/6} \right] \\ \quad + 4 \left[\frac{6}{17}(1-s)^{13/12} - \frac{6}{17}\left(\frac{3}{4}\right)^{17/6}(1-s)^{13/12} \right. \\ \quad \left. \left. - \frac{6}{17}(1-s)^{17/6} + \frac{6}{17}\left(\frac{3}{4}-s\right)^{17/6} \right] \right\}, \ \frac{1}{4} \le s < \frac{1}{2}, \\[4pt] \frac{1}{\Gamma(10/3)}(1-s)^{13/12}(1-(1-s)^{5/4}) \\ \quad + \frac{1}{\Delta_1 \Gamma(17/6)} \left\{ \frac{1}{3}\left(\frac{1}{4}\right)^{11/6}(1-s)^{13/12} \right. \\ \quad + \frac{2}{3}\left(\frac{1}{2}\right)^{11/6}(1-s)^{13/12} \\ \quad + 4 \left[\frac{6}{17}(1-s)^{13/12} - \frac{6}{17}\left(\frac{3}{4}\right)^{17/6}(1-s)^{13/12} - \frac{6}{17}(1-s)^{17/6} \right. \\ \quad \left. \left. + \frac{6}{17}\left(\frac{3}{4}-s\right)^{17/6} \right] \right\}, \ \frac{1}{2} \le s < \frac{3}{4}, \\[4pt] \frac{1}{\Gamma(10/3)}(1-s)^{13/12}(1-(1-s)^{5/4}) \\ \quad + \frac{1}{\Delta_1 \Gamma(17/6)} \left\{ \frac{1}{3}\left(\frac{1}{4}\right)^{11/6}(1-s)^{13/12} \right. \\ \quad + \frac{2}{3}\left(\frac{1}{2}\right)^{11/6}(1-s)^{13/12} \\ \quad + 4 \left[\frac{6}{17}(1-s)^{13/12} - \frac{6}{17}\left(\frac{3}{4}\right)^{17/6}(1-s)^{13/12} \right. \\ \quad \left. \left. - \frac{6}{17}(1-s)^{17/6} \right] \right\}, \ \frac{3}{4} \le s \le 1. \end{cases}$$

For $\gamma_1 = 1$, we obtain

$$\varphi(t,1) = \frac{1}{8\sqrt[4]{t(1-t)^3}}\max\left\{\left(x^3 + \frac{1}{\sqrt[5]{x}}\right),\ (1-\sigma_1\|r\|_1)t^{7/3} \le x \le 1\right\}$$

$$\le \frac{1}{8\sqrt[4]{t(1-t)^3}}\left(1 + \frac{1}{\sqrt[5]{(1-\sigma_1\|r\|_1)t^{7/3}}}\right),\quad t \in (0,1),$$

and for $\gamma_2 = 15$, we get

$$\psi(t,15) = \frac{1}{8\sqrt[4]{t(1-t)^3}}\min\left\{\left(x^3 + \frac{1}{\sqrt[5]{x}}\right),\ (15-\sigma_1\|r\|_1)t^{7/3} \le x \le 15\right\}$$

$$\ge \frac{1}{8\sqrt[4]{t(1-t)^3}}\left(((15-\sigma_1\|r\|_1)t^{7/3})^3 + \frac{1}{\sqrt[5]{15}}\right),\quad t \in (0,1).$$

Then after some computations, we deduce

$$\int_0^1 J_1(s)\varphi(s,1)\,ds \le \int_0^1 J_1(s)\frac{1}{8\sqrt[4]{s(1-s)^3}}$$

$$\times \left(1 + \frac{1}{\sqrt[5]{(1-\sigma_1\|r\|_1)s^{7/3}}}\right)ds \approx 0.1315 \le 1,$$

and

$$\int_0^1 J_1(s)\psi(s,15)\,ds \ge \int_0^1 J_1(s)\frac{1}{8\sqrt[4]{s(1-s)^3}}$$

$$\times \left(((15-\sigma_1\|r\|_1)s^{7/3})^3 + \frac{1}{\sqrt[5]{15}}\right)ds \approx 19.2383 \ge 15.$$

Therefore, all assumptions (L1)–(L5) are satisfied, and so by Theorem 2.3.1, we conclude that the boundary value problem (2.43), (2.44) has at least one positive solution $u(t)$, $t \in [0,1]$ which satisfies the inequality $u(t) \ge (1 - 0.2\sigma_1)t^{7/3}$ for all $t \in [0,1]$.

Remark 2.3.1. The results presented in this section were published in [78].

2.4 Singular Fractional Differential Equations with General Integral Boundary Conditions

In this section, we investigate the nonlinear fractional differential equation

$$D_{0+}^{\alpha} u(t) + f(t, u(t)) = 0, \quad t \in (0,1), \tag{2.45}$$

with the general integral boundary conditions

$$u(0) = u'(0) = \cdots = u^{(n-2)}(0) = 0, \quad D_{0+}^{\beta_0} u(1) = \sum_{i=1}^{m} \int_0^1 D_{0+}^{\beta_i} u(t) \, dH_i(t), \tag{2.46}$$

where $\alpha \in \mathbb{R}$, $\alpha \in (n-1, n]$, $n, m \in \mathbb{N}$, $n \geq 3$, $\beta_i \in \mathbb{R}$ for all $i = 0, \ldots, m$, $0 \leq \beta_1 < \beta_2 < \cdots < \beta_m \leq \beta_0 < \alpha - 1$, $\beta_0 \geq 1$, D_{0+}^{k} denotes the Riemann–Liouville derivative of order k (for $k = \alpha, \beta_0, \beta_1, \ldots, \beta_m$), the integrals from the boundary condition (2.46) are Riemann–Stieltjes integrals with H_i, $i = 1, \ldots, m$, functions of bounded variation, the nonlinearity $f(t, u)$ may change sign and may be singular at the points $t = 0$, $t = 1$ and/or $u = 0$.

We present conditions on the functions f and H_i, $i = 1, \ldots, m$, such that problem (2.45), (2.46) has multiple positive solutions $u \in C([0,1], \mathbb{R}_+)$, $u(t) > 0$ for all $t \in (0,1]$. The boundary conditions (2.46) cover various cases, such as multi-point boundary conditions when the functions H_i are step functions, or classical integral boundary conditions, or a combination of them. In our main results, we use various height functions of the nonlinearity of equation defined on special bounded sets, an approximation method, and two theorems from the fixed point index theory (Theorems 1.2.12 and 1.2.13). In [118], the authors proved the existence of at least three positive solutions for equation (2.45) with the boundary conditions

$$u(0) = u'(0) = \cdots = u^{(n-2)}(0) = 0, \quad D_{0+}^{\beta} u(1) = \lambda \int_0^{\eta} h(t) D_{0+}^{\beta} u(t) \, dt, \tag{2.47}$$

where $\beta \geq 1$, $\alpha - \beta - 1 > 0$, $0 < \eta \leq 1$, $0 \leq \lambda \int_0^{\eta} h(t) t^{\alpha - \beta - 1} \, dt < 1$, $h \in L^1(0,1)$ is nonnegative and may be singular at $t = 0$ and $t = 1$, and the function f is nonnegative and may be singular at the points $t = 0$, $t = 1$ and $u = 0$. Our boundary conditions (2.46) are more general than the above boundary conditions (2.47). Indeed, the last relation from (2.47) can be written as $D_{0+}^{\beta} u(1) = \int_0^1 D_{0+}^{\beta} u(t) \, dH(t)$, with

$H(t) = \left\{ \lambda \int_0^t h(s)\,ds,\ t \in [0,\eta];\ \lambda \int_0^\eta h(s)\,ds,\ t \in (\eta, 1] \right\}$, and in the right-hand side of the last condition in (2.46) we have a sum of Riemann-Stieltjes integrals from Riemann-Liouville derivatives of various orders. In [118], the authors used different height functions of the nonlinear term on special bounded sets and the Krasnosel'skii and the Leggett-Williams fixed point index theorems.

2.4.1 Auxiliary results

We consider the fractional differential equation

$$D_{0+}^\alpha u(t) + \widetilde{x}(t) = 0, \quad t \in (0,1), \tag{2.48}$$

with the boundary conditions (2.46), where $\widetilde{x} \in C(0,1) \cap L^1(0,1)$. We denote by $\Delta_2 = \frac{\Gamma(\alpha)}{\Gamma(\alpha - \beta_0)} - \sum_{i=1}^m \frac{\Gamma(\alpha)}{\Gamma(\alpha - \beta_i)} \int_0^1 s^{\alpha - \beta_i - 1}\,dH_i(s)$.

Lemma 2.4.1. *If $\Delta_2 \neq 0$, then the unique solution $u \in C[0,1]$ of problem* (2.48), (2.46) *is given by*

$$
\begin{aligned}
u(t) = {}&-\frac{1}{\Gamma(\alpha)} \int_0^t (t-s)^{\alpha-1}\widetilde{x}(s)\,ds + \frac{t^{\alpha-1}}{\Delta_2 \Gamma(\alpha - \beta_0)} \\
&\times \int_0^1 (1-s)^{\alpha-\beta_0-1}\widetilde{x}(s)\,ds - \frac{t^{\alpha-1}}{\Delta_2} \sum_{i=1}^m \frac{1}{\Gamma(\alpha - \beta_i)} \\
&\times \int_0^1 \left(\int_0^s (s-\tau)^{\alpha-\beta_i-1}\widetilde{x}(\tau)\,d\tau \right) dH_i(s), \quad t \in [0,1]. \tag{2.49}
\end{aligned}
$$

Proof. By Lemma 1.1.3, we deduce that the solutions $u \in C(0,1) \cap L^1(0,1)$ of the fractional differential equation (2.48) are given by

$$
\begin{aligned}
u(t) &= -I_{0+}^\alpha \widetilde{x}(t) + c_1 t^{\alpha-1} + \cdots + c_n t^{\alpha-n} \\
&= -\frac{1}{\Gamma(\alpha)} \int_0^t (t-s)^{\alpha-1}\widetilde{x}(s)\,ds + c_1 t^{\alpha-1} + \cdots + c_n t^{\alpha-n},
\end{aligned}
$$

where $c_1, c_2, \ldots, c_n \in \mathbb{R}$. By using the conditions $u(0) = u'(0) = \cdots = u^{(n-2)}(0) = 0$, we obtain $c_2 = \cdots = c_n = 0$. Then we conclude

$$u(t) = c_1 t^{\alpha-1} - \frac{1}{\Gamma(\alpha)} \int_0^t (t-s)^{\alpha-1}\widetilde{x}(s)\,ds, \quad t \in [0,1]. \tag{2.50}$$

For the obtained function (2.50), we find

$$D_{0+}^{\beta_i} u(t) = c_1 \frac{\Gamma(\alpha)}{\Gamma(\alpha - \beta_i)} t^{\alpha-\beta_i-1} - I_{0+}^{\alpha-\beta_i}\widetilde{x}(t), \quad i = 0, 1, \ldots, m.$$

Then the condition $D_{0+}^{\beta_0} u(1) = \sum_{i=1}^{m} \int_0^1 D_{0+}^{\beta_i} u(t) \, dH_i(t)$ gives us

$$c_1 \frac{\Gamma(\alpha)}{\Gamma(\alpha - \beta_0)} - I_{0+}^{\alpha-\beta_0} \widetilde{x}(1)$$

$$= \sum_{i=1}^{m} \int_0^1 \left(c_1 \frac{\Gamma(\alpha)}{\Gamma(\alpha - \beta_i)} t^{\alpha-\beta_i-1} - I_{0+}^{\alpha-\beta_i} \widetilde{x}(t) \right) dH_i(t),$$

or

$$c_1 \left(\frac{\Gamma(\alpha)}{\Gamma(\alpha - \beta_0)} - \sum_{i=1}^{m} \frac{\Gamma(\alpha)}{\Gamma(\alpha - \beta_i)} \int_0^1 s^{\alpha-\beta_i-1} \, dH_i(s) \right)$$

$$= \frac{1}{\Gamma(\alpha - \beta_0)} \int_0^1 (1-s)^{\alpha-\beta_0-1} \widetilde{x}(s) \, ds$$

$$- \sum_{i=1}^{m} \frac{1}{\Gamma(\alpha - \beta_i)} \int_0^1 \left(\int_0^s (s-\tau)^{\alpha-\beta_i-1} \widetilde{x}(\tau) \, d\tau \right) dH_i(s).$$

So, we deduce

$$c_1 = \frac{1}{\Delta_2 \Gamma(\alpha - \beta_0)} \int_0^1 (1-s)^{\alpha-\beta_0-1} \widetilde{x}(s) \, ds$$

$$- \frac{1}{\Delta_2} \sum_{i=1}^{m} \int_0^1 \frac{1}{\Gamma(\alpha - \beta_i)} \left(\int_0^s (s-\tau)^{\alpha-\beta_i-1} \widetilde{x}(\tau) \, d\tau \right) dH_i(s).$$

Replacing the above constant c_1 in (2.50), we obtain the expression (2.49) for the solution $u \in C[0,1]$ of problem (2.48), (2.46). Conversely, one easily verifies that $u \in C[0,1]$ given by (2.49) satisfies equation (2.48) and the boundary conditions (2.46). $\qquad\square$

Lemma 2.4.2. *If $\Delta_2 \neq 0$, then the solution u of problem (2.48), (2.46) given by (2.49) can be written as*

$$u(t) = \int_0^1 G_2(t,s) \widetilde{x}(s) \, ds, \quad t \in [0,1], \tag{2.51}$$

where the Green function G_2 is

$$G_2(t,s) = g_0(t,s) + \frac{t^{\alpha-1}}{\Delta_2} \sum_{i=1}^{m} \left(\int_0^1 g_{1i}(\tau,s) \, dH_i(\tau) \right), \tag{2.52}$$

and

$$g_0(t,s) = \frac{1}{\Gamma(\alpha)} \begin{cases} t^{\alpha-1}(1-s)^{\alpha-\beta_0-1} - (t-s)^{\alpha-1}, & 0 \le s \le t \le 1, \\ t^{\alpha-1}(1-s)^{\alpha-\beta_0-1}, & 0 \le t \le s \le 1, \end{cases}$$

$$g_{1i}(t,s) = \frac{1}{\Gamma(\alpha-\beta_i)} \begin{cases} t^{\alpha-\beta_i-1}(1-s)^{\alpha-\beta_0-1} - (t-s)^{\alpha-\beta_i-1}, \\ \qquad\qquad 0 \le s \le t \le 1, \\ t^{\alpha-\beta_i-1}(1-s)^{\alpha-\beta_0-1}, & 0 \le t \le s \le 1, \end{cases} \qquad (2.53)$$

for all $(t,s) \in [0,1] \times [0,1]$, $i = 1, \ldots, m$.

Proof. By using Lemma 2.4.1 and relation (2.49), we have

$$u(t) = \frac{1}{\Gamma(\alpha)} \left\{ \int_0^t \left[t^{\alpha-1}(1-s)^{\alpha-\beta_0-1} - (t-s)^{\alpha-1} \right] \tilde{x}(s) \, ds \right.$$
$$\left. + \int_t^1 t^{\alpha-1}(1-s)^{\alpha-\beta_0-1} \tilde{x}(s) \, ds \right\}$$
$$- \frac{1}{\Gamma(\alpha)} \int_0^1 t^{\alpha-1}(1-s)^{\alpha-\beta_0-1} \tilde{x}(s) \, ds$$
$$+ \frac{t^{\alpha-1}}{\Delta_2 \Gamma(\alpha-\beta_0)} \int_0^1 (1-s)^{\alpha-\beta_0-1} \tilde{x}(s) \, ds$$
$$- \frac{t^{\alpha-1}}{\Delta_2} \sum_{i=1}^m \frac{1}{\Gamma(\alpha-\beta_i)} \int_0^1 \left(\int_0^s (s-\tau)^{\alpha-\beta_i-1} \tilde{x}(\tau) \, d\tau \right) dH_i(s)$$
$$= \int_0^1 g_0(t,s) \tilde{x}(s) \, ds - \frac{1}{\Delta_2 \Gamma(\alpha-\beta_0)} \int_0^1 t^{\alpha-1}(1-s)^{\alpha-\beta_0-1} \tilde{x}(s) \, ds$$
$$+ \frac{1}{\Delta_2 \Gamma(\alpha)} \left(\int_0^1 t^{\alpha-1}(1-s)^{\alpha-\beta_0-1} \tilde{x}(s) \, ds \right)$$
$$\times \left(\sum_{i=1}^m \frac{\Gamma(\alpha)}{\Gamma(\alpha-\beta_i)} \int_0^1 t^{\alpha-\beta_i-1} \, dH_i(t) \right)$$
$$+ \frac{t^{\alpha-1}}{\Delta_2 \Gamma(\alpha-\beta_0)} \int_0^1 (1-s)^{\alpha-\beta_0-1} \tilde{x}(s) \, ds$$
$$- \frac{t^{\alpha-1}}{\Delta_2} \sum_{i=1}^m \frac{1}{\Gamma(\alpha-\beta_i)} \int_0^1 \left(\int_0^s (s-\tau)^{\alpha-\beta_i-1} \tilde{x}(\tau) \, d\tau \right) dH_i(s)$$

$$= \int_0^1 g_0(t,s)\widetilde{x}(s)\,ds + \frac{t^{\alpha-1}}{\Delta_2}\left[\left(\int_0^1 (1-s)^{\alpha-\beta_0-1}\widetilde{x}(s)\,ds\right)\right.$$

$$\times \left(\sum_{i=1}^m \frac{1}{\Gamma(\alpha-\beta_i)}\int_0^1 \tau^{\alpha-\beta_i-1}\,dH_i(\tau)\right)$$

$$\left.- \sum_{i=1}^m \frac{1}{\Gamma(\alpha-\beta_i)}\int_0^1 \left(\int_0^s (s-\tau)^{\alpha-\beta_i-1}\widetilde{x}(\tau)\,d\tau\right)dH_i(s)\right]$$

$$= \int_0^1 g_0(t,s)\widetilde{x}(s)\,ds + \frac{t^{\alpha-1}}{\Delta_2}\int_0^1 \left(\sum_{i=1}^m \int_0^1 g_{2i}(\tau,s)\,dH_i(\tau)\right)\widetilde{x}(s)\,ds$$

$$= \int_0^1 G_2(t,s)\widetilde{x}(s)\,ds.$$

Therefore, we obtain the expression (2.51) for the solution u of problem (2.48), (2.46) given by (2.49). $\qquad\square$

Based on Lemma 2.1.3, we obtain some properties of the functions g_0 and g_{1i}, $i = 1,\ldots,m$ in the following lemma.

Lemma 2.4.3. *The functions g_0 and g_{1i}, $i = 1,\ldots,m$ given by (2.53) have the following properties:*

(a) $g_0(t,s) \le h_0(s)$ *for all $t,\,s \in [0,1]$, where*

$$h_0(s) = \frac{1}{\Gamma(\alpha)}(1-s)^{\alpha-\beta_0-1}(1-(1-s)^{\beta_0}),\, s \in [0,1];$$

(b) $g_0(t,s) \ge t^{\alpha-1}h_0(s)$ *for all $t,\,s \in [0,1]$;*
(c) $g_0(t,s) \le \frac{t^{\alpha-1}}{\Gamma(\alpha)}$, *for all $t,\,s \in [0,1]$;*
(d) $g_{1i}(t,s) \ge t^{\alpha-\beta_i-1}h_{1i}(s)$ *for all $t,\,s \in [0,1]$, where*

$$h_{1i}(s) = \frac{1}{\Gamma(\alpha-\beta_i)}(1-s)^{\alpha-\beta_0-1}(1-(1-s)^{\beta_0-\beta_i}),\, s \in [0,1],\, i = 1,\ldots,m;$$

(e) $g_{2i}(t,s) \le \frac{1}{\Gamma(\alpha-\beta_i)}t^{\alpha-\beta_i-1}$ *for all $t,\,s \in [0,1]$, $i = 1,\ldots,m$;*
(f) *The functions g_0 and g_{1i} are continuous on $[0,1] \times [0,1]$; $g_0(t,s) \ge 0$, $g_{1i}(t,s) \ge 0$ for all $t,\,s \in [0,1]$; $g_0(t,s) > 0$, $g_{1i}(t,s) > 0$ for all $t,\,s \in (0,1)$, $i = 1,\ldots,m$.*

By the definitions of the functions G_2, g_0, g_{1i}, $i = 1,\ldots,m$, we obtain, in a similar manner as we proved Lemmas 2.3.3 and 2.3.4, the following results.

Lemma 2.4.4. *Assume that $H_i : [0,1] \to \mathbb{R}$, $i = 1, \ldots, m$ are nondecreasing functions and $\Delta_2 > 0$. Then the function G_2 given by (2.52) is a continuous function on $[0,1] \times [0,1]$ and satisfies the inequalities:*

(a) $G_2(t,s) \le J_2(s)$ *for all* $t, s \in [0,1]$, *where* $J_2(s) = h_0(s) + \frac{1}{\Delta_2} \sum_{i=1}^{m} \int_0^1 g_{1i}(\tau, s)\, dH_i(\tau)$, $s \in [0,1]$;

(b) $G_2(t,s) \ge t^{\alpha-1} J_2(s)$ *for all* $t, s \in [0,1]$;

(c) $G_2(t,s) \le \sigma_2 t^{\alpha-1}$, *for all* $t, s \in [0,1]$, *where* $\sigma_2 = \frac{1}{\Gamma(\alpha)} + \frac{1}{\Delta_2} \sum_{i=1}^{m} \frac{1}{\Gamma(\alpha-\beta_i)} \int_0^1 \tau^{\alpha-\beta_i-1}\, dH_i(\tau)$.

Lemma 2.4.5. *Assume that $H_i : [0,1] \to \mathbb{R}$, $i = 1, \ldots, m$ are nondecreasing functions, $\Delta_2 > 0$, $\widetilde{x} \in C(0,1) \cap L^1(0,1)$ and $\widetilde{x}(t) \ge 0$ for all $t \in (0,1)$. Then the solution u of problem (2.48), (2.46) given by (2.51) satisfies the inequality $u(t) \ge t^{\alpha-1} u(t')$ for all $t, t' \in [0,1]$.*

2.4.2   Existence of multiple positive solutions

In this section, we prove the existence of at least two positive solutions for problem (2.45), (2.46). We introduce the first assumptions that we will use in the sequel.

(D1) $\alpha \in \mathbb{R}$, $\alpha \in (n-1, n]$, $n, m \in \mathbb{N}$, $n \ge 3$, $\beta_i \in \mathbb{R}$ for all $i = 0, \ldots, m$, $0 \le \beta_1 < \beta_2 < \cdots < \beta_m \le \beta_0 < \alpha - 1$, $\beta_0 \ge 1$, $H_i : [0,1] \to \mathbb{R}$, $i = 1, \ldots, m$ are nondecreasing functions, and $\Delta_2 = \frac{\Gamma(\alpha)}{\Gamma(\alpha-\beta_0)} - \sum_{i=1}^{m} \frac{\Gamma(\alpha)}{\Gamma(\alpha-\beta_i)} \int_0^1 s^{\alpha-\beta_i-1}\, dH_i(s) > 0$.

(D2) The function $f \in C((0,1) \times (0,\infty), \mathbb{R})$ and there exists a function $r \in C((0,1), [0,\infty)) \cap L^1(0,1)$ such that $f(t,x) \ge -r(t)$ for all $t \in (0,1)$ and $x \in (0,\infty)$, and $\|r\|_1 > 0$, $(\|r\|_1 = \int_0^1 |r(t)|\, dt)$.

(D3) For any positive numbers θ_1, θ_2 with $\theta_1 < \theta_2$, there exists a function $h_{\theta_1,\theta_2} \in C((0,1), [0,\infty)) \cap L^1(0,1)$ such that

$$f(t,x) \le h_{\theta_1,\theta_2}(t), \quad 0 < t < 1, \quad \theta_1 t^{\alpha-1} \le x \le \theta_2.$$

For $\gamma, \gamma_1, \gamma_2 > 0$ with $\sigma_2\|r\|_1 < \gamma$, $\sigma_2\|r\|_1 < \gamma_1 < \gamma_2$ we introduce the height functions:

$$\varphi(t,\gamma) = \max\{f(t,x), \ (\gamma - \sigma_2\|r\|_1)t^{\alpha-1} \le x \le \gamma\} + r(t), \quad \forall t \in (0,1),$$

$$\psi(t,\gamma) = \min\{f(t,x), \ (\gamma - \sigma_2\|r\|_1)t^{\alpha-1} \le x \le \gamma\} + r(t), \quad \forall t \in (0,1),$$

$$\Phi(t, \gamma_1, \gamma_2) = \max\{f(t, x), \ (\gamma_1 - \sigma_2\|r\|_1)t^{\alpha-1} \le x \le \gamma_2\} + r(t),$$

$$\forall t \in (0, 1),$$

$$\Psi(t, \gamma_1, \gamma_2) = \min\{f(t, x), \ (\gamma_1 - \sigma_2\|r\|_1)t^{\alpha-1} \le x \le \gamma_2\} + r(t),$$

$$\forall t \in (0, 1).$$

We approximate the problem (2.45), (2.46) by an approximating fractional differential equation where the nonlinearity is nonnegative and in the second variable of f we have a nonnegative function, namely the equation

$$D_{0+}^\alpha x(t) + f(t, \chi_n(x(t) - w(t))) + r(t) = 0, \quad 0 < t < 1, \tag{2.54}$$

with the boundary conditions

$$x(0) = x'(0) = \cdots = x^{(n-2)}(0) = 0, \quad D_{0+}^{\beta_0} x(1) = \sum_{i=1}^m \int_0^1 D_{0+}^{\beta_i} x(t) \, dH_i(t), \tag{2.55}$$

where $\chi_n(x) = \{x, \ x \ge \frac{1}{n}; \ \frac{1}{n}, \ x < \frac{1}{n}\}$, $n \in \mathbb{N}$.

Here, $w(t) = \int_0^1 G_2(t, s)r(s) \, ds$, $t \in [0, 1]$ is solution of problem

$$\begin{cases} D_{0+}^\alpha w(t) + r(t) = 0, \quad 0 < t < 1, \\ w(0) = w'(0) = \cdots = w^{(n-2)}(0) = 0, \quad D_{0+}^{\beta_0} w(1) = \sum_{i=1}^m \int_0^1 D_{0+}^{\beta_i} w(t) \, dH_i(t). \end{cases}$$

By (D1), (D2), Lemmas 2.4.4 and 2.4.5, we deduce that $w(t) \ge 0$, $w(t) \ge t^{\alpha-1} \max_{s \in [0,1]} w(s)$ for all $t \in [0, 1]$, and

$$w(t) = \int_0^1 G_2(t, s)r(s) \, ds \le \int_0^1 \sigma_2 t^{\alpha-1} r(s) \, ds = \sigma_2 t^{\alpha-1}\|r\|_1, \quad \forall t \in [0, 1].$$

By Lemma 2.4.2, the function $x \in C[0, 1]$ is solution of the integral equation

$$x(t) = \int_0^1 G_2(t, s)(f(s, \chi_n(x(s) - w(s))) + r(s)) \, ds, \quad t \in [0, 1], \tag{2.56}$$

if and only if $x \in C[0, 1]$ is solution for problem (2.54), (2.55).

We consider the Banach space $X = C[0, 1]$ with the supremum norm $\|\cdot\|$, and we define the cone

$$P = \{x \in X, \ x(t) \ge t^{\alpha-1}\|x\|, \ \forall t \in [0, 1]\}.$$

We introduce the operators $\mathcal{A}_n : X \to X$, $n \in \mathbb{N}$ defined by

$$(\mathcal{A}_n x)(t) = \int_0^1 G_2(t,s)(f(s, \chi_n(x(s) - w(s))) + r(s)) \, ds, \quad \forall t \in [0,1], \; x \in X.$$

We can easily see that if (D1)–(D3) hold, then x_n is a fixed point of operator $\mathcal{A}_n$ if and only if x_n is a solution of problem (2.54), (2.55).

Remark 2.4.1. By using standard arguments, we can show that under assumptions (D1)–(D3), for any θ_1, θ_2 with $\sigma_2 \|r\|_1 < \theta_1 < \theta_2$, the operator $\mathcal{A}_n : \overline{P}_{\theta_2} \setminus P_{\theta_1} \to P$ for $n \geq [1/\theta_1] + 1$ fixed, is completely continuous, where $P_{\theta_1} = \{u \in P, \; \|u\| < \theta_1\}$, $\overline{P}_{\theta_2} = \{u \in P, \; \|u\| \leq \theta_2\}$.

Theorem 2.4.1. *Assume that* (D1)–(D3) *hold, and* $c, d \in \mathbb{R}$, $0 < c < d \leq 1$. *In addition, there exist five positive constants* δ_i, $i = 1, \ldots, 5$ *such that* $\sigma_2 \|r\|_1 < \delta_1 < \delta_2 < \delta_3 < \delta_4 \leq \delta_5$ *with* $\delta_3 c^{1-\alpha} \leq \delta_4$ *and*

(I$_1$) $\int_0^1 J_2(s)\varphi(s, \delta_2) \, ds < \delta_2$;

(I$_2$) $\int_0^1 J_2(s)\psi(s, \delta_1) \, ds > \delta_1$;

(I$_3$) $\int_0^1 J_2(s)\Phi(s, \delta_3, \delta_5) \, ds \leq \delta_5$;

(I$_4$) $\int_c^d J_2(s)\Psi(s, \delta_3, \delta_4) \, ds > \delta_3 c^{1-\alpha}$.

Let $n \geq n_0$ *be fixed, where* $n_0 = [1/\delta_1] + 1$. *Then the operator* $\mathcal{A}_n$ *has at least three fixed points in* $\overline{P}_{\delta_5}$.

Proof. By Remark 2.4.1, for $\theta_1 = \delta_1$ and $\theta_2 = \delta_5$, we know that the operator $\mathcal{A}_n : \overline{P}_{\delta_5} \setminus P_{\delta_1} \to P$ for $n \geq [1/\delta_1] + 1$ fixed, is completely continuous. By the extension theorem, $\mathcal{A}_n$ has a completely continuous extension (also denoted by $\mathcal{A}_n$) from P into P. We define the functional $\xi(u) = \min_{t \in [c,d]} u(t)$ for any $u \in P$.

We will verify the conditions of the Leggett–Williams theorem (Theorem 1.2.12). First, we show that for $n \geq n_0$ we have $i(\mathcal{A}_n, S(\xi, \delta_3, \delta_5), \overline{P}_{\delta_5}) = 1$. We set $x_0(t) = (\delta_3 + \delta_4)/2$, $t \in [0,1]$. Then $x_0 \in P$ (that is, $x_0(t) \geq t^{\alpha-1}\|x_0\|$ for all $t \in [0,1]$), $\xi(x_0) = (\delta_3 + \delta_4)/2 > \delta_3$, and $\|x_0\| = (\delta_3 + \delta_4)/2 \leq \delta_4$. So, $x_0 \in S(\xi, \delta_3, \delta_4)$ and then $S(\xi, \delta_3, \delta_4) \neq \emptyset$. In addition, if $x \in P(\xi, \delta_3, \delta_4)$, that is $x \in P$, $\xi(x) \geq \delta_3$ and $\|x\| \leq \delta_4$, we obtain

$$\|x\| = \max_{t \in [0,1]} x(t) \geq \min_{t \in [c,d]} x(t) \geq \delta_3,$$

$$x(t) \geq t^{\alpha-1}\|x\| \geq t^{\alpha-1}\delta_3, \quad \forall t \in [0,1],$$

$$t^{\alpha-1}(\delta_3 - \sigma_2\|r\|_1) \leq x(t) - w(t) \leq \max\left\{x(t) - w(t), \frac{1}{n}\right\}$$

$$= \chi_n(x(t) - w(t)) \leq \delta_4, \quad \forall t \in [0,1].$$

From the definition of $\Psi(t, \delta_3, \delta_4)$ we deduce that

$$f(t, \chi_n(x(t) - w(t))) + r(t) \geq \Psi(t, \delta_3, \delta_4), \quad \forall t \in [c, d].$$

By using Lemma 2.4.4 and assumption (I_4), we conclude that

$$(\mathcal{A}_n x)(t) = \int_0^1 G_2(t, s)(f(s, \chi_n(x(s) - w(s))) + r(s))\, ds$$

$$\geq \int_c^d t^{\alpha-1} J_2(s) \Psi(s, \delta_3, \delta_4)\, ds$$

$$\geq c^{\alpha-1} \int_c^d J_2(s) \Psi(s, \delta_3, \delta_4)\, ds > c^{\alpha-1} c^{1-\alpha} \delta_3 = \delta_3, \quad \forall t \in [0, 1],$$

and then

$$\xi(\mathcal{A}_n x) = \min_{t \in [c,d]} (\mathcal{A}_n x)(t) > \delta_3. \tag{2.57}$$

So, we have the condition (1) from Theorem 1.2.12.

Next, we prove that if $x \in P(\xi, \delta_3, \delta_5)$, then $\mathcal{A}_n x \in \overline{P}_{\delta_5}$. So, let $x \in X$, $x(t) \geq t^{\alpha-1} \|x\|$, $\min_{t \in [c,d]} x(t) \geq \delta_3$ and $\|x\| \leq \delta_5$. We obtain

$$\delta_3 \leq \min_{t \in [c,d]} x(t) \leq \max_{t \in [0,1]} x(t) = \|x\| \leq \delta_5,$$

and as above,

$$t^{\alpha-1}(\delta_3 - \sigma_2 \|r\|_1) \leq \chi_n(x(t) - w(t)) \leq \delta_5, \quad \forall t \in [0, 1]. \tag{2.58}$$

By Lemma 2.4.4 and assumption (I_3),

$$(\mathcal{A}_n x)(t) \leq \int_0^1 J_2(s)(f(s, \chi_n(x(s) - w(s))) + r(s))\, ds$$

$$\leq \int_0^1 J_2(s) \Phi(s, \delta_3, \delta_5)\, ds \leq \delta_5, \quad \forall t \in [0, 1], \tag{2.59}$$

and so $\|\mathcal{A}_n x\| \leq \delta_5$, which means that $\mathcal{A}_n x \in \overline{P}_{\delta_5}$. Therefore, we obtain condition (2) from Theorem 1.2.12.

Finally, we consider $x \in P(\xi, \delta_3, \delta_5)$ with $\|\mathcal{A}_n x\| > \delta_4$. By using Lemma 2.4.5,

$$\xi(\mathcal{A}_n x) = \min_{t \in [c,d]} (\mathcal{A}_n x)(t) \geq \min_{t \in [c,d]} t^{\alpha-1} \|\mathcal{A}_n x\|$$

$$= c^{\alpha-1} \|\mathcal{A}_n x\| > c^{\alpha-1} \delta_4 \geq \delta_3, \tag{2.60}$$

that is, we obtain condition (3) of Theorem 1.2.12.

Therefore, by Theorem 1.2.12 with $A = \mathcal{A}_n$, $a = \delta_3$, $b = \delta_4$ and $c = \delta_5$, we conclude that

$$i(\mathcal{A}_n, S(\xi, \delta_3, \delta_5), \overline{P}_{\delta_5}) = 1. \tag{2.61}$$

Next, if $x \in \partial P_{\delta_5}$, then $x \in P$ with $\|x\| = \delta_5$ and $\delta_3 t^{\alpha-1} \le \delta_5 t^{\alpha-1} \le x(t) \le \delta_5$ for all $t \in [0,1]$. In addition, the relation (2.58) holds. Hence, by (2.58), assumption (I_3), and Lemma 2.4.4, in a similar manner as we proved relation (2.59), we obtain

$$\|\mathcal{A}_n x\| \le \delta_5, \quad \forall x \in \partial P_{\delta_5}. \tag{2.62}$$

If $u \in \partial P_{\delta_2}$, then $x \in P$, $\|x\| = \delta_2$ and $\delta_2 t^{\alpha-1} \le x(t) \le \delta_2$, for all $t \in [0,1]$. Thus,

$$(\delta_2 - \sigma_2\|r\|_1)t^{\alpha-1} \le \chi_n(x(t) - w(t)) \le \delta_2, \quad \forall t \in [0,1].$$

By assumption (I_1) and Lemma 2.4.4,

$$(\mathcal{A}_n x)(t) \le \int_0^1 J_2(s)(f(x, \chi_n(x(s) - w(s))) + r(s))\, ds$$

$$\le \int_0^1 J_2(s)\varphi(s, \delta_2)\, ds < \delta_2, \quad \forall t \in [0,1],$$

and then

$$\|\mathcal{A}_n x\| < \delta_2, \quad \forall x \in \partial P_{\delta_2}. \tag{2.63}$$

If $u \in \partial P_{\delta_1}$, then $u \in P$, $\|u\| = \delta_1$ and $\delta_1 t^{\alpha-1} \le x(t) \le \delta_1$ for all $t \in [0,1]$. Therefore,

$$(\delta_1 - \sigma_2\|r\|_1)t^{\alpha-1} \le \chi_n(x(t) - w(t)) \le \delta_1.$$

Then by assumption (I_2) and Lemma 2.4.4,

$$(\mathcal{A}_n x)(t) \ge \int_0^1 t^{\alpha-1}J_2(s)f(s, \chi_n(x(s) - w(s))) + r(s))\, ds$$

$$\ge t^{\alpha-1}\int_0^1 J_2(s)\psi(t, \delta_1)\, ds, \quad \forall t \in [0,1],$$

and therefore

$$\|\mathcal{A}_n x\| \ge \max_{t \in [0,1]} t^{\alpha-1} \int_0^1 J_2(s)\psi(t, \delta_1)\, ds$$

$$= \int_0^1 J_2(s)\psi(t, \delta_1)\, ds > \delta_1, \quad \forall x \in \partial P_{\delta_1}. \tag{2.64}$$

Then by using (2.62)–(2.64) and Theorem 1.2.13, we have

$$i(\mathcal{A}_n, \overline{P}_{\delta_5} \setminus \overline{P}_{\delta_1}, \overline{P}_{\delta_5}) = 1, \tag{2.65}$$

$$i(\mathcal{A}_n, P_{\delta_2} \setminus \overline{P}_{\delta_1}, \overline{P}_{\delta_5}) = 1. \tag{2.66}$$

By relation (2.63), we deduce that the operator $\mathcal{A}_n$ has no fixed point on ∂P_{δ_2}. Besides, for $x \in P(\xi, \delta_3, \delta_4)$ we have $\xi(\mathcal{A}_n x) > \delta_3$ (see relation (2.57)), and for $x \in P(\xi, \delta_3, \delta_5)$ with $\|\mathcal{A}_n x\| > \delta_4$ we obtain $\xi(\mathcal{A}_n x) > \delta_3$ (see relation (2.60)). Then $\mathcal{A}_n$ has no fixed point on $P(\xi, \delta_3, \delta_5) \setminus S(\xi, \delta_3, \delta_5)$.

Therefore, by (2.61), (2.65), (2.66), and the additivity property of the fixed point index, we conclude

$$i(\mathcal{A}_n, \overline{P}_{\delta_5} \setminus (P(\xi, \delta_3, \delta_5) \cup \overline{P}_{\delta_2}), \overline{P}_{\delta_5})$$
$$= i(\mathcal{A}_n, \overline{P}_{\delta_5} \setminus \overline{P}_{\delta_1}, \overline{P}_{\delta_5}) - i(\mathcal{A}_n, P_{\delta_2} \setminus \overline{P}_{\delta_1}, \overline{P}_{\delta_5}) - i(\mathcal{A}_n, S(\xi, \delta_3, \delta_5), \overline{P}_{\delta_5})$$
$$= -1. \tag{2.67}$$

Hence, by (2.61), (2.66) and (2.67), we deduce that $\mathcal{A}_n$ has at least three fixed points $x_{1n} \in P_{\delta_2} \setminus \overline{P}_{\delta_1}, x_{2n} \in S(\xi, \delta_3, \delta_5), x_{3n} \in \overline{P}_{\delta_5} \setminus (P(\xi, \delta_3, \delta_5) \cup \overline{P}_{\delta_2})$ that satisfy the conditions

$$\delta_1 \leq \|x_{1n}\| < \delta_2, \quad \|x_{2n}\| \leq \delta_5, \quad \delta_2 < \|x_{3n}\| \leq \delta_5,$$

$$\min_{t \in [c,d]} x_{2n}(t) > \delta_3, \quad \min_{t \in [c,d]} x_{3n}(t) < \delta_3. \qquad \square$$

Theorem 2.4.2. *Assume that* (D1)–(D3) *hold, and* $c, d \in \mathbb{R}$, $0 < c < d \leq 1$. *In addition, assume there exist five positive constants* δ_i, $i = 1, \ldots, 5$ *such that* $\sigma_2 \|r\|_1 < \delta_1 < \delta_2 < \delta_3 < \delta_4 \leq \delta_5$ *with* $\delta_3 c^{1-\alpha} \leq \delta_4$, *and the assumptions* $(I_1) - (I_4)$ *are satisfied. Then the boundary value problem* (2.45), (2.46) *has at least two positive solutions* u_1, u_2 *satisfying* $\delta_1 - \sigma_2 \|r\|_1 \leq \|u_1\| \leq \delta_2, \delta_3 - \sigma_2 \|r\|_1 \leq \|u_2\| \leq \delta_5$ *and* $\min_{t \in [c,d]} u_2(t) \geq \delta_3 - \sigma_2 \|r\|_1$.

Proof. By Theorem 2.4.1, we know that the operators $\mathcal{A}_n$, $n \geq n_0$ have the fixed points $x_{in} \in P$, $i = 1, 2, 3$, that is $x_{in}(t) = (\mathcal{A}_n x_{in})(t)$ for all $t \in [0, 1]$ or

$$x_{in}(t) = \int_0^1 G_2(t, s)(f(s, \chi_n(x_{in}(s) - w(s))) + r(s)) \, ds, \quad \forall t \in [0, 1],$$

$$i = 1, 2, 3. \tag{2.68}$$

In addition, the functions x_{in}, $n \geq n_0$ satisfy the inequalities

$$x_{in}(t) - w(t) \geq t^{\alpha-1}(\|x_{in}\| - \sigma_2\|r\|_1) \geq t^{\alpha-1}(\delta_1 - \sigma_2\|r\|_1) \geq 0,$$

$$\forall t \in [0,1], \ i = 1,2,3. \tag{2.69}$$

By Remark 2.4.1, the sets $\{x_{in}, \ n \geq n_0\}$, $i = 1,2,3$, are relatively compact in the space X. Then $(x_{in})_{n \geq n_0}$ contains a subsequence $(x_{in_k})_{k \geq n_0}$ which converges in X for $k \to \infty$ to some function $x_i \in X$, $i = 1,2,3$.

By taking $k \to \infty$ in the relations (2.69) and (2.68) (with $n = n_k$), we obtain

$$x_i(t) - w(t) \geq t^{\alpha-1}(\delta_1 - \sigma_2\|r\|_1), \quad \forall t \in [0,1], \tag{2.70}$$

$$x_i(t) = \int_0^1 G_2(t,s)(f(s, x_i(s) - w(s)) + r(s)) \, ds, \quad \forall t \in [0,1], \tag{2.71}$$

and

$$\delta_1 \leq \|x_1\| \leq \delta_2, \quad \|x_2\| \leq \delta_5, \quad \min_{t \in [c,d]} x_2(t) \geq \delta_3,$$

$$\delta_2 \leq \|x_3\| \leq \delta_5, \quad \min_{t \in [c,d]} x_3(t) \leq \delta_3.$$

Here, we remark that x_2 may be equal to x_3.

By the above relations, we conclude that the functions $u_i(t) = x_i(t) - w(t)$, $t \in [0,1]$, $i = 1,2$, satisfy the relations $u_i(t) \geq t^{\alpha-1}(\delta_1 - \sigma_2\|r\|_1)$ for all $t \in [0,1]$, and

$$u_i(t) = \int_0^1 G_2(t,s)f(s, u_i(s)) \, ds, \quad \forall t \in [0,1], \ i = 1,2,$$

that is, u_i, $i = 1,2$, are positive solutions of problem (2.45), (2.46). Besides, u_1, u_2 satisfy the relations $\delta_1 - \sigma_2\|r\|_1 \leq \|u_1\| \leq \delta_2$, $\delta_3 - \sigma_2\|r\|_1 \leq \|u_2\| \leq \delta_5$ and $\min_{t \in [c,d]} u_2(t) \geq \delta_3 - \sigma_2\|r\|_1$. $\qquad\square$

2.4.3 *An example*

Let us choose $\alpha = 7/2$ $(n = 4)$, $\beta_0 = 5/3$, $m = 2$, $\beta_1 = 1/4$, $\beta_2 = 6/5$, $H_1(t) = t$ for all $t \in [0,1]$, $H_2(t) = \{0, \text{ for } t \in [0,1/2); \ 2, \text{ for } t \in [1/2,1]\}$, and $r(t) = \frac{1}{10^4 \sqrt[4]{t}}$ for all $t \in (0,1)$.

We consider the fractional differential equation

$$D_{0+}^{7/2} u(t) + f(t, u(t)) = 0, \quad 0 < t < 1, \tag{2.72}$$

with the boundary conditions

$$u(0) = u'(0) = u''(0) = 0, \quad D_{0+}^{5/3} u(1) = \int_0^1 D_{0+}^{1/4} u(t)\, dt + 2 D_{0+}^{6/5} u\left(\frac{1}{2}\right), \tag{2.73}$$

where

$$f(t, x) = \frac{1}{9\sqrt[5]{t^2(1-t)^3}} \varrho(x) - \frac{1}{10^4 \sqrt[4]{t}}, \quad 0 < t < 1,\ x > 0,$$

with

$$\varrho(x) = \begin{cases} \sqrt{x} + \frac{1}{\sqrt[7]{x}}, & 0 < x \le 1, \\ x^7 + 1, & 1 < x \le 3, \\ \sqrt[3]{x} + 2188 - \sqrt[3]{3}, & x > 3. \end{cases}$$

For $\theta_1, \theta_2 > 0$, $\theta_1 < \theta_2$, we have $f(t, x) \le h_{\theta_1, \theta_2}(t)$ for $0 < t < 1$ and $\theta_1 t^{5/2} \le x \le \theta_2$, where

$$h_{\theta_1, \theta_2}(t) = \frac{1}{9\sqrt[5]{t^2(1-t)^3}} \begin{cases} \sqrt{\theta_2} + (\theta_1 t^{5/2})^{-1/7}, & \text{for } \theta_2 \le 1, \\ \max\{1 + (\theta_1 t^{5/2})^{-1/7}, \theta_2^7 + 1\}, & \text{for } 1 < \theta_2 \le 3, \\ \max\{1 + (\theta_1 t^{5/2})^{-1/7}, \sqrt[3]{\theta_2} + 2188 - \sqrt[3]{3}\}, & \\ & \text{for } \theta_2 > 3. \end{cases}$$

This function $h_{\theta_1, \theta_2} \in L^1(0, 1)$, and in addition we obtain $\Delta_2 \approx 0.81820862 > 0$ and $\|r\|_1 = \frac{4}{3 \cdot 10^4} > 0$. Thus the assumptions (D1)–(D3) are satisfied. We also deduce $\sigma_2 \approx 1.29928727$.

We take $c = 3/4$, $d = 1$, $\delta_1 = 0.01$, $\delta_2 = 1$, $\delta_3 = 5$, $\delta_4 = 15$, $\delta_5 = 1000$. The conditions $\sigma_2 \|r\|_1 < \delta_1$ and $\delta_3 c^{1-\alpha} \le \delta_4$ are satisfied because $\sigma_2 \|r\|_1 \approx 0.00017 < \delta_1$ and $\delta_3 c^{1-\alpha} \approx 10.264 < \delta_4$. In addition, we obtain

$$h_0(s) = \frac{1}{\Gamma(7/2)} (1 - s)^{5/6} (1 - (1 - s)^{5/3}),$$

$$g_{11}(t, s) = \frac{1}{\Gamma(13/4)} \begin{cases} t^{9/4}(1-s)^{5/6} - (t-s)^{9/4}, & 0 \le s \le t \le 1, \\ t^{9/4}(1-s)^{5/6}, & 0 \le t \le s \le 1, \end{cases}$$

$$g_{12}(t, s) = \frac{1}{\Gamma(23/10)} \begin{cases} t^{13/10}(1-s)^{5/6} - (t-s)^{13/10}, & 0 \le s \le t \le 1, \\ t^{13/10}(1-s)^{5/6}, & 0 \le t \le s \le 1, \end{cases}$$

$$J_2(s) = \begin{cases} \frac{1}{\Gamma(7/2)}(1-s)^{5/6}(1-(1-s)^{5/3}) \\ \quad + \frac{1}{\Delta_2\Gamma(13/4)}\left[\frac{4}{13}(1-s)^{5/6} - \frac{4}{13}(1-s)^{13/4}\right] \\ \quad + \frac{2}{\Delta_2\Gamma(23/10)}\left[\left(\frac{1}{2}\right)^{13/10}(1-s)^{5/6} - \left(\frac{1}{2}-s\right)^{13/10}\right], \\ \qquad 0 \le s < \frac{1}{2}, \\ \frac{1}{\Gamma(7/2)}(1-s)^{5/6}(1-(1-s)^{5/3}) \\ \quad + \frac{1}{\Delta_2\Gamma(13/4)}\left[\frac{4}{13}(1-s)^{5/6} - \frac{4}{13}(1-s)^{13/4}\right] \\ \quad + \frac{2}{\Delta_2\Gamma(23/10)}\left(\frac{1}{2}\right)^{13/10}(1-s)^{5/6}, \quad \frac{1}{2} \le s \le 1. \end{cases}$$

Because

$$\varphi(t,1) = \max\{f(t,x), \ (1-\sigma_2\|r\|_1)t^{5/2} \le x \le 1\} + r(t)$$

$$= \max\left\{\frac{1}{9\sqrt[5]{t^2(1-t)^3}}\left(\sqrt{x} + \frac{1}{\sqrt[7]{x}}\right), \ (1-\sigma_2\|r\|_1)t^{5/2} \le x \le 1\right\}$$

$$\le \frac{1}{9\sqrt[5]{t^2(1-t)^3}}\left(1 + \frac{1}{((1-\sigma_2\|r\|_1)t^{5/2})^{1/7}}\right), \quad \forall t \in (0,1),$$

we deduce by using a computer program

$$\int_0^1 J_2(s)\varphi(s,1)\,ds \le \int_0^1 J_2(s)\frac{1}{9\sqrt[5]{s^2(1-s)^3}}\left(1 + \frac{1}{((1-\sigma_2\|r\|_1)s^{5/2})^{1/7}}\right)ds$$

$$\approx 0.25127612 < 1,$$

so the assumption (I_1) is satisfied.

Because

$$\psi\left(t,\frac{1}{100}\right) = \min\left\{f(t,x), \ \left(\frac{1}{100} - \sigma_2\|r\|_1\right)t^{5/2} \le x \le \frac{1}{100}\right\} + r(t)$$

$$= \min\left\{\frac{1}{9\sqrt[5]{t^2(1-t)^3}}\left(\sqrt{x} + \frac{1}{\sqrt[7]{x}}\right),\right.$$

$$\left.\left(\frac{1}{100} - \sigma_2\|r\|_1\right)t^{5/2} \le x \le \frac{1}{100}\right\}$$

$$\ge \frac{1}{9\sqrt[5]{t^2(1-t)^3}}\left(\sqrt{\left(\frac{1}{100} - \sigma_2\|r\|_1\right)t^{5/2}} + \sqrt[7]{100}\right),$$

$$\forall t \in (0,1),$$

we conclude

$$\int_0^1 J_2(s)\psi\left(s, \frac{1}{100}\right) ds \geq \int_0^1 J_2(s) \frac{1}{9\sqrt[5]{s^2(1-s)^3}}$$

$$\times \left(\sqrt{\left(\frac{1}{100} - \sigma_2\|r\|_1\right)s^{5/2}} + \sqrt[7]{100}\right) ds$$

$$\approx 0.20792415 > 0.01,$$

and so the assumption (I_2) is satisfied.

Next, we obtain

$$\Phi(t, 5, 1000) = \max\{f(t, x), \ (5 - \sigma_2\|r\|_1)t^{5/2} \leq x \leq 1000\} + r(t)$$

$$\leq \frac{1}{9\sqrt[5]{t^2(1-t)^3}} \begin{cases} 1 + \frac{1}{\sqrt[7]{(5-\sigma_2\|r\|_1)t^{5/2}}}, & 0 < t \leq t_1, \\ 2198 - \sqrt[3]{3}, & t_1 < t < 1, \end{cases}$$

where $t_1 = ((5 - \sigma_2\|r\|_1)(2197 - \sqrt[3]{3})^7)^{-2/5} \approx 2.31256 \times 10^{-10}$. We find that

$$\int_0^1 J_2(s)\Phi(s, 5, 1000)\, ds = \int_0^{t_1} J_2(s)\Phi(s, 5, 1000)\, ds$$

$$+ \int_{t_1}^1 J_2(s)\Phi(s, 5, 1000)\, ds$$

$$\leq \int_0^{t_1} J_2(s) \frac{1}{9\sqrt[5]{s^2(1-s)^3}}$$

$$\times \left(1 + \frac{1}{\sqrt[7]{(5 - \sigma_2\|r\|_1)s^{5/2}}}\right) ds$$

$$+ \int_{t_1}^1 J_2(s) \frac{1}{9\sqrt[5]{s^2(1-s)^3}}(2198 - \sqrt[3]{3})\, ds$$

$$\approx 231.13 < 1000,$$

so the assumption (I_3) is satisfied.

We also deduce for $t \in [3/4, 1)$ that

$$\Psi(t, 5, 15) = \min\{f(t, x), \ (5 - \sigma_2\|r\|_1)t^{5/2} \leq x \leq 15\} + r(t)$$

$$\geq \frac{1}{9\sqrt[5]{t^2(1-t)^3}} \begin{cases} [(5 - \sigma_2\|r\|_1)t^{5/2}]^7 + 1, & 3/4 \leq t \leq t_2, \\ \sqrt[3]{(5 - \sigma_2\|r\|_1)t^{5/2}} + 2188 - \sqrt[3]{3}, & t_2 < t < 1, \end{cases}$$

where $t_2 = (3/(5 - \sigma_2 \|r\|_1))^{2/5} \approx 0.815204$. Then we get

$$
\int_{3/4}^{1} J_2(s)\Psi(s,5,15)\,ds = \int_{3/4}^{t_2} J_2(s)\Psi(s,5,15)\,ds + \int_{t_2}^{1} J_2(s)\Psi(s,5,15)\,ds
$$

$$
\geq \int_{3/4}^{t_2} J_2(s)\frac{1}{9\sqrt[5]{s^2(1-s)^3}}([(5-\sigma_2\|r\|_1)s^{5/2}]^7 + 1)\,ds
$$

$$
+ \int_{t_2}^{1} J_2(s)\frac{1}{9\sqrt[5]{s^2(1-s)^3}}\left(\sqrt[3]{(5-\sigma_2\|r\|_1)s^{5/2}} + 2188 - \sqrt[3]{3}\right)\,ds
$$

$$
\approx 35.0712 > 5 \cdot (3/4)^{-5/2} \approx 10.264.
$$

Hence, the assumption (I_4) is satisfied.

Therefore, by Theorem 2.4.2, we conclude that the boundary value problem (2.72), (2.73) has at least two positive solutions $u_1(t)$, $u_2(t)$, $t \in [0, 1]$ that satisfy the conditions $0.009826 \leq \|u_1\| \leq 1$, $4.999826 \leq \|u_2\| \leq 1000$ and $\min_{t \in [3/4,1]} u_2(t) \geq 4.999826$.

Remark 2.4.2. The results presented in this section were published in [2]. The existence of nonnegative solutions for the fractional differential equation (2.19) subject to the boundary conditions (2.46) was studied in [85].

2.5   On a Singular Fractional Boundary Value Problem with Parameters

We consider the nonlinear fractional differential equation

$$
D_{0+}^{\alpha} u(t) + \lambda h(t) f(t, u(t)) = 0, \quad t \in (0, 1), \tag{2.74}
$$

with the nonlocal boundary conditions

$$
u(0) = u'(0) = \cdots = u^{(n-2)}(0) = 0, \quad D_{0+}^{\beta_0} u(1) = \sum_{i=1}^{m} \int_{0}^{1} D_{0+}^{\beta_i} u(t)\,dH_i(t), \tag{2.75}
$$

where $\alpha \in \mathbb{R}$, $\alpha \in (n-1, n]$, $n, m \in \mathbb{N}$, $n \geq 3$, $\beta_i \in \mathbb{R}$ for all $i = 0, \ldots, m$, $0 \leq \beta_1 < \beta_2 < \cdots < \beta_m \leq \beta_0 < \alpha - 1$, $\beta_0 \geq 1$, λ is a positive parameter, D_{0+}^{k} denotes the Riemann-Liouville derivative of order k (for $k = \alpha, \beta_0, \beta_1, \ldots, \beta_m$), the integrals from the boundary conditions (2.75) are Riemann–Stieltjes integrals with H_i, $i = 1, \ldots, m$, functions of bounded variation, the nonnegative function $f(t, u)$ may have singularity at $u = 0$ and the nonnegative function $h(t)$ may be singular at $t = 0$ and/or $t = 1$.

Under some assumptions for the functions h and f, we establish intervals for the parameter λ which are dependent on the principal characteristic value of an associated linear operator, for which the problem (2.74), (2.75) has positive solutions $u \in C([0,1], \mathbb{R}_+)$, $u(t) > 0$ for all $t \in (0,1]$. In the proof of the main theorems, we use the fixed point index theory and an application of the Krein–Rutman theorem in the space of continuous functions defined on $[0,1]$ (see Theorem 2.5.1). In the case in which $h \equiv 1$ and f is a function which changes sign and has singularities at $t = 0$ and/or $t = 1$, we present two existence results for the positive solutions of this problem, in the proof of which we apply the Guo–Krasnosel'skii fixed point theorem. In [116] the authors prove the existence of positive solutions of fractional differential equation (2.74) supplemented with the boundary conditions

$$u(0) = u'(0) = \cdots = u^{(n-2)}(0) = 0, \quad D_{0+}^{\beta} u(1) = \sum_{i=1}^{\infty} \alpha_i D_{0+}^{\gamma} u(\xi_i), \quad (2.76)$$

where $\beta \in [1, n-2]$, $\gamma \in [0, \beta]$, $\alpha_i \geq 0$, $i - 1, 2, \ldots$, $0 < \xi_1 < \xi_2 < \cdots < \xi_{i-1} < \xi_i < \cdots < 1$, and $\Gamma(\alpha - \gamma) > \Gamma(\alpha - \beta) \sum_{i=1}^{\infty} \alpha_i \xi_i^{\alpha-\gamma-1}$. The last condition of the boundary conditions (2.76) can be written as $D_{0+}^{\beta} u(1) = \int_0^1 D_{0+}^{\gamma} u(t) dH(t)$, where H is the step function defined by $H(t) = \{0, t \in [0, \xi_1]; \alpha_1, t \in (\xi_1, \xi_2]; \alpha_1 + \alpha_2, t \in (\xi_2, \xi_3]; \ldots; \sum_{i=1}^{n} \alpha_i, t \in (\xi_n, \xi_{n+1}]; \ldots \}$, so this condition is a particular case of our condition from (2.75).

In this section, we will use the notations $(\Delta_2, G_2, g_0, g_{1i}, J_2, h_0, \sigma_2)$ and the auxiliary results (Lemmas 2.4.1–2.4.5) from Section 2.4.

2.5.1   *Existence of positive solutions*

In this section, we present intervals for the parameter λ such that our problem (2.74), (2.75) has at least one positive solution. We consider the Banach space $X = C[0,1]$ with the supremum norm $\|u\| = \sup_{t \in [0,1]} |u(t)|$, and we define the cones

$$P_0 = \{u \in X, \ u(t) \geq 0, \ \forall t \in [0,1]\},$$

$$P = \{u \in X, \ u(t) \geq t^{\alpha-1} \|u\|, \ \forall t \in [0,1]\} \subset P_0.$$

We define the operator $\mathcal{D} : P_0 \to P_0$ and the linear operator $\mathcal{L} : X \to X$ by

$$\mathcal{D}u(t) = \lambda \int_0^1 G_2(t,s)h(s)f(s,u(s))\,ds, \quad t \in [0,1], \ u \in P_0,$$

$$\mathcal{L}u(t) = \int_0^1 G_2(t,s)h(s)u(s)\,ds, \quad t \in [0,1], \ u \in X.$$

We see that u is a solution of problem (2.74), (2.75) if and only if u is a fixed point of operator $\mathcal{D}$. As in Section 2.4, for $r > 0$, we denote by $P_r = \{x \in P, \ \|x\| < r\}$ and $\overline{P}_r = \{x \in P, \ \|x\| \leq r\}$.

We introduce now the assumptions that we will use in what follows.

(K1) $\alpha \in \mathbb{R}$, $\alpha \in (n-1, n]$, $n, m \in \mathbb{N}$, $n \geq 3$, $\beta_i \in \mathbb{R}$ for all $i = 0, \ldots, m$, $0 \leq \beta_1 < \beta_2 < \cdots < \beta_m \leq \beta_0 < \alpha - 1$, $\beta_0 \geq 1$, $H_i : [0,1] \to \mathbb{R}$, $i = 1, \ldots, m$, are nondecreasing functions, $\lambda > 0$, and $\Delta_2 = \frac{\Gamma(\alpha)}{\Gamma(\alpha - \beta_0)} - \sum_{i=1}^m \frac{\Gamma(\alpha)}{\Gamma(\alpha - \beta_i)} \int_0^1 s^{\alpha - \beta_i - 1}\,dH_i(s) > 0$.

(K2) The function $h \in C((0,1), [0,\infty))$ and $\int_0^1 J_2(s)h(s)\,ds < \infty$.

(K3) The function $f \in C([0,1] \times (0,\infty), [0,\infty))$ and for any $0 < r < R$ we have

$$\lim_{n \to \infty} \sup_{u \in \overline{P}_R \setminus P_r} \int_{E_n} h(s)f(s,u(s))\,ds = 0,$$

where $E_n = [0, \frac{1}{n}] \cup [\frac{n-1}{n}, 1]$.

Lemma 2.5.1. *Assume that assumptions* (K1)–(K3) *hold. Then for any* $0 < r < R$, *the operator* $\mathcal{D} : \overline{P}_R \setminus P_r \to P$ *is completely continuous.*

Proof. By $(K3)$, we deduce that there exists a natural number $n_1 \geq 3$ such that

$$\sup_{u \in \overline{P}_R \setminus P_r} \int_{E_{n_1}} h(s)f(s,u(s))\,ds < 1.$$

For $u \in \overline{P}_R \setminus P_r$, there exists $r_1 \in [r, R]$ such that $\|u\| = r_1$, and then

$$t^{\alpha-1}r \leq t^{\alpha-1}r_1 \leq u(t) \leq r_1 \leq R, \quad \forall t \in [0,1].$$

Let $L_1 = \max\left\{f(t,x),\ t \in \left[\frac{1}{n_1}, \frac{n_1-1}{n_1}\right],\ x \in \left[\frac{1}{n_1^{\alpha-1}}r, R\right]\right\}$. By Lemma 2.4.4, (K2) and (K3), we find

$$\sup_{u \in \overline{P}_R \backslash P_r} \lambda \int_0^1 G_2(t,s)h(s)f(s,u(s))\,ds$$

$$\leq \sup_{u \in \overline{P}_R \backslash P_r} \lambda \int_0^1 J_2(s)h(s)f(s,u(s))\,ds$$

$$\leq \sup_{u \in \overline{P}_R \backslash P_r} \lambda \int_{E_{n_1}} J_2(s)h(s)f(s,u(s))\,ds$$

$$+ \sup_{u \in \overline{P}_R \backslash P_r} \lambda \int_{\frac{1}{n_1}}^{\frac{n_1-1}{n_1}} J_2(s)h(s)f(s,u(s))\,ds$$

$$\leq \lambda J_0 + \lambda L_1 \int_{\frac{1}{n_1}}^{\frac{n_1-1}{n_1}} J_2(s)h(s)\,ds \leq \lambda J_0$$

$$+ \lambda L_1 \int_0^1 J_2(s)h(s)\,ds < \infty,$$

where $J_0 = \max_{t \in [0,1]} J_2(t)$. This implies that the operator $\mathcal{D}$ is well defined.

We show next that $\mathcal{D} : \overline{P}_R \backslash P_r \to P$. Indeed, for any $u \in \overline{P}_R \backslash P_r$ and $t \in [0,1]$, we have

$$(\mathcal{D}u)(t) = \lambda \int_0^1 G_2(t,s)h(s)f(s,u(s))\,ds \leq \lambda \int_0^1 J_2(s)h(s)f(s,u(s))\,ds,$$

and then

$$\|\mathcal{D}u\| \leq \lambda \int_0^1 J_2(s)h(s)f(s,u(s))\,ds.$$

On the other hand, by Lemma 2.4.4, we obtain

$$(\mathcal{D}u)(t) \geq \lambda t^{\alpha-1} \int_0^1 J_2(s)h(s)f(s,u(s))\,ds \geq t^{\alpha-1}\|\mathcal{D}u\|, \quad \forall t \in [0,1],$$

so $\mathcal{D}u \in P$. Therefore, $\mathcal{D}(\overline{P}_R \backslash P_r) \subset P$.

We prove now that $\mathcal{D} : \overline{P}_R \backslash P_r \to P$ is completely continuous. We assume that $S \subset \overline{P}_R \backslash P_r$ is an arbitrary bounded set. From the first part of the proof, we know that $\mathcal{D}(S)$ is uniformly bounded. Then we show that

$\mathcal{D}(S)$ is equicontinuous. Indeed, for $\varepsilon > 0$, there exists a natural number $n_2 \geq 3$ such that

$$\sup_{u \in \overline{P}_R \setminus P_r} \int_{E_{n_2}} h(s) f(s, u(s)) \, ds < \frac{\varepsilon}{4\lambda J_0}.$$

Since $G_2(t, s)$ is uniformly continuous on $[0, 1] \times [0, 1]$, for the above $\varepsilon > 0$ there exists $\delta > 0$ such that for any $t_1, t_2 \in [0, 1]$, with $|t_1 - t_2| < \delta$, and $s \in \left[\frac{1}{n_2}, \frac{n_2 - 1}{n_2}\right]$, we have

$$|G_2(t_1, s) - G_2(t_2, s)| < \frac{\varepsilon}{2\lambda \bar{h} L_2},$$

where $L_2 = \max\left\{1, \max\left\{f(t, x), t \in \left[\frac{1}{n_2}, \frac{n_2 - 1}{n_2}\right], x \in \left[\frac{1}{n_2^{\alpha - 1}} r, R\right]\right\}\right\}$ and $\bar{h} = \max\left\{1, \max\{h(t), t \in \left[\frac{1}{n_2}, \frac{n_2 - 1}{n_2}\right]\}\right\}$.

Then for any $u \in S$, $t_1, t_2 \in [0, 1]$, with $|t_1 - t_2| < \delta$, we deduce

$$|(\mathcal{D}u)(t_1) - (\mathcal{D}u)(t_2)| = \lambda \left| \int_0^1 (G_2(t_1, s) - G_2(t_2, s)) h(s) f(s, u(s)) \, ds \right|$$

$$\leq 2\lambda \int_{E_{n_2}} J_2(s) h(s) f(s, u(s)) \, ds$$

$$+ \lambda \sup_{u \in S} \int_{\frac{1}{n_2}}^{\frac{n_2 - 1}{n_2}} |G_2(t_1, s) - G_2(t_2, s)| h(s) f(s, u(s)) \, ds$$

$$\leq 2\lambda J_0 \frac{\varepsilon}{4\lambda J_0} + \frac{\varepsilon \lambda}{2\lambda \bar{h} L_2} \left(\int_{\frac{1}{n_2}}^{\frac{n_2 - 1}{n_2}} h(s) \, ds \right) L_2$$

$$\leq \frac{\varepsilon}{2} + \frac{\varepsilon}{2} = \varepsilon.$$

This gives us that $\mathcal{D}(S)$ is equicontinuous. By the Arzela–Ascoli theorem, we conclude that $\mathcal{D} : \overline{P}_R \setminus P_r \to P$ is compact.

Finally we prove that $\mathcal{D} : \overline{P}_R \setminus P_r \to P$ is continuous. We suppose that $u_n, u_0 \in \overline{P}_R \setminus P_r$ for all $n \geq 1$ and $\|u_n - u_0\| \to 0$ as $n \to \infty$. Then $r \leq \|u_n\| \leq R$ for all $n \geq 0$. By (K3), for $\varepsilon > 0$ there exists a natural number $n_3 \geq 3$ such that

$$\sup_{u \in \overline{P}_R \setminus P_r} \int_{E_{n_3}} h(s) f(s, u(s)) \, ds < \frac{\varepsilon}{4\lambda J_0}. \tag{2.77}$$

Because $f(t, x)$ is uniformly continuous in $\left[\frac{1}{n_3}, \frac{n_3 - 1}{n_3}\right] \times \left[\frac{1}{n_3^{\alpha - 1}} r, R\right]$, we obtain

$$\lim_{n \to \infty} |f(s, u(s)) - f(s, u_0(s))| = 0, \quad \text{uniformly for } s \in \left[\frac{1}{n_3}, \frac{n_3 - 1}{n_3}\right].$$

Then the Lebesgue dominated convergence theorem gives us

$$\int_{\frac{1}{n_3}}^{\frac{n_3-1}{n_3}} h(s)|f(s,u_n(s)) - f(s,u_0(s))|\,ds \to 0, \quad \text{as } n \to \infty.$$

Thus, for the above $\varepsilon > 0$, there exists a natural number N such that for $n > N$, we have

$$\int_{\frac{1}{n_3}}^{\frac{n_3-1}{n_3}} h(s)|f(s,u_n(s)) - f(s,u_0(s))|\,ds < \frac{\varepsilon}{2\lambda J_0}. \tag{2.78}$$

By (2.77) and (2.78), we conclude that

$$\|\mathcal{D}u_n - \mathcal{D}u_0\| \leq \sup_{u \in \overline{P}_R \backslash P_r} \lambda \int_{E_{n_3}} J_2(s)h(s)|f(s,u_n(s)) - f(s,u_0(s))|\,ds$$

$$+ \sup_{u \in \overline{P}_R \backslash P_r} \lambda \int_{\frac{1}{n_3}}^{\frac{n_3-1}{n_3}} J_2(s)h(s)|f(s,u_n(s)) - f(s,u_0(s))|\,ds$$

$$\leq \lambda J_0 \frac{\varepsilon}{4\lambda J_0} + \lambda J_0 \frac{\varepsilon}{4\lambda J_0} + \frac{\varepsilon}{2\lambda J_0}\lambda J_0 = \varepsilon.$$

This implies that $\mathcal{D} : \overline{P}_R \backslash P_r \to P$ is continuous. Hence, $\mathcal{D} : \overline{P}_R \backslash P_r \to P$ is completely continuous. $\qquad\square$

Under assumptions (K1)–(K3), by the extension theorem, the operator $\mathcal{D}$ has a completely continuous extension (also denoted by $\mathcal{D}$) from P to P.

For the next lemma, we need the following application of the Krein–Rutman theorem in the space X.

Theorem 2.5.1 (see [69, 114]). *Suppose that $A : X \to X$ is a completely continuous linear operator, and $A(P_0) \subset P_0$. If there exist $v \in X \backslash (-P_0)$ and a constant $c > 0$ such that $cAv \geq v$, then the spectral radius $r(A) \neq 0$ and A has an eigenvector $u_0 \in P_0 \backslash \{\theta\}$ corresponding to its principal characteristic value $\lambda_1 = (r(A))^{-1}$, that is $\lambda_1 Au_0 = u_0$ or $Au_0 = r(A)u_0$, and so $r(A) > 0$.*

Lemma 2.5.2. *Assume that assumptions (K1) and (K2) hold. Then the spectral radius $r(\mathcal{L}) \neq 0$ and $\mathcal{L}$ has an eigenfunction $\psi_1 \in P_0 \backslash \{\theta\}$ corresponding to the principal eigenvalue $r(\mathcal{L})$, that is $\mathcal{L}\psi_1 = r(\mathcal{L})\psi_1$. So $r(\mathcal{L}) > 0$.*

Proof. The operator $\mathcal{L} : X \to X$ is a linear completely continuous operator. By Lemma 2.4.4 we know that $G_2(t, s) > 0$ for all $t, s \in (0, 1)$. By (K2) we deduce that there exists an interval $[c, d] \subset (0, 1)$ $(0 < c < d < 1)$ such that $h(t) > 0$ for all $t \in [c, d]$. We consider a function $\varphi \in C[0, 1]$ satisfying the conditions $\varphi(t) > 0$ for $t \in (c, d)$ and $\varphi(t) = 0$ for $t \notin (c, d)$. Then for all $t \in [c, d]$ we have

$$
(\mathcal{L}\varphi)(t) = \int_0^1 G_2(t, s)h(s)\varphi(s)\, ds
$$

$$
\geq \int_c^d G_2(t, s)h(s)\varphi(s)\, ds > 0, \quad \forall t \in [c, d].
$$

Hence, there exists a constant $a > 0$ $\left(a = \frac{\max_{t \in [c,d]} \varphi(t)}{\min_{t \in [c,d]} (\mathcal{L}\varphi)(t)}\right)$ which satisfies the inequality $a(\mathcal{L}\varphi)(t) \geq \varphi(t)$, $\forall t \in [0, 1]$. By Theorem 2.5.1 we conclude that the spectral radius $r(\mathcal{L}) \neq 0$ and $\mathcal{L}$ has an eigenfunction $\psi_1 \in P_0 \setminus \{\theta\}$ corresponding to its principal characteristic value $\lambda_1 = (r(\mathcal{L}))^{-1}$ such that $\mathcal{L}\psi_1 = r(\mathcal{L})\psi_1$, and so $r(\mathcal{L}) > 0$. $\qquad \square$

Using a similar argument as that used in the proof of Lemma 2.5.1 for operator $\mathcal{L}$, we obtain that $\mathcal{L}(P) \subset P$.

Theorem 2.5.2. *Assume that assumptions* (K1)–(K3) *hold. If*

$$
0 \leq f_\infty^s := \limsup_{u \to \infty} \max_{t \in [0,1]} \frac{f(t, u)}{u} < f_0^i := \liminf_{u \to 0+} \min_{t \in [0,1]} \frac{f(t, u)}{u} \leq \infty,
$$

then for any $\lambda \in \left(\frac{1}{f_0^i r(\mathcal{L})}, \frac{1}{f_\infty^s r(\mathcal{L})}\right)$, *the problem* (2.74), (2.75) *has at least one positive solution* $u(t)$, $t \in [0, 1]$, *(with the conventions* $\frac{1}{0_+} = \infty$ *and* $\frac{1}{\infty} = 0_+$).

Proof. We consider $\lambda \in \left(\frac{1}{f_0^i r(\mathcal{L})}, \frac{1}{f_\infty^s r(\mathcal{L})}\right)$. For f_0^i we have the cases: $f_0^i \in (0, \infty)$ with $f_0^i > \frac{1}{\lambda r(\mathcal{L})}$, and $f_0^i = \infty$. In the first case, $f_0^i \in (0, \infty)$ with $f_0^i > \frac{1}{\lambda r(\mathcal{L})}$ we obtain

$$
\forall \varepsilon > 0 \; \exists \delta(\varepsilon) > 0 \;\; \text{s.t.} \;\; \frac{f(t, u)}{u} \geq f_0^i - \varepsilon, \quad \forall t \in [0, 1], \; u \in (0, \delta(\varepsilon)].
$$

By taking $\varepsilon = f_0^i - \frac{1}{\lambda r(\mathcal{L})}$ we deduce that there exists $r_1' > 0$ such that $\frac{f(t,u)}{u} \geq \frac{1}{\lambda r(\mathcal{L})}$ for all $t \in [0, 1]$ and $u \in (0, r_1']$, and so $f(t, u) \geq \frac{u}{\lambda r(\mathcal{L})}$ for all $t \in [0, 1]$ and $u \in [0, r_1']$.

In the case $f_0^i = \infty$, we have

$$\forall \varepsilon > 0 \; \exists \delta(\varepsilon) > 0 \text{ s.t. } \frac{f(t,u)}{u} \geq \varepsilon, \;\; \forall t \in [0,1], \; u \in (0, \delta(\varepsilon)].$$

So for $\varepsilon = \frac{1}{\lambda r(\mathcal{L})}$ we deduce that there exists $r_1'' > 0$ such that $f(t,u) \geq \frac{u}{\lambda r(\mathcal{L})}$ for all $t \in [0,1]$ and $u \in [0, r_1'']$.

Hence, in both the above cases, we conclude that there exists $r_1 > 0$ such that $f(t,u) \geq \frac{u}{\lambda r(\mathcal{L})}$ for all $t \in [0,1]$ and $u \in [0, r_1]$.

Then for any $u \in \partial P_{r_1}$ we find

$$\mathcal{D}u(t) = \lambda \int_0^1 G_2(t,s)h(s)f(s,u(s))\,ds$$

$$\geq \frac{1}{r(\mathcal{L})} \int_0^1 G_2(t,s)h(s)u(s)\,ds = \frac{1}{r(\mathcal{L})}\mathcal{L}u(t), \quad \forall t \in [0,1].$$

We assume that $\mathcal{D}$ has no fixed point on ∂P_{r_1}, (otherwise the proof is finished). We will prove that

$$u - \mathcal{D}u \neq \mu\psi_1, \quad \forall u \in \partial P_{r_1}, \quad \mu \geq 0, \tag{2.79}$$

where ψ_1 is given in Lemma 2.5.2. If not, there exist $u_1 \in \partial P_{r_1}$ and $\mu_1 \geq 0$ such that $u_1 - \mathcal{D}u_1 = \mu_1\psi_1$. Then $\mu_1 > 0$ and $u_1 = \mathcal{D}u_1 + \mu_1\psi_1 \geq \mu_1\psi_1$. We denote by $\mu_0 = \sup\{\mu, \; u_1 \geq \mu\psi_1\}$. Then $\mu_0 \geq \mu_1$, $u_1 \geq \mu_0\psi_1$ and

$$\mathcal{D}u_1 \geq \frac{1}{r(\mathcal{L})}\mathcal{L}u_1 \geq \frac{1}{r(\mathcal{L})}\mu_0\mathcal{L}\psi_1 = \mu_0\psi_1.$$

Hence, $u_1 = \mathcal{D}u_1 + \mu_1\psi_1 \geq \mu_0\psi_1 + \mu_1\psi_1 = (\mu_0 + \mu_1)\psi_1$, which contradicts the definition of μ_0. So, relation (2.79) holds, and by Theorem 1.2.9, we deduce that

$$i(\mathcal{D}, P_{r_1}, P) = 0. \tag{2.80}$$

For f_∞^s, we also have two cases: $f_\infty^s \in (0, \infty)$ with $f_\infty^s < \frac{1}{\lambda r(\mathcal{L})}$, and $f_\infty^s = 0$. In the first case, $f_\infty^s \in (0, \infty)$ with $f_\infty^s < \frac{1}{\lambda r(\mathcal{L})}$, we obtain

$$\forall \varepsilon > 0 \; \exists \delta(\varepsilon) > 0 \text{ s.t. } \frac{f(t,u)}{u} \leq f_\infty^s + \varepsilon, \;\; \forall t \in [0,1], \; u \geq \delta(\varepsilon).$$

By taking $\varepsilon = \frac{1}{2\lambda r(\mathcal{L})} - \frac{f_\infty^s}{2}$, we deduce that there exists $r_2' > r_1$ such that $f(t,u) \leq \frac{\theta_1}{\lambda r(\mathcal{L})}u$ for all $t \in [0,1]$ and $u \in [r_2', \infty)$, where $\theta_1 = \frac{1}{2} + \frac{f_\infty^s \lambda r(\mathcal{L})}{2} \in (0,1)$.

In the case $f_\infty^s = 0$, we have

$$\forall \varepsilon > 0 \ \exists \delta(\varepsilon) > 0 \text{ s.t. } \frac{f(t,u)}{u} \leq \varepsilon, \ \forall t \in [0,1], \ u \geq \delta(\varepsilon).$$

So, for $\varepsilon = \frac{1}{2\lambda r(\mathcal{L})}$, we deduce that there exists $r_2'' > r_1$ such that $f(t,u) \leq \frac{1}{2\lambda r(\mathcal{L})} u$ for all $t \in [0,1]$ and $u \in [r_2'', \infty)$.

Therefore, in both the above cases, we conclude that there exist $\theta \in (0,1)$ and $r_2 > r_1$ such that $f(t,u) \leq \theta \frac{1}{\lambda r(\mathcal{L})} u$ for all $t \in [0,1]$ and $u \in [r_2, \infty)$.

We define now the operator $\mathcal{L}_1 : X \to X$ by

$$\mathcal{L}_1 u = \theta \frac{1}{r(\mathcal{L})} \mathcal{L} u = \frac{\theta}{r(\mathcal{L})} \int_0^1 G_2(t,s) h(s) u(s) \, ds, \quad t \in [0,1], \ u \in X.$$

The operator $\mathcal{L}_1$ is linear and bounded, and $\mathcal{L}_1(P) \subset P$. Because $\theta \in (0,1)$ we obtain $r(\mathcal{L}_1) = \theta < 1$. We consider the set

$$Z = \{u \in P \setminus B_{r_1}, \ \mu u = \mathcal{D}u, \text{ with } \mu \geq 1\}.$$

For $u \in P$, we denote by $D(u) = \{t \in [0,1], \ u(t) \geq r_2\}$. Then for $u \in P$, we have $u(t) \geq r_2$ for all $t \in D(u)$, and so

$$f(t, u(t)) \leq \theta \frac{1}{\lambda r(\mathcal{L})} u(t), \quad \forall t \in D(u). \tag{2.81}$$

By (2.81) and the definition of operator $\mathcal{L}$, for any $u \in Z$, $\mu \geq 1$ and $t \in [0,1]$ we deduce

$$u(t) \leq \mu u(t) = (\mathcal{D}u)(t) = \lambda \int_0^1 G_2(t,s) h(s) f(s, u(s)) \, ds$$

$$= \lambda \int_{D(u)} G_2(t,s) h(s) f(s, u(s)) \, ds$$

$$+ \lambda \int_{[0,1] \setminus D(u)} G_2(t,s) h(s) f(s, u(s)) \, ds$$

$$\leq \frac{\theta}{r(\mathcal{L})} \int_{D(u)} G_2(t,s) h(s) u(s) \, ds + \lambda \int_0^1 J_2(s) h(s) f(s, \widetilde{u}(s)) \, ds$$

$$\leq \frac{\theta}{r(\mathcal{L})} \int_0^1 G_2(t,s) h(s) u(s) \, ds + \lambda J_0 M_1 = (\mathcal{L}_1 u)(t) + \lambda J_0 M_1,$$

$$\tag{2.82}$$

where $\tilde{u}(t) = \min\{u(t), r_2\}$ for all $t \in [0,1]$, (which satisfies the inequalities $r_1 t^{\alpha-1} \le \tilde{u}(t) \le r_2$ for all $t \in [0,1]$), and $M_1 = \sup_{u \in \overline{P}_{r_2} \backslash P_{r_1}} \int_0^1 h(s) f(s, u(s))\, ds$, (as in the proof of Lemma 2.5.1 we obtain that $M_1 < \infty$). By the Gelfand formula we know that $(I - \mathcal{L}_1)^{-1}$ exists and $(I - \mathcal{L}_1)^{-1} = \sum_{i=1}^{\infty} \mathcal{L}_1^i$, which implies $(I - \mathcal{L}_1)^{-1}(P) \subset P$. This together with (2.82) gives us $u(t) \le (I - \mathcal{L}_1)^{-1}(\lambda J_0 M_1)$ and so $u(t) \le \lambda J_0 M_1 \| (I - \mathcal{L}_1)^{-1} \|$ for all $t \in [0,1]$, which means that Z is bounded. Now we choose $R > \max\{r_2, \sup\{\|u\|, \ u \in Z\}\}$. Then we obtain that $\mu u \ne \mathcal{D}u$ for all $u \in \partial P_R$ and $\mu \ge 1$. By Theorem 1.2.7, we conclude that

$$i(\mathcal{D}, P_R, P) = 1. \tag{2.83}$$

By (2.80), (2.83) and the additivity property of the fixed point index, we deduce that

$$i(\mathcal{D}, P_R \backslash \overline{P}_{r_1}, P) = i(\mathcal{D}, P_R, P) - i(\mathcal{D}, P_{r_1}, P) = 1.$$

So, operator $\mathcal{D}$ has at least one fixed point on $P_R \backslash \overline{P}_{r_1}$, which is a positive solution of problem (2.74),(2.75). $\qquad\square$

By using a similar approach as that used in the proof of Theorem 2.5.2, we obtain the following result.

Theorem 2.5.3. *Assume that assumptions* (K1)–(K3) *hold. If*

$$0 \le f_0^s := \limsup_{u \to 0+} \max_{t \in [0,1]} \frac{f(t,u)}{u} < f_\infty^i := \liminf_{u \to \infty} \min_{t \in [0,1]} \frac{f(t,u)}{u} \le \infty,$$

then for any $\lambda \in \left(\frac{1}{f_\infty^i r(\mathcal{L})}, \frac{1}{f_0^s r(\mathcal{L})} \right)$, *the problem* (2.74), (2.75) *has at least one positive solution* $u(t)$, $t \in [0,1]$.

2.5.2   Some remarks on a related semipositone problem

In this section, we present two existence results for a semipositone problem associated to problem (2.74), (2.75). More precisely, we consider the fractional differential equation

$$D_{0+}^\alpha u(t) + \lambda \tilde{f}(t, u(t)) = 0, \quad t \in (0,1), \tag{2.84}$$

subject to the boundary conditions (2.75). We suppose that assumption (K1) holds and $\tilde{f}$ satisfies the following conditions:

(K2)′ The function $\tilde{f} \in C((0,1) \times [0,\infty), \mathbb{R})$ may be singular at $t = 0$ and/or $t = 1$, and there exist the functions $p, q \in C((0,1), [0,\infty))$,

$g \in C([0,1] \times [0,\infty), [0,\infty))$ such that $-p(t) \leq \widetilde{f}(t,u) \leq q(t)g(t,u)$ for all $t \in (0,1)$ and $u \in [0,\infty)$, with $0 < \int_0^1 p(t)\,dt < \infty$, $0 < \int_0^1 q(t)\,dt < \infty$.

(K3)′ There exists $\zeta \in (0,1/2)$ such that $\lim_{u\to\infty} \min_{t\in[\zeta,1-\zeta]} \frac{\widetilde{f}(t,u)}{u} = \infty$.

By using the Guo–Krasnosel'skii fixed point theorem (Theorem 1.2.2) and similar arguments as those used in the proofs of Theorem 2.1.1 and Theorem 2.1.2, we obtain the following results for the problem (2.84), (2.75).

Theorem 2.5.4. *Assume that* (K1), (K2)′ *and* (K3)′ *hold. Then there exists* $\lambda^* > 0$ *such that, for any* $\lambda \in (0,\lambda^*]$, *the boundary value problem* (2.84), (2.75) *has at least one positive solution.*

In the proof of Theorem 2.5.4 we consider $R_1 > \sigma_2 \int_0^1 p(t)\,dt > 0$, and we define $\lambda^* = \min\{1, R_1(M_2 \int_0^1 J_2(s)(q(s)+p(s))\,ds)^{-1}\}$, with $M_2 = \max\{\max_{t\in[0,1],\, u\in[0,R_1]} g(t,u), 1\}$. The solution $u(t)$, $t \in [0,1]$ satisfies the condition $u(t) \geq \Lambda_1 t^{\alpha-1}$ for all $t \in [0,1]$, where $\Lambda_1 = R_1 - \sigma_2 \int_0^1 p(s)\,ds > 0$.

Theorem 2.5.5. *Assume that* (K1), (K2)′ *and*

(K4) *There exists* $\zeta \in (0,1/2)$ *such that* $\lim_{u\to\infty} \min_{t\in[\zeta,1-\zeta]} \widetilde{f}(t,u) = \infty$ *and* $\lim_{u\to\infty} \max_{t\in[0,1]} \frac{g(t,u)}{u} = 0$,

hold. Then there exists $\lambda_* > 0$ *such that, for any* $\lambda \geq \lambda_*$, *the boundary value problem* (2.84), (2.75) *has at least one positive solution.*

By (K4), we know that for $\zeta \in (0,1/2)$ and for a fixed number $L_0 > 0$ there exists $M_3 > 0$ such that $\widetilde{f}(t,u) \geq L_0$ for all $t \in [\zeta, 1-\zeta]$ and $u \geq M_3$. In the proof of Theorem 2.5.5, we define $\lambda_* = M_3(\zeta^{\alpha-1}\sigma_2 \int_0^1 p(s)\,ds)^{-1}$. The solution $u(t)$, $t \in [0,1]$ satisfies the condition $u(t) \geq \widetilde{\Lambda}_1 t^{\alpha-1}$ for all $t \in [0,1]$, where $\widetilde{\Lambda}_1 = \frac{M_3}{\zeta^{\alpha-1}}$.

2.5.3   *Examples*

Let $\alpha = 10/3$ $(n = 4)$, $\beta_0 = 11/5$, $m = 2$, $\beta_1 = 1/2$, $\beta_2 = 5/4$, $H_1(t) = t$ for all $t \in [0,1]$, $H_2(t) = \{0$, for $t \in [0,1/2)$; 1, for $t \in [1/2,1]\}$.

We consider the fractional differential equations

$$D_{0+}^{10/3}u(t) + \lambda h(t)f(t, u(t)) = 0, \quad t \in (0,1), \tag{2.85}$$

$$D_{0+}^{10/3}u(t) + \lambda \widetilde{f}(t, u(t)) = 0, \quad t \in (0,1), \tag{2.86}$$

subject to the boundary conditions

$$u(0) = u'(0) = u''(0) = 0, \quad D_{0+}^{11/5}u(1) = \int_0^1 D_{0+}^{1/2}u(t)\,dt + D_{0+}^{5/4}u\left(\frac{1}{2}\right). \tag{2.87}$$

We have $\Delta_2 \approx 1.12792427 > 0$ and $\sigma_2 \approx 0.94443688$. So assumption (K1) is satisfied. In addition we obtain

$$g_{11}(t,s) = \frac{1}{\Gamma(17/6)} \begin{cases} t^{11/6}(1-s)^{2/15} - (t-s)^{11/6}, & 0 \le s \le t \le 1, \\ t^{11/6}(1-s)^{2/15}, & 0 \le t \le s \le 1, \end{cases}$$

$$g_{12}(t,s) = \frac{1}{\Gamma(25/12)} \begin{cases} t^{13/12}(1-s)^{2/15} - (t-s)^{13/12}, & 0 \le s \le t \le 1, \\ t^{13/12}(1-s)^{2/15}, & 0 \le t \le s \le 1, \end{cases}$$

$$h_0(s) = \frac{1}{\Gamma(10/3)}(1-s)^{2/15}(1 - (1-s)^{11/5}), \quad s \in [0,1],$$

$$J_2(s) = \begin{cases} h_0(s) + \frac{1}{\Delta_2}\left\{ \frac{1}{\Gamma(23/6)}(1-s)^{2/15} - \frac{1}{\Gamma(23/6)}(1-s)^{17/6} \right. \\ \qquad \left. + \frac{1}{\Gamma(25/12)}\left[(\frac{1}{2})^{13/12}(1-s)^{2/15} - (\frac{1}{2}-s)^{13/12}\right]\right\}, \\ \qquad 0 \le s \le 1/2, \\[4pt] h_0(s) + \frac{1}{\Delta_2}\left\{ \frac{1}{\Gamma(23/6)}(1-s)^{2/15} - \frac{1}{\Gamma(23/6)}(1-s)^{17/6} \right. \\ \qquad \left. + \frac{1}{\Gamma(25/12)}(\frac{1}{2})^{13/12}(1-s)^{2/15}\right\}, \quad 1/2 < s \le 1. \end{cases}$$

Example 1. We consider the functions

$$h(t) = \frac{1}{\sqrt[3]{t(1-t)^2}}, \quad t \in (0,1); \quad f(t,u) = \sqrt{u} + t + \frac{1}{\sqrt[4]{u}}, \quad t \in [0,1], \ u > 0.$$

The cone P from Section 2.5.1 is here $P = \{u \in C[0,1], u(t) \ge t^{7/3}\|u\|, \forall t \in [0,1]\}$. For $0 < r < R$ and $u \in \overline{P}_R \setminus P_r$ we deduce

$$f(t, u(t)) \le \sqrt{R} + 1 + \frac{1}{\sqrt[4]{t^{7/3}r}}, \quad \forall t \in (0,1].$$

Besides, we obtain $\int_0^1 J_2(s)h(s)\,ds \leq J_0\Gamma(2/3)\Gamma(1/3) < \infty$, where $J_0 = \max_{s\in[0,1]} J_2(s) \approx 0.781$. Hence, assumption (K2) is satisfied.

For $u \in \overline{P}_R \setminus P_r$ and $E_n = [0, \frac{1}{n}] \cup [\frac{n-1}{n}, 1]$, we find

$$C_n = \int_{E_n} h(s)f(s, u(s))\,ds = \int_{E_n} \frac{1}{\sqrt[3]{s(1-s)^2}}\left(\sqrt{u(s)} + s + \frac{1}{\sqrt[4]{u(s)}}\right)ds$$

$$\leq \int_{E_n} \frac{1}{\sqrt[3]{s(1-s)^2}}\left(\sqrt{R} + 1 + \frac{1}{\sqrt[4]{s^{7/3}r}}\right)ds$$

$$= (\sqrt{R}+1)\int_{E_n} \frac{ds}{\sqrt[3]{s(1-s)^2}} + \frac{1}{\sqrt[4]{r}}\int_{E_n} \frac{1}{s^{11/12}(1-s)^{2/3}}\,ds,$$

and then $\lim_{n\to\infty} \sup_{u\in\overline{P}_R\setminus P_r} C_n = 0$, because $f_1(s) = \frac{1}{\sqrt[3]{s(1-s)^2}} \in L^1(0,1)$ and $f_2(s) = \frac{1}{s^{11/12}(1-s)^{2/3}} \in L^1(0,1)$. Hence, assumption (K3) is satisfied. We also have $f_\infty^s = 0$ and $f_0^i = \infty$. Then by using Theorem 2.5.2, we deduce that, for any $\lambda \in (0,\infty)$, the problem (2.85), (2.87) has at least one positive solution $u(t)$, $t \in [0,1]$, which satisfies the condition $u(t) \geq t^{7/3}\|u\|$ for all $t \in [0,1]$.

Example 2. We consider the function

$$\widetilde{f}(t, u) = \frac{u^3 + u + 1}{\sqrt[4]{t(1-t)^3}} + \ln t, \quad t \in (0,1), \ u \geq 0.$$

For this example, we have $p(t) = -\ln t$ and $q(t) = \frac{1}{\sqrt[4]{t(1-t)^3}}$ for all $t \in (0,1)$, $g(t,u) = u^3 + u + 1$ for all $t \in [0,1]$ and $u \geq 0$, $\int_0^1 p(t)\,dt = 1$, $\int_0^1 q(t)\,dt = \Gamma(\frac{3}{4})\Gamma(\frac{1}{4}) \approx 4.44288$. Then assumption (K2)' is satisfied. In addition, for $\zeta \in (0, 1/2)$ fixed, assumption (K3)' is also satisfied. By some computations, we obtain $\int_0^1 J_2(s)(q(s)+p(s))\,ds \approx 2.71742073$. We choose $R_1 = 2$, which satisfies the condition $R_1 > \sigma_2 \int_0^1 p(t)\,dt \approx 0.944$, and then we deduce $M_2 = 11$ and $\lambda^* \approx 0.0669084$. By Theorem 2.5.4, we conclude that, for any $\lambda \in (0, \lambda^*]$, the problem (2.86), (2.87) has at least one positive solution $u(t)$, $t \in [0,1]$, which satisfies the condition $u(t) \geq \Lambda_1 t^{7/3}$ for all $t \in [0,1]$, where $\Lambda_1 \approx 1.05556$.

Example 3. We consider the function

$$\widetilde{f}(t, u) = \frac{\sqrt{u + 1/3}}{\sqrt[5]{t^3(1-t)^2}} - \frac{1}{\sqrt[3]{t}}, \quad t \in (0,1), \ u \geq 0.$$

Here, we have $p(t) = \frac{1}{\sqrt[3]{t}}$ and $q(t) = \frac{1}{\sqrt[5]{t^3(1-t)^2}}$ for all $t \in (0,1)$, $g(t,u) = \sqrt{u+1/3}$ for all $t \in [0,1]$ and $u \geq 0$. Because $\int_0^1 p(t)\,dt = 3/2$, $\int_0^1 q(t)\,dt \approx 3.30327$, the assumption (K2)' is satisfied. In addition, for $\zeta \in (0,1/2)$, we obtain $\lim_{u\to\infty} \min_{t\in[\zeta,1-\zeta]} \widetilde{f}(t,u) = \infty$ and $\lim_{u\to\infty} \max_{t\in[0,1]} g(t,u)/u = 0$, and then assumption (K4) is also satisfied. We choose $\zeta = 1/4$ and $L_0 = 100$, and then we find $M_3 = 5805$ and $\lambda_* \approx 104075$. Then by Theorem 2.5.5, we deduce that, for any $\lambda \geq \lambda_*$, the problem (2.86), (2.87) has at least one positive solution $u(t)$, $t \in [0,1]$, which satisfies the inequality $u(t) \geq \widetilde{\Lambda}_1 t^{7/3}$ for all $t \in [0,1]$, where $\widetilde{\Lambda}_1 \approx 147438$.

Remark 2.5.1. The results presented in this section will be published in [102].

2.6 A Singular Fractional Differential Equation with Integral Boundary Conditions

We consider the nonlinear fractional differential equation

$$D_{0+}^\alpha u(t) + f(t, u(t)) = 0, \quad t \in (0,1), \tag{2.88}$$

with the boundary conditions

$$\begin{cases} u(0) = u'(0) = \cdots = u^{(n-2)}(0) = 0, \\ D_{0+}^{\beta_0} u(1) = \sum_{i=1}^{m} \int_0^1 a_i(t) D_{0+}^{\beta_i} u(t)\,dH_i(t), \end{cases} \tag{2.89}$$

where $\alpha \in \mathbb{R}$, $\alpha \in (n-1, n]$, $n, m \in \mathbb{N}$, $n \geq 3$, $\beta_i \in \mathbb{R}$ for all $i = 0, \ldots, m$, $0 \leq \beta_1 < \beta_2 < \cdots < \beta_m < \alpha-1, 1 \leq \beta_0 < \alpha-1$, D_{0+}^k denotes the Riemann–Liouville derivative of order k (for $k = \alpha, \beta_0, \beta_1, \ldots, \beta_m$), the integrals from the boundary conditions (2.89) are Riemann-Stieltjes integrals with H_i, $i = 1, \ldots, m$, functions of bounded variation, the functions $a_i \in C(0,1) \cap L^1(0,1)$, $i = 1, \ldots, m$, and the nonlinearity $f(t,u)$ is nonnegative and it may be singular at the points $t = 0$, $t = 1$ and/or $u = 0$.

We will present conditions for the data of problem (2.88), (2.89) connected to the spectral radii of some associated linear operators such that this problem has at least one or two positive solutions $u \in C([0,1], \mathbb{R}_+)$, $u(t) > 0$ for all $t \in (0,1]$. In the proof of the main existence theorems, we use an application of the Krein-Rutman theorem in the space $C[0,1]$ (Theorem 2.5.1) and the fixed point index theory. In [103], the author investigates the fractional differential equation (2.88) supplemented with the boundary conditions

$$u(0) = u'(0) = \cdots = u^{(n-2)}(0) = 0, \quad D_{0+}^{\beta} u(1) = \int_0^{\eta} a(t) D_{0+}^{\gamma} u(t) \, dV(t),$$

where $\beta \in (0,1)$, $\gamma \in [0, \alpha - 1)$, $\eta \in (0,1]$, $a(t) \in L^1(0,1) \cap C(0,1)$, and the function $f(t,u)$ is nonnegative and it may be singular at $t = 0$, $t = 1$ and $x = 0$. The author proves in [103] some existence and multiplicity results which are closely associated with the relationship between 1 and the spectral radii corresponding to the relevant linear operators.

2.6.1   *Preliminary results*

We consider the fractional differential equation

$$D_{0+}^{\alpha} u(t) + \widetilde{x}(t) = 0, \quad t \in (0,1), \tag{2.90}$$

with the boundary conditions (2.89), where $\widetilde{x} \in C(0,1) \cap L^1(0,1)$. We denote by $\Delta_3 = \frac{\Gamma(\alpha)}{\Gamma(\alpha - \beta_0)} - \sum_{i=1}^{m} \frac{\Gamma(\alpha)}{\Gamma(\alpha - \beta_i)} \int_0^1 s^{\alpha - \beta_i - 1} a_i(s) \, dH_i(s)$. In a similar manner as we proved Lemma 2.4.2, we obtain the following result.

Lemma 2.6.1. *If $\Delta_3 \neq 0$, then the unique solution $u \in C[0,1]$ of problem (2.90), (2.89) is given by*

$$u(t) = \int_0^1 G_3(t,s)\widetilde{x}(s) \, ds, \quad t \in [0,1], \tag{2.91}$$

where the Green function G_3 is

$$G_3(t,s) = g_0(t,s) + \frac{t^{\alpha-1}}{\Delta_3} \sum_{i=1}^{m} \left(\int_0^1 a_i(\tau) g_{1i}(\tau,s) \, dH_i(\tau) \right), \tag{2.92}$$

and

$$g_0(t,s) = \frac{1}{\Gamma(\alpha)} \begin{cases} t^{\alpha-1}(1-s)^{\alpha-\beta_0-1} - (t-s)^{\alpha-1}, & 0 \leq s \leq t \leq 1, \\ t^{\alpha-1}(1-s)^{\alpha-\beta_0-1}, & 0 \leq t \leq s \leq 1, \end{cases}$$

$$g_{1i}(t,s) = \frac{1}{\Gamma(\alpha - \beta_i)} \begin{cases} t^{\alpha-\beta_i-1}(1-s)^{\alpha-\beta_0-1} - (t-s)^{\alpha-\beta_i-1}, \\ \qquad 0 \leq s \leq t \leq 1, \\ t^{\alpha-\beta_i-1}(1-s)^{\alpha-\beta_0-1}, \quad 0 \leq t \leq s \leq 1, \end{cases} \tag{2.93}$$

for all $(t,s) \in [0,1] \times [0,1]$, $i = 1, \ldots, m$.

Here, the functions g_{1i} may have negative values, because we did not impose a relation between β_0 and β_i, $i = 1, \ldots, m$. In Sections 2.4 and 2.5, we used the condition $\beta_m \leq \beta_0$ which implies that g_{1i}, $i = 1, \ldots, m$ are nonnegative functions.

Based on some properties of the function g_0 given by (2.93) from Lemma 2.4.3, we obtain the following lemma.

Lemma 2.6.2. *We suppose that* $\Delta_3 \neq 0$ *and* $F(s) := \frac{1}{\Delta_3} \sum_{i=1}^{m} \int_0^1 a_i(\tau) g_{1i}(\tau, s) \, dH_i(\tau) \geq 0$, *for all* $s \in [0, 1]$. *Then the Green function* G_3 *given by (2.92) is a continuous function on* $[0, 1] \times [0, 1]$ *and satisfies the inequalities:*

(a) $G_3(t, s) \leq J_3(s)$ *for all* $t, s \in [0, 1]$, *where* $J_3(s) = h_0(s) + F(s)$, $s \in [0, 1]$, *and* $h_0(s) = \frac{1}{\Gamma(\alpha)}(1 - s)^{\alpha - \beta_0 - 1}(1 - (1 - s)^{\beta_0})$, $s \in [0, 1]$;
(b) $G_3(t, s) \geq t^{\alpha - 1} J_3(s)$ *for all* $t, s \in [0, 1]$;
(c) $G_3(t, s) \leq t^{\alpha - 1} K(s)$ *for all* $t, s \in [0, 1]$, *where* $K(s) = \frac{1}{\Gamma(\alpha)}(1 - s)^{\alpha - \beta_0 - 1} + F(s)$, $s \in [0, 1]$.

In a similar manner as we proved Lemma 2.4.5, we deduce here the following result.

Lemma 2.6.3. *We suppose that* $\Delta_3 \neq 0$, $F(s) \geq 0$ *for all* $s \in [0, 1]$, $\widetilde{x} \in C(0, 1) \cap L^1(0, 1)$ *and* $\widetilde{x}(t) \geq 0$ *for all* $t \in (0, 1)$. *Then the solution* u *of problem (2.90), (2.89) given by (2.91) satisfies the inequality* $u(t) \geq t^{\alpha - 1} \|u\|$ *for all* $t \in [0, 1]$, *where* $\|u\| = \sup_{t \in [0,1]} |u(t)|$, *and so* $u(t) \geq 0$ *for all* $t \in [0, 1]$.

2.6.2 Existence and multiplicity of positive solutions

We give in this section some theorems for the existence of at least one or two positive solutions for problem (2.88), (2.89). We present firstly the assumptions that we will use in the sequel.

(V1) $\alpha \in \mathbb{R}$, $\alpha \in (n - 1, n]$, $n, m \in \mathbb{N}$, $n \geq 3$, $\beta_i \in \mathbb{R}$ for all $i = 0, \ldots, m$, $0 \leq \beta_1 < \beta_2 < \cdots < \beta_m < \alpha - 1$, $1 \leq \beta_0 < \alpha - 1$.
(V2) $a_i \in C(0, 1) \cap L^1(0, 1)$ for all $i = 1, \ldots, m$, and $H_i : [0, 1] \to \mathbb{R}$, $i = 1, \ldots, m$ are functions of bounded variation.
(V3) $\Delta_3 = \frac{\Gamma(\alpha)}{\Gamma(\alpha - \beta_0)} - \sum_{i=1}^{m} \frac{\Gamma(\alpha)}{\Gamma(\alpha - \beta_i)} \int_0^1 s^{\alpha - \beta_i - 1} a_i(s) \, dH_i(s) \neq 0$, and $F(s) = \frac{1}{\Delta_3} \sum_{i=1}^{m} \int_0^1 a_i(\tau) g_{2i}(\tau, s) \, dH_i(\tau) \geq 0$ for all $s \in [0, 1]$.

(V4) The function $f : (0,1) \times (0,\infty) \to [0,\infty)$ is continuous. Besides, for any $0 < r < R$, there exists $\phi_{r,R} \in C((0,1),[0,\infty)) \cap L^1(0,1)$ such that $f(t,u) \le \phi_{r,R}(t)$ for all $t \in (0,1)$ and $u \in [rt^{\alpha-1}, R]$.

(V5) There exist $R_1 > 0$ and a function $p_1 \in C((0,1),[0,\infty)) \cap L^1(0,1)$ with $\int_0^1 p_1(t)\, dt > 0$ such that $f(t,u) \ge p_1(t)u$ for all $(t,u) \in (0,1) \times (0,R_1]$.

(V6) There exist $R_2 > 0$ and a function $p_2 \in C((0,1),[0,\infty)) \cap L^1(0,1)$ with $\int_0^1 p_2(t)\, dt > 0$ such that $f(t,u) \le p_2(t)u$ for all $(t,u) \in (0,1) \times [R_2,\infty)$.

(V7) There exist $R_3 > 0$ and a function $p_3 \in C((0,1),[0,\infty)) \cap L^1(0,1)$ with $\int_0^1 p_3(t)\, dt > 0$ such that $f(t,u) \le p_3(t)u$ for all $(t,u) \in (0,1) \times (0,R_3]$.

(V8) There exist $R_4 > 0$ and a function $p_4 \in C((0,1),[0,\infty)) \cap L^1(0,1)$ with $\int_0^1 p_4(t)\, dt > 0$ such that $f(t,u) \ge p_4(t)u$ for all $(t,u) \in (0,1) \times [R_4,\infty)$.

We consider $X = C[0,1]$ the space of continuous functions defined on $[0,1]$ with the supremum norm $\|u\| = \sup_{t \in [0,1]} |u(t)|$, and the cone $P = \{u \in X,\ u(t) \ge t^{\alpha-1}\|u\|,\ \forall t \in [0,1]\}$.

We define the operators

$$\mathcal{G}u(t) = \int_0^1 G_3(t,s)f(s,u(s))\, ds,$$

$$\mathcal{P}_i u(t) = \int_0^1 G_3(t,s)p_i(s)u(s)\, ds, \quad i = 1,\dots,4.$$

Lemma 2.6.4. *Assume that* (V1)–(V4) *hold. Then for any $r > 0$, the operator $\mathcal{G} : P \setminus B_r \to P$ is completely continuous, $(B_r = \{u \in X,\ \|u\| < r\})$.*

Proof. For any $u \in P \setminus B_r$, we have $rt^{\alpha-1} \le u(t) \le \|u\|$. Let $R = \|u\|$. By (V4) it follows that there exists a function $\phi_{r,R} \in C((0,1),[0,\infty)) \cap L^1(0,1)$ such that $f(t,u(t)) \le \phi_{r,R}(t)$ for all $t \in (0,1)$. Then by using Lemma 2.6.2, we find

$$\mathcal{G}u(t) = \int_0^1 G_3(t,s)f(s,u(s))\, ds \le \int_0^1 J_3(s)f(s,u(s))\, ds$$

$$\le \int_0^1 J_3(s)\phi_{r,R}(s)\, ds \le \tilde{J}_0 \int_0^1 \phi_{r,R}(s)\, ds < \infty, \quad \forall t \in [0,1],$$

where $\tilde{J}_0 = \max_{t \in [0,1]} J_3(t)$, and then $\mathcal{G}u$ is well defined.

On the other hand, we obtain

$$\mathcal{G}u(t) = \int_0^1 G_3(t,s)f(s,u(s))\,ds$$

$$\geq t^{\alpha-1}\int_0^1 J_3(s)f(s,u(s))\,ds \geq t^{\alpha-1}\mathcal{G}u(t_1),$$

for all $t, t_1 \in [0,1]$, and so $\mathcal{G}u(t) \geq t^{\alpha-1}\|\mathcal{G}u\|$ for all $t \in [0,1]$. Therefore, $\mathcal{G}(P \setminus B_r) \subset P$.

We will prove next that $\mathcal{G}$ is completely continuous. Firstly, we show that $\mathcal{G}$ is continuous. Let $\{u_n\}_{n\geq 1} \subset P \setminus B_r$ and $\|u_n - u_0\| \to 0$ as $n \to \infty$, with $u_0 \in P \setminus B_r$. Then there exists $R > r$ such that $r \leq \|u_n\| \leq R$ for all $n = 0, 1, \ldots$. By (V4), for the above r, R we deduce (by the absolute continuity of the integral function) that for any $\varepsilon > 0$ there exists $\theta \in (0, 1/2)$ such that $\int_0^\theta \phi_{r,R}(s)\,ds < \varepsilon/(6\tilde{J}_0)$ and $\int_{1-\theta}^1 \phi_{r,R}(s)\,ds < \varepsilon/(6\tilde{J}_0)$. Because $f(t,u)$ is uniformly continuous on $[\theta, 1-\theta] \times [\theta^{\alpha-1}r, R]$, then there exists $N > 0$ such that for any $n > N$, we have

$$|f(t, u_n(t)) - f(t, u_0(t))| < \frac{\varepsilon}{3\int_0^1 J_3(s)\,ds}, \quad \forall t \in [\theta, 1-\theta].$$

Therefore, for any $n > N$, we find

$$\|\mathcal{G}u_n - \mathcal{G}u_0\| \leq \max_{t\in[0,1]}\int_0^1 G_3(t,s)|f(s, u_n(s)) - f(s, u_0(s))|\,ds$$

$$\leq \int_0^1 J_3(s)|f(s, u_n(s)) - f(s, u_0(s))|\,ds$$

$$\leq 2\int_0^\theta J_3(s)\phi_{r,R}(s)\,ds + \int_\theta^{1-\theta} J_3(s)|f(s, u_n(s))$$

$$- f(s, u_0(s))|\,ds + 2\int_{1-\theta}^1 J_3(s)\phi_{r,R}(s)\,ds$$

$$< 2\tilde{J}_0\int_0^\theta \phi_{r,R}(s)\,ds + \int_\theta^{1-\theta} J_3(s)|f(s, u_n(s))$$

$$- f(s, u_0(s))|\,ds + 2\tilde{J}_0\int_{1-\theta}^1 \phi_{r,R}(s)\,ds < \varepsilon.$$

Hence, $\|\mathcal{G}u_n - \mathcal{G}u_0\| \to 0$ as $n \to \infty$, and so $\mathcal{G}$ is a continuous operator.

Next, we will show that $\mathcal{G}$ is a compact operator, that is, it maps bounded sets into relatively compact sets. For this, let $E \subset P \setminus B_r$ be a bounded set. Then there exists $R_1 > r$ such that $r \leq \|u\| \leq R_1$ for all $u \in E$. By the above proof, we obtain

$$\mathcal{G}u(t) \leq \int_0^1 J_3(s)\phi_{r,R_1}(s)\,ds \leq \tilde{J}_0 \int_0^1 \phi_{r,R_1}(s)\,ds, \quad \forall t \in [0,1],\ u \in E,$$

which implies that $\mathcal{G}(E)$ is uniformly bounded.

The function $G_3(t,s)$ is uniformly continuous on $[0,1] \times [0,1]$. So, for any $\varepsilon > 0$, there exists $\zeta_1 > 0$ such that for any $t_1,\, t_2 \in [0,1]$ with $|t_1 - t_2| < \zeta_1$, and for any $s \in [0,1]$ we have

$$|G_3(t_1,s) - G_3(t_2,s)| < \frac{\varepsilon}{2\max\{\int_0^1 \phi_{r,R_1}(s)\,ds, 1\}}.$$

Therefore, for any $u \in E$, we deduce

$$
\begin{aligned}
|\mathcal{G}u(t_1) - \mathcal{G}u(t_2)| &\leq \int_0^1 |G_3(t_1,s) - G_3(t_2,s)| f(s,u(s))\,ds \\
&\leq \int_0^1 |G_3(t_1,s) - G_3(t_2,s)| \phi_{r,R_1}(s)\,ds \\
&\leq \int_0^1 \frac{\varepsilon}{2\max\{\int_0^1 \phi_{r,R_1}(s)\,ds, 1\}} \phi_{r,R_1}(s)\,ds < \varepsilon.
\end{aligned}
$$

This implies that $\mathcal{G}(E)$ is equicontinuous. By using the Arzela–Ascoli theorem, we conclude that $\mathcal{G}(E)$ is relatively compact, and then $\mathcal{G} : P \setminus B_r \to P$ is a compact operator. $\qquad\square$

Under assumptions (V1)–(V4), by the extension theorem, for any $r > 0$, the operator $\mathcal{G}$ has a completely continuous extension (also denoted by $\mathcal{G}$) from P to P.

By using Theorem 2.5.1 and similar arguments as those used in the proof of Lemma 2.5.2, we obtain the following result.

Lemma 2.6.5. *Assume that* (V1)–(V3) *hold, and* $p_i \in C((0,1),[0,\infty)) \cap L^1(0,1)$ *with* $\int_0^1 p_i(t)\,dt > 0$, $i = 1,\ldots,4$. *Then the operators* $\mathcal{P}_i : P \to P$ *are linear and completely continuous. Besides, the spectral radius* $r(\mathcal{P}_i) > 0$ *and* $\mathcal{P}_i$ *has an eigenfunction* $\psi_i \in P \setminus \{\theta\}$ *corresponding to the eigenvalue* $r(\mathcal{P}_i)$, *that is,* $\mathcal{P}_i\psi_i = r(\mathcal{P}_i)\psi_i$, $i = 1,\ldots,4$.

Theorem 2.6.1. *We assume that* (V1)–(V4) *hold, and there exist* $R_2 > R_1 > 0$ *such that* (V5) *and* (V6) *are satisfied. Besides, we suppose that* $r(\mathcal{P}_1) \geq 1 > r(\mathcal{P}_2) > 0$. *Then the boundary value problem* (2.88), (2.89) *has at least one positive solution.*

Proof. By (V5), for any $u \in \partial B_{R_1} \cap P$, we obtain

$$\mathcal{G}u(t) = \int_0^1 G_3(t,s)f(s,u(s))\,ds \geq \int_0^1 G_3(t,s)p_1(s)u(s)\,ds$$
$$= \mathcal{P}_1 u(t), \quad \forall t \in [0,1].$$

We assume that $\mathcal{G}$ has no fixed points on $\partial B_{R_1} \cap P$ (otherwise, the theorem is proved). We will show that

$$u - \mathcal{G}u \neq \mu\psi_1, \quad \forall u \in \partial B_{R_1} \cap P, \ \mu \geq 0, \tag{2.94}$$

where ψ_1 is given in Lemma 2.6.5. In fact, if not, there exist $u_1 \in \partial B_{R_1} \cap P$ and $\mu_1 \geq 0$ such that $u_1 - \mathcal{G}u_1 = \mu_1 \psi_1$. Then $\mu_1 > 0$ and $u_1 = \mathcal{G}u_1 + \mu_1 \psi_1 \geq \mu_1 \psi_1$. We denote by $\mu^0 = \sup\{\mu, \ u_1 \geq \mu\psi_1\}$. Then $\mu^0 \geq \mu_1$, $u_1 \geq \mu^0 \psi_1$ and

$$\mathcal{G}u_1 \geq \mathcal{P}_1 u_1 \geq \mu^0 \mathcal{P}_1 \psi_1 = \mu^0 r(\mathcal{P}_1)\psi_1 \geq \mu^0 \psi_1.$$

Hence, $u_1 = \mathcal{G}u_1 + \mu_1 \psi_1 \geq \mu^0 \psi_1 + \mu_1 \psi_1 = (\mu^0 + \mu_1)\psi_1$, which contradicts the definition of μ^0. We deduce that relation (2.94) holds, and by Theorem 1.2.9 we deduce

$$i(\mathcal{G}, B_{R_1} \cap P, P) = 0. \tag{2.95}$$

Now, we consider the set

$$V = \{u \in P \setminus B_{R_1}, \ \mathcal{G}u = \mu u \text{ with } \mu \geq 1\}.$$

We will prove next that the set V is bounded. For any $u \in V$, we find

$$u(t) \leq \mu u(t) = \mathcal{G}u(t) = \int_0^1 G_3(t,s)f(s,u(s))\,ds$$
$$= \int_{D_1} G_3(t,s)f(s,u(s))\,ds + \int_{D_2} G_3(t,s)f(s,u(s))\,ds$$
$$\leq \int_{D_1} G_3(t,s)p_2(s)u(s)\,ds + \int_0^1 G_3(t,s)f(s,\tilde{u}(s))\,ds$$
$$\leq \int_0^1 G_3(t,s)p_2(s)u(s)\,ds + \int_0^1 G_3(t,s)f(s,\tilde{u}(s))\,ds$$
$$= \mathcal{P}_2 u(t) + \mathcal{G}\tilde{u}(t) \leq \mathcal{P}_2 u(t) + \tilde{J}_0 M,$$

where $D_1 = \{s, \ u(s) \geq R_2\}$, $D_2 = \{s, \ u(s) < R_2\}$, $\tilde{u}(s) = \min\{u(s), R_2\}$, $M = \int_0^1 \phi_{R_1,R_2}(s)\,ds < \infty$ (by (V4)). Then we obtain $(I - \mathcal{P}_2)u(t) \leq M$ for all $t \in [0,1]$. Because $r(\mathcal{P}_2) < 1$, we deduce that the inverse operator of $(I - \mathcal{P}_2)$ exists and $(I - \mathcal{P}_2)^{-1} = \sum_{i=1}^{\infty} \mathcal{P}_2^i$. Therefore we find $u(t) \leq (I - \mathcal{P}_2)^{-1}(M)$ and $u(t) \leq M\|(I - \mathcal{P}_2)^{-1}\|$ for all $t \in [0, 1]$, which means that V is bounded. We choose $\widetilde{R}_2 > \max\{R_2, \sup\{\|u\|, u \in V\}\}$. Then $\mathcal{G}u \neq \mu u$ for all $\mu \geq 1$, $u \in \partial B_{\widetilde{R}_2} \cap P$, and by Theorem 1.2.7, we obtain

$$i(\mathcal{G}, B_{\widetilde{R}_2} \cap P, P) = 1. \tag{2.96}$$

By (2.95) and (2.96), we conclude

$$i(\mathcal{G}, (B_{\widetilde{R}_2} \setminus \overline{B}_{R_1}) \cap P, P) = i(\mathcal{G}, B_{\widetilde{R}_2} \cap P, P) - i(\mathcal{G}, B_{R_1} \cap P, P) = 1.$$

Then we deduce that $\mathcal{G}$ has at least one fixed point $\widetilde{u}$ on $(B_{\widetilde{R}_2} \setminus \overline{B}_{R_1}) \cap P$, which is a positive solution of problem (2.88), (2.89). Taking into account the remark from the beginning of the proof (that $\mathcal{G}$ may have fixed points on $\partial B_{R_1} \cap P$), we conclude that the solution $\widetilde{u}$ of problem (2.88), (2.89) satisfies $R_1 \leq \|\widetilde{u}\| < \widetilde{R}_2$. $\qquad \square$

Theorem 2.6.2. *We assume that* (V1)–(V4) *hold and there exist* $R_4 > R_3 > 0$ *such that* (V7) *and* (V8) *are satisfied. Besides, we suppose that* $r(\mathcal{P}_4) > 1 \geq r(\mathcal{P}_3) > 0$. *Then the boundary value problem* (2.88), (2.89) *has at least one positive solution.*

Proof. We suppose that $\mathcal{G}$ has no fixed points on $\partial B_{R_3} \cap P$ (otherwise, the theorem is proved). We will show that

$$\mathcal{G}u \neq \mu u, \quad \forall u \in \partial B_{R_3} \cap P, \quad \mu > 1. \tag{2.97}$$

If not, there exist $u_1 \in \partial B_{R_3} \cap P$ and $\mu_1 > 1$ such that $\mathcal{G}u_1 = \mu_1 u_1$. By using (V7) we obtain

$$\mu_1 u_1(t) = \mathcal{G}u_1(t) = \int_0^1 G_3(t, s) f(s, u_1(s)) \, ds$$

$$\leq \int_0^1 G_3(t, s) p_3(s) u_1(s) \, ds = \mathcal{P}_3 u_1(t), \quad \forall t \in [0, 1].$$

Because $\mathcal{P}_3$ is a nondecreasing operator, we deduce

$$\mu_1^2 u_1(t) \leq \mu_1 \mathcal{P}_3 u_1(t) = \mathcal{P}_3(\mu_1 u_1)(t) \leq \mathcal{P}_3(\mathcal{P}_3 u_1(t)) = \mathcal{P}_3^2 u_1(t), \quad \forall t \in [0, 1].$$

Repeating the process, we find

$$\mu_1^n u_1(t) \leq \mathcal{P}_3^n u_1(t), \quad \forall t \in [0, 1], \ n \geq 1,$$

and so

$$\mu_1^n \|u_1\| = \|\mu_1^n u_1\| \leq \|\mathcal{P}_3^n u_1\| \leq \|\mathcal{P}_3^n\| \|u_1\|, \quad \forall n \geq 1.$$

We conclude that $\|\mathcal{P}_3^n\| \geq \mu_1^n$ for all $n \geq 1$, and then $r(\mathcal{P}_3) = \lim_{n\to\infty} \sqrt[n]{\|\mathcal{P}_3^n\|} \geq \mu_1 > 1$, which is a contradiction, because $r(\mathcal{P}_3) \leq 1$. Therefore, the relation (2.97) is satisfied, and by Theorem 1.2.7, we deduce that

$$i(\mathcal{G}, B_{R_3} \cap P, P) = 1. \tag{2.98}$$

Now, we consider a decreasing sequence $(c_n)_{n\geq 1}$, with $0 < c_n < 1$, for all $n \geq 1$, convergent to 0, and we define the operators

$$\mathcal{F}_n u(t) = \int_{c_n}^1 G_3(t,s) p_4(s) u(s) \, ds.$$

By Theorem 3.7 from [107], the sequence of spectral radii $(r(\mathcal{F}_n))_n$ is increasing and converges to $r(\mathcal{P}_4)$. Then we can choose n_0 sufficiently large such that $r(\mathcal{F}_{n_0}) > 1$. We define $R_{n_0} = R_4 c_{n_0}^{1-\alpha}$. Then for any $u \in \partial B_{R_{n_0}} \cap P$, we have

$$u(t) \geq t^{\alpha-1} \|u\| = t^{\alpha-1} R_{n_0} \geq c_{n_0}^{\alpha-1} R_{n_0} = R_4, \quad \forall t \in [c_{n_0}, 1]. \tag{2.99}$$

In a similar manner as we obtained Lemma 2.6.5, we deduce that $\mathcal{F}_{n_0}$ has an eigenfunction $\psi_0 \in P \setminus \{\theta\}$ corresponding to the eigenvalue $r(\mathcal{F}_{n_0})$, that is $\mathcal{F}_{n_0} \psi_0 = r(\mathcal{F}_{n_0}) \psi_0$. Let $u \in \partial B_{R_{n_0}} \cap P$. By (V8) and (2.99), we find

$$\mathcal{G}u(t) = \int_0^1 G_3(t,s) f(s, u(s)) \, ds \geq \int_{c_{n_0}}^1 G_3(t,s) f(s, u(s)) \, ds$$

$$\geq \int_{c_{n_0}}^1 G_3(t,s) p_4(s) u(s) \, ds = \mathcal{F}_{n_0} u(t), \quad \forall t \in [0,1].$$

We assume that $\mathcal{G}$ has no fixed points on $\partial B_{R_{n_0}} \cap P$ (otherwise, the theorem is proved). We will show that

$$u - \mathcal{G}u \neq \mu \psi_0, \quad \forall u \in \partial B_{R_{n_0}} \cap P, \ \mu > 0. \tag{2.100}$$

In fact, if not, there exist $u_2 \in \partial B_{R_{n_0}} \cap P$ and $\mu_2 > 0$ such that $u_2 - \mathcal{G}u_2 = \mu_2 \psi_0$. We denote by $\mu_0 = \sup\{\mu, \ u_2 \geq \mu \psi_0\}$. Then $\mu_0 \geq \mu_2$ and $u_2 \geq \mu_0 \psi_0$. In addition, we have

$$u_2(t) = \mathcal{G}u_2(t) + \mu_2 \psi_0(t) \geq \mathcal{F}_{n_0} u_2(t) + \mu_2 \psi_0(t) \geq \mu_0 \mathcal{F}_{n_0} \psi_0(t) + \mu_2 \psi_0(t)$$

$$= \mu_0 r(\mathcal{F}_{n_0}) \psi_0(t) + \mu_2 \psi_0(t) \geq \mu_0 \psi_0(t) + \mu_2 \psi_0(t)$$

$$= (\mu_0 + \mu_2) \psi_0(t), \quad \forall t \in [0,1].$$

The obtained inequality contradicts the definition of μ_0. Therefore, the relation (2.100) is satisfied, and by Theorem 1.2.9, we deduce

$$i(\mathcal{G}, B_{R_{n_0}} \cap P, P) = 0. \tag{2.101}$$

Then by (2.98) and (2.101), we conclude that

$$i(\mathcal{G}, (B_{R_{n_0}} \setminus \overline{B}_{R_3}) \cap P, P) = i(\mathcal{G}, B_{R_{n_0}} \cap P, P) - i(\mathcal{G}, B_{R_3} \cap P, P) = -1.$$

This means that $\mathcal{G}$ has at least one fixed point $\widetilde{u}$ on $(B_{R_{n_0}} \setminus \overline{B}_{R_3}) \cap P$, which is a positive solution of problem (2.88), (2.89). Taking into account that $\mathcal{G}$ may have fixed points on $(\partial B_{R_{n_0}} \cup \partial B_{R_3}) \cap P$, then the solution $\widetilde{u}$ satisfies $R_3 \leq \|\widetilde{u}\| \leq R_{n_0}$.   □

We can also obtain existence results for multiple positive solutions by imposing various conditions similar to (V5)–(V8).

Theorem 2.6.3. *We assume that* (V1)–(V4) *hold and there exist* $R_4 > R_5 > R_1 > 0$ *such that* (V5), (V8) *and*

(V9) *There exists a function* $p_5 \in C((0,1), [0,\infty)) \cap L^1(0,1)$ *with* $\int_0^1 p_5(t)\,dt > 0$ *such that* $f(t,u) \leq p_5(t)R_5$ *for all* $u \in [R_1 t^{\alpha-1}, R_5]$ *and* $t \in (0,1)$,

hold. Besides, we suppose that $r(\mathcal{P}_1) \geq 1$, $r(\mathcal{P}_4) > 1$ *and* $\|\mathcal{P}_5\| < 1$, *where* $\mathcal{P}_5 u(t) = \int_0^1 G_3(t,s)p_5(s)u(s)\,ds$, *for* $t \in [0,1]$ *and* $u \in P$. *Then the boundary value problem* (2.88), (2.89) *has at least two positive solutions* u_1 *and* u_2 *with* $R_1 \leq \|u_1\| < R_5 < \|u_2\|$.

Proof. We will show that for any $u \in \partial B_{R_5} \cap P$, we have $\mathcal{G}u \neq \lambda u$ for all $\lambda \geq 1$. If not, there exist $u_0 \in \partial B_{R_5} \cap P$ and $\lambda_0 \geq 1$ such that $\mathcal{G}u_0 = \lambda_0 u_0$. Then we obtain

$$u_0(t) \leq \lambda_0 u_0(t) = \mathcal{G}u_0(t) = \int_0^1 G_3(t,s)f(s,u_0(s))\,ds$$

$$\leq \int_0^1 G_3(t,s)p_5(s)R_5\,ds = R_5(\mathcal{P}_5 I_d)(t) \leq R_5\|\mathcal{P}_5\| < R_5, \quad \forall t \in [0,1],$$

where $I_d(t) = t$ for all $t \in [0,1]$. So we deduce $\|u_0\| < R_5$, which is a contradiction, because $u_0 \in \partial B_{R_5} \cap P$. Therefore, by Theorem 1.2.7, we conclude

$$i(\mathcal{G}, B_{R_5} \cap P, P) = 1. \tag{2.102}$$

By the proof of Theorem 2.6.1, we obtain that

$$i(\mathcal{G}, B_{R_1} \cap P, P) = 0, \tag{2.103}$$

or $\mathcal{G}$ has a fixed point on $\partial B_{R_1} \cap P$.

By the proof of Theorem 2.6.2, we deduce that there exists $R_{n_0} > R_4$ such that

$$i(\mathcal{G}, B_{R_{n_0}} \cap P, P) = 0, \tag{2.104}$$

or $\mathcal{G}$ has a fixed point on $\partial B_{R_{n_0}} \cap P$.

If $\mathcal{G}$ has no fixed points on $(\partial B_{R_1} \cup \partial B_{R_{n_0}}) \cap P$, then by the relations (2.102)–(2.104) we conclude

$$i(\mathcal{G}, (B_{R_{n_0}} \setminus \overline{B}_{R_5}) \cap P, P) = -1, \text{ and } i(\mathcal{G}, (B_{R_5} \setminus \overline{B}_{R_1}) \cap P, P) = 1.$$

Therefore, the operator $\mathcal{G}$ has at least two fixed points $u_1 \in (B_{R_{n_0}} \setminus \overline{B}_{R_5}) \cap P$ and $u_2 \in (B_{R_5} \setminus \overline{B}_{R_1}) \cap P$, which are positive solutions for problem (2.88), (2.89). Because $\mathcal{G}$ may have fixed points on $(\partial B_{R_{n_0}} \cup \partial B_{R_1}) \cap P$, we deduce that the solutions u_1, u_2 of problem (2.88), (2.89) satisfy $R_1 \leq \|u_1\| < R_5 < \|u_2\| \leq R_{n_0}$. $\qquad \square$

Theorem 2.6.4. *We assume that* (V1)–(V4) *hold and there exist* $R_2 > R_6 > R_3 > 0$ *such that* (V6), (V7) *and*

(V10) *There exist* $c \in (0,1)$ *and a function* $p_6 \in C((0,1), [0,\infty)) \cap L^1(0,1)$
  with $\int_0^1 p_6(t)\,dt > 0$ *such that* $f(t,u) \geq p_6(t)R_6$ *for all* $t \in [c,1)$ *and*
  $u \in [c^{\alpha-1}R_6, R_6]$,

hold. Besides, we suppose that $r(\mathcal{P}_2) < 1$, $r(\mathcal{P}_3) \leq 1$ *and* $\int_c^1 J_3(s) p_6(s)\,ds > 1$. *Then the boundary value problem* (2.88), (2.89) *has at least two positive solutions* u_1 *and* u_2 *with* $R_3 \leq \|u_1\| < R_6 < \|u_2\|$.

Proof. For any $u \in \partial B_{R_6} \cap P$, we have $u(t) \geq t^{\alpha-1}\|u\| = t^{\alpha-1}R_6 \geq c^{\alpha-1}R_6$ for all $t \in [c,1]$. Then we deduce

$$\|\mathcal{G}u\| = \max_{t \in [0,1]} \int_0^1 G_3(t,s)f(s,u(s))\,ds \geq \max_{t \in [0,1]} \int_0^1 t^{\alpha-1}J_3(s)f(s,u(s))\,ds$$

$$\geq \max_{t \in [0,1]} t^{\alpha-1} \int_c^1 J_3(s)f(s,u(s))\,ds = \int_c^1 J_3(s)f(s,u(s))\,ds$$

$$\geq \int_c^1 J_3(s)p_6(s)R_6\,ds > R_6 = \|u\|.$$

Hence, $\|\mathcal{G}u\| > \|u\|$ for all $u \in \partial B_{R_6} \cap P$. This last inequality implies that $\mathcal{G}u \not\le u$ for all $u \in \partial B_{R_6} \cap P$, and then we obtain (see [17])

$$i(\mathcal{G}, B_{R_6} \cap P, P) = 0. \tag{2.105}$$

By the proof of Theorem 2.6.1, we deduce that there exists $\widetilde{R}_2 > R_2$ such that

$$i(\mathcal{G}, B_{\widetilde{R}_2} \cap P, P) = 1. \tag{2.106}$$

By the proof of Theorem 2.6.2, we conclude that

$$i(\mathcal{G}, B_{R_3} \cap P, P) = 1, \tag{2.107}$$

or $\mathcal{G}$ has a fixed point on $\partial B_{R_3} \cap P$.

If $\mathcal{G}$ has no fixed points on $\partial B_{R_3} \cap P$, then by the relations (2.105)–(2.107), we obtain

$$i(\mathcal{G}, (B_{\widetilde{R}_2} \setminus \overline{B}_{R_6}) \cap P, P) = 1, \text{ and } i(\mathcal{G}, (B_{R_6} \setminus \overline{B}_{R_3}) \cap P, P) = -1.$$

Hence, the operator $\mathcal{G}$ has at least two fixed points $u_1 \in (B_{\widetilde{R}_2} \setminus \overline{B}_{R_6}) \cap P$ and $u_2 \in (B_{R_6} \setminus \overline{B}_{R_3}) \cap P$, which are positive solutions for problem (2.88), (2.89). Because $\mathcal{G}$ may have fixed points on $\partial B_{R_3} \cap P$, we deduce that the solutions u_1, u_2 of problem (2.88), (2.89) satisfy $R_3 \le \|u_1\| < R_6 < \|u_2\| < \widetilde{R}_2$. $\square$

2.6.3 *An example*

Let $\alpha = 5/2$ $(n = 3)$, $m = 2$, $\beta_0 = 5/4$, $\beta_1 = 1/2$, $\beta_2 = 4/3$, $H_1(t) = \{0, \ t \in [0, 1/3); 1, \ t \in [1/3, 1]\}$, $H_2(t) = t$ for all $t \in [0, 1]$, $a_1 = 1$, $a_2 = 1/2$.

We consider the fractional differential equation

$$D_{0+}^{5/2}u(t) + f(t, u(t)) = 0, \quad t \in (0, 1), \tag{2.108}$$

with the boundary conditions

$$u(0) = u'(0) = 0, \quad D_{0+}^{5/4}u(1) = D_{0+}^{1/2}u\left(\frac{1}{3}\right) + \frac{1}{2}\int_0^1 D_{0+}^{4/3}u(t)\,dt. \tag{2.109}$$

We obtain $\Delta_3 \approx 0.40939289 \neq 0$. We also deduce

$$g_0(t, s) = \frac{1}{\Gamma(5/2)} \begin{cases} t^{3/2}(1-s)^{1/4} - (t-s)^{3/2}, & 0 \le s \le t \le 1, \\ t^{3/2}(1-s)^{1/4}, & 0 \le t \le s \le 1, \end{cases}$$

$$g_{11}(t,s) = \begin{cases} t(1-s)^{1/4} - (t-s), & 0 \le s \le t \le 1, \\ t(1-s)^{1/4}, & 0 \le t \le s \le 1, \end{cases}$$

$$g_{12}(t,s) = \frac{1}{\Gamma(7/6)} \begin{cases} t^{1/6}(1-s)^{1/4} - (t-s)^{1/6}, & 0 \le s \le t \le 1, \\ t^{1/6}(1-s)^{1/4}, & 0 \le t \le s \le 1, \end{cases}$$

$$F(s) = \frac{1}{\Delta_3} \begin{cases} \frac{1}{3}(1-s)^{1/4} - \left(\frac{1}{3} - s\right) + \frac{1}{2\Gamma(13/6)}[(1-s)^{1/4} - (1-s)^{7/6}], \\ \qquad 0 \le s < \frac{1}{3}, \\ \frac{1}{3}(1-s)^{1/4} + \frac{1}{2\Gamma(13/6)}[(1-s)^{1/4} - (1-s)^{7/6}], \\ \qquad \frac{1}{3} \le s \le 1, \end{cases}$$

$$G_3(t,s) = g_0(t,s) + t^{3/2}F(s), \quad t, s \in [0,1],$$

$$J_3(s) = \frac{1}{\Gamma(5/2)}(1-s)^{1/4}(1 - (1-s)^{5/4}) + F(s), \quad s \in [0,1].$$

We have $F(s) \ge 0$, for all $s \in [0,1]$. We mention here that the function g_{12} has negative values in the vicinity of $t = 1$.

We consider the function

$$f(t,u) = \begin{cases} 10(t - \frac{1}{4})^2 t^{-1/3} u^{-1/5} + (t - \frac{1}{3})^2(1-t)^{-1/4}u, \\ \qquad (t,u) \in (0,1) \times (0,1], \\ \left[10(t - \frac{1}{4})^2 t^{-1/3} + (t - \frac{1}{3})^2(1-t)^{-1/4}\right] \cos^2 \frac{\pi(u-1)}{7}, \\ \qquad (t,u) \in (0,1) \times (1,64], \\ \frac{5}{2}(t - \frac{1}{4})^2 t^{-1/3} u^{1/3} + \frac{1}{8}(t - \frac{1}{3})^2(1-t)^{-1/4} u^{1/2}, \\ \qquad (t,u) \in (0,1) \times (64,\infty). \end{cases}$$

For any $0 < r < R$, we obtain the inequality $f(t,u) \le \phi_{r,R}(t)$ for all $(t,u) \in (0,1) \times [rt^{3/2}, R]$, where $\phi_{r,R}(t)$, $t \in (0,1)$ is defined by

$$\phi_{r,R}(t) = \begin{cases} 10(t - \frac{1}{4})^2 t^{-1/3}(rt^{3/2})^{-1/5} + (t - \frac{1}{3})^2(1-t)^{-1/4}R, \quad 0 < R \le 1, \\ \max\left\{10(t - \frac{1}{4})^2 t^{-1/3} + (t - \frac{1}{3})^2(1-t)^{-1/4}; \right. \\ \quad \left. 10(t - \frac{1}{4})^2 t^{-1/3}(rt^{3/2})^{-1/5} + (t - \frac{1}{3})^2(1-t)^{-1/4}\right\}, \\ \qquad 1 < R \le 64, \\ \max\left\{\frac{5}{2}(t - \frac{1}{4})^2 t^{-1/3}R^{1/3} + \frac{1}{8}(t - \frac{1}{3})^2(1-t)^{-1/4}R^{1/2}, \right. \\ \quad \left. 10(t - \frac{1}{4})^2 t^{-1/3}(rt^{3/2})^{-1/5} + (t - \frac{1}{3})^2(1-t)^{-1/4}\right\}, \quad R > 64. \end{cases}$$

The above function $\phi_{r,R} \in C((0,1),[0,\infty)) \cap L^1(0,1)$, and so the assumptions (V1)–(V4) are satisfied.

We choose $R_1 = 1$, $R_2 = 64$ and the functions

$$p_1(t) = 10\left(t - \frac{1}{4}\right)^2 t^{-1/3} + \left(t - \frac{1}{3}\right)^2 (1-t)^{-1/4}, \quad t \in (0,1),$$

$$p_2(t) = \frac{p_1(t)}{3}, \quad t \in (0,1).$$

We have the inequalities $f(t,u) \geq p_1(t)u$ for all $(t,u) \in (0,1) \times (0,1]$, and $f(t,u) \leq p_2(t)u$ for all $(t,u) \in (0,1) \times [64,\infty)$. Evidently, $p_1, p_2 \in C((0,1),[0,\infty)) \cap L^1(0,1)$, $\int_0^1 p_1(t)\,dt \approx 1.88182 > 0$, $\int_0^1 p_2(t)\,dt \approx 0.62727 > 0$, and so assumptions (V5) and (V6) are also satisfied.

We define the linear operators $\mathcal{P}_1, \mathcal{P}_2 : P \to P$, where $P = \{u \in C[0,1], \ u(t) \geq t^{3/2}\|u\|, \ \forall t \in [0,1]\}$, by

$$\mathcal{P}_1 u(t) = \int_0^1 G_3(t,s)p_1(s)u(s)\,ds,$$

$$\mathcal{P}_2 u(t) = \int_0^1 G_3(t,s)p_2(s)u(s)\,ds = \frac{1}{3}\mathcal{P}_1 x(t), \quad t \in [0,1], \ u \in P.$$

We will show that $r(\mathcal{P}_1) \geq 1$ and $r(\mathcal{P}_2) < 1$. We denote by $I_d(t) = t$, $\forall t \in [0,1]$, and $\zeta(t) = t^{3/2}$, $\forall t \in [0,1]$. Then we find

$$\mathcal{P}_1 \zeta(t) = \int_0^1 G_3(t,s)p_1(s)\zeta(s)\,ds = \int_0^1 G_3(t,s)p_1(s)s^{3/2}\,ds$$

$$\geq \int_0^1 t^{3/2}J_3(s)p_1(s)s^{3/2}\,ds = \left(\int_0^1 J_3(s)p_1(s)s^{3/2}\,ds\right)\zeta(t),$$

and therefore

$$\mathcal{P}_1^n \zeta(t) = \mathcal{P}_1(\mathcal{P}_1^{n-1}\zeta)(t) \geq \left(\int_0^1 J_3(s)p_1(s)s^{3/2}\,ds\right)^n \zeta(t).$$

The last inequality gives us

$$r(\mathcal{P}_1) = \lim_{n\to\infty} \sqrt[n]{\|\mathcal{P}_1^n\|} \geq \lim_{n\to\infty} \sqrt[n]{\left(\int_0^1 J_3(s)p_1(s)s^{3/2}\,ds\right)^n} \max_{t\in[0,1]} \zeta(t)$$

$$= \int_0^1 J_3(s)p_1(s)s^{3/2}\,ds \approx 1.70534,$$

and then $r(\mathcal{P}_1) > 1$.

On the other hand, we obtain

$$(\mathcal{P}_1 I_d)(t) = \int_0^1 G_3(t,s) p_1(s)\, ds \leq \int_0^1 J_3(s) p_1(s)\, ds \approx 2.41727, \quad \forall\, t \in [0,1],$$

and so $\|\mathcal{P}_1 I_d\| < 2.418$. Therefore, we find

$$r(\mathcal{P}_2) = \frac{1}{3} r(\mathcal{P}_1) \leq \frac{1}{3}\|\mathcal{P}_1\| = \frac{1}{3}\|\mathcal{P}_1 I_d\| < 1.$$

Then $0 < \frac{1}{3} < r(\mathcal{P}_2) < 1 < r(\mathcal{P}_1)$. By Theorem 2.6.1, we conclude that problem (2.108), (2.109) has at least one positive solution $u(t)$, $t \in [0,1]$, which satisfies the inequality $u(t) \geq t^{3/2}\|u\|$ for all $t \in [0,1]$.

Remark 2.6.1. The results presented in this section will be published in [83].

Chapter 3

Systems of Two Riemann–Liouville Fractional Differential Equations with Multi-Point Boundary Conditions

In this chapter, we study the existence and multiplicity of positive solutions for systems of two Riemann-Liouville fractional differential equations, subject to uncoupled or coupled multi-point boundary conditions which contain fractional derivatives, and the nonlinearities of systems are nonsingular or singular functions, with nonnegative values.

3.1 Systems of Fractional Differential Equations with Uncoupled Multi-Point Boundary Conditions

We consider the system of nonlinear fractional differential equations

$$\begin{cases} D_{0+}^{\alpha} u(t) + f(t, u(t), v(t)) = 0, & t \in (0,1), \\ D_{0+}^{\beta} v(t) + g(t, u(t), v(t)) = 0, & t \in (0,1), \end{cases} \tag{3.1}$$

with the uncoupled multi-point boundary conditions

$$\begin{cases} u^{(j)}(0) = 0, & j = 0, \ldots, n-2, \quad D_{0+}^{p_1} u(1) = \sum_{i=1}^{N} a_i D_{0+}^{q_1} u(\xi_i), \\ v^{(k)}(0) = 0, & k = 0, \ldots, m-2, \quad D_{0+}^{p_2} v(1) = \sum_{i=1}^{M} b_i D_{0+}^{q_2} v(\eta_i), \end{cases} \tag{3.2}$$

where $\alpha, \beta \in \mathbb{R}$, $\alpha \in (n-1, n]$, $\beta \in (m-1, m]$, $n, m \in \mathbb{N}$, $n, m \geq 3$, $p_1, p_2, q_1, q_2 \in \mathbb{R}$, $p_1 \in [1, \alpha-1)$, $p_2 \in [1, \beta-1)$, $q_1 \in [0, p_1]$, $q_2 \in [0, p_2]$,

$\xi_i, a_i \in \mathbb{R}$ for all $i = 1, \ldots, N$ ($N \in \mathbb{N}$), $0 < \xi_1 < \cdots < \xi_N \leq 1$, $\eta_i, b_i \in \mathbb{R}$ for all $i = 1, \ldots, M$ ($M \in \mathbb{N}$), $0 < \eta_1 < \cdots < \eta_M \leq 1$, and D_{0+}^ζ denotes the Riemann-Liouville derivative of order ζ (for $\zeta = \alpha, \beta, p_1, q_1, p_2, q_2$).

Under sufficient conditions on the functions f and g, we study the existence and multiplicity of positive solutions of problem (3.1), (3.2) by using some theorems from the fixed point index theory. By a positive solution of problem (3.1), (3.2), we mean a pair of functions $(u, v) \in (C([0, 1], \mathbb{R}_+))^2$ satisfying (3.1) and (3.2) with $u(t) > 0$ for all $t \in (0, 1]$ or $v(t) > 0$ for all $t \in (0, 1]$. The system (3.1) with two positive parameters λ_1 and λ_2, and with the boundary conditions

$$\begin{cases} u^{(i)}(0) = v^{(i)}(0) = 0, & i = 0, \ldots, n - 2, \\ D_{0+}^\mu u(1) = \eta_1 D_{0+}^\mu u(\xi_1), & D_{0+}^\nu v(1) = \eta_2 D_{0+}^\nu v(\xi_2), \end{cases} \tag{3.3}$$

where $\alpha, \beta \in (n - 1, n]$, $n \in \mathbb{N}$, $n \geq 3$, $1 \leq \mu, \nu \leq n - 2$, $\xi_1, \xi_2 \in (0, 1)$, $0 < \eta_1 \xi_1^{\alpha - \mu - 1} < 1$, $0 < \eta_2 \xi_2^{\beta - \nu - 1} < 1$, was investigated in [108].

3.1.1   *Auxiliary results*

In this section, we present some auxiliary results that can be proved in a similar manner as the auxiliary results from Section 2.1.1.

We consider the fractional differential equation

$$D_{0+}^\alpha u(t) + x(t) = 0, \quad 0 < t < 1, \tag{3.4}$$

with the multi-point boundary conditions

$$u^{(j)}(0) = 0, \quad j = 0, \ldots, n - 2; \quad D_{0+}^{p_1} u(1) = \sum_{i=1}^N a_i D_{0+}^{q_1} u(\xi_i), \tag{3.5}$$

where $\alpha \in (n - 1, n]$, $n \in \mathbb{N}$, $n \geq 3$, $a_i, \xi_i \in \mathbb{R}$, $i = 1, \ldots, N$ ($N \in \mathbb{N}$), $0 < \xi_1 < \cdots < \xi_N \leq 1$, $p_1, q_1 \in \mathbb{R}$, $p_1 \in [1, \alpha - 1)$, $q_1 \in [0, p_1]$, and $x \in C(0, 1) \cap L^1(0, 1)$. We denote by $\Delta_1 = \frac{\Gamma(\alpha)}{\Gamma(\alpha - p_1)} - \frac{\Gamma(\alpha)}{\Gamma(\alpha - q_1)} \sum_{i=1}^N a_i \xi_i^{\alpha - q_1 - 1}$.

Lemma 3.1.1. *If $\Delta_1 \neq 0$, then the unique solution $u \in C[0, 1]$ of problem (3.4), (3.5) is given by*

$$u(t) = \int_0^1 G_1(t, s) x(s) \, ds, \quad t \in [0, 1], \tag{3.6}$$

where the Green function G_1 is

$$G_1(t, s) = g_1(t, s) + \frac{t^{\alpha - 1}}{\Delta_1} \sum_{i=1}^N a_i g_2(\xi_i, s), \quad \forall (t, s) \in [0, 1] \times [0, 1], \tag{3.7}$$

and

$$g_1(t,s) = \frac{1}{\Gamma(\alpha)} \begin{cases} t^{\alpha-1}(1-s)^{\alpha-p_1-1} - (t-s)^{\alpha-1}, & 0 \leq s \leq t \leq 1, \\ t^{\alpha-1}(1-s)^{\alpha-p_1-1}, & 0 \leq t \leq s \leq 1, \end{cases}$$

$$g_2(t,s) = \frac{1}{\Gamma(\alpha-q_1)} \begin{cases} t^{\alpha-q_1-1}(1-s)^{\alpha-p_1-1} - (t-s)^{\alpha-q_1-1}, \\ \qquad 0 \leq s \leq t \leq 1, \\ t^{\alpha-q_1-1}(1-s)^{\alpha-p_1-1}, & 0 \leq t \leq s \leq 1. \end{cases} \tag{3.8}$$

Lemma 3.1.2. *The functions g_1 and g_2 given by (3.8) have the properties:*

(a) $g_1(t,s) \leq h_1(s)$ *for all $t, s \in [0,1]$, where*

$$h_1(s) = \frac{1}{\Gamma(\alpha)}(1-s)^{\alpha-p_1-1}(1-(1-s)^{p_1}), s \in [0,1];$$

(b) $g_1(t,s) \geq t^{\alpha-1}h_1(s)$ *for all $t, s \in [0,1]$;*
(c) $g_1(t,s) \leq \frac{t^{\alpha-1}}{\Gamma(\alpha)}$, *for all $t, s \in [0,1]$;*
(d) $g_2(t,s) \geq t^{\alpha-q_1-1}h_2(s)$ *for all $t, s \in [0,1]$, where*

$$h_2(s) = \frac{1}{\Gamma(\alpha-q_1)}(1-s)^{\alpha-p_1-1}(1-(1-s)^{p_1-q_1}), s \in [0,1];$$

(e) $g_2(t,s) \leq \frac{1}{\Gamma(\alpha-q_1)}t^{\alpha-q_1-1}$ *for all $t, s \in [0,1]$;*
(f) *The functions g_1 and g_2 are continuous on $[0,1] \times [0,1]$; $g_1(t,s) \geq 0$, $g_2(t,s) \geq 0$ for all $t, s \in [0,1]$; $g_1(t,s) > 0$, $g_2(t,s) > 0$ for all $t, s \in (0,1)$.*

Lemma 3.1.3. *Assume that $a_i \geq 0$ for all $i = 1,\ldots,N$ and $\Delta_1 > 0$. Then the function G_1 given by (3.7) is a nonnegative continuous function on $[0,1] \times [0,1]$ and satisfies the inequalities:*

(a) $G_1(t,s) \leq J_1(s)$ *for all $t, s \in [0,1]$, where $J_1(s) = h_1(s) + \frac{1}{\Delta_1}\sum_{i=1}^{N}a_ig_2(\xi_i,s)$, $s \in [0,1]$;*
(b) $G_1(t,s) \geq t^{\alpha-1}J_1(s)$ *for all $t, s \in [0,1]$;*
(c) $G_1(t,s) \leq \sigma_1 t^{\alpha-1}$, *for all $t, s \in [0,1]$, where $\sigma_1 = \frac{1}{\Gamma(\alpha)} + \frac{1}{\Delta_1\Gamma(\alpha-q_1)}\sum_{i=1}^{N}a_i\xi_i^{\alpha-q_1-1}$.*

Lemma 3.1.4. *Assume that $a_i \geq 0$ for all $i = 1,\ldots,N$, $\Delta_1 > 0$, $x \in C(0,1) \cap L^1(0,1)$ and $x(t) \geq 0$ for all $t \in (0,1)$. Then the solution u of problem (3.4), (3.5) given by (3.6) satisfies the inequality $u(t) \geq t^{\alpha-1}u(t')$ for all $t, t' \in [0,1]$.*

We consider now the fractional differential equation

$$D_{0+}^{\beta} v(t) + y(t) = 0, \quad 0 < t < 1, \tag{3.9}$$

with the multi-point boundary conditions

$$v^{(j)}(0) = 0, \quad j = 0, \ldots, m-2; \quad D_{0+}^{p_2} v(1) = \sum_{i=1}^{M} b_i D_{0+}^{q_2} v(\eta_i), \tag{3.10}$$

where $\beta \in (m-1, m]$, $m \in \mathbb{N}$, $m \geq 3$, b_i, $\eta_i \in \mathbb{R}$, $i = 1, \ldots, M$ ($M \in \mathbb{N}$), $0 < \eta_1 < \cdots < \eta_M \leq 1$, p_2, $q_2 \in \mathbb{R}$, $p_2 \in [1, \beta - 1)$, $q_2 \in [0, p_2]$, and $y \in C(0,1) \cap L^1(0,1)$. We denote by $\Delta_2 = \frac{\Gamma(\beta)}{\Gamma(\beta-p_2)} - \frac{\Gamma(\beta)}{\Gamma(\beta-q_2)} \sum_{i=1}^{M} b_i \eta_i^{\beta-q_2-1}$.

Lemma 3.1.5. *If $\Delta_2 \neq 0$, then the unique solution of problem (3.9), (3.10) is given by*

$$v(t) = \int_0^1 G_2(t,s) y(s) \, ds, \quad t \in [0,1], \tag{3.11}$$

where the Green function G_2 is

$$G_2(t,s) = g_3(t,s) + \frac{t^{\beta-1}}{\Delta_2} \sum_{i=1}^{M} b_i g_4(\eta_i, s), \quad \forall (t,s) \in [0,1] \times [0,1], \tag{3.12}$$

and

$$g_3(t,s) = \frac{1}{\Gamma(\beta)} \begin{cases} t^{\beta-1}(1-s)^{\beta-p_2-1} - (t-s)^{\beta-1}, & 0 \leq s \leq t \leq 1, \\ t^{\beta-1}(1-s)^{\beta-p_2-1}, & 0 \leq t \leq s \leq 1, \end{cases}$$

$$g_4(t,s) = \frac{1}{\Gamma(\beta-q_2)} \begin{cases} t^{\beta-q_2-1}(1-s)^{\beta-p_2-1} - (t-s)^{\beta-q_2-1}, \\ \quad 0 \leq s \leq t \leq 1, \\ t^{\beta-q_2-1}(1-s)^{\beta-p_2-1}, & 0 \leq t \leq s \leq 1. \end{cases} \tag{3.13}$$

Lemma 3.1.6. *The functions g_3 and g_4 given by (3.13) have the properties:*

(a) $g_3(t,s) \leq h_3(s)$ *for all t, $s \in [0,1]$, where*

$$h_3(s) = \frac{1}{\Gamma(\beta)}(1-s)^{\beta-p_2-1}(1-(1-s)^{p_2}), \quad s \in [0,1];$$

(b) $g_3(t,s) \geq t^{\beta-1} h_3(s)$ *for all t, $s \in [0,1]$;*

(c) $g_3(t,s) \leq \frac{t^{\beta-1}}{\Gamma(\beta)}$, *for all t, $s \in [0,1]$;*

(d) $g_4(t,s) \geq t^{\beta-q_2-1} h_4(s)$ *for all t, $s \in [0,1]$, where*

$$h_4(s) = \frac{1}{\Gamma(\beta-q_2)}(1-s)^{\beta-p_2-1}(1-(1-s)^{p_2-q_2}), \quad s \in [0,1];$$

(e) $g_4(t,s) \le \frac{1}{\Gamma(\beta-q_2)} t^{\beta-q_2-1}$ for all t, $s \in [0,1]$;

(f) The functions g_3 and g_4 are continuous on $[0,1] \times [0,1]$; $g_3(t,s) \ge 0$, $g_4(t,s) \ge 0$ for all t, $s \in [0,1]$; $g_3(t,s) > 0$, $g_4(t,s) > 0$ for all t, $s \in (0,1)$.

Lemma 3.1.7. *Assume that $b_i \ge 0$ for all $i = 1,\ldots,M$ and $\Delta_2 > 0$. Then the function G_2 given by (3.12) is a nonnegative continuous function on $[0,1] \times [0,1]$ and satisfies the inequalities:*

(a) $G_2(t,s) \le J_2(s)$ for all t, $s \in [0,1]$, where $J_2(s) = h_3(s) + \frac{1}{\Delta_2} \sum_{i=1}^{M} b_i g_4(\eta_i, s)$, $s \in [0,1]$;

(b) $G_2(t,s) \ge t^{\beta-1} J_2(s)$ for all t, $s \in [0,1]$;

(c) $G_2(t,s) \le \sigma_2 t^{\beta-1}$, for all t, $s \in [0,1]$, where $\sigma_2 = \frac{1}{\Gamma(\beta)} + \frac{1}{\Delta_2 \Gamma(\beta-q_2)} \sum_{i=1}^{M} b_i \eta_i^{\beta-q_2-1}$.

Lemma 3.1.8. *Assume that $b_i \ge 0$ for all $i = 1,\ldots,M$, $\Delta_2 > 0$, $y \in C(0,1) \cap L^1(0,1)$ and $y(t) \ge 0$ for all $t \in (0,1)$. Then the solution v of problem (3.9), (3.10) given by (3.11) satisfies the inequality $v(t) \ge t^{\beta-1} v(t')$ for all t, $t' \in [0,1]$.*

3.1.2 Existence and multiplicity of positive solutions

In this section, we give sufficient conditions on f and g such that positive solutions with respect to a cone for our problem (3.1), (3.2) exist.

We present the assumptions that we shall use in the sequel.

(H1) $\alpha, \beta \in \mathbb{R}$, $\alpha \in (n-1, n]$, $\beta \in (m-1, m]$, $n, m \in \mathbb{N}$; $n, m \ge 3$; $\xi_i \in \mathbb{R}$ for all $i = 1,\ldots,N$ ($N \in \mathbb{N}$), $0 < \xi_1 < \cdots < \xi_N \le 1$, $\eta_i \in \mathbb{R}$ for all $i = 1,\ldots,M$ ($M \in \mathbb{N}$), $0 < \eta_1 < \cdots < \eta_M \le 1$, $p_1 \in [1, \alpha-1)$, $p_2 \in [1, \beta-1)$, $q_1 \in [0,p_1]$, $q_2 \in [0,p_2]$, $a_i \ge 0$ for all $i = 1,\ldots,N$, $b_i \ge 0$ for all $i = 1,\ldots,M$, $\Delta_1 = \frac{\Gamma(\alpha)}{\Gamma(\alpha-p_1)} - \frac{\Gamma(\alpha)}{\Gamma(\alpha-q_1)} \sum_{i=1}^{N} a_i \xi_i^{\alpha-q_1-1} > 0$, $\Delta_2 = \frac{\Gamma(\beta)}{\Gamma(\beta-p_2)} - \frac{\Gamma(\beta)}{\Gamma(\beta-q_2)} \sum_{i=1}^{M} b_i \eta_i^{\beta-q_2-1} > 0$.

(H2) The functions $f, g : [0,1] \times \mathbb{R}_+ \times \mathbb{R}_+ \to \mathbb{R}_+$ are continuous.

(H3) There exist the functions $a, b \in C(\mathbb{R}_+, \mathbb{R}_+)$ such that

(a) $a(\cdot)$ is concave and strictly increasing on $\mathbb{R}_+$ with $a(0) = 0$;

(b) There exists $\sigma \in (0,1)$ such that

$$f_0^i = \liminf_{v \to 0+} \frac{f(t,u,v)}{a(v)} \in (0,\infty], \text{ uniformly with respect to}$$

$$(t,u) \in [\sigma, 1] \times \mathbb{R}_+,$$

$$g_0^i = \liminf_{u \to 0+} \frac{g(t, u, v)}{b(u)} \in (0, \infty], \text{ uniformly with respect to}$$

$$(t, v) \in [\sigma, 1] \times \mathbb{R}_+;$$

(c) $\lim_{u \to 0+} \dfrac{a(Cb(u))}{u} = \infty$ exists for any constant $C > 0$.

(H4) There exist $\alpha_1, \alpha_2 > 0$ with $\alpha_1 \alpha_2 \leq 1$ such that

$$f_\infty^s = \limsup_{v \to \infty} \frac{f(t, u, v)}{v^{\alpha_1}} \in [0, \infty), \text{ uniformly with respect to}$$

$$(t, u) \in [0, 1] \times \mathbb{R}_+,$$

$$g_\infty^s = \lim_{u \to \infty} \frac{g(t, u, v)}{u^{\alpha_2}} = 0 \text{ exists uniformly with respect to}$$

$$(t, v) \in [0, 1] \times \mathbb{R}_+.$$

(H5) There exist the functions $c, d \in C(\mathbb{R}_+, \mathbb{R}_+)$ such that

(a) $c(\cdot)$ is concave and strictly increasing on $\mathbb{R}_+$;
(b) There exists $\sigma \in (0, 1)$ such that

$$f_\infty^i = \liminf_{v \to \infty} \frac{f(t, u, v)}{c(v)} \in (0, \infty], \text{ uniformly with respect to}$$

$$(t, u) \in [\sigma, 1] \times \mathbb{R}_+,$$

$$g_\infty^i = \liminf_{u \to \infty} \frac{g(t, u, v)}{d(u)} \in (0, \infty], \text{ uniformly with respect to}$$

$$(t, v) \in [\sigma, 1] \times \mathbb{R}_+;$$

(c) $\lim\limits_{u \to \infty} \dfrac{c(Cd(u))}{u} = \infty$ exists for any constant $C > 0$.

(H6) There exist $\beta_1, \beta_2 > 0$ with $\beta_1 \beta_2 \geq 1$ such that

$$f_0^s = \limsup_{v \to 0+} \frac{f(t, u, v)}{v^{\beta_1}} \in [0, \infty), \text{ uniformly with respect to}$$

$$(t, u) \in [0, 1] \times \mathbb{R}_+,$$

$$g_0^s = \lim_{u \to 0+} \frac{g(t, u, v)}{u^{\beta_2}} = 0 \text{ exists uniformly with respect to}$$

$$(t, v) \in [0, 1] \times \mathbb{R}_+.$$

(H7) The functions $f(t, u, v)$ and $g(t, u, v)$ are nondecreasing with respect to u and v, and there exists $N_0 > 0$ such that

$$f(t, N_0, N_0) < \frac{N_0}{2m_0}, \quad g(t, N_0, N_0) < \frac{N_0}{2m_0}, \quad \forall t \in [0, 1],$$

where $m_0 = \max\{\int_0^1 J_1(s)\, ds, \int_0^1 J_2(s)\, ds\}$, and J_1, J_2 are defined in Section 3.1.1.

Remark 3.1.1. The assumptions (H3) and (H4) with $\alpha_1 = p$, $\alpha_2 = 1/p$, $p > 0$, and the assumptions (H5) and (H6) with $\beta_1 = r$, $\beta_2 = 1/r$, $r > 0$ were used in Theorems 3.1 and 3.2 from [108] for problem (3.1) with two positive parameters λ_1 and λ_2, and with the boundary conditions (3.3).

By using the Green functions G_1 and G_2 from Section 3.1.1, we consider the following nonlinear system of integral equations:

$$\begin{cases} u(t) = \displaystyle\int_0^1 G_1(t, s) f(s, u(s), v(s))\, ds, & 0 \leq t \leq 1, \\ v(t) = \displaystyle\int_0^1 G_2(t, s) g(s, u(s), v(s))\, ds, & 0 \leq t \leq 1. \end{cases} \tag{3.14}$$

By Lemmas 3.1.1 and 3.1.5, (u, v) is solution of problem (3.1), (3.2) if and only if (u, v) is solution of the system (3.14).

We consider the Banach space $X = C[0, 1]$ with supremum norm $\| \cdot \|$ and the Banach space $Y = X \times X$ with the norm $\|(u, v)\|_Y = \|u\| + \|v\|$. We define the cones

$$P_1 = \{x \in X, \ x(t) \geq t^{\alpha - 1}\|x\|, \ \forall t \in [0, 1]\} \subset X,$$
$$P_2 = \{y \in X, \ y(t) \geq t^{\beta - 1}\|y\|, \ \forall t \in [0, 1]\} \subset X,$$

and $P = P_1 \times P_2 \subset Y$.

We introduce the operator $Q : P \to Y$ by $Q(u, v) = (Q_1(u, v), Q_2(u, v))$, $(u, v) \in P$, with $Q_1, Q_2 : P \to X$ defined by

$$Q_1(u, v)(t) = \int_0^1 G_1(t, s) f(s, u(s), v(s))\, ds, \quad t \in [0, 1], \quad (u, v) \in P,$$

$$Q_2(u, v)(t) = \int_0^1 G_2(t, s) g(s, u(s), v(s))\, ds, \quad t \in [0, 1], \quad (u, v) \in P.$$

Lemma 3.1.9. *If* (H1)–(H2) *hold, then* $Q : P \to P$ *is a completely continuous operator.*

Proof. Let $(u,v) \in P$ be an arbitrary element. Because $Q_1(u,v)$ and $Q_2(u,v)$ satisfy the problem (3.4), (3.5) for $x(t) = f(t, u(t), v(t))$, $t \in [0,1]$, and the problem (3.9), (3.10) for $y(t) = g(t, u(t), v(t))$, $t \in [0,1]$, respectively, then by Lemmas 3.1.4 and 3.1.8, we obtain

$$Q_1(u,v)(t) \geq t^{\alpha-1} Q_1(u,v)(t'), \quad Q_2(u,v)(t) \geq t^{\beta-1} Q_2(u,v)(t'),$$

$$\forall t, t' \in [0,1], \quad (u,v) \in P,$$

and so

$$Q_1(u,v)(t) \geq t^{\alpha-1} \max_{t' \in [0,1]} Q_1(u,v)(t') = t^{\alpha-1} \|Q_1(u,v)\|,$$

$$\forall t \in [0,1], \quad (u,v) \in P,$$

$$Q_2(u,v)(t) \geq t^{\beta-1} \max_{t' \in [0,1]} Q_2(u,v)(t') = t^{\beta-1} \|Q_2(u,v)\|,$$

$$\forall t \in [0,1], \quad (u,v) \in P.$$

Therefore, $Q(u,v) = (Q_1(u,v), Q_2(u,v)) \in P$ and then $Q(P) \subset P$. By using standard arguments, we can easily show that Q_1 and Q_2 are completely continuous, and then Q is a completely continuous operator. $\square$

The pair of functions (u,v) is a solution of problem (3.14) if and only if $(u,v) \in P$ is a fixed point of operator Q. So, we will investigate the existence of fixed points of operator Q.

Theorem 3.1.1. *Assume that* (H1)–(H4) *hold. Then the problem* (3.1), (3.2) *has at least one positive solution.*

Proof. By (H3), there exist $C_1 > 0$, $C_2 > 0$ and a sufficiently small $r_1 > 0$ such that

$$f(t,u,v) \geq C_1 a(v), \quad \forall\, (t,u) \in [\sigma, 1] \times \mathbb{R}_+, \quad v \in [0, r_1],$$

$$g(t,u,v) \geq C_2 b(u), \quad \forall\, (t,v) \in [\sigma, 1] \times \mathbb{R}_+, \quad u \in [0, r_1],$$

$$(3.15)$$

and

$$a(C_3 b(u)) \geq \frac{2C_3 u}{C_1 C_2 m_1 m_2 \gamma \sigma^{\alpha+\beta-2}}, \quad \forall\, u \in [0, r_1], \qquad (3.16)$$

where $C_3 = \max\{C_2 \sigma^{\beta-1} J_2(s), \ s \in [0,1]\} > 0$, $\gamma = \min\{\sigma^{\alpha-1}, \sigma^{\beta-1}\}$, $m_1 = \int_\sigma^1 J_1(s)\, ds$, $m_2 = \int_\sigma^1 J_2(s)\, ds$.

We will show that $(Q_1(u,v), Q_2(u,v)) \not\leq (u,v)$ for all $(u,v) \in \partial B_{r_1} \cap P$. We suppose that there exists $(u,v) \in \partial B_{r_1} \cap P$, that is, $\|(u,v)\|_Y = r_1$, such that $(Q_1(u,v), Q_2(u,v)) \leq (u,v)$. Then $u \geq Q_1(u,v)$ and $v \geq Q_2(u,v)$. By using the monotonicity and concavity of $a(\cdot)$, the Jensen inequality, Lemma 3.1.3, the relations (3.15) and (3.16), we obtain

$$u(t) \geq Q_1(u,v)(t) = \int_0^1 G_1(t,s)f(s,u(s),v(s))\,ds$$

$$\geq \int_\sigma^1 G_1(t,s)f(s,u(s),v(s))\,ds \geq C_1 \int_\sigma^1 t^{\alpha-1}J_1(s)a(v(s))\,ds$$

$$\geq C_1\sigma^{\alpha-1}\int_\sigma^1 J_1(s)a\left(\int_0^1 G_2(s,\tau)g(\tau,u(\tau),v(\tau))\,d\tau\right)ds$$

$$\geq C_1\sigma^{\alpha-1}\int_\sigma^1 J_1(s)a\left(\int_\sigma^1 G_2(s,\tau)g(\tau,u(\tau),v(\tau))\,d\tau\right)ds$$

$$\geq C_1\sigma^{\alpha-1}\int_\sigma^1 J_1(s)a\left(\int_\sigma^1 C_2 G_2(s,\tau)b(u(\tau))\,d\tau\right)ds$$

$$\geq C_1\sigma^{\alpha-1}\int_\sigma^1 J_1(s)a\left(\int_\sigma^1 C_2\sigma^{\beta-1}J_2(\tau)b(u(\tau))\,d\tau\right)ds$$

$$\geq C_1\sigma^{\alpha-1}\left(\int_\sigma^1 J_1(s)\,ds\right)\left(\int_\sigma^1 a\left(C_2\sigma^{\beta-1}J_2(\tau)b(u(\tau))\right)d\tau\right)$$

$$= C_1\sigma^{\alpha-1}m_1\int_\sigma^1 a\left(\left(C_3^{-1}C_2\sigma^{\beta-1}J_2(\tau)\right)C_3 b(u(\tau))\right)d\tau$$

$$\geq C_1 C_2 C_3^{-1}\sigma^{\alpha+\beta-2}m_1\int_\sigma^1 J_2(\tau)a\left(C_3 b(u(\tau))\right)d\tau$$

$$\geq C_1 C_2 C_3^{-1}\sigma^{\alpha+\beta-2}m_1\int_\sigma^1 J_2(\tau)\frac{2C_3 u(\tau)}{C_1 C_2 m_1 m_2\gamma\sigma^{\alpha+\beta-2}}\,d\tau$$

$$= \frac{2}{m_2\gamma}\int_\sigma^1 J_2(\tau)u(\tau)\,d\tau \geq \frac{2}{m_2\gamma}\sigma^{\alpha-1}m_2\|u\| \geq 2\|u\|, \quad \forall t \in [\sigma,1].$$

So, $\|u\| \geq \max_{t\in[\sigma,1]}|u(t)| \geq 2\|u\|$, and then

$$\|u\| = 0. \tag{3.17}$$

In a similar manner, by Lemma 3.1.7, we deduce

$$a(v(t)) \geq a(Q_2(u,v)(t)) = a\left(\int_0^1 G_2(t,\tau)g(\tau,u(\tau),v(\tau))\,d\tau\right)$$

$$\geq \int_0^1 a\left(G_2(t,\tau)g(\tau,u(\tau),v(\tau))\right)d\tau$$

$$\geq \int_\sigma^1 a\left(G_2(t,\tau)g(\tau,u(\tau),v(\tau))\right)d\tau$$

$$\geq \int_\sigma^1 a\left(\sigma^{\beta-1}J_2(\tau)C_2b(u(\tau))\right)d\tau$$

$$= \int_\sigma^1 a\left((C_3^{-1}C_2\sigma^{\beta-1}J_2(\tau))C_3b(u(\tau))\right)d\tau$$

$$\geq C_2C_3^{-1}\sigma^{\beta-1}\int_\sigma^1 J_2(\tau)a(C_3b(u(\tau)))\,d\tau$$

$$\geq C_2C_3^{-1}\sigma^{\beta-1}\int_\sigma^1 J_2(\tau)\frac{2C_3u(\tau)}{C_1C_2m_1m_2\gamma\sigma^{\alpha+\beta-2}}\,d\tau$$

$$= \frac{2}{C_1m_1m_2\gamma\sigma^{\alpha-1}}\int_\sigma^1 J_2(\tau)u(\tau)\,d\tau$$

$$\geq \frac{2}{C_1m_1m_2\gamma\sigma^{\alpha-1}}\int_\sigma^1 J_2(\tau)\left(\int_0^1 G_1(\tau,z)f(z,u(z),v(z))\,dz\right)d\tau$$

$$\geq \frac{2}{C_1m_1m_2\gamma\sigma^{\alpha-1}}\int_\sigma^1 J_2(\tau)\left(\int_\sigma^1 \sigma^{\alpha-1}J_1(z)C_1a(v(z))\,dz\right)d\tau$$

$$= \frac{2}{m_1m_2\gamma}\left(\int_\sigma^1 J_2(\tau)\,d\tau\right)\left(\int_\sigma^1 J_1(z)a(v(z))\,dz\right)$$

$$\geq \frac{2}{m_1\gamma}\int_\sigma^1 J_1(z)a(\sigma^{\beta-1}\|v\|)\,dz \geq \frac{2}{\gamma}\sigma^{\beta-1}a(\|v\|) \geq 2a(\|v\|),$$

$$\forall t\in[\sigma,1].$$

Then we conclude that $a(\|v\|) = a(\sup_{t\in[0,1]}v(t)) \geq a(v(\sigma)) \geq 2a(\|v\|)$, and hence $a(\|v\|) = 0$. By (H3)(a), we obtain

$$\|v\| = 0. \tag{3.18}$$

Therefore, by (3.17) and (3.18), we deduce that $\|(u,v)\|_Y = 0$, which is a contradiction. Hence, $(Q_1(u,v),Q_2(u,v)) \not\leq (u,v)$ for all $(u,v)\in \partial B_{r_1}\cap P$. By Theorem 1.2.11 (b), we conclude that the fixed point index

$$i(Q,B_{r_1}\cap P,P) = 0. \tag{3.19}$$

On the other hand, by (H4), we deduce that there exist $C_4 > 0$, $C_5 > 0$ and $C_6 > 0$ such that

$$f(t,u,v) \leq C_4v^{\alpha_1} + C_5, \quad \forall (t,u,v)\in[0,1]\times\mathbb{R}_+\times\mathbb{R}_+,$$

$$g(t,u,v) \leq \varepsilon_1u^{\alpha_2} + C_6, \quad \forall (t,u,v)\in[0,1]\times\mathbb{R}_+\times\mathbb{R}_+, \tag{3.20}$$

with $\varepsilon_1 = \min\left\{\frac{1}{M_2(8C_4M_1)^{\alpha_2}}, \frac{1}{8M_2(C_4M_1)^{\alpha_2}}\right\} > 0$, $M_1 = \int_0^1 J_1(s)\,ds$, and $M_2 = \int_0^1 J_2(s)\,ds$.

Then, by (3.20), we have

$$Q_1(u,v)(t) = \int_0^1 G_1(t,s)f(s,u(s),v(s))\,ds \leq \int_0^1 J_1(s)(C_4(v(s))^{\alpha_1} + C_5)\,ds$$

$$= C_4 \int_0^1 J_1(s)(v(s))^{\alpha_1}\,ds + C_5M_1, \quad \forall t \in [0,1],$$

$$Q_2(u,v)(t) = \int_0^1 G_2(t,s)g(s,u(s),v(s))\,ds \leq \int_0^1 J_2(s)(\varepsilon_1(u(s))^{\alpha_2} + C_6)\,ds$$

$$= \varepsilon_1 \int_0^1 J_2(s)(u(s))^{\alpha_2}\,ds + C_6M_2, \quad \forall t \in [0,1]. \tag{3.21}$$

We consider the functions p, $q : \mathbb{R}_+ \to \mathbb{R}_+$ given by

$$p(w) = C_4M_1\left[\left(\frac{w}{8C_4M_1}\right)^{\alpha_2} + C_6M_2\right]^{\alpha_1} + C_5M_1, \quad w \in \mathbb{R}_+,$$

$$q(w) = \frac{1}{8(C_4M_1)^{\alpha_2}}(C_4M_1w^{\alpha_1} + C_5M_1)^{\alpha_2} + C_6M_2, \quad w \in \mathbb{R}_+.$$

Because

$$\lim_{w\to\infty} \frac{p(w)}{w} = \lim_{w\to\infty} \frac{q(w)}{w} = \begin{cases} 0, & \text{if } \alpha_1\alpha_2 < 1, \\ 1/8, & \text{if } \alpha_1\alpha_2 = 1, \end{cases}$$

we conclude that there exists $R_1 > r_1$ such that

$$p(w) \leq \frac{1}{4}w, \ q(w) \leq \frac{1}{4}w, \ \forall w \geq R_1. \tag{3.22}$$

We will show that $(u,v) \not\leq (Q_1(u,v), Q_2(u,v))$ for all $(u,v) \in \partial B_{R_1} \cap P$. We suppose that there exists $(u,v) \in \partial B_{R_1} \cap P$, that is, $\|(u,v)\|_Y = R_1$, such that $(u,v) \leq (Q_1(u,v), Q_2(u,v))$. So, by (3.21), we obtain

$$u(t) \leq Q_1(u,v)(t) \leq C_4 \int_0^1 J_1(s)(v(s))^{\alpha_1}\,ds + C_5M_1, \quad \forall t \in [0,1],$$

$$v(t) \leq Q_2(u,v)(t) \leq \varepsilon_1 \int_0^1 J_2(s)(u(s))^{\alpha_2}\,ds + C_6M_2, \quad \forall t \in [0,1].$$

Then, we deduce

$$u(t) \leq C_4 \int_0^1 J_1(s) \left(\varepsilon_1 \int_0^1 J_2(\tau)(u(\tau))^{\alpha_2} \, d\tau + C_6 M_2 \right)^{\alpha_1} ds + C_5 M_1$$

$$= C_4 M_1 \left(\varepsilon_1 \int_0^1 J_2(\tau)(u(\tau))^{\alpha_2} \, d\tau + C_6 M_2 \right)^{\alpha_1} + C_5 M_1$$

$$\leq C_4 M_1 \left(\varepsilon_1 \int_0^1 J_2(\tau) \|u\|^{\alpha_2} \, d\tau + C_6 M_2 \right)^{\alpha_1} + C_5 M_1$$

$$= C_4 M_1 \left(\varepsilon_1 M_2 \|u\|^{\alpha_2} + C_6 M_2 \right)^{\alpha_1} + C_5 M_1$$

$$\leq C_4 M_1 \left[\left(\frac{\|u\|}{8 C_4 M_1} \right)^{\alpha_2} + C_6 M_2 \right]^{\alpha_1} + C_5 M_1$$

$$\leq C_4 M_1 \left[\left(\frac{\|(u,v)\|_Y}{8 C_4 M_1} \right)^{\alpha_2} + C_6 M_2 \right]^{\alpha_1} + C_5 M_1, \quad \forall t \in [0,1]$$

$$(3.23)$$

and

$$v(t) \leq \varepsilon_1 \int_0^1 J_2(s)(u(s))^{\alpha_2} \, ds + C_6 M_2$$

$$\leq \varepsilon_1 \int_0^1 J_2(s) \left(C_4 \int_0^1 J_1(\tau)(v(\tau))^{\alpha_1} \, d\tau + C_5 M_1 \right)^{\alpha_2} ds + C_6 M_2$$

$$\leq \varepsilon_1 M_2 \left(C_4 M_1 \|v\|^{\alpha_1} + C_5 M_1 \right)^{\alpha_2} + C_6 M_2$$

$$\leq \frac{1}{8(C_4 M_1)^{\alpha_2}} \left(C_4 M_1 \|v\|^{\alpha_1} + C_5 M_1 \right)^{\alpha_2} + C_6 M_2$$

$$\leq \frac{1}{8(C_4 M_1)^{\alpha_2}} \left(C_4 M_1 \|(u,v)\|_Y^{\alpha_1} + C_5 M_1 \right)^{\alpha_2} + C_6 M_2, \quad \forall t \in [0,1].$$

$$(3.24)$$

By using (3.23), (3.24) and (3.22), we conclude that $u(t) \leq \frac{1}{4}\|(u,v)\|_Y$ and $v(t) \leq \frac{1}{4}\|(u,v)\|_Y$ for all $t \in [0,1]$. Therefore, we obtain that $\|(u,v)\|_Y \leq \frac{1}{2}\|(u,v)\|_Y$, and so $\|(u,v)\|_Y = 0$, which is a contradiction, because $\|(u,v)\|_Y = R_1 > 0$.

So, $(u,v) \not\leq (Q_1(u,v), Q_2(u,v))$ for all $(u,v) \in \partial B_{R_1} \cap P$. By Theorem 1.2.11 (a), we deduce that the fixed point index

$$i(Q, B_{R_1} \cap P, P) = 1. \tag{3.25}$$

Because Q has no fixed points on $\partial B_{r_1} \cup \partial B_{R_1}$, by (3.19) and (3.25), we conclude that

$$i(Q, (B_{R_1} \setminus \overline{B}_{r_1}) \cap P, P) = i(Q, B_{R_1} \cap P, P) - i(Q, B_{r_1} \cap P, P) = 1.$$

Therefore, the operator Q has at least one fixed point $(u_1, v_1) \in (B_{R_1} \setminus \overline{B}_{r_1}) \cap P$, with $r_1 < \|(u_1, v_1)\|_Y < R_1$, that is $\|u_1\| > 0$ or $\|v_1\| > 0$. Because $u_1 \in P_1$ and $v_1 \in P_2$, we obtain $u_1(t) > 0$ for all $t \in (0, 1]$ or $v_1(t) > 0$ for all $t \in (0, 1]$. $\qquad \square$

Theorem 3.1.2. *Assume that* (H1), (H2), (H5) *and* (H6) *hold. Then the problem* (3.1), (3.2) *has at least one positive solution.*

Proof. By (H5) there exist $C_i > 0$, $i = 7, \ldots, 11$ such that

$$f(t, u, v) \geq C_7 c(v) - C_8, \quad g(t, u, v) \geq C_9 d(u) - C_{10},$$

$$\forall (t, u, v) \in [\sigma, 1] \times \mathbb{R}_+ \times \mathbb{R}_+, \tag{3.26}$$

and

$$c(C_{12} d(u)) \geq \frac{2 C_{12} u}{C_7 C_9 m_1 m_2 \gamma \sigma^{\alpha + \beta - 2}} - C_{11}, \quad \forall u \in \mathbb{R}_+, \tag{3.27}$$

where $C_{12} = \max\{C_9 \sigma^{\beta - 1} J_2(s), \ s \in [0, 1]\} > 0$.

Then we have

$$Q_1(u, v)(t) = \int_0^1 G_1(t, s) f(s, u(s), v(s)) \, ds$$

$$\geq \int_\sigma^1 G_1(t, s)(C_7 c(v(s)) - C_8) \, ds$$

$$\geq \sigma^{\alpha - 1} \int_\sigma^1 J_1(s)(C_7 c(v(s)) - C_8) \, ds, \quad \forall t \in [\sigma, 1],$$

$$Q_2(u, v)(t) = \int_0^1 G_2(t, s) g(s, u(s), v(s)) \, ds$$

$$\geq \int_\sigma^1 G_2(t, s)(C_9 d(u(s)) - C_{10}) \, ds$$

$$\geq \sigma^{\beta - 1} \int_\sigma^1 J_2(s)(C_9 d(u(s)) - C_{10}) \, ds, \quad \forall t \in [\sigma, 1]. \tag{3.28}$$

We will prove that the set $U = \{(u, v) \in P, \ (u, v) = Q(u, v) + \lambda(\varphi_1, \varphi_2), \ \lambda \geq 0\}$ is bounded, where $(\varphi_1, \varphi_2) \in P \setminus \{(0, 0)\}$. Indeed,

$(u, v) \in U$ implies that $u \geq Q_1(u, v)$, $v \geq Q_2(u, v)$ for some φ_1, $\varphi_2 \geq 0$. By (3.28), we obtain

$$u(t) \geq Q_1(u, v)(t) \geq \sigma^{\alpha-1} C_7 \int_\sigma^1 J_1(s) c(v(s)) \, ds - \sigma^{\alpha-1} C_8 \int_\sigma^1 J_1(s) \, ds$$

$$= \sigma^{\alpha-1} C_7 \int_\sigma^1 J_1(s) c(v(s)) \, ds - C_{13}, \quad \forall t \in [\sigma, 1], \tag{3.29}$$

$$v(t) \geq Q_2(u, v)(t) \geq \sigma^{\beta-1} C_9 \int_\sigma^1 J_2(s) d(u(s)) \, ds - \sigma^{\beta-1} C_{10} \int_\sigma^1 J_2(s) \, ds$$

$$= \sigma^{\beta-1} C_9 \int_\sigma^1 J_2(s) d(u(s)) \, ds - C_{14}, \quad \forall t \in [\sigma, 1], \tag{3.30}$$

where $C_{13} = m_1 \sigma^{\alpha-1} C_8$, $C_{14} = m_2 \sigma^{\beta-1} C_{10}$.

By the monotonicity and concavity of $c(\cdot)$ and the Jensen inequality, the inequality (3.30) implies that

$$c(v(t) + C_{14}) \geq c \left(\sigma^{\beta-1} C_9 \int_\sigma^1 J_2(s) d(u(s)) \, ds \right)$$

$$\geq \int_\sigma^1 c \left(\sigma^{\beta-1} C_9 J_2(s) d(u(s)) \right) ds$$

$$= \int_\sigma^1 c \left((\sigma^{\beta-1} C_{12}^{-1} C_9 J_2(s)) C_{12} d(u(s)) \right) ds$$

$$\geq \sigma^{\beta-1} C_{12}^{-1} C_9 \int_\sigma^1 J_2(s) c(C_{12} d(u(s))) \, ds, \quad \forall t \in [\sigma, 1]. \tag{3.31}$$

Since $c(v(t)) \geq c(v(t) + C_{14}) - c(C_{14})$, by the relations (3.27), (3.29), (3.31), we deduce

$$u(t) \geq \sigma^{\alpha-1} C_7 \int_\sigma^1 J_1(s) c(v(s)) \, ds - C_{13}$$

$$\geq \sigma^{\alpha-1} C_7 \int_\sigma^1 J_1(s) [c(v(s) + C_{14}) - c(C_{14})] \, ds - C_{13}$$

$$= \sigma^{\alpha-1} C_7 \int_\sigma^1 J_1(s) c(v(s) + C_{14}) \, ds - C_{15}$$

$$\geq \sigma^{\alpha-1} C_7 \int_\sigma^1 J_1(s) \left[\sigma^{\beta-1} C_{12}^{-1} C_9 \int_\sigma^1 J_2(\tau) c(C_{12} d(u(\tau))) \, d\tau \right] ds - C_{15}$$

$$= \sigma^{\alpha+\beta-2} C_7 C_9 C_{12}^{-1} \left(\int_\sigma^1 J_1(s) \, ds \right) \left(\int_\sigma^1 J_2(\tau) c(C_{12} d(u(\tau))) \, d\tau \right) - C_{15}$$

$$= \sigma^{\alpha+\beta-2} C_7 C_9 C_{12}^{-1} m_1 \int_\sigma^1 J_2(\tau) c(C_{12} d(u(\tau))) \, d\tau - C_{15}$$

$$\geq \sigma^{\alpha+\beta-2} C_7 C_9 C_{12}^{-1} m_1 \int_\sigma^1 J_2(\tau)$$

$$\times \left(\frac{2 C_{12} u(\tau)}{C_7 C_9 m_1 m_2 \gamma \sigma^{\alpha+\beta-2}} - C_{11} \right) d\tau - C_{15}$$

$$= \frac{2}{m_2 \gamma} \int_\sigma^1 J_2(\tau) u(\tau) \, d\tau - C_{16} \geq \frac{2 \sigma^{\alpha-1}}{m_2 \gamma} \int_\sigma^1 J_2(\tau) \|u\| \, d\tau - C_{16}$$

$$\geq 2\|u\| - C_{16}, \quad \forall t \in [\sigma, 1],$$

where $C_{15} = \sigma^{\alpha-1} C_7 c(C_{14}) m_1 + C_{13}$, $C_{16} = \sigma^{\alpha+\beta-2} C_7 C_9 C_{12}^{-1} m_1 m_2 C_{11} + C_{15}$.

Therefore, $\|u\| \geq u(\sigma) \geq 2\|u\| - C_{16}$, and then

$$\|u\| \leq C_{16}. \tag{3.32}$$

Since $c(v(t)) \geq c(\sigma^{\beta-1}\|v\|) \geq \sigma^{\beta-1} c(\|v\|)$ for all $t \in [\sigma, 1]$ and $v \in P_2$, then by the relations (3.27), (3.29)–(3.31), we obtain

$$c(v(t)) \geq c(v(t) + C_{14}) - c(C_{14})$$

$$\geq \sigma^{\beta-1} C_{12}^{-1} C_9 \int_\sigma^1 J_2(s) c(C_{12} d(u(s))) \, ds - c(C_{14})$$

$$\geq \sigma^{\beta-1} C_{12}^{-1} C_9 \int_\sigma^1 J_2(s) \left(\frac{2 C_{12} u(s)}{C_7 C_9 m_1 m_2 \gamma \sigma^{\alpha+\beta-2}} - C_{11} \right) ds - c(C_{14})$$

$$= \frac{2}{C_7 m_1 m_2 \gamma \sigma^{\alpha-1}} \int_\sigma^1 J_2(s) u(s) \, ds - C_{17}$$

$$\geq \frac{2}{C_7 m_1 m_2 \gamma \sigma^{\alpha-1}} \int_\sigma^1 J_2(s)$$

$$\times \left(\sigma^{\alpha-1} C_7 \int_\sigma^1 J_1(\tau) c(v(\tau)) \, d\tau - C_{13} \right) ds - C_{17}$$

$$= \frac{2}{m_1\gamma} \int_\sigma^1 J_1(\tau)c(v(\tau))\, d\tau - C_{18}$$

$$\geq \frac{2}{m_1\gamma} \int_\sigma^1 J_1(\tau)c(\sigma^{\beta-1}\|v\|)\, d\tau - C_{18}$$

$$\geq \frac{2\sigma^{\beta-1}}{\gamma} c(\|v\|) - C_{18} \geq 2c(\|v\|) - C_{18}, \quad \forall t \in [\sigma, 1],$$

where $C_{17} = \sigma^{\beta-1}C_{12}^{-1}C_9C_{11}m_2 + c(C_{14})$, $C_{18} = \frac{2C_{13}}{C_7 m_1 \gamma \sigma^{\alpha-1}} + C_{17}$.

Then $c(\|v\|) \geq c(v(\sigma)) \geq 2c(\|v\|) - C_{18}$, and so $c(\|v\|) \leq C_{18}$. By (H5) (a) and (c), we deduce that $\lim_{v\to\infty} c(v) = \infty$, thus there exists $C_{19} > 0$, such that

$$\|v\| \leq C_{19}. \tag{3.33}$$

By (3.32) and (3.33), we conclude that $\|(u,v)\|_Y \leq C_{16} + C_{19}$ for all $(u,v) \in U$, that is, the set U is bounded. Then there exists a sufficiently large $R_2 > 0$ such that $(u,v) \neq Q(u,v) + \lambda(\varphi_1, \varphi_2)$ for all $(u,v) \in \partial B_{R_2} \cap P$ and $\lambda \geq 0$. By Theorem 1.2.9, we deduce that

$$i(Q, B_{R_2} \cap P, P) = 0. \tag{3.34}$$

On the other hand, by (H6), there exist $C_{20} > 0$ and a sufficiently small $r_2 > 0$, $(r_2 < R_2, r_2 \leq 1)$ such that

$$f(t,u,v) \leq C_{20}v^{\beta_1}, \quad \forall (t,u) \in [0,1] \times \mathbb{R}_+, \quad v \in [0, r_2],$$
$$g(t,u,v) \leq \varepsilon_2 u^{\beta_2}, \quad \forall (t,v) \in [0,1] \times \mathbb{R}_+, \quad u \in [0, r_2], \tag{3.35}$$

where $\varepsilon_2 = (2C_{20}M_1 M_2^{\beta_1})^{-1/\beta_1} > 0$.

We will show that $(u,v) \not\leq Q(u,v)$ for all $(u,v) \in \partial B_{r_2} \cap P$. We suppose that there exists $(u,v) \in \partial B_{r_2} \cap P$, that is $\|(u,v)\|_Y = r_2 \leq 1$, such that $(u,v) \leq (Q_1(u,v), Q_2(u,v))$, or $u \leq Q_1(u,v)$ and $v \leq Q_2(u,v)$. Then by (3.35), we obtain

$$u(t) \leq Q_1(u,v)(t) = \int_0^1 G_1(t,s)f(s,u(s),v(s))ds \leq C_{20}\int_0^1 J_1(s)(v(s))^{\beta_1}\, ds$$

$$\leq C_{20}\int_0^1 J_1(s)\left(\int_0^1 G_2(s,\tau)g(\tau,u(\tau),v(\tau))\, d\tau\right)^{\beta_1} ds$$

$$\leq C_{20}\int_0^1 J_1(s)\left(\int_0^1 J_2(\tau)\varepsilon_2(u(\tau))^{\beta_2}\, d\tau\right)^{\beta_1} ds$$

$$= C_{20} M_1 \varepsilon_2^{\beta_1} \left(\int_0^1 J_2(\tau)(u(\tau))^{\beta_2} \, d\tau \right)^{\beta_1}$$

$$\leq C_{20} M_1 \varepsilon_2^{\beta_1} \left(\int_0^1 J_2(\tau) \, d\tau \right)^{\beta_1} \|u\|^{\beta_1 \beta_2}$$

$$= C_{20} M_1 \varepsilon_2^{\beta_1} M_2^{\beta_1} \|u\|^{\beta_1 \beta_2} \leq C_{20} \varepsilon_2^{\beta_1} M_1 M_2^{\beta_1} \|u\| = \frac{1}{2} \|u\|, \quad \forall t \in [0,1].$$

Therefore, $\|u\| \leq \frac{1}{2}\|u\|$, so

$$\|u\| = 0. \tag{3.36}$$

In addition,

$$v(t) \leq Q_2(u,v)(t) = \int_0^1 G_2(t,s)g(s,u(s),v(s)) \, ds \leq \int_0^1 J_2(s)\varepsilon_2(u(s))^{\beta_2} \, ds$$

$$\leq \varepsilon_2 M_2 \|u\|^{\beta_2}, \quad \forall t \in [0,1]. \tag{3.37}$$

By (3.36) and (3.37), we deduce that $\|v\| = 0$, and then $\|(u,v)\|_Y = 0$, which is a contradiction, because $\|(u,v)\|_Y = r_2 > 0$.

Then $(u,v) \nleq Q(u,v)$ for all $(u,v) \in \partial B_{r_2} \cap P$. By Theorem 1.2.11 (a), we conclude that

$$i(Q, B_{r_2} \cap P, P) = 1. \tag{3.38}$$

Because Q has no fixed points on $\partial B_{r_2} \cup \partial B_{R_2}$, by (3.34) and (3.38), we deduce that

$$i(Q, (B_{R_2} \setminus \overline{B}_{r_2}) \cap P, P) = i(Q, B_{R_2} \cap P, P) - i(Q, B_{r_2} \cap P, P) = -1.$$

So, the operator Q has at least one fixed point $(u_1, v_1) \in (B_{R_2} \setminus \overline{B}_{r_2}) \cap P$, with $r_2 < \|(u_1, v_1)\|_Y < R_2$, which is a positive solution for our problem (3.1), (3.2). $\qquad\square$

Remark 3.1.2. In a similar manner, we can prove that Theorem 3.1.1 remains valid if assumption (H3) is replaced by

(H3)$'$ There exist the functions $a, b \in C(\mathbb{R}_+, \mathbb{R}_+)$ such that

    (a) $b(\cdot)$ is concave and strictly increasing on $\mathbb{R}_+$ with $b(0) = 0$;

    (b) There exists $\sigma \in (0,1)$ such that

$$f_0^i = \liminf_{v \to 0+} \frac{f(t,u,v)}{a(v)} \in (0,\infty], \quad \text{uniformly with respect to}$$

$$(t,u) \in [\sigma, 1] \times \mathbb{R}_+,$$

$$g_0^i = \liminf_{u \to 0+} \frac{g(t, u, v)}{b(u)} \in (0, \infty], \text{ uniformly with respect to}$$

$$(t, v) \in [\sigma, 1] \times \mathbb{R}_+;$$

(c) $\lim\limits_{v \to 0+} \dfrac{b(Ca(v))}{v} = \infty$ exists for any constant $C > 0$.

Remark 3.1.3. Theorem 3.1.2 remains valid if assumption (H5) is replaced by

(H5)$'$ There exist the functions $c, d \in C(\mathbb{R}_+, \mathbb{R}_+)$ such that

(a) $d(\cdot)$ is concave and strictly increasing on $\mathbb{R}_+$;
(b) There exists $\sigma \in (0, 1)$ such that

$$f_\infty^i = \liminf_{v \to \infty} \frac{f(t, u, v)}{c(v)} \in (0, \infty], \text{ uniformly with respect to}$$

$$(t, u) \in [\sigma, 1] \times \mathbb{R}_+,$$

$$g_\infty^i = \liminf_{u \to \infty} \frac{g(t, u, v)}{d(u)} \in (0, \infty], \text{ uniformly with respect to}$$

$$(t, v) \in [\sigma, 1] \times \mathbb{R}_+;$$

(c) $\lim\limits_{v \to \infty} \dfrac{d(Cc(v))}{v} = \infty$ exists for any constant $C > 0$.

Remark 3.1.4. Theorem 3.1.1 remains valid if assumption (H4) is replaced by

(H4)$'$ There exist $\alpha_1, \alpha_2 > 0$ with $\alpha_1 \alpha_2 \le 1$ such that

$$f_\infty^s = \lim_{v \to \infty} \frac{f(t, u, v)}{v^{\alpha_1}} = 0, \text{ exists uniformly with respect to}$$

$$(t, u) \in [0, 1] \times \mathbb{R}_+,$$

$$g_\infty^s = \limsup_{u \to \infty} \frac{g(t, u, v)}{u^{\alpha_2}} \in [0, \infty), \text{ uniformly with respect to}$$

$$(t, v) \in [0, 1] \times \mathbb{R}_+.$$

Remark 3.1.5. Theorem 3.1.2 remains valid if assumption (H6) is replaced by

(H6)$'$ There exist β_1, $\beta_2 > 0$ with $\beta_1\beta_2 \geq 1$ such that

$$f_0^s = \lim_{v \to 0+} \frac{f(t, u, v)}{v^{\beta_1}} = 0, \text{ exists uniformly with respect to}$$

$$(t, u) \in [0, 1] \times \mathbb{R}_+,$$

$$g_0^s = \limsup_{u \to 0+} \frac{g(t, u, v)}{u^{\beta_2}} \in [0, \infty), \text{ uniformly with respect to}$$

$$(t, v) \in [0, 1] \times \mathbb{R}_+.$$

Theorem 3.1.3. *Assume that assumptions* (H1)–(H3), (H5) *and* (H7) *hold. Then the problem* (3.1), (3.2) *has at least two positive solutions.*

Proof. By using (H7), for any $(u, v) \in \partial B_{N_0} \cap P$, we obtain

$$Q_1(u, v)(t) \leq \int_0^1 J_1(s) f(s, N_0, N_0)\, ds < \frac{N_0}{2m_0} \int_0^1 J_1(s)\, ds$$

$$= \frac{N_0 M_1}{2m_0} \leq \frac{N_0}{2}, \quad \forall t \in [0, 1],$$

$$Q_2(u, v)(t) \leq \int_0^1 J_2(s) g(s, N_0, N_0)\, ds < \frac{N_0}{2m_0} \int_0^1 J_2(s)\, ds$$

$$= \frac{N_0 M_2}{2m_0} \leq \frac{N_0}{2}, \quad \forall t \in [0, 1].$$

Then we deduce

$$\|Q(u, v)\|_Y = \|Q_1(u, v)\| + \|Q_2(u, v)\| < N_0$$

$$= \|(u, v)\|_Y, \quad \forall (u, v) \in \partial B_{N_0} \cap P.$$

Because Q has no fixed points on ∂B_{N_0}, by Theorem 1.2.8 we conclude that

$$i(Q, B_{N_0} \cap P, P) = 1. \tag{3.39}$$

On the other hand, from (H3) and (H5), and the proofs of Theorems 3.1.1 and 3.1.2, we know that there exist a sufficiently small $r_1 > 0$ ($r_1 < N_0$) and a sufficiently large $R_2 > N_0$ such that

$$i(Q, B_{r_1} \cap P, P) = 0, \quad i(Q, B_{R_2} \cap P, P) = 0. \tag{3.40}$$

Because Q has no fixed points on $\partial B_{r_1} \cup \partial B_{R_2} \cup \partial B_{N_0}$, by the relations (3.39) and (3.40), we obtain

$$i(Q, (B_{R_2} \setminus \overline{B}_{N_0}) \cap P, P) = i(Q, B_{R_2} \cap P, P) - i(Q, B_{N_0} \cap P, P) = -1,$$

$$i(Q, (B_{N_0} \setminus \overline{B}_{r_1}) \cap P, P) = i(Q, B_{N_0} \cap P, P) - i(Q, B_{r_1} \cap P, P) = 1.$$

Then $\mathcal{Q}$ has at least one fixed point $(u_1, v_1) \in (B_{R_2} \setminus \overline{B}_{N_0}) \cap P$ and has at least one fixed point $(u_2, v_2) \in (B_{N_0} \setminus \overline{B}_{r_1}) \cap P$. Therefore, problem (3.1), (3.2) has two distinct positive solutions (u_1, v_1), (u_2, v_2). $\qquad\square$

Remark 3.1.6. Theorem 3.1.3 remains valid if (H3) is replaced by (H3)′ or (H5) is replaced by (H5)′.

Remark 3.1.7. In (H3), if $a(v) = v^p$ with $p \le 1$, and $b(u) = u^q$ with $q > 0$, the condition from (H3)(c) is satisfied if $pq < 1$. In (H5), if $c(v) = v^p$ with $p \le 1$, and $d(u) = u^q$ with $q > 0$, the condition from (H5) (c) is satisfied if $pq > 1$.

Examples

(1) We consider $f(t, u, v) = e^t(1 + e^{-(u+v)})$ and $g(t, u, v) = (1 + e^{-t})u^\theta$, for $(t, u, v) \in [0, 1] \times \mathbb{R}_+ \times \mathbb{R}_+$. For $a(v) = v^p$ with $p \le 1$, and $b(u) = u^q$ for $q > 0$ and $pq < 1$, then the assumptions (H3) and (H4) are satisfied if $q > \theta$ and $\alpha_2 > \theta$. For example, if $\theta = \frac{3}{2}$, $p = \frac{1}{3}$, $q = 2$, $\alpha_1 = \frac{1}{4}$ and $\alpha_2 = 3$, we can apply Theorem 3.1.1, and we deduce that the problem (3.1), (3.2) has at least one positive solution.

(2) We consider $f(t, u, v) = (1 + e^{-u})v^{\theta_1}$ and $g(t, u, v) = (1 + e^{-v})u^{\theta_2}$, for $(t, u, v) \in [0, 1] \times \mathbb{R}_+ \times \mathbb{R}_+$. For $c(v) = v^p$ with $p \le 1$, and $d(u) = u^q$ for $q > 0$ and $pq > 1$, then the assumptions (H5) and (H6) are satisfied if $p < \theta_1$, $q < \theta_2$, $\beta_1 < \theta_1$, and $\beta_2 < \theta_2$. For example, if $\theta_1 = 3$, $\theta_2 = \frac{2}{3}$, $p = 2$, $q = \frac{7}{12}$, $\beta_1 = \frac{5}{2}$ and $\beta_2 = \frac{1}{2}$, we can apply Theorem 3.1.2, and we conclude that the problem (3.1), (3.2) has at least one positive solution.

Remark 3.1.8. The results presented in this section under the assumptions $p_1 \in [1, n-2]$ and $p_2 \in [1, m-2]$ instead of $p_1 \in [1, \alpha-1)$ and $p_2 \in [1, \beta-1)$ were published in [50].

Remark 3.1.9. Under different assumptions on the nonlinearities f and g as those used in this section, the existence and multiplicity of positive solutions of the system (3.1) with $f(t, u, v) = \widetilde{f}(t, v)$ and $g(t, u, v) = \widetilde{g}(t, u)$ which can be nonsingular or singular functions at the points $t = 0$ and/or $t = 1$, subject to the boundary conditions (3.2) were investigated in [49]. The existence and nonexistence of positive solutions of the system (3.1) with two positive parameters supplemented with the boundary conditions (3.2) were studied in the paper [52].

3.2 Systems of Fractional Differential Equations with Coupled Multi-Point Boundary Conditions

We consider the system of nonlinear ordinary fractional differential equations

$$\begin{cases} D_{0+}^{\alpha} u(t) + f(t, u(t), v(t)) = 0, & t \in (0,1), \\ D_{0+}^{\beta} v(t) + g(t, u(t), v(t)) = 0, & t \in (0,1), \end{cases} \tag{3.41}$$

with the coupled multi-point boundary conditions

$$\begin{cases} u(0) = u'(0) = \cdots = u^{(n-2)}(0) = 0, & D_{0+}^{p_1} u(1) = \sum_{i=1}^{N} a_i D_{0+}^{q_1} v(\xi_i), \\ v(0) = v'(0) = \cdots = v^{(m-2)}(0) = 0, & D_{0+}^{p_2} v(1) = \sum_{i=1}^{M} b_i D_{0+}^{q_2} u(\eta_i), \end{cases} \tag{3.42}$$

where $\alpha \in (n-1, n]$, $\beta \in (m-1, m]$, $n, m \geq 3$, $p_1, p_2, q_1, q_2 \in \mathbb{R}$, $p_1 \in [1, \alpha - 1)$, $p_2 \in [1, \beta - 1)$, $q_1 \in [0, p_2]$, $q_2 \in [0, p_1]$, $\xi_i, a_i \in \mathbb{R}$ for all $i = 1, \ldots, N$ ($N \in \mathbb{N}$), $0 < \xi_1 < \cdots < \xi_N \leq 1$, $\eta_i, b_i \in \mathbb{R}$ for all $i = 1, \ldots, M$ ($M \in \mathbb{N}$), $0 < \eta_1 < \cdots < \eta_M \leq 1$, and D_{0+}^{ζ} denotes the Riemann-Liouville derivative of order ζ (for $\zeta = \alpha, \beta, p_1, q_1, p_2, q_2$).

Under sufficient conditions on the functions f and g, which can be non-singular or singular at the points $t = 0$ and/or $t = 1$, we study the existence and multiplicity of positive solutions of problem (3.41), (3.42). We use some theorems from the fixed point index theory and the Guo–Krasnosel'skii fixed point theorem (Theorem 1.2.2). By a positive solution of problem (3.41), (3.42) we mean a pair of functions $(u, v) \in (C([0,1], \mathbb{R}_+))^2$ satisfying (3.41) and (3.42) with $u(t) > 0$ for all $t \in (0,1]$ or $v(t) > 0$ for all $t \in (0,1]$.

3.2.1 *Preliminary results*

We consider the fractional differential system

$$\begin{cases} D_{0+}^{\alpha} u(t) + x(t) = 0, & t \in (0,1), \\ D_{0+}^{\beta} v(t) + y(t) = 0, & t \in (0,1), \end{cases} \tag{3.43}$$

with the coupled multi-point boundary conditions (3.42), where $x, y \in C(0,1) \cap L^1(0,1)$.

We denote by Δ the constant

$$\Delta = \frac{\Gamma(\alpha)\Gamma(\beta)}{\Gamma(\alpha - p_1)\Gamma(\beta - p_2)} - \frac{\Gamma(\alpha)\Gamma(\beta)}{\Gamma(\alpha - q_2)\Gamma(\beta - q_1)}$$

$$\times \left(\sum_{i=1}^{N} a_i \xi_i^{\beta - q_1 - 1} \right) \left(\sum_{i=1}^{M} b_i \eta_i^{\alpha - q_2 - 1} \right).$$

Lemma 3.2.1. *If $\Delta \neq 0$ and $x, y \in C(0,1) \cap L^1(0,1)$, then the unique solution $(u, v) \in C[0,1] \times C[0,1]$ of problem (3.43), (3.42) is given by*

$$u(t) = -\frac{1}{\Gamma(\alpha)} \int_0^t (t-s)^{\alpha-1} x(s)\, ds$$

$$+ \frac{t^{\alpha-1}}{\Delta} \left[\frac{\Gamma(\beta)}{\Gamma(\alpha - p_1)\Gamma(\beta - p_2)} \int_0^1 (1-s)^{\alpha - p_1 - 1} x(s)\, ds \right.$$

$$- \frac{\Gamma(\beta)}{\Gamma(\beta - q_1)\Gamma(\beta - p_2)} \sum_{i=1}^{N} a_i \left(\int_0^{\xi_i} (\xi_i - s)^{\beta - q_1 - 1} y(s)\, ds \right)$$

$$+ \frac{\Gamma(\beta)}{\Gamma(\beta - q_1)\Gamma(\beta - p_2)} \left(\sum_{i=1}^{N} a_i \xi_i^{\beta - q_1 - 1} \right) \int_0^1 (1-s)^{\beta - p_2 - 1} y(s)\, ds$$

$$- \frac{\Gamma(\beta)}{\Gamma(\alpha - q_2)\Gamma(\beta - q_1)} \left(\sum_{i=1}^{N} a_i \xi_i^{\beta - q_1 - 1} \right)$$

$$\left. \times \sum_{i=1}^{M} b_i \left(\int_0^{\eta_i} (\eta_i - s)^{\alpha - q_2 - 1} x(s)\, ds \right) \right], \quad t \in [0,1],$$

$$v(t) = -\frac{1}{\Gamma(\beta)} \int_0^t (t-s)^{\beta-1} y(s)\, ds$$

$$+ \frac{t^{\beta-1}}{\Delta} \left[\frac{\Gamma(\alpha)}{\Gamma(\alpha - p_1)\Gamma(\beta - p_2)} \int_0^1 (1-s)^{\beta - p_2 - 1} y(s)\, ds \right.$$

$$- \frac{\Gamma(\alpha)}{\Gamma(\alpha - p_1)\Gamma(\alpha - q_2)} \sum_{i=1}^{M} b_i \left(\int_0^{\eta_i} (\eta_i - s)^{\alpha - q_2 - 1} x(s)\, ds \right)$$

$$+ \frac{\Gamma(\alpha)}{\Gamma(\alpha - p_1)\Gamma(\alpha - q_2)} \left(\sum_{i=1}^{M} b_i \eta_i^{\alpha - q_2 - 1} \right) \int_0^1 (1-s)^{\alpha - p_1 - 1} x(s)\, ds$$

$$- \frac{\Gamma(\alpha)}{\Gamma(\alpha - q_2)\Gamma(\beta - q_1)} \left(\sum_{i=1}^{M} b_i \eta_i^{\alpha - q_2 - 1} \right)$$

$$\times \sum_{i=1}^{N} a_i \left(\int_0^{\xi_i} (\xi_i - s)^{\beta - q_1 - 1} y(s) \, ds \right) \Bigg], \quad t \in [0, 1]. \tag{3.44}$$

Proof. By Lemma 1.1.3, we deduce that the solutions $(u, v) \in (C(0, 1) \cap L^1(0, 1))^2$ of the fractional differential system (3.43) are given by

$$u(t) = -I_{0+}^{\alpha} x(t) + c_1 t^{\alpha - 1} + \cdots + c_n t^{\alpha - n}$$

$$= -\frac{1}{\Gamma(\alpha)} \int_0^t (t - s)^{\alpha - 1} x(s) \, ds + c_1 t^{\alpha - 1} + \cdots + c_n t^{\alpha - n},$$

$$v(t) = -I_{0+}^{\beta} y(t) + d_1 t^{\beta - 1} + \cdots + d_n t^{\beta - m}$$

$$= -\frac{1}{\Gamma(\beta)} \int_0^t (t - s)^{\beta - 1} y(s) \, ds + d_1 t^{\beta - 1} + \cdots + d_m t^{\beta - m},$$

where $c_1, c_2, \ldots, c_n, d_1, d_2, \ldots, d_m \in \mathbb{R}$.

By using the conditions $u(0) = u'(0) = \cdots = u^{(n-2)}(0) = 0$ and $v(0) = v'(0) = \cdots = v^{(m-2)}(0) = 0$, we obtain $c_2 = \cdots = c_n = 0$ and $d_2 = \cdots = d_m = 0$. Then we conclude

$$(t) = c_1 t^{\alpha - 1} - \frac{1}{\Gamma(\alpha)} \int_0^t (t - s)^{\alpha - 1} x(s) \, ds, \ t \in [0, 1],$$

$$v(t) = d_1 t^{\beta - 1} - \frac{1}{\Gamma(\beta)} \int_0^t (t - s)^{\beta - 1} y(s) \, ds, \ t \in [0, 1]. \tag{3.45}$$

By using some properties of the fractional integrals and fractional derivatives from Lemmas 1.1.1 and 1.1.2, we obtain

$$D_{0+}^{p_1} u(t) = c_1 \frac{\Gamma(\alpha)}{\Gamma(\alpha - p_1)} t^{\alpha - p_1 - 1} - I_{0+}^{\alpha - p_1} x(t),$$

$$D_{0+}^{q_2} u(t) = c_1 \frac{\Gamma(\alpha)}{\Gamma(\alpha - q_2)} t^{\alpha - q_2 - 1} - I_{0+}^{\alpha - q_2} x(t),$$

$$D_{0+}^{p_2} v(t) = d_1 \frac{\Gamma(\beta)}{\Gamma(\beta - p_2)} t^{\beta - p_2 - 1} - I_{0+}^{\beta - p_2} y(t),$$

$$D_{0+}^{q_1} v(t) = d_1 \frac{\Gamma(\beta)}{\Gamma(\beta - q_1)} t^{\beta - q_1 - 1} - I_{0+}^{\beta - q_1} y(t).$$

Then the conditions $D_{0+}^{p_1} u(1) = \sum_{i=1}^{N} a_i D_{0+}^{q_1} v(\xi_i)$ and $D_{0+}^{p_2} v(1) = \sum_{i=1}^{M} b_i D_{0+}^{q_2} u(\eta_i)$ give us

$$c_1 \frac{\Gamma(\alpha)}{\Gamma(\alpha - p_1)} - \frac{1}{\Gamma(\alpha - p_1)} \int_0^1 (1-s)^{\alpha - p_1 - 1} x(s) \, ds$$

$$= \sum_{i=1}^{N} a_i \left(d_1 \frac{\Gamma(\beta)}{\Gamma(\beta - q_1)} \xi_i^{\beta - q_1 - 1} \right.$$

$$\left. - \frac{1}{\Gamma(\beta - q_1)} \int_0^{\xi_i} (\xi_i - s)^{\beta - q_1 - 1} y(s) \, ds \right),$$

$$d_1 \frac{\Gamma(\beta)}{\Gamma(\beta - p_2)} - \frac{1}{\Gamma(\beta - p_2)} \int_0^1 (1-s)^{\beta - p_2 - 1} y(s) \, ds$$

$$= \sum_{i=1}^{M} b_i \left(c_1 \frac{\Gamma(\alpha)}{\Gamma(\alpha - q_2)} \eta_i^{\alpha - q_2 - 1} \right.$$

$$\left. - \frac{1}{\Gamma(\alpha - q_2)} \int_0^{\eta_i} (\eta_i - s)^{\alpha - q_2 - 1} x(s) \, ds \right). \tag{3.46}$$

The above system in c_1 and d_1 has the determinant Δ, which by the assumption of this lemma is different from 0. So, the system (3.46) has the unique solution

$$c_1 = \frac{1}{\Delta} \left[\frac{\Gamma(\beta)}{\Gamma(\alpha - p_1)\Gamma(\beta - p_2)} \int_0^1 (1-s)^{\alpha - p_1 - 1} x(s) \, ds \right.$$

$$- \frac{\Gamma(\beta)}{\Gamma(\beta - q_1)\Gamma(\beta - p_2)} \sum_{i=1}^{N} a_i \left(\int_0^{\xi_i} (\xi_i - s)^{\beta - q_1 - 1} y(s) \, ds \right)$$

$$+ \frac{\Gamma(\beta)}{\Gamma(\beta - q_1)\Gamma(\beta - p_2)} \left(\sum_{i=1}^{N} a_i \xi_i^{\beta - q_1 - 1} \right) \int_0^1 (1-s)^{\beta - p_2 - 1} y(s) \, ds$$

$$- \frac{\Gamma(\beta)}{\Gamma(\alpha - q_2)\Gamma(\beta - q_1)} \left(\sum_{i=1}^{N} a_i \xi_i^{\beta - q_1 - 1} \right)$$

$$\left. \times \sum_{i=1}^{M} b_i \left(\int_0^{\eta_i} (\eta_i - s)^{\alpha - q_2 - 1} x(s) \, ds \right) \right],$$

$$d_1 = \frac{1}{\Delta} \left[\frac{\Gamma(\alpha)}{\Gamma(\alpha - p_1)\Gamma(\beta - p_2)} \int_0^1 (1 - s)^{\beta - p_2 - 1} y(s) \, ds \right.$$

$$- \frac{\Gamma(\alpha)}{\Gamma(\alpha - p_1)\Gamma(\alpha - q_2)} \sum_{i=1}^M b_i \left(\int_0^{\eta_i} (\eta_i - s)^{\alpha - q_2 - 1} x(s) \, ds \right)$$

$$+ \frac{\Gamma(\alpha)}{\Gamma(\alpha - p_1)\Gamma(\alpha - q_2)} \left(\sum_{i=1}^M b_i \eta_i^{\alpha - q_2 - 1} \right) \int_0^1 (1 - s)^{\alpha - p_1 - 1} x(s) \, ds$$

$$- \frac{\Gamma(\alpha)}{\Gamma(\alpha - q_2)\Gamma(\beta - q_1)} \left(\sum_{i=1}^M b_i \eta_i^{\alpha - q_2 - 1} \right)$$

$$\left. \times \sum_{i=1}^N a_i \left(\int_0^{\xi_i} (\xi_i - s)^{\beta - q_1 - 1} y(s) \, ds \right) \right].$$

Replacing the above constants c_1 and d_1 in (3.45), we obtain the expression (3.44) for the solution $(u, v) \in C[0,1] \times C[0,1]$ of problem (3.43), (3.42). Conversely, one easily verifies that $(u, v) \in C[0,1] \times C[0,1]$ given by (3.44) satisfies the system (3.43) and the boundary conditions (3.42). $\square$

Lemma 3.2.2. *If $\Delta \neq 0$ and $x, y \in C(0,1) \cap L^1(0,1)$, the solution (u, v) of problem (3.43), (3.42) given by (3.44) can be written as*

$$\begin{cases} u(t) = \int_0^1 \widetilde{G}_1(t, s) x(s) \, ds + \int_0^1 \widetilde{G}_2(t, s) y(s) \, ds, & t \in [0, 1], \\ v(t) = \int_0^1 \widetilde{G}_3(t, s) y(s) \, ds + \int_0^1 \widetilde{G}_4(t, s) x(s) \, ds, & t \in [0, 1], \end{cases} \tag{3.47}$$

where the Green functions $\widetilde{G}_i, i = 1, \ldots, 4$ are

$$\begin{cases} \widetilde{G}_1(t, s) = \widetilde{g}_1(t, s) + \frac{t^{\alpha - 1}\Gamma(\beta)}{\Delta\Gamma(\beta - q_1)} \left(\sum_{i=1}^N a_i \xi_i^{\beta - q_1 - 1} \right) \left(\sum_{i=1}^M b_i \widetilde{g}_2(\eta_i, s) \right), \\ \widetilde{G}_2(t, s) = \frac{t^{\alpha - 1}\Gamma(\beta)}{\Delta\Gamma(\beta - p_2)} \left(\sum_{i=1}^N a_i \widetilde{g}_3(\xi_i, s) \right), \\ \widetilde{G}_3(t, s) = \widetilde{g}_4(t, s) + \frac{t^{\beta - 1}\Gamma(\alpha)}{\Delta\Gamma(\alpha - q_2)} \left(\sum_{i=1}^M b_i \eta_i^{\alpha - q_2 - 1} \right) \left(\sum_{i=1}^N a_i \widetilde{g}_3(\xi_i, s) \right), \\ \widetilde{G}_4(t, s) = \frac{t^{\beta - 1}\Gamma(\alpha)}{\Delta\Gamma(\alpha - p_1)} \left(\sum_{i=1}^M b_i \widetilde{g}_2(\eta_i, s) \right), \quad \forall t, s \in [0, 1], \end{cases}$$

$$\tag{3.48}$$

and

$$
\begin{cases}
\tilde{g}_1(t,s) = \dfrac{1}{\Gamma(\alpha)} \begin{cases} t^{\alpha-1}(1-s)^{\alpha-p_1-1} - (t-s)^{\alpha-1}, & 0 \le s \le t \le 1, \\ t^{\alpha-1}(1-s)^{\alpha-p_1-1}, & 0 \le t \le s \le 1, \end{cases} \\[2mm]
\tilde{g}_2(t,s) = \dfrac{1}{\Gamma(\alpha-q_2)} \begin{cases} t^{\alpha-q_2-1}(1-s)^{\alpha-p_1-1} - (t-s)^{\alpha-q_2-1}, \\ \qquad 0 \le s \le t \le 1, \\ t^{\alpha-q_2-1}(1-s)^{\alpha-p_1-1}, \\ \qquad 0 \le t \le s \le 1. \end{cases} \\[2mm]
\tilde{g}_3(t,s) = \dfrac{1}{\Gamma(\beta-q_1)} \begin{cases} t^{\beta-q_1-1}(1-s)^{\beta-p_2-1} - (t-s)^{\beta-q_1-1}, \\ \qquad 0 \le s \le t \le 1, \\ t^{\beta-q_1-1}(1-s)^{\beta-p_2-1}, & 0 \le t \le s \le 1. \end{cases} \\[2mm]
\tilde{g}_4(t,s) = \dfrac{1}{\Gamma(\beta)} \begin{cases} t^{\beta-1}(1-s)^{\beta-p_2-1} - (t-s)^{\beta-1}, & 0 \le s \le t \le 1, \\ t^{\beta-1}(1-s)^{\beta-p_2-1}, & 0 \le t \le s \le 1. \end{cases}
\end{cases}
$$

$$(3.49)$$

Proof. By Lemma 3.2.1 and relation (3.44), we conclude

$$
\begin{aligned}
u(t) = {} & \frac{1}{\Gamma(\alpha)} \left\{ \int_0^t [t^{\alpha-1}(1-s)^{\alpha-p_1-1} - (t-s)^{\alpha-1}] x(s)\,ds \right. \\
& \left. + \int_t^1 t^{\alpha-1}(1-s)^{\alpha-p_1-1} x(s)\,ds \right\} \\
& - \frac{1}{\Gamma(\alpha)} \int_0^1 t^{\alpha-1}(1-s)^{\alpha-p_1-1} x(s)\,ds \\
& + \frac{t^{\alpha-1}\Gamma(\beta)}{\Delta\Gamma(\alpha-p_1)\Gamma(\beta-p_2)} \int_0^1 (1-s)^{\alpha-p_1-1} x(s)\,ds \\
& - \frac{t^{\alpha-1}\Gamma(\beta)}{\Delta\Gamma(\alpha-q_2)\Gamma(\beta-q_1)} \left(\sum_{i=1}^N a_i \xi_i^{\beta-q_1-1} \right) \\
& \times \sum_{i=1}^M b_i \left(\int_0^{\eta_i} (\eta_i-s)^{\alpha-q_2-1} x(s)\,ds \right) \\
& - \frac{t^{\alpha-1}\Gamma(\beta)}{\Delta\Gamma(\beta-q_1)\Gamma(\beta-p_2)} \sum_{i=1}^N a_i \left(\int_0^{\xi_i} (\xi_i-s)^{\beta-q_1-1} y(s)\,ds \right) \\
& + \frac{t^{\alpha-1}\Gamma(\beta)}{\Delta\Gamma(\beta-p_2)\Gamma(\beta-q_1)} \left(\sum_{i=1}^N a_i \xi_i^{\beta-q_1-1} \right) \int_0^1 (1-s)^{\beta-p_2-1} y(s)\,ds
\end{aligned}
$$

$$= \frac{1}{\Gamma(\alpha)} \left\{ \int_0^t [t^{\alpha-1}(1-s)^{\alpha-p_1-1} - (t-s)^{\alpha-1}]x(s)\,ds \right.$$

$$\left. + \int_t^1 t^{\alpha-1}(1-s)^{\alpha-p_1-1}x(s)\,ds \right\}$$

$$- \frac{\Gamma(\beta)}{\Delta\Gamma(\alpha-p_1)\Gamma(\beta-p_2)} \int_0^1 t^{\alpha-1}(1-s)^{\alpha-p_1-1}x(s)\,ds$$

$$+ \frac{\Gamma(\beta)}{\Delta\Gamma(\alpha-q_2)\Gamma(\beta-q_1)} \left(\sum_{i=1}^N a_i \xi_i^{\beta-q_1-1} \right)$$

$$\times \left(\sum_{i=1}^M b_i \eta_i^{\alpha-q_2-1} \right) \int_0^1 t^{\alpha-1}(1-s)^{\alpha-p_1-1}x(s)\,ds$$

$$+ \frac{t^{\alpha-1}\Gamma(\beta)}{\Delta\Gamma(\alpha-p_1)\Gamma(\beta-p_2)} \int_0^1 (1-s)^{\alpha-p_1-1}x(s)\,ds$$

$$- \frac{t^{\alpha-1}\Gamma(\beta)}{\Delta\Gamma(\alpha-q_2)\Gamma(\beta-q_1)} \left(\sum_{i=1}^N a_i \xi_i^{\beta-q_1-1} \right)$$

$$\times \sum_{i=1}^M b_i \left(\int_0^{\eta_i} (\eta_i-s)^{\alpha-q_2-1}x(s)\,ds \right)$$

$$+ \frac{t^{\alpha-1}\Gamma(\beta)}{\Delta\Gamma(\beta-q_1)\Gamma(\beta-p_2)}$$

$$\times \sum_{i=1}^N a_i \left[\int_0^1 \xi_i^{\beta-q_1-1}(1-s)^{\beta-p_2-1}y(s)\,ds - \int_0^{\xi_i} (\xi_i-s)^{\beta-q_1-1}y(s)\,ds \right]$$

$$= \frac{1}{\Gamma(\alpha)} \left\{ \int_0^t [t^{\alpha-1}(1-s)^{\alpha-p_1-1} - (t-s)^{\alpha-1}]x(s)\,ds \right.$$

$$\left. + \int_t^1 t^{\alpha-1}(1-s)^{\alpha-p_1-1}x(s)\,ds \right\}$$

$$+ \frac{t^{\alpha-1}\Gamma(\beta)}{\Delta\Gamma(\alpha-q_2)\Gamma(\beta-q_1)} \left(\sum_{i=1}^N a_i \xi_i^{\beta-q_1-1} \right)$$

$$\times \sum_{i=1}^M b_i \left[\int_0^1 \eta_i^{\alpha-q_2-1}(1-s)^{\alpha-p_1-1}x(s)\,ds - \int_0^{\eta_i} (\eta_i-s)^{\alpha-q_2-1}x(s)\,ds \right]$$

$$+ \frac{t^{\alpha-1}\Gamma(\beta)}{\Delta\Gamma(\beta-q_1)\Gamma(\beta-p_2)}$$

$$\times \sum_{i=1}^{N} a_i \left[\int_0^1 \xi_i^{\beta-q_1-1}(1-s)^{\beta-p_2-1}y(s)\,ds - \int_0^{\xi_i} (\xi_i-s)^{\beta-q_1-1}y(s)\,ds \right].$$

Therefore, we obtain

$$u(t) = \frac{1}{\Gamma(\alpha)} \left\{ \int_0^t [t^{\alpha-1}(1-s)^{\alpha-p_1-1} - (t-s)^{\alpha-1}]x(s)\,ds \right.$$

$$+ \left. \int_t^1 t^{\alpha-1}(1-s)^{\alpha-p_1-1}x(s)\,ds \right\}$$

$$+ \frac{t^{\alpha-1}\Gamma(\beta)}{\Delta\Gamma(\alpha-q_2)\Gamma(\beta-q_1)} \left(\sum_{i=1}^{N} a_i\xi_i^{\beta-q_1-1} \right)$$

$$\times \sum_{i=1}^{M} b_i \left\{ \int_0^{\eta_i} \left[\eta_i^{\alpha-q_2-1}(1-s)^{\alpha-p_1-1} \right. \right.$$

$$\left. - (\eta_i-s)^{\alpha-q_2-1} \right] x(s)\,ds + \left. \int_{\eta_i}^1 \eta_i^{\alpha-q_2-1}(1-s)^{\alpha-p_1-1}x(s)\,ds \right\}$$

$$+ \frac{t^{\alpha-1}\Gamma(\beta)}{\Delta\Gamma(\beta-q_1)\Gamma(\beta-p_2)}$$

$$\times \sum_{i=1}^{N} a_i \left\{ \int_0^{\xi_i} \left[\xi_i^{\beta-q_1-1}(1-s)^{\beta-p_2-1} - (\xi_i-s)^{\beta-q_1-1} \right] y(s)\,ds \right.$$

$$+ \left. \int_{\xi_i}^1 \xi_i^{\beta-q_1-1}(1-s)^{\beta-p_2-1}y(s)\,ds \right\}$$

$$= \int_0^1 \widetilde{g}_1(t,s)x(s)\,ds + \frac{t^{\alpha-1}\Gamma(\beta)}{\Delta\Gamma(\beta-q_1)} \left(\sum_{i=1}^{N} a_i\xi_i^{\beta-q_1-1} \right)$$

$$\times \sum_{i=1}^{M} b_i \int_0^1 \widetilde{g}_2(\eta_i,s)x(s)\,ds$$

$$+ \frac{t^{\alpha-1}\Gamma(\beta)}{\Delta\Gamma(\beta-p_2)} \sum_{i=1}^{N} a_i \int_0^1 \widetilde{g}_3(\xi_i,s)y(s)\,ds$$

$$= \int_0^1 \widetilde{G}_1(t,s)x(s)\,ds + \int_0^1 \widetilde{G}_2(t,s)y(s)\,ds.$$

In a similar manner, we deduce

$$v(t) = \int_0^1 \widetilde{g}_4(t,s) y(s)\, ds + \frac{t^{\beta-1}\Gamma(\alpha)}{\Delta\Gamma(\alpha-q_2)} \left(\sum_{i=1}^M b_i \eta_i^{\alpha-q_2-1} \right)$$

$$\times \sum_{i=1}^N a_i \int_0^1 \widetilde{g}_3(\xi_i, s) y(s)\, ds$$

$$+ \frac{t^{\beta-1}\Gamma(\alpha)}{\Delta\Gamma(\alpha-p_1)} \sum_{i=1}^M b_i \int_0^1 \widetilde{g}_2(\eta_i, s) x(s)\, ds$$

$$= \int_0^1 \widetilde{G}_3(t,s) y(s)\, ds + \int_0^1 \widetilde{G}_4(t,s) x(s)\, ds,$$

where $\widetilde{G}_i$, $\widetilde{g}_i$, $i = 1,\ldots,4$, are given in (3.48) and (3.49).

Therefore, we obtain the expression (3.47) for the solution (u, v) of problem (3.43), (3.42) given by (3.44). $\qquad\square$

Using similar arguments as those used in the proof of Lemma 2.1.3 from Section 2.1, we obtain the following properties of the functions $\widetilde{g}_i$, $i = 1,\ldots,4$.

Lemma 3.2.3. *The functions $\widetilde{g}_i$, $i = 1,\ldots,4$ given by (3.49) have the properties:*

(a_1) $\widetilde{g}_1(t,s) \leq \widetilde{h}_1(s)$ *for all $t, s \in [0,1]$, where*

$$\widetilde{h}_1(s) = \frac{1}{\Gamma(\alpha)}(1-s)^{\alpha-p_1-1}(1-(1-s)^{p_1}), s \in [0,1];$$

(a_2) $\widetilde{g}_1(t,s) \geq t^{\alpha-1}\widetilde{h}_1(s)$ *for all $t, s \in [0,1]$;*

(a_3) $\widetilde{g}_1(t,s) \leq \frac{1}{\Gamma(\alpha)}t^{\alpha-1}$, *for all $t, s \in [0,1]$;*

(b_1) $\widetilde{g}_2(t,s) \geq t^{\alpha-q_2-1}\widetilde{h}_2(s)$ *for all $t, s \in [0,1]$, where*

$$\widetilde{h}_2(s) = \frac{1}{\Gamma(\alpha-q_2)}(1-s)^{\alpha-p_1-1}(1-(1-s)^{p_1-q_2}), s \in [0,1];$$

(b_2) $\widetilde{g}_2(t,s) \leq \frac{1}{\Gamma(\alpha-q_2)}t^{\alpha-q_2-1}(1-s)^{\alpha-p_1-1}$ *for all $t, s \in [0,1]$;*

(b_3) $\widetilde{g}_2(t,s) \leq \frac{1}{\Gamma(\alpha-q_2)}t^{\alpha-q_2-1}$ *for all $t, s \in [0,1]$;*

(c_1) $\widetilde{g}_3(t,s) \geq t^{\beta-q_1-1}\widetilde{h}_3(s)$ *for all $t, s \in [0,1]$, where*

$$\widetilde{h}_3(s) = \frac{1}{\Gamma(\beta-q_1)}(1-s)^{\beta-p_2-1}(1-(1-s)^{p_2-q_1}), s \in [0,1];$$

(c$_2$) $\widetilde{g}_3(t,s) \le \frac{1}{\Gamma(\beta-q_1)} t^{\beta-q_1-1}(1-s)^{\beta-p_2-1}$ *for all t, $s \in [0,1]$;*

(c$_3$) $\widetilde{g}_3(t,s) \le \frac{1}{\Gamma(\beta-q_1)} t^{\beta-q_1-1}$ *for all t, $s \in [0,1]$;*

(d$_1$) $\widetilde{g}_4(t,s) \le \widetilde{h}_4(s)$ *for all t, $s \in [0,1]$, where*

$$\widetilde{h}_4(s) = \frac{1}{\Gamma(\beta)}(1-s)^{\beta-p_2-1}(1-(1-s)^{p_2}), s \in [0,1];$$

(d$_2$) $\widetilde{g}_4(t,s) \ge t^{\beta-1}\widetilde{h}_4(s)$ *for all t, $s \in [0,1]$;*

(d$_3$) $\widetilde{g}_4(t,s) \le \frac{1}{\Gamma(\beta)} t^{\beta-1}$, *for all t, $s \in [0,1]$;*

(e) *The functions $\widetilde{g}_i$, $i = 1,\ldots,4$, are continuous on $[0,1] \times [0,1]$; $\widetilde{g}_i(t,s) \ge 0$ for all t, $s \in [0,1]$; $\widetilde{g}_i(t,s) > 0$ for all t, $s \in (0,1)$, $i = 1,\ldots,4$.*

Now, from the definitions of the Green functions $\widetilde{G}_i$, $i = 1,\ldots,4$, and the properties of functions $\widetilde{g}_i$, $i = 1,\ldots,4$, we obtain the following lemma.

Lemma 3.2.4. *If $\Delta > 0$, $a_i \ge 0$ for all $i = 1,\ldots,N$, and $b_i \ge 0$ for all $i = 1,\ldots,M$, then the functions $\widetilde{G}_i$, $i = 1,\ldots,4$, have the properties*

(a$_1$) $\widetilde{G}_1(t,s) \le \widetilde{J}_1(s)$, $\forall (t,s) \in [0,1] \times [0,1]$, *where*

$$\widetilde{J}_1(s) = \widetilde{h}_1(s) + \frac{\Gamma(\beta)}{\Delta\Gamma(\beta-q_1)}\left(\sum_{i=1}^{N} a_i\xi_i^{\beta-q_1-1}\right)\left(\sum_{i=1}^{M} b_i\widetilde{g}_2(\eta_i,s)\right),$$

$$\forall s \in [0,1];$$

(a$_2$) $\widetilde{G}_1(t,s) \ge t^{\alpha-1}\widetilde{J}_1(s)$, $\forall (t,s) \in [0,1] \times [0,1]$;

(a$_3$) $\widetilde{G}_1(t,s) \le \delta_1 t^{\alpha-1}$, $\forall (t,s) \in [0,1] \times [0,1]$, *where*

$$\delta_1 = \frac{1}{\Gamma(\alpha)} + \frac{\Gamma(\beta)}{\Delta\Gamma(\beta-q_1)\Gamma(\alpha-q_2)}\left(\sum_{i=1}^{N} a_i\xi_i^{\beta-q_1-1}\right)\left(\sum_{i=1}^{M} b_i\eta_i^{\alpha-q_2-1}\right);$$

(b$_1$) $\widetilde{G}_2(t,s) \le \widetilde{J}_2(s)$, $\forall (t,s) \in [0,1] \times [0,1]$, *where*

$$\widetilde{J}_2(s) = \frac{\Gamma(\beta)}{\Delta\Gamma(\beta-p_2)}\sum_{i=1}^{N} a_i\widetilde{g}_3(\xi_i,s), \quad \forall s \in [0,1];$$

(b$_2$) $\widetilde{G}_2(t,s) = t^{\alpha-1}\widetilde{J}_2(s)$, $\forall (t,s) \in [0,1] \times [0,1]$;

(b$_3$) $\widetilde{G}_2(t,s) \le \delta_2 t^{\alpha-1}$, $\forall (t,s) \in [0,1] \times [0,1]$, *where*

$$\delta_2 = \frac{\Gamma(\beta)}{\Delta\Gamma(\beta-p_2)\Gamma(\beta-q_1)}\sum_{i=1}^{N} a_i\xi_i^{\beta-q_1-1};$$

(c₁) $\widetilde{G}_3(t,s) \le \widetilde{J}_3(s)$, $\forall (t,s) \in [0,1] \times [0,1]$, *where*

$$\widetilde{J}_3(s) = \widetilde{h}_4(s) + \frac{\Gamma(\alpha)}{\Delta\Gamma(\alpha - q_2)}\left(\sum_{i=1}^{M} b_i \eta_i^{\alpha - q_2 - 1}\right)\left(\sum_{i=1}^{N} a_i \widetilde{g}_3(\xi_i, s)\right),$$

$$\forall s \in [0,1];$$

(c₂) $\widetilde{G}_3(t,s) \ge t^{\beta-1}\widetilde{J}_3(s)$, $\forall (t,s) \in [0,1] \times [0,1]$;
(c₃) $\widetilde{G}_3(t,s) \le \delta_3 t^{\beta-1}$, $\forall (t,s) \in [0,1] \times [0,1]$, *where*

$$\delta_3 = \frac{1}{\Gamma(\beta)} + \frac{\Gamma(\alpha)}{\Delta\Gamma(\alpha - q_2)\Gamma(\beta - q_1)}\left(\sum_{i=1}^{M} b_i \eta_i^{\alpha - q_2 - 1}\right)\left(\sum_{i=1}^{N} a_i \xi_i^{\beta - q_1 - 1}\right);$$

(d₁) $\widetilde{G}_4(t,s) \le \widetilde{J}_4(s)$, $\forall (t,s) \in [0,1] \times [0,1]$, *where*

$$\widetilde{J}_4(s) = \frac{\Gamma(\alpha)}{\Delta\Gamma(\alpha - p_1)}\sum_{i=1}^{M} b_i \widetilde{g}_2(\eta_i, s), \quad \forall s \in [0,1];$$

(d₂) $\widetilde{G}_4(t,s) = t^{\beta-1}\widetilde{J}_4(s)$, $\forall (t,s) \in [0,1] \times [0,1]$;
(d₃) $\widetilde{G}_4(t,s) \le \delta_4 t^{\beta-1}$, $\forall (t,s) \in [0,1] \times [0,1]$, *where*

$$\delta_4 = \frac{\Gamma(\alpha)}{\Delta\Gamma(\alpha - p_1)\Gamma(\alpha - q_2)}\sum_{i=1}^{M} b_i \eta_i^{\alpha - q_2 - 1}.$$

(e) *The functions* $\widetilde{G}_i$, $i = 1, \ldots, 4$, *are continuous on* $[0,1] \times [0,1]$, *and* $\widetilde{G}_i(t,s) \ge 0$ *for all* $t, s \in [0,1]$, $i = 1, \ldots, 4$.

Lemma 3.2.5. *If* $\Delta > 0$, $a_i \ge 0$ *for all* $i = 1, \ldots, N$, $b_i \ge 0$ *for all* $i = 1, \ldots, M$, *and* $x, y \in C(0,1) \cap L^1(0,1)$ *with* $x(t) \ge 0$, $y(t) \ge 0$ *for all* $t \in (0,1)$, *then the solution* (u,v) *of problem* (3.43), (3.42) *given by* (3.44) *satisfies the inequalities* $u(t) \ge 0$, $v(t) \ge 0$ *for all* $t \in [0,1]$. *Moreover, we have the inequalities* $u(t) \ge t^{\alpha-1}u(t')$ *and* $v(t) \ge t^{\beta-1}v(t')$ *for all* $t, t' \in [0,1]$.

Proof. Under the assumptions of this lemma, by using relations (3.47) and Lemma 3.2.4, we deduce that $u(t) \ge 0$ and $v(t) \ge 0$ for all $t \in [0,1]$. Besides, for all t, $t' \in [0,1]$, we obtain the following inequalities:

$$u(t) = \int_0^1 \widetilde{G}_1(t,s)x(s)\,ds + \int_0^1 \widetilde{G}_2(t,s)y(s)\,ds$$

$$\geq t^{\alpha-1}\left(\int_0^1 \widetilde{J}_1(s)x(s)\,ds + \int_0^1 \widetilde{J}_2(s)y(s)\,ds\right)$$

$$\geq t^{\alpha-1}\left(\int_0^1 \widetilde{G}_1(t',s)x(s)\,ds + \int_0^1 \widetilde{G}_2(t',s)y(s)\,ds\right) = t^{\alpha-1}u(t'),$$

$$v(t) = \int_0^1 \widetilde{G}_3(t,s)y(s)\,ds + \int_0^1 \widetilde{G}_4(t,s)x(s)\,ds$$

$$\geq t^{\beta-1}\left(\int_0^1 \widetilde{J}_3(s)y(s)\,ds + \int_0^1 \widetilde{J}_4(s)x(s)\,ds\right)$$

$$\geq t^{\beta-1}\left(\int_0^1 \widetilde{G}_3(t',s)y(s)\,ds + \int_0^1 \widetilde{G}_4(t',s)x(s)\,ds\right) = t^{\beta-1}v(t').$$

$\square$

Remark 3.2.1. Under the assumptions of Lemma 3.2.5, for $c \in (0,1)$, the solution (u,v) of problem (3.43), (3.42) given by (3.44) satisfies the inequalities $\min_{t\in[c,1]} u(t) \geq c^{\alpha-1}\max_{t'\in[0,1]} u(t')$ and $\min_{t\in[c,1]} v(t) \geq c^{\beta-1}\max_{t'\in[0,1]} v(t')$.

3.2.2 *Nonsingular nonlinearities*

In this section, we investigate the existence and multiplicity of positive solutions for our problem (3.41), (3.42) under various assumptions on the nonsingular functions f and g.

We present the basic assumptions that we shall use in the sequel.

(I1) $\alpha \in (n-1,n]$, $\beta \in (m-1,m]$, $n, m \geq 3$, $p_1, p_2, q_1, q_2 \in \mathbb{R}$, $p_1 \in [1, \alpha-1)$, $p_2 \in [1, \beta-1)$, $q_1 \in [0, p_2]$, $q_2 \in [0, p_1]$, $\xi_i \in \mathbb{R}$, $a_i \geq 0$ for all $i = 1, \ldots, N$ ($N \in \mathbb{N}$), $\sum_{i=1}^N a_i > 0$, $0 < \xi_1 < \cdots < \xi_N \leq 1$, $\eta_i \in \mathbb{R}$, $b_i \geq 0$ for all $i = 1, \ldots, M$ ($M \in \mathbb{N}$), $\sum_{i=1}^M b_i > 0$, $0 < \eta_1 < \cdots < \eta_M \leq 1$, and $\Delta = \frac{\Gamma(\alpha)\Gamma(\beta)}{\Gamma(\alpha-p_1)\Gamma(\beta-p_2)} - \frac{\Gamma(\alpha)\Gamma(\beta)}{\Gamma(\alpha-q_2)\Gamma(\beta-q_1)}\left(\sum_{i=1}^N a_i\xi_i^{\beta-q_1-1}\right)\left(\sum_{i=1}^M b_i\eta_i^{\alpha-q_2-1}\right) > 0$.

(I2) The functions $f, g : [0,1] \times [0, \infty) \times [0, \infty) \to [0, \infty)$ are continuous.

By using Lemma 3.2.2, (u,v) is solution of problem (3.41), (3.42) if and only if (u,v) is solution of the following nonlinear system of integral equations

$$\begin{cases} u(t) = \int_0^1 \widetilde{G}_1(t,s) f(s, u(s), v(s))\, ds + \int_0^1 \widetilde{G}_2(t,s) g(s, u(s), v(s))\, ds, \\ \qquad t \in [0,1], \\ v(t) = \int_0^1 \widetilde{G}_3(t,s) g(s, u(s), v(s))\, ds + \int_0^1 \widetilde{G}_4(t,s) f(s, u(s), v(s))\, ds, \\ \qquad t \in [0,1]. \end{cases}$$

$$(3.50)$$

We consider the Banach space $X = C[0,1]$ with supremum norm $\|\cdot\|$ and the Banach space $Y = X \times X$ with the norm $\|(u,v)\|_Y = \|u\| + \|v\|$. We define the cone $\widetilde{P} \subset Y$ by $\widetilde{P} = \{(u,v) \in Y, \ u(t) \geq 0, \ v(t) \geq 0 \text{ for all } t \in [0,1]\}$.

We introduce the operators $\mathcal{Q}_1, \mathcal{Q}_2 : Y \to X$ and $\mathcal{Q} : Y \to Y$ defined by

$$\mathcal{Q}_1(u,v)(t) = \int_0^1 \widetilde{G}_1(t,s) f(s, u(s), v(s))\, ds + \int_0^1 \widetilde{G}_2(t,s) g(s, u(s), v(s))\, ds,$$

$$0 \leq t \leq 1,$$

$$\mathcal{Q}_2(u,v)(t) = \int_0^1 \widetilde{G}_3(t,s) g(s, u(s), v(s))\, ds + \int_0^1 \widetilde{G}_4(t,s) f(s, u(s), v(s))\, ds,$$

$$0 \leq t \leq 1,$$

and $\mathcal{Q}(u,v) = (\mathcal{Q}_1(u,v), \mathcal{Q}_2(u,v))$, $(u,v) \in Y$.

Under the assumptions (I1) and (I2), it is easy to see that operator $\mathcal{Q} : \widetilde{P} \to \widetilde{P}$ is completely continuous. It is clear that (u,v) is a solution of the system (3.50) if and only if (u,v) is a fixed point of operator $\mathcal{Q}$. Therefore, we will investigate the existence and multiplicity of fixed points of operator $\mathcal{Q}$.

Theorem 3.2.1. *Assume that* (I1) *and* (I2) *hold. If the functions f and g also satisfy the conditions*

(I3) *There exist $p \geq 1$ and $q \geq 1$ such that*

$$f_0^s = \lim_{\substack{u+v \to 0 \\ u,v \geq 0}} \sup_{t \in [0,1]} \frac{f(t,u,v)}{(u+v)^p} = 0 \quad \text{and}$$

$$g_0^s = \lim_{\substack{u+v \to 0 \\ u,v \geq 0}} \sup_{t \in [0,1]} \frac{g(t,u,v)}{(u+v)^q} = 0;$$

(I4) *There exists $c \in (0, 1)$ such that*

$$f_\infty^i = \lim_{\substack{u+v \to \infty \\ u, v \geq 0}} \inf_{t \in [c,1]} \frac{f(t, u, v)}{u + v} = \infty \quad or$$

$$g_\infty^i = \lim_{\substack{u+v \to \infty \\ u, v \geq 0}} \inf_{t \in [c,1]} \frac{g(t, u, v)}{u + v} = \infty,$$

then problem (3.41), (3.42) *has at least one positive solution* $(u(t), v(t))$, $t \in [0, 1]$.

Proof. For c given in (I4), we define the cone

$$P_0 = \{(u, v) \in \widetilde{P}, \ \min_{t \in [c,1]} u(t) \geq c^{\alpha-1} \|u\|, \ \min_{t \in [c,1]} v(t) \geq c^{\beta-1} \|v\|\}.$$

From our assumptions and Remark 3.2.1, for any $(u, v) \in \widetilde{P}$, we deduce that $\mathcal{Q}(u, v) = (\mathcal{Q}_1(u, v), \mathcal{Q}_2(u, v)) \in P_0$, that is $\mathcal{Q}(\widetilde{P}) \subset P_0$.

We consider the functions $u_0, v_0 : [0, 1] \to \mathbb{R}$ defined by

$$\begin{cases} u_0(t) = \int_0^1 \widetilde{G}_1(t, s) \, ds + \int_0^1 \widetilde{G}_2(t, s) \, ds, \ 0 \leq t \leq 1, \\ v_0(t) = \int_0^1 \widetilde{G}_3(t, s) \, ds + \int_0^1 \widetilde{G}_4(t, s) \, ds, \ 0 \leq t \leq 1, \end{cases}$$

that is, (u_0, v_0) is solution of problem (3.43), (3.42) with $x(t) = x_0(t)$, $y(t) = y_0(t)$, $x_0(t) = 1$, $y_0(t) = 1$ for all $t \in [0, 1]$. Hence, $(u_0, v_0) = \mathcal{Q}(x_0, y_0) \in P_0$.
We define the set

$$\widetilde{M} = \{(u, v) \in \widetilde{P}, \ \text{there exists } \lambda \geq 0 \text{ such that } (u, v)$$

$$= \mathcal{Q}(u, v) + \lambda(u_0, v_0)\}.$$

We will show that $\widetilde{M} \subset P_0$ and $\widetilde{M}$ is a bounded set of Y. If $(u, v) \in \widetilde{M}$, then there exists $\lambda \geq 0$ such that $(u, v) = \mathcal{Q}(u, v) + \lambda(u_0, v_0)$ or equivalently

$$\begin{cases} u(t) = \int_0^1 \widetilde{G}_1(t, s)(f(s, u(s), v(s)) + \lambda) \, ds \\ \qquad + \int_0^1 \widetilde{G}_2(t, s)(g(s, u(s), v(s)) + \lambda) \, ds, \quad 0 \leq t \leq 1, \\ v(t) = \int_0^1 \widetilde{G}_3(t, s)(g(s, u(s), v(s)) + \lambda) \, ds \\ \qquad + \int_0^1 \widetilde{G}_4(t, s)(f(s, u(s), v(s)) + \lambda) \, ds, \quad 0 \leq t \leq 1. \end{cases}$$

By Remark 3.2.1, we obtain $(u, v) \in P_0$, so $\widetilde{M} \subset P_0$, and

$$\|u\| \leq \frac{1}{c^{\alpha-1}} \min_{t \in [c,1]} u(t), \quad \|v\| \leq \frac{1}{c^{\beta-1}} \min_{t \in [c,1]} v(t), \quad \forall (u, v) \in \widetilde{M}. \tag{3.51}$$

From (I4), we suppose that $f_\infty^i = \infty$, (in a similar manner, we can study the case $g_\infty^i = \infty$). Then for $\varepsilon_1 = \max\{\frac{2}{c^{\alpha-1}m_1}, \frac{2}{c^{\beta-1}m_4}\} > 0$, there exists $C_1 > 0$ such that

$$f(t, u, v) \geq \varepsilon_1(u + v) - C_1, \quad \forall (t, u, v) \in [c, 1] \times [0, \infty) \times [0, \infty), \quad (3.52)$$

where $m_i = \int_c^1 \widetilde{J}_i(s)\, ds$ and $\widetilde{J}_i$, $i = 1, 4$ are defined in Lemma 3.2.4.

For $(u, v) \in \overline{M}$ and $t \in [c, 1]$, by using Lemma 3.2.4 and relation (3.52), it follows that

$$u(t) = \mathcal{Q}_1(u, v)(t) + \lambda u_0(t) \geq \mathcal{Q}_1(u, v)(t)$$

$$= \int_0^1 \widetilde{G}_1(t, s)f(s, u(s), v(s))\, ds + \int_0^1 \widetilde{G}_2(t, s)g(s, u(s), v(s))\, ds$$

$$\geq \int_0^1 \widetilde{G}_1(t, s)f(s, u(s), v(s))\, ds \geq \int_c^1 t^{\alpha-1}\widetilde{J}_1(s)f(s, u(s), v(s))\, ds$$

$$\geq c^{\alpha-1}\int_c^1 \widetilde{J}_1(s)[\varepsilon_1(u(s) + v(s)) - C_1]\, ds$$

$$\geq c^{\alpha-1}\varepsilon_1 \int_c^1 \widetilde{J}_1(s)u(s)\, ds - c^{\alpha-1}m_1 C_1$$

$$\geq c^{\alpha-1}\varepsilon_1 m_1 \min_{s \in [c,1]} u(s) - c^{\alpha-1}m_1 C_1$$

$$\geq 2 \min_{s \in [c,1]} u(s) - C_2, \quad C_2 = c^{\alpha-1}m_1 C_1,$$

$$v(t) = \mathcal{Q}_2(u, v)(t) + \lambda v_0(t) \geq \mathcal{Q}_2(u, v)(t)$$

$$= \int_0^1 \widetilde{G}_3(t, s)g(s, u(s), v(s))\, ds + \int_0^1 \widetilde{G}_4(t, s)f(s, u(s), v(s))\, ds$$

$$\geq \int_0^1 \widetilde{G}_4(t, s)f(s, u(s), v(s))\, ds \geq \int_c^1 t^{\beta-1}\widetilde{J}_4(s)f(s, u(s), v(s))\, ds$$

$$\geq c^{\beta-1}\int_c^1 \widetilde{J}_4(s)[\varepsilon_1(u(s) + v(s)) - C_1]\, ds$$

$$\geq c^{\beta-1}\varepsilon_1 \int_c^1 \widetilde{J}_4(s)v(s)\, ds - c^{\beta-1}m_4 C_1$$

$$\geq c^{\beta-1}\varepsilon_1 m_4 \min_{s \in [c,1]} v(s) - c^{\beta-1}m_4 C_1$$

$$\geq 2 \min_{s \in [c,1]} v(s) - C_3, \quad C_3 = c^{\beta-1}m_4 C_1.$$

Therefore, we deduce

$$\min_{t\in[c,1]} u(t) \le C_2, \quad \min_{t\in[c,1]} v(t) \le C_3, \quad \forall\, (u,v) \in \widetilde{M}. \tag{3.53}$$

Now from relations (3.51) and (3.53), we obtain

$$\|u\| \le \frac{C_2}{c^{\alpha-1}}, \quad \|v\| \le \frac{C_3}{c^{\beta-1}}, \quad \text{and}$$

$$\|(u,v)\|_Y = \|u\| + \|v\| \le \frac{C_2}{c^{\alpha-1}} + \frac{C_3}{c^{\beta-1}} = C_4,$$

for all $(u,v) \in \widetilde{M}$, that is $\widetilde{M}$ is a bounded set of Y. Besides, there exists a sufficiently large $R_1 > 1$ such that

$$(u,v) \ne \mathcal{Q}(u,v) + \lambda(u_0,v_0), \quad \forall\, (u,v) \in \partial B_{R_1} \cap \widetilde{P}, \ \forall \lambda \ge 0.$$

From Theorem 1.2.9, we deduce that the fixed point index of operator $\mathcal{Q}$ over $B_{R_1} \cap \widetilde{P}$ with respect to $\widetilde{P}$ is

$$i(\mathcal{Q}, B_{R_1} \cap \widetilde{P}, \widetilde{P}) = 0. \tag{3.54}$$

Next, from assumption (I3), we conclude that for $\varepsilon_2 = \min\left\{\frac{1}{8M_1}, \frac{1}{8M_4}\right\}$ and $\varepsilon_3 = \min\left\{\frac{1}{8M_2}, \frac{1}{8M_3}\right\}$, there exists $r_1 \in (0,1]$ such that

$$f(t,u,v) \le \varepsilon_2(u+v)^p, \quad g(t,u,v) \le \varepsilon_3(u+v)^q,$$

$$\forall t \in [0,1], \ u,v \ge 0, \ u+v \le r_1, \tag{3.55}$$

where $M_i = \int_0^1 \widetilde{J}_i(s)\,ds$, $i = 1,\dots,4$.

By using (3.55), we deduce that for all $(u,v) \in \overline{B}_{r_1} \cap \widetilde{P}$ and $t \in [0,1]$

$$\mathcal{Q}_1(u,v)(t) \le \int_0^1 \widetilde{J}_1(s)\varepsilon_2(u(s)+v(s))^p\,ds + \int_0^1 \widetilde{J}_2(s)\varepsilon_3(u(s)+v(s))^q\,ds$$

$$\le \varepsilon_2 M_1\|(u,v)\|_Y^p + \varepsilon_3 M_2\|(u,v)\|_Y^q \le \frac{1}{8}\|(u,v)\|_Y + \frac{1}{8}\|(u,v)\|_Y$$

$$= \frac{1}{4}\|(u,v)\|_Y,$$

$$\mathcal{Q}_2(u,v)(t) \le \int_0^1 \widetilde{J}_3(s)\varepsilon_3(u(s)+v(s))^q\,ds + \int_0^1 \widetilde{J}_4(s)\varepsilon_2(u(s)+v(s))^p\,ds$$

$$\le \varepsilon_3 M_3\|(u,v)\|_Y^q + \varepsilon_2 M_4\|(u,v)\|_Y^p \le \frac{1}{8}\|(u,v)\|_Y + \frac{1}{8}\|(u,v)\|_Y$$

$$= \frac{1}{4}\|(u,v)\|_Y.$$

These imply that

$$\|\mathcal{Q}_1(u,v)\| \le \frac{1}{4}\|(u,v)\|_Y, \quad \|\mathcal{Q}_2(u,v)\| \le \frac{1}{4}\|(u,v)\|_Y,$$

$$\|\mathcal{Q}(u,v)\|_Y = \|\mathcal{Q}_1(u,v)\| + \|\mathcal{Q}_2(u,v)\| \le \frac{1}{2}\|(u,v)\|_Y,$$

$$\forall\,(u,v) \in \partial B_{r_1} \cap \widetilde{P}.$$

From Theorem 1.2.8, we conclude that the fixed point index of operator $\mathcal{Q}$ over $B_{r_1} \cap \widetilde{P}$ with respect to $\widetilde{P}$ is

$$i(\mathcal{Q}, B_{r_1} \cap \widetilde{P}, \widetilde{P}) = 1. \tag{3.56}$$

Combining (3.54) and (3.56), we obtain

$$i(\mathcal{Q}, (B_{R_1} \setminus \overline{B}_{r_1}) \cap \widetilde{P}, \widetilde{P}) = i(\mathcal{Q}, B_{R_1} \cap \widetilde{P}, \widetilde{P}) - i(\mathcal{Q}, B_{r_1} \cap \widetilde{P}, \widetilde{P}) = -1.$$

We deduce that $\mathcal{Q}$ has at least one fixed point $(u,v) \in (B_{R_1} \setminus \overline{B}_{r_1}) \cap \widetilde{P}$, that is $r_1 < \|(u,v)\|_Y < R_1$ or $r_1 < \|u\| + \|v\| < R_1$. By Lemma 3.2.5, we obtain that $u(t) > 0$ for all $t \in (0,1]$ or $v(t) > 0$ for all $t \in (0,1]$. The proof of the theorem is completed. $\qquad\square$

Theorem 3.2.2. *Assume that* (I1) *and* (I2) *hold. If the functions f and g also satisfy the conditions*

(I5) $f_\infty^s = \displaystyle\lim_{\substack{u+v\to\infty \\ u,\,v\ge 0}} \sup_{t\in[0,1]} \frac{f(t,u,v)}{u+v} = 0$ *and* $g_\infty^s = \displaystyle\lim_{\substack{u+v\to\infty \\ u,\,v\ge 0}} \sup_{t\in[0,1]} \frac{g(t,u,v)}{u+v} = 0;$

(I6) *There exist $c \in (0,1)$, $\hat{p} \in (0,1]$ and $\hat{q} \in (0,1]$ such that*

$$f_0^i = \lim_{\substack{u+v\to 0 \\ u,\,v\ge 0}} \inf_{t\in[c,1]} \frac{f(t,u,v)}{(u+v)^{\hat{p}}} = \infty \quad or \quad g_0^i = \lim_{\substack{u+v\to 0 \\ u,\,v\ge 0}} \inf_{t\in[c,1]} \frac{g(t,u,v)}{(u+v)^{\hat{q}}} = \infty,$$

then problem (3.41), (3.42) *has at least one positive solution* $(u(t), v(t))$, $t \in [0,1]$.

Proof. From the assumption (I5), we deduce that for $\varepsilon_4 = \min\left\{\frac{1}{8M_1}, \frac{1}{8M_4}\right\}$ and $\varepsilon_5 = \min\left\{\frac{1}{8M_2}, \frac{1}{8M_3}\right\}$ there exist $C_5, C_6 > 0$ such that

$$f(t, u, v) \leq \varepsilon_4(u + v) + C_5, \quad g(t, u, v) \leq \varepsilon_5(u + v) + C_6,$$

$$\forall (t, u, v) \in [0, 1] \times [0, \infty) \times [0, \infty). \tag{3.57}$$

Hence, for $(u, v) \in \widetilde{P}$, by using (3.57), we obtain

$$\mathcal{Q}_1(u, v)(t) \leq \int_0^1 \widetilde{J}_1(s)(\varepsilon_4(u(s) + v(s)) + C_5) \, ds$$

$$+ \int_0^1 \widetilde{J}_2(s)(\varepsilon_5(u(s) + v(s)) + C_6) \, ds$$

$$\leq \varepsilon_4(\|u\| + \|v\|) \int_0^1 \widetilde{J}_1(s) \, ds + C_5 \int_0^1 \widetilde{J}_1(s) \, ds + \varepsilon_5(\|u\| + \|v\|)$$

$$\times \int_0^1 \widetilde{J}_2(s) \, ds + C_6 \int_0^1 \widetilde{J}_2(s) \, ds$$

$$= \varepsilon_4 \|(u, v)\|_Y M_1 + C_5 M_1 + \varepsilon_5 \|(u, v)\|_Y M_2 + C_6 M_2$$

$$\leq \frac{1}{8} \|(u, v)\|_Y + \frac{1}{8} \|(u, v)\|_Y + C_7 = \frac{1}{4} \|(u, v)\|_Y + C_7,$$

$$\forall t \in [0, 1], \quad C_7 = C_5 M_1 + C_6 M_2,$$

$$\mathcal{Q}_2(u, v)(t) \leq \int_0^1 \widetilde{J}_3(s)(\varepsilon_5(u(s) + v(s)) + C_6) \, ds$$

$$+ \int_0^1 \widetilde{J}_4(s)(\varepsilon_4(u(s) + v(s)) + C_5) \, ds$$

$$\leq \varepsilon_5(\|u\| + \|v\|) \int_0^1 \widetilde{J}_3(s) \, ds + C_6 \int_0^1 \widetilde{J}_3(s) \, ds + \varepsilon_4(\|u\| + \|v\|)$$

$$\times \int_0^1 \widetilde{J}_4(s) \, ds + C_5 \int_0^1 \widetilde{J}_4(s) \, ds$$

$$= \varepsilon_5 \|(u, v)\|_Y M_3 + C_6 M_3 + \varepsilon_4 \|(u, v)\|_Y M_4 + C_5 M_4$$

$$\leq \frac{1}{8} \|(u, v)\|_Y + \frac{1}{8} \|(u, v)\|_Y + C_8 = \frac{1}{4} \|(u, v)\|_Y + C_8,$$

$$\forall t \in [0, 1], \quad C_8 = C_6 M_3 + C_5 M_4,$$

and so

$$\|\mathcal{Q}(u, v)\|_Y = \|\mathcal{Q}_1(u, v)\| + \|\mathcal{Q}_2(u, v)\| \leq \frac{1}{2} \|(u, v)\|_Y + C_9,$$

$$C_9 = C_7 + C_8.$$

Then there exists a sufficiently large $R_2 \geq \max\{4C_9, 1\}$ such that

$$\|Q(u,v)\|_Y \leq \frac{3}{4}\|(u,v)\|_Y, \quad \forall (u,v) \in \widetilde{P}, \quad \|(u,v)\|_Y \geq R_2.$$

Hence, $\|Q(u,v)\|_Y < \|(u,v)\|_Y$ for all $(u,v) \in \partial B_{R_2} \cap \widetilde{P}$ and from Theorem 1.2.8 we have

$$i(Q, B_{R_2} \cap \widetilde{P}, \widetilde{P}) = 1. \tag{3.58}$$

On the other hand, from (I6), we suppose that $f_0^i = \infty$, (in a similar manner we can study the case $g_0^i = \infty$). We conclude that for $\varepsilon_6 = \max\{\frac{1}{c^{\alpha-1}m_1}, \frac{1}{c^{\beta-1}m_4}\}$, there exists $r_2 \in (0,1)$ such that

$$f(t,u,v) \geq \varepsilon_6(u+v)^{\hat{p}}, \quad \forall t \in [c,1], \ u, v \geq 0, \ u+v \leq r_2. \tag{3.59}$$

From (3.59), we deduce that for any $(u,v) \in \overline{B}_{r_2} \cap \widetilde{P}$

$$Q_1(u,v)(t) \geq \int_c^1 \widetilde{G}_1(t,s)f(s,u(s),v(s))\,ds + \int_c^1 \widetilde{G}_2(t,s)g(s,u(s),v(s))\,ds$$

$$\geq \int_c^1 \widetilde{G}_1(t,s)f(s,u(s),v(s))\,ds$$

$$\geq \varepsilon_6 \int_c^1 \widetilde{G}_1(t,s)(u(s)+v(s))^{\hat{p}}\,ds$$

$$\geq \varepsilon_6 \int_c^1 \widetilde{G}_1(t,s)(u(s)+v(s))\,ds =: L_1(u,v)(t), \quad \forall t \in [0,1],$$

$$Q_2(u,v)(t) \geq \int_c^1 \widetilde{G}_3(t,s)g(s,u(s),v(s))\,ds + \int_c^1 \widetilde{G}_4(t,s)f(s,u(s),v(s))\,ds$$

$$\geq \int_c^1 \widetilde{G}_4(t,s)f(s,u(s),v(s))\,ds$$

$$\geq \varepsilon_6 \int_c^1 \widetilde{G}_4(t,s)(u(s)+v(s))^{\hat{p}}\,ds$$

$$\geq \varepsilon_6 \int_c^1 \widetilde{G}_4(t,s)(u(s)+v(s))\,ds =: L_2(u,v)(t), \quad \forall t \in [0,1],$$

Hence,

$$Q(u,v) \geq L(u,v), \quad \forall (u,v) \in \partial B_{r_2} \cap \widetilde{P}, \tag{3.60}$$

where the linear operator $L : \tilde{P} \to \tilde{P}$ is defined by $L(u, v) = (L_1(u, v), L_2(u, v))$. For $(\tilde{u}_0, \tilde{v}_0) \in \tilde{P} \setminus \{(0, 0)\}$ defined by

$$\tilde{u}_0(t) = \int_c^1 \tilde{G}_1(t, s) \, ds, \quad \tilde{v}_0(t) = \int_c^1 \tilde{G}_4(t, s) \, ds, \quad t \in [0, 1],$$

we have $L(\tilde{u}_0, \tilde{v}_0) = (L_1(\tilde{u}_0, \tilde{v}_0), L_2(\tilde{u}_0, \tilde{v}_0))$ with

$$L_1(\tilde{u}_0, \tilde{v}_0)(t) = \varepsilon_6 \int_c^1 \tilde{G}_1(t, s) \left(\int_c^1 \tilde{G}_1(s, \tau) \, d\tau + \int_c^1 \tilde{G}_4(s, \tau) \, d\tau \right) ds$$

$$\geq \varepsilon_6 \int_c^1 \tilde{G}_1(t, s) \left(\int_c^1 \tilde{G}_1(s, \tau) \, d\tau \right) ds$$

$$\geq \varepsilon_6 \int_c^1 \tilde{G}_1(t, s) \left(\int_c^1 c^{\alpha-1} \tilde{J}_1(\tau) \, d\tau \right) ds$$

$$= \varepsilon_6 c^{\alpha-1} m_1 \int_c^1 \tilde{G}_1(t, s) \, ds \geq \int_c^1 \tilde{G}_1(t, s) \, ds$$

$$= \tilde{u}_0(t), \quad \forall t \in [0, 1],$$

$$L_2(\tilde{u}_0, \tilde{v}_0)(t) = \varepsilon_6 \int_c^1 \tilde{G}_4(t, s) \left(\int_c^1 \tilde{G}_1(s, \tau) \, d\tau + \int_c^1 \tilde{G}_4(s, \tau) \, d\tau \right) ds$$

$$\geq \varepsilon_6 \int_c^1 \tilde{G}_4(t, s) \left(\int_c^1 \tilde{G}_4(s, \tau) \, d\tau \right) ds$$

$$\geq \varepsilon_6 \int_c^1 \tilde{G}_4(t, s) \left(\int_c^1 c^{\beta-1} \tilde{J}_4(\tau) \, d\tau \right) ds$$

$$= \varepsilon_6 c^{\beta-1} m_4 \int_c^1 \tilde{G}_4(t, s) \, ds \geq \int_c^1 \tilde{G}_4(t, s) \, ds$$

$$= \tilde{v}_0(t), \quad \forall t \in [0, 1],$$

So,

$$L(\tilde{u}_0, \tilde{v}_0) \geq (\tilde{u}_0, \tilde{v}_0). \tag{3.61}$$

We may suppose that $\mathcal{Q}$ has no fixed point on $\partial B_{r_2} \cap \tilde{P}$ (otherwise the proof is finished). From (3.60), (3.61) and Theorem 1.2.10, we conclude that

$$i(\mathcal{Q}, B_{r_2} \cap \tilde{P}, \tilde{P}) = 0. \tag{3.62}$$

Therefore, from (3.58) and (3.62), we have

$$i(\mathcal{Q}, (B_{R_2} \setminus \overline{B}_{r_2}) \cap \tilde{P}, \tilde{P}) = i(\mathcal{Q}, B_{R_2} \cap \tilde{P}, \tilde{P}) - i(\mathcal{Q}, B_{r_2} \cap \tilde{P}, \tilde{P}) = 1.$$

Then $\mathcal{Q}$ has at least one fixed point in $(B_{R_2} \setminus \overline{B}_{r_2}) \cap \widetilde{P}$, that is $r_2 < \|(u,v)\|_Y < R_2$. Because $\mathcal{Q}$ may have a fixed point on $\partial B_{r_2} \cap \widetilde{P}$, we obtain for the fixed point of operator $\mathcal{Q}$ the inequality $\|(u,v)\|_Y \geq r_2$. Thus, problem (3.41), (3.42) has at least one positive solution $(u,v) \in \widetilde{P}$. This completes the proof of the theorem. $\square$

Theorem 3.2.3. *Assume that* (I1), (I2), (I4) *and* (I6) *hold. If the functions* f *and* g *also satisfy the condition*

(I7) *For each* $t \in [0,1]$, $f(t,u,v)$ *and* $g(t,u,v)$ *are nondecreasing with respect to* u *and* v, *and there exists a constant* $N_0 > 0$ *such that*

$$f(t,N_0,N_0) < \frac{N_0}{4m_0}, \quad g(t,N_0,N_0) < \frac{N_0}{4m_0}, \quad \forall t \in [0,1],$$

where $m_0 = \max\{M_i, \ i=1,\ldots,4\}$ $(M_i = \int_0^1 \widetilde{J}_i(s)\, ds, \ i=1,\ldots,4)$, *then problem* (3.41), (3.42) *has at least two positive solutions* $(u_1(t),v_1(t))$, $(u_2(t),v_2(t))$, $t \in [0,1]$.

Proof. By using (I7), for any $(u,v) \in \partial B_{N_0} \cap \widetilde{P}$, we obtain

$$\mathcal{Q}_1(u,v)(t) \leq \int_0^1 \widetilde{G}_1(t,s)f(s,N_0,N_0)\, ds + \int_0^1 \widetilde{G}_2(t,s)g(s,N_0,N_0)\, ds$$

$$\leq \int_0^1 \widetilde{J}_1(s)f(s,N_0,N_0)\, ds + \int_0^1 \widetilde{J}_2(s)g(s,N_0,N_0)\, ds$$

$$< \frac{N_0}{4m_0} \int_0^1 \widetilde{J}_1(s)\, ds + \frac{N_0}{4m_0} \int_0^1 \widetilde{J}_2(s)\, ds$$

$$= \frac{N_0 M_1}{4m_0} + \frac{N_0 M_2}{4m_0} \leq \frac{N_0}{2}, \quad \forall t \in [0,1],$$

$$\mathcal{Q}_2(u,v)(t) \leq \int_0^1 \widetilde{G}_3(t,s)g(s,N_0,N_0)\, ds + \int_0^1 \widetilde{G}_4(t,s)f(s,N_0,N_0)\, ds$$

$$\leq \int_0^1 \widetilde{J}_3(s)g(s,N_0,N_0)\, ds + \int_0^1 \widetilde{J}_4(s)f(s,N_0,N_0)\, ds$$

$$< \frac{N_0}{4m_0} \int_0^1 \widetilde{J}_3(s)\, ds + \frac{N_0}{4m_0} \int_0^1 \widetilde{J}_4(s)\, ds$$

$$= \frac{N_0 M_3}{4m_0} + \frac{N_0 M_4}{4m_0} \leq \frac{N_0}{2}, \quad \forall t \in [0,1].$$

Then we deduce

$$\|Q(u,v)\|_Y = \|Q_1(u,v)\| + \|Q_2(u,v)\|$$
$$< N_0 = \|(u,v)\|_Y, \quad \forall (u,v) \in \partial B_{N_0} \cap \tilde{P}.$$

By Theorem 1.2.8, we conclude that

$$i(Q, B_{N_0} \cap \tilde{P}, \tilde{P}) = 1. \tag{3.63}$$

On the other hand, from (I4), (I6) and the proofs of Theorem 3.2.1 and Theorem 3.2.2, we know that there exists a sufficiently large $R_1 > N_0$ and a sufficiently small $r_2 \in (0, N_0)$ such that

$$i(Q, B_{R_1} \cap \tilde{P}, \tilde{P}) = 0, \quad i(Q, B_{r_2} \cap \tilde{P}, \tilde{P}) = 0. \tag{3.64}$$

From the relations (3.63) and (3.64), we obtain

$$i(Q, (B_{R_1} \setminus \overline{B}_{N_0}) \cap \tilde{P}, \tilde{P}) = i(Q, B_{R_1} \cap \tilde{P}, \tilde{P}) - i(Q, B_{N_0} \cap \tilde{P}, \tilde{P}) = -1,$$
$$i(Q, (B_{N_0} \setminus \overline{B}_{r_2}) \cap \tilde{P}, \tilde{P}) = i(Q, B_{N_0} \cap \tilde{P}, \tilde{P}) - i(Q, B_{r_2} \cap \tilde{P}, \tilde{P}) = 1.$$
$$\tag{3.65}$$

Then Q has at least one fixed point $(u_1, v_1) \in (B_{R_1} \setminus \overline{B}_{N_0}) \cap \tilde{P}$ and has at least one fixed point $(u_2, v_2) \in (B_{N_0} \setminus \overline{B}_{r_2}) \cap \tilde{P}$. If in Theorem 3.2.2, the operator Q has at least one fixed point on $\partial B_{r_2} \cap \tilde{P}$, then by using the first relation from (3.65), we deduce that Q has at least one fixed point $(u_1, v_1) \in (B_{R_1} \setminus \overline{B}_{N_0}) \cap \tilde{P}$ and has at least one fixed point (u_2, v_2) on $\partial B_{r_2} \cap \tilde{P}$. Therefore, problem (3.41), (3.42) has two distinct positive solutions (u_1, v_1), (u_2, v_2). The proof of the theorem is completed. □

3.2.3 *Singular nonlinearities*

In this section, we investigate the existence of positive solutions for our problem (3.41), (3.42) under various assumptions on functions f and g which may be singular at $t = 0$ and/or $t = 1$.

The basic assumptions used here are the following:

$(\widetilde{I1}) \equiv (I1)$.

$(\widetilde{I2})$ The functions $f, g \in C((0,1) \times \mathbb{R}_+ \times \mathbb{R}_+, \mathbb{R}_+)$, and there exist the functions $\tilde{p}_i \in C((0,1), \mathbb{R}_+)$ and $\tilde{q}_i \in C([0,1] \times \mathbb{R}_+ \times \mathbb{R}_+, \mathbb{R}_+)$ with $0 < \int_0^1 (1-s)^{\alpha-p_1-1} \tilde{p}_1(s)\, ds < \infty$, $0 < \int_0^1 (1-s)^{\beta-p_2-1} \tilde{p}_2(s)\, ds < \infty$, such that

$$f(t,u,v) \le \tilde{p}_1(t)\tilde{q}_1(t,u,v),$$
$$g(t,u,v) \le \tilde{p}_2(t)\tilde{q}_2(t,u,v), \quad \forall t \in (0,1), \ u, v \in \mathbb{R}_+.$$

We consider again the Banach space $X = C[0,1]$ with supremum norm $\|\cdot\|$ and the Banach space $Y = X \times X$ with the norm $\|(u,v)\|_Y = \|u\| + \|v\|$. We also define the cone $\widetilde{P} \subset Y$ by

$$\widetilde{P} = \{(u,v) \in Y, \; u(t) \geq 0, \; v(t) \geq 0, \; \forall t \in [0,1]\}.$$

We introduce the operators $\widetilde{\mathcal{Q}}_1, \widetilde{\mathcal{Q}}_2 : Y \to X$ and $\widetilde{\mathcal{Q}} : Y \to Y$ defined by

$$\widetilde{\mathcal{Q}}_1(u,v)(t) = \int_0^1 \widetilde{G}_1(t,s) f(s,u(s),v(s)) \, ds$$

$$+ \int_0^1 \widetilde{G}_2(t,s) g(s,u(s),v(s)) \, ds, \quad 0 \leq t \leq 1,$$

$$\widetilde{\mathcal{Q}}_2(u,v)(t) = \int_0^1 \widetilde{G}_3(t,s) g(s,u(s),v(s)) \, ds$$

$$+ \int_0^1 \widetilde{G}_4(t,s) f(s,u(s),v(s)) \, ds, \quad 0 \leq t \leq 1,$$

and $\widetilde{\mathcal{Q}}(u,v) = (\widetilde{\mathcal{Q}}_1(u,v), \widetilde{\mathcal{Q}}_2(u,v))$, $(u,v) \in Y$.

The pair of functions (u,v) is a solution of problem (3.41), (3.42) if and only if (u,v) is a fixed point of operator $\widetilde{\mathcal{Q}}$.

Lemma 3.2.6. *Assume that* $(\widetilde{I1})$ *and* $(\widetilde{I2})$ *hold. Then* $\widetilde{\mathcal{Q}} : \widetilde{P} \to \widetilde{P}$ *is completely continuous.*

Proof. We denote by $\widetilde{\alpha} = \int_0^1 \widetilde{J}_1(s)\widetilde{p}_1(s)\,ds$, $\widetilde{\beta} = \int_0^1 \widetilde{J}_2(s)\widetilde{p}_2(s)\,ds$, $\widetilde{\gamma} = \int_0^1 \widetilde{J}_3(s)\widetilde{p}_2(s)\,ds$ and $\widetilde{\delta} = \int_0^1 \widetilde{J}_4(s)\widetilde{p}_1(s)\,ds$, where $\widetilde{J}_i$, $i = 1,\ldots,4$, are defined in Lemma 3.2.4. Using $(\widetilde{I2})$ and Lemma 3.2.3, we deduce that $\widetilde{\alpha}, \widetilde{\beta}, \widetilde{\gamma}, \widetilde{\delta} > 0$ and

$$\widetilde{\alpha} \leq \int_0^1 \left[\frac{(1-s)^{\alpha-p_1-1}}{\Gamma(\alpha)} + \frac{\Gamma(\beta)}{\Delta\Gamma(\beta-q_1)} \left(\sum_{i=1}^N a_i \xi_i^{\beta-q_1-1} \right) \right.$$

$$\left. \times \sum_{i=1}^M \frac{b_i \eta_i^{\alpha-q_2-1}}{\Gamma(\alpha-q_2)} (1-s)^{\alpha-p_1-1} \right] \widetilde{p}_1(s)\,ds$$

$$= \left[\frac{1}{\Gamma(\alpha)} + \frac{\Gamma(\beta)}{\Delta\Gamma(\beta-q_1)} \left(\sum_{i=1}^N a_i \xi_i^{\beta-q_1-1} \right) \sum_{i=1}^M \frac{b_i \eta_i^{\alpha-q_2-1}}{\Gamma(\alpha-q_2)} \right]$$

$$\times \int_0^1 (1-s)^{\alpha-p_1-1} \widetilde{p}_1(s)\,ds < \infty,$$

$$\widetilde{\beta} \le \frac{\Gamma(\beta)}{\Delta\Gamma(\beta - p_2)} \sum_{i=1}^{N} \frac{a_i \xi_i^{\beta - q_1 - 1}}{\Gamma(\beta - q_1)} \int_0^1 (1 - s)^{\beta - p_2 - 1} \widetilde{p}_2(s) \, ds < \infty,$$

$$\widetilde{\gamma} \le \int_0^1 \left[\frac{(1 - s)^{\beta - p_2 - 1}}{\Gamma(\beta)} + \frac{\Gamma(\alpha)}{\Delta\Gamma(\alpha - q_2)} \left(\sum_{i=1}^{M} b_i \eta_i^{\alpha - q_2 - 1} \right) \right.$$
$$\left. \times \sum_{i=1}^{N} \frac{a_i \xi_i^{\beta - q_1 - 1}}{\Gamma(\beta - q_1)} (1 - s)^{\beta - p_2 - 1} \right] \widetilde{p}_2(s) ds$$

$$= \left[\frac{1}{\Gamma(\beta)} + \frac{\Gamma(\alpha)}{\Delta\Gamma(\alpha - q_2)} \left(\sum_{i=1}^{M} b_i \eta_i^{\alpha - q_2 - 1} \right) \sum_{i=1}^{N} \frac{a_i \xi_i^{\beta - q_1 - 1}}{\Gamma(\beta - q_1)} \right]$$
$$\times \int_0^1 (1 - s)^{\beta - p_2 - 1} \widetilde{p}_2(s) \, ds < \infty,$$

$$\widetilde{\delta} \le \frac{\Gamma(\alpha)}{\Delta\Gamma(\alpha - p_1)} \sum_{i=1}^{M} \frac{b_i \eta_i^{\alpha - q_2 - 1}}{\Gamma(\alpha - q_2)} \int_0^1 (1 - s)^{\alpha - p_1 - 1} \widetilde{p}_1(s) \, ds < \infty.$$

By Lemma 3.2.4, we also conclude that $\widetilde{Q}$ maps $\widetilde{P}$ into $\widetilde{P}$.

We shall prove that $\widetilde{Q}$ maps bounded sets into relatively compact sets. Suppose $D \subset \widetilde{P}$ is an arbitrary bounded set. Then there exists $\widetilde{M}_1 > 0$ such that $\|(u, v)\|_Y \le \widetilde{M}_1$ for all $(u, v) \in D$. By the continuity of $\widetilde{q}_1$ and $\widetilde{q}_2$, there exists $\widetilde{M}_2 > 0$ such that $\widetilde{M}_2 = \max\left\{ \sup_{t \in [0,1], \, u,v \in [0, \widetilde{M}_1]} \widetilde{q}_1(t, u, v), \sup_{t \in [0,1], \, u,v \in [0, \widetilde{M}_1]} \widetilde{q}_2(t, u, v) \right\}$. By using Lemma 3.2.4, for any $(u, v) \in D$ and $t \in [0, 1]$, we obtain

$$\widetilde{Q}_1(u, v)(t) \le \int_0^1 \widetilde{J}_1(s) f(s, u(s), v(s)) \, ds$$

$$+ \int_0^1 \widetilde{J}_2(s) g(s, u(s), v(s)) \, ds$$

$$\le \int_0^1 \widetilde{J}_1(s) \widetilde{p}_1(s) \widetilde{q}_1(s, u(s), v(s)) \, ds$$

$$+ \int_0^1 \widetilde{J}_2(s) \widetilde{p}_2(s) \widetilde{q}_2(s, u(s), v(s)) \, ds$$

$$\le \widetilde{M}_2 \int_0^1 \widetilde{J}_1(s) \widetilde{p}_1(s) \, ds$$

$$+ \widetilde{M}_2 \int_0^1 \widetilde{J}_2(s) \widetilde{p}_2(s) \, ds = \widetilde{M}_2(\widetilde{\alpha} + \widetilde{\beta}),$$

$$\widetilde{\mathcal{Q}}_2(u,v)(t) \le \int_0^1 \widetilde{J}_3(s)g(s,u(s),v(s))\,ds + \int_0^1 \widetilde{J}_4(s)f(s,u(s),v(s))\,ds$$

$$\le \int_0^1 \widetilde{J}_3(s)\widetilde{p}_2(s)\widetilde{q}_2(s,u(s),v(s))\,ds$$

$$+ \int_0^1 \widetilde{J}_4(s)\widetilde{p}_1(s)\widetilde{q}_1(s,u(s),v(s))\,ds$$

$$\le \widetilde{M}_2 \int_0^1 \widetilde{J}_3(s)\widetilde{p}_2(s)\,ds$$

$$+ \widetilde{M}_2 \int_0^1 \widetilde{J}_4(s)\widetilde{p}_1(s)\,ds = \widetilde{M}_2(\widetilde{\gamma}+\widetilde{\delta}).$$

Therefore, $\|\widetilde{\mathcal{Q}}_1(u,v)\| \le \widetilde{M}_2(\widetilde{\alpha}+\widetilde{\beta})$, $\|\widetilde{\mathcal{Q}}_2(u,v)\| \le \widetilde{M}_2(\widetilde{\gamma}+\widetilde{\delta})$, for all $(u,v) \in D$, and so $\widetilde{\mathcal{Q}}_1(D)$, $\widetilde{\mathcal{Q}}_2(D)$ and $\widetilde{\mathcal{Q}}(D)$ are bounded.

In what follows, we shall prove that $\widetilde{\mathcal{Q}}(D)$ is equicontinuous. By using Lemma 3.2.2, we have for $(u,v) \in D$ and $t \in [0,1]$

$$\widetilde{\mathcal{Q}}_1(u,v)(t) = \int_0^1 \left[\widetilde{g}_1(t,s) + \frac{t^{\alpha-1}\Gamma(\beta)}{\Delta\Gamma(\beta-q_1)} \left(\sum_{i=1}^N a_i\xi_i^{\beta-q_1-1} \right) \sum_{i=1}^M b_i\widetilde{g}_2(\eta_i,s) \right]$$

$$\times f(s,u(s),v(s))\,ds$$

$$+ \int_0^1 \frac{t^{\alpha-1}\Gamma(\beta)}{\Delta\Gamma(\beta-p_2)} \sum_{i=1}^N a_i\widetilde{g}_3(\xi_i,s)g(s,u(s),v(s))\,ds$$

$$= \int_0^t \frac{1}{\Gamma(\alpha)}[t^{\alpha-1}(1-s)^{\alpha-p_1-1} - (t-s)^{\alpha-1}]f(s,u(s),v(s))\,ds$$

$$+ \int_t^1 \frac{1}{\Gamma(\alpha)}t^{\alpha-1}(1-s)^{\alpha-p_1-1}f(s,u(s),v(s))\,ds$$

$$+ \frac{t^{\alpha-1}\Gamma(\beta)}{\Delta\Gamma(\beta-q_1)} \left(\sum_{i=1}^N a_i\xi_i^{\beta-q_1-1} \right)$$

$$\times \int_0^1 \left(\sum_{i=1}^M b_i\widetilde{g}_2(\eta_i,s) \right) f(s,u(s),v(s))\,ds$$

$$+ \frac{t^{\alpha-1}\Gamma(\beta)}{\Delta\Gamma(\beta-p_2)} \int_0^1 \left(\sum_{i=1}^N a_i\widetilde{g}_3(\xi_i,s) \right) g(s,u(s),v(s))\,ds.$$

Therefore, for any $t \in (0, 1)$, we conclude

$$(\widetilde{\mathcal{Q}}_1(u,v))'(t) = \int_0^t \frac{1}{\Gamma(\alpha)} \left[(\alpha-1)t^{\alpha-2}(1-s)^{\alpha-p_1-1} - (\alpha-1)(t-s)^{\alpha-2} \right]$$
$$\times f(s, u(s), v(s)) \, ds$$
$$+ \int_t^1 \frac{1}{\Gamma(\alpha)} (\alpha-1)t^{\alpha-2}(1-s)^{\alpha-p_1-1} f(s, u(s), v(s)) \, ds$$
$$+ \frac{(\alpha-1)t^{\alpha-2}\Gamma(\beta)}{\Delta\Gamma(\beta-q_1)} \left(\sum_{i=1}^N a_i \xi_i^{\beta-q_1-1} \right)$$
$$\times \int_0^1 \left(\sum_{i=1}^M b_i \widetilde{g}_2(\eta_i, s) \right) f(s, u(s), v(s)) \, ds$$
$$+ \frac{(\alpha-1)t^{\alpha-2}\Gamma(\beta)}{\Delta\Gamma(\beta-p_2)} \int_0^1 \left(\sum_{i=1}^N a_i \widetilde{g}_3(\xi_i, s) \right)$$
$$\times g(s, u(s), v(s)) \, ds.$$

So, for any $t \in (0, 1)$, we deduce

$$|(\widetilde{\mathcal{Q}}_1(u,v))'(t)| \leq \frac{1}{\Gamma(\alpha-1)} \int_0^t [t^{\alpha-2}(1-s)^{\alpha-p_1-1} + (t-s)^{\alpha-2}]$$
$$\times \widetilde{p}_1(s)\widetilde{q}_1(s, u(s), v(s)) \, ds$$
$$+ \frac{1}{\Gamma(\alpha-1)} \int_t^1 t^{\alpha-2}(1-s)^{\alpha-p_1-1}$$
$$\times \widetilde{p}_1(s)\widetilde{q}_1(s, u(s), v(s)) \, ds$$
$$+ \frac{(\alpha-1)t^{\alpha-2}\Gamma(\beta)}{\Delta\Gamma(\beta-q_1)} \left(\sum_{i=1}^N a_i \xi_i^{\beta-q_1-1} \right)$$
$$\times \int_0^1 \left(\sum_{i=1}^M b_i \widetilde{g}_2(\eta_i, s) \right) \widetilde{p}_1(s)\widetilde{q}_1(s, u(s), v(s)) \, ds$$
$$+ \frac{(\alpha-1)t^{\alpha-2}\Gamma(\beta)}{\Delta\Gamma(\beta-p_2)} \int_0^1 \left(\sum_{i=1}^N a_i \widetilde{g}_3(\xi_i, s) \right)$$
$$\times \widetilde{p}_2(s)\widetilde{q}_2(s, u(s), v(s)) \, ds.$$

Then we obtain

$$|(\widetilde{\mathcal{Q}}_1(u,v))'(t)|$$

$$\leq \widetilde{M}_2 \left[\frac{1}{\Gamma(\alpha-1)} \int_0^t [t^{\alpha-2}(1-s)^{\alpha-p_1-1} + (t-s)^{\alpha-2}]\widetilde{p}_1(s)\,ds \right.$$

$$+ \frac{1}{\Gamma(\alpha-1)} \int_t^1 t^{\alpha-2}(1-s)^{\alpha-p_1-1}\widetilde{p}_1(s)\,ds$$

$$+ \frac{(\alpha-1)t^{\alpha-2}\Gamma(\beta)}{\Delta\Gamma(\beta-q_1)} \left(\sum_{i=1}^N a_i \xi_i^{\beta-q_1-1} \right) \int_0^1 \left(\sum_{i=1}^M b_i \widetilde{g}_2(\eta_i,s) \right) \widetilde{p}_1(s)\,ds$$

$$+ \left. \frac{(\alpha-1)t^{\alpha-2}\Gamma(\beta)}{\Delta\Gamma(\beta-p_2)} \int_0^1 \left(\sum_{i=1}^N a_i \widetilde{g}_3(\xi_i,s) \right) \widetilde{p}_2(s)\,ds \right]. \tag{3.66}$$

We denote

$$h(t) = \frac{1}{\Gamma(\alpha-1)} \int_0^t [t^{\alpha-2}(1-s)^{\alpha-p_1-1} + (t-s)^{\alpha-2}]\widetilde{p}_1(s)\,ds$$

$$+ \frac{1}{\Gamma(\alpha-1)} \int_t^1 t^{\alpha-2}(1-s)^{\alpha-p_1-1}\widetilde{p}_1(s)\,ds,$$

$$\mu(t) = h(t) + \frac{(\alpha-1)t^{\alpha-2}\Gamma(\beta)}{\Delta\Gamma(\beta-q_1)} \left(\sum_{i=1}^N a_i \xi_i^{\beta-q_1-1} \right)$$

$$\times \int_0^1 \left(\sum_{i=1}^M b_i \widetilde{g}_2(\eta_i,s) \right) \widetilde{p}_1(s)\,ds$$

$$+ \frac{(\alpha-1)t^{\alpha-2}\Gamma(\beta)}{\Delta\Gamma(\beta-p_2)} \int_0^1 \left(\sum_{i=1}^N a_i \widetilde{g}_3(\xi_i,s) \right) \widetilde{p}_2(s)\,ds.$$

For the integral of the function h, by exchanging the order of integration, we obtain

$$\int_0^1 h(t)\,dt = \frac{1}{\Gamma(\alpha-1)} \int_0^1 \left(\int_0^t [t^{\alpha-2}(1-s)^{\alpha-p_1-1} + (t-s)^{\alpha-2}]\widetilde{p}_1(s)\,ds \right) dt$$

$$+ \frac{1}{\Gamma(\alpha-1)} \int_0^1 \left(\int_t^1 t^{\alpha-2}(1-s)^{\alpha-p_1-1}\widetilde{p}_1(s)\,ds \right) dt$$

$$= \frac{1}{\Gamma(\alpha-1)} \int_0^1 \left(\int_s^1 [t^{\alpha-2}(1-s)^{\alpha-p_1-1} + (t-s)^{\alpha-2}] dt \right) \widetilde{p}_1(s) ds$$

$$+ \frac{1}{\Gamma(\alpha-1)} \int_0^1 \left(\int_0^s t^{\alpha-2}(1-s)^{\alpha-p_1-1} dt \right) \widetilde{p}_1(s) ds$$

$$= \frac{1}{\Gamma(\alpha-1)} \int_0^1 \left[\frac{(1-s)^{\alpha-p_1-1}}{\alpha-1} - \frac{s^{\alpha-1}(1-s)^{\alpha-p_1-1}}{\alpha-1} \right.$$

$$\left. + \frac{(1-s)^{\alpha-1}}{\alpha-1} \right] \widetilde{p}_1(s) ds$$

$$+ \frac{1}{\Gamma(\alpha-1)} \int_0^1 \frac{s^{\alpha-1}(1-s)^{\alpha-p_1-1}}{\alpha-1} \widetilde{p}_1(s) ds$$

$$= \frac{1}{\Gamma(\alpha)} \int_0^1 (1-s)^{\alpha-p_1-1}(1+(1-s)^{p_1})\widetilde{p}_1(s) ds$$

$$\leq \frac{2}{\Gamma(\alpha)} \int_0^1 (1-s)^{\alpha-p_1-1}\widetilde{p}_1(s) ds < \infty.$$

For the integral of the function μ, we have

$$\int_0^1 \mu(t) dt = \int_0^1 h(t) dt + \frac{(\alpha-1)\Gamma(\beta)}{\Delta\Gamma(\beta-q_1)} \left(\sum_{i=1}^N a_i \xi_i^{\beta-q_1-1} \right) \left(\int_0^1 t^{\alpha-2} dt \right)$$

$$\times \left(\int_0^1 \left(\sum_{i=1}^M b_i \widetilde{g}_2(\eta_i, s) \right) \widetilde{p}_1(s) ds \right)$$

$$+ \frac{(\alpha-1)\Gamma(\beta)}{\Delta\Gamma(\beta-p_2)} \left(\int_0^1 t^{\alpha-2} dt \right) \left(\int_0^1 \left(\sum_{i=1}^N a_i \widetilde{g}_3(\xi_i, s) \right) \widetilde{p}_2(s) ds \right)$$

$$\leq \frac{2}{\Gamma(\alpha)} \int_0^1 (1-s)^{\alpha-p_1-1}\widetilde{p}_1(s) ds$$

$$+ \frac{\Gamma(\beta)}{\Delta\Gamma(\beta-q_1)} \left(\sum_{i=1}^N a_i \xi_i^{\beta-q_1-1} \right)$$

$$\times \left(\int_0^1 \left(\sum_{i=1}^M \frac{b_i \eta_i^{\alpha-q_2-1}}{\Gamma(\alpha-q_2)} \right) (1-s)^{\alpha-p_1-1}\widetilde{p}_1(s) ds \right)$$

$$+ \frac{\Gamma(\beta)}{\Delta\Gamma(\beta-p_2)} \left(\int_0^1 \left(\sum_{i=1}^N \frac{a_i \xi_i^{\beta-q_1-1}}{\Gamma(\beta-q_1)} \right) (1-s)^{\beta-p_2-1}\widetilde{p}_2(s) ds \right)$$

$$= \frac{2}{\Gamma(\alpha)} \int_0^1 (1-s)^{\alpha-p_1-1} \widetilde{p}_1(s) \, ds$$

$$+ \frac{\Gamma(\beta)}{\Delta\Gamma(\beta-q_1)\Gamma(\alpha-q_2)} \left(\sum_{i=1}^N a_i \xi_i^{\beta-q_1-1} \right)$$

$$\times \left(\sum_{i=1}^M b_i \eta_i^{\alpha-q_2-1} \right) \int_0^1 (1-s)^{\alpha-p_1-1} \widetilde{p}_1(s) \, ds$$

$$+ \frac{\Gamma(\beta)}{\Delta\Gamma(\beta-p_2)\Gamma(\beta-q_1)} \left(\sum_{i=1}^N a_i \xi_i^{\beta-q_1-1} \right)$$

$$\times \int_0^1 (1-s)^{\beta-p_2-1} \widetilde{p}_2(s) \, ds.$$

Therefore, we obtain

$$\int_0^1 \mu(t) \, dt \le \left[\frac{2}{\Gamma(\alpha)} + \frac{\Gamma(\beta)}{\Delta\Gamma(\beta-q_1)\Gamma(\alpha-q_2)} \left(\sum_{i=1}^N a_i \xi_i^{\beta-q_1-1} \right) \right.$$

$$\left. \times \left(\sum_{i=1}^M b_i \eta_i^{\alpha-q_2-1} \right) \right]$$

$$\times \int_0^1 (1-s)^{\alpha-p_1-1} \widetilde{p}_1(s) \, ds + \frac{\Gamma(\beta)}{\Delta\Gamma(\beta-p_2)\Gamma(\beta-q_1)} \quad (3.67)$$

$$\times \left(\sum_{i=1}^N a_i \xi_i^{\beta-q_1-1} \right)$$

$$\times \int_0^1 (1-s)^{\beta-p_2-1} \widetilde{p}_2(s) \, ds < \infty.$$

We deduce that $\mu \in L^1(0,1)$. Thus, for any $t_1, t_2 \in [0,1]$ with $t_1 \le t_2$ and $(u,v) \in D$, by (3.66) and (3.67), we conclude

$$|\widetilde{\mathcal{Q}}_1(u,v)(t_1) - \widetilde{\mathcal{Q}}_1(u,v)(t_2)| = \left| \int_{t_1}^{t_2} (\widetilde{\mathcal{Q}}_1(u,v))'(t) \, dt \right| \le \widetilde{M}_2 \int_{t_1}^{t_2} \mu(t) \, dt.$$
$$(3.68)$$

From (3.67), (3.68) and the absolute continuity of the integral function, we obtain that $\widetilde{\mathcal{Q}}_1(D)$ is equicontinuous. In a similar manner, we deduce that $\widetilde{\mathcal{Q}}_2(D)$ is also equicontinuous. By the Arzela–Ascoli theorem, we conclude that $\widetilde{\mathcal{Q}}_1(D)$ and $\widetilde{\mathcal{Q}}_2(D)$ are relatively compact sets, and so $\widetilde{\mathcal{Q}}(D)$ is also relatively compact. Therefore, $\widetilde{\mathcal{Q}}$ is a compact operator. Besides, we can show that $\widetilde{Q}$ is continuous on $\widetilde{P}$, (see the proof of Lemma 1.4.1 from [43]). Hence $\widetilde{Q} : \widetilde{P} \to \widetilde{P}$ is completely continuous. $\qquad\qquad\qquad\square$

For $c \in (0,1)$ we define the cone

$$P_0 = \left\{ (u,v) \in \widetilde{P}, \ \min_{t \in [c,1]} u(t) \geq c^{\alpha-1}\|u\|, \ \min_{t \in [c,1]} v(t) \geq c^{\beta-1}\|v\| \right\}.$$

Under assumptions $(\widetilde{I1})$, $(\widetilde{I2})$ and Remark 3.2.1, we have $\widetilde{Q}(\widetilde{P}) \subset P_0$ and so $\widetilde{Q}|_{P_0} : P_0 \to P_0$ (denoted again by $\widetilde{Q}$) is also a completely continuous operator.

Theorem 3.2.4. *Assume that $(\widetilde{I1})$ and $(\widetilde{I2})$ hold. If the functions $\widetilde{q}_1$, $\widetilde{q}_2$, f and g also satisfy the conditions*

$(\widetilde{I3})$ *There exist $a \geq 1$ and $b \geq 1$ such that*

$$q_{10} = \lim_{\substack{u+v\to 0 \\ u,v\geq 0}} \sup_{t\in[0,1]} \frac{\widetilde{q}_1(t,u,v)}{(u+v)^a} = 0 \quad and$$

$$q_{20} = \lim_{\substack{u+v\to 0 \\ u,v\geq 0}} \sup_{t\in[0,1]} \frac{\widetilde{q}_2(t,u,v)}{(u+v)^b} = 0;$$

$(\widetilde{I4})$ *There exists $c \in (0,1/2)$ such that*

$$\widetilde{f}^i_\infty = \lim_{\substack{u+v\to\infty \\ u,v\geq 0}} \inf_{t\in[c,1-c]} \frac{f(t,u,v)}{u+v} = \infty \quad or$$

$$\widetilde{g}^i_\infty = \lim_{\substack{u+v\to\infty \\ u,v\geq 0}} \inf_{t\in[c,1-c]} \frac{g(t,u,v)}{u+v} = \infty,$$

then problem (3.41), (3.42) has at least one positive solution $(u(t), v(t))$, $t \in [0,1]$.

Proof. We consider the above cone P_0 with c given in $(\widetilde{I4})$. From $(\widetilde{I3})$, we deduce that for $\varepsilon_7 = \min\left\{\frac{1}{4\widetilde{\alpha}}, \frac{1}{4\widetilde{\delta}}\right\} > 0$, $\varepsilon_8 = \min\left\{\frac{1}{4\widetilde{\beta}}, \frac{1}{4\widetilde{\gamma}}\right\} > 0$ there exists $r_3 \in (0,1)$ such that

$$\widetilde{q}_1(t,u,v) \leq \varepsilon_7(u+v)^a, \quad \widetilde{q}_2(t,u,v) \leq \varepsilon_8(u+v)^b,$$

$$\forall t \in [0,1], \ u, v \geq 0, \ u+v \leq r_3. \qquad\qquad (3.69)$$

Then by (3.69) and Lemma 3.2.4, for any $(u,v) \in \partial B_{r_3} \cap P_0$ and $t \in [0,1]$, we obtain

$$\widetilde{\mathcal{Q}}_1(u,v)(t) \leq \int_0^1 \widetilde{J}_1(s)\widetilde{p}_1(s)\widetilde{q}_1(s,u(s),v(s))\,ds$$

$$+ \int_0^1 \widetilde{J}_2(s)\widetilde{p}_2(s)\widetilde{q}_2(s,u(s),v(s))\,ds$$

$$\leq \varepsilon_7 \int_0^1 \widetilde{J}_1(s)\widetilde{p}_1(s)(u(s)+v(s))^a\,ds$$

$$+ \varepsilon_8 \int_0^1 \widetilde{J}_2(s)\widetilde{p}_2(s)(u(s)+v(s))^b\,ds$$

$$\leq \varepsilon_7\widetilde{\alpha}\|(u,v)\|_Y^a + \varepsilon_8\widetilde{\beta}\|(u,v)\|_Y^b \leq \varepsilon_7\widetilde{\alpha}\|(u,v)\|_Y + \varepsilon_8\widetilde{\beta}\|(u,v)\|_Y$$

$$\leq \frac{1}{4}\|(u,v)\|_Y + \frac{1}{4}\|(u,v)\|_Y = \frac{1}{2}\|(u,v)\|_Y,$$

$$\widetilde{\mathcal{Q}}_2(u,v)(t) \leq \int_0^1 \widetilde{J}_3(s)\widetilde{p}_2(s)\widetilde{q}_2(s,u(s),v(s))\,ds$$

$$+ \int_0^1 \widetilde{J}_4(s)\widetilde{p}_1(s)\widetilde{q}_1(s,u(s),v(s))\,ds$$

$$\leq \varepsilon_8 \int_0^1 \widetilde{J}_3(s)\widetilde{p}_2(s)(u(s)+v(s))^b\,ds$$

$$+ \varepsilon_7 \int_0^1 \widetilde{J}_4(s)\widetilde{p}_1(s)(u(s)+v(s))^a\,ds$$

$$\leq \varepsilon_8\widetilde{\gamma}\|(u,v)\|_Y^b + \varepsilon_7\widetilde{\delta}\|(u,v)\|_Y^a \leq \varepsilon_8\widetilde{\gamma}\|(u,v)\|_Y + \varepsilon_7\widetilde{\delta}\|(u,v)\|_Y$$

$$\leq \frac{1}{4}\|(u,v)\|_Y + \frac{1}{4}\|(u,v)\|_Y = \frac{1}{2}\|(u,v)\|_Y.$$

Therefore, we deduce $\|\widetilde{\mathcal{Q}}_1(u,v)\| \leq \frac{1}{2}\|(u,v)\|_Y$, $\|\widetilde{\mathcal{Q}}_2(u,v)\| \leq \frac{1}{2}\|(u,v)\|_Y$ for all $(u,v) \in \partial B_{r_3} \cap P_0$, and so

$$\|\widetilde{\mathcal{Q}}(u,v)\|_Y \leq \|(u,v)\|_Y, \quad \forall\,(u,v) \in \partial B_{r_3} \cap P_0. \tag{3.70}$$

From $(\widetilde{I}4)$, we suppose that $\widetilde{f}^i_\infty = \infty$, (in a similar manner, we can study the case $\widetilde{g}^i_\infty = \infty$). So, for $\varepsilon_9 = 2(c^{\alpha-1}\widetilde{m}_1\min\{c^{\alpha-1},c^{\beta-1}\})^{-1}$, $(\widetilde{m}_1 = \int_c^{1-c} J_1(s)\,ds)$, there exists $C_{10} > 0$ such that

$$f(t,u,v) \geq \varepsilon_9(u+v) - C_{10}, \quad \forall\,t \in [c,1-c],\ u,v \geq 0. \tag{3.71}$$

Then by using (3.71), for any $(u,v) \in P_0$ and $t \in [c,1]$, we have

$$
\begin{aligned}
\widetilde{\mathcal{Q}}_1(u,v)(t) &\geq \int_c^{1-c} \widetilde{G}_1(t,s) f(s,u(s),v(s))\, ds \\
&\quad + \int_c^{1-c} \widetilde{G}_2(t,s) g(s,u(s),v(s))\, ds \\
&\geq c^{\alpha-1} \int_c^{1-c} \widetilde{J}_1(s) f(s,u(s),v(s))\, ds \\
&\geq c^{\alpha-1} \int_c^{1-c} \widetilde{J}_1(s)(\varepsilon_9(u(s)+v(s)) - C_{10})\, ds \\
&\geq c^{\alpha-1}\varepsilon_9 \widetilde{m}_1 \min_{s \in [c,1-c]}(u(s)+v(s)) - c^{\alpha-1}\widetilde{m}_1 C_{10} \\
&\geq c^{\alpha-1}\varepsilon_9 \widetilde{m}_1 \min_{s \in [c,1]}(u(s)+v(s)) - c^{\alpha-1}\widetilde{m}_1 C_{10} \\
&\geq c^{\alpha-1}\varepsilon_9 \widetilde{m}_1 \left(\min_{s \in [c,1]} u(s) + \min_{s \in [c,1]} v(s) \right) - c^{\alpha-1}\widetilde{m}_1 C_{10} \\
&\geq c^{\alpha-1}\varepsilon_9 \widetilde{m}_1 (c^{\alpha-1}\|u\| + c^{\beta-1}\|v\|) - c^{\alpha-1}\widetilde{m}_1 C_{10} \\
&\geq c^{\alpha-1}\varepsilon_9 \widetilde{m}_1 \min\{c^{\alpha-1}, c^{\beta-1}\}\|(u,v)\|_Y - C_{11} \\
&= 2\|(u,v)\|_Y - C_{11}, \quad C_{11} = c^{\alpha-1}\widetilde{m}_1 C_{10}.
\end{aligned}
$$

Hence, we obtain $\|\widetilde{\mathcal{Q}}_1(u,v)\| \geq 2\|(u,v)\|_Y - C_{11}$ for all $(u,v) \in P_0$. We can choose $R_3 \geq \max\{C_{11}, 1\}$, and then we deduce

$$
\|\widetilde{\mathcal{Q}}(u,v)\|_Y \geq \|\widetilde{\mathcal{Q}}_1(u,v)\| \geq \|(u,v)\|_Y, \quad \forall\, (u,v) \in \partial B_{R_3} \cap P_0. \tag{3.72}
$$

By (3.70), (3.72) and the Guo–Krasnosel'skii fixed point theorem (Theorem 1.2.2), we conclude that $\widetilde{\mathcal{Q}}$ has a fixed point $(u,v) \in (\overline{B}_{R_3} \setminus B_{r_3}) \cap P_0$, that is $r_3 \leq \|(u,v)\|_Y \leq R_3$. By Lemma 3.2.5, we obtain that $u(t) > 0$ for all $t \in (0,1]$ or $v(t) > 0$ for all $t \in (0,1]$. The proof of the theorem is completed. $\qquad \square$

Theorem 3.2.5. *Assume that* $(\widetilde{I}1)$ *and* $(\widetilde{I}2)$ *hold. If the functions* $\widetilde{q}_1$, $\widetilde{q}_2$, f *and* g *also satisfy the conditions*

$(\widetilde{I}5)$
$$
q_{1\infty} = \lim_{\substack{u+v \to \infty \\ u,v \geq 0}} \sup_{t \in [0,1]} \frac{\widetilde{q}_1(t,u,v)}{u+v} = 0 \quad and
$$

$$
q_{2\infty} = \lim_{\substack{u+v \to \infty \\ u,v \geq 0}} \sup_{t \in [0,1]} \frac{\widetilde{q}_2(t,u,v)}{u+v} = 0;
$$

($\widetilde{I}6$) *There exist* $c \in (0, 1/2)$, $\hat{a} \in (0, 1]$ *and* $\hat{b} \in (0, 1]$ *such that*

$$\widetilde{f}_0^i = \lim_{\substack{u+v \to 0 \\ u,v \geq 0}} \inf_{t \in [c, 1-c]} \frac{f(t, u, v)}{(u+v)^{\hat{a}}} = \infty \quad or$$

$$\widetilde{g}_0^i = \lim_{\substack{u+v \to 0 \\ u,v \geq 0}} \inf_{t \in [c, 1-c]} \frac{g(t, u, v)}{(u+v)^{\hat{b}}} = \infty,$$

then problem (3.41), (3.42) *has at least one positive solution* $(u(t), v(t))$, $t \in [0, 1]$.

Proof. We consider again the cone P_0 with c given in ($\widetilde{I}6$). From ($\widetilde{I}5$) we deduce that for $\varepsilon_{10} \in \left(0, \frac{1}{2(\widetilde{\alpha}+\widetilde{\delta})} \right)$ and $\varepsilon_{11} \in \left(0, \frac{1}{2(\widetilde{\beta}+\widetilde{\gamma})} \right)$, there exist $C_{12}, C_{13} > 0$ such that

$$\widetilde{q}_1(t, u, v) \leq \varepsilon_{10}(u+v) + C_{12},$$
$$\widetilde{q}_2(t, u, v) \leq \varepsilon_{11}(u+v) + C_{13}, \quad \forall t \in [0, 1], \ u, \ v \geq 0. \quad (3.73)$$

By using (3.73) and ($\widetilde{I}2$), for any $(u, v) \in P_0$, we conclude

$$\widetilde{\mathcal{Q}}_1(u, v)(t) \leq \int_0^1 \widetilde{J}_1(s)\widetilde{p}_1(s)\widetilde{q}_1(s, u(s), v(s)) \, ds$$

$$+ \int_0^1 \widetilde{J}_2(s)\widetilde{p}_2(s)\widetilde{q}_2(s, u(s), v(s)) \, ds$$

$$\leq \int_0^1 \widetilde{J}_1(s)\widetilde{p}_1(s)(\varepsilon_{10}(u(s) + v(s)) + C_{12}) \, ds$$

$$+ \int_0^1 \widetilde{J}_2(s)\widetilde{p}_2(s)(\varepsilon_{11}(u(s) + v(s)) + C_{13}) \, ds$$

$$\leq (\varepsilon_{10}\|(u, v)\|_Y + C_{12})\widetilde{\alpha} + (\varepsilon_{11}\|(u, v)\|_Y + C_{13})\widetilde{\beta}$$

$$= (\varepsilon_{10}\widetilde{\alpha} + \varepsilon_{11}\widetilde{\beta})\|(u, v)\|_Y + C_{12}\widetilde{\alpha} + C_{13}\widetilde{\beta}, \quad \forall t \in [0, 1],$$

$$\widetilde{\mathcal{Q}}_2(u, v)(t) \leq \int_0^1 \widetilde{J}_3(s)\widetilde{p}_2(s)\widetilde{q}_2(s, u(s), v(s)) \, ds$$

$$+ \int_0^1 \widetilde{J}_4(s)\widetilde{p}_1(s)\widetilde{q}_1(s, u(s), v(s)) \, ds$$

$$\leq \int_0^1 \widetilde{J}_3(s)\widetilde{p}_2(s)(\varepsilon_{11}(u(s) + v(s)) + C_{13})\, ds$$

$$+ \int_0^1 \widetilde{J}_4(s)\widetilde{p}_1(s)(\varepsilon_{10}(u(s) + v(s)) + C_{12})\, ds$$

$$\leq (\varepsilon_{11}\|(u, v)\|_Y + C_{13})\widetilde{\gamma} + (\varepsilon_{10}\|(u, v)\|_Y + C_{12})\widetilde{\delta}$$

$$= (\varepsilon_{11}\widetilde{\gamma} + \varepsilon_{10}\widetilde{\delta})\|(u, v)\|_Y + C_{13}\widetilde{\gamma} + C_{12}\widetilde{\delta}, \ \forall t \in [0, 1].$$

Therefore,

$$\|\widetilde{\mathcal{Q}}_1(u, v)\| \leq (\varepsilon_{10}\widetilde{\alpha} + \varepsilon_{11}\widetilde{\beta})\|(u, v)\|_Y + C_{12}\widetilde{\alpha} + C_{13}\widetilde{\beta},$$

$$\|\widetilde{\mathcal{Q}}_2(u, v)\| \leq (\varepsilon_{11}\widetilde{\gamma} + \varepsilon_{10}\widetilde{\delta})\|(u, v)\|_Y + C_{13}\widetilde{\gamma} + C_{12}\widetilde{\delta},$$

and so

$$\|\widetilde{\mathcal{Q}}(u, v)\|_Y \leq [\varepsilon_{10}(\widetilde{\alpha} + \widetilde{\delta}) + \varepsilon_{11}(\widetilde{\beta} + \widetilde{\gamma})]\|(u, v)\|_Y + C_{12}(\widetilde{\alpha} + \widetilde{\delta}) + C_{13}(\widetilde{\beta} + \widetilde{\gamma})$$

$$< \|(u, v)\|_Y + C_{14}, \ \ C_{14} = C_{12}(\widetilde{\alpha} + \widetilde{\delta}) + C_{13}(\widetilde{\beta} + \widetilde{\gamma}).$$

We can choose large $R_4 > 1$ such that

$$\|\widetilde{\mathcal{Q}}(u, v)\|_Y \leq \|(u, v)\|_Y, \ \ \forall (u, v) \in \partial B_{R_4} \cap P_0. \tag{3.74}$$

From $(\widetilde{I}6)$, we suppose that $\widetilde{f}_0^i = \infty$, (in a similar manner we can study the case $\widetilde{g}_0^i = \infty$). So, we deduce that for $\varepsilon_{12} = (c^{\alpha-1}\widetilde{m}_1 \min\{c^{\alpha-1}, c^{\beta-1}\})^{-1} > 0$, $(\widetilde{m}_1 = \int_c^{1-c} \widetilde{J}_1(s)\, ds)$, there exists $r_4 \in (0, 1]$ such that

$$f(t, u, v) \geq \varepsilon_{12}(u + v)^{\hat{a}}, \ \ \forall t \in [c, 1 - c], \ u, v \geq 0, \ u + v \leq r_4. \tag{3.75}$$

Then by using (3.75), for any $(u, v) \in \partial B_{r_4} \cap P_0$ and $t \in [c, 1]$, we have

$$\widetilde{\mathcal{Q}}_1(u, v)(t) \geq \int_c^{1-c} \widetilde{G}_1(t, s)f(s, u(s), v(s))\, ds$$

$$+ \int_c^{1-c} \widetilde{G}_2(t, s)g(s, u(s), v(s))\, ds$$

$$\geq c^{\alpha-1} \int_c^{1-c} \widetilde{J}_1(s)\varepsilon_{12}(u(s) + v(s))^{\hat{a}}\, ds$$

$$\geq c^{\alpha-1} \int_c^{1-c} \widetilde{J}_1(s)\varepsilon_{12}(u(s) + v(s))\, ds$$

$$\geq c^{\alpha-1}\varepsilon_{12} \int_c^{1-c} \widetilde{J}_1(s) \left(\min_{s\in[c,1-c]}(u(s)+v(s)) \right) ds$$

$$\geq c^{\alpha-1}\varepsilon_{12} \min_{s\in[c,1]}(u(s)+v(s)) \int_c^{1-c} \widetilde{J}_1(s)\, ds$$

$$\geq c^{\alpha-1}\varepsilon_{12}\widetilde{m}_1 \left(\min_{s\in[c,1]} u(s) + \min_{s\in[c,1]} v(s) \right)$$

$$\geq c^{\alpha-1}\varepsilon_{12}\widetilde{m}_1(c^{\alpha-1}\|u\| + c^{\beta-1}\|v\|)$$

$$\geq c^{\alpha-1}\varepsilon_{12}\widetilde{m}_1 \min\{c^{\alpha-1}, c^{\beta-1}\}(\|u\| + \|v\|) = \|(u,v)\|_Y.$$

Therefore, $\|\widetilde{Q}_1(u,v)\| \geq \|(u,v)\|_Y$ for all $(u,v) \in \partial B_{r_4} \cap P_0$, and so

$$\|\widetilde{Q}(u,v)\|_Y \geq \|\widetilde{Q}_1(u,v)\| \geq \|(u,v)\|_Y, \quad \forall\, (u,v) \in \partial B_{r_4} \cap P_0. \tag{3.76}$$

By (3.74), (3.76) and the Guo–Krasnosel'skii fixed point theorem (Theorem 1.2.2), we deduce that $\widetilde{Q}$ has at least one fixed point $(u,v) \in (\overline{B}_{R_4} \setminus B_{r_4}) \cap P_0$, that is $r_4 \leq \|(u,v)\|_Y \leq R_4$. The proof of the theorem is complete. $\qquad\square$

3.2.4 Examples

Let $\alpha = 9/2$ $(n = 5)$, $\beta = 8/3$ $(m = 3)$, $p_1 = 4/3$, $p_2 = 1$, $q_1 = 1/2$, $q_2 = 2/3$, $N = 1$, $M = 2$, $\xi_1 = 1/2$, $a_1 = 2$, $\eta_1 = 1/3$, $\eta_2 = 2/3$, $b_1 = 1$ and $b_2 = 1/2$.

We consider the system of fractional differential equations

$$\begin{cases} D_{0+}^{9/2}u(t) + f(t, u(t), v(t)) = 0, & t \in (0,1), \\ D_{0+}^{8/3}v(t) + g(t, u(t), v(t)) = 0, & t \in (0,1), \end{cases} \tag{3.77}$$

with the multi-point boundary conditions

$$\begin{cases} u(0) = u'(0) = u''(0) = u'''(0) = 0, & D_{0+}^{4/3}u(1) = 2D_{0+}^{1/2}v\left(\frac{1}{2}\right), \\ v(0) = v'(0) = 0, & v'(1) = D_{0+}^{2/3}u\left(\frac{1}{3}\right) + \frac{1}{2}D_{0+}^{2/3}u\left(\frac{2}{3}\right). \end{cases} \tag{3.78}$$

Then we obtain $\Delta = \frac{5\Gamma(9/2)}{3\Gamma(19/6)} - \frac{\Gamma(9/2)\Gamma(8/3)(1+2^{11/6})}{2^{1/6}3^{17/6}\Gamma(13/6)\Gamma(23/6)} \approx 7.6683666$. We also deduce

$$\widetilde{g}_1(t,s) = \frac{1}{\Gamma(9/2)} \begin{cases} t^{7/2}(1-s)^{13/6} - (t-s)^{7/2}, & 0 \leq s \leq t \leq 1, \\ t^{7/2}(1-s)^{13/6}, & 0 \leq t \leq s \leq 1, \end{cases}$$

$$\widetilde{g}_2(t,s) = \frac{1}{\Gamma(23/6)} \begin{cases} t^{17/6}(1-s)^{13/6} - (t-s)^{17/6}, & 0 \le s \le t \le 1, \\ t^{17/6}(1-s)^{13/6}, & 0 \le t \le s \le 1, \end{cases}$$

$$\widetilde{g}_3(t,s) = \frac{1}{\Gamma(13/6)} \begin{cases} t^{7/6}(1-s)^{2/3} - (t-s)^{7/6}, & 0 \le s \le t \le 1, \\ t^{7/6}(1-s)^{2/3}, & 0 \le t \le s \le 1, \end{cases}$$

$$\widetilde{g}_4(t,s) = \frac{1}{\Gamma(8/3)} \begin{cases} t^{5/3}(1-s)^{2/3} - (t-s)^{5/3}, & 0 \le s \le t \le 1, \\ t^{5/3}(1-s)^{2/3}, & 0 \le t \le s \le 1, \end{cases}$$

$\widetilde{h}_1(s) = \frac{1}{\Gamma(9/2)}(1-s)^{13/6}(1-(1-s)^{4/3})$, $\widetilde{h}_2(s) = \frac{1}{\Gamma(23/6)}(1-s)^{13/6}(1-(1-s)^{2/3})$, $\widetilde{h}_3(s) = \frac{1}{\Gamma(13/6)}(1-s)^{2/3}(1-(1-s)^{1/2})$, $\widetilde{h}_4(s) = \frac{1}{\Gamma(8/3)}s(1-s)^{2/3}$ for all $s \in [0,1]$. For the functions $\widetilde{J}_i$, $i = 1, \ldots, 4$, we obtain

$$\widetilde{J}_1(s) = \begin{cases} \frac{1}{\Gamma(9/2)}(1-s)^{13/6}(1-(1-s)^{4/3}) \\ \quad + \frac{\Gamma(8/3)}{2^{7/6}3^{17/6}\Delta\Gamma(13/6)\Gamma(23/6)}\big[2(1-s)^{13/6} - 2(1-3s)^{17/6} \\ \quad + 2^{17/6}(1-s)^{13/6} - (2-3s)^{17/6}\big], 0 \le s < \frac{1}{3}, \\[4pt] \frac{1}{\Gamma(9/2)}(1-s)^{13/6}(1-(1-s)^{4/3}) \\ \quad + \frac{\Gamma(8/3)}{2^{7/6}3^{17/6}\Delta\Gamma(13/6)\Gamma(23/6)}\big[2(1-s)^{13/6} \\ \quad + 2^{17/6}(1-s)^{13/6} - (2-3s)^{17/6}\big], \ \frac{1}{3} \le s < \frac{2}{3}, \\[4pt] \frac{1}{\Gamma(9/2)}(1-s)^{13/6}(1-(1-s)^{4/3}) \\ \quad + \frac{\Gamma(8/3)}{2^{1/6}3^{17/6}\Delta\Gamma(13/6)\Gamma(23/6)}\big[(1-s)^{13/6} \\ \quad + 2^{11/6}(1-s)^{13/6}\big], \ \frac{2}{3} \le s \le 1, \end{cases}$$

$$\widetilde{J}_2(s) = \frac{5}{3 \cdot 2^{1/6}\Delta\Gamma(13/6)} \begin{cases} (1-s)^{2/3} - (1-2s)^{7/6}, & 0 \le s < \frac{1}{2}, \\ (1-s)^{2/3}, & \frac{1}{2} \le s \le 1, \end{cases}$$

$$\widetilde{J}_3(s) = \begin{cases} \frac{1}{\Gamma(8/3)}s(1-s)^{2/3} + \frac{(1+2^{11/6})\Gamma(9/2)}{2^{1/6}3^{17/6}\Delta\Gamma(13/6)\Gamma(23/6)}\big[(1-s)^{2/3} \\ \quad -(1-2s)^{7/6}\big], \ 0 \le s < \frac{1}{2}, \\[4pt] \frac{1}{\Gamma(8/3)}s(1-s)^{2/3} + \frac{(1+2^{11/6})\Gamma(9/2)}{2^{1/6}3^{17/6}\Delta\Gamma(13/6)\Gamma(23/6)}(1-s)^{2/3}, \\ \quad \frac{1}{2} \le s \le 1, \end{cases}$$

$$\tilde{J}_4(s) = \frac{\Gamma(9/2)}{2 \cdot 3^{17/6} \Delta \Gamma(19/6) \Gamma(23/6)}$$

$$\times \begin{cases} 2(1-s)^{13/6} - 2(1-3s)^{17/6} + 2^{17/6}(1-s)^{13/6} \\ \quad -(2-3s)^{17/6}, \ 0 \le s < \frac{1}{3}, \\ 2(1-s)^{13/6} + 2^{17/6}(1-s)^{13/6} \\ \quad -(2-3s)^{17/6}, \ \frac{1}{3} \le s < \frac{2}{3}, \\ 2(1-s)^{13/6} + 2^{17/6}(1-s)^{13/6}, \ \frac{2}{3} \le s \le 1. \end{cases}$$

We also deduce $M_1 = \int_0^1 \tilde{J}_1(s)\,ds \approx 0.00912385$, $M_2 = \int_0^1 \tilde{J}_2(s)\,ds \approx 0.06605546$, $M_3 = \int_0^1 \tilde{J}_3(s)\,ds \approx 0.16869513$ and $M_4 = \int_0^1 \tilde{J}_4(s)\,ds \approx 0.00432433$.

Example 1. We consider the functions

$$f(t,u,v) = a_0 t^{\gamma_0}(u+v)^{\alpha_0},$$

$$g(t,u,v) = b_0(t+1)^{\delta_0}(u+v)^{\beta_0}, \ t \in [0,1], \quad u,v \ge 0,$$

where $\alpha_0 > 1$, $\beta_0 \in (0,1)$, $a_0, b_0, \gamma_0, \delta_0 > 0$. In addition, we have $m_0 = \max\{M_i, \ i = 1,\dots,4\} = M_3$. The functions $f(t,u,v)$ and $g(t,u,v)$ are nondecreasing with respect to u and v for any $t \in [0,1]$, and for $\hat{q} = 1$ and $c \in (0,1)$ the assumptions (I4) and (I6) are satisfied; indeed we obtain $f_\infty^i = \infty$ and $g_0^i = \infty$. We take $N_0 = 1$ and then $f(t,N_0,N_0) \le 2^{\alpha_0} a_0$, $g(t,N_0,N_0) \le 2^{\beta_0+\delta_0} b_0$ for all $t \in [0,1]$. If $a_0 < \frac{1}{2^{\alpha_0+2}m_0}$ and $b_0 < \frac{1}{2^{\beta_0+\delta_0+2}m_0}$, then the assumption (I7) is satisfied. For example, for $\alpha_0 = 2$, $\beta_0 = \frac{1}{2}$ and $\delta_0 = 1$, if $a_0 \le 0.37$ and $b_0 \le 0.52$, then by Theorem 3.2.3, we deduce that problem (3.77), (3.78) has at least two positive solutions.

Example 2. We consider the functions

$$f t,u,v) = \frac{(u+v)^{a_0}}{t^{\zeta_1}}, \quad g(t,u,v) = \frac{(u+v)^{b_0}}{(1-t)^{\zeta_2}}, \ t \in (0,1), \ u,v \ge 0,$$

with $a_0, b_0 > 1$ and $\zeta_1, \zeta_2 \in (0,1)$. Here, $f(t,u,v) = \tilde{p}_1(t)\tilde{q}_1(t,u,v)$, $g(t,u,v) = \tilde{p}_2(t)\tilde{q}_2(t,u,v)$, where $\tilde{p}_1(t) = \frac{1}{t^{\zeta_1}}$, $\tilde{p}_2(t) = \frac{1}{(1-t)^{\zeta_2}}$ for all $t \in (0,1)$, and $\tilde{q}_1(t,u,v) = (u+v)^{a_0}$, $\tilde{q}_2(t,u,v) = (u+v)^{b_0}$ for all $t \in [0,1]$, $u,v \ge 0$. We have $0 < \int_0^1 (1-s)^{13/6}\tilde{p}_1(s)\,ds < \infty$, $0 < \int_0^1 (1-s)^{2/3}\tilde{p}_2(s)\,ds < \infty$.

In (I3), for $a = b = 1$, we obtain $q_{10} = 0$ and $q_{20} = 0$. In (I4), for $c \in (0,1/2)$, we have $\tilde{f}_\infty^i = \infty$. Then, by Theorem 3.2.4, we deduce that problem (3.77), (3.78) has at least one positive solution.

Remark 3.2.2. The results presented in this section under the assumptions $p_1 \in [1, n-2]$ and $p_2 \in [1, m-2]$ instead of $p_1 \in [1, \alpha-1)$ and $p_2 \in [1, \beta-1)$ were published in [48]. The existence and nonexistence of positive solutions for the system (3.41) with two positive parameters, subject to the boundary conditions (3.42) were investigated in [53].

Chapter 4

Systems of Two Riemann–Liouville Fractional Differential Equations with p-Laplacian Operators, Parameters and Multi-Point Boundary Conditions

In this chapter, we investigate the existence and nonexistence of positive solutions for systems of two Riemann–Liouville fractional differential equations with p-Laplacian operators and positive parameters, subject to uncoupled or coupled multi-point boundary conditions which contain fractional derivatives, and the nonlinearities of systems are nonnegative functions.

4.1 Systems of Fractional Differential Equations with Uncoupled Multi-Point Boundary Conditions

We consider the system of nonlinear ordinary fractional differential equations with r_1-Laplacian and r_2-Laplacian operators

$$\begin{cases} D_{0+}^{\alpha_1}(\varphi_{r_1}(D_{0+}^{\beta_1}u(t))) + \lambda f(t, u(t), v(t)) = 0, & t \in (0, 1), \\ D_{0+}^{\alpha_2}(\varphi_{r_2}(D_{0+}^{\beta_2}v(t))) + \mu g(t, u(t), v(t)) = 0, & t \in (0, 1), \end{cases} \tag{4.1}$$

with the multi-point boundary conditions

$$
\begin{cases}
u^{(j)}(0) = 0, \ j = 0, \ldots, n - 2; \quad D_{0+}^{\beta_1} u(0) = 0, \\[2mm]
D_{0+}^{p_1} u(1) = \sum_{i=1}^{N} a_i D_{0+}^{q_1} u(\xi_i), \\[2mm]
v^{(j)}(0) = 0, \ j = 0, \ldots, m - 2; \quad D_{0+}^{\beta_2} v(0) = 0, \\[2mm]
D_{0+}^{p_2} v(1) = \sum_{i=1}^{M} b_i D_{0+}^{q_2} v(\eta_i),
\end{cases}
\tag{4.2}
$$

where $\alpha_1, \alpha_2 \in (0,1]$, $\beta_1 \in (n-1,n]$, $\beta_2 \in (m-1,m]$, $n, m \in \mathbb{N}$, $n, m \geq 3$, $p_1, p_2, q_1, q_2 \in \mathbb{R}$, $p_1 \in [1, \beta_1 - 1)$, $p_2 \in [1, \beta_2 - 1)$, $q_1 \in [0, p_1]$, $q_2 \in [0, p_2]$, $\xi_i, a_i \in \mathbb{R}$ for all $i = 1, \ldots, N$ ($N \in \mathbb{N}$), $0 < \xi_1 < \cdots < \xi_N \leq 1$, $\eta_i, b_i \in \mathbb{R}$ for all $i = 1, \ldots, M$ ($M \in \mathbb{N}$), $0 < \eta_1 < \cdots < \eta_M \leq 1$, $r_1, r_2 > 1$, $\varphi_{r_i}(s) = |s|^{r_i - 2} s$, $\varphi_{r_i}^{-1} = \varphi_{\varrho_i}$, $\frac{1}{r_i} + \frac{1}{\varrho_i} = 1$, $i = 1, 2$, $\lambda, \mu > 0$, $f, g \in C([0,1] \times [0, \infty) \times [0, \infty), [0, \infty))$, and D_{0+}^{k} denotes the Riemann-Liouville derivative of order k (for $k = \alpha_1, \beta_1, \alpha_2, \beta_2, p_1, q_1, p_2, q_2$).

Under some assumptions on the functions f and g, we give intervals for the positive parameters λ and μ such that positive solutions of (4.1), (4.2) exist. By a positive solution of problem (4.1), (4.2), we mean a pair of functions $(u, v) \in (C([0,1], \mathbb{R}_+))^2$, satisfying (4.1) and (4.2) with $u(t) > 0$ for all $t \in (0,1]$, or $v(t) > 0$ for all $t \in (0,1]$. The nonexistence of positive solutions for the above problem is also studied.

4.1.1   *Auxiliary results*

We consider first the nonlinear fractional differential equation

$$
D_{0+}^{\alpha_1}(\varphi_{r_1}(D_{0+}^{\beta_1} u(t))) + h(t) = 0, \quad t \in (0,1),
\tag{4.3}
$$

with the boundary conditions

$$
\begin{cases}
u^{(j)}(0) = 0, \ j = 0, \ldots, n - 2; \\[2mm]
D_{0+}^{\beta_1} u(0) = 0, \quad D_{0+}^{p_1} u(1) = \sum_{i=1}^{N} a_i D_{0+}^{q_1} u(\xi_i),
\end{cases}
\tag{4.4}
$$

where $\alpha_1 \in (0,1]$, $\beta_1 \in (n-1,n]$, $n \in \mathbb{N}$, $n \geq 3$, $p_1, q_1 \in \mathbb{R}$, $p_1 \in [1, \beta_1 - 1)$, $q_1 \in [0, p_1]$, $\xi_i, a_i \in \mathbb{R}$ for all $i = 1, \ldots, N$ ($N \in \mathbb{N}$), $0 < \xi_1 < \cdots < \xi_N \leq 1$, and $h \in C[0,1]$.

If we denote by $\varphi_{r_1}(D_{0+}^{\beta_1}u(t)) = x(t)$, then problem (4.3), (4.4) is equivalent to the following two boundary value problems:

$$D_{0+}^{\alpha_1}x(t) + h(t) = 0, \quad 0 < t < 1, \tag{4.5}$$

with the boundary condition

$$x(0) = 0, \tag{4.6}$$

and

$$D_{0+}^{\beta_1}u(t) = \varphi_{\varrho_1}x(t), \quad 0 < t < 1, \tag{4.7}$$

with the boundary conditions

$$u^{(j)}(0) = 0, \ j = 0, \ldots, n-2; \quad D_{0+}^{p_1}u(1) = \sum_{i=1}^{N} a_i D_{0+}^{q_1}u(\xi_i). \tag{4.8}$$

For the first problem (4.5), (4.6), the function

$$x(t) = -I_{0+}^{\alpha_1}h(t) = -\frac{1}{\Gamma(\alpha_1)} \int_0^t (t-s)^{\alpha_1-1}h(s)\,ds, \quad t \in [0,1] \tag{4.9}$$

is the unique solution $x \in C[0,1]$ of (4.5), (4.6).

For the second problem (4.7), (4.8), if $\Delta_1 = \frac{\Gamma(\beta_1)}{\Gamma(\beta_1-p_1)} - \frac{\Gamma(\beta_1)}{\Gamma(\beta_1-q_1)}$
$\times \sum_{i=1}^{N} a_i\xi_i^{\beta_1-q_1-1} \neq 0$, then by Lemma 2.1.2 from Section 2.1, we deduce that the function

$$u(t) = -\int_0^1 G_1(t,s)\varphi_{\varrho_1}x(s)\,ds, \quad t \in [0,1], \tag{4.10}$$

is the unique solution $u \in C[0,1]$ of (4.7), (4.8). Here, the Green function G_1 is given by

$$G_1(t,s) = g_1(t,s) + \frac{t^{\beta_1-1}}{\Delta_1}\sum_{i=1}^{N} a_i g_2(\xi_i,s), \quad (t,s) \in [0,1] \times [0,1], \tag{4.11}$$

with

$$g_1(t,s) = \frac{1}{\Gamma(\beta_1)} \begin{cases} t^{\beta_1-1}(1-s)^{\beta_1-p_1-1} - (t-s)^{\beta_1-1}, & 0 \le s \le t \le 1, \\ t^{\beta_1-1}(1-s)^{\beta_1-p_1-1}, & 0 \le t \le s \le 1, \end{cases}$$

$$g_2(t,s) = \frac{1}{\Gamma(\beta_1-q_1)} \begin{cases} t^{\beta_1-q_1-1}(1-s)^{\beta_1-p_1-1} - (t-s)^{\beta_1-q_1-1}, \\ \qquad 0 \le s \le t \le 1, \\ t^{\beta_1-q_1-1}(1-s)^{\beta_1-p_1-1}, \quad 0 \le t \le s \le 1. \end{cases}$$

$$\tag{4.12}$$

Therefore by (4.9) and (4.10) we obtain the following lemma.

Lemma 4.1.1. *If $\Delta_1 \neq 0$, then the function*

$$u(t) = \int_0^1 G_1(t,s)\varphi_{\varrho_1}(I_{0+}^{\alpha_1}h(s))\,ds, \quad t \in [0,1] \qquad (4.13)$$

is the unique solution $u \in C[0,1]$ of problem (4.3), (4.4).

Next, we consider the nonlinear fractional differential equation

$$D_{0+}^{\alpha_2}(\varphi_{r_2}(D_{0+}^{\beta_2}v(t))) + k(t) = 0, \quad t \in (0,1), \qquad (4.14)$$

with the boundary conditions

$$\begin{cases} v^{(j)}(0) = 0, \quad j = 0,\ldots,m-2; \\ D_{0+}^{\beta_2}v(0) = 0, \quad D_{0+}^{p_2}v(1) = \sum_{i=1}^{M} b_i D_{0+}^{q_2}v(\eta_i), \end{cases} \qquad (4.15)$$

where $\alpha_2 \in (0,1]$, $\beta_2 \in (m-1,m]$, $m \in \mathbb{N}$, $m \geq 3$, $p_2, q_2 \in \mathbb{R}$, $p_2 \in [1,\beta_2-1)$, $q_2 \in [0,p_2]$, $\eta_i, b_i \in \mathbb{R}$ for all $i = 1,\ldots,M$ ($M \in \mathbb{N}$), $0 < \eta_1 < \cdots < \eta_M \leq 1$, and $k \in C[0,1]$.

We denote by $\Delta_2 = \frac{\Gamma(\beta_2)}{\Gamma(\beta_2-p_2)} - \frac{\Gamma(\beta_2)}{\Gamma(\beta_2-q_2)}\sum_{i=1}^{M} b_i\eta_i^{\beta_2-q_2-1}$, and by G_2, g_3, g_4 the following functions:

$$G_2(t,s) = g_3(t,s) + \frac{t^{\beta_2-1}}{\Delta_2}\sum_{i=1}^{M} b_i g_4(\eta_i,s), \quad (t,s) \in [0,1] \times [0,1], \quad (4.16)$$

$$g_3(t,s) = \frac{1}{\Gamma(\beta_2)}\begin{cases} t^{\beta_2-1}(1-s)^{\beta_2-p_2-1} - (t-s)^{\beta_2-1}, & 0 \leq s \leq t \leq 1, \\ t^{\beta_2-1}(1-s)^{\beta_2-p_2-1}, & 0 \leq t \leq s \leq 1, \end{cases}$$

$$g_4(t,s) = \frac{1}{\Gamma(\beta_2-q_2)}\begin{cases} t^{\beta_2-q_2-1}(1-s)^{\beta_2-p_2-1} - (t-s)^{\beta_2-q_2-1}, \\ \qquad 0 \leq s \leq t \leq 1, \\ t^{\beta_2-q_2-1}(1-s)^{\beta_2-p_2-1}, & 0 \leq t \leq s \leq 1. \end{cases}$$
$$\qquad (4.17)$$

In a similar manner as above, we obtain the following result.

Lemma 4.1.2. *If $\Delta_2 \neq 0$, then the function*

$$v(t) = \int_0^1 G_2(t,s)\varphi_{\varrho_2}(I_{0+}^{\alpha_2}k(s))\,ds, \quad t \in [0,1] \qquad (4.18)$$

is the unique solution $v \in C[0,1]$ of problem (4.14), (4.15).

Based on the properties of the functions g_i, $i = 1, \ldots, 4$ given by (4.12) and (4.17) (see Lemma 2.1.3 from Section 2.1), we obtain the following properties of the Green functions G_1 and G_2 that will be used in the following sections.

Lemma 4.1.3. *Assume that* $a_i, b_j \geq 0$ *for all* $i = 1, \ldots, N$ *and* $j = 1, \ldots, M$, *and* $\Delta_1, \Delta_2 > 0$. *Then the functions* G_1, G_2 *given by* (4.11) *and* (4.16), *respectively, have the properties*

(a) $G_1, G_2 : [0,1] \times [0,1] \to [0,\infty)$ *are continuous functions;*

(b) $G_1(t,s) \leq J_1(s)$ *for all* $t, s \in [0,1]$, *where* $J_1(s) = h_1(s) + \frac{1}{\Delta_1} \sum_{i=1}^{N} a_i g_2(\xi_i, s)$, $s \in [0,1]$ *and* $h_1(s) = \frac{1}{\Gamma(\beta_1)}(1-s)^{\beta_1 - p_1 - 1}(1 - (1-s)^{p_1})$, $s \in [0,1]$;

(c) $G_1(t,s) \geq t^{\beta_1 - 1} J_1(s)$ *for all* $t, s \in [0,1]$;

(d) $G_2(t,s) \leq J_2(s)$ *for all* $t, s \in [0,1]$, *where* $J_2(s) = h_3(s) + \frac{1}{\Delta_2} \sum_{i=1}^{M} b_i g_4(\eta_i, s)$, $s \in [0,1]$ *and* $h_3(s) = \frac{1}{\Gamma(\beta_2)}(1-s)^{\beta_2 - p_2 - 1}(1 - (1-s)^{p_2})$, $s \in [0,1]$;

(e) $G_2(t,s) \geq t^{\beta_2 - 1} J_2(s)$ *for all* $t, s \in [0,1]$.

4.1.2 Existence of positive solutions

In this section, we present sufficient conditions on the functions f, g, and intervals for the parameters λ, μ such that positive solutions with respect to a cone for our problem (4.1), (4.2) exist.

We present now the assumptions that we will use in the sequel.

(H1) $\alpha_1, \alpha_2 \in (0,1]$, $\beta_1 \in (n-1, n]$, $\beta_2 \in (m-1, m]$, $n, m \in \mathbb{N}$, $n, m \geq 3$, $p_1, p_2, q_1, q_2 \in \mathbb{R}$, $p_1 \in [1, \beta_1 - 1)$, $p_2 \in [1, \beta_2 - 1)$, $q_1 \in [0, p_1]$, $q_2 \in [0, p_2]$, $\xi_i \in \mathbb{R}$, $a_i \geq 0$ for all $i = 1, \ldots, N$ ($N \in \mathbb{N}$), $0 < \xi_1 < \cdots < \xi_N \leq 1$, $\eta_i \in \mathbb{R}$, $b_i \geq 0$ for all $i = 1, \ldots, M$ ($M \in \mathbb{N}$), $0 < \eta_1 < \cdots < \eta_M \leq 1$, $\lambda, \mu > 0$, $\Delta_1 = \frac{\Gamma(\beta_1)}{\Gamma(\beta_1 - p_1)} - \frac{\Gamma(\beta_1)}{\Gamma(\beta_1 - q_1)} \sum_{i=1}^{N} a_i \xi_i^{\beta_1 - q_1 - 1} > 0$, $\Delta_2 = \frac{\Gamma(\beta_2)}{\Gamma(\beta_2 - p_2)} - \frac{\Gamma(\beta_2)}{\Gamma(\beta_2 - q_2)} \sum_{i=1}^{M} b_i \eta_i^{\beta_2 - q_2 - 1} > 0$, $r_i > 1$, $\varphi_{r_i}(s) = |s|^{r_i - 2} s$, $\varphi_{r_i}^{-1} = \varphi_{\varrho_i}$, $\varrho_i = \frac{r_i}{r_i - 1}$, $i = 1, 2$.

(H2) The functions $f, g : [0,1] \times [0,\infty) \times [0,\infty) \to [0,\infty)$ are continuous.

For $[c_1, c_2] \subset [0,1]$ with $0 < c_1 < c_2 \leq 1$, we introduce the following extreme limits:

$$f_0^s = \limsup_{\substack{u+v \to 0^+ \\ u,v \geq 0}} \max_{t \in [0,1]} \frac{f(t,u,v)}{(u+v)^{r_1 - 1}}, \qquad g_0^s = \limsup_{\substack{u+v \to 0^+ \\ u,v \geq 0}} \max_{t \in [0,1]} \frac{g(t,u,v)}{(u+v)^{r_2 - 1}},$$

$$f_0^i = \liminf_{\substack{u+v\to 0^+ \\ u,v\ge 0}} \min_{t\in[c_1,c_2]} \frac{f(t,u,v)}{(u+v)^{r_1-1}}, \quad g_0^i = \liminf_{\substack{u+v\to 0^+ \\ u,v\ge 0}} \min_{t\in[c_1,c_2]} \frac{g(t,u,v)}{(u+v)^{r_2-1}},$$

$$f_\infty^s = \limsup_{\substack{u+v\to\infty \\ u,v\ge 0}} \max_{t\in[0,1]} \frac{f(t,u,v)}{(u+v)^{r_1-1}}, \quad g_\infty^s = \limsup_{\substack{u+v\to\infty \\ u,v\ge 0}} \max_{t\in[0,1]} \frac{g(t,u,v)}{(u+v)^{r_2-1}},$$

$$f_\infty^i = \liminf_{\substack{u+v\to\infty \\ u,v\ge 0}} \min_{t\in[c_1,c_2]} \frac{f(t,u,v)}{(u+v)^{r_1-1}}, \quad g_\infty^i = \liminf_{\substack{u+v\to\infty \\ u,v\ge 0}} \min_{t\in[c_1,c_2]} \frac{g(t,u,v)}{(u+v)^{r_2-1}}.$$

By using Lemmas 4.1.1 and 4.1.2 (relations (4.13) and (4.18)), (u,v) is a solution of the following nonlinear system of integral equations

$$\begin{cases} u(t) = \lambda^{\varrho_1-1} \displaystyle\int_0^1 G_1(t,s)\varphi_{\varrho_1}(I_{0+}^{\alpha_1}f(s,u(s),v(s)))\,ds, & t\in[0,1], \\[2mm] v(t) = \mu^{\varrho_2-1} \displaystyle\int_0^1 G_2(t,s)\varphi_{\varrho_2}(I_{0+}^{\alpha_2}g(s,u(s),v(s)))\,ds, & t\in[0,1], \end{cases}$$

if and only if (u,v) is solution of problem (4.1), (4.2).

We consider the Banach space $X = C[0,1]$ with the supremum norm $\|\cdot\|$, and the Banach space $Y = X\times X$ with the norm $\|(u,v)\|_Y = \|u\|+\|v\|$. We define the cones

$$P_1 = \{u\in X,\ u(t)\ge t^{\beta_1-1}\|u\|,\ \forall t\in[0,1]\} \subset X,$$

$$P_2 = \{v\in X,\ v(t)\ge t^{\beta_2-1}\|v\|,\ \forall t\in[0,1]\} \subset X,$$

and $P = P_1 \times P_2 \subset Y$.

We define now the operators $Q_1, Q_2 : Y \to X$ and $Q : Y \to Y$ by

$$Q_1(u,v)(t) = \lambda^{\varrho_1-1} \int_0^1 G_1(t,s)\varphi_{\varrho_1}(I_{0+}^{\alpha_1}f(s,u(s),v(s)))\,ds, \quad t\in[0,1],$$

$$Q_2(u,v)(t) = \mu^{\varrho_2-1} \int_0^1 G_2(t,s)\varphi_{\varrho_2}(I_{0+}^{\alpha_2}g(s,u(s),v(s)))\,ds, \quad t\in[0,1],$$

and $Q(u,v) = (Q_1(u,v), Q_2(u,v))$, $(u,v)\in Y$. Then (u,v) is a solution of problem (4.1), (4.2) if and only if (u,v) is a fixed point of operator Q.

Lemma 4.1.4. *If (H1)–(H2) hold, then $Q : P \to P$ is a completely continuous operator.*

Proof. Let $(u,v)\in P$ be an arbitrary element. Because $Q_1(u,v)$ and $Q_2(u,v)$ satisfy problem (4.3), (4.4) for $h(t) = \lambda f(t,u(t),v(t))$, $t\in[0,1]$, and problem (4.14), (4.15) for $k(t) = \mu g(t, u(t), v(t))$, $t \in [0, 1]$, respectively, then we obtain

$$Q_1(u, v)(t) \leq \lambda^{\varrho_1 - 1} \int_0^1 J_1(s) \varphi_{\varrho_1}(I_{0+}^{\alpha_1} f(s, u(s), v(s))) \, ds, \quad t \in [0, 1],$$

$$Q_2(u, v)(t) \leq \mu^{\varrho_2 - 1} \int_0^1 J_2(s) \varphi_{\varrho_2}(I_{0+}^{\alpha_2} g(s, u(s), v(s))) \, ds, \quad t \in [0, 1],$$

and so

$$\|Q_1(u, v)\| \leq \lambda^{\varrho_1 - 1} \int_0^1 J_1(s) \varphi_{\varrho_1}(I_{0+}^{\alpha_1} f(s, u(s), v(s))) \, ds,$$

$$\|Q_2(u, v)\| \leq \mu^{\varrho_2 - 1} \int_0^1 J_2(s) \varphi_{\varrho_2}(I_{0+}^{\alpha_2} g(s, u(s), v(s))) \, ds.$$

Therefore, we conclude that

$$Q_1(u, v)(t) \geq \lambda^{\varrho_1 - 1} \int_0^1 t^{\beta_1 - 1} J_1(s) \varphi_{\varrho_1}(I_{0+}^{\alpha_1} f(s, u(s), v(s))) \, ds$$

$$\geq t^{\beta_1 - 1} \|Q_1(u, v)\|, \quad \forall t \in [0, 1],$$

$$Q_2(u, v)(t) \geq \mu^{\varrho_2 - 1} \int_0^1 t^{\beta_2 - 1} J_2(s) \varphi_{\varrho_2}(I_{0+}^{\alpha_2} g(s, u(s), v(s))) \, ds$$

$$\geq t^{\beta_2 - 1} \|Q_2(u, v)\|, \quad \forall t \in [0, 1].$$

Hence, $Q(u, v) = (Q_1(u, v), Q_2(u, v)) \in P$, and then $Q(P) \subset P$. By the continuity of the functions f, g, G_1, G_2 and the Arzela–Ascoli theorem, we can show that Q_1 and Q_2 are completely continuous operators, and then Q is a completely continuous operator. $\qquad\square$

For $[c_1, c_2] \subset [0, 1]$ with $0 < c_1 < c_2 \leq 1$, we denote by

$$A = \frac{1}{(\Gamma(\alpha_1 + 1))^{\varrho_1 - 1}} \int_{c_1}^{c_2} (s - c_1)^{\alpha_1(\varrho_1 - 1)} J_1(s) \, ds,$$

$$B = \frac{1}{(\Gamma(\alpha_1 + 1))^{\varrho_1 - 1}} \int_0^1 s^{\alpha_1(\varrho_1 - 1)} J_1(s) \, ds,$$

$$C = \frac{1}{(\Gamma(\alpha_2 + 1))^{\varrho_2 - 1}} \int_{c_1}^{c_2} (s - c_1)^{\alpha_2(\varrho_2 - 1)} J_2(s) \, ds,$$

$$D = \frac{1}{(\Gamma(\alpha_2 + 1))^{\varrho_2 - 1}} \int_0^1 s^{\alpha_2(\varrho_2 - 1)} J_2(s) \, ds,$$

where J_1 and J_2 are defined in Lemma 4.1.3.

First, for $f_0^s, g_0^s, f_\infty^i, g_\infty^i \in (0, \infty)$ and numbers $\alpha_1', \alpha_2' \geq 0$, $\widetilde{\alpha}_1, \widetilde{\alpha}_2 > 0$ such that $\alpha_1' + \alpha_2' = 1$ and $\widetilde{\alpha}_1 + \widetilde{\alpha}_2 = 1$, we define the numbers $L_1, L_2, L_3, L_4, L_2', L_4'$ by

$$L_1 = \frac{1}{f_\infty^i}\left(\frac{\alpha_1'}{\gamma\gamma_1 A}\right)^{r_1 - 1}, \quad L_2 = \frac{1}{f_0^s}\left(\frac{\widetilde{\alpha}_1}{B}\right)^{r_1 - 1}, \quad L_3 = \frac{1}{g_\infty^i}\left(\frac{\alpha_2'}{\gamma\gamma_2 C}\right)^{r_2 - 1},$$

$$L_4 = \frac{1}{g_0^s}\left(\frac{\widetilde{\alpha}_2}{D}\right)^{r_2 - 1}, \quad L_2' = \frac{1}{f_0^s B^{r_1 - 1}}, \quad L_4' = \frac{1}{g_0^s D^{r_2 - 1}},$$

where $\gamma_1 = c_1^{\beta_1 - 1}$, $\gamma_2 = c_1^{\beta_2 - 1}$, $\gamma = \min\{\gamma_1, \gamma_2\}$.

Theorem 4.1.1. *Assume that* (H1) *and* (H2) *hold,* $[c_1, c_2] \subset [0, 1]$ *with* $0 < c_1 < c_2 \leq 1$, $\alpha_1', \alpha_2' \geq 0$, $\widetilde{\alpha}_1, \widetilde{\alpha}_2 > 0$ *such that* $\alpha_1' + \alpha_2' = 1$, $\widetilde{\alpha}_1 + \widetilde{\alpha}_2 = 1$.

(1) *If* $f_0^s, g_0^s, f_\infty^i, g_\infty^i \in (0, \infty)$, $L_1 < L_2$ *and* $L_3 < L_4$, *then for each* $\lambda \in (L_1, L_2)$ *and* $\mu \in (L_3, L_4)$ *there exists a positive solution* $(u(t), v(t))$, $t \in [0, 1]$ *for* (4.1), (4.2).

(2) *If* $f_0^s = 0$, $g_0^s, f_\infty^i, g_\infty^i \in (0, \infty)$ *and* $L_3 < L_4'$, *then for each* $\lambda \in (L_1, \infty)$ *and* $\mu \in (L_3, L_4')$ *there exists a positive solution* $(u(t), v(t))$, $t \in [0, 1]$ *for* (4.1), (4.2).

(3) *If* $g_0^s = 0$, $f_0^s, f_\infty^i, g_\infty^i \in (0, \infty)$ *and* $L_1 < L_2'$, *then for each* $\lambda \in (L_1, L_2')$ *and* $\mu \in (L_3, \infty)$ *there exists a positive solution* $(u(t), v(t))$, $t \in [0, 1]$ *for* (4.1), (4.2).

(4) *If* $f_0^s = g_0^s = 0$, $f_\infty^i, g_\infty^i \in (0, \infty)$, *then for each* $\lambda \in (L_1, \infty)$ *and* $\mu \in (L_3, \infty)$ *there exists a positive solution* $(u(t), v(t))$, $t \in [0, 1]$ *for* (4.1), (4.2).

(5) *If* $f_0^s, g_0^s \in (0, \infty)$ *and at least one of* f_∞^i, g_∞^i *is* ∞, *then for each* $\lambda \in (0, L_2)$ *and* $\mu \in (0, L_4)$ *there exists a positive solution* $(u(t), v(t))$, $t \in [0, 1]$ *for* (4.1), (4.2).

(6) *If* $f_0^s = 0$, $g_0^s \in (0, \infty)$ *and at least one of* f_∞^i, g_∞^i *is* ∞, *then for each* $\lambda \in (0, \infty)$ *and* $\mu \in (0, L_4')$ *there exists a positive solution* $(u(t), v(t))$, $t \in [0, 1]$ *for* (4.1), (4.2).

(7) *If* $f_0^s \in (0, \infty)$, $g_0^s = 0$ *and at least one of* f_∞^i, g_∞^i *is* ∞, *then for each* $\lambda \in (0, L_2')$ *and* $\mu \in (0, \infty)$ *there exists a positive solution* $(u(t), v(t))$, $t \in [0, 1]$ *for* (4.1), (4.2).

(8) *If* $f_0^s = g_0^s = 0$ *and at least one of* f_∞^i, g_∞^i *is* ∞, *then for each* $\lambda \in (0, \infty)$ *and* $\mu \in (0, \infty)$ *there exists a positive solution* $(u(t), v(t))$, $t \in [0, 1]$ *for* (4.1), (4.2).

Proof. We consider the above cone $P \subset Y$ and the operators Q_1, Q_2 and Q. Because the proofs of the above cases are similar, in what follows, we will prove some representative cases.

Case (1). We have $f_0^s, g_0^s, f_\infty^i, g_\infty^i \in (0, \infty)$, $L_1 < L_2$ and $L_3 < L_4$. Let $\lambda \in (L_1, L_2)$ and $\mu \in (L_3, L_4)$. We consider $\varepsilon > 0$ such that $\varepsilon < f_\infty^i$, $\varepsilon < g_\infty^i$ and

$$\frac{1}{f_\infty^i - \varepsilon} \left(\frac{\alpha_1'}{\gamma \gamma_1 A} \right)^{r_1 - 1} \le \lambda \le \frac{1}{f_0^s + \varepsilon} \left(\frac{\tilde{\alpha}_1}{B} \right)^{r_1 - 1},$$

$$\frac{1}{g_\infty^i - \varepsilon} \left(\frac{\alpha_2'}{\gamma \gamma_2 C} \right)^{r_2 - 1} \le \mu \le \frac{1}{g_0^s + \varepsilon} \left(\frac{\tilde{\alpha}_2}{D} \right)^{r_2 - 1}.$$

By using (H2) and the definitions of f_0^s and g_0^s, we deduce that there exists $R_1 > 0$ such that

$$f(t, u, v) \le (f_0^s + \varepsilon)(u + v)^{r_1 - 1}, \quad g(t, u, v) \le (g_0^s + \varepsilon)(u + v)^{r_2 - 1},$$

for all $t \in [0, 1]$ and $u, v \ge 0$, $u + v \le R_1$.

We define the set $\Omega_1 = \{(u, v) \in Y, \|(u, v)\|_Y < R_1\}$. Now, let $(u, v) \in P \cap \partial\Omega_1$, that is $(u, v) \in P$ with $\|(u, v)\|_Y = R_1$, or equivalently $\|u\| + \|v\| = R_1$. Then $u(t) + v(t) \le R_1$ for all $t \in [0, 1]$ and by Lemma 4.1.3, we obtain

$$Q_1(u, v)(t) \le \lambda^{\varrho_1 - 1} \int_0^1 J_1(s) \varphi_{\varrho_1} \left(\frac{1}{\Gamma(\alpha_1)} \int_0^s (s - \tau)^{\alpha_1 - 1} \right.$$

$$\times f(\tau, u(\tau), v(\tau)) \, d\tau \bigg) ds$$

$$\le \lambda^{\varrho_1 - 1} \int_0^1 J_1(s) \varphi_{\varrho_1} \left(\frac{1}{\Gamma(\alpha_1)} \int_0^s (s - \tau)^{\alpha_1 - 1} (f_0^s + \varepsilon) \right.$$

$$\times (u(\tau) + v(\tau))^{r_1 - 1} \, d\tau \bigg) ds$$

$$\le \lambda^{\varrho_1 - 1} (f_0^s + \varepsilon)^{\varrho_1 - 1} \int_0^1 J_1(s) \varphi_{\varrho_1} \left(\frac{1}{\Gamma(\alpha_1)} \int_0^s (s - \tau)^{\alpha_1 - 1} \right.$$

$$\times (\|u\| + \|v\|)^{r_1 - 1} \, d\tau \bigg) ds$$

$$= \lambda^{\varrho_1 - 1}(f_0^s + \varepsilon)^{\varrho_1 - 1}(\|u\| + \|v\|) \int_0^1 J_1(s)$$

$$\times \varphi_{\varrho_1} \left(\frac{1}{\Gamma(\alpha_1)} \int_0^s (s - \tau)^{\alpha_1 - 1} \, d\tau \right) ds$$

$$= \lambda^{\varrho_1 - 1}(f_0^s + \varepsilon)^{\varrho_1 - 1} \|(u,v)\|_Y \int_0^1 J_1(s)\varphi_{\varrho_1} \left(\frac{s^{\alpha_1}}{\alpha_1 \Gamma(\alpha_1)} \right) ds$$

$$= \lambda^{\varrho_1 - 1}(f_0^s + \varepsilon)^{\varrho_1 - 1} \|(u,v)\|_Y$$

$$\times \int_0^1 J_1(s) \frac{1}{(\Gamma(\alpha_1 + 1))^{\varrho_1 - 1}} s^{\alpha_1(\varrho_1 - 1)} \, ds$$

$$= \lambda^{\varrho_1 - 1}(f_0^s + \varepsilon)^{\varrho_1 - 1} B \|(u,v)\|_Y \le \tilde{\alpha}_1 \|(u,v)\|_Y, \quad \forall t \in [0,1].$$

Therefore, we have $\|Q_1(u,v)\| \le \tilde{\alpha}_1 \|(u,v)\|_Y$.

In a similar manner, we conclude

$$Q_2(u,v)(t) \le \mu^{\varrho_2 - 1} \int_0^1 J_2(s)\varphi_{\varrho_2} \left(\frac{1}{\Gamma(\alpha_2)} \int_0^s (s - \tau)^{\alpha_2 - 1} \right.$$

$$\times g(\tau, u(\tau), v(\tau)) \, d\tau \Bigg) ds$$

$$\le \mu^{\varrho_2 - 1} \int_0^1 J_2(s)\varphi_{\varrho_2} \left(\frac{1}{\Gamma(\alpha_2)} \int_0^s (s - \tau)^{\alpha_2 - 1} (g_0^s + \varepsilon) \right.$$

$$\times (u(\tau) + v(\tau))^{r_2 - 1} \, d\tau \Bigg) ds$$

$$\le \mu^{\varrho_2 - 1}(g_0^s + \varepsilon)^{\varrho_2 - 1} \int_0^1 J_2(s)\varphi_{\varrho_2} \left(\frac{1}{\Gamma(\alpha_2)} \int_0^s (s - \tau)^{\alpha_2 - 1} \right.$$

$$\times (\|u\| + \|v\|)^{r_2 - 1} \, d\tau \Bigg) ds$$

$$= \mu^{\varrho_2 - 1}(g_0^s + \varepsilon)^{\varrho_2 - 1}(\|u\| + \|v\|) \int_0^1 J_2(s)$$

$$\times \varphi_{\varrho_2} \left(\frac{1}{\Gamma(\alpha_2)} \int_0^s (s - \tau)^{\alpha_2 - 1} \, d\tau \right) ds$$

$$= \mu^{\varrho_2 - 1}(g_0^s + \varepsilon)^{\varrho_2 - 1} \|(u,v)\|_Y \int_0^1 J_2(s)\varphi_{\varrho_2} \left(\frac{s^{\alpha_2}}{\alpha_2 \Gamma(\alpha_2)} \right) ds$$

$$= \mu^{\varrho_2 - 1} (g_0^s + \varepsilon)^{\varrho_2 - 1} \|(u, v)\|_Y$$

$$\times \int_0^1 J_2(s) \frac{1}{(\Gamma(\alpha_2 + 1))^{\varrho_2 - 1}} s^{\alpha_2(\varrho_2 - 1)} \, ds$$

$$= \mu^{\varrho_2 - 1} (g_0^s + \varepsilon)^{\varrho_2 - 1} D \|(u, v)\|_Y \leq \widetilde{\alpha}_2 \|(u, v)\|_Y, \quad \forall t \in [0, 1].$$

Hence, we get $\|Q_2(u, v)\| \leq \widetilde{\alpha}_2 \|(u, v)\|_Y$.

Therefore, for $(u, v) \in P \cap \partial\Omega_1$, we deduce

$$\|Q(u, v)\|_Y = \|Q_1(u, v)\| + \|Q_2(u, v)\|$$

$$\leq \widetilde{\alpha}_1 \|(u, v)\|_Y + \widetilde{\alpha}_2 \|(u, v)\|_Y = \|(u, v)\|_Y. \tag{4.19}$$

Next, by the definitions of f_∞^i and g_∞^i there exists $\overline{R}_2 > 0$ such that

$$f(t, u, v) \geq (f_\infty^i - \varepsilon)(u + v)^{r_1 - 1}, \quad g(t, u, v) \geq (g_\infty^i - \varepsilon)(u + v)^{r_2 - 1},$$

for all $t \in [c_1, c_2]$ and $u, v \geq 0$, $u + v \geq \overline{R}_2$.

We consider $R_2 = \max\{2R_1, \overline{R}_2/\gamma\}$ and we define the set $\Omega_2 = \{(u, v) \in Y, \|(u, v)\|_Y < R_2\}$. Then for $(u, v) \in P \cap \partial\Omega_2$, we obtain

$$u(t) + v(t) \geq \min_{t \in [c_1, c_2]} t^{\beta_1 - 1} \|u\| + \min_{t \in [c_1, c_2]} t^{\beta_2 - 1} \|v\| = c_1^{\beta_1 - 1} \|u\| + c_1^{\beta_2 - 1} \|v\|$$

$$= \gamma_1 \|u\| + \gamma_2 \|v\| \geq \gamma \|(u, v)\|_Y = \gamma R_2 \geq \overline{R}_2, \quad \forall t \in [c_1, c_2].$$

Then, by Lemma 4.1.3, we conclude

$$Q_1(u, v)(c_1) \geq \lambda^{\varrho_1 - 1} \int_0^1 c_1^{\beta_1 - 1} J_1(s) \varphi_{\varrho_1}(I_{0+}^{\alpha_1} f(s, u(s), v(s))) \, ds$$

$$\geq \lambda^{\varrho_1 - 1} c_1^{\beta_1 - 1} \int_{c_1}^{c_2} J_1(s) \varphi_{\varrho_1} \left(\frac{1}{\Gamma(\alpha_1)} \int_{c_1}^s (s - \tau)^{\alpha_1 - 1} \right.$$

$$\times f(\tau, u(\tau), v(\tau)) \, d\tau \bigg) \, ds$$

$$\geq \lambda^{\varrho_1 - 1} c_1^{\beta_1 - 1} \int_{c_1}^{c_2} J_1(s) \varphi_{\varrho_1} \left(\frac{1}{\Gamma(\alpha_1)} \int_{c_1}^s (s - \tau)^{\alpha_1 - 1} \right.$$

$$\times (f_\infty^i - \varepsilon)(u(\tau) + v(\tau))^{r_1 - 1} \, d\tau \bigg) \, ds$$

$$\geq \lambda^{\varrho_1-1}c_1^{\beta_1-1}\int_{c_1}^{c_2} J_1(s)\varphi_{\varrho_1}\left(\frac{1}{\Gamma(\alpha_1)}\int_{c_1}^{s}(s-\tau)^{\alpha_1-1}\right.$$

$$\times (f_\infty^i - \varepsilon)(\gamma\|(u,v)\|_Y)^{r_1-1}\,d\tau\Big)\,ds$$

$$= \lambda^{\varrho_1-1}c_1^{\beta_1-1}(f_\infty^i - \varepsilon)^{\varrho_1-1}\gamma\|(u,v)\|_Y$$

$$\times \int_{c_1}^{c_2} J_1(s)\varphi_{\varrho_1}\left(\frac{1}{\Gamma(\alpha_1)}\int_{c_1}^{s}(s-\tau)^{\alpha_1-1}\,d\tau\right)\,ds$$

$$= \lambda^{\varrho_1-1}c_1^{\beta_1-1}(f_\infty^i - \varepsilon)^{\varrho_1-1}\gamma\|(u,v)\|_Y$$

$$\times \int_{c_1}^{c_2} J_1(s)\varphi_{\varrho_1}\left(\frac{(s-c_1)^{\alpha_1}}{\alpha_1\Gamma(\alpha_1)}\right)\,ds$$

$$= \gamma\gamma_1\lambda^{\varrho_1-1}(f_\infty^i - \varepsilon)^{\varrho_1-1}\|(u,v)\|_Y$$

$$\times \int_{c_1}^{c_2} J_1(s)\frac{1}{(\Gamma(\alpha_1+1))^{\varrho_1-1}}(s-c_1)^{\alpha_1(\varrho_1-1)}\,ds$$

$$= \gamma\gamma_1\lambda^{\varrho_1-1}(f_\infty^i - \varepsilon)^{\varrho_1-1}A\|(u,v)\|_Y \geq \alpha_1'\|(u,v)\|_Y.$$

Therefore, we obtain $\|Q_1(u,v)\| \geq Q_1(u,v)(c_1) \geq \alpha_1'\|(u,v)\|_Y$.
In a similar manner, we deduce

$$Q_2(u,v)(c_1) \geq \mu^{\varrho_2-1}\int_0^1 c_1^{\beta_2-1}J_2(s)\varphi_{\varrho_2}(I_{0+}^{\alpha_2}g(s,u(s),v(s)))\,ds$$

$$\geq \mu^{\varrho_2-1}c_1^{\beta_2-1}\int_{c_1}^{c_2} J_2(s)\varphi_{\varrho_2}\left(\frac{1}{\Gamma(\alpha_2)}\int_{c_1}^{s}(s-\tau)^{\alpha_2-1}\right.$$

$$\times g(\tau,u(\tau),v(\tau))\,d\tau\Big)\,ds$$

$$\geq \mu^{\varrho_2-1}c_1^{\beta_2-1}\int_{c_1}^{c_2} J_2(s)\varphi_{\varrho_2}\left(\frac{1}{\Gamma(\alpha_2)}\int_{c_1}^{s}(s-\tau)^{\alpha_2-1}\right.$$

$$\times (g_\infty^i - \varepsilon)(u(\tau)+v(\tau))^{r_2-1}\,d\tau\Big)\,ds$$

$$\geq \mu^{\varrho_2-1}c_1^{\beta_2-1}\int_{c_1}^{c_2} J_2(s)\varphi_{\varrho_2}\left(\frac{1}{\Gamma(\alpha_2)}\int_{c_1}^{s}(s-\tau)^{\alpha_2-1}\right.$$

$$\times (g_\infty^i - \varepsilon)(\gamma\|(u,v)\|_Y)^{r_2-1}\,d\tau\Big)\,ds$$

$$= \mu^{\varrho_2 - 1} c_1^{\beta_2 - 1} (g_\infty^i - \varepsilon)^{\varrho_2 - 1} \gamma \|(u, v)\|_Y$$

$$\times \int_{c_1}^{c_2} J_2(s) \varphi_{\varrho_2} \left(\frac{1}{\Gamma(\alpha_2)} \int_{c_1}^{s} (s - \tau)^{\alpha_2 - 1} d\tau \right) ds$$

$$= \mu^{\varrho_2 - 1} c_1^{\beta_2 - 1} (g_\infty^i - \varepsilon)^{\varrho_2 - 1} \gamma \|(u, v)\|_Y$$

$$\times \int_{c_1}^{c_2} J_2(s) \varphi_{\varrho_2} \left(\frac{(s - c_1)^{\alpha_2}}{\alpha_2 \Gamma(\alpha_2)} \right) ds$$

$$= \gamma \gamma_2 \mu^{\varrho_2 - 1} (g_\infty^i - \varepsilon)^{\varrho_2 - 1} \|(u, v)\|_Y$$

$$\times \int_{c_1}^{c_2} J_2(s) \frac{1}{(\Gamma(\alpha_2 + 1))^{\varrho_2 - 1}} (s - c_1)^{\alpha_2 (\varrho_2 - 1)} ds$$

$$= \gamma \gamma_2 \mu^{\varrho_2 - 1} (g_\infty^i - \varepsilon)^{\varrho_2 - 1} C \|(u, v)\|_Y \geq \alpha_2' \|(u, v)\|_Y.$$

Hence, we get $\|Q_2(u, v)\| \geq Q_2(u, v)(c_1) \geq \alpha_2' \|(u, v)\|_Y$.
Then for $(u, v) \in P \cap \partial\Omega_2$, we obtain

$$\|Q(u, v)\|_Y = \|Q_1(u, v)\| + \|Q_2(u, v)\| \geq (\alpha_1' + \alpha_2') \|(u, v)\|_Y = \|(u, v)\|_Y.$$
$$\text{(4.20)}$$

By using Lemma 4.1.4, (4.19), (4.20) and the Guo–Krasnosel'skii fixed point theorem (Theorem 1.2.2), we conclude that the operator Q has a fixed point $(u, v) \in P \cap (\overline{\Omega}_2 \setminus \Omega_1)$, so $u(t) \geq t^{\beta_1 - 1} \|u\|$, $v(t) \geq t^{\beta_2 - 1} \|v\|$ for all $t \in [0, 1]$ and $R_1 \leq \|u\| + \|v\| \leq R_2$. If $\|u\| > 0$ then $u(t) > 0$ for all $t \in (0, 1]$, and if $\|v\| > 0$ then $v(t) > 0$ for all $t \in (0, 1]$.

Case (3). We have $g_0^s = 0$, $f_0^s, f_\infty^i, g_\infty^i \in (0, \infty)$ and $L_1 < L_2'$. Let $\lambda \in (L_1, L_2')$ and $\mu \in (L_3, \infty)$. Instead of the numbers $\widetilde{\alpha}_1$ and $\widetilde{\alpha}_2$ used in the first case, we choose $\widetilde{\alpha}_1'$ such that $\widetilde{\alpha}_1' \in (B(\lambda f_0^s)^{\varrho_1 - 1}, 1)$ and $\widetilde{\alpha}_2' = 1 - \widetilde{\alpha}_1'$. The choice of the $\widetilde{\alpha}_1'$ is possible because $\lambda < 1/(f_0^s B^{r_1 - 1})$. Then let $\varepsilon > 0$ with $\varepsilon < f_\infty^i$, $\varepsilon < g_\infty^i$ and

$$\frac{1}{f_\infty^i - \varepsilon} \left(\frac{\alpha_1'}{\gamma \gamma_1 A} \right)^{r_1 - 1} \leq \lambda \leq \frac{1}{f_0^s + \varepsilon} \left(\frac{\widetilde{\alpha}_1'}{B} \right)^{r_1 - 1},$$

$$\frac{1}{g_\infty^i - \varepsilon} \left(\frac{\alpha_2'}{\gamma \gamma_2 C} \right)^{r_2 - 1} \leq \mu \leq \frac{1}{\varepsilon} \left(\frac{\widetilde{\alpha}_2'}{D} \right)^{r_2 - 1}.$$

By using (H2) and the definitions of f_0^s and g_0^s, we deduce that there exists $R_1 > 0$ such that

$$f(t, u, v) \leq (f_0^s + \varepsilon)(u + v)^{r_1 - 1}, \quad g(t, u, v) \leq \varepsilon (u + v)^{r_2 - 1},$$

for all $t \in [0, 1]$ and $u, v \geq 0$, $u + v \leq R_1$.

We define the set $\Omega_1 = \{(u,v) \in Y, \|(u,v)\|_Y < R_1\}$. In a similar manner as in the proof of Case (1), for any $(u,v) \in P \cap \partial\Omega_1$, we obtain

$$Q_1(u,v)(t) \le \lambda^{\varrho_1-1}(f_0^s + \varepsilon)^{\varrho_1-1}B\|(u,v)\|_Y \le \tilde{\alpha}_1'\|(u,v)\|_Y, \quad \forall t \in [0,1],$$

$$Q_2(u,v)(t) \le \mu^{\varrho_2-1}\varepsilon^{\varrho_2-1}D\|(u,v)\|_Y \le \tilde{\alpha}_2'\|(u,v)\|_Y, \quad \forall t \in [0,1],$$

and so $\|Q(u,v)\|_Y \le (\tilde{\alpha}_1' + \tilde{\alpha}_2')\|(u,v)\|_Y = \|(u,v)\|_Y$.

The second part of the proof is the same as the corresponding one from Case (1). For Ω_2 defined in Case (1) and for any $(u,v) \in P \cap \partial\Omega_2$, we conclude

$$Q_1(u,v)(c_1) \ge \gamma\gamma_1\lambda^{\varrho_1-1}(f_\infty^i - \varepsilon)^{\varrho_1-1}A\|(u,v)\|_Y \ge \alpha_1'\|(u,v)\|_Y,$$

$$Q_2(u,v)(c_1) \ge \gamma\gamma_2\mu^{\varrho_2-1}(g_\infty^i - \varepsilon)^{\varrho_2-1}C\|(u,v)\|_Y \ge \alpha_2'\|(u,v)\|_Y,$$

and then $\|Q(u,v)\|_Y \ge (\alpha_1' + \alpha_2')\|(u,v)\|_Y = \|(u,v)\|_Y$.

Therefore, we deduce the conclusion of the theorem.

Case (6). We consider here $f_0^s = 0$, $g_0^s \in (0,\infty)$ and $f_\infty^i = \infty$. Let $\lambda \in (0,\infty)$ and $\mu \in (0, L_4')$. Instead of the numbers $\tilde{\alpha}_1$ and $\tilde{\alpha}_2$ used in the first case, we choose $\tilde{\alpha}_2'$ such that $\tilde{\alpha}_2' \in (D(\mu g_0^s)^{\varrho_2-1}, 1)$ and $\tilde{\alpha}_1' = 1 - \tilde{\alpha}_2'$. The choice of the $\tilde{\alpha}_2'$ is possible because $\mu < 1/(g_0^s D^{r_2-1})$. Then let $\varepsilon > 0$ such that

$$\varepsilon\left(\frac{1}{\gamma\gamma_1 A}\right)^{r_1-1} \le \lambda \le \frac{1}{\varepsilon}\left(\frac{\tilde{\alpha}_1'}{B}\right)^{r_1-1}, \quad \mu \le \frac{1}{g_0^s + \varepsilon}\left(\frac{\tilde{\alpha}_2'}{D}\right)^{r_2-1}.$$

By using (H2) and the definitions of f_0^s and g_0^s, we deduce that there exists $R_1 > 0$ such that

$$f(t,u,v) \le \varepsilon(u+v)^{r_1-1}, \quad g(t,u,v) \le (g_0^s + \varepsilon)(u+v)^{r_2-1},$$

for all $t \in [0,1]$ and $u,v \ge 0$, $u + v \le R_1$.

We define the set $\Omega_1 = \{(u,v) \in Y, \|(u,v)\|_Y < R_1\}$. In a similar manner as in the proof of Case (1), for any $(u,v) \in P \cap \partial\Omega_1$ we obtain

$$Q_1(u,v)(t) \le \lambda^{\varrho_1-1}\varepsilon^{\varrho_1-1}B\|(u,v)\|_Y \le \tilde{\alpha}_1'\|(u,v)\|_Y, \quad \forall t \in [0,1],$$

$$Q_2(u,v)(t) \le \mu^{\varrho_2-1}(g_0^s + \varepsilon)^{\varrho_2-1}D\|(u,v)\|_Y \le \tilde{\alpha}_2'\|(u,v)\|_Y, \quad \forall t \in [0,1],$$

and so $\|Q(u,v)\|_Y \le (\tilde{\alpha}_1' + \tilde{\alpha}_2')\|(u,v)\|_Y = \|(u,v)\|_Y$.

For the second part of the proof, by the definition of f_∞^i, there exists $\overline{R}_2 > 0$ such that

$$f(t, u, v) \geq \frac{1}{\varepsilon}(u + v)^{r_1 - 1}, \quad \forall t \in [c_1, c_2], \quad u, v \geq 0, \quad u + v \geq \overline{R}_2.$$

We consider $R_2 = \max\{2R_1, \overline{R}_2/\gamma\}$ and we define $\Omega_2 = \{(u, v) \in Y, \|(u, v)\|_Y < R_2\}$. Then for $(u, v) \in P \cap \partial\Omega_2$ we deduce as in Case (1) that $u(t) + v(t) \geq \gamma R_2 \geq \overline{R}_2$ for all $t \in [c_1, c_2]$.

Then by Lemma 4.1.3, we have

$$Q_1(u, v)(c_1) \geq \lambda^{\varrho_1 - 1} \int_0^1 c_1^{\beta_1 - 1} J_1(s)\varphi_{\varrho_1}(I_{0+}^{\alpha_1} f(s, u(s), v(s)))\, ds$$

$$\geq \lambda^{\varrho_1 - 1} c_1^{\beta_1 - 1} \int_{c_1}^{c_2} J_1(s)\varphi_{\varrho_1}\left(\frac{1}{\Gamma(\alpha_1)} \int_{c_1}^s (s - \tau)^{\alpha_1 - 1}\right.$$
$$\times f(\tau, u(\tau), v(\tau))\, d\tau\Bigg)\, ds$$

$$\geq \lambda^{\varrho_1 - 1} c_1^{\beta_1 - 1} \int_{c_1}^{c_2} J_1(s)\varphi_{\varrho_1}\left(\frac{1}{\Gamma(\alpha_1)} \int_{c_1}^s (s - \tau)^{\alpha_1 - 1}\right.$$
$$\times \frac{1}{\varepsilon}(u(\tau) + v(\tau))^{r_1 - 1}\, d\tau\Bigg)\, ds$$

$$\geq \lambda^{\varrho_1 - 1} c_1^{\beta_1 - 1} \int_{c_1}^{c_2} J_1(s)\varphi_{\varrho_1}\left(\frac{1}{\Gamma(\alpha_1)} \int_{c_1}^s (s - \tau)^{\alpha_1 - 1}\right.$$
$$\times \frac{1}{\varepsilon}(\gamma\|(u, v)\|_Y)^{r_1 - 1}\, d\tau\Bigg)\, ds$$

$$= \lambda^{\varrho_1 - 1} c_1^{\beta_1 - 1}\left(\frac{1}{\varepsilon}\right)^{\varrho_1 - 1} \gamma\|(u, v)\|_Y$$
$$\times \int_{c_1}^{c_2} J_1(s)\varphi_{\varrho_1}\left(\frac{(s - c_1)^{\alpha_1}}{\Gamma(\alpha_1 + 1)}\right)\, ds$$

$$= \lambda^{\varrho_1 - 1} c_1^{\beta_1 - 1}\left(\frac{1}{\varepsilon}\right)^{\varrho_1 - 1} \gamma\|(u, v)\|_Y$$
$$\times \int_{c_1}^{c_2} J_1(s)\frac{1}{(\Gamma(\alpha_1 + 1))^{\varrho_1 - 1}}(s - c_1)^{\alpha_1(\varrho_1 - 1)}\, ds$$

$$= \gamma\gamma_1 \lambda^{\varrho_1 - 1}\left(\frac{1}{\varepsilon}\right)^{\varrho_1 - 1} A\|(u, v)\|_Y \geq \|(u, v)\|_Y.$$

So, we conclude that $\|Q_1(u,v)\| \geq Q_1(u,v)(c_1) \geq \|(u,v)\|_Y$ and $\|Q(u,v)\|_Y \geq \|Q_1(u,v)\| \geq \|(u,v)\|_Y$.

Therefore, we deduce the conclusion of the theorem.

Case (8). We consider $f_0^s = g_0^s = 0$ and $g_\infty^i = \infty$. Let $\lambda \in (0,\infty)$ and $\mu \in (0,\infty)$, and let $\varepsilon > 0$ such that

$$\lambda \leq \frac{1}{\varepsilon(2B)^{r_1-1}}, \quad \varepsilon\left(\frac{1}{\gamma\gamma_2 C}\right)^{r_2-1} \leq \mu \leq \frac{1}{\varepsilon(2D)^{r_2-1}}.$$

By using (H2) and the definition of f_0^s and g_0^s, we deduce that there exists $R_1 > 0$ such that

$$f(t,u,v) \leq \varepsilon(u+v)^{r_1-1}, \quad g(t,u,v) \leq \varepsilon(u+v)^{r_2-1},$$

for all $t \in [0,1]$, $u,v \geq 0$, $u+v \leq R_1$.

We define the set $\Omega_1 = \{(u,v) \in Y, \|(u,v)\|_Y < R_1\}$. In a similar manner as in the proof of Case (1), for any $(u,v) \in P \cap \partial\Omega_1$, we obtain

$$Q_1(u,v)(t) \leq \lambda^{\varrho_1-1}\varepsilon^{\varrho_1-1}B\|(u,v)\|_Y \leq \frac{1}{2}\|(u,v)\|_Y, \quad \forall t \in [0,1],$$

$$Q_2(u,v)(t) \leq \mu^{\varrho_2-1}\varepsilon^{\varrho_2-1}D\|(u,v)\|_Y \leq \frac{1}{2}\|(u,v)\|_Y, \quad \forall t \in [0,1],$$

and so $\|Q(u,v)\|_Y \leq (\frac{1}{2}+\frac{1}{2})\|(u,v)\|_Y = \|(u,v)\|_Y$.

For the second part of the proof, by the definition of g_∞^i, there exists $\overline{R}_2 > 0$ such that

$$g(t,u,v) \geq \frac{1}{\varepsilon}(u+v)^{r_2-1}, \quad \forall t \in [c_1,c_2], \ u,v \geq 0, \ u+v \geq \overline{R}_2.$$

We consider $R_2 = \max\{2R_1, \overline{R}_2/\gamma\}$ and we define $\Omega_2 = \{(u,v) \in Y, \|(u,v)\|_Y < R_2\}$. Then for $(u,v) \in P \cap \partial\Omega_2$ we deduce as in Case (1) that $u(t) + v(t) \geq \gamma R_2 \geq \overline{R}_2$ for all $t \in [c_1,c_2]$.

Then by Lemma 4.1.3, we have

$$Q_2(u,v)(c_1) \geq \mu^{\varrho_2-1}\int_0^1 c_1^{\beta_2-1}J_2(s)\varphi_{\varrho_2}(I_{0+}^{\alpha_2}g(s,u(s),v(s)))\,ds$$

$$\geq \mu^{\varrho_2-1}c_1^{\beta_2-1}\int_{c_1}^{c_2} J_2(s)\varphi_{\varrho_2}\left(\frac{1}{\Gamma(\alpha_2)}\int_{c_1}^s (s-\tau)^{\alpha_2-1}\right.$$

$$\left. \times\, g(\tau,u(\tau),v(\tau))\,d\tau\right)ds$$

$$\geq \mu^{\varrho_2-1}c_1^{\beta_2-1}\int_{c_1}^{c_2} J_2(s)\varphi_{\varrho_2}\left(\frac{1}{\Gamma(\alpha_2)}\int_{c_1}^s (s-\tau)^{\alpha_2-1}\right.$$

$$\times \frac{1}{\varepsilon}(u(\tau) + v(\tau))^{r_2-1} d\tau \bigg) ds$$

$$\geq \mu^{\varrho_2-1} c_1^{\beta_2-1} \int_{c_1}^{c_2} J_2(s)\varphi_{\varrho_2} \left(\frac{1}{\Gamma(\alpha_2)} \int_{c_1}^{s} (s-\tau)^{\alpha_2-1} \right.$$

$$\times \frac{1}{\varepsilon}(\gamma\|(u,v)\|_Y)^{r_2-1} d\tau \bigg) ds$$

$$= \mu^{\varrho_2-1} c_1^{\beta_2-1} \left(\frac{1}{\varepsilon} \right)^{\varrho_2-1} \gamma\|(u,v)\|_Y$$

$$\times \int_{c_1}^{c_2} J_2(s)\varphi_{\varrho_2} \left(\frac{(s-c_1)^{\alpha_2}}{\Gamma(\alpha_2+1)} \right) ds$$

$$= \mu^{\varrho_2-1} c_1^{\beta_2-1} \left(\frac{1}{\varepsilon} \right)^{\varrho_2-1} \gamma\|(u,v)\|_Y$$

$$\times \int_{c_1}^{c_2} J_2(s) \frac{1}{(\Gamma(\alpha_2+1))^{\varrho_2-1}} (s-c_1)^{\alpha_2(\varrho_2-1)} ds$$

$$= \gamma\gamma_2\mu^{\varrho_2-1} \left(\frac{1}{\varepsilon} \right)^{\varrho_2-1} C\|(u,v)\|_Y \geq \|(u,v)\|_Y.$$

Then we conclude that $\|Q_2(u,v)\| \geq Q_2(u,v)(c_1) \geq \|(u,v)\|_Y$ and $\|Q(u,v)\|_Y \geq \|Q_2(u,v)\| \geq \|(u,v)\|_Y$.

Therefore, we deduce the conclusion of the theorem. $\qquad\square$

In what follows, for f_0^i, g_0^i, f_∞^s, $g_\infty^s \in (0,\infty)$ and numbers α_1', $\alpha_2' \geq 0$, $\widetilde{\alpha}_1$, $\widetilde{\alpha}_2 > 0$ such that $\alpha_1' + \alpha_2' = 1$ and $\widetilde{\alpha}_1 + \widetilde{\alpha}_2 = 1$, we define the numbers $\widetilde{L}_1, \widetilde{L}_2, \widetilde{L}_3, \widetilde{L}_4, \widetilde{L}_2'$ and $\widetilde{L}_4'$ by

$$\widetilde{L}_1 = \frac{1}{f_0^i} \left(\frac{\alpha_1'}{\gamma\gamma_1 A} \right)^{r_1-1}, \quad \widetilde{L}_2 = \frac{1}{f_\infty^s} \left(\frac{\widetilde{\alpha}_1}{B} \right)^{r_1-1}, \quad \widetilde{L}_3 = \frac{1}{g_0^i} \left(\frac{\alpha_2'}{\gamma\gamma_2 C} \right)^{r_2-1},$$

$$\widetilde{L}_4 = \frac{1}{g_\infty^s} \left(\frac{\widetilde{\alpha}_2}{D} \right)^{r_2-1}, \quad \widetilde{L}_2' = \frac{1}{f_\infty^s B^{r_1-1}}, \quad \widetilde{L}_4' = \frac{1}{g_\infty^s D^{r_2-1}}.$$

Theorem 4.1.2. *Assume that (H1) and (H2) hold, $[c_1, c_2] \subset [0,1]$ with $0 < c_1 < c_2 \leq 1$, $\alpha_1', \alpha_2' \geq 0$, $\widetilde{\alpha}_1, \widetilde{\alpha}_2 > 0$ such that $\alpha_1' + \alpha_2' = 1$, $\widetilde{\alpha}_1 + \widetilde{\alpha}_2 = 1$.*

(1) *If $f_0^i, g_0^i, f_\infty^s, g_\infty^s \in (0,\infty)$, $\widetilde{L}_1 < \widetilde{L}_2$ and $\widetilde{L}_3 < \widetilde{L}_4$, then for each $\lambda \in (\widetilde{L}_1, \widetilde{L}_2)$ and $\mu \in (\widetilde{L}_3, \widetilde{L}_4)$ there exists a positive solution $(u(t), v(t))$, $t \in [0,1]$ for (4.1), (4.2).*

(2) If f_0^i, g_0^i, $f_\infty^s \in (0, \infty)$, $g_\infty^s = 0$ and $\widetilde{L}_1 < \widetilde{L}_2'$, then for each $\lambda \in (\widetilde{L}_1, \widetilde{L}_2')$ and $\mu \in (\widetilde{L}_3, \infty)$ there exists a positive solution $(u(t), v(t))$, $t \in [0, 1]$ for (4.1), (4.2).

(3) If f_0^i, g_0^i, $g_\infty^s \in (0, \infty)$, $f_\infty^s = 0$ and $\widetilde{L}_3 < \widetilde{L}_4'$, then for each $\lambda \in (\widetilde{L}_1, \infty)$ and $\mu \in (\widetilde{L}_3, \widetilde{L}_4')$ there exists a positive solution $(u(t), v(t))$, $t \in [0, 1]$ for (4.1), (4.2).

(4) If f_0^i, $g_0^i \in (0, \infty)$, $f_\infty^s = g_\infty^s = 0$, then for each $\lambda \in (\widetilde{L}_1, \infty)$ and $\mu \in (\widetilde{L}_3, \infty)$ there exists a positive solution $(u(t), v(t))$, $t \in [0, 1]$ for (4.1), (4.2).

(5) If f_∞^s, $g_\infty^s \in (0, \infty)$ and at least one of f_0^i, g_0^i is ∞, then for each $\lambda \in (0, \widetilde{L}_2)$ and $\mu \in (0, \widetilde{L}_4)$ there exists a positive solution $(u(t), v(t))$, $t \in [0, 1]$ for (4.1), (4.2).

(6) If $f_\infty^s \in (0, \infty)$, $g_\infty^s = 0$ and at least one of f_0^i, g_0^i is ∞, then for each $\lambda \in (0, \widetilde{L}_2')$ and $\mu \in (0, \infty)$ there exists a positive solution $(u(t), v(t))$, $t \in [0, 1]$ for (4.1), (4.2).

(7) If $f_\infty^s = 0$, $g_\infty^s \in (0, \infty)$ and at least one of f_0^i, g_0^i is ∞, then for each $\lambda \in (0, \infty)$ and $\mu \in (0, \widetilde{L}_4')$ there exists a positive solution $(u(t), v(t))$, $t \in [0, 1]$ for (4.1), (4.2).

(8) If $f_\infty^s = g_\infty^s = 0$ and at least one of f_0^i, g_0^i is ∞, then for each $\lambda \in (0, \infty)$ and $\mu \in (0, \infty)$ there exists a positive solution $(u(t), v(t))$, $t \in [0, 1]$ for (4.1), (4.2).

Proof. We consider the cone $P \subset Y$ and the operators Q_1, Q_2 and Q defined at the beginning of this section. Because the proofs of the above cases are similar, in what follows, we will prove some representative cases.

Case (1). We have $f_0^i, g_0^i, f_\infty^s, g_\infty^s \in (0, \infty)$, $\widetilde{L}_1 < \widetilde{L}_2$ and $\widetilde{L}_3 < \widetilde{L}_4$. Let $\lambda \in (\widetilde{L}_1, \widetilde{L}_2)$ and $\mu \in (\widetilde{L}_3, \widetilde{L}_4)$. We consider $\varepsilon > 0$ such that $\varepsilon < f_0^i$, $\varepsilon < g_0^i$ and

$$\frac{1}{f_0^i - \varepsilon} \left(\frac{\alpha_1'}{\gamma \gamma_1 A} \right)^{r_1 - 1} \le \lambda \le \frac{1}{f_\infty^s + \varepsilon} \left(\frac{\widetilde{\alpha}_1}{B} \right)^{r_1 - 1},$$

$$\frac{1}{g_0^i - \varepsilon} \left(\frac{\alpha_2'}{\gamma \gamma_2 C} \right)^{r_2 - 1} \le \mu \le \frac{1}{g_\infty^s + \varepsilon} \left(\frac{\widetilde{\alpha}_2}{D} \right)^{r_2 - 1}.$$

By using (H2) and the definitions of f_0^i and g_0^i, we deduce that there exists $R_3 > 0$ such that

$$f(t, u, v) \ge (f_0^i - \varepsilon)(u + v)^{r_1 - 1}, \quad g(t, u, v) \ge (g_0^i - \varepsilon)(u + v)^{r_2 - 1},$$

for all $t \in [c_1, c_2]$, $u, v \ge 0$, $u + v \le R_3$.

We denote by $\Omega_3 = \{(u,v) \in Y, \|(u,v)\|_Y < R_3\}$. Let $(u,v) \in P$ with $\|(u,v)\|_Y = R_3$, that is $\|u\| + \|v\| = R_3$. Because $u(t) + v(t) \leq \|u\| + \|v\| = R_3$ for all $t \in [0,1]$, then by Lemma 4.1.3, we obtain

$$Q_1(u,v)(c_1) \geq \lambda^{\varrho_1 - 1} \int_0^1 c_1^{\beta_1 - 1} J_1(s) \varphi_{\varrho_1} (I_{0+}^{\alpha_1} f(s, u(s), v(s)))\, ds$$

$$\geq \lambda^{\varrho_1 - 1} c_1^{\beta_1 - 1} \int_{c_1}^{c_2} J_1(s) \varphi_{\varrho_1} \left(\frac{1}{\Gamma(\alpha_1)} \int_{c_1}^s (s - \tau)^{\alpha_1 - 1} \right.$$

$$\left. \times f(\tau, u(\tau), v(\tau))\, d\tau \right) ds$$

$$\geq \lambda^{\varrho_1 - 1} c_1^{\beta_1 - 1} \int_{c_1}^{c_2} J_1(s) \varphi_{\varrho_1} \left(\frac{1}{\Gamma(\alpha_1)} \int_{c_1}^s (s - \tau)^{\alpha_1 - 1} \right.$$

$$\left. \times (f_0^i - \varepsilon)(u(\tau) + v(\tau))^{r_1 - 1}\, d\tau \right) ds$$

$$\geq \lambda^{\varrho_1 - 1} c_1^{\beta_1 - 1} \int_{c_1}^{c_2} J_1(s) \varphi_{\varrho_1} \left(\frac{1}{\Gamma(\alpha_1)} \int_{c_1}^s (s - \tau)^{\alpha_1 - 1} \right.$$

$$\left. \times (f_0^i - \varepsilon)(\gamma \|(u,v)\|_Y)^{r_1 - 1}\, d\tau \right) ds$$

$$= \lambda^{\varrho_1 - 1} c_1^{\beta_1 - 1} (f_0^i - \varepsilon)^{\varrho_1 - 1} \gamma \|(u,v)\|_Y$$

$$\times \int_{c_1}^{c_2} J_1(s) \varphi_{\varrho_1} \left(\frac{1}{\Gamma(\alpha_1)} \int_{c_1}^s (s - \tau)^{\alpha_1 - 1}\, d\tau \right) ds$$

$$= \gamma \gamma_1 \lambda^{\varrho_1 - 1} (f_0^i - \varepsilon)^{\varrho_1 - 1} A \|(u,v)\|_Y \geq \alpha_1' \|(u,v)\|_Y.$$

Therefore, we conclude $\|Q_1(u,v)\| \geq Q_1(u,v)(c_1) \geq \alpha_1' \|(u,v)\|_Y$. In a similar manner, we deduce

$$Q_2(u,v)(c_1) \geq \mu^{\varrho_2 - 1} \int_0^1 c_1^{\beta_2 - 1} J_2(s) \varphi_{\varrho_2} (I_{0+}^{\alpha_2} g(s, u(s), v(s)))\, ds$$

$$\geq \mu^{\varrho_2 - 1} c_1^{\beta_2 - 1} \int_{c_1}^{c_2} J_2(s) \varphi_{\varrho_2} \left(\frac{1}{\Gamma(\alpha_2)} \int_{c_1}^s (s - \tau)^{\alpha_2 - 1} \right.$$

$$\left. \times g(\tau, u(\tau), v(\tau))\, d\tau \right) ds$$

$$\geq \mu^{\varrho_2 - 1} c_1^{\beta_2 - 1} \int_{c_1}^{c_2} J_2(s) \varphi_{\varrho_2} \left(\frac{1}{\Gamma(\alpha_2)} \int_{c_1}^s (s - \tau)^{\alpha_2 - 1} \right.$$

$$\left. \times (g_0^i - \varepsilon)(u(\tau) + v(\tau))^{r_2 - 1}\, d\tau \right) ds$$

$$\geq \mu^{\varrho_2-1} c_1^{\beta_2-1} \int_{c_1}^{c_2} J_2(s) \varphi_{\varrho_2} \left(\frac{1}{\Gamma(\alpha_2)} \int_{c_1}^{s} (s-\tau)^{\alpha_2-1} \right.$$

$$\times \left. (g_0^i - \varepsilon)(\gamma \|(u,v)\|_Y)^{r_2-1} d\tau \right) ds$$

$$= \mu^{\varrho_2-1} c_1^{\beta_2-1} (g_0^i - \varepsilon)^{\varrho_2-1} \gamma \|(u,v)\|_Y$$

$$\times \int_{c_1}^{c_2} J_2(s) \varphi_{\varrho_2} \left(\frac{1}{\Gamma(\alpha_2)} \int_{c_1}^{s} (s-\tau)^{\alpha_2-1} d\tau \right) ds$$

$$= \gamma \gamma_2 \mu^{\varrho_2-1} (g_0^i - \varepsilon)^{\varrho_2-1} C \|(u,v)\|_Y \geq \alpha_2' \|(u,v)\|_Y.$$

Hence, we get $\|Q_2(u,v)\| \geq Q_2(u,v)(c_1) \geq \alpha_2' \|(u,v)\|_Y$.
Then for $(u,v) \in P \cap \partial \Omega_3$, we obtain

$$\|Q(u,v)\|_Y = \|Q_1(u,v)\| + \|Q_2(u,v)\| \geq (\alpha_1' + \alpha_2') \|(u,v)\|_Y = \|(u,v)\|_Y. \tag{4.21}$$

Now, we define the functions $f^*, g^* : [0,1] \times [0,\infty) \to [0,\infty)$ by $f^*(t,x) = \max_{0 \leq u+v \leq x} f(t,u,v)$, $g^*(t,x) = \max_{0 \leq u+v \leq x} g(t,u,v)$, for all $t \in [0,1]$ and $x \in [0,\infty)$. Then

$$f(t,u,v) \leq f^*(t,x), \quad g(t,u,v) \leq g^*(t,x), \quad \forall t \in [0,1], \ u,v \geq 0, \ u+v \leq x.$$

The functions $f^*(t,\cdot)$, $g^*(t,\cdot)$ are nondecreasing for every $t \in [0,1]$ and they satisfy the conditions (see Section 4.1.5)

$$\limsup_{x \to \infty} \max_{t \in [0,1]} \frac{f^*(t,x)}{x^{r_1-1}} = f_\infty^s, \quad \limsup_{x \to \infty} \max_{t \in [0,1]} \frac{g^*(t,x)}{x^{r_2-1}} = g_\infty^s.$$

Therefore, for $\varepsilon > 0$ there exists $\overline{R}_4 > 0$ such that for all $x \geq \overline{R}_4$ and $t \in [0,1]$, we have

$$\frac{f^*(t,x)}{x^{r_1-1}} \leq \limsup_{x \to \infty} \max_{t \in [0,1]} \frac{f^*(t,x)}{x^{r_1-1}} + \varepsilon = f_\infty^s + \varepsilon,$$

$$\frac{g^*(t,x)}{x^{r_2-1}} \leq \limsup_{x \to \infty} \max_{t \in [0,1]} \frac{g^*(t,x)}{x^{r_2-1}} + \varepsilon = g_\infty^s + \varepsilon,$$

and so $f^*(t,x) \leq (f_\infty^s + \varepsilon) x^{r_1-1}$ and $g^*(t,x) \leq (g_\infty^s + \varepsilon) x^{r_2-1}$.

We consider $R_4 = \max\{2R_3, \overline{R}_4\}$ and we denote by $\Omega_4 = \{(u,v) \in Y, \|(u,v)\|_Y < R_4\}$. Let $(u,v) \in P \cap \partial\Omega_4$. By the definitions of f^* and g^*, we conclude

$$f(t, u(t), v(t)) \leq f^*(t, \|(u,v)\|_Y),$$

$$g(t, u(t), v(t)) \leq g^*(t, \|(u,v)\|_Y), \quad \forall t \in [0,1]. \tag{4.22}$$

Then for all $t \in [0,1]$, we obtain

$$Q_1(u,v)(t) \leq \lambda^{\varrho_1 - 1} \int_0^1 J_1(s)\varphi_{\varrho_1} \left(\frac{1}{\Gamma(\alpha_1)} \int_0^s (s - \tau)^{\alpha_1 - 1} \right.$$

$$\times f(\tau, u(\tau), v(\tau)) \, d\tau \bigg) ds$$

$$\leq \lambda^{\varrho_1 - 1} \int_0^1 J_1(s)\varphi_{\varrho_1} \left(\frac{1}{\Gamma(\alpha_1)} \int_0^s (s - \tau)^{\alpha_1 - 1} \right.$$

$$\times f^*(\tau, \|(u,v)\|_Y) \, d\tau \bigg) ds$$

$$\leq \lambda^{\varrho_1 - 1} \int_0^1 J_1(s)\varphi_{\varrho_1} \left(\frac{1}{\Gamma(\alpha_1)} (f_\infty^s + \varepsilon)\|(u,v)\|_Y^{r_1 - 1} \right.$$

$$\times \left(\int_0^s (s - \tau)^{\alpha_1 - 1} \, d\tau \right) \bigg) ds$$

$$= \lambda^{\varrho_1 - 1}(f_\infty^s + \varepsilon)^{\varrho_1 - 1} B\|(u,v)\|_Y \leq \widetilde{\alpha}_1 \|(u,v)\|_Y, \quad \forall t \in [0,1],$$

and so $\|Q_1(u,v)\| \leq \widetilde{\alpha}_1 \|(u,v)\|_Y$.

In a similar manner, we deduce

$$Q_2(u,v)(t) \leq \mu^{\varrho_2 - 1} \int_0^1 J_2(s)\varphi_{\varrho_2} \left(\frac{1}{\Gamma(\alpha_2)} \int_0^s (s - \tau)^{\alpha_2 - 1} \right.$$

$$\times g(\tau, u(\tau), v(\tau)) \, d\tau \bigg) ds$$

$$\leq \mu^{\varrho_2 - 1} \int_0^1 J_2(s)\varphi_{\varrho_2} \left(\frac{1}{\Gamma(\alpha_2)} \int_0^s (s - \tau)^{\alpha_2 - 1} \right.$$

$$\times g^*(\tau, \|(u,v)\|_Y) \, d\tau \bigg) ds$$

$$\leq \mu^{\varrho_2 - 1} \int_0^1 J_2(s) \varphi_{\varrho_2} \left(\frac{1}{\Gamma(\alpha_2)} (g_\infty^s + \varepsilon) \|(u, v)\|_Y^{r_2 - 1} \right.$$

$$\times \left. \left(\int_0^s (s - \tau)^{\alpha_2 - 1} \, d\tau \right) \right) ds$$

$$= \mu^{\varrho_2 - 1} (g_\infty^s + \varepsilon)^{\varrho_2 - 1} D \|(u, v)\|_Y \leq \tilde{\alpha}_2 \|(u, v)\|_Y, \quad \forall t \in [0, 1],$$

and then $\|Q_2(u, v)\| \leq \tilde{\alpha}_2 \|(u, v)\|_Y$.

Therefore, for $(u, v) \in P \cap \partial \Omega_4$, it follows that

$$\|Q(u, v)\|_Y = \|Q_1(u, v)\| + \|Q_2(u, v)\| \leq (\tilde{\alpha}_1 + \tilde{\alpha}_2) \|(u, v)\|_Y = \|(u, v)\|_Y. \tag{4.23}$$

By using Lemma 4.1.4, (4.21), (4.23) and the Guo–Krasnosel'skii fixed point theorem (Theorem 1.2.2), we conclude that Q has a fixed point $(u, v) \in P \cap (\overline{\Omega}_4 \setminus \Omega_3)$.

Case (3). We have $f_\infty^s = 0$, $f_0^i, g_0^i, g_\infty^s \in (0, \infty)$ and $\tilde{L}_3 < \tilde{L}_4'$. Let $\lambda \in (\tilde{L}_1, \infty)$ and $\mu \in (\tilde{L}_3, \tilde{L}_4')$. We choose $\tilde{\alpha}_2' \in (D(\mu g_\infty^s)^{\varrho_2 - 1}, 1)$ and $\tilde{\alpha}_1' = 1 - \tilde{\alpha}_2'$. Let $\varepsilon > 0$ with $\varepsilon < f_0^i$, $\varepsilon < g_0^i$ and

$$\frac{1}{f_0^i - \varepsilon} \left(\frac{\alpha_1'}{\gamma \gamma_1 A} \right)^{r_1 - 1} \leq \lambda \leq \frac{1}{\varepsilon} \left(\frac{\tilde{\alpha}_1'}{B} \right)^{r_1 - 1},$$

$$\frac{1}{g_0^i - \varepsilon} \left(\frac{\alpha_2'}{\gamma \gamma_2 C} \right)^{r_2 - 1} \leq \mu \leq \frac{1}{g_\infty^s + \varepsilon} \left(\frac{\tilde{\alpha}_2'}{D} \right)^{r_2 - 1}.$$

The first part of the proof is the same as the corresponding one from Case (1). For Ω_3 defined in Case (1), for $(u, v) \in P \cap \partial \Omega_3$, we obtain

$$Q_1(u, v)(c_1) \geq \gamma \gamma_1 \lambda^{\varrho_1 - 1} (f_0^i - \varepsilon)^{\varrho_1 - 1} A \|(u, v)\|_Y \geq \alpha_1' \|(u, v)\|_Y,$$

$$Q_2(u, v)(c_1) \geq \gamma \gamma_2 \mu^{\varrho_2 - 1} (g_0^i - \varepsilon)^{\varrho_2 - 1} C \|(u, v)\|_Y \geq \alpha_2' \|(u, v)\|_Y$$

and so $\|Q(u, v)\|_Y \geq (\alpha_1' + \alpha_2') \|(u, v)\|_Y = \|(u, v)\|_Y$.

For the second part, we use the same functions f^* and g^* from Case (1) which satisfy in this case the conditions

$$\lim_{x \to \infty} \max_{t \in [0,1]} \frac{f^*(t, x)}{x^{r_1 - 1}} = 0, \quad \limsup_{x \to \infty} \max_{t \in [0,1]} \frac{g^*(t, x)}{x^{r_2 - 1}} = g_\infty^s.$$

Therefore, for $\varepsilon > 0$, there exists $\overline{R}_4 > 0$ such that for all $x \geq \overline{R}_4$ and $t \in [0,1]$, we have

$$\frac{f^*(t,x)}{x^{r_1-1}} \leq \lim_{x \to \infty} \max_{t \in [0,1]} \frac{f^*(t,x)}{x^{r_1-1}} + \varepsilon = \varepsilon,$$

$$\frac{g^*(t,x)}{x^{r_2-1}} \leq \limsup_{x \to \infty} \max_{t \in [0,1]} \frac{g^*(t,x)}{x^{r_2-1}} + \varepsilon = g_\infty^s + \varepsilon,$$

and so $f^*(t,x) \leq \varepsilon x^{r_1-1}$ and $g^*(t,x) \leq (g_\infty^s + \varepsilon)x^{r_2-1}$.

We consider $R_4 = \max\{2R_3, \overline{R}_4\}$ and we denote by $\Omega_4 = \{(u,v) \in Y, \|(u,v)\|_Y < R_4\}$. Let $(u,v) \in P \cap \partial\Omega_4$. By the definitions of f^* and g^*, we obtain the relations (4.22). In addition, in a similar manner as in the proof of Case (1), we conclude

$$Q_1(u,v)(t) \leq \lambda^{\varrho_1-1}\varepsilon^{\varrho_1-1}\|(u,v\|_Y B \leq \widetilde{\alpha}_1'\|(u,v)\|_Y, \quad \forall t \in [0,1],$$

$$Q_2(u,v)(t) \leq \mu^{\varrho_2-1}(g_\infty^s + \varepsilon)^{\varrho_2-1}\|(u,v)\|_Y D \leq \widetilde{\alpha}_2'\|(u,v)\|_Y, \quad \forall t \in [0,1],$$

and so $\|Q(u,v)\|_Y \leq (\widetilde{\alpha}_1' + \widetilde{\alpha}_2')\|(u,v\|_Y = \|(u,v)\|_Y$.

Therefore, we deduce the conclusion of the theorem.

Case (6). We consider here $g_\infty^s = 0$, $f_\infty^s \in (0,\infty)$ and $g_0^i = \infty$. Let $\lambda \in (0, \widetilde{L}_2')$ and $\mu \in (0,\infty)$. We choose $\widetilde{\alpha}_1' \in (B(\lambda f_\infty^s)^{\varrho_1-1}, 1)$ and $\widetilde{\alpha}_2' = 1 - \widetilde{\alpha}_1'$, and let $\varepsilon > 0$ such that

$$\lambda \leq \frac{1}{f_\infty^s + \varepsilon}\left(\frac{\widetilde{\alpha}_1'}{B}\right)^{r_1-1}, \quad \varepsilon\left(\frac{1}{\gamma\gamma_2 C}\right)^{r_2-1} \leq \mu \leq \frac{1}{\varepsilon}\left(\frac{\widetilde{\alpha}_2'}{D}\right)^{r_2-1}.$$

By (H2) and the definition of g_0^i, we deduce that there exists $R_3 > 0$ such that

$$g(t,u,v) \geq \frac{1}{\varepsilon}(u+v)^{r_2-1}, \quad \forall t \in [c_1, c_2], \ u, v \geq 0, \ u + v \leq R_3.$$

We denote by $\Omega_3 = \{(u,v) \in Y, \ \|(u,v)\|_Y < R_3\}$. Let $(u,v) \in P$ with $\|(u,v)\|_Y = R_3$, that is $\|u\| + \|v\| = R_3$. Because $u(t) + v(t) \leq \|u\| + \|v\| = R_3$

for all $t \in [0, 1]$, then by Lemma 4.1.3, we obtain

$$Q_2(u, v)(c_1) \geq \mu^{\varrho_2 - 1} \int_0^1 c_1^{\beta_2 - 1} J_2(s) \varphi_{\varrho_2} (I_{0+}^{\alpha_2} g(s, u(s), v(s))) \, ds$$

$$\geq \mu^{\varrho_2 - 1} c_1^{\beta_2 - 1} \int_{c_1}^{c_2} J_2(s) \varphi_{\varrho_2} \left(\frac{1}{\Gamma(\alpha_2)} \int_{c_1}^s (s - \tau)^{\alpha_2 - 1} \right.$$

$$\times \, g(\tau, u(\tau), v(\tau)) \, d\tau \bigg) \, ds$$

$$\geq \mu^{\varrho_2 - 1} c_1^{\beta_2 - 1} \int_{c_1}^{c_2} J_2(s) \varphi_{\varrho_2} \left(\frac{1}{\Gamma(\alpha_2)} \int_{c_1}^s (s - \tau)^{\alpha_2 - 1} \right.$$

$$\times \, \frac{1}{\varepsilon} (u(\tau) + v(\tau))^{r_2 - 1} \, d\tau \bigg) \, ds$$

$$\geq \mu^{\varrho_2 - 1} c_1^{\beta_2 - 1} \int_{c_1}^{c_2} J_2(s) \varphi_{\varrho_2} \left(\frac{1}{\Gamma(\alpha_2)} \int_{c_1}^s (s - \tau)^{\alpha_2 - 1} \right.$$

$$\times \, \frac{1}{\varepsilon} (\gamma \| (u, v) \|_Y)^{r_2 - 1} \, d\tau \bigg) \, ds$$

$$= \mu^{\varrho_2 - 1} c_1^{\beta_2 - 1} \left(\frac{1}{\varepsilon} \right)^{\varrho_2 - 1} \gamma \| (u, v) \|_Y$$

$$\times \int_{c_1}^{c_2} J_2(s) \frac{1}{(\Gamma(\alpha_2 + 1))^{\varrho_2 - 1}} (s - c_1)^{\alpha_2 (\varrho_2 - 1)} \, ds$$

$$= \gamma \gamma_2 \mu^{\varrho_2 - 1} \left(\frac{1}{\varepsilon} \right)^{\varrho_2 - 1} C \| (u, v) \|_Y \geq \| (u, v) \|_Y.$$

Hence, we get $\| Q_2(u, v) \| \geq Q_2(u, v)(c_1) \geq \| (u, v) \|_Y$ and $\| Q(u, v) \|_Y \geq \| Q_2(u, v) \| \geq \| (u, v) \|_Y$.

For the second part of the proof, we consider the functions f^* and g^* from Case (1), which satisfy in this case the conditions

$$\limsup_{x \to \infty} \max_{t \in [0,1]} \frac{f^*(t, x)}{x^{r_1 - 1}} = f_\infty^s, \quad \lim_{x \to \infty} \max_{t \in [0,1]} \frac{g^*(t, x)}{x^{r_2 - 1}} = 0.$$

Then for $\varepsilon > 0$, there exists $\overline{R}_4 > 0$ such that for all $x \geq \overline{R}_4$ and $t \in [0, 1]$ we have

$$\frac{f^*(t, x)}{x^{r_1-1}} \leq \limsup_{x \to \infty} \max_{t \in [0,1]} \frac{f^*(t, x)}{x^{r_1-1}} + \varepsilon = f_\infty^s + \varepsilon,$$

$$\frac{g^*(t, x)}{x^{r_2-1}} \leq \lim_{x \to \infty} \max_{t \in [0,1]} \frac{g^*(t, x)}{x^{r_2-1}} + \varepsilon = \varepsilon,$$

and so $f^*(t, x) \leq (f_\infty^s + \varepsilon)x^{r_1-1}$ and $g^*(t, x) \leq \varepsilon x^{r_2-1}$.

We consider $R_4 = \max\{2R_3, \overline{R}_4\}$ and we denote by $\Omega_4 = \{(u, v) \in Y, \|(u, v)\|_Y < R_4\}$. Let $(u, v) \in P \cap \partial\Omega_4$. By the definitions of f^* and g^*, we deduce the relations (4.22). In addition, in a similar manner as in the proof of Case (1), we conclude

$$Q_1(u, v)(t) \leq \lambda^{\varrho_1-1}(f_\infty^s + \varepsilon)^{\varrho_1-1}\|(u, v)\|_Y B \leq \widetilde{\alpha}_1'\|(u, v)\|_Y, \quad \forall t \in [0, 1],$$

$$Q_2(u, v)(t) \leq \mu^{\varrho_2-1}\varepsilon^{\varrho_2-1}\|(u, v)\|_Y D \leq \widetilde{\alpha}_2'\|(u, v)\|_Y, \quad \forall t \in [0, 1],$$

and so $\|Q(u, v)\|_Y \leq (\widetilde{\alpha}_1' + \widetilde{\alpha}_2')\|(u, v\|_Y = \|(u, v)\|_Y$.

Therefore, we obtain the conclusion of the theorem.

Case (8). We consider $f_\infty^s = g_\infty^s = 0$ and $f_0^i = \infty$. Let $\lambda \in (0, \infty)$ and $\mu \in (0, \infty)$. We choose $\varepsilon > 0$ such that

$$\varepsilon\left(\frac{1}{\gamma\gamma_1 A}\right)^{r_1-1} \leq \lambda \leq \frac{1}{\varepsilon(2B)^{r_1-1}}, \quad \mu \leq \frac{1}{\varepsilon(2D)^{r_2-1}}.$$

By using (H2) and the definition of f_0^i, we deduce that there exists $R_3 > 0$ such that

$$f(t, u, v) \geq \frac{1}{\varepsilon}(u + v)^{r_1-1}, \quad \forall t \in [c_1, c_2], \ u, v \geq 0, \ u + v \leq R_3.$$

We denote by $\Omega_3 = \{(u, v) \in Y, \|(u, v)\|_Y < R_3\}$. Let $(u, v) \in P$ with $\|(u, v)\|_Y = R_3$, that is $\|u\| + \|v\| = R_3$. Because $u(t) + v(t) \leq \|u\| + \|v\| = R_3$

for all $t \in [0,1]$, then by Lemma 4.1.3, we obtain

$$Q_1(u,v)(c_1) \geq \lambda^{\varrho_1-1} \int_0^1 c_1^{\beta_1-1} J_1(s)\varphi_{\varrho_1}(I_{0+}^{\alpha_1} f(s,u(s),v(s)))\, ds$$

$$\geq \lambda^{\varrho_1-1} c_1^{\beta_1-1} \int_{c_1}^{c_2} J_1(s)\varphi_{\varrho_1}\left(\frac{1}{\Gamma(\alpha_1)} \int_{c_1}^s (s-\tau)^{\alpha_1-1}\right.$$

$$\times f(\tau,u(\tau),v(\tau))\, d\tau \Bigg)\, ds$$

$$\geq \lambda^{\varrho_1-1} c_1^{\beta_1-1} \int_{c_1}^{c_2} J_1(s)\varphi_{\varrho_1}\left(\frac{1}{\Gamma(\alpha_1)} \int_{c_1}^s (s-\tau)^{\alpha_1-1}\right.$$

$$\times \frac{1}{\varepsilon}(u(\tau)+v(\tau))^{r_1-1}\, d\tau \Bigg)\, ds$$

$$\geq \lambda^{\varrho_1-1} c_1^{\beta_1-1} \int_{c_1}^{c_2} J_1(s)\varphi_{\varrho_1}\left(\frac{1}{\Gamma(\alpha_1)} \int_{c_1}^s (s-\tau)^{\alpha_1-1}\right.$$

$$\times \frac{1}{\varepsilon}(\gamma\|(u,v)\|_Y)^{r_1-1}\, d\tau \Bigg)\, ds$$

$$= \lambda^{\varrho_1-1} c_1^{\beta_1-1}\left(\frac{1}{\varepsilon}\right)^{\varrho_1-1}\gamma\|(u,v)\|_Y$$

$$\times \int_{c_1}^{c_2} J_1(s)\frac{1}{(\Gamma(\alpha_1+1))^{\varrho_1-1}}(s-c_1)^{\alpha_1(\varrho_1-1)}\, ds$$

$$= \gamma\gamma_1\lambda^{\varrho_1-1}\left(\frac{1}{\varepsilon}\right)^{\varrho_1-1} A\|(u,v)\|_Y \geq \|(u,v)\|_Y.$$

Hence, we get $\|Q_1(u,v)\| \geq Q_1(u,v)(c_1) \geq \|(u,v)\|_Y$ and $\|Q(u,v)\|_Y \geq \|Q_1(u,v)\| \geq \|(u,v)\|_Y$.

For the second part of the proof, we consider the functions f^* and g^* from Case (1), which satisfy in this case the conditions

$$\lim_{x\to\infty} \max_{t\in[0,1]} \frac{f^*(t,x)}{x^{r_1-1}} = 0, \quad \lim_{x\to\infty} \max_{t\in[0,1]} \frac{g^*(t,x)}{x^{r_2-1}} = 0.$$

Then for $\varepsilon > 0$ there exists $\overline{R}_4 > 0$ such that for all $x \geq \overline{R}_4$ and $t \in [0,1]$ we have

$$\frac{f^*(t,x)}{x^{r_1-1}} \leq \lim_{x\to\infty} \max_{t\in[0,1]} \frac{f^*(t,x)}{x^{r_1-1}} + \varepsilon = \varepsilon,$$

$$\frac{g^*(t,x)}{x^{r_2-1}} \leq \lim_{x \to \infty} \max_{t \in [0,1]} \frac{g^*(t,x)}{x^{r_2-1}} + \varepsilon = \varepsilon,$$

and so $f^*(t,x) \leq \varepsilon x^{r_1-1}$ and $g^*(t,x) \leq \varepsilon x^{r_2-1}$.

We consider $R_4 = \max\{2R_3, \overline{R}_4\}$ and we denote by $\Omega_4 = \{(u,v) \in Y, \ \|(u,v)\|_Y < R_4\}$. Let $(u,v) \in P \cap \partial\Omega_4$. By the definitions of f^* and g^*, we obtain the relations (4.22). In addition, in a similar manner as in the proof of Case (1), we deduce

$$Q_1(u,v)(t) \leq \lambda^{\varrho_1-1}\varepsilon^{\varrho_1-1}\|(u,v)\|_Y B \leq \frac{1}{2}\|(u,v)\|_Y, \quad \forall t \in [0,1],$$

$$Q_2(u,v)(t) \leq \mu^{\varrho_2-1}\varepsilon^{\varrho_2-1}\|(u,v)\|_Y D \leq \frac{1}{2}\|(u,v)\|_Y, \quad \forall t \in [0,1],$$

and so $\|Q(u,v)\|_Y \leq \left(\frac{1}{2} + \frac{1}{2}\right)\|(u,v\|_Y = \|(u,v)\|_Y$.

Therefore, we obtain the conclusion of the theorem. $\qquad\square$

4.1.3 *Nonexistence of positive solutions*

In this section, we present intervals for λ and μ for which there exist no positive solutions of problem (4.1), (4.2).

Theorem 4.1.3. *Assume that* (H1) *and* (H2) *hold. If there exist positive numbers* M_1, M_2 *such that*

$$\begin{aligned} f(t,u,v) &\leq M_1(u+v)^{r_1-1}, \\ g(t,u,v) &\leq M_2(u+v)^{r_2-1}, \quad \forall t \in [0,1], \ u,v \geq 0, \end{aligned} \tag{4.24}$$

then there exist positive constants λ_0 *and* μ_0 *such that for every* $\lambda \in (0, \lambda_0)$ *and* $\mu \in (0, \mu_0)$ *the boundary value problem* (4.1), (4.2) *has no positive solution.*

Proof. We define $\lambda_0 = \frac{1}{M_1(2B)^{r_1-1}}$ and $\mu_0 = \frac{1}{M_2(2D)^{r_2-1}}$, where

$$B = \frac{1}{(\Gamma(\alpha_1+1))^{\varrho_1-1}} \int_0^1 s^{\alpha_1(\varrho_1-1)} J_1(s) \, ds,$$

$$D = \frac{1}{(\Gamma(\alpha_2+1))^{\varrho_2-1}} \int_0^1 s^{\alpha_2(\varrho_2-1)} J_2(s) \, ds.$$

We will prove that for every $\lambda \in (0, \lambda_0)$ and $\mu \in (0, \mu_0)$, the problem (4.1), (4.2) has no positive solution.

Let $\lambda \in (0, \lambda_0)$ and $\mu \in (0, \mu_0)$. We suppose that the problem (4.1), (4.2) has a positive solution $(u(t), v(t))$, $t \in [0, 1]$. Then we obtain

$$u(t) = Q_1(u, v)(t) = \lambda^{\varrho_1 - 1} \int_0^1 G_1(t, s) \varphi_{\varrho_1} (I_{0+}^{\alpha_1} f(s, u(s), v(s))) \, ds$$

$$\leq \lambda^{\varrho_1 - 1} \int_0^1 J_1(s) \varphi_{\varrho_1} \left(\frac{1}{\Gamma(\alpha_1)} \int_0^s (s - \tau)^{\alpha_1 - 1} f(\tau, u(\tau), v(\tau)) \, d\tau \right) ds$$

$$\leq \lambda^{\varrho_1 - 1} \int_0^1 J_1(s) \varphi_{\varrho_1} \left(\frac{1}{\Gamma(\alpha_1)} \int_0^s (s - \tau)^{\alpha_1 - 1} \right.$$
$$\left. \times M_1 (u(\tau) + v(\tau))^{r_1 - 1} \, d\tau \right) ds$$

$$\leq \lambda^{\varrho_1 - 1} \int_0^1 J_1(s) \varphi_{\varrho_1} (M_1) \varphi_{\varrho_1} \left(\frac{1}{\Gamma(\alpha_1)} \int_0^s (s - \tau)^{\alpha_1 - 1} \right.$$
$$\left. \times (\|u\| + \|v\|)^{r_1 - 1} \, d\tau \right) ds$$

$$= \lambda^{\varrho_1 - 1} M_1^{\varrho_1 - 1} B \|(u, v)\|_Y, \ \forall t \in [0, 1]$$

and

$$v(t) = Q_2(u, v)(t) = \mu^{\varrho_2 - 1} \int_0^1 G_2(t, s) \varphi_{\varrho_2} (I_{0+}^{\alpha_2} g(s, u(s), v(s))) \, ds$$

$$\leq \mu^{\varrho_2 - 1} \int_0^1 J_2(s) \varphi_{\varrho_2} \left(\frac{1}{\Gamma(\alpha_2)} \int_0^s (s - \tau)^{\alpha_2 - 1} g(\tau, u(\tau), v(\tau)) \, d\tau \right) ds$$

$$\leq \mu^{\varrho_2 - 1} \int_0^1 J_2(s) \varphi_{\varrho_2} \left(\frac{1}{\Gamma(\alpha_2)} \int_0^s (s - \tau)^{\alpha_2 - 1} \right.$$
$$\left. \times M_2 (u(\tau) + v(\tau))^{r_2 - 1} \, d\tau \right) ds$$

$$\leq \mu^{\varrho_2 - 1} \int_0^1 J_2(s) \varphi_{\varrho_2} (M_2) \varphi_{\varrho_2} \left(\frac{1}{\Gamma(\alpha_2)} \int_0^s (s - \tau)^{\alpha_2 - 1} \right.$$
$$\left. \times (\|u\| + \|v\|)^{r_2 - 1} \, d\tau \right) ds$$

$$= \mu^{\varrho_2 - 1} M_2^{\varrho_2 - 1} D \|(u, v)\|_Y, \quad \forall t \in [0, 1].$$

Then we deduce

$$\|u\| \le \lambda^{\varrho_1-1} M_1^{\varrho_1-1} B\|(u,v)\|_Y < \lambda_0^{\varrho_1-1} M_1^{\varrho_1-1} B\|(u,v)\|_Y = \frac{1}{2}\|(u,v)\|_Y,$$

$$\|v\| \le \mu^{\varrho_2-1} M_2^{\varrho_2-1} D\|(u,v)\|_Y < \mu_0^{\varrho_2-1} M_2^{\varrho_2-1} D\|(u,v)\|_Y = \frac{1}{2}\|(u,v)\|_Y,$$

and so $\|(u,v)\|_Y = \|u\| + \|v\| < \|(u,v)\|_Y$, which is a contradiction.

Therefore, the boundary value problem (4.1), (4.2) has no positive solution. $\qquad\square$

Remark 4.1.1. In the proof of Theorem 4.1.3, we can also define $\lambda_0 = \frac{1}{M_1}\left(\frac{\alpha_1}{B}\right)^{r_1-1}$ and $\mu_0 = \frac{1}{M_2}\left(\frac{\alpha_2}{D}\right)^{r_2-1}$ with α_1, $\alpha_2 > 0$ and $\alpha_1 + \alpha_2 = 1$.

Remark 4.1.2. If f_0^s, g_0^s, f_∞^s, $g_\infty^s < \infty$, then there exist positive constants M_1, M_2 such that relation (4.24) holds, and then we obtain the conclusion of Theorem 4.1.3.

Theorem 4.1.4. *Assume that* (H1) *and* (H2) *hold. If there exist positive numbers* c_1, c_2 *with* $0 < c_1 < c_2 \le 1$ *and* $m_1 > 0$ *such that*

$$f(t,u,v) \ge m_1(u+v)^{r_1-1}, \quad \forall t \in [c_1,c_2], \ u,v \ge 0, \qquad (4.25)$$

then there exists a positive constant $\widetilde{\lambda}_0$ *such that for every* $\lambda > \widetilde{\lambda}_0$ *and* $\mu > 0$, *the boundary value problem* (4.1), (4.2) *has no positive solution.*

Proof. We define $\widetilde{\lambda}_0 = \frac{1}{m_1(\gamma\gamma_1 A)^{r_1-1}}$, where $A = \frac{1}{(\Gamma(\alpha_1+1))^{\varrho_1-1}}\int_{c_1}^{c_2}(s - c_1)^{\alpha_1(\varrho_1-1)}J_1(s)\,ds$.

We will show that for every $\lambda > \widetilde{\lambda}_0$ and $\mu > 0$, the problem (4.1), (4.2) has no positive solution.

Let $\lambda > \widetilde{\lambda}_0$ and $\mu > 0$. We suppose that the problem (4.1), (4.2) has a positive solution $(u(t),v(t))$, $t \in [0,1]$. Then we obtain

$$u(c_1) = Q_1(u,v)(c_1) = \lambda^{\varrho_1-1}\int_0^1 G_1(c_1,s)\varphi_{\varrho_1}(I_{0+}^{\alpha_1}f(s,u(s),v(s)))\,ds$$

$$\ge \lambda^{\varrho_1-1}\int_0^1 c_1^{\beta_1-1}J_1(s)\varphi_{\varrho_1}(I_{0+}^{\alpha_1}f(s,u(s),v(s)))\,ds$$

$$\ge \lambda^{\varrho_1-1}c_1^{\beta_1-1}\int_{c_1}^{c_2} J_1(s)\varphi_{\varrho_1}\left(\frac{1}{\Gamma(\alpha_1)}\int_{c_1}^s (s-\tau)^{\alpha_1-1}\right.$$

$$\times\ f(\tau,u(\tau),v(\tau))\,d\tau\Big)\,ds$$

$$\geq \lambda^{\varrho_1-1} c_1^{\beta_1-1} \int_{c_1}^{c_2} J_1(s) \varphi_{\varrho_1} \left(\frac{1}{\Gamma(\alpha_1)} \int_{c_1}^{s} (s-\tau)^{\alpha_1-1} \right.$$

$$\times m_1 (u(\tau) + v(\tau))^{r_1-1} d\tau \bigg) ds$$

$$\geq \lambda^{\varrho_1-1} c_1^{\beta_1-1} \int_{c_1}^{c_2} J_1(s) \varphi_{\varrho_1} \left(\frac{1}{\Gamma(\alpha_1)} \int_{c_1}^{s} (s-\tau)^{\alpha_1-1} \right.$$

$$\times m_1 \gamma^{r_1-1} \|(u,v)\|_Y^{r_1-1} d\tau \bigg) ds$$

$$= \lambda^{\varrho_1-1} \gamma\gamma_1 m_1^{\varrho_1-1} \|(u,v)\|_Y$$

$$\times \int_{c_1}^{c_2} J_1(s) \frac{1}{(\Gamma(\alpha_1+1))^{\varrho_1-1}} (s-c_1)^{\alpha_1(\varrho_1-1)} ds$$

$$= \gamma\gamma_1 \lambda^{\varrho_1-1} m_1^{\varrho_1-1} A \|(u,v)\|_Y.$$

Then we conclude

$$\|u\| \geq u(c_1) \geq \gamma\gamma_1 A (\lambda m_1)^{\varrho_1-1} \|(u,v)\|_Y$$
$$> \gamma\gamma_1 A (\widetilde{\lambda}_0 m_1)^{\varrho_1-1} \|(u,v)\|_Y = \|(u,v)\|_Y,$$

and so $\|(u,v)\|_Y = \|u\| + \|v\| \geq \|u\| > \|(u,v)\|_Y$, which is a contradiction.

Therefore, the boundary value problem (4.1), (4.2) has no positive solution. $\qquad\square$

Theorem 4.1.5. *Assume that* (H1) *and* (H2) *hold. If there exist positive numbers* c_1, c_2 *with* $0 < c_1 < c_2 \leq 1$ *and* $m_2 > 0$ *such that*

$$g(t,u,v) \geq m_2 (u+v)^{r_2-1}, \quad \forall t \in [c_1, c_2], \ u, v \geq 0, \qquad (4.26)$$

then there exists a positive constant $\widetilde{\mu}_0$ *such that for every* $\mu > \widetilde{\mu}_0$ *and* $\lambda > 0$, *the boundary value problem* (4.1), (4.2) *has no positive solution.*

Proof. We define $\widetilde{\mu}_0 = \frac{1}{m_2(\gamma\gamma_2 C)^{r_2-1}}$, where $C = \frac{1}{(\Gamma(\alpha_2+1))^{\varrho_2-1}} \int_{c_1}^{c_2} (s - c_1)^{\alpha_2(\varrho_2-1)} J_2(s) \, ds$.

We will show that for every $\mu > \widetilde{\mu}_0$ and $\lambda > 0$, the problem (4.1), (4.2) has no positive solution.

Let $\mu > \tilde{\mu}_0$ and $\lambda > 0$. We suppose that the problem (4.1), (4.2) has a positive solution $(u(t), v(t)), t \in [0, 1]$. Then we obtain

$$v(c_1) = Q_2(u, v)(c_1) = \mu^{\varrho_2 - 1} \int_0^1 G_2(c_1, s)\varphi_{\varrho_2}(I_{0+}^{\alpha_1} g(s, u(s), v(s))) \, ds$$

$$\geq \mu^{\varrho_2 - 1} \int_0^1 c_1^{\beta_2 - 1} J_2(s)\varphi_{\varrho_2}(I_{0+}^{\alpha_2} g(s, u(s), v(s))) \, ds$$

$$\geq \mu^{\varrho_2 - 1} c_1^{\beta_2 - 1} \int_{c_1}^{c_2} J_2(s)\varphi_{\varrho_2} \left(\frac{1}{\Gamma(\alpha_2)} \int_{c_1}^s (s - \tau)^{\alpha_2 - 1} \right.$$

$$\times g(\tau, u(\tau), v(\tau)) \, d\tau \bigg) \, ds$$

$$\geq \mu^{\varrho_2 - 1} c_1^{\beta_2 - 1} \int_{c_1}^{c_2} J_2(s)\varphi_{\varrho_2} \left(\frac{1}{\Gamma(\alpha_2)} \int_{c_1}^s (s - \tau)^{\alpha_2 - 1} \right.$$

$$\times m_2(u(\tau) + v(\tau))^{r_2 - 1} \, d\tau \bigg) \, ds$$

$$\geq \mu^{\varrho_2 - 1} c_1^{\beta_2 - 1} \int_{c_1}^{c_2} J_2(s)\varphi_{\varrho_2} \left(\frac{1}{\Gamma(\alpha_2)} \int_{c_1}^s (s - \tau)^{\alpha_2 - 1} \right.$$

$$\times m_2 \gamma^{r_2 - 1} \|(u, v)\|_Y^{r_2 - 1} \, d\tau \bigg) \, ds$$

$$= \mu^{\varrho_2 - 1} \gamma \gamma_2 m_2^{\varrho_2 - 1} \|(u, v)\|_Y$$

$$\times \int_{c_1}^{c_2} J_2(s) \frac{1}{(\Gamma(\alpha_2 + 1))^{\varrho_2 - 1}} (s - c_1)^{\alpha_2(\varrho_2 - 1)} \, ds$$

$$= \gamma \gamma_2 \mu^{\varrho_2 - 1} m_2^{\varrho_2 - 1} C \|(u, v)\|_Y.$$

Then we deduce

$$\|v\| \geq v(c_1) \geq \gamma \gamma_2 C(\mu m_2)^{\varrho_2 - 1} \|(u, v)\|_Y$$

$$> \gamma \gamma_2 C(\tilde{\mu}_0 m_2)^{\varrho_2 - 1} \|(u, v)\|_Y = \|(u, v)\|_Y,$$

and so $\|(u, v)\|_Y = \|u\| + \|v\| \geq \|v\| > \|(u, v)\|_Y$, which is a contradiction.

Therefore, the boundary value problem (4.1), (4.2) has no positive solution. $\square$

Theorem 4.1.6. *Assume that* (H1) *and* (H2) *hold. If there exist positive numbers* c_1, c_2 *with* $0 < c_1 < c_2 \leq 1$ *and* $m_1, m_2 > 0$ *such that*

$$f(t, u, v) \geq m_1 (u + v)^{r_1 - 1},$$
$$g(t, u, v) \geq m_2 (u + v)^{r_2 - 1}, \quad \forall t \in [c_1, c_2], \ u, v \geq 0, \tag{4.27}$$

then there exist positive constants $\hat{\lambda}_0$ *and* $\hat{\mu}_0$ *such that for every* $\lambda > \hat{\lambda}_0$ *and* $\mu > \hat{\mu}_0$, *the boundary value problem* (4.1), (4.2) *has no positive solution.*

Proof. We define $\hat{\lambda}_0 = \frac{1}{m_1 (2\gamma\gamma_1 A)^{r_1 - 1}}$ and $\hat{\mu}_0 = \frac{1}{m_2 (2\gamma\gamma_2 C)^{r_2 - 1}}$. Then for every $\lambda > \hat{\lambda}_0$ and $\mu > \hat{\mu}_0$, the problem (4.1), (4.2) has no positive solution. Indeed, let $\lambda > \hat{\lambda}_0$ and $\mu > \hat{\mu}_0$. We suppose that the problem (4.1), (4.2) has a positive solution $(u(t), v(t))$, $t \in [0, 1]$. In a similar manner as that used in the proofs of Theorems 4.1.4 and 4.1.5 we obtain

$$\|u\| \geq u(c_1) \geq \gamma\gamma_1 A (\lambda m_1)^{\varrho_1 - 1} \|(u, v)\|_Y,$$
$$\|v\| \geq v(c_1) \geq \gamma\gamma_2 C (\mu m_2)^{\varrho_2 - 1} \|(u, v)\|_Y,$$

and so

$$\|(u, v)\|_Y = \|u\| + \|v\| \geq \gamma\gamma_1 A (\lambda m_1)^{\varrho_1 - 1} \|(u, v)\|_Y$$
$$+ \gamma\gamma_2 C (\mu m_2)^{\varrho_2 - 1} \|(u, v)\|_Y$$
$$> \gamma\gamma_1 A (\hat{\lambda}_0 m_1)^{\varrho_1 - 1} \|(u, v)\|_Y + \gamma\gamma_2 C (\hat{\mu}_0 m_2)^{\varrho_2 - 1} \|(u, v)\|_Y$$
$$= \frac{1}{2} \|(u, v)\|_Y + \frac{1}{2} \|(u, v)\|_Y = \|(u, v)\|_Y,$$

which is a contradiction. Therefore, the boundary value problem (4.1), (4.2) has no positive solution. $\square$

Remark 4.1.3. (i) If for c_1, c_2 with $0 < c_1 < c_2 \leq 1$, we have $f_0^i, f_\infty^i > 0$ and $f(t, u, v) > 0$ for all $t \in [c_1, c_2]$ and $u, v \geq 0$ with $u + v > 0$, then the relation (4.25) holds and we obtain the conclusion of Theorem 4.1.4.

(ii) If for c_1, c_2 with $0 < c_1 < c_2 \leq 1$, we have $g_0^i, g_\infty^i > 0$ and $g(t, u, v) > 0$ for all $t \in [c_1, c_2]$ and $u, v \geq 0$ with $u + v > 0$, then the relation (4.26) holds and we obtain the conclusion of Theorem 4.1.5.

(iii) If for c_1, c_2 with $0 < c_1 < c_2 \leq 1$, we have $f_0^i, f_\infty^i, g_0^i, g_\infty^i > 0$ and $f(t, u, v) > 0$, $g(t, u, v) > 0$ for all $t \in [c_1, c_2]$ and $u, v \geq 0$ with $u + v > 0$, then the relation (4.27) holds and we obtain the conclusion of Theorem 4.1.6.

4.1.4 *An example*

Let $\alpha_1 = 1/2$, $\alpha_2 = 1/3$, $n = 3$, $\beta_1 = 7/3$, $m = 4$, $\beta_2 = 15/4$, $p_1 = 1$, $q_1 = 1/3$, $p_2 = 3/2$, $q_2 = 6/5$, $N = 2$, $M = 1$, $\xi_1 = 1/4$, $\xi_2 = 3/4$, $a_1 = 3$, $a_2 = 1/4$, $\eta_1 = 1/3$, $b_1 = 2$, $r_1 = 4$, $\varrho_1 = 4/3$, $\varphi_{r_1}(s) = s^3$, $\varphi_{\varrho_1}(s) = s^{1/3}$, $r_2 = 3$, $\varrho_2 = 3/2$, $\varphi_{r_2}(s) = s|s|$, $\varphi_{\varrho_2}(s) = s|s|^{-1/2}$.

We consider the system of fractional differential equations

$$\begin{cases} D_{0+}^{1/2}(\varphi_4(D_{0+}^{7/3}u(t))) + \lambda(t+1)^a\left(e^{(u(t)+v(t))^3} - 1\right) = 0, & t \in (0,1), \\ D_{0+}^{1/3}(\varphi_3(D_{0+}^{15/4}v(t))) + \mu(2-t)^b(u^3(t) + v^3(t)) = 0, & t \in (0,1), \end{cases}$$

$$(4.28)$$

with the multi-point boundary conditions

$$\begin{cases} u(0) = u'(0) = 0, \ D_{0+}^{7/3}u(0) = 0, \ u'(1) = 3D_{0+}^{1/3}u\left(\frac{1}{4}\right) + \frac{1}{4}D_{0+}^{1/3}u\left(\frac{3}{4}\right), \\ v(0) = v'(0) = v''(0) = 0, \ D_{0+}^{15/4}v(0) = 0, \ D_{0+}^{3/2}v(1) = 2D_{0+}^{6/5}v\left(\frac{1}{3}\right), \end{cases}$$

$$(4.29)$$

where $a, b > 0$.

Here, we have $f(t, u, v) = (t + 1)^a(e^{(u+v)^3} - 1)$, $g(t, u, v) = (2 - t)^b(u^3 + v^3)$, $\forall t \in [0, 1]$, $u, v \geq 0$. Then we obtain $\Delta_1 \approx 0.21710894 > 0$, $\Delta_2 \approx 2.73417069 > 0$, and so the assumptions (H1) and (H2) are satisfied. In addition, we deduce

$$g_1(t, s) = \frac{1}{\Gamma(7/3)} \begin{cases} t^{4/3}(1-s)^{1/3} - (t-s)^{4/3}, & 0 \leq s \leq t \leq 1, \\ t^{4/3}(1-s)^{1/3}, & 0 \leq t \leq s \leq 1, \end{cases}$$

$$g_2(t, s) = \begin{cases} t(1-s)^{1/3} - t + s, & 0 \leq s \leq t \leq 1, \\ t(1-s)^{1/3}, & 0 \leq t \leq s \leq 1, \end{cases}$$

$$g_3(t, s) = \frac{1}{\Gamma(15/4)} \begin{cases} t^{11/4}(1-s)^{5/4} - (t-s)^{11/4}, & 0 \leq s \leq t \leq 1, \\ t^{11/4}(1-s)^{5/4}, & 0 \leq t \leq s \leq 1, \end{cases}$$

$$g_4(t, s) = \frac{1}{\Gamma(51/20)} \begin{cases} t^{31/20}(1-s)^{5/4} - (t-s)^{31/20}, & 0 \leq s \leq t \leq 1, \\ t^{31/20}(1-s)^{5/4}, & 0 \leq t \leq s \leq 1, \end{cases}$$

$$G_1(t, s) = g_1(t, s) + \frac{t^{4/3}}{\Delta_1}\left(3g_2\left(\frac{1}{4}, s\right) + \frac{1}{4}g_2\left(\frac{3}{4}, s\right)\right),$$

$$G_2(t,s) = g_3(t,s) + \frac{2t^{11/4}}{\Delta_2} g_4\left(\frac{1}{3}, s\right),$$

$$h_1(s) = \frac{1}{\Gamma(7/3)} s(1-s)^{1/3}, \quad h_3(s) = \frac{1}{\Gamma(15/4)}(1-s)^{5/4}(1-(1-s)^{3/2}).$$

For the functions J_1 and J_2, we obtain

$$J_1(s) = \begin{cases} \dfrac{1}{\Gamma(7/3)} s(1-s)^{1/3} + \dfrac{1}{\Delta_1}\left[\dfrac{15}{16}(1-s)^{1/3} + \dfrac{13s}{4} - \dfrac{15}{16}\right], \\ \quad 0 \le s < \dfrac{1}{4}, \\[2mm] \dfrac{1}{\Gamma(7/3)} s(1-s)^{1/3} + \dfrac{1}{\Delta_1}\left[\dfrac{15}{16}(1-s)^{1/3} + \dfrac{s}{4} - \dfrac{3}{16}\right], \\ \quad \dfrac{1}{4} \le s < \dfrac{3}{4}, \\[2mm] \dfrac{1}{\Gamma(7/3)} s(1-s)^{1/3} + \dfrac{15}{16\Delta_1}(1-s)^{1/3}, \quad \dfrac{3}{4} \le s \le 1, \end{cases}$$

$$J_2(s) = \begin{cases} \dfrac{1}{\Gamma(15/4)}(1-s)^{5/4}(1-(1-s)^{3/2}) + \dfrac{2}{3^{31/20}\Delta_2\Gamma(51/20)} \\ \quad \times [(1-s)^{5/4} - (1-3s)^{31/20}], \quad 0 \le s < \dfrac{1}{3}, \\[2mm] \dfrac{1}{\Gamma(15/4)}(1-s)^{5/4}(1-(1-s)^{3/2}) \\ \quad + \dfrac{2}{3^{31/20}\Delta_2\Gamma(51/20)}(1-s)^{5/4}, \quad \dfrac{1}{3} \le s \le 1. \end{cases}$$

Now, we choose $c_1 = 1/4$ and $c_2 = 3/4$, and then we deduce $\gamma_1 = (1/4)^{4/3}$, $\gamma_2 = (1/4)^{11/4}$, $\gamma = \gamma_2$. In addition, we have $f_0^s = 2^a$, $f_\infty^i = \infty$, $g_0^s = 0$, $g_\infty^i = \infty$, $A \approx 1.35668478$, $B \approx 2.51926854$.

By Theorem 4.1.1(7), for any $\lambda \in (0, L_2')$ and $\mu \in (0, \infty)$ with $L_2' = 1/(f_0^s B^3)$, the problem (4.28), (4.29) has a positive solution $(u(t), v(t))$, $t \in [0,1]$. For example, if $a = 2$, we obtain $L_2' \approx 0.0156357$.

We can also use Theorem 4.1.4, because $f(t,u,v) \ge (5/4)^a(u+v)^3$ for all $t \in [1/4, 3/4]$ and $u, v \ge 0$, that is $m_1 = (5/4)^a$. If $a = 2$, we deduce $\tilde{\lambda}_0 = 1/(m_1(\gamma\gamma_1 A)^3) \approx 6.0810421 \times 10^6$, and then we conclude that for every $\lambda > \tilde{\lambda}_0$ and $\mu > 0$, the boundary value problem (4.28), (4.29) has no positive solution.

4.1.5 *A relation between two supremum limits*

In this section, we will prove that if

$$\limsup_{\substack{u+v\to\infty \\ u,v\geq 0}} \max_{t\in[0,1]} \frac{f(t,u,v)}{(u+v)^{r_1-1}} = f_\infty^s,$$

then

$$\limsup_{x\to\infty} \max_{t\in[0,1]} \frac{f^*(t,x)}{x^{r_1-1}} = f_\infty^s,$$

where $f : [0,1]\times[0,\infty)\times[0,\infty) \to [0,\infty)$ is a continuous function, $f^*(t,x) = \max\{f(t,u,v),\ u+v\leq x,\ u,v\geq 0\}$ for $t\in[0,1]$ and $x\geq 0$, and $r_1 > 1$.

(I) In the case $f_\infty^s \in (0,\infty)$, from the characterization theorem of supremum limit, we have

(a) $\forall\varepsilon > 0$, $\exists M(\varepsilon) > 0$ such that $\forall u,v\geq 0$, $u+v > M(\varepsilon)$ we have

$$\max_{t\in[0,1]} \frac{f(t,u,v)}{(u+v)^{r_1-1}} < f_\infty^s + \varepsilon,$$

(b) $\forall\varepsilon > 0$ $\forall M' > 0$ then $\exists(u_0,v_0)$, $u_0\geq 0$, $v_0\geq 0$, $u_0+v_0 > M'$ such that

$$\max_{t\in[0,1]} \frac{f(t,u_0,v_0)}{(u_0+v_0)^{r_1-1}} > f_\infty^s - \varepsilon.$$

The second relation (b) is verified for an arbitrary (u,v) with $u+v > M'$ if $\varepsilon > f_\infty^s$, because f has nonnegative values.

From (a), for $\varepsilon > 0$ arbitrary, but fixed for the moment, there exists $M_1 = M(\varepsilon/2) > 0$ such that for all $u,v\geq 0$, $u+v > M_1$ we have

$$\max_{t\in[0,1]} \frac{f(t,u,v)}{(u+v)^{r_1-1}} < f_\infty^s + \frac{\varepsilon}{2},$$

and then $f(t,u,v) < \left(f_\infty^s + \frac{\varepsilon}{2}\right)(u+v)^{r_1-1}$ for all $t\in[0,1]$.

Then for $\varepsilon > 0$, there exists $M_1 > 0$ such that for all $x > M_1$ and $t\in[0,1]$, we obtain

$$f^*(t,x) = \max_{0\leq u+v\leq x} f(t,u,v) \leq \max_{0\leq u+v\leq M_1} f(t,u,v) + \sup_{M_1<u+v\leq x} f(t,u,v)$$

$$= f^*(t,M_1) + \sup_{M_1<u+v\leq x} f(t,u,v) \leq \max_{t\in[0,1]} f^*(t,M_1)$$

$$+ \sup_{M_1 < u+v \leq x} \left(f_\infty^s + \frac{\varepsilon}{2} \right) (u+v)^{r_1-1}$$

$$\leq K_{M_1} + \left(f_\infty^s + \frac{\varepsilon}{2} \right) x^{r_1-1},$$

where $K_{M_1} = \max_{t \in [0,1]} f^*(t, M_1)$.

Therefore, for $\varepsilon > 0$ there exist $M_1 > 0$ and $K_{M_1} > 0$ such that

$$\frac{f^*(t,x)}{x^{r_1-1}} \leq \frac{K_{M_1}}{x^{r_1-1}} + f_\infty^s + \frac{\varepsilon}{2}, \quad \forall x > M_1, \ \forall t \in [0,1],$$

and so

$$\max_{t \in [0,1]} \frac{f^*(t,x)}{x^{r_1-1}} \leq \frac{K_{M_1}}{x^{r_1-1}} + f_\infty^s + \frac{\varepsilon}{2}, \quad \forall x > M_1.$$

Because $\lim_{x \to \infty} \frac{1}{x^{r_1-1}} = 0$, then for $\varepsilon > 0$ there exists $M_2 \geq M_1$ such that $\frac{1}{x^{r_1-1}} < \frac{\varepsilon}{2K_{M_1}}$ for all $x > M_2$.

So, we conclude

$$\forall \varepsilon > 0, \exists M_2 > 0 \text{ such that } \max_{t \in [0,1]} \frac{f^*(t,x)}{x^{r_1-1}} < \frac{\varepsilon}{2} + f_\infty^s + \frac{\varepsilon}{2}$$

$$= f_\infty^s + \varepsilon, \quad \forall x > M_2. \tag{4.30}$$

From relation (b) we deduce that for any $\varepsilon > 0$ and any $M' > 0$ there exists $x_0 = u_0 + v_0 > 0$ such that

$$\max_{t \in [0,1]} \frac{f^*(t,x_0)}{x_0^{r_1-1}} \geq \max_{t \in [0,1]} \frac{f(t,u_0,v_0)}{x_0^{r_1-1}} = \max_{t \in [0,1]} \frac{f(t,u_0,v_0)}{(u_0+v_0)^{r_1-1}} > f_\infty^s - \varepsilon.$$

Then we obtain

$$\forall \varepsilon > 0, \ \forall M' > 0, \quad \exists x_0 > M' \text{ such that } \max_{t \in [0,1]} \frac{f^*(t,x_0)}{x_0^{r_1-1}} > f_\infty^s - \varepsilon. \tag{4.31}$$

By relations (4.30), (4.31) and the characterization theorem for supremum limit, we conclude that $\limsup_{x \to \infty} \max_{t \in [0,1]} \frac{f^*(t,x)}{x^{r_1-1}} = f_\infty^s$.

(II) If $f_\infty^s = 0$, then $\limsup_{\substack{u+v \to \infty \\ u,v \geq 0}} \max_{t \in [0,1]} \frac{f(t,u,v)}{(u+v)^{r_1-1}} = 0$ is equivalent to $\lim_{\substack{u+v \to \infty \\ u,v \geq 0}} \max_{t \in [0,1]} \frac{f(t,u,v)}{(u+v)^{r_1-1}} = 0$, because f has nonnegative values. Also, $\limsup_{x \to \infty} \max_{t \in [0,1]} \frac{f^*(t,x)}{x^{r_1-1}} = 0$ is equivalent to $\lim_{x \to \infty} \max_{t \in [0,1]} \frac{f^*(t,x)}{x^{r_1-1}} = 0$.

In the same manner as used in the case (I) (for the implication $(a) \Rightarrow$ (4.30)), we can show that relation

$$\forall \varepsilon > 0 \ \exists M > 0 \text{ such that } \forall u, v \geq 0,$$

$$u + v > M \text{ we have } 0 \leq \max_{t \in [0,1]} \frac{f(t, u, v)}{(u + v)^{r_1 - 1}} < \varepsilon,$$

implies the relation

$$\forall \varepsilon > 0 \ \exists \widetilde{M} > 0 \text{ such that } \forall x > \widetilde{M} \text{ we have } 0 \leq \max_{t \in [0,1]} \frac{f^*(t, x)}{x^{r_1 - 1}} < \varepsilon,$$

that is, $\lim_{x \to \infty} \max_{t \in [0,1]} \frac{f^*(t,x)}{x^{r_1-1}} = 0$.

(III) If $f_\infty^s = \infty$, then by the characterization theorem, we have

$$\forall M_1 > 0 \quad \forall M_2 > 0 \ \exists (u, v), \ u \geq 0, \ v \geq 0,$$

$$u + v > M_1 \text{ such that } \max_{t \in [0,1]} \frac{f(t, u, v)}{(u + v)^{r_1 - 1}} > M_2.$$

Then we deduce that for any $M_1 > 0$ and any $M_2 > 0$ there exists $x = u + v > M_1$ such that

$$\max_{t \in [0,1]} \frac{f^*(t, x)}{x^{r_1 - 1}} \geq \max_{t \in [0,1]} \frac{f(t, u, v)}{x^{r_1 - 1}} = \max_{t \in [0,1]} \frac{f(t, u, v)}{(u + v)^{r_1 - 1}} > M_2.$$

So we obtain that $\limsup_{x \to \infty} \max_{t \in [0,1]} \frac{f^*(t,x)}{x^{r_1-1}} = \infty$.

Remark 4.1.4. The results presented in this section under the assumptions $p_1 \in [1, n-2]$ and $p_2 \in [1, m-2]$ instead of $p_1 \in [1, \beta_1 - 1)$ and $p_2 \in [1, \beta_2 - 1)$ were published in [79].

4.2 Systems of Fractional Differential Equations with Coupled Multi-Point Boundary Conditions

We consider the system of nonlinear ordinary fractional differential equations with r_1-Laplacian and r_2-Laplacian operators

$$\begin{cases} D_{0+}^{\alpha_1} (\varphi_{r_1}(D_{0+}^{\beta_1} u(t))) + \lambda f(t, u(t), v(t)) = 0, & t \in (0, 1), \\ D_{0+}^{\alpha_2} (\varphi_{r_2}(D_{0+}^{\beta_2} v(t))) + \mu g(t, u(t), v(t)) = 0, & t \in (0, 1), \end{cases} \tag{4.32}$$

with the coupled multi-point boundary conditions

$$
\begin{cases}
u^{(j)}(0) = 0, \quad j = 0, \ldots, n-2; \quad D_{0+}^{\beta_1} u(0) = 0, \\
\displaystyle D_{0+}^{p_1} u(1) = \sum_{i=1}^{N} a_i D_{0+}^{q_1} v(\xi_i), \\
v^{(j)}(0) = 0, \quad j = 0, \ldots, m-2; \quad D_{0+}^{\beta_2} v(0) = 0, \\
\displaystyle D_{0+}^{p_2} v(1) = \sum_{i=1}^{M} b_i D_{0+}^{q_2} u(\eta_i),
\end{cases}
\tag{4.33}
$$

where $\alpha_1, \alpha_2 \in (0, 1]$, $\beta_1 \in (n-1, n]$, $\beta_2 \in (m-1, m]$, $n, m \in \mathbb{N}$, $n, m \geq 3$, $p_1, p_2, q_1, q_2 \in \mathbb{R}$, $p_1 \in [1, \beta_1 - 1)$, $p_2 \in [1, \beta_2 - 1)$, $q_1 \in [0, p_2]$, $q_2 \in [0, p_1]$, $\xi_i, a_i \in \mathbb{R}$ for all $i = 1, \ldots, N$ ($N \in \mathbb{N}$), $0 < \xi_1 < \cdots < \xi_N \leq 1$, η_i, $b_i \in \mathbb{R}$ for all $i = 1, \ldots, M$ ($M \in \mathbb{N}$), $0 < \eta_1 < \cdots < \eta_M \leq 1$, $r_1, r_2 > 1$, $\varphi_{r_i}(s) = |s|^{r_i - 2} s$, $\varphi_{r_i}^{-1} = \varphi_{\varrho_i}$, $\frac{1}{r_i} + \frac{1}{\varrho_i} = 1$, $i = 1, 2$, $\lambda, \mu > 0$, $f, g \in C([0, 1] \times [0, \infty) \times [0, \infty), [0, \infty))$, and D_{0+}^k denotes the Riemann-Liouville derivative of order k (for $k = \alpha_1, \beta_1, \alpha_2, \beta_2, p_1, q_1, p_2, q_2$).

Under sufficient conditions on the functions f and g, we present intervals for the parameters λ and μ such that problem (4.32), (4.33) has positive solutions. By a positive solution of problem (4.32), (4.33) we mean a pair of functions $(u, v) \in (C([0, 1], \mathbb{R}_+))^2$, satisfying (S) and (BC) with $u(t) > 0$ for all $t \in (0, 1]$, or $v(t) > 0$ for all $t \in (0, 1]$. We also investigate the nonexistence of positive solutions for the above problem.

4.2.1   *Preliminary results*

We consider the system of fractional differential equations

$$
\begin{cases}
D_{0+}^{\alpha_1} (\varphi_{r_1}(D_{0+}^{\beta_1} u(t))) + h(t) = 0, \quad t \in (0, 1), \\
D_{0+}^{\alpha_2} (\varphi_{r_2}(D_{0+}^{\beta_2} v(t))) + k(t) = 0, \quad t \in (0, 1),
\end{cases}
\tag{4.34}
$$

with the coupled multi-point boundary conditions (4.33), where $h, k \in C[0, 1]$.

If we denote by $\varphi_{r_1}(D_{0+}^{\beta_1} u(t)) = x(t)$ and $\varphi_{r_2}(D_{0+}^{\beta_2} v(t)) = y(t)$, then the problem (4.34), (4.33) is equivalent to the following three problems:

(I)
$$
\begin{cases}
D_{0+}^{\alpha_1} x(t) + h(t) = 0, \quad 0 < t < 1, \\
x(0) = 0,
\end{cases}
$$

(II)
$$
\begin{cases}
D_{0+}^{\alpha_2} y(t) + k(t) = 0, \quad 0 < t < 1, \\
y(0) = 0,
\end{cases}
$$

and

(III)
$$\begin{cases} D_{0+}^{\beta_1} u(t) = \varphi_{\varrho_1}(x(t)), & t \in (0,1), \\ D_{0+}^{\beta_2} v(t) = \varphi_{\varrho_2}(y(t)), & t \in (0,1), \end{cases}$$

with the boundary conditions

$$\begin{cases} u^{(j)}(0) = 0, & j = 0,\ldots, n-2; \quad D_{0+}^{p_1} u(1) = \sum_{i=1}^{N} a_i D_{0+}^{q_1} v(\xi_i), \\ v^{(j)}(0) = 0, & j = 0,\ldots, m-2; \quad D_{0+}^{p_2} v(1) = \sum_{i=1}^{M} b_i D_{0+}^{q_2} u(\eta_i). \end{cases}$$

For the first two problems (I) and (II), the functions

$$x(t) = -I_{0+}^{\alpha_1} h(t) = -\frac{1}{\Gamma(\alpha_1)} \int_0^t (t-s)^{\alpha_1 - 1} h(s)\, ds, \quad t \in [0,1], \qquad (4.35)$$

and

$$y(t) = -I_{0+}^{\alpha_2} k(t) = -\frac{1}{\Gamma(\alpha_2)} \int_0^t (t-s)^{\alpha_2 - 1} k(s)\, ds, \quad t \in [0,1], \qquad (4.36)$$

are the unique solutions $x, y \in C[0,1]$ for (I) and (II), respectively.

For the third problem (III), if

$$\Delta := \frac{\Gamma(\beta_1)\Gamma(\beta_2)}{\Gamma(\beta_1 - p_1)\Gamma(\beta_2 - p_2)} - \frac{\Gamma(\beta_1)\Gamma(\beta_2)}{\Gamma(\beta_1 - q_2)\Gamma(\beta_2 - q_1)} \left(\sum_{i=1}^{N} a_i \xi_i^{\beta_2 - q_1 - 1} \right)$$

$$\times \left(\sum_{i=1}^{M} b_i \eta_i^{\beta_1 - q_2 - 1} \right) \neq 0,$$

and $x, y \in C[0,1]$, then by Lemma 3.2.2 from Section 3.2, we deduce that the pair of functions

$$\begin{cases} u(t) = -\int_0^1 \widetilde{G}_1(t,s)\varphi_{\varrho_1}(x(s))\, ds - \int_0^1 \widetilde{G}_2(t,s)\varphi_{\varrho_2}(y(s))\, ds, & t \in [0,1], \\ v(t) = -\int_0^1 \widetilde{G}_3(t,s)\varphi_{\varrho_2}(y(s))\, ds - \int_0^1 \widetilde{G}_4(t,s)\varphi_{\varrho_1}(x(s))\, ds, & t \in [0,1] \end{cases}$$

$$(4.37)$$

is the unique solution $(u, v) \in C[0, 1] \times C[0, 1]$ of problem (III). Here, the Green functions $\widetilde{G}_i$, $i = 1, \ldots, 4$ (see Lemma 3.2.2) are defined by

$$\widetilde{G}_1(t, s) = \widetilde{g}_1(t, s) + \frac{t^{\beta_1 - 1}\Gamma(\beta_2)}{\Delta\Gamma(\beta_2 - q_1)} \left(\sum_{i=1}^{N} a_i \xi_i^{\beta_2 - q_1 - 1} \right) \left(\sum_{i=1}^{M} b_i \widetilde{g}_2(\eta_i, s) \right),$$

$$\widetilde{G}_2(t, s) = \frac{t^{\beta_1 - 1}\Gamma(\beta_2)}{\Delta\Gamma(\beta_2 - p_2)} \left(\sum_{i=1}^{N} a_i \widetilde{g}_3(\xi_i, s) \right),$$

$$\widetilde{G}_3(t, s) = \widetilde{g}_4(t, s) + \frac{t^{\beta_2 - 1}\Gamma(\beta_1)}{\Delta\Gamma(\beta_1 - q_2)} \left(\sum_{i=1}^{M} b_i \eta_i^{\beta_1 - q_2 - 1} \right) \left(\sum_{i=1}^{N} a_i \widetilde{g}_3(\xi_i, s) \right),$$

$$\widetilde{G}_4(t, s) = \frac{t^{\beta_2 - 1}\Gamma(\beta_1)}{\Delta\Gamma(\beta_1 - p_1)} \left(\sum_{i=1}^{M} b_i \widetilde{g}_2(\eta_i, s) \right), \quad \forall t, \ s \in [0, 1],$$

$$(4.38)$$

where

$$\widetilde{g}_1(t, s) = \frac{1}{\Gamma(\beta_1)} \begin{cases} t^{\beta_1 - 1}(1 - s)^{\beta_1 - p_1 - 1} - (t - s)^{\beta_1 - 1}, & 0 \leq s \leq t \leq 1, \\ t^{\beta_1 - 1}(1 - s)^{\beta_1 - p_1 - 1}, & 0 \leq t \leq s \leq 1, \end{cases}$$

$$\widetilde{g}_2(t, s) = \frac{1}{\Gamma(\beta_1 - q_2)} \begin{cases} t^{\beta_1 - q_2 - 1}(1 - s)^{\beta_1 - p_1 - 1} - (t - s)^{\beta_1 - q_2 - 1}, \\ \qquad 0 \leq s \leq t \leq 1, \\ t^{\beta_1 - q_2 - 1}(1 - s)^{\beta_1 - p_1 - 1}, & 0 \leq t \leq s \leq 1. \end{cases}$$

$$\widetilde{g}_3(t, s) = \frac{1}{\Gamma(\beta_2 - q_1)} \begin{cases} t^{\beta_2 - q_1 - 1}(1 - s)^{\beta_2 - p_2 - 1} - (t - s)^{\beta_2 - q_1 - 1}, \\ \qquad 0 \leq s \leq t \leq 1, \\ t^{\beta_2 - q_1 - 1}(1 - s)^{\beta_2 - p_2 - 1}, & 0 \leq t \leq s \leq 1. \end{cases}$$

$$\widetilde{g}_4(t, s) = \frac{1}{\Gamma(\beta_2)} \begin{cases} t^{\beta_2 - 1}(1 - s)^{\beta_2 - p_2 - 1} - (t - s)^{\beta_2 - 1}, & 0 \leq s \leq t \leq 1, \\ t^{\beta_2 - 1}(1 - s)^{\beta_2 - p_2 - 1}, & 0 \leq t \leq s \leq 1. \end{cases}$$

$$(4.39)$$

Therefore, by (4.35)–(4.37), we obtain the following lemma.

Lemma 4.2.1. *If $\Delta \neq 0$, then the unique solution $(u, v) \in C[0, 1] \times C[0, 1]$ of problem (4.34), (4.33) is given by*

$$
\begin{cases}
u(t) = \int_0^1 \widetilde{G}_1(t, s)\varphi_{\varrho_1}(I_{0+}^{\alpha_1} h(s))\, ds \\
\qquad + \int_0^1 \widetilde{G}_2(t, s)\varphi_{\varrho_2}(I_{0+}^{\alpha_2} k(s))\, ds, \quad t \in [0, 1], \\
v(t) = \int_0^1 \widetilde{G}_3(t, s)\varphi_{\varrho_2}(I_{0+}^{\alpha_2} k(s))\, ds \\
\qquad + \int_0^1 \widetilde{G}_4(t, s)\varphi_{\varrho_1}(I_{0+}^{\alpha_1} h(s))\, ds, \quad t \in [0, 1].
\end{cases}
\tag{4.40}
$$

Based on the properties of the functions $\widetilde{g}_i$, $i = 1, \ldots, 4$ given by (4.39) (see Lemma 3.2.3) we obtain the following properties of the Green functions $\widetilde{G}_i$, $i = 1, \ldots, 4$ that will be used in the next sections.

Lemma 4.2.2. *Assume that $\Delta > 0$, $a_i \geq 0$ for all $i = 1, \ldots, N$, and $b_i \geq 0$ for all $i = 1, \ldots, M$. Then the functions $\widetilde{G}_i$, $i = 1, \ldots, 4$, given by (4.38) have the properties*

(a) $\widetilde{G}_i : [0, 1] \times [0, 1] \to [0, \infty)$, $i = 1, \ldots, 4$ *are continuous functions;*
(b) $\widetilde{G}_1(t, s) \leq \widetilde{J}_1(s)$, $\forall (t, s) \in [0, 1] \times [0, 1]$, *where*

$$
\widetilde{J}_1(s) = \widetilde{h}_1(s) + \frac{\Gamma(\beta_2)}{\Delta\Gamma(\beta_2 - q_1)} \left(\sum_{i=1}^N a_i \xi_i^{\beta_2 - q_1 - 1} \right)
$$

$$
\times \left(\sum_{i=1}^M b_i \widetilde{g}_2(\eta_i, s) \right), \quad \forall s \in [0, 1],
$$

and $\widetilde{h}_1(s) = \frac{1}{\Gamma(\beta_1)}(1 - s)^{\beta_1 - p_1 - 1}(1 - (1 - s)^{p_1})$, $s \in [0, 1]$;

(c) $\widetilde{G}_1(t, s) \geq t^{\beta_1 - 1} \widetilde{J}_1(s)$, $\forall (t, s) \in [0, 1] \times [0, 1]$;
(d) $\widetilde{G}_2(t, s) \leq \widetilde{J}_2(s)$, $\forall (t, s) \in [0, 1] \times [0, 1]$, *where*

$$
\widetilde{J}_2(s) = \frac{\Gamma(\beta_2)}{\Delta\Gamma(\beta_2 - p_2)} \sum_{i=1}^N a_i \widetilde{g}_3(\xi_i, s), \quad \forall s \in [0, 1];
$$

(e) $\widetilde{G}_2(t,s) = t^{\beta_1-1}\widetilde{J}_2(s), \ \forall(t,s) \in [0,1] \times [0,1];$

(f) $\widetilde{G}_3(t,s) \leq \widetilde{J}_3(s), \ \forall(t,s) \in [0,1] \times [0,1],$ *where*

$$\widetilde{J}_3(s) = \widetilde{h}_4(s) + \frac{\Gamma(\beta_1)}{\Delta\Gamma(\beta_1-q_2)}\left(\sum_{i=1}^{M}b_i\eta_i^{\beta_1-q_2-1}\right)\left(\sum_{i=1}^{N}a_i\widetilde{g}_3(\xi_i,s)\right),$$

and $\widetilde{h}_4(s) = \frac{1}{\Gamma(\beta_2)}(1-s)^{\beta_2-p_2-1}(1-(1-s)^{p_2}), \ s \in [0,1];$

(g) $\widetilde{G}_3(t,s) \geq t^{\beta_2-1}\widetilde{J}_3(s), \ \forall(t,s) \in [0,1] \times [0,1];$

(h) $\widetilde{G}_4(t,s) \leq \widetilde{J}_4(s), \ \forall(t,s) \in [0,1] \times [0,1],$ *where*

$$\widetilde{J}_4(s) = \frac{\Gamma(\beta_1)}{\Delta\Gamma(\beta_1-p_1)}\sum_{i=1}^{M}b_i\widetilde{g}_2(\eta_i,s), \ \forall s \in [0,1];$$

(i) $\widetilde{G}_4(t,s) = t^{\beta_2-1}\widetilde{J}_4(s), \ \forall(t,s) \in [0,1] \times [0,1].$

4.2.2   *Existence of positive solutions*

In this section, we investigate the existence of positive solutions of problem (4.32), (4.33) under some assumptions on the functions f and g, by establishing in the same time various intervals for the positive parameters λ and μ.

We present the assumptions that we will use in the sequel.

(I1) $\alpha_1, \ \alpha_2 \ \in \ (0,1], \ \beta_1 \ \in \ (n-1,n], \ \beta_2 \ \in \ (m-1,m], \ n,m \ \geq \ 3,$
$p_1,p_2,q_1,q_2 \ \in \ \mathbb{R}, \ p_1 \ \in \ [1,\beta_1-1), \ p_2 \ \in \ [1,\beta_2-1), \ q_1 \ \in \ [0,p_2],$
$q_2 \ \in \ [0,p_1], \ \xi_i \ \in \ \mathbb{R}, \ a_i \geq 0$ for all $i=1,\ldots,N \ (N \in \mathbb{N}), \ \sum_{i=1}^{N}a_i > 0,$
$0 < \xi_1 < \cdots < \xi_N \leq 1, \ \eta_i \in \mathbb{R}, \ b_i \geq 0$ for all $i=1,\ldots,M \ (M \in \mathbb{N}),$
$\sum_{i=1}^{M}b_i > 0, 0 < \eta_1 < \cdots < \eta_M \leq 1, \ \lambda,\mu > 0, \ \Delta = \frac{\Gamma(\beta_1)\Gamma(\beta_2)}{\Gamma(\beta_1-p_1)\Gamma(\beta_2-p_2)} -$
$\frac{\Gamma(\beta_1)\Gamma(\beta_2)}{\Gamma(\beta_1-q_2)\Gamma(\beta_2-q_1)}\left(\sum_{i=1}^{N}a_i\xi_i^{\beta_2-q_1-1}\right)\left(\sum_{i=1}^{M}b_i\eta_i^{\beta_1-q_2-1}\right) > 0, \ r_i > 1,$
$\varphi_{r_i}(s) = |s|^{r_i-2}s, \ \varphi_{r_i}^{-1} = \varphi_{\varrho_i}, \ \varrho_i = \frac{r_i}{r_i-1}, \ i=1,2.$

(I2) The functions $f, \ g : [0,1] \times [0,\infty) \times [0,\infty) \to [0,\infty)$ are continuous.

For $[c_1,c_2] \subset [0,1]$ with $0 < c_1 < c_2 \leq 1$, we introduce the following extreme limits:

$$f_0^s = \limsup_{\substack{u+v\to 0+ \\ u,v\geq 0}} \max_{t\in[0,1]} \frac{f(t,u,v)}{(u+v)^{r_1-1}}, \quad g_0^s = \limsup_{\substack{u+v\to 0+ \\ u,v\geq 0}} \max_{t\in[0,1]} \frac{g(t,u,v)}{(u+v)^{r_2-1}},$$

$$f_0^i = \liminf_{\substack{u+v\to 0+ \\ u,v\geq 0}} \min_{t\in[c_1,c_2]} \frac{f(t,u,v)}{(u+v)^{r_1-1}}, \quad g_0^i = \liminf_{\substack{u+v\to 0+ \\ u,v\geq 0}} \min_{t\in[c_1,c_2]} \frac{g(t,u,v)}{(u+v)^{r_2-1}},$$

$$f_\infty^s = \limsup_{\substack{u+v\to\infty \\ u,v\geq 0}} \max_{t\in[0,1]} \frac{f(t,u,v)}{(u+v)^{r_1-1}}, \quad g_\infty^s = \limsup_{\substack{u+v\to\infty \\ u,v\geq 0}} \max_{t\in[0,1]} \frac{g(t,u,v)}{(u+v)^{r_2-1}},$$

$$f_\infty^i = \liminf_{\substack{u+v\to\infty \\ u,v\geq 0}} \min_{t\in[c_1,c_2]} \frac{f(t,u,v)}{(u+v)^{r_1-1}}, \quad g_\infty^i = \liminf_{\substack{u+v\to\infty \\ u,v\geq 0}} \min_{t\in[c_1,c_2]} \frac{g(t,u,v)}{(u+v)^{r_2-1}}.$$

By using Lemma 4.2.1 (relations (4.40)), (u,v) is solution of problem (4.32), (4.33) if and only if (u,v) is solution of the nonlinear system of integral equations

$$\begin{cases} u(t) = \lambda^{\varrho_1-1} \displaystyle\int_0^1 \widetilde{G}_1(t,s)\varphi_{\varrho_1}(I_{0+}^{\alpha_1} f(s,u(s),v(s)))\,ds \\[2mm] \qquad + \mu^{\varrho_2-1} \displaystyle\int_0^1 \widetilde{G}_2(t,s)\varphi_{\varrho_2}(I_{0+}^{\alpha_2} g(s,u(s),v(s)))\,ds, \quad t\in[0,1], \\[4mm] v(t) = \mu^{\varrho_2-1} \displaystyle\int_0^1 \widetilde{G}_3(t,s)\varphi_{\varrho_2}(I_{0+}^{\alpha_2} g(s,u(s),v(s)))\,ds \\[2mm] \qquad + \lambda^{\varrho_1-1} \displaystyle\int_0^1 \widetilde{G}_4(t,s)\varphi_{\varrho_1}(I_{0+}^{\alpha_1} f(s,u(s),v(s)))\,ds, \quad t\in[0,1]. \end{cases}$$

We consider the Banach space $X = C[0,1]$ with the supremum norm $\|\cdot\|$, and the Banach space $Y = X \times X$ with the norm $\|(u,v)\|_Y = \|u\| + \|v\|$. We define the cones

$$P_1 = \{u \in X, \; u(t) \geq t^{\beta_1-1}\|u\|, \; \forall t \in [0,1]\} \subset X,$$
$$P_2 = \{v \in X, \; v(t) \geq t^{\beta_2-1}\|v\|, \; \forall t \in [0,1]\} \subset X,$$

and $P = P_1 \times P_2 \subset Y$.

We define now the operators $\mathcal{Q}_1, \mathcal{Q}_2 : Y \to X$ and $\mathcal{Q} : Y \to Y$ by

$$\mathcal{Q}_1(u,v)(t) = \lambda^{\varrho_1-1} \int_0^1 \widetilde{G}_1(t,s)\varphi_{\varrho_1}(I_{0+}^{\alpha_1} f(s,u(s),v(s)))\,ds$$

$$\qquad + \mu^{\varrho_2-1} \int_0^1 \widetilde{G}_2(t,s)\varphi_{\varrho_2}(I_{0+}^{\alpha_2} g(s,u(s),v(s)))\,ds, \quad t\in[0,1],$$

$$\mathcal{Q}_2(u,v)(t) = \mu^{\varrho_2-1} \int_0^1 \widetilde{G}_3(t,s)\varphi_{\varrho_2}(I_{0+}^{\alpha_2} g(s,u(s),v(s)))\,ds$$

$$\qquad + \lambda^{\varrho_1-1} \int_0^1 \widetilde{G}_4(t,s)\varphi_{\varrho_1}(I_{0+}^{\alpha_1} f(s,u(s),v(s)))\,ds, \quad t\in[0,1],$$

and $\mathcal{Q}(u,v) = (\mathcal{Q}_1(u,v), \mathcal{Q}_2(u,v))$, $(u,v) \in Y$. Then (u,v) is a solution of problem (4.32), (4.33) if and only if (u,v) is a fixed point of operator $\mathcal{Q}$.

Lemma 4.2.3. *If* (I1)–(I2) *hold, then* $\mathcal{Q} : P \to P$ *is a completely continuous operator.*

Proof. Let $(u,v) \in P$ be an arbitrary element. Because $\mathcal{Q}_1(u,v)$ and $\mathcal{Q}_2(u,v)$ satisfy the problem (4.34), (4.33) for $h(t) = \lambda f(t, u(t), v(t))$ and $k(t) = \mu g(t, u(t), v(t))$, $t \in [0,1]$, then by Lemma 4.2.2, we obtain

$$\mathcal{Q}_1(u,v)(t) \leq \lambda^{\varrho_1 - 1} \int_0^1 \tilde{J}_1(s)\varphi_{\varrho_1}(I_{0+}^{\alpha_1} f(s, u(s), v(s)))\, ds$$

$$+ \mu^{\varrho_2 - 1} \int_0^1 \tilde{J}_2(s)\varphi_{\varrho_2}(I_{0+}^{\alpha_2} g(s, u(s), v(s)))\, ds, \quad \forall t \in [0,1],$$

$$\mathcal{Q}_2(u,v)(t) \leq \mu^{\varrho_2 - 1} \int_0^1 \tilde{J}_3(s)\varphi_{\varrho_2}(I_{0+}^{\alpha_2} g(s, u(s), v(s)))\, ds$$

$$+ \lambda^{\varrho_1 - 1} \int_0^1 \tilde{J}_4(s)\varphi_{\varrho_1}(I_{0+}^{\alpha_1} f(s, u(s), v(s)))\, ds, \quad \forall t \in [0,1],$$

and so

$$\|\mathcal{Q}_1(u,v)\| \leq \lambda^{\varrho_1 - 1} \int_0^1 \tilde{J}_1(s)\varphi_{\varrho_1}(I_{0+}^{\alpha_1} f(s, u(s), v(s)))\, ds$$

$$+ \mu^{\varrho_2 - 1} \int_0^1 \tilde{J}_2(s)\varphi_{\varrho_2}(I_{0+}^{\alpha_2} g(s, u(s), v(s)))\, ds,$$

$$\|\mathcal{Q}_2(u,v)\| \leq \mu^{\varrho_2 - 1} \int_0^1 \tilde{J}_3(s)\varphi_{\varrho_2}(I_{0+}^{\alpha_2} g(s, u(s), v(s)))\, ds$$

$$+ \lambda^{\varrho_1 - 1} \int_0^1 \tilde{J}_4(s)\varphi_{\varrho_1}(I_{0+}^{\alpha_1} f(s, u(s), v(s)))\, ds.$$

Therefore, we conclude that

$$\mathcal{Q}_1(u,v)(t) \geq \lambda^{\varrho_1 - 1} \int_0^1 t^{\beta_1 - 1} \tilde{J}_1(s)\varphi_{\varrho_1}(I_{0+}^{\alpha_1} f(s, u(s), v(s)))\, ds$$

$$+ \mu^{\varrho_2 - 1} \int_0^1 t^{\beta_1 - 1} \tilde{J}_2(s)\varphi_{\varrho_2}(I_{0+}^{\alpha_2} g(s, u(s), v(s)))\, ds$$

$$\geq t^{\beta_1-1}\|\mathcal{Q}_1(u,v)\|, \quad \forall t \in [0,1],$$

$$\mathcal{Q}_2(u,v)(t) \geq \mu^{\varrho_2-1} \int_0^1 t^{\beta_2-1} \widetilde{J}_3(s)\varphi_{\varrho_2}(I_{0+}^{\alpha_2}g(s,u(s),v(s)))\,ds$$

$$+ \lambda^{\varrho_1-1} \int_0^1 t^{\beta_2-1} \widetilde{J}_4(s)\varphi_{\varrho_1}(I_{0+}^{\alpha_1}f(s,u(s),v(s)))\,ds$$

$$\geq t^{\beta_2-1}\|\mathcal{Q}_2(u,v)\|, \quad \forall t \in [0,1].$$

Hence, $\mathcal{Q}(u,v) = (\mathcal{Q}_1(u,v), \mathcal{Q}_2(u,v)) \in P$, and then $\mathcal{Q}(P) \subset P$. By the continuity of the functions $f, g, \widetilde{G}_i,\ i = 1,\dots,4$, and the Arzela–Ascoli theorem, we can show that $\mathcal{Q}_1$ and $\mathcal{Q}_2$ are completely continuous operators, and then $\mathcal{Q}$ is a completely continuous operator. $\square$

For $[c_1, c_2] \subset [0,1]$ with $0 < c_1 < c_2 \leq 1$, we denote by

$$A = \frac{1}{(\Gamma(\alpha_1+1))^{\varrho_1-1}} \int_0^1 s^{\alpha_1(\varrho_1-1)} \widetilde{J}_1(s)\,ds,$$

$$B = \frac{1}{(\Gamma(\alpha_2+1))^{\varrho_2-1}} \int_0^1 s^{\alpha_2(\varrho_2-1)} \widetilde{J}_2(s)\,ds,$$

$$C = \frac{1}{(\Gamma(\alpha_2+1))^{\varrho_2-1}} \int_0^1 s^{\alpha_2(\varrho_2-1)} \widetilde{J}_3(s)\,ds,$$

$$D = \frac{1}{(\Gamma(\alpha_1+1))^{\varrho_1-1}} \int_0^1 s^{\alpha_1(\varrho_1-1)} \widetilde{J}_4(s)\,ds,$$

$$\widetilde{A} = \frac{1}{(\Gamma(\alpha_1+1))^{\varrho_1-1}} \int_{c_1}^{c_2} (s-c_1)^{\alpha_1(\varrho_1-1)} \widetilde{J}_1(s)\,ds, \qquad (4.41)$$

$$\widetilde{B} = \frac{1}{(\Gamma(\alpha_2+1))^{\varrho_2-1}} \int_{c_1}^{c_2} (s-c_1)^{\alpha_2(\varrho_2-1)} \widetilde{J}_2(s)\,ds,$$

$$\widetilde{C} = \frac{1}{(\Gamma(\alpha_2+1))^{\varrho_2-1}} \int_{c_1}^{c_2} (s-c_1)^{\alpha_2(\varrho_2-1)} \widetilde{J}_3(s)\,ds,$$

$$\widetilde{D} = \frac{1}{(\Gamma(\alpha_1+1))^{\varrho_1-1}} \int_{c_1}^{c_2} (s-c_1)^{\alpha_1(\varrho_1-1)} \widetilde{J}_4(s)\,ds,$$

where $\widetilde{J}_i,\ i = 1,\dots,4$ are defined in Lemma 4.2.2.

First, for $f_0^s, g_0^s, f_\infty^i, g_\infty^i \in (0, \infty)$ and numbers $\gamma_1, \gamma_2 \in [0,1]$, γ_3, $\gamma_4 \in (0,1)$, $a \in [0,1]$ and $b \in (0,1)$, we define the numbers

$$L_1 = \max\left\{ \frac{1}{f_\infty^i}\left(\frac{a\gamma_1}{\theta\theta_1\widetilde{A}}\right)^{r_1-1}, \frac{1}{f_\infty^i}\left(\frac{(1-a)\gamma_2}{\theta\theta_2\widetilde{D}}\right)^{r_1-1} \right\},$$

$$L_2 = \min\left\{ \frac{1}{f_0^s}\left(\frac{b\gamma_3}{A}\right)^{r_1-1}, \frac{1}{f_0^s}\left(\frac{(1-b)\gamma_4}{D}\right)^{r_1-1} \right\},$$

$$L_3 = \max\left\{ \frac{1}{g_\infty^i}\left(\frac{a(1-\gamma_1)}{\theta\theta_1\widetilde{B}}\right)^{r_2-1}, \frac{1}{g_\infty^i}\left(\frac{(1-a)(1-\gamma_2)}{\theta\theta_2\widetilde{C}}\right)^{r_2-1} \right\},$$

$$L_4 = \min\left\{ \frac{1}{g_0^s}\left(\frac{b(1-\gamma_3)}{B}\right)^{r_2-1}, \frac{1}{g_0^s}\left(\frac{(1-b)(1-\gamma_4)}{C}\right)^{r_2-1} \right\},$$

$$L_2' = \min\left\{ \frac{1}{f_0^s}\left(\frac{b}{A}\right)^{r_1-1}, \frac{1}{f_0^s}\left(\frac{1-b}{D}\right)^{r_1-1} \right\},$$

$$L_4' = \min\left\{ \frac{1}{g_0^s}\left(\frac{b}{B}\right)^{r_2-1}, \frac{1}{g_0^s}\left(\frac{1-b}{C}\right)^{r_2-1} \right\},$$

where $\theta_1 = c_1^{\beta_1-1}$, $\theta_2 = c_1^{\beta_2-1}$, $\theta = \min\{\theta_1, \theta_2\}$.

Theorem 4.2.1. *Assume that* (I1) *and* (I2) *hold*, $[c_1, c_2] \subset [0,1]$ *with* $0 < c_1 < c_2 \leq 1$, $\gamma_1, \gamma_2 \in [0,1]$, $\gamma_3, \gamma_4 \in (0,1)$, $a \in [0,1]$ *and* $b \in (0,1)$.

(1) *If* $f_0^s, g_0^s, f_\infty^i, g_\infty^i \in (0, \infty)$, $L_1 < L_2$ *and* $L_3 < L_4$, *then for each* $\lambda \in (L_1, L_2)$ *and* $\mu \in (L_3, L_4)$ *there exists a positive solution* $(u(t), v(t))$, $t \in [0,1]$ *for* (4.32), (4.33).

(2) *If* $f_0^s = 0$, $g_0^s, f_\infty^i, g_\infty^i \in (0, \infty)$ *and* $L_3 < L_4'$, *then for each* $\lambda \in (L_1, \infty)$ *and* $\mu \in (L_3, L_4')$ *there exists a positive solution* $(u(t), v(t))$, $t \in [0,1]$ *for* (4.32), (4.33).

(3) *If* $g_0^s = 0$, $f_0^s, f_\infty^i, g_\infty^i \in (0, \infty)$ *and* $L_1 < L_2'$, *then for each* $\lambda \in (L_1, L_2')$ *and* $\mu \in (L_3, \infty)$ *there exists a positive solution* $(u(t), v(t))$, $t \in [0,1]$ *for* (4.32), (4.33).

(4) *If* $f_0^s = g_0^s = 0$, $f_\infty^i, g_\infty^i \in (0, \infty)$, *then for each* $\lambda \in (L_1, \infty)$ *and* $\mu \in (L_3, \infty)$ *there exists a positive solution* $(u(t), v(t))$, $t \in [0,1]$ *for* (4.32), (4.33).

(5) *If $f_0^s, g_0^s \in (0, \infty)$ and at least one of f_∞^i, g_∞^i is ∞, then for each $\lambda \in (0, L_2)$ and $\mu \in (0, L_4)$ there exists a positive solution $(u(t), v(t))$, $t \in [0, 1]$ for (4.32), (4.33).*

(6) *If $f_0^s = 0$, $g_0^s \in (0, \infty)$ and at least one of f_∞^i, g_∞^i is ∞, then for each $\lambda \in (0, \infty)$ and $\mu \in (0, L_4')$ there exists a positive solution $(u(t), v(t))$, $t \in [0, 1]$ for (4.32), (4.33).*

(7) *If $f_0^s \in (0, \infty)$, $g_0^s = 0$ and at least one of f_∞^i, g_∞^i is ∞, then for each $\lambda \in (0, L_2')$ and $\mu \in (0, \infty)$ there exists a positive solution $(u(t), v(t))$, $t \in [0, 1]$ for (4.32), (4.33).*

(8) *If $f_0^s = g_0^s = 0$ and at least one of f_∞^i, g_∞^i is ∞, then for each $\lambda \in (0, \infty)$ and $\mu \in (0, \infty)$ there exists a positive solution $(u(t), v(t))$, $t \in [0, 1]$ for (4.32), (4.33).*

Proof. We consider the above cone $P \subset Y$ and the operators $\mathcal{Q}_1$, $\mathcal{Q}_2$ and $\mathcal{Q}$. Because the proofs of the above cases are similar, in what follows, we will prove two of them, namely Cases (1) and (7).

Case (1). We have $f_0^s, g_0^s, f_\infty^i, g_\infty^i \in (0, \infty)$, $L_1 < L_2$ and $L_3 < L_4$. Let $\lambda \in (L_1, L_2)$ and $\mu \in (L_3, L_4)$. We consider $\varepsilon > 0$ such that $\varepsilon < f_\infty^i$, $\varepsilon < g_\infty^i$ and

$$
\max \left\{ \frac{1}{f_\infty^i - \varepsilon} \left(\frac{a\gamma_1}{\theta\theta_1 \widetilde{A}} \right)^{r_1 - 1}, \frac{1}{f_\infty^i - \varepsilon} \left(\frac{(1-a)\gamma_2}{\theta\theta_2 \widetilde{D}} \right)^{r_1 - 1} \right\} \leq \lambda
$$

$$
\leq \min \left\{ \frac{1}{f_0^s + \varepsilon} \left(\frac{b\gamma_3}{A} \right)^{r_1 - 1}, \frac{1}{f_0^s + \varepsilon} \left(\frac{(1-b)\gamma_4}{D} \right)^{r_1 - 1} \right\},
$$

$$
\max \left\{ \frac{1}{g_\infty^i - \varepsilon} \left(\frac{a(1-\gamma_1)}{\theta\theta_1 \widetilde{B}} \right)^{r_2 - 1}, \frac{1}{g_\infty^i - \varepsilon} \left(\frac{(1-a)(1-\gamma_2)}{\theta\theta_2 \widetilde{C}} \right)^{r_2 - 1} \right\} \leq \mu
$$

$$
\leq \min \left\{ \frac{1}{g_0^s + \varepsilon} \left(\frac{b(1-\gamma_3)}{B} \right)^{r_2 - 1}, \frac{1}{g_0^s + \varepsilon} \left(\frac{(1-b)(1-\gamma_4)}{C} \right)^{r_2 - 1} \right\}.
$$

By using (I2) and the definition of f_0^s and g_0^s, we deduce that there exists $R_1 > 0$ such that

$$
f(t, u, v) \leq (f_0^s + \varepsilon)(u + v)^{r_1 - 1}, \quad g(t, u, v) \leq (g_0^s + \varepsilon)(u + v)^{r_2 - 1},
$$

for all $t \in [0, 1]$ and $u, v \geq 0$, $u + v \leq R_1$.

We define the set $\Omega_1 = \{(u, v) \in Y, \|(u, v)\|_Y < R_1\}$. Now, let $(u, v) \in P \cap \partial\Omega_1$, that is, $(u, v) \in P$ with $\|(u, v)\|_Y = R_1$, or equivalently $\|u\| + \|v\| = R_1$. Then $u(t) + v(t) \leq R_1$ for all $t \in [0, 1]$, and by

Lemma 4.2.2, we obtain

$$\mathcal{Q}_1(u,v)(t) \leq \lambda^{\varrho_1-1} \int_0^1 \tilde{J}_1(s)\varphi_{\varrho_1} \left(\frac{1}{\Gamma(\alpha_1)} \int_0^s (s-\tau)^{\alpha_1-1} \right.$$

$$\left. \times f(\tau, u(\tau), v(\tau))\, d\tau \right) ds$$

$$+ \mu^{\varrho_2-1} \int_0^1 \tilde{J}_2(s)\varphi_{\varrho_2} \left(\frac{1}{\Gamma(\alpha_2)} \int_0^s (s-\tau)^{\alpha_2-1} \right.$$

$$\left. \times g(\tau, u(\tau), v(\tau))\, d\tau \right) ds$$

$$\leq \lambda^{\varrho_1-1} \int_0^1 \tilde{J}_1(s)\varphi_{\varrho_1} \left(\frac{1}{\Gamma(\alpha_1)} \int_0^s (s-\tau)^{\alpha_1-1}(f_0^s + \varepsilon) \right.$$

$$\left. \times (u(\tau)+v(\tau))^{r_1-1}\, d\tau \right) ds$$

$$+ \mu^{\varrho_2-1} \int_0^1 \tilde{J}_2(s)\varphi_{\varrho_2} \left(\frac{1}{\Gamma(\alpha_2)} \int_0^s (s-\tau)^{\alpha_2-1}(g_0^s + \varepsilon) \right.$$

$$\left. \times (u(\tau)+v(\tau))^{r_2-1}\, d\tau \right) ds$$

$$\leq \lambda^{\varrho_1-1}(f_0^s + \varepsilon)^{\varrho_1-1} \int_0^1 \tilde{J}_1(s)\varphi_{\varrho_1} \left(\frac{1}{\Gamma(\alpha_1)} \int_0^s (s-\tau)^{\alpha_1-1} \right.$$

$$\left. \times (\|u\|+\|v\|)^{r_1-1}\, d\tau \right) ds$$

$$+ \mu^{\varrho_2-1}(g_0^s + \varepsilon)^{\varrho_2-1} \int_0^1 \tilde{J}_2(s)\varphi_{\varrho_2} \left(\frac{1}{\Gamma(\alpha_2)} \int_0^s (s-\tau)^{\alpha_2-1} \right.$$

$$\left. \times (\|u\|+\|v\|)^{r_2-1}\, d\tau \right) ds$$

$$= \lambda^{\varrho_1-1}(f_0^s + \varepsilon)^{\varrho_1-1}\|(u,v)\|_Y$$

$$\times \int_0^1 \tilde{J}_1(s)\frac{1}{(\Gamma(\alpha_1+1))^{\varrho_1-1}} s^{\alpha_1(\varrho_1-1)}\, ds$$

$$+ \mu^{\varrho_2-1}(g_0^s + \varepsilon)^{\varrho_2-1}\|(u,v)\|_Y$$

$$\times \int_0^1 \widetilde{J}_2(s) \frac{1}{(\Gamma(\alpha_2+1))^{\varrho_2-1}} s^{\alpha_2(\varrho_2-1)} \, ds$$

$$= [\lambda^{\varrho_1-1}(f_0^s+\varepsilon)^{\varrho_1-1}A + \mu^{\varrho_2-1}(g_0^s+\varepsilon)^{\varrho_2-1}B]\|(u,v)\|_Y$$

$$\leq [b\gamma_3 + b(1-\gamma_3)]\|(u,v)\|_Y = b\|(u,v)\|_Y, \quad \forall t \in [0,1].$$

Therefore, $\|\mathcal{Q}_1(u,v)\| \leq b\|(u,v)\|_Y$.

In a similar manner, we conclude

$$\mathcal{Q}_2(u,v)(t) \leq \mu^{\varrho_2-1} \int_0^1 \widetilde{J}_3(s)\varphi_{\varrho_2}\left(\frac{1}{\Gamma(\alpha_2)}\int_0^s (s-\tau)^{\alpha_2-1}\right.$$

$$\times g(\tau, u(\tau), v(\tau)) \, d\tau \bigg) \, ds$$

$$+ \lambda^{\varrho_1-1} \int_0^1 \widetilde{J}_4(s)\varphi_{\varrho_1}\left(\frac{1}{\Gamma(\alpha_1)}\int_0^s (s-\tau)^{\alpha_1-1}\right.$$

$$\times f(\tau, u(\tau), v(\tau)) \, d\tau \bigg) \, ds$$

$$\leq \mu^{\varrho_2-1} \int_0^1 \widetilde{J}_3(s)\varphi_{\varrho_2}\left(\frac{1}{\Gamma(\alpha_2)}\int_0^s (s-\tau)^{\alpha_2-1}(g_0^s+\varepsilon)\right.$$

$$\times (u(\tau)+v(\tau))^{r_2-1} \, d\tau \bigg) \, ds$$

$$+ \lambda^{\varrho_1-1} \int_0^1 \widetilde{J}_4(s)\varphi_{\varrho_1}\left(\frac{1}{\Gamma(\alpha_1)}\int_0^s (s-\tau)^{\alpha_1-1}(f_0^s+\varepsilon)\right.$$

$$\times (u(\tau)+v(\tau))^{r_1-1} \, d\tau \bigg) \, ds$$

$$\leq \mu^{\varrho_2-1}(g_0^s+\varepsilon)^{\varrho_2-1} \int_0^1 \widetilde{J}_3(s)\varphi_{\varrho_2}\left(\frac{1}{\Gamma(\alpha_2)}\int_0^s (s-\tau)^{\alpha_2-1}\right.$$

$$\times (\|u\|+\|v\|)^{r_2-1} \, d\tau \bigg) \, ds$$

$$+ \lambda^{\varrho_1-1}(f_0^s+\varepsilon)^{\varrho_1-1} \int_0^1 \widetilde{J}_4(s)\varphi_{\varrho_1}\left(\frac{1}{\Gamma(\alpha_1)}\int_0^s (s-\tau)^{\alpha_1-1}\right.$$

$$\times (\|u\|+\|v\|)^{r_1-1} \, d\tau \bigg) \, ds$$

$$= \mu^{\varrho_2-1}(g_0^s+\varepsilon)^{\varrho_2-1}\|(u,v)\|_Y$$

$$\times \int_0^1 \widetilde{J_3}(s)\frac{1}{(\Gamma(\alpha_2+1))^{\varrho_2-1}}s^{\alpha_2(\varrho_2-1)}\,ds$$

$$+\lambda^{\varrho_1-1}(f_0^s+\varepsilon)^{\varrho_1-1}\|(u,v)\|_Y$$

$$\times \int_0^1 \widetilde{J_4}(s)\frac{1}{(\Gamma(\alpha_1+1))^{\varrho_1-1}}s^{\alpha_1(\varrho_1-1)}\,ds$$

$$=[\mu^{\varrho_2-1}(g_0^s+\varepsilon)^{\varrho_2-1}C+\lambda^{\varrho_1-1}(f_0^s+\varepsilon)^{\varrho_1-1}D]\|(u,v)\|_Y$$

$$\leq [(1-b)(1-\gamma_4)+(1-b)\gamma_4]\|(u,v)\|_Y$$

$$=(1-b)\|(u,v)\|_Y, \quad \forall t \in [0,1].$$

Hence, $\|\mathcal{Q}_2(u,v)\| \leq (1-b)\|(u,v)\|_Y$.

Then for $(u,v) \in P \cap \partial\Omega_1$, we deduce

$$\|\mathcal{Q}(u,v)\|_Y = \|\mathcal{Q}_1(u,v)\| + \|\mathcal{Q}_2(u,v)\|$$

$$\leq b\|(u,v)\|_Y + (1-b)\|(u,v)\|_Y = \|(u,v)\|_Y. \qquad (4.42)$$

By the definition of f_∞^i and g_∞^i, there exists $\overline{R}_2 > 0$ such that

$$f(t,u,v) \geq (f_\infty^i - \varepsilon)(u+v)^{r_1-1}, \quad g(t,u,v) \geq (g_\infty^i - \varepsilon)(u+v)^{r_2-1},$$

for all $t \in [c_1,c_2]$ and $u,v \geq 0$, $u+v \geq \overline{R}_2$.

We consider $R_2 = \max\{2R_1, \overline{R}_2/\theta\}$ and we define the set $\Omega_2 = \{(u,v) \in Y, \|(u,v)\|_Y < R_2\}$. Then for $(u,v) \in P \cap \partial\Omega_2$, we obtain

$$u(t)+v(t) \geq \min_{t\in[c_1,c_2]} t^{\beta_1-1}\|u\| + \min_{t\in[c_1,c_2]} t^{\beta_2-1}\|v\| = c_1^{\beta_1-1}\|u\| + c_1^{\beta_2-1}\|v\|$$

$$= \theta_1\|u\| + \theta_2\|v\| \geq \theta\|(u,v)\|_Y = \theta R_2 \geq \overline{R}_2, \quad \forall t \in [c_1,c_2].$$

Therefore, by Lemma 4.2.2, we conclude

$$\mathcal{Q}_1(u,v)(c_1) \geq \lambda^{\varrho_1-1}\int_0^1 c_1^{\beta_1-1}\widetilde{J_1}(s)\varphi_{\varrho_1}(I_{0+}^{\alpha_1}f(s,u(s),v(s)))\,ds$$

$$+\mu^{\varrho_2-1}\int_0^1 c_1^{\beta_1-1}\widetilde{J_2}(s)\varphi_{\varrho_2}(I_{0+}^{\alpha_2}g(s,u(s),v(s)))\,ds$$

$$\geq \lambda^{\varrho_1-1}c_1^{\beta_1-1}\int_{c_1}^{c_2}\widetilde{J_1}(s)\varphi_{\varrho_1}\left(\frac{1}{\Gamma(\alpha_1)}\int_{c_1}^s (s-\tau)^{\alpha_1-1}\right.$$

$$\left.\times f(\tau,u(\tau),v(\tau))\,d\tau\right)ds$$

$$+ \mu^{\varrho_2 - 1} c_1^{\beta_1 - 1} \int_{c_1}^{c_2} \tilde{J}_2(s) \varphi_{\varrho_2} \left(\frac{1}{\Gamma(\alpha_2)} \int_{c_1}^{s} (s - \tau)^{\alpha_2 - 1} \right.$$

$$\left. \times \, g(\tau, u(\tau), v(\tau)) \, d\tau \right) ds$$

$$\geq \lambda^{\varrho_1 - 1} c_1^{\beta_1 - 1} \int_{c_1}^{c_2} \tilde{J}_1(s) \varphi_{\varrho_1} \left(\frac{1}{\Gamma(\alpha_1)} \int_{c_1}^{s} (s - \tau)^{\alpha_1 - 1} \right.$$

$$\left. \times \, (f_\infty^i - \varepsilon)(u(\tau) + v(\tau))^{r_1 - 1} \, d\tau \right) ds$$

$$+ \mu^{\varrho_2 - 1} c_1^{\beta_1 - 1} \int_{c_1}^{c_2} \tilde{J}_2(s) \varphi_{\varrho_2} \left(\frac{1}{\Gamma(\alpha_2)} \int_{c_1}^{s} (s - \tau)^{\alpha_2 - 1} \right.$$

$$\left. \times \, (g_\infty^i - \varepsilon)(u(\tau) + v(\tau))^{r_2 - 1} \, d\tau \right) ds$$

$$\geq \lambda^{\varrho_1 - 1} c_1^{\beta_1 - 1} \int_{c_1}^{c_2} \tilde{J}_1(s) \varphi_{\varrho_1} \left(\frac{1}{\Gamma(\alpha_1)} \int_{c_1}^{s} (s - \tau)^{\alpha_1 - 1} \right.$$

$$\left. \times \, (f_\infty^i - \varepsilon)(\theta \| (u, v) \|_Y)^{r_1 - 1} \, d\tau \right) ds$$

$$+ \mu^{\varrho_2 - 1} c_1^{\beta_1 - 1} \int_{c_1}^{c_2} \tilde{J}_2(s) \varphi_{\varrho_2} \left(\frac{1}{\Gamma(\alpha_2)} \int_{c_1}^{s} (s - \tau)^{\alpha_2 - 1} \right.$$

$$\left. \times \, (g_\infty^i - \varepsilon)(\theta \| (u, v) \|_Y)^{r_2 - 1} \, d\tau \right) ds$$

$$= \theta \theta_1 \lambda^{\varrho_1 - 1} (f_\infty^i - \varepsilon)^{\varrho_1 - 1} \| (u, v) \|_Y$$

$$\times \int_{c_1}^{c_2} \tilde{J}_1(s) \frac{1}{(\Gamma(\alpha_1 + 1))^{\varrho_1 - 1}} (s - c_1)^{\alpha_1(\varrho_1 - 1)} ds$$

$$+ \theta \theta_1 \mu^{\varrho_2 - 1} (g_\infty^i - \varepsilon)^{\varrho_2 - 1} \| (u, v) \|_Y$$

$$\times \int_{c_1}^{c_2} \tilde{J}_2(s) \frac{1}{(\Gamma(\alpha_2 + 1))^{\varrho_2 - 1}} (s - c_1)^{\alpha_2(\varrho_2 - 1)} ds$$

$$= [\theta \theta_1 \lambda^{\varrho_1 - 1} (f_\infty^i - \varepsilon)^{\varrho_1 - 1} \tilde{A} + \theta \theta_1 \mu^{\varrho_2 - 1}$$

$$\times (g_\infty^i - \varepsilon)^{\varrho_2 - 1} \tilde{B}] \| (u, v) \|_Y$$

$$\geq [a \gamma_1 + a(1 - \gamma_1)] \| (u, v) \|_Y = a \| (u, v) \|_Y.$$

So, $\| \mathcal{Q}_1(u, v) \| \geq \mathcal{Q}_1(u, v)(c_1) \geq a \| (u, v) \|_Y$.

In a similar manner, we deduce

$$\mathcal{Q}_2(u,v)(c_1) \geq \mu^{\varrho_2 - 1} \int_0^1 c_1^{\beta_2 - 1} \widetilde{J}_3(s) \varphi_{\varrho_2} (I_{0+}^{\alpha_2} g(s, u(s), v(s))) \, ds$$

$$+ \lambda^{\varrho_1 - 1} \int_0^1 c_1^{\beta_2 - 1} \widetilde{J}_4(s) \varphi_{\varrho_1} (I_{0+}^{\alpha_1} f(s, u(s), v(s))) \, ds$$

$$\geq \mu^{\varrho_2 - 1} c_1^{\beta_2 - 1} \int_{c_1}^{c_2} \widetilde{J}_3(s) \varphi_{\varrho_2} \left(\frac{1}{\Gamma(\alpha_2)} \int_{c_1}^{s} (s - \tau)^{\alpha_2 - 1} \right.$$

$$\times \, g(\tau, u(\tau), v(\tau)) \, d\tau \Bigg) \, ds$$

$$+ \lambda^{\varrho_1 - 1} c_1^{\beta_2 - 1} \int_{c_1}^{c_2} \widetilde{J}_4(s) \varphi_{\varrho_1} \left(\frac{1}{\Gamma(\alpha_1)} \int_{c_1}^{s} (s - \tau)^{\alpha_1 - 1} \right.$$

$$\times \, f(\tau, u(\tau), v(\tau)) \, d\tau \Bigg) \, ds$$

$$\geq \mu^{\varrho_2 - 1} c_1^{\beta_2 - 1} \int_{c_1}^{c_2} \widetilde{J}_3(s) \varphi_{\varrho_2} \left(\frac{1}{\Gamma(\alpha_2)} \int_{c_1}^{s} (s - \tau)^{\alpha_2 - 1} \right.$$

$$\times \, (g_\infty^i - \varepsilon)(u(\tau) + v(\tau))^{r_2 - 1} \, d\tau \Bigg) \, ds$$

$$+ \lambda^{\varrho_1 - 1} c_1^{\beta_2 - 1} \int_{c_1}^{c_2} \widetilde{J}_4(s) \varphi_{\varrho_1} \left(\frac{1}{\Gamma(\alpha_1)} \int_{c_1}^{s} (s - \tau)^{\alpha_1 - 1} \right.$$

$$\times \, (f_\infty^i - \varepsilon)(u(\tau) + v(\tau))^{r_1 - 1} \, d\tau \Bigg) \, ds$$

$$\geq \mu^{\varrho_2 - 1} c_1^{\beta_2 - 1} \int_{c_1}^{c_2} \widetilde{J}_3(s) \varphi_{\varrho_2} \left(\frac{1}{\Gamma(\alpha_2)} \int_{c_1}^{s} (s - \tau)^{\alpha_2 - 1} \right.$$

$$\times \, (g_\infty^i - \varepsilon)(\theta \| (u, v) \|_Y)^{r_2 - 1} \, d\tau \Bigg) \, ds$$

$$+ \lambda^{\varrho_1 - 1} c_1^{\beta_2 - 1} \int_{c_1}^{c_2} \widetilde{J}_4(s) \varphi_{\varrho_1} \left(\frac{1}{\Gamma(\alpha_1)} \int_{c_1}^{s} (s - \tau)^{\alpha_1 - 1} \right.$$

$$\times \, (f_\infty^i - \varepsilon)(\theta \| (u, v) \|_Y)^{r_1 - 1} \, d\tau \Bigg) \, ds$$

$$= \theta \theta_2 \mu^{\varrho_2 - 1} (g_\infty^i - \varepsilon)^{\varrho_2 - 1} \| (u, v) \|_Y$$

$$\times \int_{c_1}^{c_2} \widetilde{J}_3(s) \frac{1}{(\Gamma(\alpha_2 + 1))^{\varrho_2 - 1}} (s - c_1)^{\alpha_2 (\varrho_2 - 1)} \, ds$$

$$+ \theta \theta_2 \lambda^{\varrho_1 - 1} (f_\infty^i - \varepsilon)^{\varrho_1 - 1} \| (u, v) \|_Y$$

$$\times \int_{c_1}^{c_2} \widetilde{J}_4(s) \frac{1}{(\Gamma(\alpha_1+1))^{\varrho_1-1}} (s-c_1)^{\alpha_1(\varrho_1-1)} ds$$

$$= [\theta\theta_2\mu^{\varrho_2-1}(g_\infty^i - \varepsilon)^{\varrho_2-1}\widetilde{C} + \theta\theta_2\lambda^{\varrho_1-1}$$

$$\times (f_\infty^i - \varepsilon)^{\varrho_1-1}\widetilde{D}] \|(u,v)\|_Y$$

$$\geq [(1-a)(1-\gamma_2) + (1-a)\gamma_2]\|(u,v)\|_Y = (1-a)\|(u,v)\|_Y.$$

So, $\|Q_2(u,v)\| \geq Q_2(u,v)(c_1) \geq (1-a)\|(u,v)\|_Y$.

Hence, for $(u,v) \in P \cap \partial\Omega_2$ we obtain

$$\|Q(u,v)\|_Y = \|Q_1(u,v)\| + \|Q_2(u,v)\|$$

$$\geq a\|(u,v)\|_Y + (1-a)\|(u,v)\|_Y = \|(u,v)\|_Y. \qquad (4.43)$$

By using (4.42), (4.43), Lemma 4.2.3 and the Guo–Krasnosel'skii fixed point theorem (Theorem 1.2.2), we deduce that Q has a fixed point $(u,v) \in P \cap (\overline{\Omega}_2 \setminus \Omega_1)$ such that $R_1 \leq \|u\| + \|v\| \leq R_2$, $u(t) \geq t^{\beta_1-1}\|u\|$, $v(t) \geq t^{\beta_2-1}\|v\|$ for all $t \in [0,1]$. If $\|u\| > 0$ then $u(t) > 0$ for all $t \in (0,1]$ and if $\|v\| > 0$ then $v(t) > 0$ for all $t \in (0,1]$. So (u,v) is a positive solution for problem (4.32), (4.33).

Case (7). We consider here $g_0^s = 0$, $f_0^s \in (0,\infty)$ and $g_\infty^i = \infty$. Let $\lambda \in (0, L_2')$ and $\mu \in (0, \infty)$. Instead of the numbers $\gamma_3, \gamma_4 \in (0,1)$ used in the first case, we choose $\widetilde{\gamma}_3 \in ((\lambda f_0^s)^{\varrho_1-1}\frac{A}{b}, 1)$ and $\widetilde{\gamma}_4 \in ((\lambda f_0^s)^{\varrho_1-1}\frac{D}{1-b}, 1)$. The choice of $\widetilde{\gamma}_3$ and $\widetilde{\gamma}_4$ is possible because $\lambda < \frac{1}{f_0^s}(\frac{b}{A})^{r_1-1}$ and $\lambda < \frac{1}{f_0^s}(\frac{1-b}{D})^{r_1-1}$. Let $\varepsilon > 0$ be such that

$$\lambda \leq \min\left\{ \frac{1}{f_0^s + \varepsilon}\left(\frac{b\widetilde{\gamma}_3}{A}\right)^{r_1-1}, \frac{1}{f_0^s + \varepsilon}\left(\frac{(1-b)\widetilde{\gamma}_4}{D}\right)^{r_1-1} \right\},$$

$$\varepsilon\left(\frac{1}{\theta\theta_1\widetilde{B}}\right)^{r_2-1} \leq \mu \leq \min\left\{ \frac{1}{\varepsilon}\left(\frac{b(1-\widetilde{\gamma}_3)}{B}\right)^{r_2-1}, \right.$$

$$\left. \frac{1}{\varepsilon}\left(\frac{(1-b)(1-\widetilde{\gamma}_4)}{C}\right)^{r_2-1} \right\}.$$

By using (I2) and the definition of f_0^s and g_0^s, we deduce that there exists $R_1 > 0$ such that

$$f(t,u,v) \leq (f_0^s + \varepsilon)(u+v)^{r_1-1}, \quad g(t,u,v) \leq \varepsilon(u+v)^{r_2-1}$$

for all $t \in [0,1]$ and $u,v \geq 0$, $u+v \leq R_1$.

We define the set $\Omega_1 = \{(u,v) \in Y, \ \|(u,v)\|_Y < R_1\}$. In a similar manner as in the proof of Case (1), for any $(u,v) \in P \cap \partial\Omega_1$, we obtain

$$\mathcal{Q}_1(u,v)(t) \leq [\lambda^{\varrho_1-1}(f_0^s + \varepsilon)^{\varrho_1-1}A + \mu^{\varrho_2-1}\varepsilon^{\varrho_2-1}B]\|(u,v)\|_Y$$
$$\leq [b\tilde{\gamma}_3 + b(1-\tilde{\gamma}_3)]\|(u,v)\|_Y = b\|(u,v)\|_Y, \quad \forall t \in [0,1],$$

$$\mathcal{Q}_2(u,v)(t) \leq [\mu^{\varrho_2-1}\varepsilon^{\varrho_2-1}C + \lambda^{\varrho_1-1}(f_0^s + \varepsilon)^{\varrho_1-1}D]\|(u,v)\|_Y$$
$$\leq [(1-b)(1-\tilde{\gamma}_4) + (1-b)\tilde{\gamma}_4]\|(u,v)\|_Y$$
$$= (1-b)\|(u,v)\|_Y, \quad \forall t \in [0,1],$$

and so $\|\mathcal{Q}(u,v)\|_Y \leq \|(u,v)\|_Y$.

For the second part of the proof, by the definition of g_∞^i, there exists $\overline{R}_2 > 0$ such that

$$g(t,u,v) \geq \frac{1}{\varepsilon}(u+v)^{r_2-1}, \quad \forall t \in [c_1,c_2], \ u,v \geq 0, \ u+v \geq \overline{R}_2.$$

We consider $R_2 = \max\{2R_1, \overline{R}_2/\theta\}$ and we define $\Omega_2 = \{(u,v) \in Y, \ \|(u,v)\|_Y < R_2\}$. Then for $(u,v) \in P \cap \partial\Omega_2$, we deduce as in Case (1) that $u(t) + v(t) \geq \theta R_2 \geq \overline{R}_2$ for all $t \in [c_1,c_2]$.

Then by Lemma 4.2.2, we have

$$\mathcal{Q}_1(u,v)(c_1) \geq \lambda^{\varrho_1-1}\int_0^1 c_1^{\beta_1-1}\tilde{J}_1(s)\varphi_{\varrho_1}(I_{0+}^{\alpha_1}f(s,u(s),v(s)))\,ds$$

$$+ \mu^{\varrho_2-1}\int_0^1 c_1^{\beta_1-1}\tilde{J}_2(s)\varphi_{\varrho_2}(I_{0+}^{\alpha_2}g(s,u(s),v(s)))\,ds$$

$$\geq \mu^{\varrho_2-1}\int_0^1 c_1^{\beta_1-1}\tilde{J}_2(s)\varphi_{\varrho_2}(I_{0+}^{\alpha_2}g(s,u(s),v(s)))\,ds$$

$$\geq \mu^{\varrho_2-1}c_1^{\beta_1-1}\int_{c_1}^{c_2}\tilde{J}_2(s)\varphi_{\varrho_2}\left(\frac{1}{\Gamma(\alpha_2)}\int_{c_1}^s (s-\tau)^{\alpha_2-1}\right.$$
$$\left. \times\, g(\tau,u(\tau),v(\tau))\,d\tau\right)ds$$

$$\geq \mu^{\varrho_2-1}c_1^{\beta_1-1}\int_{c_1}^{c_2}\tilde{J}_2(s)\varphi_{\varrho_2}\left(\frac{1}{\Gamma(\alpha_2)}\int_{c_1}^s (s-\tau)^{\alpha_2-1}\right.$$
$$\left. \times\, \frac{1}{\varepsilon}(u(\tau)+v(\tau))^{r_2-1}d\tau\right)ds$$

$$\geq \mu^{\varrho_2-1}c_1^{\beta_1-1}\int_{c_1}^{c_2}\tilde{J}_2(s)\varphi_{\varrho_2}\left(\frac{1}{\Gamma(\alpha_2)}\int_{c_1}^s (s-\tau)^{\alpha_2-1}\right.$$

$$\times \frac{1}{\varepsilon}(\theta\|(u,v)\|_Y)^{r_2-1}\,d\tau\Big)\,ds$$

$$= \theta\theta_1\mu^{\varrho_2-1}\left(\frac{1}{\varepsilon}\right)^{\varrho_2-1}\|(u,v)\|_Y$$

$$\times \int_{c_1}^{c_2}\widetilde{J}_2(s)\frac{1}{(\Gamma(\alpha_2+1))^{\varrho_2-1}}(s-c_1)^{\alpha_2(\varrho_2-1)}ds$$

$$= \theta\theta_1\mu^{\varrho_2-1}\left(\frac{1}{\varepsilon}\right)^{\varrho_2-1}\|(u,v)\|_Y\widetilde{B} \geq \|(u,v)\|_Y.$$

So, we conclude that $\|\mathcal{Q}_1(u,v)\| \geq \mathcal{Q}_1(u,v)(c_1) \geq \|(u,v)\|_Y$ and $\|\mathcal{Q}(u,v)\|_Y \geq \|\mathcal{Q}_1(u,v)\| \geq \|(u,v)\|_Y$.

Therefore, we deduce the conclusion of the theorem.   □

In what follows, for $f_0^i, g_0^i, f_\infty^s, g_\infty^s \in (0,\infty)$ and numbers $\gamma_1, \gamma_2 \in [0,1]$, $\gamma_3, \gamma_4 \in (0,1)$, $a \in [0,1]$ and $b \in (0,1)$, we define the numbers

$$\widetilde{L}_1 = \max\left\{\frac{1}{f_0^i}\left(\frac{a\gamma_1}{\theta\theta_1\widetilde{A}}\right)^{r_1-1}, \frac{1}{f_0^i}\left(\frac{(1-a)\gamma_2}{\theta\theta_2\widetilde{D}}\right)^{r_1-1}\right\},$$

$$\widetilde{L}_2 = \min\left\{\frac{1}{f_\infty^s}\left(\frac{b\gamma_3}{A}\right)^{r_1-1}, \frac{1}{f_\infty^s}\left(\frac{(1-b)\gamma_4}{D}\right)^{r_1-1}\right\},$$

$$\widetilde{L}_3 = \max\left\{\frac{1}{g_0^i}\left(\frac{a(1-\gamma_1)}{\theta\theta_1\widetilde{B}}\right)^{r_2-1}, \frac{1}{g_0^i}\left(\frac{(1-a)(1-\gamma_2)}{\theta\theta_2\widetilde{C}}\right)^{r_2-1}\right\},$$

$$\widetilde{L}_4 = \min\left\{\frac{1}{g_\infty^s}\left(\frac{b(1-\gamma_3)}{B}\right)^{r_2-1}, \frac{1}{g_\infty^s}\left(\frac{(1-b)(1-\gamma_4)}{C}\right)^{r_2-1}\right\},$$

$$\widetilde{L}_2' = \min\left\{\frac{1}{f_\infty^s}\left(\frac{b}{A}\right)^{r_1-1}, \frac{1}{f_\infty^s}\left(\frac{1-b}{D}\right)^{r_1-1}\right\},$$

$$\widetilde{L}_4' = \min\left\{\frac{1}{g_\infty^s}\left(\frac{b}{B}\right)^{r_2-1}, \frac{1}{g_\infty^s}\left(\frac{1-b}{C}\right)^{r_2-1}\right\}.$$

Theorem 4.2.2. *Assume that* (I1) *and* (I2) *hold,* $[c_1,c_2] \subset [0,1]$ *with* $0 < c_1 < c_2 \leq 1$, $\gamma_1, \gamma_2 \in [0,1]$, $\gamma_3, \gamma_4 \in (0,1)$, $a \in [0,1]$ *and* $b \in (0,1)$.

(1) *If* $f_0^i, g_0^i, f_\infty^s, g_\infty^s \in (0,\infty)$, $\widetilde{L}_1 < \widetilde{L}_2$ *and* $\widetilde{L}_3 < \widetilde{L}_4$, *then for each* $\lambda \in (\widetilde{L}_1, \widetilde{L}_2)$ *and* $\mu \in (\widetilde{L}_3, \widetilde{L}_4)$ *there exists a positive solution* $(u(t), v(t))$, $t \in [0,1]$ *for* (4.32), (4.33).

(2) If $f_0^i, g_0^i, f_\infty^s \in (0, \infty)$, $g_\infty^s = 0$ and $\widetilde{L}_1 < \widetilde{L}_2'$, then for each $\lambda \in (\widetilde{L}_1, \widetilde{L}_2')$ and $\mu \in (\widetilde{L}_3, \infty)$ there exists a positive solution $(u(t), v(t))$, $t \in [0, 1]$ for (4.32), (4.33).

(3) If $f_0^i, g_0^i, g_\infty^s \in (0, \infty)$, $f_\infty^s = 0$ and $\widetilde{L}_3 < \widetilde{L}_4'$, then for each $\lambda \in (\widetilde{L}_1, \infty)$ and $\mu \in (\widetilde{L}_3, \widetilde{L}_4')$ there exists a positive solution $(u(t), v(t))$, $t \in [0, 1]$ for (4.32), (4.33).

(4) If $f_0^i, g_0^i \in (0, \infty)$, $f_\infty^s = g_\infty^s = 0$, then for each $\lambda \in (\widetilde{L}_1, \infty)$ and $\mu \in (\widetilde{L}_3, \infty)$ there exists a positive solution $(u(t), v(t))$, $t \in [0, 1]$ for (4.32), (4.33).

(5) If $f_\infty^s, g_\infty^s \in (0, \infty)$ and at least one of f_0^i, g_0^i is ∞, then for each $\lambda \in (0, \widetilde{L}_2)$ and $\mu \in (0, \widetilde{L}_4)$ there exists a positive solution $(u(t), v(t))$, $t \in [0, 1]$ for (4.32), (4.33).

(6) If $f_\infty^s \in (0, \infty)$, $g_\infty^s = 0$ and at least one of f_0^i, g_0^i is ∞, then for each $\lambda \in (0, \widetilde{L}_2')$ and $\mu \in (0, \infty)$ there exists a positive solution $(u(t), v(t))$, $t \in [0, 1]$ for (4.32), (4.33).

(7) If $f_\infty^s = 0$, $g_\infty^s \in (0, \infty)$ and at least one of f_0^i, g_0^i is ∞, then for each $\lambda \in (0, \infty)$ and $\mu \in (0, \widetilde{L}_4')$ there exists a positive solution $(u(t), v(t))$, $t \in [0, 1]$ for (4.32), (4.33).

(8) If $f_\infty^s = g_\infty^s = 0$ and at least one of f_0^i, g_0^i is ∞, then for each $\lambda \in (0, \infty)$ and $\mu \in (0, \infty)$ there exists a positive solution $(u(t), v(t))$, $t \in [0, 1]$ for (4.32), (4.33).

Proof. We consider the cone $P \subset Y$ and the operators $\mathcal{Q}_1$, $\mathcal{Q}_2$ and $\mathcal{Q}$ defined at the beginning of this section. Because the proofs of the above cases are similar, in what follows, we will prove two of them, namely Cases (1) and (7).

Case (1). We have $f_0^i, g_0^i, f_\infty^s, g_\infty^s \in (0, \infty)$, $\widetilde{L}_1 < \widetilde{L}_2$ and $\widetilde{L}_3 < \widetilde{L}_4$. Let $\lambda \in (\widetilde{L}_1, \widetilde{L}_2)$ and $\mu \in (\widetilde{L}_3, \widetilde{L}_4)$. We consider $\varepsilon > 0$ such that $\varepsilon < f_0^i$, $\varepsilon < g_0^i$ and

$$
\max \left\{ \frac{1}{f_0^i - \varepsilon} \left(\frac{a\gamma_1}{\theta\theta_1 \widetilde{A}} \right)^{r_1-1}, \frac{1}{f_0^i - \varepsilon} \left(\frac{(1-a)\gamma_2}{\theta\theta_2 \widetilde{D}} \right)^{r_1-1} \right\} \le \lambda
$$

$$
\le \min \left\{ \frac{1}{f_\infty^s + \varepsilon} \left(\frac{b\gamma_3}{A} \right)^{r_1-1}, \frac{1}{f_\infty^s + \varepsilon} \left(\frac{(1-b)\gamma_4}{D} \right)^{r_1-1} \right\},
$$

$$\max\left\{\frac{1}{g_0^i-\varepsilon}\left(\frac{a(1-\gamma_1)}{\theta\theta_1\widetilde{B}}\right)^{r_2-1},\frac{1}{g_0^i-\varepsilon}\left(\frac{(1-a)(1-\gamma_2)}{\theta\theta_2\widetilde{C}}\right)^{r_2-1}\right\}\leq\mu$$

$$\leq\min\left\{\frac{1}{g_\infty^s+\varepsilon}\left(\frac{b(1-\gamma_3)}{B}\right)^{r_2-1},\frac{1}{g_\infty^s+\varepsilon}\left(\frac{(1-b)(1-\gamma_4)}{C}\right)^{r_2-1}\right\}.$$

By using (I2) and the definition of f_0^i and g_0^i, we deduce that there exists $R_3 > 0$ such that

$$f(t,u,v)\geq(f_0^i-\varepsilon)(u+v)^{r_1-1},\quad g(t,u,v)\geq(g_0^i-\varepsilon)(u+v)^{r_2-1},$$

for all $t\in[c_1,c_2]$, $u,v\geq0$, $u+v\leq R_3$.

We denote by $\Omega_3=\{(u,v)\in Y,\ \|(u,v)\|_Y<R_3\}$. Let $(u,v)\in P$ with $\|(u,v)\|_Y=R_3$, that is $\|u\|+\|v\|=R_3$. Because $u(t)+v(t)\leq\|u\|+\|v\|=R_3$ for all $t\in[0,1]$, then by Lemma 4.2.2, we obtain

$$\mathcal{Q}_1(u,v)(c_1)\geq\lambda^{\varrho_1-1}c_1^{\beta_1-1}\int_{c_1}^{c_2}\widetilde{J}_1(s)\varphi_{\varrho_1}\left(\frac{1}{\Gamma(\alpha_1)}\int_{c_1}^{s}(s-\tau)^{\alpha_1-1}\right.$$

$$\times\, f(\tau,u(\tau),v(\tau))\,d\tau\Bigg)\,ds$$

$$+\mu^{\varrho_2-1}c_1^{\beta_1-1}\int_{c_1}^{c_2}\widetilde{J}_2(s)\varphi_{\varrho_2}\left(\frac{1}{\Gamma(\alpha_2)}\int_{c_1}^{s}(s-\tau)^{\alpha_2-1}\right.$$

$$\times\, g(\tau,u(\tau),v(\tau))\,d\tau\Bigg)\,ds$$

$$\geq\lambda^{\varrho_1-1}c_1^{\beta_1-1}\int_{c_1}^{c_2}\widetilde{J}_1(s)\varphi_{\varrho_1}\left(\frac{1}{\Gamma(\alpha_1)}\int_{c_1}^{s}(s-\tau)^{\alpha_1-1}\right.$$

$$\times\,(f_0^i-\varepsilon)(u(\tau)+v(\tau))^{r_1-1}\,d\tau\Bigg)\,ds$$

$$+\mu^{\varrho_2-1}c_1^{\beta_1-1}\int_{c_1}^{c_2}\widetilde{J}_2(s)\varphi_{\varrho_2}\left(\frac{1}{\Gamma(\alpha_2)}\int_{c_1}^{s}(s-\tau)^{\alpha_2-1}\right.$$

$$\times\,(g_0^i-\varepsilon)(u(\tau)+v(\tau))^{r_2-1}\,d\tau\Bigg)\,ds$$

$$\geq \lambda^{\varrho_1-1} c_1^{\beta_1-1} \int_{c_1}^{c_2} \tilde{J}_1(s) \varphi_{\varrho_1} \left(\frac{1}{\Gamma(\alpha_1)} \int_{c_1}^{s} (s-\tau)^{\alpha_1-1} \right.$$

$$\left. \times (f_0^i - \varepsilon)(\theta \|(u,v)\|_Y)^{r_1-1} \, d\tau \right) ds$$

$$+ \mu^{\varrho_2-1} c_1^{\beta_1-1} \int_{c_1}^{c_2} \tilde{J}_2(s) \varphi_{\varrho_2} \left(\frac{1}{\Gamma(\alpha_2)} \int_{c_1}^{s} (s-\tau)^{\alpha_2-1} \right.$$

$$\left. \times (g_0^i - \varepsilon)(\theta \|(u,v)\|_Y)^{r_2-1} \, d\tau \right) ds$$

$$= \theta \theta_1 \lambda^{\varrho_1-1} (f_0^i - \varepsilon)^{\varrho_1-1} \|(u,v)\|_Y$$

$$\times \int_{c_1}^{c_2} \tilde{J}_1(s) \frac{1}{(\Gamma(\alpha_1+1))^{\varrho_1-1}} (s-c_1)^{\alpha_1(\varrho_1-1)} \, ds$$

$$+ \theta \theta_1 \mu^{\varrho_2-1} (g_0^i - \varepsilon)^{\varrho_2-1} \|(u,v)\|_Y$$

$$\times \int_{c_1}^{c_2} \tilde{J}_2(s) \frac{1}{(\Gamma(\alpha_2+1))^{\varrho_2-1}} (s-c_1)^{\alpha_2(\varrho_2-1)} \, ds$$

$$= [\theta \theta_1 \lambda^{\varrho_1-1} (f_0^i - \varepsilon)^{\varrho_1-1} \tilde{A} + \theta \theta_1 \mu^{\varrho_2-1} (g_0^i - \varepsilon)^{\varrho_2-1} \tilde{B}] \|(u,v)\|_Y$$

$$\geq [a\gamma_1 + a(1-\gamma_1)] \|(u,v)\|_Y = a \|(u,v)\|_Y.$$

Then, $\|\mathcal{Q}_1(u,v)\| \geq \mathcal{Q}_1(u,v)(c_1) \geq a \|(u,v)\|_Y$.
In a similar manner, we deduce

$$\mathcal{Q}_2(u,v)(c_1) \geq \mu^{\varrho_2-1} c_1^{\beta_2-1} \int_{c_1}^{c_2} \tilde{J}_3(s) \varphi_{\varrho_2} \left(\frac{1}{\Gamma(\alpha_2)} \int_{c_1}^{s} (s-\tau)^{\alpha_2-1} \right.$$

$$\left. \times g(\tau, u(\tau), v(\tau)) \, d\tau \right) ds$$

$$+ \lambda^{\varrho_1-1} c_1^{\beta_2-1} \int_{c_1}^{c_2} \tilde{J}_4(s) \varphi_{\varrho_1} \left(\frac{1}{\Gamma(\alpha_1)} \int_{c_1}^{s} (s-\tau)^{\alpha_1-1} \right.$$

$$\left. \times f(\tau, u(\tau), v(\tau)) \, d\tau \right) ds$$

$$\geq \mu^{\varrho_2-1} c_1^{\beta_2-1} \int_{c_1}^{c_2} \tilde{J}_3(s) \varphi_{\varrho_2} \left(\frac{1}{\Gamma(\alpha_2)} \int_{c_1}^{s} (s-\tau)^{\alpha_2-1} \right.$$

$$\left. \times (g_0^i - \varepsilon)(u(\tau) + v(\tau))^{r_2-1} \, d\tau \right) ds$$

$$+ \lambda^{\varrho_1 - 1} c_1^{\beta_2 - 1} \int_{c_1}^{c_2} \tilde{J}_4(s) \varphi_{\varrho_1} \left(\frac{1}{\Gamma(\alpha_1)} \int_{c_1}^{s} (s - \tau)^{\alpha_1 - 1} \right.$$

$$\times (f_0^i - \varepsilon)(u(\tau) + v(\tau))^{r_1 - 1} \, d\tau \bigg) \, ds$$

$$\geq \mu^{\varrho_2 - 1} c_1^{\beta_2 - 1} \int_{c_1}^{c_2} \tilde{J}_3(s) \varphi_{\varrho_2} \left(\frac{1}{\Gamma(\alpha_2)} \int_{c_1}^{s} (s - \tau)^{\alpha_2 - 1} \right.$$

$$\times (g_0^i - \varepsilon)(\theta \| (u, v) \|_Y)^{r_2 - 1} \, d\tau \bigg) \, ds$$

$$+ \lambda^{\varrho_1 - 1} c_1^{\beta_2 - 1} \int_{c_1}^{c_2} \tilde{J}_4(s) \varphi_{\varrho_1} \left(\frac{1}{\Gamma(\alpha_1)} \int_{c_1}^{s} (s - \tau)^{\alpha_1 - 1} \right.$$

$$\times (f_0^i - \varepsilon)(\theta \| (u, v) \|_Y)^{r_1 - 1} \, d\tau \bigg) \, ds$$

$$= \theta \theta_2 \mu^{\varrho_2 - 1} (g_0^i - \varepsilon)^{\varrho_2 - 1} \| (u, v) \|_Y$$

$$\times \int_{c_1}^{c_2} \tilde{J}_3(s) \frac{1}{(\Gamma(\alpha_2 + 1))^{\varrho_2 - 1}} (s - c_1)^{\alpha_2 (\varrho_2 - 1)} \, ds$$

$$+ \theta \theta_2 \lambda^{\varrho_1 - 1} (f_0^i - \varepsilon)^{\varrho_1 - 1} \| (u, v) \|_Y$$

$$\times \int_{c_1}^{c_2} \tilde{J}_4(s) \frac{1}{(\Gamma(\alpha_1 + 1))^{\varrho_1 - 1}} (s - c_1)^{\alpha_1 (\varrho_1 - 1)} \, ds$$

$$= [\theta \theta_2 \mu^{\varrho_2 - 1} (g_0^i - \varepsilon)^{\varrho_2 - 1} \tilde{C} + \theta \theta_2 \lambda^{\varrho_1 - 1} (f_0^i - \varepsilon)^{\varrho_1 - 1} \tilde{D}] \| (u, v) \|_Y$$

$$\geq [(1 - a)(1 - \gamma_2) + (1 - a)\gamma_2] \| (u, v) \|_Y = (1 - a) \| (u, v) \|_Y.$$

So, $\| \mathcal{Q}_2(u, v) \| \geq \mathcal{Q}_2(u, v)(c_1) \geq (1 - a) \| (u, v) \|_Y$.
Hence, for $(u, v) \in P \cap \partial \Omega_3$, we obtain

$$\| \mathcal{Q}(u, v) \|_Y = \| \mathcal{Q}_1(u, v) \| + \| \mathcal{Q}_2(u, v) \|$$
$$\geq a \| (u, v) \|_Y + (1 - a) \| (u, v) \|_Y = \| (u, v) \|_Y. \quad (4.44)$$

Now, we define the functions $f^*, g^* : [0, 1] \times [0, \infty) \to [0, \infty)$ by $f^*(t, x) = \max_{0 \leq u + v \leq x} f(t, u, v)$, $g^*(t, x) = \max_{0 \leq u + v \leq x} g(t, u, v)$, for all $t \in [0, 1]$ and $x \in [0, \infty)$. Then

$$f(t, u, v) \leq f^*(t, x), \quad g(t, u, v) \leq g^*(t, x), \quad \forall t \in [0, 1], \ u, v \geq 0, \ u + v \leq x.$$

The functions $f^*(t, \cdot)$, $g^*(t, \cdot)$ are nondecreasing for every $t \in [0, 1]$ and they satisfy the conditions

$$\limsup_{x \to \infty} \max_{t \in [0,1]} \frac{f^*(t, x)}{x^{r_1 - 1}} = f_\infty^s, \quad \limsup_{x \to \infty} \max_{t \in [0,1]} \frac{g^*(t, x)}{x^{r_2 - 1}} = g_\infty^s.$$

Therefore, for $\varepsilon > 0$ there exists $\overline{R}_4 > 0$ such that for all $x \geq \overline{R}_4$ and $t \in [0, 1]$ we have

$$\frac{f^*(t, x)}{x^{r_1 - 1}} \leq \limsup_{x \to \infty} \max_{t \in [0,1]} \frac{f^*(t, x)}{x^{r_1 - 1}} + \varepsilon = f_\infty^s + \varepsilon,$$

$$\frac{g^*(t, x)}{x^{r_2 - 1}} \leq \limsup_{x \to \infty} \max_{t \in [0,1]} \frac{g^*(t, x)}{x^{r_2 - 1}} + \varepsilon = g_\infty^s + \varepsilon,$$

and so $f^*(t, x) \leq (f_\infty^s + \varepsilon) x^{r_1 - 1}$ and $g^*(t, x) \leq (g_\infty^s + \varepsilon) x^{r_2 - 1}$.

We consider $R_4 = \max\{2R_3, \overline{R}_4\}$ and we denote by $\Omega_4 = \{(u, v) \in Y, \|(u, v)\|_Y < R_4\}$. Let $(u, v) \in P \cap \partial \Omega_4$. By the definition of f^* and g^* we conclude

$$f(t, u(t), v(t)) \leq f^*(t, \|(u, v)\|_Y),$$

$$g(t, u(t), v(t)) \leq g^*(t, \|(u, v)\|_Y), \quad \forall t \in [0, 1]. \tag{4.45}$$

Then for all $t \in [0, 1]$ we obtain

$$\mathcal{Q}_1(u, v)(t) \leq \lambda^{\varrho_1 - 1} \int_0^1 \tilde{J}_1(s) \varphi_{\varrho_1} \left(\frac{1}{\Gamma(\alpha_1)} \int_0^s (s - \tau)^{\alpha_1 - 1} \right.$$

$$\times f(\tau, u(\tau), v(\tau)) \, d\tau \bigg) \, ds$$

$$+ \mu^{\varrho_2 - 1} \int_0^1 \tilde{J}_2(s) \varphi_{\varrho_2} \left(\frac{1}{\Gamma(\alpha_2)} \int_0^s (s - \tau)^{\alpha_2 - 1} \right.$$

$$\times g(\tau, u(\tau), v(\tau)) \, d\tau \bigg) \, ds$$

$$\leq \lambda^{\varrho_1 - 1} \int_0^1 \tilde{J}_1(s) \varphi_{\varrho_1} \left(\frac{1}{\Gamma(\alpha_1)} \int_0^s (s - \tau)^{\alpha_1 - 1} \right.$$

$$\times f^*(\tau, \|(u, v)\|_Y) \, d\tau \bigg) \, ds$$

$$+ \mu^{\varrho_2 - 1} \int_0^1 \tilde{J}_2(s) \varphi_{\varrho_2} \left(\frac{1}{\Gamma(\alpha_2)} \int_0^s (s - \tau)^{\alpha_2 - 1} \right.$$

$$\times g^*(\tau, \|(u, v)\|_Y) \, d\tau \bigg) \, ds$$

$$\leq \lambda^{\varrho_1-1} \int_0^1 \widetilde{J}_1(s)\varphi_{\varrho_1}\left(\frac{1}{\Gamma(\alpha_1)}\int_0^s (s-\tau)^{\alpha_1-1}\right.$$

$$\left. \times (f_\infty^s + \varepsilon)\|(u,v)\|_Y^{r_1-1}\,d\tau\right)ds$$

$$+ \mu^{\varrho_2-1}\int_0^1 \widetilde{J}_2(s)\varphi_{\varrho_2}\left(\frac{1}{\Gamma(\alpha_2)}\int_0^s (s-\tau)^{\alpha_2-1}\right.$$

$$\left. \times (g_\infty^s + \varepsilon)\|(u,v)\|_Y^{r_2-1}\,d\tau\right)ds$$

$$= \lambda^{\varrho_1-1}(f_\infty^s + \varepsilon)^{\varrho_1-1}\|(u,v)\|_Y$$

$$\times \int_0^1 \frac{1}{(\Gamma(\alpha_1+1))^{\varrho_1-1}}s^{\alpha_1(\varrho_1-1)}\widetilde{J}_1(s)\,ds$$

$$+ \mu^{\varrho_2-1}(g_\infty^s + \varepsilon)^{\varrho_2-1}\|(u,v)\|_Y$$

$$\times \int_0^1 \frac{1}{(\Gamma(\alpha_2+1))^{\varrho_2-1}}s^{\alpha_2(\varrho_2-1)}\widetilde{J}_2(s)\,ds$$

$$= [\lambda^{\varrho_1-1}(f_\infty^s + \varepsilon)^{\varrho_1-1}A + \mu^{\varrho_2-1}(g_\infty^s + \varepsilon)^{\varrho_2-1}B]\|(u,v)\|_Y$$

$$\leq [b\gamma_3 + b(1-\gamma_3)]\|(u,v)\|_Y = b\|(u,v)\|_Y,$$

and so $\|Q_1(u,v)\| \leq b\|(u,v)\|_Y$.

In a similar manner, we conclude

$$Q_2(u,v)(t) \leq \mu^{\varrho_2-1}\int_0^1 \widetilde{J}_3(s)\varphi_{\varrho_2}\left(\frac{1}{\Gamma(\alpha_2)}\int_0^s (s-\tau)^{\alpha_2-1}\right.$$

$$\left. \times g(\tau, u(\tau), v(\tau))\,d\tau\right)ds$$

$$+ \lambda^{\varrho_1-1}\int_0^1 \widetilde{J}_4(s)\varphi_{\varrho_1}\left(\frac{1}{\Gamma(\alpha_1)}\int_0^s (s-\tau)^{\alpha_1-1}\right.$$

$$\left. \times f(\tau, u(\tau), v(\tau))\,d\tau\right)ds$$

$$\leq \mu^{\varrho_2-1}\int_0^1 \widetilde{J}_3(s)\varphi_{\varrho_2}\left(\frac{1}{\Gamma(\alpha_2)}\int_0^s (s-\tau)^{\alpha_2-1}\right.$$

$$\left. \times g^*(\tau, \|(u,v)\|_Y)\,d\tau\right)ds$$

$$+ \lambda^{\varrho_1 - 1} \int_0^1 \tilde{J}_4(s) \varphi_{\varrho_1} \left(\frac{1}{\Gamma(\alpha_1)} \int_0^s (s - \tau)^{\alpha_1 - 1} \right.$$

$$\left. \times f^*(\tau, \|(u, v)\|_Y) \, d\tau \right) ds$$

$$\leq \mu^{\varrho_2 - 1} \int_0^1 \tilde{J}_3(s) \varphi_{\varrho_2} \left(\frac{1}{\Gamma(\alpha_2)} \int_0^s (s - \tau)^{\alpha_2 - 1} \right.$$

$$\left. \times (g_\infty^s + \varepsilon) \|(u, v)\|_Y^{r_2 - 1} \, d\tau \right) ds$$

$$+ \lambda^{\varrho_1 - 1} \int_0^1 \tilde{J}_4(s) \varphi_{\varrho_1} \left(\frac{1}{\Gamma(\alpha_1)} \int_0^s (s - \tau)^{\alpha_1 - 1} \right.$$

$$\left. \times (f_\infty^s + \varepsilon) \|(u, v)\|_Y^{r_1 - 1} \, d\tau \right) ds$$

$$= \mu^{\varrho_2 - 1} (g_\infty^s + \varepsilon)^{\varrho_2 - 1} \|(u, v)\|_Y$$

$$\times \int_0^1 \frac{1}{(\Gamma(\alpha_2 + 1))^{\varrho_2 - 1}} s^{\alpha_2(\varrho_2 - 1)} \tilde{J}_3(s) \, ds$$

$$+ \lambda^{\varrho_1 - 1} (f_\infty^s + \varepsilon)^{\varrho_1 - 1} \|(u, v)\|_Y$$

$$\times \int_0^1 \frac{1}{(\Gamma(\alpha_1 + 1))^{\varrho_1 - 1}} s^{\alpha_1(\varrho_1 - 1)} \tilde{J}_4(s) \, ds$$

$$= [\mu^{\varrho_2 - 1} (g_\infty^s + \varepsilon)^{\varrho_2 - 1} C + \lambda^{\varrho_1 - 1} (f_\infty^s + \varepsilon)^{\varrho_1 - 1} D] \|(u, v)\|_Y$$

$$\leq [(1 - b)(1 - \gamma_4) + (1 - b)\gamma_4] \|(u, v)\|_Y$$

$$= (1 - b) \|(u, v)\|_Y, \quad \forall t \in [0, 1].$$

Hence, $\|Q_2(u, v)\| \leq (1 - b) \|(u, v)\|_Y$.

Then for $(u, v) \in P \cap \partial \Omega_4$, we deduce

$$\|Q(u, v)\|_Y = \|Q_1(u, v)\| + \|Q_2(u, v)\|$$

$$\leq b \|(u, v)\|_Y + (1 - b) \|(u, v)\|_Y = \|(u, v)\|_Y. \quad (4.46)$$

By using (4.44), (4.46), Lemma 4.2.3 and the Guo–Krasnosel'skii fixed point theorem (Theorem 1.2.2), we conclude that Q has a fixed point $(u, v) \in P \cap (\overline{\Omega}_4 \setminus \Omega_3)$ which is a positive solution of problem (4.32), (4.33).

Case (7). We consider $f_\infty^s = 0$, $g_\infty^s \in (0, \infty)$ and $f_0^i = \infty$. Let $\lambda \in (0, \infty)$ and $\mu \in (0, \tilde{L}_4')$. Instead of the numbers $\gamma_3, \gamma_4 \in (0, 1)$ used in the first case, we choose $\tilde{\gamma}_3 \in \left(0, 1 - (\mu g_\infty^s)^{\varrho_2 - 1} \frac{B}{b} \right)$ and $\tilde{\gamma}_4 \in \left(0, 1 - (\mu g_\infty^s)^{\varrho_2 - 1} \frac{C}{1 - b} \right)$.

The choice of $\tilde{\gamma}_3$ and $\tilde{\gamma}_4$ is possible because $\mu < \frac{1}{g_\infty^s}\left(\frac{b}{B}\right)^{r_2-1}$ and $\mu < \frac{1}{g_\infty^s}\left(\frac{1-b}{C}\right)^{r_2-1}$. Let $\varepsilon > 0$ such that

$$\varepsilon\left(\frac{1}{\theta\theta_1\tilde{A}}\right)^{r_1-1} \leq \lambda \leq \min\left\{\frac{1}{\varepsilon}\left(\frac{b\tilde{\gamma}_3}{A}\right)^{r_1-1}, \frac{1}{\varepsilon}\left(\frac{(1-b)\tilde{\gamma}_4}{D}\right)^{r_1-1}\right\},$$

$$\mu \leq \min\left\{\frac{1}{g_\infty^s+\varepsilon}\left(\frac{b(1-\tilde{\gamma}_3)}{B}\right)^{r_2-1}, \frac{1}{g_\infty^s+\varepsilon}\left(\frac{(1-b)(1-\tilde{\gamma}_4)}{C}\right)^{r_2-1}\right\}.$$

By (I2) and the definition of f_0^i, we deduce that there exists $R_3 > 0$ such that

$$f(t,u,v) \geq \frac{1}{\varepsilon}(u+v)^{r_1-1}, \quad \forall t \in [c_1,c_2], \ u,v \geq 0, \ u+v \leq R_3.$$

We denote by $\Omega_3 = \{(u,v) \in Y, \|(u,v)\|_Y < R_3\}$. Let $(u,v) \in P \cap \partial\Omega_3$, that is $\|u\| + \|v\| = R_3$. Because $u(t) + v(t) \leq \|u\| + \|v\| = R_3$ for all $t \in [0,1]$, then by Lemma 4.2.2, we obtain

$$\mathcal{Q}_1(u,v)(c_1) \geq \lambda^{\varrho_1-1}\int_0^1 c_1^{\beta_1-1}\tilde{J}_1(s)\varphi_{\varrho_1}(I_{0+}^{\alpha_1}f(s,u(s),v(s)))\,ds$$

$$+ \mu^{\varrho_2-1}\int_0^1 c_1^{\beta_1-1}\tilde{J}_2(s)\varphi_{\varrho_2}(I_{0+}^{\alpha_2}g(s,u(s),v(s)))\,ds$$

$$\geq \lambda^{\varrho_1-1}c_1^{\beta_1-1}\int_{c_1}^{c_2}\tilde{J}_1(s)\varphi_{\varrho_1}\left(\frac{1}{\Gamma(\alpha_1)}\int_{c_1}^s(s-\tau)^{\alpha_1-1}\right.$$

$$\left. \times\, f(\tau,u(\tau),v(\tau))\,d\tau\right)ds$$

$$\geq \lambda^{\varrho_1-1}c_1^{\beta_1-1}\int_{c_1}^{c_2}\tilde{J}_1(s)\varphi_{\varrho_1}\left(\frac{1}{\Gamma(\alpha_1)}\int_{c_1}^s(s-\tau)^{\alpha_1-1}\right.$$

$$\left. \times\, \frac{1}{\varepsilon}(u(\tau)+v(\tau))^{r_1-1}\,d\tau\right)ds$$

$$\geq \lambda^{\varrho_1-1}c_1^{\beta_1-1}\int_{c_1}^{c_2}\tilde{J}_1(s)\varphi_{\varrho_1}\left(\frac{1}{\Gamma(\alpha_1)}\int_{c_1}^s(s-\tau)^{\alpha_1-1}\right.$$

$$\left. \times\, \frac{1}{\varepsilon}(\theta\|(u,v)\|_Y)^{r_1-1}\,d\tau\right)ds$$

$$= \theta\theta_1\lambda^{\varrho_1-1}\left(\frac{1}{\varepsilon}\right)^{\varrho_1-1}\|(u,v)\|_Y\tilde{A} \geq \|(u,v)\|_Y.$$

Hence, we get $\|Q_1(u,v)\| \geq Q_1(u,v)(c_1) \geq \|(u,v)\|_Y$ and $\|Q(u,v)\|_Y \geq \|Q_1(u,v)\| \geq \|(u,v)\|_Y$.

For the second part of the proof, we consider the functions f^* and g^* from Case (1), which satisfy in this case the conditions

$$\lim_{x\to\infty} \max_{t\in[0,1]} \frac{f^*(t,x)}{x^{r_1-1}} = 0, \quad \limsup_{x\to\infty} \max_{t\in[0,1]} \frac{g^*(t,x)}{x^{r_2-1}} = g_\infty^s.$$

Then for $\varepsilon > 0$ there exists $\overline{R}_4 > 0$ such that for all $x \geq \overline{R}_4$ and $t \in [0,1]$ we have

$$\frac{f^*(t,x)}{x^{r_1-1}} \leq \lim_{x\to\infty} \max_{t\in[0,1]} \frac{f^*(t,x)}{x^{r_1-1}} + \varepsilon = \varepsilon,$$

$$\frac{g^*(t,x)}{x^{r_2-1}} \leq \limsup_{x\to\infty} \max_{t\in[0,1]} \frac{g^*(t,x)}{x^{r_2-1}} + \varepsilon = g_\infty^s + \varepsilon,$$

and so $f^*(t,x) \leq \varepsilon x^{r_1-1}$ and $g^*(t,x) \leq (g_\infty^s + \varepsilon)x^{r_2-1}$.

We consider $R_4 = \max\{2R_3, \overline{R}_4\}$ and we denote by $\Omega_4 = \{(u,v) \in Y, \|(u,v)\|_Y < R_4\}$. Let $(u,v) \in P \cap \partial\Omega_4$. By the definition of f^* and g^*, we deduce the relation (4.45). In addition, in a similar manner as in the proof of Case (1), we conclude

$$Q_1(u,v)(t) \leq \lambda^{\varrho_1-1}\varepsilon^{\varrho_1-1}A\|(u,v)\|_Y + \mu^{\varrho_2-1}(g_\infty^s + \varepsilon)^{\varrho_2-1}B\|(u,v)\|_Y$$

$$\leq [b\tilde{\gamma}_3 + b(1-\tilde{\gamma}_3)]\|(u,v)\|_Y = b\|(u,v)\|_Y, \quad \forall t \in [0,1],$$

$$Q_2(u,v)(t) \leq \mu^{\varrho_2-1}(g_\infty^s + \varepsilon)^{\varrho_2-1}C\|(u,v)\|_Y + \lambda^{\varrho_1-1}\varepsilon^{\varrho_1-1}D\|(u,v)\|_Y$$

$$\leq [(1-b)(1-\tilde{\gamma}_4) + (1-b)\tilde{\gamma}_4]\|(u,v)\|_Y$$

$$= (1-b)\|(u,v)\|_Y, \quad \forall t \in [0,1],$$

and so $\|Q(u,v)\|_Y \leq b\|(u,v)\|_Y + (1-b)\|(u,v)\|_Y = \|(u,v)\|_Y$.

Therefore, we obtain the conclusion of the theorem. $\square$

4.2.3   *Nonexistence of positive solutions*

In this section, we present intervals for λ and μ for which our problem (4.32), (4.33) has no positive solutions.

Theorem 4.2.3. *Assume that* (I1) *and* (I2) *hold. If there exist positive numbers* M_1, M_2 *such that*

$$f(t, u, v) \leq M_1(u + v)^{r_1 - 1},$$

$$g(t, u, v) \leq M_2(u + v)^{r_2 - 1}, \quad \forall t \in [0, 1], \ u, v \geq 0,$$

(4.47)

then there exist positive constants λ_0 *and* μ_0 *such that for every* $\lambda \in (0, \lambda_0)$ *and* $\mu \in (0, \mu_0)$ *the boundary value problem* (4.32), (4.33) *has no positive solution.*

Proof. We define $\lambda_0 = \min\left\{\frac{1}{M_1(4A)^{r_1-1}}, \frac{1}{M_1(4D)^{r_1-1}}\right\}$ and $\mu_0 = \min\left\{\frac{1}{M_2(4B)^{r_2-1}}, \frac{1}{M_2(4C)^{r_2-1}}\right\}$, where A, B, C, D are given in (4.41).

We will prove that for every $\lambda \in (0, \lambda_0)$ and $\mu \in (0, \mu_0)$, the problem (4.32), (4.33) has no positive solution.

Let $\lambda \in (0, \lambda_0)$ and $\mu \in (0, \mu_0)$. We suppose that the problem (4.32), (4.33) has a positive solution $(u(t), v(t))$, $t \in [0, 1]$. Then we obtain

$$u(t) = \mathcal{Q}_1(u, v)(t) \leq \lambda^{\varrho_1 - 1} \int_0^1 \tilde{J}_1(s) \varphi_{\varrho_1} \left(\frac{1}{\Gamma(\alpha_1)} \int_0^s (s - \tau)^{\alpha_1 - 1}\right.$$

$$\times \left. f(\tau, u(\tau), v(\tau)) \, d\tau\right) ds$$

$$+ \mu^{\varrho_2 - 1} \int_0^1 \tilde{J}_2(s) \varphi_{\varrho_2} \left(\frac{1}{\Gamma(\alpha_2)} \int_0^s (s - \tau)^{\alpha_2 - 1}\right.$$

$$\times \left. g(\tau, u(\tau), v(\tau)) \, d\tau\right) ds$$

$$\leq \lambda^{\varrho_1 - 1} \int_0^1 \tilde{J}_1(s) \varphi_{\varrho_1} \left(\frac{1}{\Gamma(\alpha_1)} \int_0^s (s - \tau)^{\alpha_1 - 1}\right.$$

$$\times \left. M_1(u(\tau) + v(\tau))^{r_1 - 1} \, d\tau\right) ds$$

$$+ \mu^{\varrho_2 - 1} \int_0^1 \tilde{J}_2(s) \varphi_{\varrho_2} \left(\frac{1}{\Gamma(\alpha_2)} \int_0^s (s - \tau)^{\alpha_2 - 1}\right.$$

$$\times \left. M_2(u(\tau) + v(\tau))^{r_2 - 1} \, d\tau\right) ds$$

$$\leq \lambda^{\varrho_1-1} M_1^{\varrho_1-1} \int_0^1 \tilde{J}_1(s) \varphi_{\varrho_1} \left(\frac{1}{\Gamma(\alpha_1)} \int_0^s (s-\tau)^{\alpha_1-1} \right.$$

$$\left. \times \ (\|u\| + \|v\|)^{r_1-1} \, d\tau \right) ds$$

$$+ \mu^{\varrho_2-1} M_2^{\varrho_2-1} \int_0^1 \tilde{J}_2(s) \varphi_{\varrho_2} \left(\frac{1}{\Gamma(\alpha_2)} \int_0^s (s-\tau)^{\alpha_2-1} \right.$$

$$\left. \times \ (\|u\| + \|v\|)^{r_2-1} \, d\tau \right) ds$$

$$= \lambda^{\varrho_1-1} M_1^{\varrho_1-1} A \|(u,v)\|_Y + \mu^{\varrho_2-1} M_2^{\varrho_2-1} B \|(u,v)\|_Y, \quad \forall t \in [0,1],$$

and

$$v(t) = \mathcal{Q}_2(u,v)(t) \leq \mu^{\varrho_2-1} \int_0^1 \tilde{J}_3(s) \varphi_{\varrho_2} \left(\frac{1}{\Gamma(\alpha_2)} \int_0^s (s-\tau)^{\alpha_2-1} \right.$$

$$\left. \times \ g(\tau, u(\tau), v(\tau)) \, d\tau \right) ds$$

$$+ \lambda^{\varrho_1-1} \int_0^1 \tilde{J}_4(s) \varphi_{\varrho_1} \left(\frac{1}{\Gamma(\alpha_1)} \int_0^s (s-\tau)^{\alpha_1-1} f(\tau, u(\tau), v(\tau)) \, d\tau \right) ds$$

$$\leq \mu^{\varrho_2-1} \int_0^1 \tilde{J}_3(s) \varphi_{\varrho_2} \left(\frac{1}{\Gamma(\alpha_2)} \int_0^s (s-\tau)^{\alpha_2-1} \right.$$

$$\left. \times \ M_2(u(\tau) + v(\tau))^{r_2-1} \, d\tau \right) ds$$

$$+ \lambda^{\varrho_1-1} \int_0^1 \tilde{J}_4(s) \varphi_{\varrho_1} \left(\frac{1}{\Gamma(\alpha_1)} \int_0^s (s-\tau)^{\alpha_1-1} \right.$$

$$\left. \times \ M_1(u(\tau) + v(\tau))^{r_1-1} \, d\tau \right) ds$$

$$\leq \mu^{\varrho_2-1} M_2^{\varrho_2-1} \int_0^1 \tilde{J}_3(s) \varphi_{\varrho_2} \left(\frac{1}{\Gamma(\alpha_2)} \int_0^s (s-\tau)^{\alpha_2-1} \right.$$

$$\left. \times \ (\|u\| + \|v\|)^{r_2-1} \, d\tau \right) ds$$

$$+ \lambda^{\varrho_1 - 1} M_1^{\varrho_1 - 1} \int_0^1 \tilde{J}_4(s) \varphi_{\varrho_1} \left(\frac{1}{\Gamma(\alpha_1)} \int_0^s (s - \tau)^{\alpha_1 - 1} \right.$$

$$\times \left. (\|u\| + \|v\|)^{r_1 - 1} \, d\tau \right) ds$$

$$= \mu^{\varrho_2 - 1} M_2^{\varrho_2 - 1} C \|(u, v)\|_Y + \lambda^{\varrho_1 - 1} M_1^{\varrho_1 - 1} D \|(u, v)\|_Y, \quad \forall t \in [0, 1].$$

Then we deduce

$$\|u\| \leq \lambda^{\varrho_1 - 1} M_1^{\varrho_1 - 1} A \|(u, v)\|_Y + \mu^{\varrho_2 - 1} M_2^{\varrho_2 - 1} B \|(u, v)\|_Y$$

$$< \lambda_0^{\varrho_1 - 1} M_1^{\varrho_1 - 1} A \|(u, v)\|_Y + \mu_0^{\varrho_2 - 1} M_2^{\varrho_2 - 1} B \|(u, v)\|_Y$$

$$\leq \frac{1}{4} \|(u, v)\|_Y + \frac{1}{4} \|(u, v)\|_Y = \frac{1}{2} \|(u, v)\|_Y,$$

$$\|v\| \leq \mu^{\varrho_2 - 1} M_2^{\varrho_2 - 1} C \|(u, v)\|_Y + \lambda^{\varrho_1 - 1} M_1^{\varrho_1 - 1} D \|(u, v)\|_Y$$

$$< \mu_0^{\varrho_2 - 1} M_2^{\varrho_2 - 1} C \|(u, v)\|_Y + \lambda_0^{\varrho_1 - 1} M_1^{\varrho_1 - 1} D \|(u, v)\|_Y$$

$$\leq \frac{1}{4} \|(u, v)\|_Y + \frac{1}{4} \|(u, v)\|_Y = \frac{1}{2} \|(u, v)\|_Y,$$

and so $\|(u, v)\|_Y = \|u\| + \|v\| < \|(u, v)\|_Y$, which is a contradiction.

Therefore, the boundary value problem (4.32), (4.33) has no positive solution. $\square$

Remark 4.2.1. If $f_0^s, g_0^s, f_\infty^s, g_\infty^s < \infty$, then there exist positive constants M_1, M_2 such that relation (4.47) holds, and then we obtain the conclusion of Theorem 4.2.3.

Theorem 4.2.4. *Assume that* (I1) *and* (I2) *hold. If there exist positive numbers* c_1, c_2 *with* $0 < c_1 < c_2 \leq 1$ *and* $m_1 > 0$ *such that*

$$f(t, u, v) \geq m_1 (u + v)^{r_1 - 1}, \quad \forall t \in [c_1, c_2], \ u, v \geq 0, \tag{4.48}$$

then there exists a positive constant $\tilde{\lambda}_0$ *such that for every* $\lambda > \tilde{\lambda}_0$ *and* $\mu > 0$, *the boundary value problem* (4.32), (4.33) *has no positive solution.*

Proof. We define $\tilde{\lambda}_0 = \min \left\{ \frac{1}{m_1 (\theta \theta_1 \tilde{A})^{r_1 - 1}}, \frac{1}{m_1 (\theta \theta_2 \tilde{D})^{r_1 - 1}} \right\}$, where $\tilde{A}$ and $\tilde{D}$ are given by (4.41).

We will show that for every $\lambda > \tilde{\lambda}_0$ and $\mu > 0$ the problem (4.32), (4.33) has no positive solution. Let $\lambda > \tilde{\lambda}_0$ and $\mu > 0$. We suppose that the problem (4.32), (4.33) has a positive solution $(u(t), v(t))$, $t \in [0, 1]$.

If $\theta_1 \widetilde{A} \geq \theta_2 \widetilde{D}$, then $\widetilde{\lambda}_0 = \frac{1}{m_1(\theta\theta_1\widetilde{A})^{r_1-1}}$, and therefore, we obtain

$$u(c_1) = \mathcal{Q}_1(u,v)(c_1) \geq \lambda^{\varrho_1-1} \int_0^1 c_1^{\beta_1-1} \widetilde{J}_1(s)\varphi_{\varrho_1}(I_{0+}^{\alpha_1}f(s,u(s),v(s)))\,ds$$

$$+\, \mu^{\varrho_2-1} \int_0^1 c_1^{\beta_1-1} \widetilde{J}_2(s)\varphi_{\varrho_2}(I_{0+}^{\alpha_2}g(s,u(s),v(s)))\,ds$$

$$\geq \lambda^{\varrho_1-1} c_1^{\beta_1-1} \int_{c_1}^{c_2} \widetilde{J}_1(s)\varphi_{\varrho_1}\left(\frac{1}{\Gamma(\alpha_1)}\int_{c_1}^s (s-\tau)^{\alpha_1-1}\right.$$

$$\left. \times\, f(\tau,u(\tau),v(\tau))\,d\tau\right)ds$$

$$\geq \lambda^{\varrho_1-1} c_1^{\beta_1-1} \int_{c_1}^{c_2} \widetilde{J}_1(s)\varphi_{\varrho_1}\left(\frac{1}{\Gamma(\alpha_1)}\int_{c_1}^s (s-\tau)^{\alpha_1-1}\right.$$

$$\left. \times\, m_1(u(\tau)+v(\tau))^{r_1-1}\,d\tau\right)ds$$

$$\geq \lambda^{\varrho_1-1} c_1^{\beta_1-1} m_1^{\varrho_1-1} \int_{c_1}^{c_2} \widetilde{J}_1(s)\varphi_{\varrho_1}\left(\frac{1}{\Gamma(\alpha_1)}\int_{c_1}^s (s-\tau)^{\alpha_1-1}\right.$$

$$\left. \times\, (\theta\|(u,v)\|_Y)^{r_1-1}\,d\tau\right)ds$$

$$= (\lambda m_1)^{\varrho_1-1}\theta\theta_1\widetilde{A}\|(u,v)\|_Y.$$

Then we conclude

$$\|u\| \geq u(c_1) \geq (\lambda m_1)^{\varrho_1-1}\theta\theta_1\widetilde{A}\|(u,v)\|_Y$$

$$> (\widetilde{\lambda}_0 m_1)^{\varrho_1-1}\theta\theta_1\widetilde{A}\|(u,v)\|_Y = \|(u,v)\|_Y,$$

and so $\|(u,v)\|_Y = \|u\| + \|v\| \geq \|u\| > \|(u,v)\|_Y$, which is a contradiction.
If $\theta_1 \widetilde{A} < \theta_2 \widetilde{D}$, then $\widetilde{\lambda}_0 = \frac{1}{m_1(\theta\theta_2\widetilde{D})^{r_1-1}}$, and therefore, we deduce

$$v(c_1) = \mathcal{Q}_2(u,v)(c_1) \geq \mu^{\varrho_2-1} \int_0^1 c_1^{\beta_2-1} \widetilde{J}_3(s)\varphi_{\varrho_2}(I_{0+}^{\alpha_2}g(s,u(s),v(s)))\,ds$$

$$+\, \lambda^{\varrho_1-1} \int_0^1 c_1^{\beta_2-1} \widetilde{J}_4(s)\varphi_{\varrho_1}(I_{0+}^{\alpha_1}f(s,u(s),v(s)))\,ds$$

$$\geq \lambda^{\varrho_1-1} c_1^{\beta_2-1} \int_{c_1}^{c_2} \widetilde{J}_4(s)\varphi_{\varrho_1}\left(\frac{1}{\Gamma(\alpha_1)}\int_{c_1}^s (s-\tau)^{\alpha_1-1}\right.$$

$$\left. \times\, f(\tau,u(\tau),v(\tau))\,d\tau\right)ds$$

$$\geq \lambda^{\varrho_1 - 1} c_1^{\beta_2 - 1} \int_{c_1}^{c_2} \tilde{J}_4(s) \varphi_{\varrho_1} \left(\frac{1}{\Gamma(\alpha_1)} \int_{c_1}^{s} (s - \tau)^{\alpha_1 - 1} \right.$$

$$\left. \times \, m_1 (u(\tau) + v(\tau))^{r_1 - 1} \, d\tau \right) ds$$

$$\geq \lambda^{\varrho_1 - 1} c_1^{\beta_2 - 1} m_1^{\varrho_1 - 1} \int_{c_1}^{c_2} \tilde{J}_4(s) \varphi_{\varrho_1} \left(\frac{1}{\Gamma(\alpha_1)} \int_{c_1}^{s} (s - \tau)^{\alpha_1 - 1} \right.$$

$$\left. \times \, (\theta \|(u, v)\|_Y)^{r_1 - 1} \, d\tau \right) ds$$

$$= (\lambda m_1)^{\varrho_1 - 1} \theta \theta_2 \tilde{D} \|(u, v)\|_Y.$$

Then we conclude

$$\|v\| \geq v(c_1) \geq (\lambda m_1)^{\varrho_1 - 1} \theta \theta_2 \tilde{D} \|(u, v)\|_Y$$
$$> (\tilde{\lambda}_0 m_1)^{\varrho_1 - 1} \theta \theta_2 \tilde{D} \|(u, v)\|_Y = \|(u, v)\|_Y,$$

and so $\|(u, v)\|_Y = \|u\| + \|v\| \geq \|v\| > \|(u, v)\|_Y$, which is a contradiction.

Therefore, the boundary value problem (4.32), (4.33) has no positive solution. $\qquad\square$

Remark 4.2.2. If for c_1, c_2 with $0 < c_1 < c_2 \leq 1$, we have f_0^i, $f_\infty^i > 0$ and $f(t, u, v) > 0$ for all $t \in [c_1, c_2]$ and $u, v \geq 0$ with $u + v > 0$, then the relation (4.48) holds, and we obtain the conclusion of Theorem 4.2.4.

Theorem 4.2.5. *Assume that* (I1) *and* (I2) *hold. If there exist positive numbers* c_1, c_2 *with* $0 < c_1 < c_2 \leq 1$ *and* $m_2 > 0$ *such that*

$$g(t, u, v) \geq m_2 (u + v)^{r_2 - 1}, \quad \forall t \in [c_1, c_2], \ u, v \geq 0, \tag{4.49}$$

then there exists a positive constant $\tilde{\mu}_0$ *such that for every* $\mu > \tilde{\mu}_0$ *and* $\lambda > 0$, *the boundary value problem* (4.32), (4.33) *has no positive solution.*

Proof. We define $\tilde{\mu}_0 = \min \left\{ \frac{1}{m_2(\theta \theta_1 \tilde{B})^{r_2 - 1}}, \frac{1}{m_2(\theta \theta_2 \tilde{C})^{r_2 - 1}} \right\}$, where $\tilde{B}$ and $\tilde{C}$ are given by (4.41).

We will show that for every $\mu > \tilde{\mu}_0$ and $\lambda > 0$, the problem (4.32), (4.33) has no positive solution. Let $\mu > \tilde{\mu}_0$ and $\lambda > 0$. We suppose that the problem (4.32), (4.33) has a positive solution $(u(t), v(t))$, $t \in [0, 1]$.

If $\theta_1\widetilde{B} \geq \theta_2\widetilde{C}$, then $\widetilde{\mu}_0 = \frac{1}{m_2(\theta\theta_1\widetilde{B})^{r_2-1}}$, and therefore, we obtain

$$u(c_1) = \mathcal{Q}_1(u,v)(c_1) \geq \lambda^{\varrho_1-1} \int_0^1 c_1^{\beta_1-1}\widetilde{J}_1(s)\varphi_{\varrho_1}\left(I_{0+}^{\alpha_1}f(s,u(s),v(s))\right)ds$$

$$+ \mu^{\varrho_2-1}\int_0^1 c_1^{\beta_1-1}\widetilde{J}_2(s)\varphi_{\varrho_2}\left(I_{0+}^{\alpha_2}g(s,u(s),v(s))\right)ds$$

$$\geq \mu^{\varrho_2-1}c_1^{\beta_1-1}\int_{c_1}^{c_2}\widetilde{J}_2(s)\varphi_{\varrho_2}\left(\frac{1}{\Gamma(\alpha_2)}\int_{c_1}^s (s-\tau)^{\alpha_2-1}\right.$$

$$\times\; g(\tau,u(\tau),v(\tau))\,d\tau\Bigg)\,ds$$

$$\geq \mu^{\varrho_2-1}c_1^{\beta_1-1}\int_{c_1}^{c_2}\widetilde{J}_2(s)\varphi_{\varrho_2}\left(\frac{1}{\Gamma(\alpha_2)}\int_{c_1}^s (s-\tau)^{\alpha_2-1}\right.$$

$$\times\; m_2(u(\tau)+v(\tau))^{r_2-1}\,d\tau\Bigg)\,ds$$

$$\geq \mu^{\varrho_2-1}c_1^{\beta_1-1}m_2^{\varrho_2-1}\int_{c_1}^{c_2}\widetilde{J}_2(s)\varphi_{\varrho_2}\left(\frac{1}{\Gamma(\alpha_2)}\int_{c_1}^s (s-\tau)^{\alpha_2-1}\right.$$

$$\times\; (\theta\|(u,v)\|_Y)^{r_2-1}\,d\tau\Bigg)\,ds$$

$$= (\mu m_2)^{\varrho_2-1}\theta\theta_1\widetilde{B}\|(u,v)\|_Y.$$

Then we conclude

$$\|u\| \geq u(c_1) \geq (\mu m_2)^{\varrho_2-1}\theta\theta_1\widetilde{B}\|(u,v)\|_Y$$
$$> (\widetilde{\mu}_0 m_2)^{\varrho_2-1}\theta\theta_1\widetilde{B}\|(u,v)\|_Y = \|(u,v)\|_Y,$$

and so $\|(u,v)\|_Y = \|u\| + \|v\| \geq \|u\| > \|(u,v)\|_Y$, which is a contradiction.
If $\theta_1\widetilde{B} < \theta_2\widetilde{C}$, then $\widetilde{\mu}_0 = \frac{1}{m_2(\theta\theta_2\widetilde{C})^{r_2-1}}$, and therefore, we deduce

$$v(c_1) = \mathcal{Q}_2(u,v)(c_1) \geq \mu^{\varrho_2-1}\int_0^1 c_1^{\beta_2-1}\widetilde{J}_3(s)\varphi_{\varrho_2}\left(I_{0+}^{\alpha_2}g(s,u(s),v(s))\right)ds$$

$$+ \lambda^{\varrho_1-1}\int_0^1 c_1^{\beta_2-1}\widetilde{J}_4(s)\varphi_{\varrho_1}\left(I_{0+}^{\alpha_1}f(s,u(s),v(s))\right)ds$$

$$\geq \mu^{\varrho_2-1}c_1^{\beta_2-1}\int_{c_1}^{c_2}\widetilde{J}_3(s)\varphi_{\varrho_2}\left(\frac{1}{\Gamma(\alpha_2)}\int_{c_1}^s (s-\tau)^{\alpha_2-1}\right.$$

$$\times\; g(\tau,u(\tau),v(\tau))\,d\tau\Bigg)\,ds$$

$$\geq \mu^{\varrho_2 - 1} c_1^{\beta_2 - 1} \int_{c_1}^{c_2} \widetilde{J}_3(s) \varphi_{\varrho_2} \left(\frac{1}{\Gamma(\alpha_2)} \int_{c_1}^{s} (s - \tau)^{\alpha_2 - 1} \right.$$

$$\left. \times\, m_2(u(\tau) + v(\tau))^{r_2 - 1}\, d\tau \right) ds$$

$$\geq \mu^{\varrho_2 - 1} c_1^{\beta_2 - 1} m_2^{\varrho_2 - 1} \int_{c_1}^{c_2} \widetilde{J}_3(s) \varphi_{\varrho_2} \left(\frac{1}{\Gamma(\alpha_2)} \int_{c_1}^{s} (s - \tau)^{\alpha_2 - 1} \right.$$

$$\left. \times\, (\theta \| (u, v) \|_Y)^{r_2 - 1}\, d\tau \right) ds$$

$$= (\mu m_2)^{\varrho_2 - 1} \theta \theta_2 \widetilde{C} \| (u, v) \|_Y.$$

Then we conclude

$$\| v \| \geq v(c_1) \geq (\mu m_2)^{\varrho_2 - 1} \theta \theta_2 \widetilde{C} \| (u, v) \|_Y$$

$$> (\widetilde{\mu}_0 m_2)^{\varrho_2 - 1} \theta \theta_2 \widetilde{C} \| (u, v) \|_Y = \| (u, v) \|_Y,$$

and so $\| (u, v) \|_Y = \| u \| + \| v \| \geq \| v \| > \| (u, v) \|_Y$, which is a contradiction.

Therefore, the boundary value problem (4.32), (4.33) has no positive solution. $\square$

Remark 4.2.3. If for c_1, c_2 with $0 < c_1 < c_2 \leq 1$, we have $g_0^i, g_\infty^i > 0$ and $g(t, u, v) > 0$ for all $t \in [c_1, c_2]$ and $u, v \geq 0$ with $u + v > 0$, then the relation (4.49) holds, and we obtain the conclusion of Theorem 4.2.5.

Theorem 4.2.6. *Assume that* (I1) *and* (I2) *hold. If there exist positive numbers* c_1, c_2 *with* $0 < c_1 < c_2 \leq 1$ *and* $m_1, m_2 > 0$ *such that*

$$f(t, u, v) \geq m_1 (u + v)^{r_1 - 1},$$

$$g(t, u, v) \geq m_2 (u + v)^{r_2 - 1}, \quad \forall t \in [c_1, c_2], \ u, v \geq 0, \tag{4.50}$$

then there exist positive constants $\hat{\lambda}_0$ *and* $\hat{\mu}_0$ *such that for every* $\lambda > \hat{\lambda}_0$ *and* $\mu > \hat{\mu}_0$, *the boundary value problem* (4.32), (4.33) *has no positive solution.*

Proof. We define $\hat{\lambda}_0 = \frac{1}{m_1 (2\theta \theta_1 \widetilde{A})^{r_1 - 1}}$ and $\widetilde{\mu}_0 = \frac{1}{m_2 (2\theta \theta_2 \widetilde{C})^{r_2 - 1}}$, where $\widetilde{A}$ and $\widetilde{C}$ are given by (4.41). Then for every $\lambda > \hat{\lambda}_0$ and $\mu > \hat{\mu}_0$, the problem (4.32), (4.33) has no positive solution. Indeed, let $\lambda > \hat{\lambda}_0$ and $\mu > \hat{\mu}_0$. We suppose that the problem (4.32), (4.33) has a positive solution $(u(t), v(t))$, $t \in [0, 1]$. In a similar manner as that used in the proofs of

Theorems 4.2.4 and 4.2.5, we obtain

$$\|u\| \geq u(c_1) \geq (\lambda m_1)^{\varrho_1 - 1} \theta \theta_1 \widetilde{A} \|(u,v)\|_Y,$$

$$\|v\| \geq v(c_1) \geq (\mu m_2)^{\varrho_2 - 1} \theta \theta_2 \widetilde{C} \|(u,v)\|_Y,$$

and so

$$\|(u,v)\|_Y = \|u\| + \|v\| \geq (\lambda m_1)^{\varrho_1 - 1} \theta \theta_1 \widetilde{A} \|(u,v)\|_Y$$

$$+ (\mu m_2)^{\varrho_2 - 1} \theta \theta_2 \widetilde{C} \|(u,v)\|_Y$$

$$> (\hat{\lambda}_0 m_1)^{\varrho_1 - 1} \theta \theta_1 \widetilde{A} \|(u,v)\|_Y + (\hat{\mu}_0 m_2)^{\varrho_2 - 1} \theta \theta_2 \widetilde{C} \|(u,v)\|_Y$$

$$= \frac{1}{2} \|(u,v)\|_Y + \frac{1}{2} \|(u,v)\|_Y = \|(u,v)\|_Y,$$

which is a contradiction. Therefore, the boundary value problem (4.32), (4.33) has no positive solution.

We can also define $\hat{\lambda}_0' = \frac{1}{m_1(2\theta\theta_2\widetilde{D})^{r_1 - 1}}$ and $\widetilde{\mu}_0' = \frac{1}{m_2(2\theta\theta_1\widetilde{B})^{r_2 - 1}}$, where $\widetilde{B}$ and $\widetilde{D}$ are given by (4.41). Then for every $\lambda > \hat{\lambda}_0'$ and $\mu > \hat{\mu}_0'$, the problem (4.32), (4.33) has no positive solution. Indeed, let $\lambda > \hat{\lambda}_0'$ and $\mu > \hat{\mu}_0'$. We suppose that the problem (4.32), (4.33) has a positive solution $(u(t), v(t))$, $t \in [0,1]$. In a similar manner as that used in the proofs of Theorems 4.2.4 and 4.2.5, we obtain

$$\|v\| \geq v(c_1) \geq (\lambda m_1)^{\varrho_1 - 1} \theta \theta_2 \widetilde{D} \|(u,v)\|_Y,$$

$$\|u\| \geq u(c_1) \geq (\mu m_2)^{\varrho_2 - 1} \theta \theta_1 \widetilde{B} \|(u,v)\|_Y,$$

and so

$$\|(u,v)\|_Y = \|u\| + \|v\| \geq (\lambda m_1)^{\varrho_1 - 1} \theta \theta_2 \widetilde{D} \|(u,v)\|_Y$$

$$+ (\mu m_2)^{\varrho_2 - 1} \theta \theta_1 \widetilde{B} \|(u,v)\|_Y$$

$$> (\hat{\lambda}_0' m_1)^{\varrho_1 - 1} \theta \theta_2 \widetilde{D} \|(u,v)\|_Y + (\hat{\mu}_0' m_2)^{\varrho_2 - 1} \theta \theta_1 \widetilde{B} \|(u,v)\|_Y$$

$$= \frac{1}{2} \|(u,v)\|_Y + \frac{1}{2} \|(u,v)\|_Y = \|(u,v)\|_Y,$$

which is a contradiction. Therefore, the boundary value problem (4.32), (4.33) has no positive solution. $\qquad \square$

Remark 4.2.4. If for c_1, c_2 with $0 < c_1 < c_2 \leq 1$, we have f_0^i, f_∞^i, g_0^i, $g_\infty^i > 0$ and $f(t,u,v) > 0$, $g(t,u,v) > 0$ for all $t \in [c_1, c_2]$ and $u, v \geq 0$ with $u + v > 0$, then the relation (4.50) holds, and we obtain the conclusion of Theorem 4.2.6.

4.2.4 *An example*

Let $\alpha_1 = 1/3$, $\alpha_2 = 1/4$, $\beta_1 = 7/2$, $n = 4$, $\beta_2 = 14/3$, $m = 5$, $p_1 = 4/3$, $p_2 = 5/2$, $q_1 = 5/4$, $q_2 = 2/3$, $N = 2$, $\xi_1 = 1/4$, $\xi_2 = 3/5$, $a_1 = 2$, $a_2 = 1/3$, $M = 1$, $\eta_1 = 1/2$, $b_1 = 4$, $r_1 = 5$, $\varrho_1 = 5/4$, $\varphi_{r_1}(s) = s|s|^3$, $\varphi_{\varrho_1}(s) = s|s|^{-3/4}$, $r_2 = 3$, $\varrho_2 = 3/2$, $\varphi_{r_2}(s) = s|s|$, $\varphi_{\varrho_2}(s) = s|s|^{-1/2}$.

We consider the system of fractional differential equations

$$
\begin{cases}
D_{0+}^{1/3}(\varphi_5(D_{0+}^{7/2}u(t))) + \lambda(t+1)^{\widetilde{a}}(u^5(t) + v^5(t)) = 0, & t \in (0,1), \\
D_{0+}^{1/4}(\varphi_3(D_{0+}^{14/3}v(t))) + \mu(2-t)^{\widetilde{b}}(e^{(u(t)+v(t))^2} - 1) = 0, & t \in (0,1),
\end{cases}
\tag{4.51}
$$

with the coupled multi-point boundary conditions

$$
\begin{cases}
u(0) = u'(0) = u''(0) = 0, \quad D_{0+}^{7/2}u(0) = 0, \\
D_{0+}^{4/3}u(1) = 2D_{0+}^{5/4}v\left(\dfrac{1}{4}\right) + \dfrac{1}{3}D_{0+}^{5/4}v\left(\dfrac{3}{5}\right), \\
v(0) = v'(0) = v''(0) = v'''(0) = 0, \quad D_{0+}^{14/3}v(0) = 0, \\
D_{0+}^{5/2}u(1) = 4D_{0+}^{2/3}v\left(\dfrac{1}{2}\right),
\end{cases}
\tag{4.52}
$$

where $\widetilde{a}$, $\widetilde{b} > 0$.

Here, we have $f(t,u,v) = (t+1)^{\widetilde{a}}(u^5+v^5)$, $g(t,u,v) = (2-t)^{\widetilde{b}}(e^{(u+v)^2}-1)$ for all $t \in [0,1]$ and $u, v \geq 0$. Then we obtain $\Delta \approx 39.98272963 > 0$, and so the assumptions (I1) and (I2) are satisfied. In addition, we deduce

$$
\widetilde{g}_1(t,s) = \frac{1}{\Gamma(7/2)}
\begin{cases}
t^{5/2}(1-s)^{7/6} - (t-s)^{5/2}, & 0 \leq s \leq t \leq 1, \\
t^{5/2}(1-s)^{7/6}, & 0 \leq t \leq s \leq 1,
\end{cases}
$$

$$
\widetilde{g}_2(t,s) = \frac{1}{\Gamma(17/6)}
\begin{cases}
t^{11/6}(1-s)^{7/6} - (t-s)^{11/6}, & 0 \leq s \leq t \leq 1, \\
t^{11/6}(1-s)^{7/6}, & 0 \leq t \leq s \leq 1,
\end{cases}
$$

$$\tilde{g}_3(t,s) = \frac{1}{\Gamma(41/12)} \begin{cases} t^{29/12}(1-s)^{7/6} - (t-s)^{29/12}, & 0 \le s \le t \le 1, \\ t^{29/12}(1-s)^{7/6}, & 0 \le t \le s \le 1, \end{cases}$$

$$\tilde{g}_4(t,s) = \frac{1}{\Gamma(14/3)} \begin{cases} t^{11/3}(1-s)^{7/6} - (t-s)^{11/3}, & 0 \le s \le t \le 1, \\ t^{11/3}(1-s)^{7/6}, & 0 \le t \le s \le 1, \end{cases}$$

$$\tilde{G}_1(t,s) = \tilde{g}_1(t,s) + \frac{4t^{5/2}\Gamma(14/3)}{\Delta\Gamma(41/12)} \left[2\left(\frac{1}{4}\right)^{29/12} + \frac{1}{3}\left(\frac{3}{5}\right)^{29/12} \right] \tilde{g}_2\left(\frac{1}{2}, s\right),$$

$$\tilde{G}_2(t,s) = \frac{t^{5/2}\Gamma(14/3)}{\Delta\Gamma(13/6)} \left[2\tilde{g}_3\left(\frac{1}{4}, s\right) + \frac{1}{3}\tilde{g}_3\left(\frac{3}{5}, s\right) \right],$$

$$\tilde{G}_3(t,s) = \tilde{g}_4(t,s) + \frac{4t^{11/3}\Gamma(7/2)(1/2)^{11/6}}{\Delta\Gamma(17/6)} \left[2\tilde{g}_3\left(\frac{1}{4}, s\right) + \frac{1}{3}\tilde{g}_3\left(\frac{3}{5}, s\right) \right],$$

$$\tilde{G}_4(t,s) = \frac{4t^{11/3}\Gamma(7/2)}{\Delta\Gamma(13/6)} \tilde{g}_2\left(\frac{1}{2}, s\right), \quad \forall t, s \in [0,1].$$

For the functions $\tilde{h}_1$, $\tilde{h}_4$ and $\tilde{J}_i$, $i = 1, \ldots, 4$, we obtain

$$\tilde{h}_1(s) = \frac{1}{\Gamma(7/2)}(1-s)^{7/6}(1 - (1-s)^{4/3}),$$

$$\tilde{h}_4(s) = \frac{1}{\Gamma(14/3)}(1-s)^{7/6}(1 - (1-s)^{5/2}),$$

$$\tilde{J}_1(s) = \begin{cases} \dfrac{1}{\Gamma(7/2)}(1-s)^{7/6}(1 - (1-s)^{4/3}) \\[2mm] \quad + \dfrac{\Gamma(14/3)}{\Delta\Gamma(41/12)} \left[2\left(\dfrac{1}{4}\right)^{29/12} + \dfrac{1}{3}\left(\dfrac{3}{5}\right)^{29/12} \right] \\[2mm] \quad \times \dfrac{4}{\Gamma(17/6)} \left[\left(\dfrac{1}{2}\right)^{11/6}(1-s)^{7/6} - \left(\dfrac{1}{2} - s\right)^{11/6} \right], & 0 \le s < \dfrac{1}{2}, \\[4mm] \dfrac{1}{\Gamma(7/2)}(1-s)^{7/6}(1 - (1-s)^{4/3}) \\[2mm] \quad + \dfrac{\Gamma(14/3)}{\Delta\Gamma(41/12)} \left[2\left(\dfrac{1}{4}\right)^{29/12} + \dfrac{1}{3}\left(\dfrac{3}{5}\right)^{29/12} \right] \\[2mm] \quad \times \dfrac{4}{\Gamma(17/6)} \left(\dfrac{1}{2}\right)^{11/6}(1-s)^{7/6}, & \dfrac{1}{2} \le s \le 1, \end{cases}$$

$$\tilde{J}_2(s) = \begin{cases} \dfrac{\Gamma(14/3)}{\Delta\Gamma(13/6)\Gamma(41/12)} \left\{ \dfrac{2}{4^{29/12}} \left[(1-s)^{7/6} - (1-4s)^{29/12} \right] \right. \\ \qquad \left. + \dfrac{1}{3 \cdot 5^{29/12}} \left[3^{29/12}(1-s)^{7/6} - (3-5s)^{29/12} \right] \right\}, \quad 0 \le s < \dfrac{1}{4}, \\[2mm] \dfrac{\Gamma(14/3)}{\Delta\Gamma(13/6)\Gamma(41/12)} \left\{ \dfrac{2}{4^{29/12}}(1-s)^{7/6} + \dfrac{1}{3 \cdot 5^{29/12}} \right. \\ \qquad \left. \times [3^{29/12}(1-s)^{7/6} - (3-5s)^{29/12}] \right\}, \quad \dfrac{1}{4} \le s < \dfrac{3}{5}, \\[2mm] \dfrac{\Gamma(14/3)}{\Delta\Gamma(13/6)\Gamma(41/12)} \left[2\left(\dfrac{1}{4}\right)^{29/12} + \dfrac{1}{3}\left(\dfrac{3}{5}\right)^{29/12} \right] \\ \qquad \times (1-s)^{7/6}, \quad \dfrac{3}{5} \le s \le 1, \end{cases}$$

$$\tilde{J}_3(s) = \begin{cases} \dfrac{1}{\Gamma(14/3)}(1-s)^{7/6}(1-(1-s)^{5/2}) + \dfrac{\Gamma(7/2)2^{1/6}}{\Delta\Gamma(17/6)\Gamma(41/12)} \\ \qquad \times \left\{ \dfrac{2}{4^{29/12}}[(1-s)^{7/6} - (1-4s)^{29/12}] + \dfrac{1}{3 \cdot 5^{29/12}} \right. \\ \qquad \left. \times [3^{29/12}(1-s)^{7/6} - (3-5s)^{29/12}] \right\}, \quad 0 \le s < \dfrac{1}{4}, \\[2mm] \dfrac{1}{\Gamma(14/3)}(1-s)^{7/6}(1-(1-s)^{5/2}) + \dfrac{\Gamma(7/2)2^{1/6}}{\Delta\Gamma(17/6)\Gamma(41/12)} \\ \qquad \times \left\{ \dfrac{2}{4^{29/12}}(1-s)^{7/6} + \dfrac{1}{3 \cdot 5^{29/12}}[3^{29/12}(1-s)^{7/6} \right. \\ \qquad \left. -(3-5s)^{29/12}] \right\}, \quad \dfrac{1}{4} \le s < \dfrac{3}{5}, \\[2mm] \dfrac{1}{\Gamma(14/3)}(1-s)^{7/6}(1-(1-s)^{5/2}) + \dfrac{\Gamma(7/2)2^{1/6}}{\Delta\Gamma(17/6)\Gamma(41/12)} \\ \qquad \times \left(\dfrac{2}{4^{29/12}} + \dfrac{1}{3}\left(\dfrac{3}{5}\right)^{29/12} \right) (1-s)^{7/6}, \quad \dfrac{3}{5} \le s \le 1, \end{cases}$$

$$\tilde{J}_4(s) = \begin{cases} \dfrac{\Gamma(7/2)2^{1/6}}{\Delta\Gamma(13/6)\Gamma(17/6)} \left[(1-s)^{7/6} - (1-2s)^{11/6} \right], \quad 0 \le s < \dfrac{1}{2}, \\[2mm] \dfrac{\Gamma(7/2)2^{1/6}}{\Delta\Gamma(13/6)\Gamma(17/6)}(1-s)^{7/6}, \quad \dfrac{1}{2} \le s \le 1. \end{cases}$$

Now, we choose $c_1 = 1/4$ and $c_2 = 3/4$, and then we deduce $\theta_1 = (1/4)^{5/2}$, $\theta_2 = (1/4)^{11/3}$ and $\theta = \theta_2$. In addition, we have $f_0^s = 0$, $f_\infty^i = \infty$, $g_0^s = 2^{\tilde{b}}$, $g_\infty^i = \infty$, $B \approx 0.00564278$, $\tilde{B} \approx 0.00325593$, $C \approx 0.01653798$, $\tilde{C} \approx 0.0102111$.

By Theorem 4.2.1(6), if we consider $b = 1/2$, then for any $\lambda \in (0, \infty)$ and $\mu \in (0, L_4')$ with $L_4' = \frac{1}{2^b}\left(\frac{1}{2C}\right)^2$, the problem (4.51), (4.52) has a positive solution $(u(t), v(t))$, $t \in [0, 1]$. For example, if $\tilde{b} = 1$ we obtain $L_4' \approx 457.0303$.

We can also use Theorem 4.2.5, because $g(t, u, v) \geq m_2(u + v)^2$ for all $t \in [1/4, 3/4]$ and $u, v \geq 0$, with $m_2 = (5/4)^{\tilde{b}}$. If $\tilde{b} = 1$, we deduce $\tilde{\mu}_0 = \frac{4}{5(\theta\theta_1\tilde{B})^2} \approx 2.0097701 \times 10^{12}$, and then we conclude that for every $\lambda > 0$ and $\mu > \tilde{\mu}_0$, the boundary value problem (4.51), (4.52) has no positive solution.

Remark 4.2.5. The results presented in this section under the assumptions $p_1 \in [1, n-2]$ and $p_2 \in [1, m-2]$ instead of $p_1 \in [1, \beta_1-1)$ and $p_2 \in [1, \beta_2-1)$ were published in [80].

Remark 4.2.6. The existence of positive solutions for the system of Riemann-Liouville fractional differential equations

$$\begin{cases} D_{0+}^\alpha u(t) + \lambda f(t, u(t), v(t)) = 0, & t \in (0, 1), \\ D_{0+}^\beta v(t) + \mu g(t, u(t), v(t)) = 0, & t \in (0, 1), \end{cases} \tag{4.53}$$

subject to the coupled integral boundary conditions

$$\begin{cases} u(0) = u'(0) = \cdots = u^{(n-2)}(0) = 0, & u'(1) = \int_0^1 v(s)\, dH(s), \\ v(0) = v'(0) = \cdots = v^{(m-2)}(0) = 0, & v'(1) = \int_0^1 u(s)\, dK(s), \end{cases}$$

where $\alpha \in (n-1, n]$, $\beta \in (m-1, m]$, $n, m \in \mathbb{N}$, $n, m \geq 3$, λ, μ are positive parameters, and H, K are nondecreasing functions, was investigated in [51] (in the case where f, g are nonnegative functions) and in [44] (in the case where f, g are sign-changing functions). The system (4.53) with f, g sign-changing functions, supplemented with the coupled integral boundary conditions

$$
\begin{cases}
u(0) = u'(0) = \cdots = u^{(n-2)}(0) = 0, & D_{0+}^{p} u(1) = \displaystyle\int_{0}^{1} v(s)\, dH(s), \\[3mm]
v(0) = v'(0) = \cdots = v^{(m-2)}(0) = 0, & D_{0+}^{q} v(1) = \displaystyle\int_{0}^{1} u(s)\, dK(s),
\end{cases}
$$

where $p, q \in \mathbb{R}$, $p \in [1, n-2]$, $q \in [1, m-2]$, and H, K are nondecreasing functions, was studied in [45].

Chapter 5

Systems of Three Riemann–Liouville Fractional Differential Equations with Parameters and Multi-Point Boundary Conditions

In this chapter, we study the existence and nonexistence of positive solutions for systems of three Riemann-Liouville fractional differential equations with positive parameters subject to uncoupled multi-point boundary conditions, which contain fractional derivatives, and the nonlinearities of systems are nonnegative functions.

5.1 Systems of Fractional Differential Equations with Uncoupled Multi-Point Boundary Conditions

We consider the system of nonlinear ordinary fractional differential equations

$$
\begin{cases}
D_{0+}^{\alpha} u(t) + \lambda f(t, u(t), v(t), w(t)) = 0, & t \in (0,1), \\
D_{0+}^{\beta} v(t) + \mu g(t, u(t), v(t), w(t)) = 0, & t \in (0,1), \\
D_{0+}^{\gamma} w(t) + \nu h(t, u(t), v(t), w(t)) = 0, & t \in (0,1),
\end{cases}
\tag{5.1}
$$

with the multi-point boundary conditions

$$
\begin{cases}
u^{(j)}(0) = 0, \quad j = 0, \dots, n-2; \quad D_{0+}^{p_1} u(1) = \sum_{i=1}^{N} a_i D_{0+}^{q_1} u(\xi_i), \\[2mm]
v^{(j)}(0) = 0, \quad j = 0, \dots, m-2; \quad D_{0+}^{p_2} v(1) = \sum_{i=1}^{M} b_i D_{0+}^{q_2} v(\eta_i), \quad (5.2) \\[2mm]
w^{(j)}(0) = 0, \quad j = 0, \dots, l-2; \quad D_{0+}^{p_3} w(1) = \sum_{i=1}^{L} c_i D_{0+}^{q_3} w(\zeta_i),
\end{cases}
$$

where $\lambda, \mu, \nu > 0$, $\alpha, \beta, \gamma \in \mathbb{R}$, $\alpha \in (n-1, n]$, $\beta \in (m-1, m]$, $\gamma \in (l-1, l]$, $n, m, l \in \mathbb{N}$, $n, m, l \geq 3$, $p_1, p_2, p_3, q_1, q_2, q_3 \in \mathbb{R}$, $p_1 \in [1, \alpha - 1)$, $p_2 \in [1, \beta - 1)$, $p_3 \in [1, \gamma - 1)$, $q_1 \in [0, p_1]$, $q_2 \in [0, p_2]$, $q_3 \in [0, p_3]$, $\xi_i, a_i \in \mathbb{R}$ for all $i = 1, \dots, N$ ($N \in \mathbb{N}$), $0 < \xi_1 < \cdots < \xi_N \leq 1$, $\eta_i, b_i \in \mathbb{R}$ for all $i = 1, \dots, M$ ($M \in \mathbb{N}$), $0 < \eta_1 < \cdots < \eta_M \leq 1$, $\zeta_i, c_i \in \mathbb{R}$ for all $i = 1, \dots, L$ ($L \in \mathbb{N}$), $0 < \zeta_1 < \cdots < \zeta_L \leq 1$, and D_{0+}^k denotes the Riemann-Liouville derivative of order k (for $k = \alpha, \beta, \gamma, p_1, q_1, p_2, q_2, p_3, q_3$).

Under some assumptions on f, g and h, we give intervals for the parameters λ, μ and ν such that positive solutions of (5.1), (5.2) exist. By a positive solution of problem (5.1), (5.2) we mean a triplet of functions $(u, v, w) \in (C([0, 1], \mathbb{R}_+))^3$, satisfying (5.1) and (5.2) with $u(t) > 0$ for all $t \in (0, 1]$, or $v(t) > 0$ for all $t \in (0, 1]$, or $w(t) > 0$ for all $t \in (0, 1]$. The nonexistence of positive solutions for the above problem is also studied. The results obtained in this section improve and extend the results from [98], where only a few cases are presented for the existence of positive solutions for a system of integral equations, and, as an application, for a system with three fractional equations subject to some boundary conditions in points $t = 0$ and $t = 1$ (Application 4.3 from [98]).

5.1.1   *Auxiliary results*

We present firstly some auxiliary results from Section 2.1.1 that will be used to prove our main results.

We consider the fractional differential equation

$$
D_{0+}^{\alpha} u(t) + x(t) = 0, \quad 0 < t < 1, \quad (5.3)
$$

with the multi-point boundary conditions

$$u^{(j)}(0) = 0, \quad j = 0, \ldots, n-2; \quad D_{0+}^{p_1} u(1) = \sum_{i=1}^{N} a_i D_{0+}^{q_1} u(\xi_i), \qquad (5.4)$$

where $\alpha \in (n-1, n]$, $n \in \mathbb{N}$, $n \geq 3$, $a_i, \xi_i \in \mathbb{R}$, $i = 1, \ldots, N$ ($N \in \mathbb{N}$), $0 < \xi_1 < \cdots < \xi_N \leq 1$, $p_1, q_1 \in \mathbb{R}$, $p_1 \in [1, \alpha - 1)$, $q_1 \in [0, p_1]$, and $x \in C[0, 1]$. We denote by $\Delta_1 = \frac{\Gamma(\alpha)}{\Gamma(\alpha-p_1)} - \frac{\Gamma(\alpha)}{\Gamma(\alpha-q_1)} \sum_{i=1}^{N} a_i \xi_i^{\alpha-q_1-1}$.

Lemma 5.1.1. *If* $\Delta_1 \neq 0$, *then the unique solution* $u \in C[0, 1]$ *of problem* (5.3), (5.4) *is*

$$u(t) = \int_0^1 G_1(t, s) x(s) \, ds, \quad t \in [0, 1], \qquad (5.5)$$

where the Green function G_1 *is given by*

$$G_1(t, s) = g_1(t, s) + \frac{t^{\alpha-1}}{\Delta_1} \sum_{i=1}^{N} a_i g_2(\xi_i, s), \quad \forall (t, s) \in [0, 1] \times [0, 1], \qquad (5.6)$$

and

$$g_1(t, s) = \frac{1}{\Gamma(\alpha)} \begin{cases} t^{\alpha-1}(1-s)^{\alpha-p_1-1} - (t-s)^{\alpha-1}, & 0 \leq s \leq t \leq 1, \\ t^{\alpha-1}(1-s)^{\alpha-p_1-1}, & 0 \leq t \leq s \leq 1, \end{cases}$$

$$g_2(t, s) = \frac{1}{\Gamma(\alpha-q_1)} \begin{cases} t^{\alpha-q_1-1}(1-s)^{\alpha-p_1-1} - (t-s)^{\alpha-q_1-1}, \\ \qquad 0 \leq s \leq t \leq 1, \\ t^{\alpha-q_1-1}(1-s)^{\alpha-p_1-1}, \quad 0 \leq t \leq s \leq 1. \end{cases} \qquad (5.7)$$

Lemma 5.1.2. *The functions* g_1 *and* g_2 *given by* (5.7) *have the properties:*

(a) $g_1(t, s) \leq h_1(s)$ *for all* $t, s \in [0, 1]$, *where*

$$h_1(s) = \frac{1}{\Gamma(\alpha)} (1-s)^{\alpha-p_1-1}(1 - (1-s)^{p_1}), s \in [0, 1];$$

(b) $g_1(t, s) \geq t^{\alpha-1} h_1(s)$ *for all* $t, s \in [0, 1]$;

(c) $g_1(t, s) \leq \frac{t^{\alpha-1}}{\Gamma(\alpha)}$, *for all* $t, s \in [0, 1]$;

(d) $g_2(t, s) \geq t^{\alpha-q_1-1} h_2(s)$ *for all* $t, s \in [0, 1]$, *where*

$$h_2(s) = \frac{1}{\Gamma(\alpha-q_1)} (1-s)^{\alpha-p_1-1}(1 - (1-s)^{p_1-q_1}), s \in [0, 1];$$

(e) $g_2(t, s) \leq \frac{1}{\Gamma(\alpha-q_1)} t^{\alpha-q_1-1}$ *for all* $t, s \in [0, 1]$;

(f) *The functions g_1 and g_2 are continuous on $[0,1] \times [0,1]$; $g_1(t,s) \geq 0$, $g_2(t,s) \geq 0$ for all $t, s \in [0,1]$; $g_1(t,s) > 0$, $g_2(t,s) > 0$ for all $t, s \in (0,1)$.*

Lemma 5.1.3. *Assume that $a_i \geq 0$ for all $i = 1, \ldots, N$ and $\Delta_1 > 0$. Then the function G_1 given by (5.6) is a nonnegative continuous function on $[0,1] \times [0,1]$ and satisfies the inequalities:*

(a) $G_1(t,s) \leq J_1(s)$ *for all* $t, s \in [0,1]$, *where* $J_1(s) = h_1(s) + \frac{1}{\Delta_1} \sum_{i=1}^{N} a_i g_2(\xi_i, s)$, $s \in [0,1]$;

(b) $G_1(t,s) \geq t^{\alpha-1} J_1(s)$ *for all* $t, s \in [0,1]$;

(c) $G_1(t,s) \leq \sigma_1 t^{\alpha-1}$, *for all* $t, s \in [0,1]$, *where* $\sigma_1 = \frac{1}{\Gamma(\alpha)} + \frac{1}{\Delta_1 \Gamma(\alpha - q_1)}$ $\times \sum_{i=1}^{N} a_i \xi_i^{\alpha - q_1 - 1}$.

Lemma 5.1.4. *Assume that $a_i \geq 0$ for all $i = 1, \ldots, N$, $\Delta_1 > 0$, $x \in C[0,1]$ and $x(t) \geq 0$ for all $t \in [0,1]$. Then the solution u of problem (5.3), (5.4) given by (5.5) satisfies the inequality $u(t) \geq t^{\alpha-1} u(t')$ for all $t, t' \in [0,1]$.*

We can also formulate similar results as Lemmas 5.1.1–5.1.4 for the fractional boundary value problems

$$D_{0+}^{\beta} v(t) + y(t) = 0, \quad 0 < t < 1, \tag{5.8}$$

$$v^{(j)}(0) = 0, \quad j = 0, \ldots, m-2; \quad D_{0+}^{p_2} v(1) = \sum_{i=1}^{M} b_i D_{0+}^{q_2} v(\eta_i), \tag{5.9}$$

and

$$D_{0+}^{\gamma} w(t) + z(t) = 0, \quad 0 < t < 1, \tag{5.10}$$

$$w^{(j)}(0) = 0, \quad j = 0, \ldots, l-2; \quad D_{0+}^{p_3} w(1) = \sum_{i=1}^{L} c_i D_{0+}^{q_3} w(\zeta_i), \tag{5.11}$$

where $\beta \in (m-1, m]$, $\gamma \in (l-1, l]$, $m, l \in \mathbb{N}$, $m, l \geq 3$, $b_i, \eta_i \in \mathbb{R}$, $i = 1, \ldots, M$ ($M \in \mathbb{N}$), $0 < \eta_1 < \cdots < \eta_M \leq 1$, $c_i, \zeta_i \in \mathbb{R}$, $i = 1, \ldots, L$ ($L \in \mathbb{N}$), $0 < \zeta_1 < \cdots < \zeta_L \leq 1$, $p_2, q_2, p_3, q_3 \in \mathbb{R}$, $p_2 \in [1, \beta - 1)$, $q_2 \in [0, p_2]$, $p_3 \in [1, \gamma - 1)$, $q_3 \in [0, p_3]$, and $y, z \in C[0,1]$.

We denote by Δ_2, g_3, g_4, G_2, h_3, h_4, J_2 and σ_2, and Δ_3, g_5, g_6, G_3, h_5, h_6, J_3 and σ_3 the corresponding constants and functions for problem (5.8),

(5.9) and problem (5.10), (5.11), respectively, defined in a similar manner as Δ_1, g_1, g_2, G_1, h_1, h_2, J_1 and σ_1, respectively. More precisely, we have

$$\Delta_2 = \frac{\Gamma(\beta)}{\Gamma(\beta - p_2)} - \frac{\Gamma(\beta)}{\Gamma(\beta - q_2)} \sum_{i=1}^{M} b_i \eta_i^{\beta - q_2 - 1},$$

$$g_3(t, s) = \frac{1}{\Gamma(\beta)} \begin{cases} t^{\beta-1}(1-s)^{\beta-p_2-1} - (t-s)^{\beta-1}, & 0 \leq s \leq t \leq 1, \\ t^{\beta-1}(1-s)^{\beta-p_2-1}, & 0 \leq t \leq s \leq 1, \end{cases}$$

$$g_4(t, s) = \frac{1}{\Gamma(\beta - q_2)} \begin{cases} t^{\beta-q_2-1}(1-s)^{\beta-p_2-1} - (t-s)^{\beta-q_2-1}, \\ \quad 0 \leq s \leq t \leq 1, \\ t^{\beta-q_2-1}(1-s)^{\beta-p_2-1}, \quad 0 \leq t \leq s \leq 1, \end{cases}$$

$$G_2(t, s) = g_3(t, s) + \frac{t^{\beta-1}}{\Delta_2} \sum_{i=1}^{M} b_i g_4(\eta_i, s), \quad \forall (t, s) \in [0, 1] \times [0, 1],$$

$$h_3(s) = \frac{1}{\Gamma(\beta)}(1-s)^{\beta-p_2-1}(1 - (1-s)^{p_2}), \quad s \in [0, 1],$$

$$h_4(s) = \frac{1}{\Gamma(\beta - q_2)}(1-s)^{\beta-p_2-1}(1 - (1-s)^{p_2-q_2}), \quad s \in [0, 1],$$

$$J_2(s) = h_3(s) + \frac{1}{\Delta_2} \sum_{i=1}^{M} b_i g_4(\eta_i, s), \quad s \in [0, 1],$$

$$\sigma_2 = \frac{1}{\Gamma(\beta)} + \frac{1}{\Delta_2 \Gamma(\beta - q_2)} \sum_{i=1}^{M} b_i \eta_i^{\beta - q_2 - 1},$$

and

$$\Delta_3 = \frac{\Gamma(\gamma)}{\Gamma(\gamma - p_3)} - \frac{\Gamma(\gamma)}{\Gamma(\gamma - q_3)} \sum_{i=1}^{L} c_i \zeta_i^{\gamma - q_3 - 1},$$

$$g_5(t, s) = \frac{1}{\Gamma(\gamma)} \begin{cases} t^{\gamma-1}(1-s)^{\gamma-p_3-1} - (t-s)^{\gamma-1}, & 0 \leq s \leq t \leq 1, \\ t^{\gamma-1}(1-s)^{\gamma-p_3-1}, & 0 \leq t \leq s \leq 1, \end{cases}$$

$$g_6(t, s) = \frac{1}{\Gamma(\gamma - q_3)} \begin{cases} t^{\gamma-q_3-1}(1-s)^{\gamma-p_3-1} - (t-s)^{\gamma-q_3-1}, \\ \quad 0 \leq s \leq t \leq 1, \\ t^{\gamma-q_3-1}(1-s)^{\gamma-p_3-1}, \quad 0 \leq t \leq s \leq 1, \end{cases}$$

$$G_3(t, s) = g_5(t, s) + \frac{t^{\gamma-1}}{\Delta_3} \sum_{i=1}^{L} c_i g_6(\zeta_i, s), \quad \forall (t, s) \in [0, 1] \times [0, 1],$$

$$h_5(s) = \frac{1}{\Gamma(\gamma)}(1-s)^{\gamma-p_3-1}(1-(1-s)^{p_3}), \ s \in [0,1],$$

$$h_6(s) = \frac{1}{\Gamma(\gamma-q_3)}(1-s)^{\gamma-p_3-1}(1-(1-s)^{p_3-q_3}), \ s \in [0,1],$$

$$J_3(s) = h_5(s) + \frac{1}{\Delta_3}\sum_{i=1}^{L} c_i g_6(\zeta_i, s), \ s \in [0,1],$$

$$\sigma_3 = \frac{1}{\Gamma(\gamma)} + \frac{1}{\Delta_3\Gamma(\gamma-q_3)}\sum_{i=1}^{L} c_i \zeta_i^{\gamma-q_3-1}.$$

The inequalities from Lemmas 5.1.3 and 5.1.4 for the functions G_2, G_3, v and w are the following $G_2(t,s) \le J_2(s)$, $G_2(t,s) \ge t^{\beta-1}J_2(s)$, $G_2(t,s) \le \sigma_2 t^{\beta-1}$, $G_3(t,s) \le J_3(s)$, $G_3(t,s) \ge t^{\gamma-1}J_3(s)$, $G_3(t,s) \le \sigma_3 t^{\gamma-1}$ for all $t, s \in [0,1]$, and $v(t) \ge t^{\beta-1}v(t')$, $w(t) \ge t^{\gamma-1}w(t')$ for all $t, t' \in [0,1]$.

5.1.2   *Existence of positive solutions*

In this section, we give sufficient conditions on λ, μ, ν, f, g and h such that positive solutions with respect to a cone for our problem (5.1), (5.2) exist. We present the assumptions that we shall use in the sequel.

(H1) $\alpha, \beta, \gamma \in \mathbb{R}$, $\alpha \in (n-1, n]$, $\beta \in (m-1, m]$, $\gamma \in (l-1, l]$, $n, m, l \in \mathbb{N}$, $n, m, l \ge 3$, $p_1, p_2, p_3, q_1, q_2, q_3 \in \mathbb{R}$, $p_1 \in [1, \alpha-1)$, $p_2 \in [1, \beta-1)$, $p_3 \in [1, \gamma-1)$, $q_1 \in [0, p_1]$, $q_2 \in [0, p_2]$, $q_3 \in [0, p_3]$, $\xi_i \in \mathbb{R}$, $a_i \ge 0$ for all $i = 1, \ldots, N$ ($N \in \mathbb{N}$), $0 < \xi_1 < \cdots < \xi_N \le 1$, $\eta_i \in \mathbb{R}$, $b_i \ge 0$ for all $i = 1, \ldots, M$ ($M \in \mathbb{N}$), $0 < \eta_1 < \cdots < \eta_M \le 1$, and $\zeta_i \in \mathbb{R}$, $c_i \ge 0$ for all $i = 1, \ldots, L$ ($L \in \mathbb{N}$), $0 < \zeta_1 < \cdots < \zeta_L \le 1$; $\lambda, \mu, \nu > 0$, $\Delta_1 = \frac{\Gamma(\alpha)}{\Gamma(\alpha-p_1)} - \frac{\Gamma(\alpha)}{\Gamma(\alpha-q_1)}\sum_{i=1}^{N} a_i \xi_i^{\alpha-q_1-1} > 0$, $\Delta_2 = \frac{\Gamma(\beta)}{\Gamma(\beta-p_2)} - \frac{\Gamma(\beta)}{\Gamma(\beta-q_2)}\sum_{i=1}^{M} b_i \eta_i^{\beta-q_2-1} > 0$, $\Delta_3 = \frac{\Gamma(\gamma)}{\Gamma(\gamma-p_3)} - \frac{\Gamma(\gamma)}{\Gamma(\gamma-q_3)}\sum_{i=1}^{L} c_i \zeta_i^{\gamma-q_3-1} > 0$.

(H2) The functions $f, g, h : [0,1] \times \mathbb{R}_+ \times \mathbb{R}_+ \times \mathbb{R}_+ \to \mathbb{R}_+$ are continuous.

For $\sigma \in (0,1)$, we introduce the following extreme limits:

$$f_0^s = \limsup_{u+v+w\to 0+} \max_{t\in[0,1]} \frac{f(t,u,v,w)}{u+v+w}, \quad g_0^s = \limsup_{u+v+w\to 0+} \max_{t\in[0,1]} \frac{g(t,u,v,w)}{u+v+w},$$

$$h_0^s = \limsup_{u+v+w\to 0+} \max_{t\in[0,1]} \frac{h(t,u,v,w)}{u+v+w}, \quad f_0^i = \liminf_{u+v+w\to 0+} \min_{t\in[\sigma,1]} \frac{f(t,u,v,w)}{u+v+w},$$

$$g_0^i = \liminf_{u+v+w\to0+} \min_{t\in[\sigma,1]} \frac{g(t,u,v,w)}{u+v+w}, \quad h_0^i = \liminf_{u+v+w\to0+} \min_{t\in[\sigma,1]} \frac{h(t,u,v,w)}{u+v+w},$$

$$f_\infty^s = \limsup_{u+v+w\to\infty} \max_{t\in[0,1]} \frac{f(t,u,v,w)}{u+v+w}, \quad g_\infty^s = \limsup_{u+v+w\to\infty} \max_{t\in[0,1]} \frac{g(t,u,v,w)}{u+v+w},$$

$$h_\infty^s = \limsup_{u+v+w\to\infty} \max_{t\in[0,1]} \frac{h(t,u,v,w)}{u+v+w}, \quad f_\infty^i = \liminf_{u+v+w\to\infty} \min_{t\in[\sigma,1]} \frac{f(t,u,v,w)}{u+v+w},$$

$$g_\infty^i = \liminf_{u+v+w\to\infty} \min_{t\in[\sigma,1]} \frac{g(t,u,v,w)}{u+v+w}, \quad h_\infty^i = \liminf_{u+v+w\to\infty} \min_{t\in[\sigma,1]} \frac{h(t,u,v,w)}{u+v+w},$$

In the definition of the extreme limits above, the variables u, v and w are nonnegative.

By using the Green functions G_i, $i = 1, 2, 3$ from Section 5.1.1, we consider the following nonlinear system of integral equations:

$$\begin{cases} u(t) = \lambda \int_0^1 G_1(t,s) f(s, u(s), v(s), w(s))\, ds, & t \in [0,1], \\[2mm] v(t) = \mu \int_0^1 G_2(t,s) g(s, u(s), v(s), w(s))\, ds, & t \in [0,1], \\[2mm] w(t) = \nu \int_0^1 G_3(t,s) h(s, u(s), v(s), w(s))\, ds, & t \in [0,1]. \end{cases}$$

By Lemma 5.1.1 and the corresponding lemmas for problems (5.8), (5.9) and (5.10), (5.11), we deduce that (u, v, w) is a solution of the above system if and only if (u, v, w) is a solution of problem (5.1), (5.2).

We consider the Banach space $X = C[0, 1]$ with supremum norm $\|\cdot\|$ and the Banach space $Y = X \times X \times X$ with the norm $\|(u, v, w)\|_Y = \|u\| + \|v\| + \|w\|$. We define the cones

$$P_1 = \{u \in X, \ u(t) \geq t^{\alpha-1}\|u\|, \ \forall t \in [0,1]\} \subset X,$$

$$P_2 = \{v \in X, \ v(t) \geq t^{\beta-1}\|v\|, \ \forall t \in [0,1]\} \subset X,$$

$$P_3 = \{w \in X, \ w(t) \geq t^{\gamma-1}\|w\|, \ \forall t \in [0,1]\} \subset X,$$

and $P = P_1 \times P_2 \times P_3 \subset Y$.

For λ, μ, $\nu > 0$, we define now the operator $Q : P \to Y$ by $Q(u, v, w) = (Q_1(u, v, w), Q_2(u, v, w), Q_3(u, v, w))$ with

$$Q_1(u, v, w)(t) = \lambda \int_0^1 G_1(t, s) f(s, u(s), v(s), w(s)) \, ds,$$

$$t \in [0, 1], \ (u, v, w) \in P,$$

$$Q_2(u, v, w)(t) = \mu \int_0^1 G_2(t, s) g(s, u(s), v(s), w(s)) \, ds,$$

$$t \in [0, 1], \ (u, v, w) \in P,$$

$$Q_3(u, v, w)(t) = \nu \int_0^1 G_3(t, s) h(s, u(s), v(s), w(s)) \, ds,$$

$$t \in [0, 1], \ (u, v, w) \in P.$$

Lemma 5.1.5. *If* (H1) *and* (H2) *hold, then* $Q : P \to P$ *is a completely continuous operator.*

Proof. Let $(u, v, w) \in P$ be an arbitrary element. Because $Q_1(u, v, w)$, $Q_2(u, v, w)$ and $Q_3(u, v, w)$ satisfy the problem (5.3), (5.4) for $x(t) = \lambda f(t, u(t), v(t), w(t))$, $t \in [0, 1]$, the problem (5.8), (5.9) for $y(t) = \mu g(t, u(t), v(t), w(t))$, $t \in [0, 1]$, and the problem (5.10), (5.11) for $z(t) = \nu h(t, u(t), v(t), w(t))$, $t \in [0, 1]$, respectively, then by Lemma 5.1.4 and the corresponding ones for problems (5.8), (5.9) and (5.10), (5.11), we obtain

$$Q_1(u, v, w)(t') \geq t^{\alpha-1} Q_1(u, v, w)(t'), \quad Q_2(u, v, w)(t') \geq t^{\beta-1} Q_2(u, v, w)(t'),$$

$$Q_3(u, v, w)(t') \geq t^{\gamma-1} Q_3(u, v, w)(t'), \quad \forall t, \, t' \in [0, 1], \ (u, v, w) \in P,$$

and so

$$Q_1(u, v, w)(t) \geq t^{\alpha-1} \max_{t' \in [0,1]} Q_1(u, v, w)(t') = t^{\alpha-1} \|Q_1(u, v, w)\|,$$

$$\forall t \in [0, 1], \ (u, v, w) \in P,$$

$$Q_2(u, v, w)(t) \geq t^{\beta-1} \max_{t' \in [0,1]} Q_2(u, v, w)(t') = t^{\beta-1} \|Q_2(u, v, w)\|,$$

$$\forall t \in [0, 1], \ (u, v, w) \in P,$$

$$Q_3(u, v, w)(t) \geq t^{\gamma-1} \max_{t' \in [0,1]} Q_3(u, v, w)(t') = t^{\gamma-1} \|Q_3(u, v, w)\|,$$

$$\forall t \in [0, 1], \ (u, v, w) \in P.$$

Therefore, $Q(u, v, w) = (Q_1(u, v, w), Q_2(u, v, w), Q_3(u, v, w)) \in P$, and then $Q(P) \subset P$. By using standard arguments, we can easily show that Q_1, Q_2 and Q_3 are completely continuous, and then Q is a completely continuous operator. $\qquad\square$

The triplet $(u, v, w) \in P$ is a fixed point of operator Q if and only if (u, v, w) is solution of problem (5.1), (5.2). So, we will investigate the existence of fixed points of operator Q.

For $\sigma \in (0, 1)$, we denote by $A = \int_\sigma^1 J_1(s)\, ds$, $B = \int_0^1 J_1(s)\, ds$, $C = \int_\sigma^1 J_2(s)\, ds$, $D = \int_0^1 J_2(s)\, ds$, $E = \int_\sigma^1 J_3(s)\, ds$, $F = \int_0^1 J_3(s)\, ds$, where J_1, J_2 and J_3 are defined in Section 5.1.1.

First, for f_0^s, g_0^s, h_0^s, f_∞^i, g_∞^i, $h_\infty^i \in (0, \infty)$ and numbers α_1, α_2, $\alpha_3 \geq 0$ with $\alpha_1 + \alpha_2 + \alpha_3 = 1$, $\widetilde{\alpha}_1$, $\widetilde{\alpha}_2$, $\widetilde{\alpha}_3 > 0$ with $\widetilde{\alpha}_1 + \widetilde{\alpha}_2 + \widetilde{\alpha}_3 = 1$, $\widetilde{\alpha}_2'$, $\widetilde{\alpha}_3' > 0$ with $\widetilde{\alpha}_2' + \widetilde{\alpha}_3' = 1$, $\widetilde{\alpha}_1''$, $\widetilde{\alpha}_3'' > 0$ with $\widetilde{\alpha}_1'' + \widetilde{\alpha}_3'' = 1$, $\widetilde{\alpha}_1'''$, $\widetilde{\alpha}_2''' > 0$ with $\widetilde{\alpha}_1''' + \widetilde{\alpha}_2''' = 1$, we define the numbers

$$L_1 = \frac{\alpha_1}{\theta\sigma^{\alpha-1} f_\infty^i A}, \quad L_3 = \frac{\alpha_2}{\theta\sigma^{\beta-1} g_\infty^i C}, \quad L_5 = \frac{\alpha_3}{\theta\sigma^{\gamma-1} h_\infty^i E},$$

$$L_2 = \frac{\widetilde{\alpha}_1}{f_0^s B}, \quad L_4 = \frac{\widetilde{\alpha}_2}{g_0^s D}, \quad L_6 = \frac{\widetilde{\alpha}_3}{h_0^s F}, \quad L_4' = \frac{\widetilde{\alpha}_2'}{g_0^s D},$$

$$L_6' = \frac{\widetilde{\alpha}_3'}{h_0^s F}, \quad L_2'' = \frac{\widetilde{\alpha}_1''}{f_0^s B}, \quad L_6'' = \frac{\widetilde{\alpha}_3''}{h_0^s F}, \quad L_2''' = \frac{\widetilde{\alpha}_1'''}{f_0^s B},$$

$$L_4''' = \frac{\widetilde{\alpha}_2'''}{g_0^s D}, \quad \widetilde{L}_2 = \frac{1}{f_0^s B}, \quad \widetilde{L}_4 = \frac{1}{g_0^s D}, \quad \widetilde{L}_6 = \frac{1}{h_0^s F},$$

where $\theta = \min\{\sigma^{\alpha-1}, \sigma^{\beta-1}, \sigma^{\gamma-1}\}$.

Theorem 5.1.1. *Assume that* (H1) *and* (H2) *hold,* $\sigma \in (0, 1)$, α_1, α_2, $\alpha_3 \geq 0$ *with* $\alpha_1 + \alpha_2 + \alpha_3 = 1$, $\widetilde{\alpha}_1$, $\widetilde{\alpha}_2$, $\widetilde{\alpha}_3 > 0$ *with* $\widetilde{\alpha}_1 + \widetilde{\alpha}_2 + \widetilde{\alpha}_3 = 1$, $\widetilde{\alpha}_2'$, $\widetilde{\alpha}_3' > 0$ *with* $\widetilde{\alpha}_2' + \widetilde{\alpha}_3' = 1$, $\widetilde{\alpha}_1''$, $\widetilde{\alpha}_3'' > 0$ *with* $\widetilde{\alpha}_1'' + \widetilde{\alpha}_3'' = 1$, $\widetilde{\alpha}_1'''$, $\widetilde{\alpha}_2''' > 0$ *with* $\widetilde{\alpha}_1''' + \widetilde{\alpha}_2''' = 1$.

(1) *If* f_0^s, g_0^s, h_0^s, f_∞^i, g_∞^i, $h_\infty^i \in (0, \infty)$, $L_1 < L_2$, $L_3 < L_4$ *and* $L_5 < L_6$, *then for each* $\lambda \in (L_1, L_2)$, $\mu \in (L_3, L_4)$, $\nu \in (L_5, L_6)$ *there exists a positive solution* $(u(t), v(t), w(t))$, $t \in [0, 1]$ *for problem* (5.1), (5.2).

(2) *If* $f_0^s = 0$, g_0^s, h_0^s, f_∞^i, g_∞^i, $h_\infty^i \in (0, \infty)$, $L_3 < L_4'$ *and* $L_5 < L_6'$, *then for each* $\lambda \in (L_1, \infty)$, $\mu \in (L_3, L_4')$, $\nu \in (L_5, L_6')$ *there exists a positive solution* $(u(t), v(t), w(t))$, $t \in [0, 1]$ *for problem* (5.1), (5.2).

(3) If $g_0^s = 0$, f_0^s, h_0^s, f_∞^i, g_∞^i, $h_\infty^i \in (0, \infty)$, $L_1 < L_2''$ and $L_5 < L_6''$, then for each $\lambda \in (L_1, L_2'')$, $\mu \in (L_3, \infty)$, $\nu \in (L_5, L_6'')$ there exists a positive solution $(u(t), v(t), w(t))$, $t \in [0, 1]$ for problem (5.1), (5.2).

(4) If $h_0^s = 0$, f_0^s, g_0^s, f_∞^i, g_∞^i, $h_\infty^i \in (0, \infty)$, $L_1 < L_2'''$ and $L_3 < L_4'''$, then for each $\lambda \in (L_1, L_2''')$, $\mu \in (L_3, L_4''')$, $\nu \in (L_5, \infty)$, there exists a positive solution $(u(t), v(t), w(t))$, $t \in [0, 1]$ for problem (5.1), (5.2).

(5) If $f_0^s = g_0^s = 0$, h_0^s, f_∞^i, g_∞^i, $h_\infty^i \in (0, \infty)$, $L_5 < \widetilde{L}_6$, then for each $\lambda \in (L_1, \infty)$, $\mu \in (L_3, \infty)$, $\nu \in (L_5, \widetilde{L}_6)$, there exists a positive solution $(u(t), v(t), w(t))$, $t \in [0, 1]$ for problem (5.1), (5.2).

(6) If $f_0^s = h_0^s = 0$, g_0^s, f_∞^i, g_∞^i, $h_\infty^i \in (0, \infty)$, $L_3 < \widetilde{L}_4$, then for each $\lambda \in (L_1, \infty)$, $\mu \in (L_3, \widetilde{L}_4)$, $\nu \in (L_5, \infty)$, there exists a positive solution $(u(t), v(t), w(t))$, $t \in [0, 1]$ for problem (5.1), (5.2).

(7) If $g_0^s = h_0^s = 0$, f_0^s, f_∞^i, g_∞^i, $h_\infty^i \in (0, \infty)$, $L_1 < \widetilde{L}_2$, then for each $\lambda \in (L_1, \widetilde{L}_2)$, $\mu \in (L_3, \infty)$, $\nu \in (L_5, \infty)$, there exists a positive solution $(u(t), v(t), w(t))$, $t \in [0, 1]$ for problem (5.1), (5.2).

(8) If $f_0^s = g_0^s = h_0^s = 0$, f_∞^i, g_∞^i, $h_\infty^i \in (0, \infty)$, then for each $\lambda \in (L_1, \infty)$, $\mu \in (L_3, \infty)$, $\nu \in (L_5, \infty)$, there exists a positive solution $(u(t), v(t), w(t))$, $t \in [0, 1]$ for problem (5.1), (5.2).

(9) If f_0^s, g_0^s, $h_0^s \in (0, \infty)$ and at least one of f_∞^i, g_∞^i, h_∞^i is ∞, then for each $\lambda \in (0, L_2)$, $\mu \in (0, L_4)$, $\nu \in (0, L_6)$, there exists a positive solution $(u(t), v(t), w(t))$, $t \in [0, 1]$ for problem (5.1), (5.2).

(10) If $f_0^s = 0$, g_0^s, $h_0^s \in (0, \infty)$ and at least one of f_∞^i, g_∞^i, h_∞^i is ∞, then for each $\lambda \in (0, \infty)$, $\mu \in (0, L_4')$, $\nu \in (0, L_6')$, there exists a positive solution $(u(t), v(t), w(t))$, $t \in [0, 1]$ for problem (5.1), (5.2).

(11) If $g_0^s = 0$, f_0^s, $h_0^s \in (0, \infty)$ and at least one of f_∞^i, g_∞^i, h_∞^i is ∞, then for each $\lambda \in (0, L_2'')$, $\mu \in (0, \infty)$, $\nu \in (0, L_6'')$, there exists a positive solution $(u(t), v(t), w(t))$, $t \in [0, 1]$ for problem (5.1), (5.2).

(12) If $h_0^s = 0$, f_0^s, $g_0^s \in (0, \infty)$ and at least one of f_∞^i, g_∞^i, h_∞^i is ∞, then for each $\lambda \in (0, L_2''')$, $\mu \in (0, L_4''')$, $\nu \in (0, \infty)$, there exists a positive solution $(u(t), v(t), w(t))$, $t \in [0, 1]$ for problem (5.1), (5.2).

(13) If $f_0^s = g_0^s = 0$, $h_0^s \in (0, \infty)$ and at least one of f_∞^i, g_∞^i, h_∞^i is ∞, then for each $\lambda \in (0, \infty)$, $\mu \in (0, \infty)$, $\nu \in (0, \widetilde{L}_6)$, there exists a positive solution $(u(t), v(t), w(t))$, $t \in [0, 1]$ for problem (5.1), (5.2).

(14) If $f_0^s = h_0^s = 0$, $g_0^s \in (0, \infty)$ and at least one of f_∞^i, g_∞^i, h_∞^i is ∞, then for each $\lambda \in (0, \infty)$, $\mu \in (0, \widetilde{L}_4)$, $\nu \in (0, \infty)$, there exists a positive solution $(u(t), v(t), w(t))$, $t \in [0, 1]$ for problem (5.1), (5.2).

(15) If $g_0^s = h_0^s = 0$, $f_0^s \in (0, \infty)$ and at least one of f_∞^i, g_∞^i, h_∞^i is ∞, then for each $\lambda \in (0, \widetilde{L}_2)$, $\mu \in (0, \infty)$, $\nu \in (0, \infty)$, there exists a positive solution $(u(t), v(t), w(t))$, $t \in [0, 1]$ for problem (5.1), (5.2).

(16) *If $f_0^s = g_0^s = h_0^s = 0$ and at least one of f_∞^i, g_∞^i, h_∞^i is ∞, then for each $\lambda \in (0, \infty)$, $\mu \in (0, \infty)$, $\nu \in (0, \infty)$, there exists a positive solution $(u(t), v(t), w(t))$, $t \in [0, 1]$ for problem (5.1), (5.2).*

Proof. We consider the above cone $P \subset Y$ and the operators Q_1, Q_2, Q_3 and Q. We will prove some illustrative cases of this theorem.

Case (1). We consider f_0^s, g_0^s, h_0^s, f_∞^i, g_∞^i, $h_\infty^i \in (0, \infty)$. Let $\lambda \in (L_1, L_2)$, $\mu \in (L_3, L_4)$ and $\nu \in (L_5, L_6)$. We choose $\varepsilon > 0$ a positive number such that $\varepsilon < f_\infty^i$, $\varepsilon < g_\infty^i$, $\varepsilon < h_\infty^i$ and

$$\frac{\widetilde{\alpha}_1}{(f_0^s + \varepsilon)B} \geq \lambda, \quad \frac{\widetilde{\alpha}_2}{(g_0^s + \varepsilon)D} \geq \mu, \quad \frac{\widetilde{\alpha}_3}{(h_0^s + \varepsilon)F} \geq \nu,$$

$$\frac{\alpha_1}{\theta\sigma^{\alpha-1}(f_\infty^i - \varepsilon)A} \leq \lambda, \quad \frac{\alpha_2}{\theta\sigma^{\beta-1}(g_\infty^i - \varepsilon)C} \leq \mu, \quad \frac{\alpha_3}{\theta\sigma^{\gamma-1}(h_\infty^i - \varepsilon)E} \leq \nu.$$

By using (H2) and the definition of f_0^s, g_0^s and h_0^s, we deduce that there exists $R_1 > 0$ such that $f(t, u, v, w) \leq (f_0^s + \varepsilon)(u + v + w)$, $g(t, u, v, w) \leq (g_0^s + \varepsilon)(u + v + w)$, $h(t, u, v, w) \leq (h_0^s + \varepsilon)(u + v + w)$, for all $t \in [0, 1]$ and $u, v, w \geq 0$ with $u + v + w \leq R_1$. We define the set $\Omega_1 = \{(u, v, w) \in Y, \|(u, v, w)\|_Y < R_1\}$.

Now, let $(u, v, w) \in P \cap \partial\Omega_1$, that is, $\|(u, v, w)\|_Y = R_1$ or equivalently $\|u\| + \|v\| + \|w\| = R_1$. Then $u(t) + v(t) + w(t) \leq R_1$ for all $t \in [0, 1]$, and by Lemma 5.1.3, we obtain

$$Q_1(u, v, w)(t) \leq \lambda \int_0^1 J_1(s)f(s, u(s), v(s), w(s))\, ds$$

$$\leq \lambda \int_0^1 J_1(s)(f_0^s + \varepsilon)(u(s) + v(s) + w(s))\, ds$$

$$\leq \lambda(f_0^s + \varepsilon)\int_0^1 J_1(s)(\|u\| + \|v\| + \|w\|)\, ds$$

$$= \lambda(f_0^s + \varepsilon)B\|(u, v, w)\|_Y \leq \widetilde{\alpha}_1\|(u, v, w)\|_Y, \quad \forall t \in [0, 1],$$

$$Q_2(u, v, w)(t) \leq \mu \int_0^1 J_2(s)g(s, u(s), v(s), w(s))\, ds$$

$$\leq \mu \int_0^1 J_2(s)(g_0^s + \varepsilon)(u(s) + v(s) + w(s))\, ds$$

$$\leq \mu(g_0^s + \varepsilon)\int_0^1 J_2(s)(\|u\| + \|v\| + \|w\|)\, ds$$

$$= \mu(g_0^s + \varepsilon)D\|(u, v, w)\|_Y \leq \widetilde{\alpha}_2\|(u, v, w)\|_Y, \quad \forall t \in [0, 1],$$

$$Q_3(u,v,w)(t) \le \nu \int_0^1 J_3(s)h(s,u(s),v(s),w(s))\,ds$$

$$\le \nu \int_0^1 J_3(s)(h_0^s + \varepsilon)(u(s) + v(s) + w(s))\,ds$$

$$\le \nu(h_0^s + \varepsilon) \int_0^1 J_3(s)(\|u\| + \|v\| + \|w\|)\,ds$$

$$= \nu(h_0^s + \varepsilon)F\|(u,v,w)\|_Y \le \widetilde{\alpha}_3\|(u,v,w)\|_Y, \quad \forall t \in [0,1].$$

Therefore, $\|Q_1(u,v,w)\| \le \widetilde{\alpha}_1\|(u,v,w)\|_Y$, $\|Q_2(u,v,w)\| \le \widetilde{\alpha}_2\|(u,v,w)\|_Y$, $\|Q_3(u,v,w)\| \le \widetilde{\alpha}_3\|(u,v,w)\|_Y$.

Then for $(u,v,w) \in P \cap \partial\Omega_1$ we deduce

$$\|Q(u,v,w)\|_Y = \|Q_1(u,v,w)\| + \|Q_2(u,v,w)\| + \|Q_3(u,v,w)\|$$

$$\le (\widetilde{\alpha}_1 + \widetilde{\alpha}_2 + \widetilde{\alpha}_3)\|(u,v,w)\|_Y = \|(u,v,w)\|_Y. \quad (5.12)$$

By the definition of f_∞^i, g_∞^i and h_∞^i, there exists $\overline{R}_2 > 0$ such that $f(t,u,v,w) \ge (f_\infty^i - \varepsilon)(u + v + w)$, $g(t,u,v,w) \ge (g_\infty^i - \varepsilon)(u + v + w)$, $h(t,u,v,w) \ge (h_\infty^i - \varepsilon)(u + v + w)$ for all u, v, $w \ge 0$ with $u + v + w \ge \overline{R}_2$ and $t \in [\sigma,1]$. We consider $R_2 = \max\{2R_1, \overline{R}_2/\theta\}$ and we define $\Omega_2 = \{(u,v,w) \in Y, \|(u,v,w)\|_Y < R_2\}$. Then for $(u,v,w) \in P$ with $\|(u,v,w)\|_Y = R_2$, we obtain

$$u(t) + v(t) + w(t) \ge \sigma^{\alpha-1}\|u\| + \sigma^{\beta-1}\|v\| + \sigma^{\gamma-1}\|w\| \ge \theta(\|u\| + \|v\| + \|w\|)$$

$$= \theta\|(u,v,w)\|_Y = \theta R_2 \ge \overline{R}_2, \quad \forall t \in [\sigma,1].$$

Then by Lemma 5.1.3, we conclude

$$Q_1(u,v,w)(t) \ge \lambda \int_0^1 t^{\alpha-1} J_1(s)f(s,u(s),v(s),w(s))\,ds$$

$$\ge \lambda\sigma^{\alpha-1} \int_\sigma^1 J_1(s)f(s,u(s),v(s),w(s))\,ds$$

$$\ge \lambda\sigma^{\alpha-1} \int_\sigma^1 J_1(s)(f_\infty^i - \varepsilon)(u(s) + v(s) + w(s))\,ds$$

$$\ge \lambda\sigma^{\alpha-1}\theta(f_\infty^i - \varepsilon) \int_\sigma^1 J_1(s)\|(u,v,w)\|_Y\,ds$$

$$= \lambda\sigma^{\alpha-1}\theta(f_\infty^i - \varepsilon)A\|(u,v,w)\|_Y$$

$$\ge \alpha_1\|(u,v,w)\|_Y, \quad \forall t \in [\sigma,1],$$

$$Q_2(u,v,w)(t) \geq \mu \int_0^1 t^{\beta-1} J_2(s) g(s, u(s), v(s), w(s)) \, ds$$

$$\geq \mu\sigma^{\beta-1} \int_\sigma^1 J_2(s) g(s, u(s), v(s), w(s)) \, ds$$

$$\geq \mu\sigma^{\beta-1} \int_\sigma^1 J_2(s)(g_\infty^i - \varepsilon)(u(s) + v(s) + w(s)) \, ds$$

$$\geq \mu\sigma^{\beta-1} \theta(g_\infty^i - \varepsilon) \int_\sigma^1 J_2(s) \|(u,v,w)\|_Y \, ds$$

$$= \mu\sigma^{\beta-1} \theta(g_\infty^i - \varepsilon) C \|(u,v,w)\|_Y$$

$$\geq \alpha_2 \|(u,v,w)\|_Y, \quad \forall t \in [\sigma, 1],$$

$$Q_3(u,v,w)(t) \geq \nu \int_0^1 t^{\gamma-1} J_3(s) h(s, u(s), v(s), w(s)) \, ds$$

$$\geq \nu\sigma^{\gamma-1} \int_\sigma^1 J_3(s) h(s, u(s), v(s), w(s)) \, ds$$

$$\geq \nu\sigma^{\gamma-1} \int_\sigma^1 J_3(s)(h_\infty^i - \varepsilon)(u(s) + v(s) + w(s)) \, ds$$

$$\geq \nu\sigma^{\gamma-1} \theta(h_\infty^i - \varepsilon) \int_\sigma^1 J_3(s) \|(u,v,w)\|_Y \, ds$$

$$= \nu\sigma^{\gamma-1} \theta(h_\infty^i - \varepsilon) F \|(u,v,w)\|_Y$$

$$\geq \alpha_3 \|(u,v,w)\|_Y, \quad \forall t \in [\sigma, 1].$$

So, $\|Q_1(u,v,w)\| \geq Q_1(u,v,w)(\sigma) \geq \alpha_1 \|(u,v,w)\|_Y$, $\|Q_2(u,v,w)\| \geq Q_2(u,v,w)(\sigma) \geq \alpha_2 \|(u,v,w)\|_Y$, $\|Q_3(u,v,w)\| \geq Q_3(u,v,w)(\sigma) \geq \alpha_3 \|(u,v,w)\|_Y$.

Hence, for $(u,v,w) \in P \cap \partial\Omega_2$, we obtain

$$\|Q(u,v,w)\|_Y = \|Q_1(u,v,w)\| + \|Q_2(u,v,w)\| + \|Q_3(u,v,w)\|$$

$$\geq (\alpha_1 + \alpha_2 + \alpha_3) \|(u,v,w)\|_Y = \|(u,v,w)\|_Y. \quad (5.13)$$

By using Lemma 5.1.5, Theorem 1.2.2 (i), and the relations (5.12), (5.13), we deduce that Q has a fixed point $(u,v,w) \in P \cap (\overline{\Omega}_2 \setminus \Omega_1)$, $u(t) \geq t^{\alpha-1} \|u\|$, $v(t) \geq t^{\beta-1} \|v\|$, $w(t) \geq t^{\gamma-1} \|w\|$ for all $t \in [0,1]$, and $R_1 \leq \|u\| + \|v\| + \|w\| \leq R_2$. If $\|u\| > 0$ then $u(t) > 0$ for all $t \in (0,1]$, if $\|v\| > 0$ then $v(t) > 0$ for all $t \in (0,1]$, and if $\|w\| > 0$ then $w(t) > 0$ for all $t \in (0,1]$. So, (u,v,w) is a positive solution for our problem (5.1), (5.2). $\qquad \square$

Case (10). We consider $f_0^s = 0$, $f_\infty^i = \infty$, g_0^s, h_0^s, g_∞^i, $h_\infty^i \in (0, \infty)$. Let $\lambda \in (0, \infty)$, $\mu \in (0, L_4')$ and $\nu \in (0, L_6')$. We choose $\varepsilon > 0$ a positive number such that $\varepsilon \leq \lambda \theta \sigma^{\alpha-1} A$ and

$$\varepsilon \leq \frac{1 - \mu g_0^s D - \nu h_0^s F}{2\lambda B}, \quad \varepsilon \leq \frac{\widetilde{\alpha}_2' - \mu g_0^s D}{2\mu D}, \quad \varepsilon \leq \frac{\widetilde{\alpha}_3' - \nu h_0^s F}{2\nu F}.$$

The numerators of the above fractions are positive, because $\mu < \frac{\widetilde{\alpha}_2'}{g_0^s D}$, that is, $\widetilde{\alpha}_2' > \mu g_0^s D$, $\nu < \frac{\widetilde{\alpha}_3'}{h_0^s F}$, that is, $\widetilde{\alpha}_3' > \nu h_0^s F$, and $1 - \mu g_0^s D - \nu h_0^s F = \widetilde{\alpha}_2' + \widetilde{\alpha}_3' - \mu g_0^s D - \nu h_0^s F = (\widetilde{\alpha}_2' - \mu g_0^s D) + (\widetilde{\alpha}_3' - \nu h_0^s F) > 0$.

By using (H2) and the definition of f_0^s, g_0^s, h_0^s, we deduce that there exists $R_1 > 0$ such that $f(t, u, v, w) \leq \varepsilon(u + v + w)$, $g(t, u, v, w) \leq (g_0^s + \varepsilon)(u + v + w)$, $h(t, u, v, w) \leq (h_0^s + \varepsilon)(u + v + w)$ for all $t \in [0, 1]$, $u, v, w \geq 0$ with $u + v + w \leq R_1$. We define the set $\Omega_1 = \{(u, v, w) \in Y, \|(u, v, w)\|_Y < R_1\}$.

Now, let $(u, v, w) \in P \cap \partial\Omega_1$, that is, $\|(u, v, w)\|_Y = R_1$. Then $u(t) + v(t) + w(t) \leq R_1$ for all $t \in [0, 1]$, and by Lemma 5.1.3, we obtain

$$Q_1(u, v, w)(t) \leq \lambda \int_0^1 J_1(s) f(s, u(s), v(s), w(s)) \, ds$$

$$\leq \lambda \int_0^1 J_1(s) \varepsilon(u(s) + v(s) + w(s)) \, ds$$

$$\leq \lambda \varepsilon \int_0^1 J_1(s)(\|u\| + \|v\| + \|w\|) \, ds$$

$$= \lambda \varepsilon B \|(u, v, w)\|_Y \leq \frac{1}{2}(1 - \mu g_0^s D - \nu h_0^s F)\|(u, v, w)\|_Y,$$

$$Q_2(u, v, w)(t) \leq \mu \int_0^1 J_2(s) g(s, u(s), v(s), w(s)) \, ds$$

$$\leq \mu \int_0^1 J_2(s)(g_0^s + \varepsilon)(u(s) + v(s) + w(s)) \, ds$$

$$\leq \mu(g_0^s + \varepsilon) \int_0^1 J_2(s)(\|u\| + \|v\| + \|w\|) \, ds$$

$$= \mu(g_0^s + \varepsilon) D \|(u, v, w)\|_Y$$

$$\leq \mu \left(g_0^s + \frac{\widetilde{\alpha}_2' - \mu g_0^s D}{2\mu D} \right) D \|(u, v, w)\|_Y$$

$$= \frac{1}{2}(\mu g_0^s D + \widetilde{\alpha}_2')\|(u, v, w)\|_Y,$$

$$Q_3(u,v,w)(t) \le \nu \int_0^1 J_3(s)h(s,u(s),v(s),w(s))\,ds$$

$$\le \nu \int_0^1 J_3(s)(h_0^s + \varepsilon)(u(s) + v(s) + w(s))\,ds$$

$$\le \nu(h_0^s + \varepsilon)\int_0^1 J_3(s)(\|u\| + \|v\| + \|w\|)\,ds$$

$$= \nu(h_0^s + \varepsilon)F\|(u,v,w)\|_Y$$

$$\le \nu\left(h_0^s + \frac{\widetilde{\alpha}_3' - \nu h_0^s F}{2\nu F}\right)F\|(u,v,w)\|_Y$$

$$= \frac{1}{2}(\nu h_0^s F + \widetilde{\alpha}_3')\|(u,v,w)\|_Y, \quad \forall t \in [0,1].$$

Therefore,

$$\|Q_1(u,v,w)\| \le \frac{1}{2}(1 - \mu g_0^s D - \nu h_0^s F)\|(u,v,w)\|_Y,$$

$$\|Q_2(u,v,w)\| \le \frac{1}{2}(\mu g_0^s D + \widetilde{\alpha}_2')\|(u,v,w)\|_Y,$$

$$\|Q_3(u,v,w)\| \le \frac{1}{2}(\nu h_0^s F + \widetilde{\alpha}_3')\|(u,v,w)\|_Y.$$

Then for $(u,v,w) \in P \cap \partial\Omega_1$, we conclude

$$\|Q(u,v,w)\|_Y = \|Q_1(u,v,w)\| + \|Q_2(u,v,w)\| + \|Q_3(u,v,w)\|$$

$$\le \frac{1}{2}(1 - \mu g_0^s D - \nu h_0^s F + \mu g_0^s D + \widetilde{\alpha}_2' + \nu h_0^s F + \widetilde{\alpha}_3')$$

$$\times \|(u,v,w)\|_Y = \|(u,v,w)\|_Y. \tag{5.14}$$

By the definition of f_∞^i, there exists $\overline{R}_2 > 0$ such that $f(t,u,v,w) \ge \frac{1}{\varepsilon}(u+v+w)$ for all $u, v, w \ge 0$ with $u+v+w \ge \overline{R}_2$ and $t \in [\sigma, 1]$. We consider $R_2 = \max\{2R_1, \overline{R}_2/\theta\}$ and we define $\Omega_2 = \{(u,v,w) \in Y, \|(u,v,w)\|_Y < R_2\}$. Then for $(u,v,w) \in P$ with $\|(u,v,w)\|_Y = R_2$, we obtain $u(t) + v(t) + w(t) \ge \theta\|(u,v,w)\|_Y = \theta R_2 \ge \overline{R}_2$ for all $t \in [\sigma, 1]$. Then by Lemma 5.1.3, we deduce

$$Q_1(u,v,w)(t) \ge \lambda \int_0^1 t^{\alpha-1} J_1(s)f(s,u(s),v(s),w(s))\,ds$$

$$\ge \lambda\sigma^{\alpha-1}\int_\sigma^1 J_1(s)f(s,u(s),v(s),w(s))\,ds$$

$$\geq \lambda \sigma^{\alpha-1} \int_{\sigma}^{1} J_1(s) \frac{1}{\varepsilon} (u(s) + v(s) + w(s))\, ds$$

$$\geq \lambda \sigma^{\alpha-1} \theta \frac{1}{\varepsilon} \int_{\sigma}^{1} J_1(s) \|(u,v,w)\|_Y\, ds$$

$$= \lambda \sigma^{\alpha-1} \theta \frac{1}{\varepsilon} A \|(u,v,w)\|_Y \geq \|(u,v,w)\|_Y, \quad \forall t \in [\sigma,1].$$

Then for $(u,v,w) \in P \cap \partial \Omega_2$ we find $\|Q_1(u,v,w)\| \geq Q_1(u,v,w)(\sigma) \geq \|(u,v,w)\|_Y$, and

$$\|Q(u,v,w)\|_Y \geq \|Q_1(u,v,w)\| \geq \|(u,v,w)\|_Y. \qquad (5.15)$$

By using Lemma 5.1.5, Theorem 1.2.2 (i) and the inequalities (5.14), (5.15), we conclude that Q has a fixed point $(u,v,w) \in P \cap (\overline{\Omega}_2 \setminus \Omega_1)$, which is a positive solution of problem (5.1), (5.2).

Case (15). We consider $g_0^s = h_0^s = 0$, $g_\infty^i = \infty$, $f_0^s, f_\infty^i, h_\infty^i \in (0,\infty)$. Let $\lambda \in (0, \widetilde{L}_2)$, $\mu \in (0,\infty)$, $\nu \in (0,\infty)$. We choose $\varepsilon > 0$ a positive number such that $\varepsilon \leq \mu \theta \sigma^{\beta-1} C$ and

$$\varepsilon \leq \frac{1 - \lambda f_0^s B}{2\lambda B}, \quad \varepsilon \leq \frac{1 - \lambda f_0^s B}{4\mu D}, \quad \varepsilon \leq \frac{1 - \lambda f_0^s B}{4\nu F}.$$

The numerator of the above fractions is positive because $\lambda < \frac{1}{f_0^s B}$ that is $1 - \lambda f_0^s B > 0$.

By using (H2) and the definition of f_0^s, g_0^s, h_0^s, we deduce that there exists $R_1 > 0$ such that $f(t,u,v,w) \leq (f_0^s + \varepsilon)(u+v+w)$, $g(t,u,v,w) \leq \varepsilon(u+v+w)$, $h(t,u,v,w) \leq \varepsilon(u+v+w)$ for all $t \in [0,1]$, $u, v, w \geq 0$ with $u+v+w \leq R_1$. We define the set $\Omega_1 = \{(u,v,w) \in Y, \|(u,v,w)\|_Y < R_1\}$.

Now, let $(u,v,w) \in P \cap \partial \Omega_1$, that is $\|(u,v,w)\|_Y = R_1$. Then $u(t) + v(t) + w(t) \leq R_1$ for all $t \in [0,1]$, and by Lemma 5.1.3, we obtain

$$Q_1(u,v,w)(t) \leq \lambda \int_0^1 J_1(s) f(s,u(s),v(s),w(s))\, ds$$

$$\leq \lambda \int_0^1 J_1(s)(f_0^s + \varepsilon)(u(s) + v(s) + w(s))\, ds$$

$$\leq \lambda(f_0^s + \varepsilon) \int_0^1 J_1(s)(\|u\| + \|v\| + \|w\|)\, ds$$

$$= \lambda(f_0^s + \varepsilon) B \|(u,v,w)\|_Y \leq \lambda \left(f_0^s + \frac{1 - \lambda f_0^s B}{2\lambda B} \right)$$

$$\times B \|(u,v,w)\|_Y = \frac{1}{2}(\lambda f_0^s B + 1) \|(u,v,w)\|_Y,$$

$$Q_2(u, v, w)(t) \leq \mu \int_0^1 J_2(s) g(s, u(s), v(s), w(s)) \, ds$$

$$\leq \mu \int_0^1 J_2(s) \varepsilon(u(s) + v(s) + w(s)) \, ds$$

$$\leq \mu \varepsilon \int_0^1 J_2(s)(\|u\| + \|v\| + \|w\|) \, ds = \mu \varepsilon D \|(u, v, w)\|_Y$$

$$\leq \mu \frac{1 - \lambda f_0^s B}{4 \mu D} D \|(u, v, w)\|_Y = \frac{1}{4}(1 - \lambda f_0^s B) \|(u, v, w)\|_Y,$$

$$Q_3(u, v, w)(t) \leq \nu \int_0^1 J_3(s) h(s, u(s), v(s), w(s)) \, ds$$

$$\leq \nu \int_0^1 J_3(s) \varepsilon(u(s) + v(s) + w(s)) \, ds$$

$$\leq \nu \varepsilon \int_0^1 J_3(s)(\|u\| + \|v\| + \|w\|) \, ds = \nu \varepsilon F \|(u, v, w)\|_Y$$

$$\leq \nu \frac{1 - \lambda f_0^s B}{4 \nu F} F \|(u, v, w)\|_Y$$

$$= \frac{1}{4}(1 - \lambda f_0^s B) \|(u, v, w)\|_Y, \quad \forall t \in [0, 1].$$

Therefore,

$$\|Q_1(u, v, w)\| \leq \frac{1}{2}(\lambda f_0^s B + 1) \|(u, v, w)\|_Y,$$

$$\|Q_2(u, v, w)\| \leq \frac{1}{4}(1 - \lambda f_0^s B) \|(u, v, w)\|_Y,$$

$$\|Q_3(u, v, w)\| \leq \frac{1}{4}(1 - \lambda f_0^s B) \|(u, v, w)\|_Y.$$

Then for $(u, v, w) \in P \cap \partial \Omega_1$, we deduce

$$\|Q(u, v, w)\|_Y = \|Q_1(u, v, w)\| + \|Q_2(u, v, w)\| + \|Q_3(u, v, w)\|$$

$$\leq \frac{1}{4}(2 + 2\lambda f_0^s B + 1 - \lambda f_0^s B + 1 - \lambda f_0^s B)$$

$$\times \|(u, v, w)\|_Y = \|(u, v, w)\|_Y. \tag{5.16}$$

By the definition of g^i_∞, there exists $\overline{R}_2 > 0$ such that $g(t, u, v, w) \geq \frac{1}{\varepsilon}(u+v+w)$ for all u, v, $w \geq 0$ with $u+v+w \geq \overline{R}_2$ and $t \in [\sigma, 1]$. We consider $R_2 = \max\{2R_1, \overline{R}_2/\theta\}$ and we define $\Omega_2 = \{(u, v, w) \in Y, \|(u, v, w)\|_Y < R_2\}$. Then for $(u, v, w) \in P$ with $\|(u, v, w)\|_Y = R_2$, we obtain $u(t) + v(t) + w(t) \geq \theta\|(u, v, w)\|_Y = \theta R_2 \geq \overline{R}_2$, for all $t \in [\sigma, 1]$.

Then by Lemma 5.1.3, we conclude

$$Q_2(u, v, w)(t) \geq \mu \int_0^1 t^{\beta-1} J_2(s) g(s, u(s), v(s), w(s))\, ds$$

$$\geq \mu\sigma^{\beta-1} \int_\sigma^1 J_2(s) g(s, u(s), v(s), w(s))\, ds$$

$$\geq \mu\sigma^{\beta-1} \int_\sigma^1 J_2(s) \frac{1}{\varepsilon}(u(s) + v(s) + w(s))\, ds$$

$$\geq \mu\sigma^{\beta-1}\theta\frac{1}{\varepsilon} \int_\sigma^1 J_2(s)\|(u, v, w)\|_Y\, ds$$

$$= \mu\sigma^{\beta-1}\theta\frac{1}{\varepsilon}C\|(u, v, w)\|_Y \geq \|(u, v, w)\|_Y, \quad \forall t \in [\sigma, 1].$$

Then for $(u, v, w) \in P \cap \partial\Omega_2$, we find $\|Q_2(u, v, w)\| \geq Q_2(u, v, w)(\sigma) \geq \|(u, v, w)\|_Y$, and

$$\|Q(u, v, w)\|_Y \geq \|Q_2(u, v, w)\| \geq \|(u, v, w)\|_Y. \tag{5.17}$$

By using Lemma 5.1.5, Theorem 1.2.2 (i) and the inequalities (5.16), (5.17), we deduce that Q has a fixed point $(u, v, w) \in P \cap (\overline{\Omega}_2 \setminus \Omega_1)$ which is a positive solution of problem (5.1), (5.2).

Case (16). We consider $f^s_0 = g^s_0 = h^s_0 = 0$, $h^i_\infty = \infty$, f^i_∞, $g^i_\infty \in (0, \infty)$. Let $\lambda \in (0, \infty)$, $\mu \in (0, \infty)$ and $\nu \in (0, \infty)$. We choose $\varepsilon > 0$ such that

$$\varepsilon \leq \nu\theta\sigma^{\gamma-1}E, \quad \varepsilon \leq \frac{1}{3\lambda B}, \quad \varepsilon \leq \frac{1}{3\mu D}, \quad \varepsilon \leq \frac{1}{3\nu F}.$$

By using (H2) and the definition of f^s_0, g^s_0, h^s_0, we deduce that there exists $R_1 > 0$ such that $f(t, u, v, w) \leq \varepsilon(u + v + w)$, $g(t, u, v, w) \leq \varepsilon(u + v + w)$, $h(t, u, v, w) \leq \varepsilon(u + v + w)$ for all $t \in [0, 1]$, u, v, $w \geq 0$ with $u+v+w \leq R_1$. We define the set $\Omega_1 = \{(u, v, w) \in Y, \|(u, v, w)\|_Y < R_1\}$.

Now, let $(u, v, w) \in P \cap \partial\Omega_1$, that is $\|(u, v, w)\|_Y = R_1$. Then $u(t) + v(t) + w(t) \leq R_1$ for all $t \in [0, 1]$, and by Lemma 5.1.3, we obtain

$$Q_1(u, v, w)(t) \leq \lambda \int_0^1 J_1(s) f(s, u(s), v(s), w(s)) \, ds$$

$$\leq \lambda \int_0^1 J_1(s) \varepsilon(u(s) + v(s) + w(s)) \, ds$$

$$\leq \lambda \varepsilon \int_0^1 J_1(s) (\|u\| + \|v\| + \|w\|) \, ds$$

$$= \lambda \varepsilon B \|(u, v, w)\|_Y \leq \frac{1}{3} \|(u, v, w)\|_Y,$$

$$Q_2(u, v, w)(t) \leq \mu \int_0^1 J_2(s) g(s, u(s), v(s), w(s)) \, ds$$

$$\leq \mu \int_0^1 J_2(s) \varepsilon(u(s) + v(s) + w(s)) \, ds$$

$$\leq \mu \varepsilon \int_0^1 J_2(s) (\|u\| + \|v\| + \|w\|) \, ds$$

$$= \mu \varepsilon D \|(u, v, w)\|_Y \leq \frac{1}{3} \|(u, v, w)\|_Y,$$

$$Q_3(u, v, w)(t) \leq \nu \int_0^1 J_3(s) h(s, u(s), v(s), w(s)) \, ds$$

$$\leq \nu \int_0^1 J_3(s) \varepsilon(u(s) + v(s) + w(s)) \, ds$$

$$\leq \nu \varepsilon \int_0^1 J_3(s) (\|u\| + \|v\| + \|w\|) \, ds$$

$$= \nu \varepsilon F \|(u, v, w)\|_Y \leq \frac{1}{3} \|(u, v, w)\|_Y, \quad \forall t \in [0, 1].$$

Therefore, $\|Q_1(u, v, w)\| \leq \frac{1}{3}\|(u, v, w)\|_Y$, $\|Q_2(u, v, w)\| \leq \frac{1}{3}\|(u, v, w)\|_Y$, $\|Q_3(u, v, w)\| \leq \frac{1}{3}\|(u, v, w)\|_Y$.

Then for $(u, v, w) \in P \cap \partial\Omega_1$, we conclude

$$\|Q(u, v, w)\|_Y \leq \|(u, v, w)\|_Y. \tag{5.18}$$

By the definition of h_∞^i, there exists $\overline{R}_2 > 0$ such that $h(t, u, v, w) \geq \frac{1}{\varepsilon}(u + v + w)$ for all u, v, $w \geq 0$ with $u + v + w \geq \overline{R}_2$ and $t \in [\sigma, 1]$. We consider

$R_2 = \max\{2R_1, \overline{R}_2/\theta\}$ and we define $\Omega_2 = \{(u, v, w) \in Y, \|(u, v, w)\|_Y < R_2\}$. Then for $(u, v, w) \in P$ with $\|(u, v, w)\|_Y = R_2$, we obtain $u(t) + v(t) + w(t) \geq \theta\|(u, v, w)\|_Y = \theta R_2 \geq \overline{R}_2$, for all $t \in [\sigma, 1]$.

Then by Lemma 5.1.3, we deduce

$$Q_3(u, v, w)(t) \geq \nu \int_0^1 t^{\gamma-1} J_3(s) h(s, u(s), v(s), w(s)) \, ds$$

$$\geq \nu \sigma^{\gamma-1} \int_\sigma^1 J_3(s) h(s, u(s), v(s), w(s)) \, ds$$

$$\geq \nu \sigma^{\gamma-1} \int_\sigma^1 J_3(s) \frac{1}{\varepsilon} (u(s) + v(s) + w(s)) \, ds$$

$$\geq \nu \sigma^{\gamma-1} \theta \frac{1}{\varepsilon} \int_\sigma^1 J_3(s) \|(u, v, w)\|_Y \, ds$$

$$= \nu \sigma^{\gamma-1} \theta \frac{1}{\varepsilon} E \|(u, v, w)\|_Y \geq \|(u, v, w)\|_Y, \quad \forall t \in [\sigma, 1].$$

Then for $(u, v, w) \in P \cap \partial\Omega_2$ we find $\|Q_3(u, v, w)\| \geq Q_3(u, v, w)(\sigma) \geq \|(u, v, w)\|_Y$, and

$$\|Q(u, v, w)\|_Y \geq \|Q_3(u, v, w)\| \geq \|(u, v, w)\|_Y. \tag{5.19}$$

By using Lemma 5.1.5, Theorem 1.2.2 (i) and the inequalities (5.18), (5.19), we conclude that Q has a fixed point $(u, v, w) \in P \cap (\overline{\Omega}_2 \setminus \Omega_1)$ which is a positive solution of problem (5.1), (5.2).

Remark 5.1.1. Each of the cases (9)–(16) of Theorem 5.1.1 contains 7 cases as follows: $\{f_\infty^i = \infty, g_\infty^i, h_\infty^i \in (0, \infty)\}$, or $\{g_\infty^i = \infty, f_\infty^i, h_\infty^i \in (0, \infty)\}$, or $\{h_\infty^i = \infty, f_\infty^i, g_\infty^i \in (0, \infty)\}$, or $\{f_\infty^i = g_\infty^i = \infty, h_\infty^i \in (0, \infty)\}$, or $\{f_\infty^i = h_\infty^i = \infty, g_\infty^i \in (0, \infty)\}$, or $\{g_\infty^i = h_\infty^i = \infty, f_\infty^i \in (0, \infty)\}$, or $\{f_\infty^i = g_\infty^i = h_\infty^i = \infty\}$. So, the total number of cases from Theorem 5.1.1 is 64, which we grouped in 16 cases.

Each of the cases (1)–(8) contains 4 subcases because $\alpha_1, \alpha_2, \alpha_3 \in (0, 1)$, or $\alpha_1 = 1$ and $\alpha_2 = \alpha_3 = 0$, or $\alpha_2 = 1$ and $\alpha_1 = \alpha_3 = 0$, or $\alpha_3 = 1$ and $\alpha_1 = \alpha_2 = 0$.

Remark 5.1.2. In [98], the authors present only 15 cases (Theorems 2.1–2.15 from [98]) from 64 cases, namely the first 9 cases of our Theorem 5.1.1. They did not study the cases when some extreme limits are 0 and other are ∞. Besides, our intervals for parameters λ, μ, ν presented in Theorem 5.1.1 (our cases (2)–(7) and (9)) are better than the corresponding ones from [98, Theorems 2.2–2.7 and 2.9–2.15]. In addition, the cone used in [98] implies the existence of nonnegative solutions, which satisfy the condition $\inf_{t\in[\xi,\eta]}(u(t) + v(t) + w(t)) > 0$, which is different than our definition of positive solutions.

Remark 5.1.3. One can formulate existence results for the general case of the system of n fractional differential equations

$$D_{0+}^{\alpha_j} u_j(t) + \lambda_j f_j(t, u_1(t), \ldots, u_n(t)) = 0, \quad j = 1, \ldots, n, \qquad (5.20)$$

with the boundary conditions

$$\begin{cases} u_j^{(k)}(0) = 0, \quad k = 0, \ldots, m_j - 2, \quad j = 1, \ldots, n, \\ D_{0+}^{p_j} u_j(1) = \sum_{k=1}^{N_j} a_{jk} D_{0+}^{q_j} u_j(\xi_{jk}), \quad j = 1, \ldots, n, \end{cases} \qquad (5.21)$$

where $\alpha_j \in (m_j - 1, m_j]$, $m_j \in \mathbb{N}$, $m_j \geq 3$; ξ_{jk}, $a_{jk} \in \mathbb{R}$ for all $k = 1, \ldots, N_j$, $(N_j \in \mathbb{N})$; $0 < \xi_{j1} < \xi_{j2} < \cdots \leq \xi_{jN_j}$, $p_j \in [1, \alpha_j - 1)$, $q_j \in [0, p_j]$, $j = 1, \ldots, N$.

According to the values of $f_{j0}^s = \limsup_{u_1+\cdots+u_n\to 0+} \sup_{t\in[0,1]}$ $\frac{f_j(t,u_1,\ldots,u_n)}{u_1+\cdots+u_n} \in [0, \infty)$, and $f_{j\infty}^i = \liminf_{u_1+\cdots+u_n\to\infty} \inf_{t\in[\sigma,1]} \frac{f_j(t,u_1,\ldots,u_n)}{u_1+\cdots+u_n} \in (0, \infty]$, $j = 1, \ldots, n$ we have 2^{2n} cases, which can be grouped in 2^{n+1} cases.

In what follows, for f_0^i, g_0^i, h_0^i, f_∞^s, g_∞^s, $h_\infty^s \in (0, \infty)$ and numbers α_1, α_2, $\alpha_3 \geq 0$ with $\alpha_1 + \alpha_2 + \alpha_3 = 1$, $\widetilde{\alpha}_1$, $\widetilde{\alpha}_2$, $\widetilde{\alpha}_3 > 0$ with $\widetilde{\alpha}_1 + \widetilde{\alpha}_2 + \widetilde{\alpha}_3 = 1$, $\widetilde{\alpha}_2'$, $\widetilde{\alpha}_3' > 0$ with $\widetilde{\alpha}_2' + \widetilde{\alpha}_3' = 1$, $\widetilde{\alpha}_1''$, $\widetilde{\alpha}_3'' > 0$ with $\widetilde{\alpha}_1'' + \widetilde{\alpha}_3'' = 1$, $\widetilde{\alpha}_1'''$, $\widetilde{\alpha}_2''' > 0$ with $\widetilde{\alpha}_1''' + \widetilde{\alpha}_2''' = 1$, we define the numbers

$$M_1 = \frac{\alpha_1}{\theta\sigma^{\alpha-1}f_0^i A}, \quad M_3 = \frac{\alpha_2}{\theta\sigma^{\beta-1}g_0^i C}, \quad M_5 = \frac{\alpha_3}{\theta\sigma^{\gamma-1}h_0^i E},$$

$$M_2 = \frac{\widetilde{\alpha}_1}{f_\infty^s B}, \quad M_4 = \frac{\widetilde{\alpha}_2}{g_\infty^s D}, \quad M_6 = \frac{\widetilde{\alpha}_3}{h_\infty^s F}, \quad M_4' = \frac{\widetilde{\alpha}_2'}{g_\infty^s D},$$

$$M_6' = \frac{\widetilde{\alpha}_3'}{h_\infty^s F}, \quad M_2'' = \frac{\widetilde{\alpha}_1''}{f_\infty^s B}, \quad M_6'' = \frac{\widetilde{\alpha}_3''}{h_\infty^s F}, \quad M_2''' = \frac{\widetilde{\alpha}_1'''}{f_\infty^s B},$$

$$M_4''' = \frac{\widetilde{\alpha}_2'''}{g_\infty^s D}, \quad \widetilde{M}_2 = \frac{1}{f_\infty^s B}, \quad \widetilde{M}_4 = \frac{1}{g_\infty^s D}, \quad \widetilde{M}_6 = \frac{1}{h_\infty^s F},$$

where $\theta = \min\{\sigma^{\alpha-1}, \sigma^{\beta-1}, \sigma^{\gamma-1}\}$.

Theorem 5.1.2. *Assume that* (H1) *and* (H2) *hold,* $\sigma \in (0, 1)$, α_1, α_2, $\alpha_3 \geq 0$ *with* $\alpha_1 + \alpha_2 + \alpha_3 = 1$, $\widetilde{\alpha}_1$, $\widetilde{\alpha}_2$, $\widetilde{\alpha}_3 > 0$ *with* $\widetilde{\alpha}_1 + \widetilde{\alpha}_2 + \widetilde{\alpha}_3 = 1$, $\widetilde{\alpha}_2'$, $\widetilde{\alpha}_3' > 0$ *with* $\widetilde{\alpha}_2' + \widetilde{\alpha}_3' = 1$, $\widetilde{\alpha}_1''$, $\widetilde{\alpha}_3'' > 0$ *with* $\widetilde{\alpha}_1'' + \widetilde{\alpha}_3'' = 1$, $\widetilde{\alpha}_1'''$, $\widetilde{\alpha}_2''' > 0$ *with* $\widetilde{\alpha}_1''' + \widetilde{\alpha}_2''' = 1$.

(1) If $f_0^i, g_0^i, h_0^i, f_\infty^s, g_\infty^s, h_\infty^s \in (0, \infty)$, $M_1 < M_2$, $M_3 < M_4$ and $M_5 < M_6$, then for each $\lambda \in (M_1, M_2)$, $\mu \in (M_3, M_4)$, $\nu \in (M_5, M_6)$, there exists a positive solution $(u(t), v(t), w(t))$, $t \in [0, 1]$ for problem (5.1), (5.2).

(2) If $f_\infty^s = 0$, $g_\infty^s, h_\infty^s, f_0^i, g_0^i, h_0^i \in (0, \infty)$, $M_3 < M_4'$ and $M_5 < M_6'$, then for each $\lambda \in (M_1, \infty)$, $\mu \in (M_3, M_4')$, $\nu \in (M_5, M_6')$, there exists a positive solution $(u(t), v(t), w(t))$, $t \in [0, 1]$ for problem (5.1), (5.2).

(3) If $g_\infty^s = 0$, $f_\infty^s, h_\infty^s, f_0^i, g_0^i, h_0^i \in (0, \infty)$, $M_1 < M_2''$ and $M_5 < M_6''$, then for each $\lambda \in (M_1, M_2'')$, $\mu \in (M_3, \infty)$, $\nu \in (M_5, M_6'')$, there exists a positive solution $(u(t), v(t), w(t))$, $t \in [0, 1]$ for problem (5.1), (5.2).

(4) If $h_\infty^s = 0$, $f_\infty^s, g_\infty^s, f_0^i, g_0^i, h_0^i \in (0, \infty)$, $M_1 < M_2'''$ and $M_3 < M_4'''$, then for each $\lambda \in (M_1, M_2''')$, $\mu \in (M_3, M_4''')$, $\nu \in (M_5, \infty)$, there exists a positive solution $(u(t), v(t), w(t))$, $t \in [0, 1]$ for problem (5.1), (5.2).

(5) If $f_\infty^s = g_\infty^s = 0$, $h_\infty^s, f_0^i, g_0^i, h_0^i \in (0, \infty)$, $M_5 < \widetilde{M_6}$, then for each $\lambda \in (M_1, \infty)$, $\mu \in (M_3, \infty)$, $\nu \in (M_5, \widetilde{M_6})$, there exists a positive solution $(u(t), v(t), w(t))$, $t \in [0, 1]$ for problem (5.1), (5.2).

(6) If $f_\infty^s = h_\infty^s = 0$, $g_\infty^s, f_0^i, g_0^i, h_0^i \in (0, \infty)$, $M_3 < \widetilde{M_4}$, then for each $\lambda \in (M_1, \infty)$, $\mu \in (M_3, \widetilde{M_4})$, $\nu \in (M_5, \infty)$, there exists a positive solution $(u(t), v(t), w(t))$, $t \in [0, 1]$ for problem (5.1), (5.2).

(7) If $g_\infty^s = h_\infty^s = 0$, $f_\infty^s, f_0^i, g_0^i, h_0^i \in (0, \infty)$, $M_1 < \widetilde{M_2}$, then for each $\lambda \in (M_1, \widetilde{M_2})$, $\mu \in (M_3, \infty)$, $\nu \in (M_5, \infty)$, there exists a positive solution $(u(t), v(t), w(t))$, $t \in [0, 1]$ for problem (5.1), (5.2).

(8) If $f_\infty^s = g_\infty^s = h_\infty^s = 0$, $f_0^i, g_0^i, h_0^i \in (0, \infty)$, then for each $\lambda \in (M_1, \infty)$, $\mu \in (M_3, \infty)$, $\nu \in (M_5, \infty)$, there exists a positive solution $(u(t), v(t), w(t))$, $t \in [0, 1]$ for problem (5.1), (5.2).

(9) If $f_\infty^s, g_\infty^s, h_\infty^s \in (0, \infty)$ and at least one of f_0^i, g_0^i, h_0^i is ∞, then for each $\lambda \in (0, M_2)$, $\mu \in (0, M_4)$, $\nu \in (0, M_6)$, there exists a positive solution $(u(t), v(t), w(t))$, $t \in [0, 1]$ for problem (5.1), (5.2).

(10) If $f_\infty^s = 0$, $g_\infty^s, h_\infty^s \in (0, \infty)$ and at least one of f_0^i, g_0^i, h_0^i is ∞, then for each $\lambda \in (0, \infty)$, $\mu \in (0, M_4')$, $\nu \in (0, M_6')$, there exists a positive solution $(u(t), v(t), w(t))$, $t \in [0, 1]$ for problem (5.1), (5.2).

(11) If $g_\infty^s = 0$, $f_\infty^s, h_\infty^s \in (0, \infty)$ and at least one of f_0^i, g_0^i, h_0^i is ∞, then for each $\lambda \in (0, M_2'')$, $\mu \in (0, \infty)$, $\nu \in (0, M_6'')$, there exists a positive solution $(u(t), v(t), w(t))$, $t \in [0, 1]$ for problem (5.1), (5.2).

(12) If $h_\infty^s = 0$, $f_\infty^s, g_\infty^s \in (0, \infty)$ and at least one of f_0^i, g_0^i, h_0^i is ∞, then for each $\lambda \in (0, M_2''')$, $\mu \in (0, M_4''')$, $\nu \in (0, \infty)$, there exists a positive solution $(u(t), v(t), w(t))$, $t \in [0, 1]$ for problem (5.1), (5.2).

(13) If $f_\infty^s = g_\infty^s = 0$, $h_\infty^s \in (0, \infty)$ and at least one of f_0^i, g_0^i, h_0^i is ∞, then for each $\lambda \in (0, \infty)$, $\mu \in (0, \infty)$, $\nu \in (0, \widetilde{M_6})$, there exists a positive solution $(u(t), v(t), w(t))$, $t \in [0, 1]$ for problem (5.1), (5.2).

(14) If $f_\infty^s = h_\infty^s = 0$, $g_\infty^s \in (0, \infty)$ and at least one of f_0^i, g_0^i, h_0^i is ∞, then for each $\lambda \in (0, \infty)$, $\mu \in (0, \widetilde{M_4})$, $\nu \in (0, \infty)$, there exists a positive solution $(u(t), v(t), w(t))$, $t \in [0, 1]$ for problem (5.1), (5.2).

(15) If $g_\infty^s = h_\infty^s = 0$, $\underline{f_\infty^s} \in (0, \infty)$ and at least one of f_0^i, g_0^i, h_0^i is ∞, then for each $\lambda \in (0, \widetilde{M_2})$, $\mu \in (0, \infty)$, $\nu \in (0, \infty)$, there exists a positive solution $(u(t), v(t), w(t))$, $t \in [0, 1]$ for problem (5.1), (5.2).

(16) If $f_\infty^s = g_\infty^s = h_\infty^s = 0$ and at least one of f_0^i, g_0^i, h_0^i is ∞, then for each $\lambda \in (0, \infty)$, $\mu \in (0, \infty)$, $\nu \in (0, \infty)$, there exists a positive solution $(u(t), v(t), w(t))$, $t \in [0, 1]$ for problem (5.1), (5.2).

Proof. We consider again the above cone $P \subset Y$ and the operators Q_1, Q_2, Q_3 and Q. We will also prove for this theorem some illustrative cases.

Case (1). We consider $f_0^i, g_0^i, h_0^i, f_\infty^s, g_\infty^s, h_\infty^s \in (0, \infty)$. Let $\lambda \in (M_1, M_2)$, $\mu \in (M_3, M_4)$, $\nu \in (M_5, M_6)$. We choose $\varepsilon > 0$ a positive number such that $\varepsilon < f_0^i$, $\varepsilon < g_0^i$, $\varepsilon < h_0^i$ and

$$\frac{\alpha_1}{\theta \sigma^{\alpha-1}(f_0^i - \varepsilon)A} \le \lambda, \quad \frac{\alpha_2}{\theta \sigma^{\beta-1}(g_0^i - \varepsilon)C} \le \mu, \quad \frac{\alpha_3}{\theta \sigma^{\gamma-1}(h_0^i - \varepsilon)E} \le \nu,$$

$$\frac{\widetilde{\alpha_1}}{(f_\infty^s + \varepsilon)B} \ge \lambda, \quad \frac{\widetilde{\alpha_2}}{(g_\infty^s + \varepsilon)D} \ge \mu, \quad \frac{\widetilde{\alpha_3}}{(h_\infty^s + \varepsilon)F} \ge \nu.$$

By using (H2) and the definition of f_0^i, g_0^i, h_0^i, we deduce that there exists $R_3 > 0$ such that $f(t, u, v, w) \ge (f_0^i - \varepsilon)(u + v + w)$, $g(t, u, v, w) \ge (g_0^i - \varepsilon)(u + v + w)$, $h(t, u, v, w) \ge (h_0^i - \varepsilon)(u + v + w)$ for all $u, v, w \ge 0$ with $u + v + w \le R_3$ and $t \in [\sigma, 1]$. We denote by $\Omega_3 = \{(u, v, w) \in Y, \|(u, v, w)\|_Y < R_3\}$.

Let $(u, v, w) \in P \cap \partial\Omega_3$, that is $\|(u, v, w)\|_Y = R_3$ or equivalently $\|u\| + \|v\| + \|w\| = R_3$. Because $u(t) + v(t) + w(t) \le R_3$ for all $t \in [0, 1]$, then by Lemma 5.1.3 we obtain for all $t \in [\sigma, 1]$

$$Q_1(u, v, w)(t) \ge \lambda \int_0^1 t^{\alpha-1} J_1(s) f(s, u(s), v(s), w(s)) \, ds$$

$$\ge \lambda \sigma^{\alpha-1} \int_\sigma^1 J_1(s) f(s, u(s), v(s), w(s)) \, ds$$

$$\geq \lambda \sigma^{\alpha-1} \int_\sigma^1 J_1(s)(f_0^i - \varepsilon)(u(s) + v(s) + w(s)) \, ds$$

$$\geq \lambda \sigma^{\alpha-1} \theta(f_0^i - \varepsilon) \int_\sigma^1 J_1(s) \|(u, v, w)\|_Y \, ds$$

$$= \lambda \sigma^{\alpha-1} \theta(f_0^i - \varepsilon) A \|(u, v, w)\|_Y \geq \alpha_1 \|(u, v, w)\|_Y,$$

$$Q_2(u, v, w)(t) \geq \mu \int_0^1 t^{\beta-1} J_2(s) g(s, u(s), v(s), w(s)) \, ds$$

$$\geq \mu \sigma^{\beta-1} \int_\sigma^1 J_2(s) g(s, u(s), v(s), w(s)) \, ds$$

$$\geq \mu \sigma^{\beta-1} \int_\sigma^1 J_2(s)(g_0^i - \varepsilon)(u(s) + v(s) + w(s)) \, ds$$

$$\geq \mu \sigma^{\beta-1} \theta(g_0^i - \varepsilon) \int_\sigma^1 J_2(s) \|(u, v, w)\|_Y \, ds$$

$$= \mu \sigma^{\beta-1} \theta(g_0^i - \varepsilon) C \|(u, v, w)\|_Y \geq \alpha_2 \|(u, v, w)\|_Y,$$

$$Q_3(u, v, w)(t) \geq \nu \int_0^1 t^{\gamma-1} J_3(s) h(s, u(s), v(s), w(s)) \, ds$$

$$\geq \nu \sigma^{\gamma-1} \int_\sigma^1 J_3(s) h(s, u(s), v(s), w(s)) \, ds$$

$$\geq \nu \sigma^{\gamma-1} \int_\sigma^1 J_3(s)(h_0^i - \varepsilon)(u(s) + v(s) + w(s)) \, ds$$

$$\geq \nu \sigma^{\gamma-1} \theta(h_0^i - \varepsilon) \int_\sigma^1 J_3(s) \|(u, v, w)\|_Y \, ds$$

$$= \nu \sigma^{\gamma-1} \theta(h_0^i - \varepsilon) E \|(u, v, w)\|_Y \geq \alpha_3 \|(u, v, w)\|_Y.$$

So

$$\|Q_1(u, v, w)\| \geq Q_1(u, v, w)(\sigma) \geq \alpha_1 \|(u, v, w)\|_Y,$$

$$\|Q_2(u, v, w)\| \geq Q_2(u, v, w)(\sigma) \geq \alpha_2 \|(u, v, w)\|_Y,$$

$$\|Q_3(u, v, w)\| \geq Q_3(u, v, w)(\sigma) \geq \alpha_3 \|(u, v, w)\|_Y.$$

Then for an arbitrary element $(u, v, w) \in P \cap \partial\Omega_3$, we deduce

$$\|Q(u, v, w)\|_Y \geq (\alpha_1 + \alpha_2 + \alpha_3) \|(u, v, w)\|_Y = \|(u, v, w)\|_Y. \tag{5.22}$$

Now, we define the functions $f^*, g^*, h^* : [0,1] \times \mathbb{R}_+ \to \mathbb{R}_+$, $f^*(t,x) = \max_{0 \le u+v+w \le x} f(t,u,v,w)$, $g^*(t,x) = \max_{0 \le u+v+w \le x} g(t,u,v,w)$, $h^*(t,x) = \max_{0 \le u+v+w \le x} h(t,u,v,w)$, $t \in [0,1]$, $x \in \mathbb{R}_+$. Then $f(t,u,v,w) \le f^*(t,x)$, $g(t,u,v,w) \le g^*(t,x)$, $h(t,u,v,w) \le h^*(t,x)$ for all $t \in [0,1]$, $u, v, w \ge 0$ and $u + v + w \le x$. The functions $f^*(t,\cdot)$, $g^*(t,\cdot)$, $h^*(t,\cdot)$ are nondecreasing for every $t \in [0,1]$, and they satisfy the conditions

$$\limsup_{x \to \infty} \max_{t \in [0,1]} \frac{f^*(t,x)}{x} = f_\infty^s, \quad \limsup_{x \to \infty} \max_{t \in [0,1]} \frac{g^*(t,x)}{x} = g_\infty^s,$$

$$\limsup_{x \to \infty} \max_{t \in [0,1]} \frac{h^*(t,x)}{x} = h_\infty^s.$$

Therefore, for $\varepsilon > 0$, there exists $\overline{R}_4 > 0$ such that for all $x \ge \overline{R}_4$ and $t \in [0,1]$, we have $f^*(t,x) \le (f_\infty^s + \varepsilon)x$, $g^*(t,x) \le (g_\infty^s + \varepsilon)x$, $h^*(t,x) \le (h_\infty^s + \varepsilon)x$.

We consider $R_4 = \max\{2R_3, \overline{R}_4\}$ and we denote by $\Omega_4 = \{(u,v,w) \in Y, \|(u,v,w)\|_Y < R_4\}$. Let $(u,v,w) \in P \cap \partial\Omega_4$. By the definition of f^*, g^*, h^* we conclude

$$f(t,u(t),v(t),w(t)) \le f^*(t,\|(u,v,w)\|_Y),$$

$$g(t,u(t),v(t),w(t)) \le g^*(t,\|(u,v,w)\|_Y),$$

$$h(t,u(t),v(t),w(t)) \le h^*(t,\|(u,v,w)\|_Y),$$

$$\forall t \in [0,1].$$

Then for all $t \in [0,1]$, we obtain

$$Q_1(u,v,w)(t) \le \lambda \int_0^1 J_1(s)f(s,u(s),v(s),w(s))\,ds$$

$$\le \lambda \int_0^1 J_1(s)f^*(s,\|(u,v,w)\|_Y)\,ds$$

$$\le \lambda(f_\infty^s + \varepsilon)\int_0^1 J_1(s)\|(u,v,w)\|_Y\,ds \le \widetilde{\alpha}_1\|(u,v,w)\|_Y,$$

$$Q_2(u,v,w)(t) \le \mu \int_0^1 J_2(s)g(s,u(s),v(s),w(s))\,ds$$

$$\le \mu \int_0^1 J_2(s)g^*(s,\|(u,v,w)\|_Y)\,ds$$

$$\le \mu(g_\infty^s + \varepsilon)\int_0^1 J_2(s)\|(u,v,w)\|_Y\,ds \le \widetilde{\alpha}_2\|(u,v,w)\|_Y,$$

$$Q_3(u, v, w)(t) \leq \nu \int_0^1 J_3(s) h(s, u(s), v(s), w(s)) \, ds$$

$$\leq \nu \int_0^1 J_3(s) h^*(s, \|(u, v, w)\|_Y) \, ds$$

$$\leq \nu (h_\infty^s + \varepsilon) \int_0^1 J_3(s) \|(u, v, w)\|_Y \, ds \leq \tilde{\alpha}_3 \|(u, v, w)\|_Y.$$

Therefore, we deduce $\|Q_1(u, v, w)\| \leq \tilde{\alpha}_1 \|(u, v, w)\|_Y$, $\|Q_2(u, v, w)\| \leq \tilde{\alpha}_2 \|(u, v, w)\|_Y$, $\|Q_3(u, v, w)\| \leq \tilde{\alpha}_3 \|(u, v, w)\|_Y$.

Hence, for $(u, v, w) \in P \cap \partial\Omega_4$, we conclude that

$$\|Q(u, v, w)\|_Y \leq (\tilde{\alpha}_1 + \tilde{\alpha}_2 + \tilde{\alpha}_3)\|(u, v, w)\|_Y = \|(u, v, w)\|_Y. \qquad (5.23)$$

By using Lemma 5.1.5, Theorem 1.2.2 (ii), the relations (5.22), (5.23), we deduce that Q has a fixed point $(u, v, w) \in P \cap (\overline{\Omega}_4 \setminus \Omega_3)$, which is a positive solution for our problem (5.1), (5.2).

Case (11). We consider $g_\infty^s = 0$, $h_0^i = \infty$, f_∞^s, h_∞^s, f_0^i, $g_0^i \in (0, \infty)$. Let $\lambda \in (0, M_2'')$, $\mu \in (0, \infty)$, $\nu \in (0, M_6'')$. We choose $\varepsilon > 0$ such that $\varepsilon \leq \nu \theta \sigma^{\gamma-1} E$ and

$$\varepsilon \leq \frac{\tilde{\alpha}_1'' - \lambda f_\infty^s B}{2\lambda B}, \quad \varepsilon \leq \frac{1 - \lambda f_\infty^s B - \nu h_\infty^s F}{2\mu D}, \quad \varepsilon \leq \frac{\tilde{\alpha}_3'' - \nu h_\infty^s F}{2\nu F}.$$

The numerators of the above fractions are positive, because $\lambda < \frac{\tilde{\alpha}_1''}{f_\infty^s B}$ that is $\tilde{\alpha}_1'' > \lambda f_\infty^s B$, $\nu < \frac{\tilde{\alpha}_3''}{h_\infty^s F}$ that is $\tilde{\alpha}_3'' > \nu h_\infty^s F$, and $1 - \lambda f_\infty^s B - \nu h_\infty^s F = \tilde{\alpha}_1'' + \tilde{\alpha}_3'' - \lambda f_\infty^s B - \nu h_\infty^s F = (\tilde{\alpha}_1'' - \lambda f_\infty^s B) + (\tilde{\alpha}_3'' - \nu h_\infty^s F) > 0$.

By using (H2) and the definition of h_0^i, we deduce that there exists $R_3 > 0$ such that $h(t, u, v, w) \geq \frac{1}{\varepsilon}(u + v + w)$ for all $u, v, w \geq 0$ with $u + v + w \leq R_3$ and $t \in [\sigma, 1]$. We denote by $\Omega_3 = \{(u, v, w) \in Y, \|(u, v, w)\|_Y < R_3\}$.

Let $(u, v, w) \in P \cap \partial\Omega_3$, that is $\|(u, v, w)\|_Y = R_3$. Because $u(t) + v(t) + w(t) \leq R_3$ for all $t \in [0, 1]$, then by using Lemma 5.1.3, we obtain

$$Q_3(u, v, w)(t) \geq \nu \int_0^1 t^{\gamma-1} J_3(s) h(s, u(s), v(s), w(s)) \, ds$$

$$\geq \nu \sigma^{\gamma-1} \int_\sigma^1 J_3(s) h(s, u(s), v(s), w(s)) \, ds$$

$$\geq \nu \sigma^{\gamma-1} \int_\sigma^1 J_3(s) \frac{1}{\varepsilon}(u(s) + v(s) + w(s)) \, ds$$

$$\geq \nu \sigma^{\gamma-1} \theta \frac{1}{\varepsilon} \int_\sigma^1 J_3(s) \|(u,v,w)\|_Y \, ds$$

$$= \nu \sigma^{\gamma-1} \theta \frac{1}{\varepsilon} E \|(u,v,w)\|_Y \geq \|(u,v,w)\|_Y, \quad \forall t \in [\sigma, 1].$$

Then for $(u,v,w) \in P \cap \partial\Omega_3$ we find $\|Q_3(u,v,w)\| \geq Q_3(u,v,w)(\sigma) \geq \|(u,v,w)\|_Y$, and

$$\|Q(u,v,w)\|_Y \geq \|Q_3(u,v,w)\| \geq \|(u,v,w)\|_Y. \tag{5.24}$$

Now, using the functions f^*, g^*, h^* defined in the proof of Case (1), we have

$$\limsup_{x \to \infty} \max_{t \in [0,1]} \frac{f^*(t,x)}{x} = f_\infty^s, \quad \lim_{x \to \infty} \max_{t \in [0,1]} \frac{g^*(t,x)}{x} = 0,$$

$$\limsup_{x \to \infty} \max_{t \in [0,1]} \frac{h^*(t,x)}{x} = h_\infty^s.$$

Therefore for $\varepsilon > 0$, there exists $\overline{R}_4 > 0$ such that for all $x \geq \overline{R}_4$ and $t \in [0,1]$, we deduce $f^*(t,x) \leq (f_\infty^s + \varepsilon)x$, $g^*(t,x) \leq \varepsilon x$, $h^*(t,x) \leq (h_\infty^s + \varepsilon)x$. We consider $R_4 = \max\{2R_3, \overline{R}_4\}$ and we denote by $\Omega_4 = \{(u,v,w) \in Y, \|(u,v,w)\|_Y < R_4\}$. Let $(u,v,w) \in P \cap \partial\Omega_4$. Then for all $t \in [0,1]$, we obtain

$$Q_1(u,v,w)(t) \leq \lambda \int_0^1 J_1(s) f(s, u(s), v(s), w(s)) \, ds$$

$$\leq \lambda \int_0^1 J_1(s) f^*(s, \|(u,v,w)\|_Y) \, ds$$

$$\leq \lambda(f_\infty^s + \varepsilon) \int_0^1 J_1(s) \|(u,v,w)\|_Y \, ds$$

$$\leq \lambda \left(f_\infty^s + \frac{\widetilde{\alpha}_1'' - \lambda f_\infty^s B}{2\lambda B} \right) B \|(u,v,w)\|_Y$$

$$= \frac{1}{2}(\lambda f_\infty^s B + \widetilde{\alpha}_1'') \|(u,v,w)\|_Y,$$

$$Q_2(u,v,w)(t) \leq \mu \int_0^1 J_2(s) g(s, u(s), v(s), w(s)) \, ds$$

$$\leq \mu \int_0^1 J_2(s) g^*(s, \|(u,v,w)\|_Y) \, ds$$

$$\le \mu\varepsilon \int_0^1 J_2(s)\|(u,v,w)\|_Y\, ds$$

$$\le \mu\frac{1-\lambda f_\infty^s B - \nu h_\infty^s F}{2\mu D} D\|(u,v,w)\|_Y$$

$$= \frac{1}{2}(1-\lambda f_\infty^s B - \nu h_\infty^s F)\|(u,v,w)\|_Y,$$

$$Q_3(u,v,w)(t) \le \nu \int_0^1 J_3(s)h(s,u(s),v(s),w(s))\, ds$$

$$\le \nu \int_0^1 J_3(s)h^*(s,\|(u,v,w)\|_Y)\, ds$$

$$\le \nu(h_\infty^s + \varepsilon) \int_0^1 J_3(s)\|(u,v,w)\|_Y\, ds$$

$$\le \nu\left(h_\infty^s + \frac{\tilde\alpha_3'' - \nu h_\infty^s F}{2\nu F}\right) F\|(u,v,w)\|_Y$$

$$= \frac{1}{2}(\nu h_\infty^s F + \tilde\alpha_3'')\|(u,v,w)\|_Y.$$

Therefore,

$$\|Q_1(u,v,w)\| \le \frac{1}{2}(\lambda f_\infty^s B + \tilde\alpha_1'')\|(u,v,w)\|_Y,$$

$$\|Q_2(u,v,w)\| \le \frac{1}{2}(1-\lambda f_\infty^s B - \nu h_\infty^s F)\|(u,v,w)\|_Y,$$

$$\|Q_3(u,v,w)\| \le \frac{1}{2}(\nu h_\infty^s F + \tilde\alpha_3'')\|(u,v,w)\|_Y.$$

Then for $(u,v,w) \in P \cap \partial\Omega_4$, we conclude that

$$\|Q(u,v,w)\|_Y \le \frac{1}{2}(\lambda f_\infty^s B + \tilde\alpha_1'' + 1 - \lambda f_\infty^s B - \nu h_\infty^s F + \nu h_\infty^s F + \tilde\alpha_3'')$$

$$\times \|(u,v,w)\|_Y = \|(u,v,w)\|_Y. \tag{5.25}$$

By using Lemma 5.1.5, Theorem 1.2.2 (ii) and relations (5.24), (5.25), we deduce that Q has a fixed point $(u,v,w) \in P \cap (\overline\Omega_4 \setminus \Omega_3)$, which is a positive solution for our problem (5.1), (5.2).

Case (14). We consider $f_\infty^s = h_\infty^s = 0$, $g_0^i = \infty$, g_∞^s, f_0^i, $h_0^i \in (0,\infty)$. Let $\lambda \in (0,\infty)$, $\mu \in (0,\tilde M_4)$, $\nu \in (0,\infty)$. We choose $\varepsilon > 0$ such that

$\varepsilon \leq \mu\theta\sigma^{\beta-1}C$ and

$$\varepsilon \leq \frac{1 - \mu g_\infty^s D}{4\lambda B}, \quad \varepsilon \leq \frac{1 - \mu g_\infty^s D}{2\mu D}, \quad \varepsilon \leq \frac{1 - \mu g_\infty^s D}{4\nu F}.$$

The numerator of the above fractions is positive because $\mu < \frac{1}{g_0^s D}$, that is, $1 - \mu g_\infty^s D > 0$.

By using (H2) and the definition of g_0^i, we deduce that there exists $R_3 > 0$ such that $g(t, u, v, w) \geq \frac{1}{\varepsilon}(u+v+w)$ for all u, v, $w \geq 0$ with $u+v+w \leq R_3$ and $t \in [\sigma, 1]$. We denote by $\Omega_3 = \{(u, v, w) \in Y, \|(u, v, w)\|_Y < R_3\}$.

Let $(u, v, w) \in P \cap \partial\Omega_3$, that is $\|(u, v, w)\|_Y = R_3$. Because $u(t) + v(t) + w(t) \leq R_3$ for all $t \in [0, 1]$, then by using Lemma 5.1.3, we obtain

$$Q_2(u, v, w)(t) \geq \mu \int_0^1 t^{\beta-1} J_2(s) g(s, u(s), v(s), w(s)) \, ds$$

$$\geq \mu\sigma^{\beta-1} \int_\sigma^1 J_2(s) g(s, u(s), v(s), w(s)) \, ds$$

$$\geq \mu\sigma^{\beta-1} \int_\sigma^1 J_2(s) \frac{1}{\varepsilon}(u(s) + v(s) + w(s)) \, ds$$

$$\geq \mu\sigma^{\beta-1}\theta\frac{1}{\varepsilon} \int_\sigma^1 J_2(s)\|(u, v, w)\|_Y \, ds$$

$$= \mu\sigma^{\beta-1}\theta\frac{1}{\varepsilon}C\|(u, v, w)\|_Y \geq \|(u, v, w)\|_Y, \quad \forall t \in [\sigma, 1].$$

Then for $(u, v, w) \in P \cap \partial\Omega_3$ we find $\|Q_2(u, v, w)\| \geq Q_2(u, v, w)(\sigma) \geq \|(u, v, w)\|_Y$, and

$$\|Q(u, v, w)\|_Y \geq \|Q_2(u, v, w)\| \geq \|(u, v, w)\|_Y. \tag{5.26}$$

Now, using the functions f^*, g^*, h^* defined in the proof of Case (1), we have

$$\lim_{x\to\infty} \max_{t\in[0,1]} \frac{f^*(t, x)}{x} = 0, \quad \limsup_{x\to\infty} \max_{t\in[0,1]} \frac{g^*(t, x)}{x} = g_\infty^s,$$

$$\lim_{x\to\infty} \max_{t\in[0,1]} \frac{h^*(t, x)}{x} = 0.$$

Therefore, for $\varepsilon > 0$, there exists $\overline{R}_4 > 0$ such that for all $x \geq \overline{R}_4$ and $t \in [0, 1]$ we deduce $f^*(t, x) \leq \varepsilon x$, $g^*(t, x) \leq (g_\infty^s + \varepsilon)x$, $h^*(t, x) \leq \varepsilon x$.

We consider $R_4 = \max\{2R_3, \overline{R}_4\}$, and we denote by $\Omega_4 = \{(u, v, w) \in Y,$ $\|(u, v, w)\|_Y < R_4\}$. Let $(u, v, w) \in P \cap \partial\Omega_4$. Then for all $t \in [0, 1]$ we obtain

$$Q_1(u, v, w)(t) \leq \lambda \int_0^1 J_1(s) f(s, u(s), v(s), w(s))\, ds$$

$$\leq \lambda \int_0^1 J_1(s) f^*(s, \|(u, v, w)\|_Y)\, ds$$

$$\leq \lambda\varepsilon \int_0^1 J_1(s) \|(u, v, w)\|_Y\, ds$$

$$= \lambda\varepsilon B \|(u, v, w)\|_Y \leq \lambda \frac{1 - \mu g_\infty^s D}{4\lambda B} B \|(u, v, w)\|_Y$$

$$= \frac{1}{4}(1 - \mu g_\infty^s D) \|(u, v, w)\|_Y,$$

$$Q_2(u, v, w)(t) \leq \mu \int_0^1 J_2(s) g(s, u(s), v(s), w(s))\, ds$$

$$\leq \mu \int_0^1 J_2(s) g^*(s, \|(u, v, w)\|_Y)\, ds$$

$$\leq \mu(g_\infty^s + \varepsilon) \int_0^1 J_2(s) \|(u, v, w)\|_Y\, ds$$

$$= \mu(g_\infty^s + \varepsilon) D \|(u, v, w)\|_Y$$

$$\leq \mu \left(g_\infty^s + \frac{1 - \mu g_\infty^s D}{2\mu D} \right) D \|(u, v, w)\|_Y$$

$$= \frac{1}{2}(\mu g_\infty^s D + 1) \|(u, v, w)\|_Y,$$

$$Q_3(u, v, w)(t) \leq \nu \int_0^1 J_3(s) h(s, u(s), v(s), w(s))\, ds$$

$$\leq \nu \int_0^1 J_3(s) h^*(s, \|(u, v, w)\|_Y)\, ds$$

$$\leq \nu\varepsilon \int_0^1 J_3(s) \|(u, v, w)\|_Y\, ds$$

$$= \nu\varepsilon F \|(u, v, w)\|_Y \leq \nu \frac{1 - \nu g_\infty^s D}{4\nu F} F \|(u, v, w)\|_Y$$

$$= \frac{1}{4}(1 - \mu g_\infty^s D) \|(u, v, w)\|_Y.$$

Therefore,

$$\|Q_1(u,v,w)\| \le \frac{1}{4}(1 - \mu g_\infty^s D)\|(u,v,w)\|_Y,$$

$$\|Q_2(u,v,w)\| \le \frac{1}{2}(1 + \mu g_\infty^s D)\|(u,v,w)\|_Y,$$

$$\|Q_3(u,v,w)\| \le \frac{1}{4}(1 - \mu g_\infty^s D)\|(u,v,w)\|_Y.$$

Then for $(u,v,w) \in P \cap \partial\Omega_4$, we conclude that

$$\|Q(u,v,w)\|_Y \le \frac{1}{4}(1 - \mu g_\infty^s D + 2 + 2\mu g_\infty^s D + 1 - \mu g_\infty^s D)$$

$$\times \|(u,v,w)\|_Y = \|(u,v,w)\|_Y. \tag{5.27}$$

By using Lemma 5.1.5, Theorem 1.2.2 (ii), the relations (5.26) and (5.27), we deduce that Q has a fixed point $(u,v,w) \in P \cap (\overline{\Omega}_4 \setminus \Omega_3)$, which is a positive solution for our problem (5.1), (5.2).

Case (16). We consider $f_\infty^s = g_\infty^s = h_\infty^s = 0$, $f_0^i = g_0^i = \infty$ and $h_0^i \in (0,\infty)$. Let $\lambda \in (0,\infty)$, $\mu \in (0,\infty)$, $\nu \in (0,\infty)$. We choose $\varepsilon > 0$ such that

$$\varepsilon \le \lambda\theta\sigma^{\alpha-1}A, \quad \varepsilon \le \frac{1}{3\lambda B}, \quad \varepsilon \le \frac{1}{3\mu D}, \quad \varepsilon \le \frac{1}{3\nu F}.$$

By using (H2) and the definition of f_0^i, we deduce that there exists $R_3 > 0$ such that $f(t,u,v,w) \ge \frac{1}{\varepsilon}(u+v+w)$ for all u, v, $w \ge 0$ with $u+v+w \le R_3$ and $t \in [\sigma,1]$. We denote by $\Omega_3 = \{(u,v,w) \in Y, \ \|(u,v,w)\|_Y < R_3\}$.

Let $(u,v,w) \in P \cap \partial\Omega_3$, that is $\|(u,v,w)\|_Y = R_3$. Because $u(t) + v(t) + w(t) \le R_3$ for all $t \in [0,1]$, then by using Lemma 5.1.3, we obtain

$$Q_1(u,v,w)(t) \ge \lambda \int_0^1 t^{\alpha-1} J_1(s) f(s,u(s),v(s),w(s))\,ds$$

$$\ge \lambda\sigma^{\alpha-1} \int_\sigma^1 J_1(s) f(s,u(s),v(s),w(s))\,ds$$

$$\ge \lambda\sigma^{\alpha-1} \int_\sigma^1 J_1(s)\frac{1}{\varepsilon}(u(s) + v(s) + w(s))\,ds$$

$$\ge \lambda\sigma^{\alpha-1}\theta\frac{1}{\varepsilon} \int_\sigma^1 J_1(s)\|(u,v,w)\|_Y\,ds$$

$$= \lambda\sigma^{\alpha-1}\theta\frac{1}{\varepsilon}A\|(u,v,w)\|_Y \ge \|(u,v,w)\|_Y, \quad \forall t \in [\sigma,1].$$

Then for $(u, v, w) \in P \cap \partial \Omega_3$ we find $\|Q_1(u, v, w)\| \geq Q_1(u, v, w)(\sigma) \geq \|(u, v, w)\|_Y$, and

$$\|Q(u, v, w)\|_Y \geq \|Q_1(u, v, w)\| \geq \|(u, v, w)\|_Y. \tag{5.28}$$

Now, using the functions f^*, g^*, h^* defined in the proof of Case (1), we have

$$\lim_{x \to \infty} \max_{t \in [0,1]} \frac{f^*(t, x)}{x} = 0, \quad \lim_{x \to \infty} \max_{t \in [0,1]} \frac{g^*(t, x)}{x} = 0,$$

$$\lim_{x \to \infty} \max_{t \in [0,1]} \frac{h^*(t, x)}{x} = 0.$$

Therefore, for $\varepsilon > 0$, there exists $\overline{R}_4 > 0$ such that $f^*(t, x) \leq \varepsilon x$, $g^*(t, x) \leq \varepsilon x$, $h^*(t, x) \leq \varepsilon x$ for all $x \geq \overline{R}_4$ and $t \in [0, 1]$.

We consider $R_4 = \max\{2R_3, \overline{R}_4\}$ and we denote by $\Omega_4 = \{(u, v, w) \in Y, \|(u, v, w)\|_Y < R_4\}$. Let $(u, v, w) \in P \cap \partial \Omega_4$. Then for all $t \in [0, 1]$, we obtain

$$Q_1(u, v, w)(t) \leq \lambda \int_0^1 J_1(s) f(s, u(s), v(s), w(s)) \, ds$$

$$\leq \lambda \int_0^1 J_1(s) f^*(s, \|(u, v, w)\|_Y) \, ds$$

$$\leq \lambda \varepsilon \int_0^1 J_1(s) \|(u, v, w)\|_Y \, ds$$

$$= \lambda \varepsilon B \|(u, v, w)\|_Y \leq \frac{1}{3} \|(u, v, w)\|_Y,$$

$$Q_2(u, v, w)(t) \leq \mu \int_0^1 J_2(s) g(s, u(s), v(s), w(s)) \, ds$$

$$\leq \mu \int_0^1 J_2(s) g^*(s, \|(u, v, w)\|_Y) \, ds$$

$$\leq \mu \varepsilon \int_0^1 J_2(s) \|(u, v, w)\|_Y \, ds$$

$$= \mu \varepsilon D \|(u, v, w)\|_Y \leq \frac{1}{3} \|(u, v, w)\|_Y,$$

$$Q_3(u, v, w)(t) \leq \nu \int_0^1 J_3(s) h(s, u(s), v(s), w(s)) \, ds$$

$$\leq \nu \int_0^1 J_3(s) h^*(s, \|(u, v, w)\|_Y) \, ds$$

$$\leq \nu\varepsilon \int_0^1 J_3(s)\|(u,v,w)\|_Y \, ds$$

$$= \nu\varepsilon F\|(u,v,w)\|_Y \leq \frac{1}{3}\|(u,v,w)\|_Y.$$

Therefore $\|Q_1(u,v,w)\| \leq \frac{1}{3}\|(u,v,w)\|_Y$, $\|Q_2(u,v,w)\| \leq \frac{1}{3}\|(u,v,w)\|_Y$, $\|Q_3(u,v,w)\| \leq \frac{1}{3}\|(u,v,w)\|_Y$. Then for $(u,v,w) \in P \cap \partial\Omega_4$, we conclude that

$$\|Q(u,v,w)\|_Y \leq \|(u,v,w)\|_Y. \tag{5.29}$$

By using Lemma 5.1.5, Theorem 1.2.2 (ii), the relations (5.28) and (5.29), we deduce that Q has a fixed point $(u,v,w) \in P \cap (\overline{\Omega}_4 \setminus \Omega_3)$, which is a positive solution for our problem (5.1), (5.2). $\qquad\square$

Remark 5.1.4. Each of the cases (9)–(16) of Theorem 5.1.2 contains 7 cases as follows: $\{f_0^i = \infty,\ g_0^i,\ h_0^i \in (0,\infty)\}$, or $\{g_0^i = \infty,\ f_0^i,\ h_0^i \in (0,\infty)\}$, or $\{h_0^i = \infty,\ f_0^i,\ g_0^i \in (0,\infty)\}$, or $\{f_0^i = g_0^i = \infty,\ h_0^i \in (0,\infty)\}$, or $\{f_0^i = h_0^i = \infty,\ g_0^i \in (0,\infty)\}$, or $\{g_0^i = h_0^i = \infty,\ f_0^i \in (0,\infty)\}$, or $\{f_0^i = g_0^i = h_0^i = \infty\}$. So, the total number of cases from Theorem 5.1.2 is 64, which we grouped in 16 cases.

Each of the cases (1)–(8) contains 4 subcases because $\alpha_1,\ \alpha_2,\ \alpha_3 \in (0,1)$, or $\alpha_1 = 1$ and $\alpha_2 = \alpha_3 = 0$, or $\alpha_2 = 1$ and $\alpha_1 = \alpha_3 = 0$, or $\alpha_3 = 1$ and $\alpha_1 = \alpha_2 = 0$.

Remark 5.1.5. In [98], the authors present only 15 cases (Theorems 2.16–2.30 from [98]) from 64 cases, namely the first 9 cases of our Theorem 5.1.2. They did not study the cases when some extreme limits are 0 and other are ∞. Besides, our intervals for parameters $\lambda,\ \mu,\ \nu$ presented in Theorem 5.1.2 (our cases (2)–(7) and (9)) are better than the corresponding ones from [98, Theorems 2.17–2.22 and 2.24–2.30].

Remark 5.1.6. One can formulate existence results for the general case of the system of n fractional differential equations (5.20) with the boundary conditions (5.21) from Remark 5.1.3. According to the values of $f_{j\infty}^s = \limsup_{u_1+\cdots+u_n \to \infty} \sup_{t\in[0,1]} \frac{f_j(t,u_1,\ldots,u_n)}{u_1+\cdots+u_n} \in [0,\infty)$, and $f_{j0}^i = \liminf_{u_1+\cdots+u_n \to 0} \inf_{t\in[\sigma,1]} \frac{f_j(t,u_1,\ldots,u_n)}{u_1+\cdots+u_n} \in (0,\infty]$, $j = 1,\ldots,n$ we have 2^{2n} cases, which can be grouped in 2^{n+1} cases.

5.1.3   Nonexistence of positive solutions

We present in this section intervals for λ, μ and ν for which there exist no positive solutions of problem (5.1), (5.2).

Theorem 5.1.3. *Assume that* (H1) *and* (H2) *hold. If there exist positive numbers* A_1, A_2, A_3 *such that*

$$f(t,u,v,w) \le A_1(u+v+w), \quad g(t,u,v,w) \le A_2(u+v+w),$$

$$h(t,u,v,w) \le A_3(u+v+w), \quad \forall t \in [0,1], \ u,v,w \ge 0, \qquad (5.30)$$

then there exist positive constants λ_0, μ_0, ν_0 *such that for every* $\lambda \in (0,\lambda_0)$, $\mu \in (0,\mu_0)$, $\nu \in (0,\nu_0)$ *the boundary value problem* (5.1), (5.2) *has no positive solution.*

Proof. We define $\lambda_0 = \frac{1}{3A_1 B}$, $\mu_0 = \frac{1}{3A_2 D}$, $\nu_0 = \frac{1}{3A_3 F}$, where $B = \int_0^1 J_1(s)\,ds$, $D = \int_0^1 J_2(s)\,ds$, $F = \int_0^1 J_3(s)\,ds$. We will show that for any $\lambda \in (0,\lambda_0)$, $\mu \in (0,\mu_0)$, $\nu \in (0,\nu_0)$, the problem (5.1), (5.2) has no positive solution.

Let $\lambda \in (0,\lambda_0)$, $\mu \in (0,\mu_0)$, $\nu \in (0,\nu_0)$. We suppose that the problem (5.1), (5.2) has a positive solution $(u(t),v(t),w(t))$, $t \in [0,1]$. Then, we have

$$u(t) = Q_1(u,v,w)(t) = \lambda \int_0^1 G_1(t,s)f(s,u(s),v(s),w(s))\,ds$$

$$\le \lambda \int_0^1 J_1(s)f(s,u(s),v(s),w(s))\,ds$$

$$\le \lambda A_1 \int_0^1 J_1(s)(u(s)+v(s)+w(s))\,ds$$

$$\le \lambda A_1(\|u\| + \|v\| + \|w\|) \int_0^1 J_1(s)\,ds$$

$$= \lambda A_1 B \|(u,v,w)\|_Y, \quad \forall t \in [0,1],$$

$$v(t) = Q_2(u,v,w)(t) = \mu \int_0^1 G_2(t,s)g(s,u(s),v(s),w(s))\,ds$$

$$\le \mu \int_0^1 J_2(s)g(s,u(s),v(s),w(s))\,ds$$

$$\le \mu A_2 \int_0^1 J_2(s)(u(s)+v(s)+w(s))\,ds$$

$$\leq \mu A_2(\|u\| + \|v\| + \|w\|) \int_0^1 J_2(s)\,ds$$

$$= \mu A_2 D \|(u,v,w)\|_Y, \quad \forall t \in [0,1],$$

$$w(t) = Q_3(u,v,w)(t) = \nu \int_0^1 G_3(t,s)h(s,u(s),v(s),w(s))\,ds$$

$$\leq \nu \int_0^1 J_3(s)h(s,u(s),v(s),w(s))\,ds$$

$$\leq \nu A_3 \int_0^1 J_3(s)(u(s) + v(s) + w(s))\,ds$$

$$\leq \nu A_3(\|u\| + \|v\| + \|w\|) \int_0^1 J_3(s)\,ds$$

$$= \nu A_3 F \|(u,v,w)\|_Y, \quad \forall t \in [0,1].$$

Therefore, we conclude

$$\|u\| \leq \lambda A_1 B \|(u,v,w)\|_Y < \lambda_0 A_1 B \|(u,v,w)\|_Y = \frac{1}{3}\|(u,v,w)\|_Y,$$

$$\|v\| \leq \mu A_2 D \|(u,v,w)\|_Y < \mu_0 A_2 D \|(u,v,w)\|_Y = \frac{1}{3}\|(u,v,w)\|_Y,$$

$$\|w\| \leq \nu A_3 F \|(u,v,w)\|_Y < \nu_0 A_3 F \|(u,v,w)\|_Y = \frac{1}{3}\|(u,v,w)\|_Y.$$

Hence, we deduce $\|(u,v,w)\|_Y = \|u\| + \|v\| + \|w\| < \|(u,v,w)\|_Y$, which is a contradiction. So, the boundary value problem (5.1), (5.2) has no positive solution. $\qquad\square$

Remark 5.1.7. In the proof of Theorem 5.1.3, we can also define $\lambda_0 = \frac{\alpha_1}{A_1 B}$, $\mu_0 = \frac{\alpha_2}{A_2 D}$, $\nu_0 = \frac{\alpha_3}{A_3 F}$ with $\alpha_1, \alpha_2, \alpha_3 > 0$ and $\alpha_1 + \alpha_2 + \alpha_3 = 1$.

Remark 5.1.8. If f_0^s, g_0^s, h_0^s, f_∞^s, g_∞^s, $h_\infty^s < \infty$, then there exist positive constants A_1, A_2, A_3 such that (5.30) holds, and then we obtain the conclusion of Theorem 5.1.3.

Theorem 5.1.4. *Assume that* (H1) *and* (H2) *hold. If there exist positive numbers* $\sigma \in (0,1)$ *and* $m_1 > 0$ *such that*

$$f(t,u,v,w) \geq m_1(u+v+w), \quad \forall t \in [\sigma,1], \ u, v, w \geq 0, \qquad (5.31)$$

then there exists a positive constant $\widetilde{\lambda}_0$ *such that for every* $\lambda > \widetilde{\lambda}_0$, $\mu > 0$ *and* $\nu > 0$, *the boundary value problem* (5.1), (5.2) *has no positive solution.*

Proof. We define $\widetilde{\lambda}_0 = \frac{1}{\theta\sigma^{\alpha-1}m_1 A}$ where $A = \int_\sigma^1 J_1(s)\,ds$. We will show that for every $\lambda > \widetilde{\lambda}_0$, $\mu > 0$ and $\nu > 0$, the problem (5.1), (5.2) has no positive solution.

Let $\lambda > \widetilde{\lambda}_0$, $\mu > 0$ and $\nu > 0$. We suppose that the problem (5.1), (5.2) has a positive solution $(u(t), v(t), w(t))$, $t \in [0, 1]$. Then we obtain

$$u(t) = Q_1(u, v, w)(t) = \lambda \int_0^1 G_1(t, s) f(s, u(s), v(s), w(s))\,ds$$

$$\geq \lambda t^{\alpha-1} \int_\sigma^1 J_1(s) f(s, u(s), v(s), w(s))\,ds$$

$$\geq \lambda \sigma^{\alpha-1} \int_\sigma^1 J_1(s) m_1(u(s) + v(s) + w(s))\,ds$$

$$\geq \lambda \theta \sigma^{\alpha-1} m_1 \int_\sigma^1 J_1(s)(\|u\| + \|v\| + \|w\|)\,ds$$

$$= \lambda \theta \sigma^{\alpha-1} m_1 A \|(u, v, w)\|_Y.$$

Therefore, we deduce

$$\|u\| \geq u(\sigma) \geq \lambda \theta \sigma^{\alpha-1} m_1 A \|(u, v, w)\|_Y > \widetilde{\lambda}_0 \theta \sigma^{\alpha-1} m_1 A \|(u, v, w)\|_Y$$

$$= \|(u, v, w)\|_Y,$$

and so, $\|(u, v, w)\|_Y = \|u\| + \|v\| + \|w\| > \|(u, v, w)\|_Y$, which is a contradiction. Therefore, the boundary value problem (5.1), (5.2) has no positive solution. $\qquad\square$

In a similar manner, we obtain the following theorems.

Theorem 5.1.5. *Assume that* (H1) *and* (H2) *hold. If there exist positive numbers* $\sigma \in (0, 1)$ *and* $m_2 > 0$ *such that*

$$g(t, u, v, w) \geq m_2(u + v + w), \quad \forall t \in [\sigma, 1], \ u, v, w \geq 0, \qquad (5.32)$$

then there exists a positive constant $\widetilde{\mu}_0$ *such that for every* $\lambda > 0$, $\mu > \widetilde{\mu}_0$ *and* $\nu > 0$, *the boundary value problem* (5.1), (5.2) *has no positive solution.*

In Theorem 5.1.5, we define $\widetilde{\mu}_0 = \frac{1}{\theta\sigma^{\beta-1}m_2 C}$, where $C = \int_\sigma^1 J_2(s)\,ds$.

Theorem 5.1.6. *Assume that* (H1) *and* (H2) *hold. If there exist positive numbers* $\sigma \in (0, 1)$ *and* $m_3 > 0$ *such that*

$$h(t, u, v, w) \geq m_3(u + v + w), \quad \forall t \in [\sigma, 1], \ u, v, w \geq 0, \qquad (5.33)$$

then there exists a positive constant $\widetilde{\nu}_0$ such that for every $\lambda > 0$, $\mu > 0$ and $\nu > \widetilde{\nu}_0$, the boundary value problem (5.1), (5.2) has no positive solution.

In Theorem 5.1.6, we define $\widetilde{\nu}_0 = \frac{1}{\theta \sigma^{\gamma-1} m_3 E}$, where $E = \int_\sigma^1 J_3(s)\, ds$.

Remark 5.1.9.

(a) If for $\sigma \in (0,1)$, f_0^i, $f_\infty^i > 0$ and $f(t,u,v,w) > 0$ for all $t \in [\sigma,1]$ and u, v, $w \geq 0$ with $u+v+w > 0$, then relation (5.31) holds and we obtain the conclusion of Theorem 5.1.4.

(b) If for $\sigma \in (0,1)$, g_0^i, $g_\infty^i > 0$ and $g(t,u,v,w) > 0$ for all $t \in [\sigma,1]$ and u, v, $w \geq 0$ with $u+v+w > 0$, then relation (5.32) holds and we obtain the conclusion of Theorem 5.1.5.

(c) If for $\sigma \in (0,1)$, h_0^i, $h_\infty^i > 0$ and $h(t,u,v,w) > 0$ for all $t \in [\sigma,1]$ and u, v, $w \geq 0$ with $u+v+w > 0$, then relation (5.33) holds and we obtain the conclusion of Theorem 5.1.6.

Theorem 5.1.7. *Assume that* (H1) *and* (H2) *hold. If there exist positive numbers* $\sigma \in (0,1)$ *and* m_1, $m_2 > 0$ *such that*

$$f(t,u,v,w) \geq m_1(u+v+w), \quad g(t,u,v,w) \geq m_2(u+v+w),$$
$$\forall t \in [\sigma,1], \ u, \ v, \ w \geq 0, \tag{5.34}$$

then there exist positive constants $\widetilde{\widetilde{\lambda}}_0$ *and* $\widetilde{\widetilde{\mu}}_0$ *such that for every* $\lambda > \widetilde{\widetilde{\lambda}}_0$, $\mu > \widetilde{\widetilde{\mu}}_0$ *and* $\nu > 0$, *the boundary value problem* (5.1), (5.2) *has no positive solution.*

Proof. We define $\widetilde{\widetilde{\lambda}}_0 = \frac{1}{2\theta \sigma^{\alpha-1} m_1 A} \left(= \frac{\widetilde{\lambda}_0}{2}\right)$ and $\widetilde{\widetilde{\mu}}_0 = \frac{1}{2\theta \sigma^{\beta-1} m_2 C} \left(= \frac{\widetilde{\mu}_0}{2}\right)$.

Then for every $\lambda > \widetilde{\widetilde{\lambda}}_0$, $\mu > \widetilde{\widetilde{\mu}}_0$ and $\nu > 0$, the problem (5.1), (5.2) has no positive solution. Indeed, let $\lambda > \widetilde{\widetilde{\lambda}}_0$, $\mu > \widetilde{\widetilde{\mu}}_0$ and $\nu > 0$. We suppose that the problem (5.1), (5.2) has a positive solution $(u(t), v(t), w(t))$, $t \in [0,1]$. Then in a similar manner as in the proof of Theorem 5.1.4, we deduce

$$\|u\| \geq \lambda \theta \sigma^{\alpha-1} m_1 A \|(u,v,w)\|_Y, \quad \|v\| \geq \mu \theta \sigma^{\beta-1} m_2 C \|(u,v,w)\|_Y,$$

and so

$$\|(u,v,w)\|_Y = \|u\| + \|v\| + \|w\| \geq \|u\| + \|v\|$$
$$\geq (\lambda \theta \sigma^{\alpha-1} m_1 A + \mu \theta \sigma^{\beta-1} m_2 C) \|(u,v,w)\|_Y$$
$$> (\widetilde{\widetilde{\lambda}}_0 \theta \sigma^{\alpha-1} m_1 A + \widetilde{\widetilde{\mu}}_0 \theta \sigma^{\beta-1} m_2 C) \|(u,v,w)\|_Y$$
$$= \left(\frac{1}{2} + \frac{1}{2}\right) \|(u,v,w)\|_Y = \|(u,v,w)\|_Y,$$

which is a contradiction. Therefore, the boundary value problem (5.1), (5.2) has no positive solution. □

Remark 5.1.10. In the proof of Theorem 5.1.7, we can also define $\widetilde{\lambda}_0 = \frac{\widetilde{\alpha}_1}{\theta\sigma^{\alpha-1}m_1A}$, $\widetilde{\mu}_0 = \frac{\widetilde{\alpha}_2}{\theta\sigma^{\beta-1}m_2C}$ with $\widetilde{\alpha}_1$, $\widetilde{\alpha}_2 > 0$ with $\widetilde{\alpha}_1 + \widetilde{\alpha}_2 = 1$.

In a similar manner, we obtain the following theorems.

Theorem 5.1.8. *Assume that* (H1) *and* (H2) *hold. If there exist positive numbers* $\sigma \in (0,1)$ *and* m_1, $m_3 > 0$ *such that*

$$f(t, u, v, w) \geq m_1(u + v + w), \quad h(t, u, v, w) \geq m_3(u + v + w),$$

$$\forall \in t \in [\sigma, 1], \ u, v, w \geq 0, \tag{5.35}$$

then there exist positive constants $\widetilde{\lambda}'_0$ *and* $\widetilde{\nu}'_0$ *such that for every* $\lambda > \widetilde{\lambda}'_0$, $\mu > 0$ *and* $\nu > \widetilde{\nu}'_0$, *the boundary value problem* (5.1), (5.2) *has no positive solution.*

In Theorem 5.1.8, we define $\widetilde{\lambda}'_0 = \frac{1}{2\theta\sigma^{\alpha-1}m_1A} \left(= \frac{\widetilde{\lambda}_0}{2} \right)$ and $\widetilde{\nu}'_0 = \frac{1}{2\theta\sigma^{\gamma-1}m_3E} \left(= \frac{\widetilde{\nu}_0}{2} \right)$, or in general $\widetilde{\lambda}'_0 = \frac{\widetilde{\alpha}_1}{\theta\sigma^{\alpha-1}m_1A}$ and $\widetilde{\nu}'_0 = \frac{\widetilde{\alpha}_2}{\theta\sigma^{\gamma-1}m_3E}$ with $\widetilde{\alpha}_1, \widetilde{\alpha}_2 > 0$, $\widetilde{\alpha}_1 + \widetilde{\alpha}_2 = 1$.

Theorem 5.1.9. *Assume that* (H1) *and* (H2) *hold. If there exist positive numbers* $\sigma \in (0,1)$ *and* m_2, $m_3 > 0$ *such that*

$$g(t, u, v, w) \geq m_2(u + v + w), \quad h(t, u, v, w) \geq m_3(u + v + w),$$

$$\forall \in t \in [\sigma, 1], \ u, v, w \geq 0, \tag{5.36}$$

then there exist positive constants $\widetilde{\mu}''_0$ *and* $\widetilde{\nu}''_0$ *such that for every* $\lambda > 0$, $\mu > \widetilde{\mu}''_0$ *and* $\nu > \widetilde{\nu}''_0$, *the boundary value problem* (5.1), (5.2) *has no positive solution.*

In Theorem 5.1.9, we define $\widetilde{\mu}''_0 = \frac{1}{2\theta\sigma^{\beta-1}m_2C} \left(= \frac{\widetilde{\mu}_0}{2} \right)$ and $\widetilde{\nu}''_0 = \frac{1}{2\theta\sigma^{\gamma-1}m_3E} \left(= \frac{\widetilde{\nu}_0}{2} \right)$, or in general $\widetilde{\mu}''_0 = \frac{\widetilde{\alpha}_1}{\theta\sigma^{\beta-1}m_2C}$ and $\widetilde{\nu}''_0 = \frac{\widetilde{\alpha}_2}{\theta\sigma^{\gamma-1}m_3E}$ with $\widetilde{\alpha}_1, \widetilde{\alpha}_2 > 0$, $\widetilde{\alpha}_1 + \widetilde{\alpha}_2 = 1$.

Remark 5.1.11.

(a) If for $\sigma \in (0,1)$, f_0^i, f_∞^i, g_0^i, $g_\infty^i > 0$ and $f(t,u,v,w) > 0$, $g(t,u,v,w) > 0$ for all $t \in [\sigma,1]$ and u, v, $w \geq 0$ with $u+v+w > 0$, then the relation (5.34) holds, and we obtain the conclusion of Theorem 5.1.7.

(b) If for $\sigma \in (0,1)$, f_0^i, f_∞^i, h_0^i, $h_\infty^i > 0$ and $f(t,u,v,w) > 0$, $h(t,u,v,w) > 0$ for all $t \in [\sigma,1]$ and u, v, $w \geq 0$ with $u+v+w > 0$, then the relation (5.35) holds, and we obtain the conclusion of Theorem 5.1.8.

(c) If for $\sigma \in (0,1)$, g_0^i, g_∞^i, h_0^i, $h_\infty^i > 0$ and $g(t,u,v,w) > 0$, $h(t,u,v,w) > 0$ for all $t \in [\sigma,1]$ and u, v, $w \geq 0$ with $u+v+w > 0$, then the relation (5.36) holds, and we obtain the conclusion of Theorem 5.1.9.

Theorem 5.1.10. *Assume that* (H1) *and* (H2) *hold. If there exist positive numbers* $\sigma \in (0,1)$ *and* m_1, m_2, $m_3 > 0$ *such that*

$$f(t,u,v,w) \geq m_1(u+v+w), \quad g(t,u,v,w) \geq m_2(u+v+w),$$

$$h(t,u,v,w) \geq m_3(u+v+w), \quad \forall t \in [\sigma,1], \ u, v, w \geq 0, \tag{5.37}$$

then there exist positive constants $\hat{\lambda}_0$, $\hat{\mu}_0$ *and* $\hat{\nu}_0$ *such that for every* $\lambda > \hat{\lambda}_0$, $\mu > \hat{\mu}_0$ *and* $\nu > \hat{\nu}_0$, *the boundary value problem* (5.1), (5.2) *has no positive solution.*

Proof. We define $\hat{\lambda}_0 = \frac{1}{3\theta\sigma^{\alpha-1}m_1 A}$, $\hat{\mu}_0 = \frac{1}{3\theta\sigma^{\beta-1}m_2 C}$, $\hat{\nu}_0 = \frac{1}{3\theta\sigma^{\gamma-1}m_3 E}$. Then for every $\lambda > \hat{\lambda}_0$, $\mu > \hat{\mu}_0$, $\nu > \hat{\nu}_0$, the problem (5.1), (5.2) has no positive solution. Indeed, let $\lambda > \hat{\lambda}_0$, $\mu > \hat{\mu}_0$ and $\nu > \hat{\nu}_0$. We suppose that the problem (5.1), (5.2) has a positive solution $(u(t), v(t), w(t))$, $t \in [0,1]$. Then in a similar manner as in the proof of Theorem 5.1.7, we deduce

$$\|u\| \geq \lambda\theta\sigma^{\alpha-1}m_1 A \|(u,v,w)\|_Y,$$

$$\|v\| \geq \mu\theta\sigma^{\beta-1}m_2 C \|(u,v,w)\|_Y,$$

$$\|w\| \geq \nu\theta\sigma^{\gamma-1}m_3 E \|(u,v,w)\|_Y,$$

and so

$$\|(u,v,w)\|_Y = \|u\| + \|v\| + \|w\|$$

$$\geq (\lambda\theta\sigma^{\alpha-1}m_1 A + \mu\theta\sigma^{\beta-1}m_2 C + \nu\theta\sigma^{\gamma-1}m_3 E)\|(u,v,w)\|_Y$$

$$> \left(\hat{\lambda}_0\theta\sigma^{\alpha-1}m_1 A + \hat{\mu}_0\theta\sigma^{\beta-1}m_2 C + \hat{\nu}_0\theta\sigma^{\gamma-1}m_3 E\right)$$

$$\times \|(u,v,w)\|_Y = \|(u,v,w)\|_Y,$$

which is a contradiction. Therefore, the boundary value problem (5.1), (5.2) has no positive solution. $\square$

Remark 5.1.12. In the proof of Theorem 5.1.10, we can also define $\hat{\lambda}_0 = \frac{\alpha'_1}{\theta \sigma^{\alpha-1} m_1 A}$, $\hat{\mu}_0 = \frac{\alpha'_2}{\theta \sigma^{\beta-1} m_2 C}$, $\hat{\nu}_0 = \frac{\alpha'_3}{\theta \sigma^{\gamma-1} m_3 F}$, where $\alpha'_1, \alpha'_2, \alpha'_3 > 0$ with $\alpha'_1 + \alpha'_2 + \alpha'_3 = 1$.

Remark 5.1.13. If for $\sigma \in (0,1)$, $f_0^i, f_\infty^i, g_0^i, g_\infty^i, h_0^i, h_\infty^i > 0$ and $f(t,u,v,w) > 0$, $g(t,u,v,w) > 0$, $h(t,u,v,w) > 0$ for all $t \in [\sigma, 1]$, $u, v, w \geq 0$, $u + v + w > 0$, then the relation (5.37) holds, and we have the conclusion of Theorem 5.1.10.

Remark 5.1.14. The conclusions of Theorems 5.1.1–5.1.10 remain valid for general systems of Hammerstein integral equations of the form

$$
\begin{cases}
u(t) = \lambda \displaystyle\int_0^1 G_1(t,s) f(s, u(s), v(s), w(s)) \, ds, & t \in [0,1], \\[2mm]
v(t) = \mu \displaystyle\int_0^1 G_2(t,s) g(s, u(s), v(s), w(s)) \, ds, & t \in [0,1], \qquad (5.38) \\[2mm]
w(t) = \nu \displaystyle\int_0^1 G_3(t,s) h(s, u(s), v(s), w(s)) \, ds, & t \in [0,1],
\end{cases}
$$

with positive parameters λ, μ, ν, and instead of assumptions (H1) − (H2), the following assumptions are satisfied

$(\widetilde{H1})$ The functions G_1, G_2, $G_3 : [0,1] \times [0,1] \to \mathbb{R}$ are continuous and there exist the continuous functions J_1, J_2, $J_3 : [0,1] \to \mathbb{R}$ and $\sigma \in (0,1)$, $\alpha, \beta, \gamma > 2$ such that

  (a) $0 \leq G_i(t,s) \leq J_i(s)$, $\forall t, s \in [0,1]$, $i = 1,2,3$;
  (b) $G_1(t,s) \geq t^{\alpha-1} J_1(s)$, $G_2(t,s) \geq t^{\beta-1} J_2(s)$, $G_3(t,s) \geq t^{\gamma-1} J_3(s)$, $\forall t, s \in [0,1]$;
  (c) $\displaystyle\int_\sigma^1 J_i(s) \, ds > 0$, $i = 1,2,3$.

$(\widetilde{H2})$ The functions f, g, $h : [0,1] \times \mathbb{R}_+ \times \mathbb{R}_+ \times \mathbb{R}_+ \to \mathbb{R}_+$ are continuous.

5.1.4 Examples

Let $n = 3$, $m = 5$, $l = 4$, $\alpha = \frac{5}{2}$, $\beta = \frac{17}{4}$, $\gamma = \frac{10}{3}$, $p_1 = 1$, $q_1 = \frac{1}{2}$, $p_2 = \frac{7}{3}$, $q_2 = \frac{3}{2}$, $p_3 = \frac{7}{4}$, $q_3 = \frac{2}{3}$, $N = 2$, $M = 1$, $L = 3$, $\xi_1 = \frac{1}{3}$, $\xi_2 = \frac{2}{3}$, $a_1 = 2$, $a_2 = \frac{1}{2}$, $\eta_1 = \frac{1}{2}$, $b_1 = 4$, $\zeta_1 = \frac{1}{4}$, $\zeta_2 = \frac{1}{2}$, $\zeta_3 = \frac{3}{4}$, $c_1 = 3$, $c_2 = 2$, $c_3 = 1$.

We consider the system of fractional differential equations

$$\begin{cases} D_{0+}^{5/2}u(t) + \lambda f(t, u(t), v(t), w(t)) = 0, & t \in (0,1), \\ D_{0+}^{17/4}v(t) + \mu g(t, u(t), v(t), w(t)) = 0, & t \in (0,1), \\ D_{0+}^{10/3}w(t) + \nu h(t, u(t), v(t), w(t)) = 0, & t \in (0,1), \end{cases} \tag{5.39}$$

with the multi-point boundary conditions

$$\begin{cases} u(0) = u'(0) = 0, \ u'(1) = 2D_{0+}^{1/2}u\left(\dfrac{1}{3}\right) + \dfrac{1}{2}D_{0+}^{1/2}u\left(\dfrac{2}{3}\right), \\ v(0) = v'(0) = v''(0) = v'''(0) = 0, \ D_{0+}^{7/3}v(1) = 4D_{0+}^{3/2}v\left(\dfrac{1}{2}\right), \\ w(0) = w'(0) = w''(0) = 0, \end{cases} \tag{5.40}$$

$$D_{0+}^{7/4}w(1) = 3D_{0+}^{2/3}w\left(\dfrac{1}{4}\right) + 2D_{0+}^{2/3}w\left(\dfrac{1}{2}\right) + D_{0+}^{2/3}w\left(\dfrac{3}{4}\right).$$

We have $\Delta_1 = \frac{6-3\sqrt{\pi}}{4} \approx 0.17065961 > 0$, $\Delta_2 = \frac{\Gamma(17/4)}{\Gamma(23/12)} - \frac{2^{1/4}\Gamma(17/4)}{\Gamma(11/4)} \approx 2.43672831 > 0$, $\Delta_3 = \frac{\Gamma(10/3)}{\Gamma(19/12)} - (3+2^{8/3}+3^{5/3})\frac{\Gamma(10/3)}{4^{5/3}\Gamma(8/3)} \approx 0.25945301 > 0$. So assumption (H1) is satisfied.

Besides, we deduce

$$g_1(t,s) = \frac{1}{\Gamma(5/2)} \begin{cases} t^{3/2}(1-s)^{1/2} - (t-s)^{3/2}, & 0 \le s \le t \le 1, \\ t^{3/2}(1-s)^{1/2}, & 0 \le t \le s \le 1, \end{cases}$$

$$g_2(t,s) = \begin{cases} t(1-s)^{1/2} - (t-s), & 0 \le s \le t \le 1, \\ t(1-s)^{1/2}, & 0 \le t \le s \le 1, \end{cases}$$

$$g_3(t,s) = \frac{1}{\Gamma(17/4)} \begin{cases} t^{13/4}(1-s)^{11/12} - (t-s)^{13/4}, & 0 \le s \le t \le 1, \\ t^{13/4}(1-s)^{11/12}, & 0 \le t \le s \le 1, \end{cases}$$

$$g_4(t,s) = \frac{1}{\Gamma(11/4)} \begin{cases} t^{7/4}(1-s)^{11/12} - (t-s)^{7/4}, & 0 \le s \le t \le 1, \\ t^{7/4}(1-s)^{11/12}, & 0 \le t \le s \le 1, \end{cases}$$

$$g_5(t,s) = \frac{1}{\Gamma(10/3)} \begin{cases} t^{7/3}(1-s)^{7/12} - (t-s)^{7/3}, & 0 \le s \le t \le 1, \\ t^{7/3}(1-s)^{7/12}, & 0 \le t \le s \le 1, \end{cases}$$

$$g_6(t,s) = \frac{1}{\Gamma(8/3)} \begin{cases} t^{5/3}(1-s)^{7/12} - (t-s)^{5/3}, & 0 \le s \le t \le 1, \\ t^{5/3}(1-s)^{7/12}, & 0 \le t \le s \le 1. \end{cases}$$

Then we obtain

$$G_1(t, s) = g_1(t, s) + \frac{t^{3/2}}{\Delta_1} \left(2g_2 \left(\frac{1}{3}, s \right) + \frac{1}{2}g_2 \left(\frac{2}{3}, s \right) \right),$$

$$G_2(t, s) = g_3(t, s) + \frac{4t^{13/4}}{\Delta_2} g_4 \left(\frac{1}{2}, s \right),$$

$$G_3(t, s) = g_5(t, s) + \frac{t^{7/3}}{\Delta_3} \left(3g_6 \left(\frac{1}{4}, s \right) + 2g_6 \left(\frac{1}{2}, s \right) + g_6 \left(\frac{3}{4}, s \right) \right),$$

$$h_1(s) = \frac{4}{3\sqrt{\pi}} s(1-s)^{1/2}, \quad h_3(s) = \frac{1}{\Gamma(17/4)} (1-s)^{11/12}(1-(1-s)^{7/3}),$$

$$h_5(s) = \frac{1}{\Gamma(10/3)} (1-s)^{7/12}(1-(1-s)^{7/4}),$$

$$J_1(s) = \frac{4}{3\sqrt{\pi}} s(1-s)^{1/2} + \frac{1}{\Delta_1} \left(2g_2 \left(\frac{1}{3}, s \right) + \frac{1}{2}g_2 \left(\frac{2}{3}, s \right) \right)$$

$$= \begin{cases} \dfrac{4}{3\sqrt{\pi}} s(1-s)^{1/2} + \dfrac{1}{2\Delta_1} \left[2(1-s)^{1/2} + 5s - 2 \right], & 0 \le s < \dfrac{1}{3}, \\[2mm] \dfrac{4}{3\sqrt{\pi}} s(1-s)^{1/2} + \dfrac{1}{6\Delta_1} \left[6(1-s)^{1/2} + 3s - 2 \right], & \dfrac{1}{3} \le s < \dfrac{2}{3}, \\[2mm] \dfrac{4}{3\sqrt{\pi}} s(1-s)^{1/2} + \dfrac{1}{\Delta_1} (1-s)^{1/2}, & \dfrac{2}{3} \le s \le 1. \end{cases}$$

$$J_2(s) = \frac{1}{\Gamma(17/4)} (1-s)^{11/12}(1-(1-s)^{7/3}) + \frac{4}{\Delta_2} g_4 \left(\frac{1}{2}, s \right)$$

$$= \begin{cases} \dfrac{1}{\Gamma(17/4)} (1-s)^{11/12}(1-(1-s)^{7/3}) \\[2mm] \quad + \dfrac{2^{1/4}}{\Delta_2 \Gamma(11/4)} [(1-s)^{11/12} - (1-2s)^{7/4}], & 0 \le s < \dfrac{1}{2}, \\[3mm] \dfrac{1}{\Gamma(17/4)} (1-s)^{11/12}(1-(1-s)^{7/3}) \\[2mm] \quad + \dfrac{2^{1/4}}{\Delta_2 \Gamma(11/4)} (1-s)^{11/12}, & \dfrac{1}{2} \le s \le 1. \end{cases}$$

$$J_3(s) = \frac{1}{\Gamma(10/3)} (1-s)^{7/12}(1-(1-s)^{7/4})$$

$$+ \frac{1}{\Delta_3} \left(3g_6 \left(\frac{1}{4}, s \right) + 2g_6 \left(\frac{1}{2}, s \right) + g_6 \left(\frac{3}{4}, s \right) \right)$$

$$
= \begin{cases}
\dfrac{1}{\Gamma(10/3)}(1-s)^{7/12}(1-(1-s)^{7/4}) \\[2mm]
\quad + \dfrac{1}{2^{10/3}\Delta_3\Gamma(8/3)}\Big[(3+2^{8/3}+3^{5/3})(1-s)^{7/12} \\[2mm]
\quad - 3(1-4s)^{5/3} - 2^{8/3}(1-2s)^{5/3} - (3-4s)^{5/3}\Big], \ 0 \le s < \dfrac{1}{4}, \\[4mm]
\dfrac{1}{\Gamma(10/3)}(1-s)^{7/12}(1-(1-s)^{7/4}) \\[2mm]
\quad + \dfrac{1}{2^{10/3}\Delta_3\Gamma(8/3)}\Big[(3+2^{8/3}+3^{5/3})(1-s)^{7/12} \\[2mm]
\quad - 2^{8/3}(1-2s)^{5/3} - (3-4s)^{5/3}\Big], \ \dfrac{1}{4} \le s < \dfrac{1}{2}, \\[4mm]
\dfrac{1}{\Gamma(10/3)}(1-s)^{7/12}(1-(1-s)^{7/4}) \\[2mm]
\quad + \dfrac{1}{2^{10/3}\Delta_3\Gamma(8/3)}\Big[(3+2^{8/3}+3^{5/3})(1-s)^{7/12} \\[2mm]
\quad - (3-4s)^{5/3}\Big], \ \dfrac{1}{2} \le s < \dfrac{3}{4}, \\[4mm]
\dfrac{1}{\Gamma(10/3)}(1-s)^{7/12}(1-(1-s)^{7/4}) \\[2mm]
\quad + \dfrac{1}{2^{10/3}\Delta_3\Gamma(8/3)}(3+2^{8/3}+3^{5/3})(1-s)^{7/12}, \ \dfrac{3}{4} \le s \le 1.
\end{cases}
$$

Now, we choose $\sigma = \frac{1}{4} \in (0,1)$ and then $\theta = 2^{-13/2} \approx 0.01104854$. We also obtain $A = \int_{1/4}^{1} J_1(s)\,ds \approx 2.42142749$, $B = \int_0^1 J_1(s)\,ds \approx 2.80487506$, $C = \int_{1/4}^{1} J_2(s)\,ds \approx 0.11093116$, $D = \int_0^1 J_2(s)\,ds \approx 0.13771787$, $E = \int_{1/4}^{1} J_3(s)\,ds \approx 1.49070723$, $F = \int_0^1 J_3(s)\,ds \approx 1.80167568$.

Example 1. We consider the functions

$$
f(t,u,v,w) = \frac{(2t+1)[\widetilde{p}_1(u+v+w)+1](u+v+w)(\widetilde{q}_1+\sin v)}{u+v+w+1},
$$

$$
g(t,u,v,w) = \frac{\sqrt{t+1}[\widetilde{p}_2(u+v+w)+1](u+v+w)(\widetilde{q}_2+\cos w)}{u+v+w+1},
$$

$$
h(t,u,v,w) = \frac{t^2[\widetilde{p}_3(u+v+w)+1](u+v+w)(\widetilde{q}_3+\sin u)}{u+v+w+1},
$$

for $t \in [0,1]$, $u, v, w \ge 0$, where $\widetilde{p}_1, \widetilde{p}_2, \widetilde{p}_3 > 0$, $\widetilde{q}_1, \widetilde{q}_2, \widetilde{q}_3 > 1$.

We have $f_0^s = 3\tilde{q}_1$, $g_0^s = \sqrt{2}(\tilde{q}_2 + 1)$, $h_0^s = \tilde{q}_3$, $f_\infty^i = \frac{3}{2}\tilde{p}_1(\tilde{q}_1 - 1)$, $g_\infty^i = \frac{\sqrt{5}}{2}\tilde{p}_2(\tilde{q}_2 - 1)$, $h_\infty^i = \frac{1}{16}\tilde{p}_3(\tilde{q}_3 - 1)$. For $\alpha_1 = \alpha_2 = \alpha_3 = \tilde{\alpha}_1 = \tilde{\alpha}_2 = \tilde{\alpha}_3 = \frac{1}{3}$, we obtain $L_1 = \frac{2^{21/2}}{9\tilde{p}_1(\tilde{q}_1 - 1)A}$, $L_2 = \frac{1}{9\tilde{q}_1 B}$, $L_3 = \frac{2^{14}}{3\sqrt{5}\tilde{p}_2(\tilde{q}_2 - 1)C}$, $L_4 = \frac{1}{3\sqrt{2}(\tilde{q}_2 + 1)D}$, $L_5 = \frac{2^{91/6}}{3\tilde{p}_3(\tilde{q}_3 - 1)E}$, and $L_6 = \frac{1}{3\tilde{q}_3 F}$.

The conditions $L_1 < L_2$, $L_3 < L_4$ and $L_5 < L_6$ become

$$\frac{\tilde{p}_1(\tilde{q}_1 - 1)}{\tilde{q}_1} > \frac{2^{21/2}B}{A}, \quad \frac{\tilde{p}_2(\tilde{q}_2 - 1)}{\tilde{q}_2 + 1} > \frac{2^{29/2}D}{5^{1/2}C}, \quad \frac{\tilde{p}_3(\tilde{q}_3 - 1)}{\tilde{q}_3} > \frac{2^{91/6}F}{E}.$$

For example, if $\frac{\tilde{p}_1(\tilde{q}_1 - 1)}{\tilde{q}_1} \geq 1678$, $\frac{\tilde{p}_2(\tilde{q}_2 - 1)}{\tilde{q}_2 + 1} \geq 12865$ and $\frac{\tilde{p}_3(\tilde{q}_3 - 1)}{\tilde{q}_3} \geq 44454$, then the above conditions are satisfied.

As an example, we consider $\tilde{q}_1 = 2$, $\tilde{q}_2 = 3$, $\tilde{q}_3 = 4$, $\tilde{p}_1 = 3356$, $\tilde{p}_2 = 25730$, $\tilde{p}_3 = 59272$, and then the inequalities $L_1 < L_2$, $L_3 < L_4$ and $L_5 < L_6$ are satisfied. In this case, $L_1 \approx 0.01980063$, $L_2 \approx 0.01980678$, $L_3 \approx 0.42784885$, $L_4 \approx 0.4278716$, $L_5 \approx 0.04625271$, $L_6 \approx 0.04625324$. By Theorem 5.1.1 (1) we deduce that for every $\lambda \in (L_1, L_2)$, $\mu \in (L_3, L_4)$ and $\nu \in (L_5, L_6)$ there exists a positive solution $(u(t), v(t), w(t))$, $t \in [0, 1]$ of problem (5.39), (5.40).

Because $f_0^s = 3\tilde{q}_1$, $f_\infty^s = 3\tilde{p}_1(\tilde{q}_1 + 1)$, $g_0^s = \sqrt{2}(\tilde{q}_2 + 1)$, $g_\infty^s = \sqrt{2}\tilde{p}_2(\tilde{q}_2 + 1)$, $h_0^s = \tilde{q}_3$, $h_\infty^s = \tilde{p}_3(\tilde{q}_3 + 1)$, then by Theorem 5.1.3 and Remark 5.1.8, we conclude that for any $\lambda \in (0, \lambda_0)$, $\mu \in (0, \mu_0)$ and $\nu \in (0, \nu_0)$, the problem (5.39), (5.40) has no positive solution, where $\lambda_0 = \frac{1}{3A_1 B}$, $\mu_0 = \frac{1}{3A_2 D}$, $\nu_0 = \frac{1}{3A_3 F}$. If we consider as above $\tilde{p}_1 = 3356$, $\tilde{q}_1 = 2$, $\tilde{p}_2 = 36386$, $\tilde{q}_2 = 3$, $\tilde{p}_3 = 59272$, $\tilde{q}_3 = 4$, then $A_1 = 30204$, $A_2 = 102920\sqrt{2} \approx 145551$, $A_3 = 296360$. Therefore we obtain $\lambda_0 \approx 3.9346 \times 10^{-6}$, $\mu_0 \approx 1.6629 \times 10^{-5}$, $\nu_0 \approx 6.24284 \times 10^{-7}$.

Because f_0^i, f_∞^i, g_0^i, g_∞^i, h_0^i, $h_\infty^i > 0$ and $f(t, u, v, w) > 0$, $g(t, u, v, w) > 0$, $h(t, u, v, w) > 0$ for all $t \in [1/4, 1]$ and u, v, $w \geq 0$ with $u + v + w > 0$, we can also apply Theorem 5.1.10 and Remark 5.1.13. Here, $\hat{\lambda}_0 = \frac{1}{3\theta\sigma^{\alpha - 1}m_1 A}$, $\hat{\mu}_0 = \frac{1}{3\theta\sigma^{\beta - 1}m_2 C}$ and $\hat{\nu}_0 = \frac{1}{3\theta\sigma^{\gamma - 1}m_3 E}$. For the functions f, g, h presented above, we have $m_1 = 3$, $m_2 = 2\sqrt{5}$, $m_3 = \frac{1}{4}$, $\hat{\lambda}_0 \approx 33.22545838$, $\hat{\mu}_0 \approx 5504.275396$, $\hat{\nu}_0 \approx 2056.117822$. So, if $\lambda > 33.23$, $\mu > 5504.28$ and $\nu > 2056.12$, the problem (5.39), (5.40) has no positive solution.

Example 2. We consider the functions

$$f(t, u, v, w) = t^a(u^2 + v^2 + w^2), \quad g(t, u, v, w) = (2 - t)^b(e^{u + v + w} - 1),$$

$$h(t, u, v, w) = (u + v + w)^c, \quad t \in [0, 1], \ u, v, w \geq 0,$$

where a, $b > 0$, $c > 1$. We have $f_0^s = 0$, $f_\infty^i = \infty$, $g_0^s = 2^b$, $g_\infty^i = \infty$, $h_0^s = 0$, $h_\infty^i = \infty$.

By Theorem 5.1.1 (14), for any $\lambda \in (0, \infty)$, $\mu \in (0, \widetilde{L}_4)$ and $\nu \in (0, \infty)$, with $\widetilde{L}_4 = \frac{1}{2^b D}$, the problem (5.39),(5.40) has a positive solution. Here, $D = \int_0^1 J_2(s)\, ds \approx 0.13771787$. For example, if $b = 2$, we obtain $\widetilde{L}_4 = \frac{1}{4D} \approx 1.8153054$.

We can also use Theorem 5.1.5, because $g(t, u, v, w) \geq u + v + w$ for all $t \in [1/4, 1]$ and u, v, $w \geq 0$, that is $m_2 = 1$. Because $\widetilde{\mu}_0 = \frac{1}{\theta \sigma^{\beta - 1} m_2 C} \approx 73847.6037$, we deduce that for every $\lambda > 0$, $\mu > 73847.61$ and $\nu > 0$, the boundary value problem (5.39), (5.40) has no positive solution.

Remark 5.1.15. The results presented in this section under the assumptions $p_1 \in [1, n-2]$, $p_2 \in [1, m-2]$ and $p_3 \in [1, l-2]$ instead of $p_1 \in [1, \alpha - 1)$, $p_2 \in [1, \beta - 1)$ and $p_3 \in [1, \gamma - 1)$ were published in [77].

Chapter 6

Existence of Solutions for Riemann–Liouville Fractional Boundary Value Problems

In this chapter, we investigate the existence of solutions for Riemann–Liouville fractional differential equations and systems of Riemann–Liouville fractional differential equations with integral terms, subject to nonlocal boundary conditions which contain fractional derivatives and Riemann–Stieltjes integrals.

6.1 Riemann–Liouville Fractional Differential Equations with Nonlocal Boundary Conditions

We consider the nonlinear fractional differential equation

$$D_{0+}^{\alpha} u(t) + f(t, u(t), I_{0+}^{p} u(t))) = 0, \quad t \in (0, 1), \tag{6.1}$$

with the nonlocal boundary conditions

$$\begin{cases} u(0) = u'(0) = \cdots = u^{(n-2)}(0) = 0, \\ D_{0+}^{\beta_0} u(1) = \sum_{i=1}^{m} \int_0^1 D_{0+}^{\beta_i} u(t) \, dH_i(t), \end{cases} \tag{6.2}$$

where $\alpha \in \mathbb{R}$, $\alpha \in (n-1, n]$, $n, m \in \mathbb{N}$, $n \geq 2$, $\beta_i \in \mathbb{R}$ for all $i = 0, \ldots, m$, $0 \leq \beta_1 < \beta_2 < \cdots < \beta_m < \alpha - 1$, $\beta_0 \in [0, \alpha - 1)$, D_{0+}^{k} denotes the Riemann–Liouville derivative of order k (for $k = \alpha, \beta_0, \beta_1, \ldots, \beta_m$), $p > 0$, I_{0+}^{p} is the Riemann–Liouville integral of order p, f is a nonlinear function, and the

279

integrals from the boundary condition (6.2) are Riemann–Stieltjes integrals with H_i, $i = 1, \dots, m$ functions of bounded variation.

We present conditions for the nonlinearity f such that problem (6.1), (6.2) has at least one solution $u \in C[0,1]$. In the proofs of the main existence results, we use some theorems from the fixed point theory.

6.1.1   *Preliminary results*

We consider the fractional differential equation

$$D_{0+}^{\alpha} u(t) + y(t) = 0, \quad t \in (0,1), \tag{6.3}$$

with the boundary conditions (6.2), where $y \in C(0,1) \cap L^1(0,1)$. We denote by

$$\Delta_0 = \frac{\Gamma(\alpha)}{\Gamma(\alpha - \beta_0)} - \sum_{i=1}^{m} \frac{\Gamma(\alpha)}{\Gamma(\alpha - \beta_i)} \int_0^1 s^{\alpha - \beta_i - 1} \, dH_i(s).$$

In a similar manner as we proved Lemma 2.4.1 from Section 2.4, we obtain the following lemma.

Lemma 6.1.1. *If $\Delta_0 \neq 0$, then the unique solution $u \in C[0,1]$ of problem (6.2), (6.3) is given by*

$$
\begin{aligned}
u(t) = &-\frac{1}{\Gamma(\alpha)} \int_0^t (t-s)^{\alpha-1} y(s) \, ds \\
&+ \frac{t^{\alpha-1}}{\Delta_0 \Gamma(\alpha - \beta_0)} \int_0^1 (1-s)^{\alpha - \beta_0 - 1} y(s) \, ds \\
&- \frac{t^{\alpha-1}}{\Delta_0} \sum_{i=1}^{m} \frac{1}{\Gamma(\alpha - \beta_i)} \int_0^1 \left(\int_0^s (s-\tau)^{\alpha - \beta_i - 1} y(\tau) \, d\tau \right) dH_i(s),
\end{aligned}
\tag{6.4}
$$

$$t \in [0,1].$$

By using standard computations, we obtain the following result.

Lemma 6.1.2. *If $x \in C[0,1]$ then for $\theta > 0$, we have*

$$|I_{0+}^{\theta} x(t)| \leq \frac{\|x\|}{\Gamma(\theta + 1)}, \quad \forall t \in [0,1],$$

where $\|x\| = \sup_{t \in [0,1]} |x(t)|$.

6.1.2 Existence of solutions

We introduce firstly the assumptions that we will use in our main existence theorems for problem (6.1), (6.2).

(H1) $\alpha \in \mathbb{R}$, $\alpha \in (n-1, n]$, $n, m \in \mathbb{N}$, $n \geq 2$, $\beta_i \in \mathbb{R}$ for all $i = 0, \ldots, m$, $0 \leq \beta_1 < \beta_2 < \cdots < \beta_m < \alpha - 1$, $\beta_0 \in [0, \alpha - 1)$, $p > 0$, $H_i :$ $[0, 1] \to \mathbb{R}$, $i = 1, \ldots, m$ are functions of bounded variation, and $\Delta_0 = \frac{\Gamma(\alpha)}{\Gamma(\alpha - \beta_0)} - \sum_{i=1}^{m} \frac{\Gamma(\alpha)}{\Gamma(\alpha - \beta_i)} \int_0^1 s^{\alpha - \beta_i - 1} \, dH_i(s) \neq 0$.

(H2) The function $f : [0, 1] \times \mathbb{R}^2 \to \mathbb{R}$ is continuous and there exists $L_1 > 0$ such that
$$|f(t, x, y) - f(t, x_1, y_1)| \leq L_1(|x - x_1| + |y - y_1|),$$
for all $t \in [0, 1]$, $x, y, x_1, y_1 \in \mathbb{R}$.

(H3) There exists a function $g \in C([0, 1], [0, \infty))$ such that
$$|f(t, x, y)| \leq g(t), \quad \forall (t, x, y) \in [0, 1] \times \mathbb{R}^2.$$

(H4) The function $f : [0, 1] \times \mathbb{R}^2 \to \mathbb{R}$ is continuous and there exist real constants $a_0 > 0$, $a_1 \geq 0$, $a_2 \geq 0$ such that
$$|f(t, x, y)| \leq a_0 + a_1|x| + a_2|y|, \quad \forall t \in [0, 1], \ x, y \in \mathbb{R}.$$

(H5) The function $f : [0, 1] \times \mathbb{R}^2 \to \mathbb{R}$ is continuous and there exist the constants $b_0, b_1, b_2 \geq 0$ with at least one nonzero, and $l_1, l_2 \in (0, 1)$ such that
$$|f(t, x, y)| \leq b_0 + b_1|x|^{l_1} + b_2|y|^{l_2}, \quad \forall t \in [0, 1], \ x, y \in \mathbb{R}.$$

(H6) The function $f : [0, 1] \times \mathbb{R}^2 \to \mathbb{R}$ is continuous and there exist $c_0, c_1, c_2 \geq 0$ with at least one nonzero, and nondecreasing functions $h_1, h_2 \in C([0, \infty), [0, \infty))$ such that
$$|f(t, x, y)| \leq c_0 + c_1 h_1(|x|) + c_2 h_2(|y|), \quad \forall t \in [0, 1], \ x, y \in \mathbb{R}.$$

We denote by

$$M_1 = \frac{1}{\Gamma(\alpha + 1)} + \frac{1}{|\Delta_0|\Gamma(\alpha - \beta_0 + 1)}$$
$$+ \frac{1}{|\Delta_0|} \sum_{i=1}^{m} \frac{1}{\Gamma(\alpha - \beta_i + 1)} \left| \int_0^1 s^{\alpha - \beta_i} \, dH_i(s) \right|, \tag{6.5}$$

$$M_2 = M_1 - \frac{1}{\Gamma(\alpha + 1)}, \quad L_0 = 1 + \frac{1}{\Gamma(p + 1)}.$$

We consider the Banach space $X = C([0, 1])$ with the supremum norm $\|u\| = \sup_{t \in [0,1]} |u(t)|$, and define the operator $A : X \to X$ by

$$
(Au)(t) = -\frac{1}{\Gamma(\alpha)} \int_0^t (t - s)^{\alpha - 1} f(s, u(s), I_{0+}^p u(s))\, ds
$$

$$
+ \frac{t^{\alpha - 1}}{\Delta_0 \Gamma(\alpha - \beta_0)} \int_0^1 (1 - s)^{\alpha - \beta_0 - 1} f(s, u(s), I_{0+}^p u(s))\, ds
$$

$$
- \frac{t^{\alpha - 1}}{\Delta_0} \sum_{i=1}^m \frac{1}{\Gamma(\alpha - \beta_i)} \tag{6.6}
$$

$$
\times \int_0^1 \left(\int_0^s (s - \tau)^{\alpha - \beta_i - 1} f(\tau, u(\tau), I_{0+}^p u(\tau))\, d\tau \right) dH_i(s),
$$

$$
t \in [0, 1].
$$

By using Lemma 6.1.1, we see that u is a fixed point of operator A if and only if u is a solution of problem (6.1), (6.2). Therefore, next, we will investigate the existence of fixed points of operator A.

Theorem 6.1.1. *Assume that* (H1) *and* (H2) *hold. If* $\Xi := L_1 L_0 M_1 < 1$, *then problem* (6.1), (6.2) *has a unique solution on* $[0, 1]$, *where* L_0 *and* M_1 *are given by* (6.5).

Proof. Let us fix $r > 0$ such that $r \geq M_0 M_1 (1 - L_1 L_0 M_1)^{-1}$, where $M_0 = \sup_{t \in [0,1]} |f(t, 0, 0)|$. We consider the set $\overline{B}_r = \{u \in X, \|u\|_X \leq r\}$ and we show firstly that $A(\overline{B}_r) \subset \overline{B}_r$. Let $u \in \overline{B}_r$. By using (H2) and Lemma 6.1.2, for $f(t, u(t), I_{0+}^p u(t))$, we obtain the following inequalities

$$
|f(t, u(t), I_{0+}^p u(t))| \leq |f(t, u(t), I_{0+}^p u(t)) - f(t, 0, 0)| + |f(t, 0, 0)|
$$

$$
\leq L_1 (|u(t)| + |I_{0+}^p u(t)|)
$$

$$
+ M_0 \leq L_1 \left(\|u\| + \frac{1}{\Gamma(p+1)} \|u\| \right) + M_0
$$

$$
= L_1 \left(1 + \frac{1}{\Gamma(p+1)} \right) \|u\|
$$

$$
+ M_0 \leq L_1 L_0 r + M_0, \quad \forall t \in [0, 1].
$$

Then by the definition of operator A from (6.6), we deduce

$$|(Au)(t)| \leq \frac{1}{\Gamma(\alpha)} \int_0^t (t-s)^{\alpha-1} (L_1 L_0 r + M_0) \, ds$$

$$+ \frac{t^{\alpha-1}}{|\Delta_0| \Gamma(\alpha - \beta_0)} \int_0^1 (1-s)^{\alpha-\beta_0-1} (L_1 L_0 r + M_0) \, ds$$

$$+ \frac{t^{\alpha-1}}{|\Delta_0|} \sum_{i=1}^m \frac{1}{\Gamma(\alpha - \beta_i)}$$

$$\times \left| \int_0^1 \left(\int_0^s (s-\tau)^{\alpha-\beta_i-1} (L_1 L_0 r + M_0) \, d\tau \right) dH_i(s) \right|$$

$$= (L_1 L_0 r + M_0) \left\{ \frac{t^\alpha}{\Gamma(\alpha+1)} + \frac{t^{\alpha-1}}{|\Delta_0| \Gamma(\alpha - \beta_0 + 1)} + \frac{t^{\alpha-1}}{|\Delta_0|} \right.$$

$$\left. \times \sum_{i=1}^m \frac{1}{\Gamma(\alpha - \beta_i + 1)} \left| \int_0^1 s^{\alpha-\beta_i} \, dH_i(s) \right| \right\}, \quad \forall t \in [0,1].$$

Therefore, we conclude

$$\|Au\| \leq (L_1 L_0 r + M_0) \left[\frac{1}{\Gamma(\alpha+1)} + \frac{1}{|\Delta_0| \Gamma(\alpha - \beta_0 + 1)} \right.$$

$$\left. + \frac{1}{|\Delta_0|} \sum_{i=1}^m \frac{1}{\Gamma(\alpha - \beta_i + 1)} \left| \int_0^1 s^{\alpha-\beta_i} \, dH_i(s) \right| \right]$$

$$= (L_1 L_0 r + M_0) M_1 \leq r.$$

So, we deduce that A maps $\overline{B}_r$ into itself.

Now, for $u, v \in \overline{B}_r$ we have

$$|(Au)(t) - (Av)(t)| \leq \left| -\frac{1}{\Gamma(\alpha)} \int_0^t (t-s)^{\alpha-1} f(s, u(s), I_{0+}^p u(s)) \, ds \right.$$

$$+ \frac{1}{\Gamma(\alpha)} \int_0^t (t-s)^{\alpha-1} f(s, v(s), I_{0+}^p v(s)) \, ds \left| + \frac{t^{\alpha-1}}{|\Delta_0| \Gamma(\alpha - \beta_0)} \right.$$

$$\times \int_0^1 (1-s)^{\alpha-\beta_0-1} |f(s, u(s), I_{0+}^p u(s)) - f(s, v(s), I_{0+}^p v(s))| \, ds$$

$$+ \frac{t^{\alpha-1}}{|\Delta_0|} \sum_{i=1}^m \frac{1}{\Gamma(\alpha - \beta_i)} \left| \int_0^1 \left(\int_0^s (s-\tau)^{\alpha-\beta_i-1} |f(\tau, u(\tau), I_{0+}^p u(\tau)) \right. \right.$$

$$\left. \left. - f(\tau, v(\tau), I_{0+}^p v(\tau))| \, d\tau \right) dH_i(s) \right|$$

$$\leq \frac{L_1}{\Gamma(\alpha)} \int_0^t (t-s)^{\alpha-1} [|u(s) - v(s)|$$

$$+ |I_{0+}^p u(s) - I_{0+}^p v(s)|] \, ds + \frac{t^{\alpha-1} L_1}{|\Delta_0| \Gamma(\alpha - \beta_0)}$$

$$\times \int_0^1 (1-s)^{\alpha-\beta_0-1} [|u(s) - v(s)| + |I_{0+}^p u(s) - I_{0+}^p v(s)|] \, ds$$

$$+ \frac{t^{\alpha-1} L_1}{|\Delta_0|} \sum_{i=1}^m \frac{1}{\Gamma(\alpha - \beta_i)} \left| \int_0^1 \left(\int_0^s (s-\tau)^{\alpha-\beta_i-1} [|u(\tau) - v(\tau)| \right. \right.$$

$$\left. \left. + |I_{0+}^p u(\tau) - I_{0+}^p v(\tau)|] \, d\tau \right) dH_i(s) \right|$$

$$\leq \frac{L_1}{\Gamma(\alpha)} \int_0^t (t-s)^{\alpha-1} \left[\|u-v\| + \frac{1}{\Gamma(p+1)} \|u-v\| \right] ds$$

$$+ \frac{t^{\alpha-1} L_1}{|\Delta_0| \Gamma(\alpha - \beta_0)} \int_0^1 (1-s)^{\alpha-\beta_0-1} \left[\|u-v\| + \frac{1}{\Gamma(p+1)} \|u-v\| \right] ds$$

$$+ \frac{t^{\alpha-1} L_1}{|\Delta_0|} \sum_{i=1}^m \frac{1}{\Gamma(\alpha - \beta_i)} \left| \int_0^1 \left(\int_0^s (s-\tau)^{\alpha-\beta_i-1} \right. \right.$$

$$\left. \left. \times \left[\|u-v\| + \frac{1}{\Gamma(p+1)} \|u-v\| \right] d\tau \right) dH_i(s) \right|$$

$$= L_1 L_0 \|u-v\| \left[\frac{t^\alpha}{\Gamma(\alpha+1)} + \frac{t^{\alpha-1}}{|\Delta_0| \Gamma(\alpha - \beta_0 + 1)} \right.$$

$$\left. + \frac{t^{\alpha-1}}{|\Delta_0|} \sum_{i=1}^m \frac{1}{\Gamma(\alpha - \beta_i + 1)} \left| \int_0^1 s^{\alpha-\beta_i} dH_i(s) \right| \right], \quad \forall t \in [0, 1].$$

Hence, we obtain

$$\|Au - Av\| \leq L_1 L_0 \|u-v\| \left(\frac{1}{\Gamma(\alpha+1)} + \frac{1}{|\Delta_0| \Gamma(\alpha - \beta_0 + 1)} \right.$$

$$\left. + \frac{1}{|\Delta_0|} \sum_{i=1}^m \frac{1}{\Gamma(\alpha - \beta_i + 1)} \left| \int_0^1 s^{\alpha-\beta_i} dH_i(s) \right| \right)$$

$$= \Xi \|u-v\|.$$

By using the condition $\Xi < 1$, we deduce that operator A is a contraction. Then by Theorem 1.2.1, we conclude that operator A has a unique fixed point $u \in \overline{B}_r$, which is the unique solution of problem (6.1), (6.2) on $[0, 1]$. $\square$

Theorem 6.1.2. *Assume that* (H1)–(H3) *hold. If* $\Xi_1 := L_1 L_0 \frac{1}{\Gamma(\alpha+1)} < 1$, *then problem* (6.1), (6.2) *has at least one solution on* $[0, 1]$.

Proof. Let us fix $r_1 > 0$ such that $r_1 \geq M_1 \|g\|$. We consider the set $\overline{B}_{r_1} = \{u \in X, \|u\| \leq r_1\}$, and we introduce the operators $A_1, A_2 : \overline{B}_{r_1} \to X$ defined by

$$(A_1 u)(t) = -\frac{1}{\Gamma(\alpha)} \int_0^t (t-s)^{\alpha-1} f(s, u(s), I_{0+}^p u(s))\, ds, \quad t \in [0, 1],$$

$$(A_2 u)(t) = \frac{t^{\alpha-1}}{\Delta_0 \Gamma(\alpha - \beta_0)} \int_0^1 (1-s)^{\alpha-\beta_0-1} f(s, u(s), I_{0+}^p u(s))\, ds$$

$$-\frac{t^{\alpha-1}}{\Delta_0} \sum_{i=1}^m \frac{1}{\Gamma(\alpha - \beta_i)}$$

$$\times \int_0^1 \left(\int_0^s (s-\tau)^{\alpha-\beta_i-1} f(\tau, u(\tau), I_{0+}^p u(\tau))\, d\tau \right) dH_i(s),$$

$$\tag{6.7}$$

for all $t \in [0, 1]$ and $u \in \overline{B}_{r_1}$.

By using (H3), we obtain for all $u, v \in \overline{B}_{r_1}$

$$\|A_1 u + A_2 v\| \leq \|A_1 u\| + \|A_2 v\| \leq \frac{1}{\Gamma(\alpha+1)} \|g\|$$

$$+ \left(\frac{1}{|\Delta_0| \Gamma(\alpha - \beta_0 + 1)} + \frac{1}{|\Delta_0|} \sum_{i=1}^m \frac{1}{\Gamma(\alpha - \beta_i + 1)} \right.$$

$$\left. \times \left| \int_0^1 s^{\alpha-\beta_i}\, dH_i(s) \right| \right) \|g\| = M_1 \|g\| \leq r_1.$$

Hence, $A_1 u + A_2 v \in \overline{B}_{r_1}$ for all $u, v \in \overline{B}_{r_1}$.

The operator A_1 is a contraction, because

$$\|A_1 u - A_1 v\| \leq L_1 L_0 \frac{1}{\Gamma(\alpha+1)} \|u - v\| = \Xi_1 \|u - v\|, \quad \forall u, v \in \overline{B}_{r_1},$$

and $\Xi_1 < 1$.

The continuity of f implies that the operator A_2 is continuous on $\overline{B}_{r_1}$. We prove next that A_2 is compact. The operator A_2 is uniformly bounded on $\overline{B}_{r_1}$, because

$$\|A_2 u\| \leq \left(\frac{1}{|\Delta_0|\Gamma(\alpha - \beta_0 + 1)} \right.$$

$$\left. + \frac{1}{|\Delta_0|} \sum_{i=1}^{m} \frac{1}{\Gamma(\alpha - \beta_i + 1)} \left| \int_0^1 s^{\alpha - \beta_i} dH_i(s) \right| \right) \|g\| = M_2 \|g\|,$$

for all $u \in \overline{B}_{r_1}$. Now, we prove that A_2 is equicontinuous on $\overline{B}_{r_1}$. We denote by

$$\Lambda_{r_1} = \sup \left\{ |f(t, x, y)|, \quad t \in [0, 1], \ |x| \leq r_1, \ |y| \leq \frac{1}{\Gamma(p+1)} r_1 \right\}.$$

$$(6.8)$$

Then for $u \in \overline{B}_{r_1}$ and $t_1, t_2 \in [0, 1]$ with $t_1 < t_2$, we obtain

$$|(A_2 u)(t_2) - (A_2 u)(t_1)| \leq \frac{(t_2^{\alpha-1} - t_1^{\alpha-1})}{|\Delta_0|\Gamma(\alpha - \beta_0)} \int_0^1 (1 - s)^{\alpha - \beta_0 - 1} \Lambda_{r_1} \, ds$$

$$+ \frac{(t_2^{\alpha-1} - t_1^{\alpha-1})}{|\Delta_0|} \sum_{i=1}^{m} \frac{1}{\Gamma(\alpha - \beta_i)}$$

$$\times \left| \int_0^1 \left(\int_0^s (s - \tau)^{\alpha - \beta_i - 1} \Lambda_{r_1} \, d\tau \right) dH_i(s) \right|$$

$$\leq \Lambda_{r_1} (t_2^{\alpha-1} - t_1^{\alpha-1}) \left[\frac{1}{|\Delta_0|\Gamma(\alpha - \beta_0 + 1)} \right.$$

$$\left. + \frac{1}{|\Delta_0|} \sum_{i=1}^{m} \frac{1}{\Gamma(\alpha - \beta_i + 1)} \left| \int_0^1 s^{\alpha - \beta_i} dH_i(s) \right| \right]$$

$$= \Lambda_{r_1} M_2 (t_2^{\alpha-1} - t_1^{\alpha-1}).$$

Therefore, we conclude

$$|(A_2 u)(t_2) - (A_2 u)(t_1)| \to 0, \quad \text{as } t_2 \to t_1,$$

uniformly with respect to $u \in \overline{B}_{r_1}$.

We deduce that A_2 is equicontinuous on $\overline{B}_{r_1}$, and so, by using the Arzela-Ascoli theorem, the set $A_2(\overline{B}_{r_1})$ is relatively compact. We conclude that operator A_2 is compact on $\overline{B}_{r_1}$. Thus, all assumptions of Theorem 1.2.3 are satisfied, and then by Theorem 1.2.3 we deduce that there exists a fixed point of operator $A_1 + A_2$, which is a solution of the boundary value problem (6.1), (6.2) on $[0, 1]$. $\square$

Theorem 6.1.3. *Assume that* (H1)–(H3) *hold. If* $\Xi_2 := L_1 L_0 M_2 < 1$, *then problem* (6.1), (6.2) *has at least one solution on* $[0, 1]$.

Proof. We consider again a positive number $r_1 \geq M_1 \|g\|$ and the operators A_1, A_2 defined on $\overline{B}_{r_1}$ given by (6.7). In a similar manner as in the proof of Theorem 6.1.2, we obtain that $A_1 u + A_2 v \in \overline{B}_{r_1}$ for all $u, v \in \overline{B}_{r_1}$.

The operator A_2 is a contraction because

$$\|A_2 u - A_2 v\| \leq L_1 L_0 \left(\frac{1}{|\Delta_0| \Gamma(\alpha - \beta_0 + 1)} \right.$$

$$+ \frac{1}{|\Delta_0|} \sum_{i=1}^{m} \frac{1}{\Gamma(\alpha - \beta_i + 1)} \left| \int_0^1 s^{\alpha - \beta_i} dH_i(s) \right| \Bigg)$$

$$\times \|u - v\| = L_1 L_0 M_2 \|u - v\| = \Xi_2 \|u - v\|,$$

$$\forall u, v \in \overline{B}_{r_1},$$

with $\Xi_2 < 1$.

Then the continuity of f implies that the operator A_1 is continuous on $\overline{B}_{r_1}$. We prove now that A_1 is compact. The operator A_1 is uniformly bounded on $\overline{B}_{r_1}$ because

$$\|A_1 u\| \leq \frac{1}{\Gamma(\alpha + 1)} \|g\|, \quad \forall u \in \overline{B}_{r_1}.$$

Now, we show that A_1 is equicontinuous on $\overline{B}_{r_1}$. By using Λ_{r_1} (defined in the proof of Theorem 6.1.2), we obtain for $u \in \overline{B}_{r_1}$ and $t_1, t_2 \in [0, 1]$ with $t_1 < t_2$

$$|(A_1 u)(t_2) - (A_1 u)(t_1)|$$

$$= \left| -\frac{1}{\Gamma(\alpha)} \int_0^{t_2} (t_2 - s)^{\alpha - 1} f(s, u(s), I_{0+}^p u(s)) \, ds \right.$$

$$+ \frac{1}{\Gamma(\alpha)} \int_0^{t_1} (t_1 - s)^{\alpha - 1} f(s, u(s), I_{0+}^p u(s)) \, ds \Bigg|$$

$$= \left| -\frac{1}{\Gamma(\alpha)} \int_0^{t_1} [(t_2 - s)^{\alpha - 1} - (t_1 - s)^{\alpha - 1}] f(s, u(s), I_{0+}^p u(s)) \, ds \right.$$

$$- \frac{1}{\Gamma(\alpha)} \int_{t_1}^{t_2} (t_2 - s)^{\alpha - 1} f(s, u(s), I_{0+}^p u(s)) \, ds \Bigg|$$

$$\leq \frac{\Lambda_{r_1}}{\Gamma(\alpha)} \int_0^{t_1} [(t_2 - s)^{\alpha-1} - (t_1 - s)^{\alpha-1}] \, ds$$

$$+ \frac{\Lambda_{r_1}}{\Gamma(\alpha)} \int_{t_1}^{t_2} (t_2 - s)^{\alpha-1} \, ds$$

$$= \frac{\Lambda_{r_1}}{\Gamma(\alpha+1)} [-(t_2 - t_1)^{\alpha} + t_2^{\alpha} - t_1^{\alpha}] + \frac{\Lambda_{r_1}}{\Gamma(\alpha+1)} (t_2 - t_1)^{\alpha}$$

$$\leq \frac{\Lambda_{r_1}}{\Gamma(\alpha+1)} (t_2^{\alpha} - t_1^{\alpha}).$$

Then we deduce

$$|(A_1 u)(t_2) - (A_1 u)(t_1)| \to 0, \quad \text{as } t_2 \to t_1,$$

uniformly with respect to $u \in \overline{B}_{r_1}$.

We conclude that A_1 is equicontinuous on $\overline{B}_{r_1}$, and by using the Arzela-Ascoli theorem, the set $A_1(\overline{B}_{r_1})$ is relatively compact. We deduce that operator A_1 is compact on $\overline{B}_{r_1}$. By Theorem 1.2.3, we obtain that there exists a fixed point of operator $A_1 + A_2$, which is a solution of the boundary value problem (6.1), (6.2) on $[0, 1]$. □

Theorem 6.1.4. *Assume that* (H1) *and* (H4) *hold. If* $\Xi_3 := M_1(a_1 + \frac{a_2}{\Gamma(p+1)}) < 1$, *then the boundary value problem* (6.1), (6.2) *has at least one solution on* $[0, 1]$.

Proof. We consider the operator $A : X \to X$ defined in (6.6). We firstly prove that A is completely continuous. By the continuity of f we deduce that A is a continuous operator.

We show next that A is a compact operator. Let $\Omega \subset X$ be a bounded set. Then there exist a positive constant L_2 such that

$$|f(t, u(t), I_{0+}^p u(t))| \leq L_2, \quad \forall t \in [0, 1], \ u \in \Omega.$$

Therefore, we obtain as in the proof of Theorem 6.1.1 that $|(Au)(t)| \leq L_2 M_1$, for all $t \in [0, 1]$ and $u \in \Omega$. So, $A(\Omega)$ is uniformly bounded.

We will show next that $A(\Omega)$ is equicontinuous. Let $u \in \Omega$ and $t_1, t_2 \in [0, 1]$ with $t_1 < t_2$. Then by using the operators A_1 and A_2 defined on Ω (given by (6.7)), and based on a similar approach as that used in the proof of Theorems 6.1.2 and 6.1.3, we obtain

$$|(Au)(t_2) - (Au)(t_1)|$$

$$= |(A_1u)(t_2) - (A_1u)(t_1) + (A_2u)(t_2) - (A_2u)(t_1)|$$

$$\leq |(A_1u)(t_2) - (A_1u)(t_1)| + |(A_2u)(t_2) - (A_1u)(t_1)|$$

$$\leq \frac{L_2}{\Gamma(\alpha+1)}(t_2^\alpha - t_1^\alpha) + L_2M_2(t_2^{\alpha-1} - t_1^{\alpha-1}).$$

Then $|(Au)(t_2) - (Au)(t_1)| \to 0$ as $t_2 \to t_1$ uniformly with respect to $u \in \Omega$. Thus, $A(\Omega)$ is equicontinuous. By the Arzela-Ascoli theorem, we deduce that $A(\Omega)$ is relatively compact, and so A is compact. Therefore, A is completely continuous.

Now, we will prove that the set $F = \{u \in X, \ u = \nu A(u), \ 0 < \nu < 1\}$ is bounded. Let $u \in F$, that is $u = \nu A(u)$ for some $\nu \in (0,1)$. Then we have

$$|u(t)| = |\nu(Au)(t)| \leq |(Au)(t)|, \quad \forall t \in [0,1].$$

By (H4), we obtain

$$|u(t)| \leq |(Au)(t)|$$

$$\leq \frac{1}{\Gamma(\alpha)} \int_0^t (t-s)^{\alpha-1}[a_0 + a_1|u(s)| + a_2|I_{0+}^p u(s)|]\, ds$$

$$+ \frac{t^{\alpha-1}}{|\Delta_0|\Gamma(\alpha-\beta_0)} \int_0^1 (t-s)^{\alpha-\beta_0-1}[a_0 + a_1|u(s)| + a_2|I_{0+}^p u(s)|]\, ds$$

$$+ \frac{t^{\alpha-1}}{|\Delta_0|} \sum_{i=1}^m \frac{1}{\Gamma(\alpha-\beta_i)} \left| \int_0^1 \left(\int_0^s (s-\tau)^{\alpha-\beta_i-1}[a_0 + a_1|u(\tau)| \right. \right.$$

$$\left. \left. + a_2|I_{0+}^p u(\tau)|]\, d\tau \right) dH_i(s) \right|$$

$$\leq \left(a_0 + a_1\|u\| + \frac{a_2}{\Gamma(p+1)}\|u\| \right) \left[\frac{t^\alpha}{\Gamma(\alpha+1)} + \frac{t^{\alpha-1}}{|\Delta_0|\Gamma(\alpha-\beta_0+1)} \right.$$

$$\left. + \frac{t^{\alpha-1}}{|\Delta_0|} \sum_{i=1}^m \frac{1}{\Gamma(\alpha-\beta_i+1)} \left| \int_0^1 s^{\alpha-\beta_i}\, dH_i(s) \right| \right].$$

Therefore, we deduce

$$\|u\| \leq M_1 \left(a_0 + a_1\|u\| + \frac{a_2}{\Gamma(p+1)}\|u\| \right).$$

Because $\Xi_3 < 1$, we obtain

$$\|u\| \leq M_1 a_0 \left(1 - M_1 a_1 - \frac{M_1 a_2}{\Gamma(p+1)}\right)^{-1}.$$

Hence, we deduce that the set F is bounded.

By using Theorem 1.2.5, we conclude that the operator A has at least one fixed point, which is a solution for our problem (6.1), (6.2). $\qquad\square$

Theorem 6.1.5. *Assume that* (H1) *and* (H5) *hold. Then the problem* (6.1), (6.2) *has at least one solution.*

Proof. Let $\overline{B}_R = \{u \in X, \ \|u\| \leq R\}$, where

$$R \geq \max \left\{ 3b_0 M_1, (3b_1 M_1)^{\frac{1}{1-l_1}}, \left(\frac{3b_2 M_1}{(\Gamma(p+1))^{l_2}}\right)^{\frac{1}{1-l_2}} \right\}.$$

We prove now that $A : \overline{B}_R \to \overline{B}_R$. For $u \in \overline{B}_R$, we deduce

$$|(Au)(t)| \leq \left(b_0 + b_1 R^{l_1} + \frac{b_2}{(\Gamma(p+1))^{l_2}} R^{l_2}\right) M_1 \leq \frac{R}{3} + \frac{R}{3} + \frac{R}{3} = R,$$

$$\forall t \in [0,1],$$

and then $\|Au\| \leq R$, which implies that $A(\overline{B}_R) \subset \overline{B}_R$.

From the continuity of the function f, we can easily show that the operator A is continuous. The functions from $A(\overline{B}_R)$ are uniformly bounded and equicontinuous. Indeed, by using the notation (6.8), with r_1 replaced by R, we obtain for any $u \in \overline{B}_R$ and $t_1, t_2 \in [0,1]$, $t_1 < t_2$ that

$$|(Au)(t_2) - (Au)(t_1)| \leq \frac{\Lambda_R}{\Gamma(\alpha+1)}(t_2^{\alpha} - t_1^{\alpha}) + \Lambda_R M_2(t_2^{\alpha-1} - t_1^{\alpha-1}).$$

Therefore, $|(Au)(t_2) - (Au)(t_1)| \to 0$ as $t_2 \to t_1$ uniformly with respect to $u \in \overline{B}_R$. By the Arzela-Ascoli theorem, we conclude that $A(\overline{B}_R)$ is relatively compact, and then A is a completely continuous operator. By Theorem 1.2.4, we deduce that operator A has at least one fixed point u in $\overline{B}_R$ which is a solution of our problem (6.1), (6.2). $\qquad\square$

Theorem 6.1.6. *Assume that* (H1) *and* (H6) *hold. If there exists* $\Xi_0 > 0$ *such that*

$$\left(c_0 + c_1 h_1(\Xi_0) + c_2 h_2\left(\frac{\Xi_0}{\Gamma(p+1)}\right)\right) M_1 < \Xi_0, \tag{6.9}$$

where c_0, c_1, c_2, h_1, h_2 *are given in* (H6), *then problem* (6.1), (6.2) *has at least one solution on* $[0,1]$.

Proof. We consider the set $\overline{B}_{\Xi_0} = \{u \in X, \|u\| \leq \Xi_0\}$, where Ξ_0 is given in the assumptions of the theorem. We will show that $A : \overline{B}_{\Xi_0} \to \overline{B}_{\Xi_0}$. For $u \in \overline{B}_{\Xi_0}$ and $t \in [0,1]$, we obtain

$$|(Au)(t)| \leq \left(c_0 + c_1 h_1(\Xi_0) + c_2 h_2\left(\frac{\Xi_0}{\Gamma(p+1)}\right)\right) M_1 < \Xi_0.$$

Then $A(\overline{B}_{\Xi_0}) \subset \overline{B}_{\Xi_0}$. In a similar manner as in the proof of Theorem 6.1.5 we can show that operator A is completely continuous.

We suppose now that there exists $u \in \partial B_{\Xi_0}$ such that $u = \nu A(u)$ for some $\nu \in (0,1)$. We obtain as above that $\|u\| \leq \|Au\| < \Xi_0$, which is a contradiction, because $u \in \partial B_{\Xi_0}$. Then by Theorem 1.2.6, we conclude that operator A has a fixed point $u \in \overline{B}_{\Xi_0}$, and so problem (6.1), (6.2) has at least one solution. $\qquad\square$

6.1.3   *Examples*

Let $\alpha = \frac{5}{2}$ $(n = 3)$, $p = \frac{10}{3}$, $m = 2$, $\beta_0 = \frac{6}{5}$, $\beta_1 = \frac{1}{3}$, $\beta_2 = \frac{3}{4}$, $H_1(t) = t^2$ for all $t \in [0,1]$, $H_2(t) = \{0, \text{ if } t \in [0,1/2); 3, \text{ if } t \in [1/2,1]\}$.

We consider the fractional differential equation

$$D_{0+}^{5/2}u(t) + f(t, u(t), I_{0+}^{10/3}u(t)) = 0, \quad 0 < t < 1, \tag{6.10}$$

with the boundary conditions

$$u(0) = u'(0) = 0, \quad D_{0+}^{6/5}u(1) = 2\int_0^1 t D_{0+}^{1/3}u(t)\,dt + 3D_{0+}^{3/4}u\left(\frac{1}{2}\right). \tag{6.11}$$

We obtain here $\Delta_0 \approx -1.87462428 \neq 0$, $L_0 \approx 1.1079852$, $M_1 \approx 1.16312084$ and $M_2 \approx 0.86221973$. So, assumption (H1) is satisfied.

Example 1. We consider the function

$$f(t, x, y) = \frac{|x|}{2(t+1)^2(1+|x|)} - \frac{t}{4}\arctan y - \frac{3t}{t^2+4},$$

$$t \in [0,1], \ x, y \in \mathbb{R}.$$

Here, we have

$$|f(t, x, y) - f(t, x_1, y_1)| \leq \frac{1}{2}(|x - x_1| + |y - y_1|),$$

$$\forall t \in [0,1], \ x, y, x_1, y_1 \in \mathbb{R},$$

so $L_1 = \frac{1}{2}$, and then $\Xi \approx 0.64436034 < 1$. Therefore, assumption (H2) is satisfied, and by Theorem 6.1.1, we deduce that problem (6.10), (6.11) has a unique solution $u(t)$, $t \in [0,1]$.

Example 2. We consider the function

$$f(t,x,y) = \frac{1}{\sqrt{4+t^2}} \sin t + \frac{|x|}{3(2+|x|)} - \frac{1}{2(1+t)} \sin^2 y,$$

$$t \in [0,1], \ x, \ y \in \mathbb{R}.$$

In this case, we have

$$|f(t,x,y) - f(t,x_1,y_1)| \le |x - x_1| + |y - y_1|,$$

$$\forall t \in [0,1], \ x, \ y, \ x_1, \ y_1 \in \mathbb{R},$$

so $L_1 = 1$, and $|f(t,x,y)| \le g(t)$ for all $t \in [0,1]$ and $x, \ y \in \mathbb{R}$, where

$$g(t) = \frac{|\sin t|}{\sqrt{4+t^2}} + \frac{1}{3} + \frac{1}{2(1+t)}, \quad \forall t \in [0,1].$$

Then assumptions (H2) and (H3) are satisfied, and, in addition, we obtain $\Xi_1 \approx 0.33339398 < 1$. Therefore, by Theorem 6.1.2, we conclude that problem (6.10), (6.11) has at least one solution on $[0,1]$.

Example 3. We consider the function

$$f(t,x,y) = \frac{t}{t^2+1} \left(4\cos t + \frac{1}{2}\sin x \right) - \frac{1}{(t+1)^3} y,$$

$$\forall t \in [0,1], \ x, \ y \in \mathbb{R}.$$

Because we have

$$|f(t,x,y)| \le 2 + \frac{1}{4}|x| + |y|, \quad \forall t \in [0,1], \ x, \ y \in \mathbb{R},$$

the assumption (H4) is satisfied with $a_0 = 2$, $a_1 = \frac{1}{4}$ and $a_2 = 1$. In addition, we obtain $\Xi_3 \approx 0.41638005 < 1$, and then by Theorem 6.1.4 we deduce that problem (6.10), (6.11) has at least one solution on $[0,1]$.

Example 4. We consider the function

$$f(t,x,y) = \frac{e^{-t}}{1+t^3} - \frac{1}{3}x^{2/3} + \frac{1}{4(3+t)} \arctan y^{1/5}, \quad t \in [0,1], \ x, \ y \in \mathbb{R}.$$

Because we obtain

$$|f(t,x,y)| \le 1 + \frac{1}{3}|x|^{2/3} + \frac{1}{12}|y|^{1/5}, \quad \forall t \in [0,1], \ x, \ y \in \mathbb{R},$$

then assumption (H5) is satisfied with $b_0 = 1$, $b_1 = \frac{1}{3}$, $b_2 = \frac{1}{12}$, $l_1 = \frac{2}{3}$, $l_2 = \frac{1}{5}$. Then by Theorem 6.1.5, we deduce that problem (6.10), (6.11) has at least one solution on $[0,1]$.

Example 5. We consider the function

$$f(t, x, y) = \frac{(1-t)^2}{10} + \frac{(1-t)x^2}{15(1+x^2)} - \frac{t^4 y^3}{5}, \quad \forall t \in [0, 1], \ x, \ y \in \mathbb{R}.$$

Because we have

$$|f(t, x, y)| \leq \frac{1}{10} + \frac{1}{15}|x|^2 + \frac{1}{5}|y|^3, \quad t \in [0, 1], \ x, \ y \in \mathbb{R},$$

then assumption (H6) is satisfied with $h_1(x) = x^2$ and $h_2(x) = x^3$ for $x \in [0, \infty)$, $c_0 = \frac{1}{10}$, $c_1 = \frac{1}{15}$ and $c_2 = \frac{1}{5}$. For $\Xi_0 = 2$, the condition (6.9) is satisfied, because $\left(c_0 + c_1 h_1(2) + c_2 h_2\left(\frac{2}{\Gamma(p+1)}\right)\right) M_1 \approx 0.428821 < 2$. Therefore, by Theorem 6.1.6, we conclude that problem (6.10), (6.11) has at least one solution on $[0, 1]$.

Remark 6.1.1. The results presented in this section were published in [82].

6.2 Systems of Riemann–Liouville Fractional Differential Equations with Uncoupled Boundary Conditions

We consider the nonlinear system of fractional differential equations

$$\begin{cases} D_{0+}^{\alpha} u(t) + f(t, u(t), v(t), I_{0+}^{\theta_1} u(t), I_{0+}^{\sigma_1} v(t)) = 0, & t \in (0, 1), \\ D_{0+}^{\beta} v(t) + g(t, u(t), v(t), I_{0+}^{\theta_2} u(t), I_{0+}^{\sigma_2} v(t)) = 0, & t \in (0, 1), \end{cases} \tag{6.12}$$

with the uncoupled nonlocal boundary conditions

$$\begin{cases} u(0) = u'(0) = \cdots = u^{(n-2)}(0) = 0, \\ \\ D_{0+}^{\gamma_0} u(1) = \sum_{i=1}^{p} \int_0^1 D_{0+}^{\gamma_i} u(t) \, dH_i(t), \\ \\ v(0) = v'(0) = \cdots = v^{(m-2)}(0) = 0, \\ \\ D_{0+}^{\delta_0} v(1) = \sum_{i=1}^{q} \int_0^1 D_{0+}^{\delta_i} v(t) \, dK_i(t), \end{cases} \tag{6.13}$$

where α, $\beta \in \mathbb{R}$, $\alpha \in (n-1, n]$, $\beta \in (m-1, m]$, n, $m \in \mathbb{N}$, $n \geq 2$, $m \geq 2$, θ_1, θ_2, σ_1, $\sigma_2 > 0$, p, $q \in \mathbb{N}$, $\gamma_i \in \mathbb{R}$ for all $i = 0, \ldots, p$, $0 \leq \gamma_1 < \gamma_2 < \cdots < \gamma_p < \alpha - 1$, $\gamma_0 \in [0, \alpha - 1)$, $\delta_i \in \mathbb{R}$ for all $i = 0, \ldots, q$, $0 \leq \delta_1 < \delta_2 < \cdots < \delta_q < \beta - 1$, $\delta_0 \in [0, \beta - 1)$, D_{0+}^k denotes the Riemann–Liouville derivative of order k (for $k = \alpha, \beta, \gamma_0, \gamma_i, i = 1, \ldots, p, \delta_0, \delta_i, i = 1, \ldots, q)$, I_{0+}^ζ is the Riemann–Liouville integral of order ζ (for $\zeta = \theta_1, \sigma_1, \theta_2, \sigma_2$), f and g are nonlinear functions, and the integrals from the boundary conditions

(6.13) are Riemann–Stieltjes integrals with H_i for $i = 1, \ldots, p$ and K_i for $i = 1, \ldots, q$ functions of bounded variation.

Based on some theorems from the fixed point theory, we give conditions for the nonlinearities f and g such that problem (6.12), (6.13) has at least one solution $(u, v) \in (C[0, 1])^2$.

6.2.1 *Auxiliary results*

We consider the fractional differential equation

$$D_{0+}^{\alpha} u(t) + h(t) = 0, \quad t \in (0, 1), \tag{6.14}$$

with the boundary conditions

$$
\begin{cases}
u(0) = u'(0) = \cdots = u^{(n-2)}(0) = 0, \\
D_{0+}^{\gamma_0} u(1) = \sum_{i=1}^{p} \int_0^1 D_{0+}^{\gamma_i} u(t) \, dH_i(t),
\end{cases}
\tag{6.15}
$$

where $h \in C(0, 1) \cap L^1(0, 1)$. We denote by

$$\Delta_1 = \frac{\Gamma(\alpha)}{\Gamma(\alpha - \gamma_0)} - \sum_{i=1}^{p} \frac{\Gamma(\alpha)}{\Gamma(\alpha - \gamma_i)} \int_0^1 s^{\alpha - \gamma_i - 1} \, dH_i(s).$$

In a similar manner as we proved Lemma 2.4.1 from Section 2.4, we obtain the following lemma.

Lemma 6.2.1. *If $\Delta_1 \neq 0$, then the unique solution $u \in C[0, 1]$ of problem (6.14), (6.15) is given by*

$$u(t) = -\frac{1}{\Gamma(\alpha)} \int_0^t (t - s)^{\alpha - 1} h(s) \, ds$$

$$+ \frac{t^{\alpha - 1}}{\Delta_1 \Gamma(\alpha - \gamma_0)} \int_0^1 (1 - s)^{\alpha - \gamma_0 - 1} h(s) \, ds$$

$$- \frac{t^{\alpha - 1}}{\Delta_1} \sum_{i=1}^{p} \frac{1}{\Gamma(\alpha - \gamma_i)} \int_0^1 \left(\int_0^s (s - \tau)^{\alpha - \gamma_i - 1} h(\tau) \, d\tau \right) dH_i(s),$$

$$t \in [0, 1]. \tag{6.16}$$

We also consider the fractional differential equation

$$D_{0+}^{\beta} v(t) + k(t) = 0, \quad t \in (0, 1), \tag{6.17}$$

with the boundary conditions

$$\begin{cases} v(0) = v'(0) = \cdots = v^{(m-2)}(0) = 0, \\ D_{0+}^{\delta_0} v(1) = \sum_{i=1}^{q} \int_0^1 D_{0+}^{\delta_i} v(t) \, dK_i(t), \end{cases} \tag{6.18}$$

where $k \in C(0,1) \cap L^1(0,1)$. We denote by

$$\Delta_2 = \frac{\Gamma(\beta)}{\Gamma(\beta - \delta_0)} - \sum_{i=1}^{q} \frac{\Gamma(\beta)}{\Gamma(\beta - \delta_i)} \int_0^1 s^{\beta - \delta_i - 1} \, dK_i(s).$$

Lemma 6.2.2. *If $\Delta_2 \neq 0$, then the unique solution $v \in C[0,1]$ of problem (6.17), (6.18) is given by*

$$v(t) = -\frac{1}{\Gamma(\beta)} \int_0^t (t-s)^{\beta-1} k(s) \, ds$$

$$+ \frac{t^{\beta-1}}{\Delta_2 \Gamma(\beta - \delta_0)} \int_0^1 (1-s)^{\beta-\delta_0-1} k(s) \, ds$$

$$- \frac{t^{\beta-1}}{\Delta_2} \sum_{i=1}^{q} \frac{1}{\Gamma(\beta - \delta_i)} \int_0^1 \left(\int_0^s (s-\tau)^{\beta-\delta_i-1} k(\tau) \, d\tau \right) dK_i(s),$$

$$t \in [0,1]. \tag{6.19}$$

We denote by (I1) the following basic assumptions for problem (6.12), (6.13) that will be used in the main theorems.

(I1) $\alpha, \beta \in \mathbb{R}$, $\alpha \in (n-1, n]$, $\beta \in (m-1, m]$, $n, m \in \mathbb{N}$, $n \geq 2$, $m \geq 2$, $\theta_1, \theta_2, \sigma_1, \sigma_2 > 0$, $p, q \in \mathbb{N}$, $\gamma_i \in \mathbb{R}$ for all $i = 0, \ldots, p$, $0 \leq \gamma_1 < \gamma_2 < \cdots < \gamma_p < \alpha - 1$, $\gamma_0 \in [0, \alpha - 1)$, $\delta_i \in \mathbb{R}$ for all $i = 0, \ldots, q$, $0 \leq \delta_1 < \delta_2 < \cdots < \delta_q < \beta - 1$, $\delta_0 \in [0, \beta - 1)$, $H_i : [0,1] \to \mathbb{R}$, $i = 1, \ldots, p$ and $K_j : [0,1] \to \mathbb{R}$, $j = 1, \ldots, q$ are functions of bounded variation, $\Delta_1 \neq 0$, $\Delta_2 \neq 0$.

We introduce the following constants:

$$M_1 = 1 + \frac{1}{\Gamma(\theta_1 + 1)}, \quad M_2 = 1 + \frac{1}{\Gamma(\sigma_1 + 1)},$$

$$M_3 = 1 + \frac{1}{\Gamma(\theta_2 + 1)}, \quad M_4 = 1 + \frac{1}{\Gamma(\sigma_2 + 1)},$$

$$M_5 = \max\{M_1, M_2\}, \quad M_6 = \max\{M_3, M_4\},$$

$$M_7 = \frac{1}{\Gamma(\alpha+1)} + \frac{1}{|\Delta_1|\Gamma(\alpha-\gamma_0+1)}$$

$$+ \frac{1}{|\Delta_1|} \sum_{i=1}^{p} \frac{1}{\Gamma(\alpha-\gamma_i+1)} \left| \int_0^1 s^{\alpha-\gamma_i} \, dH_i(s) \right|, \qquad (6.20)$$

$$M_9 = \frac{1}{\Gamma(\beta+1)} + \frac{1}{|\Delta_2|\Gamma(\beta-\delta_0+1)}$$

$$+ \frac{1}{|\Delta_2|} \sum_{i=1}^{q} \frac{1}{\Gamma(\beta-\delta_i+1)} \left| \int_0^1 s^{\beta-\delta_i} \, dK_i(s) \right|,$$

$$M_8 = M_7 - \frac{1}{\Gamma(\alpha+1)}, \quad M_{10} = M_9 - \frac{1}{\Gamma(\beta+1)}.$$

We consider the Banach space $X = C[0,1]$ with supremum norm $\|u\| = \sup_{t\in[0,1]} |u(t)|$, and the Banach space $Y = X \times X$ with the norm $\|(u,v)\|_Y = \|u\| + \|v\|$. We introduce the operator $\mathcal{A} : Y \to Y$ defined by $\mathcal{A}(u,v) = (\mathcal{A}_1(u,v), \mathcal{A}_2(u,v))$ for $(u,v) \in Y$, where the operators $\mathcal{A}_1, \mathcal{A}_2 : Y \to X$ are given by

$$\mathcal{A}_1(u,v)(t) = -\frac{1}{\Gamma(\alpha)} \int_0^t (t-s)^{\alpha-1} f(s, u(s), v(s), I_{0+}^{\theta_1} u(s), I_{0+}^{\sigma_1} v(s)) \, ds$$

$$+ \frac{t^{\alpha-1}}{\Delta_1 \Gamma(\alpha-\gamma_0)} \int_0^1 (1-s)^{\alpha-\gamma_0-1}$$

$$\times f(s, u(s), v(s), I_{0+}^{\theta_1} u(s), I_{0+}^{\sigma_1} v(s)) \, ds$$

$$- \frac{t^{\alpha-1}}{\Delta_1} \sum_{i=1}^{p} \frac{1}{\Gamma(\alpha-\gamma_i)} \int_0^1 \left(\int_0^s (s-\tau)^{\alpha-\gamma_i-1} \right.$$

$$\left. \times f(\tau, u(\tau), v(\tau), I_{0+}^{\theta_1} u(\tau), I_{0+}^{\sigma_1} v(\tau)) \, d\tau \right) dH_i(s),$$

$$\mathcal{A}_2(u,v)(t) = -\frac{1}{\Gamma(\beta)} \int_0^t (t-s)^{\beta-1} g(s, u(s), v(s), I_{0+}^{\theta_2} u(s), I_{0+}^{\sigma_2} v(s)) \, ds$$

$$+ \frac{t^{\beta-1}}{\Delta_2 \Gamma(\beta-\delta_0)} \int_0^1 (1-s)^{\beta-\delta_0-1}$$

$$\times g(s, u(s), v(s), I_{0+}^{\theta_2} u(s), I_{0+}^{\sigma_2} v(s)) \, ds$$

$$- \frac{t^{\beta-1}}{\Delta_2} \sum_{i=1}^{q} \frac{1}{\Gamma(\beta-\delta_i)} \int_0^1 \left(\int_0^s (s-\tau)^{\beta-\delta_i-1} \right.$$

$$\left. \times g(\tau, u(\tau), v(\tau), I_{0+}^{\theta_2} u(\tau), I_{0+}^{\sigma_2} v(\tau)) \, d\tau \right) dK_i(s),$$

$$\forall t \in [0,1], \ (u,v) \in Y. \qquad (6.21)$$

By using Lemmas 6.2.1 and 6.2.2, we note that (u, v) is a solution of problem (6.12), (6.13) if and only if (u, v) is a fixed point of operator $\mathcal{A}$.

6.2.2 Existence of solutions

In this section, we present some conditions for the nonlinearities f and g such that operator $\mathcal{A}$ has at least one fixed point, which is a solution of problem (6.12), (6.13).

Theorem 6.2.1. *Assume that* (I1) *and*

(I2) *The functions* $f, g : [0,1] \times \mathbb{R}^4 \to \mathbb{R}$ *are continuous and there exist* L_1, $L_2 > 0$ *such that*

$$|f(t, x_1, x_2, x_3, x_4) - f(t, \widetilde{x}_1, \widetilde{x}_2, \widetilde{x}_3, \widetilde{x}_4)| \leq L_1 \sum_{i=1}^{4} |x_i - \widetilde{x}_i|,$$

$$|g(t, y_1, y_2, y_3, y_4) - g(t, \widetilde{y}_1, \widetilde{y}_2, \widetilde{y}_3, \widetilde{y}_4)| \leq L_2 \sum_{i=1}^{4} |y_i - \widetilde{y}_i|,$$

for all $t \in [0,1]$, $x_i, y_i, \widetilde{x}_i, \widetilde{y}_i \in \mathbb{R}$, $i = 1, \ldots, 4$,

hold. If $\Xi := L_1 M_5 M_7 + L_2 M_6 M_9 < 1$, *then problem* (6.12), (6.13) *has a unique solution* $(u(t), v(t))$, $t \in [0,1]$, *where* M_5, M_6, M_7, M_9 *are given by* (6.20).

Proof. We consider the positive number r given by

$$r = (M_0 M_7 + \widetilde{M}_0 M_9)(1 - L_1 M_5 M_7 - L_2 M_6 M_9)^{-1},$$

where $M_0 = \sup_{t \in [0,1]} |f(t, 0, 0, 0, 0)|$, $\widetilde{M}_0 = \sup_{t \in [0,1]} |g(t, 0, 0, 0, 0)|$. We define the set $\overline{B}_r = \{(u, v) \in Y, \|(u, v)\|_Y \leq r\}$ and first we show that $\mathcal{A}(\overline{B}_r) \subset \overline{B}_r$. Let $(u, v) \in \overline{B}_r$. By using (I2) and Lemma 6.1.2, for $f(t, u(t), v(t), I_{0+}^{\theta_1} u(t), I_{0+}^{\sigma_1} v(t))$, we deduce the following inequalities:

$$|f(t, u(t), v(t), I_{0+}^{\theta_1} u(t), I_{0+}^{\sigma_1} v(t))|$$

$$\leq |f(t, u(t), v(t), I_{0+}^{\theta_1} u(t), I_{0+}^{\sigma_1} v(t)) - f(t, 0, 0, 0, 0)|$$

$$+ |f(t, 0, 0, 0, 0)| \leq L_1(|u(t)| + |v(t)| + |I_{0+}^{\theta_1} u(t)|$$

$$+ |I_{0+}^{\sigma_1} v(t)|) + M_0$$

$$\leq L_1 \left(\|u\| + \|v\| + \frac{\|u\|}{\Gamma(\theta_1 + 1)} + \frac{\|v\|}{\Gamma(\sigma_1 + 1)} \right) + M_0$$

$$= L_1\left(\left(1 + \frac{1}{\Gamma(\theta_1 + 1)}\right)\|u\| + \left(1 + \frac{1}{\Gamma(\sigma_1 + 1)}\right)\|v\|\right) + M_0$$

$$= L_1(M_1\|u\| + M_2\|v\|) + M_0$$

$$\le L_1 M_5\|(u, v)\|_Y + M_0 \le L_1 M_5 r + M_0, \quad \forall t \in [0, 1].$$

In a similar manner, we have

$$|g(t, u(t), v(t), I_{0+}^{\theta_2} u(t), I_{0+}^{\sigma_2} v(t))|$$

$$\le |g(t, u(t), v(t), I_{0+}^{\theta_2} u(t), I_{0+}^{\sigma_2} v(t)) - g(t, 0, 0, 0, 0)|$$

$$+ |g(t, 0, 0, 0, 0)| \le L_2(|u(t)| + |v(t)| + |I_{0+}^{\theta_2} u(t)|$$

$$+ |I_{0+}^{\sigma_2} v(t)|) + \widetilde{M}_0$$

$$\le L_2\left(\|u\| + \|v\| + \frac{\|u\|}{\Gamma(\theta_2 + 1)} + \frac{\|v\|}{\Gamma(\sigma_2 + 1)}\right) + \widetilde{M}_0$$

$$= L_2\left(\left(1 + \frac{1}{\Gamma(\theta_2 + 1)}\right)\|u\| + \left(1 + \frac{1}{\Gamma(\sigma_2 + 1)}\right)\|v\|\right) + \widetilde{M}_0$$

$$= L_2(M_3\|u\| + M_4\|v\|) + \widetilde{M}_0$$

$$\le L_2 M_6\|(u, v)\|_Y + \widetilde{M}_0 \le L_2 M_6 r + \widetilde{M}_0, \quad \forall t \in [0, 1].$$

Then by (6.21) (the definition of operators $\mathcal{A}_1$ and $\mathcal{A}_2$), we obtain

$$|\mathcal{A}_1(u, v)(t)| \le \frac{1}{\Gamma(\alpha)} \int_0^t (t - s)^{\alpha - 1}(L_1 M_5 r + M_0)\, ds$$

$$+ \frac{t^{\alpha - 1}}{|\Delta_1|\Gamma(\alpha - \gamma_0)} \int_0^1 (1 - s)^{\alpha - \gamma_0 - 1}(L_1 M_5 r + M_0)\, ds$$

$$+ \frac{t^{\alpha - 1}}{|\Delta_1|} \sum_{i=1}^p \frac{1}{\Gamma(\alpha - \gamma_i)}$$

$$\times \left|\int_0^1 \left(\int_0^s (s - \tau)^{\alpha - \gamma_i - 1}(L_1 M_5 r + M_0)\, d\tau\right) dH_i(s)\right|$$

$$= (L_1 M_5 r + M_0)\left[\frac{t^\alpha}{\Gamma(\alpha + 1)} + \frac{t^{\alpha - 1}}{|\Delta_1|\Gamma(\alpha - \gamma_0 + 1)}\right.$$

$$\left.+ \frac{t^{\alpha - 1}}{|\Delta_1|} \sum_{i=1}^p \frac{1}{\Gamma(\alpha - \gamma_i + 1)}\left|\int_0^1 s^{\alpha - \gamma_i}\, dH_i(s)\right|\right],$$

$$\forall t \in [0, 1].$$

Therefore, we conclude

$$\|\mathcal{A}_1(u,v)\| \le (L_1 M_5 r + M_0) \left[\frac{1}{\Gamma(\alpha+1)} + \frac{1}{|\Delta_1|\Gamma(\alpha-\gamma_0+1)} \right.$$

$$\left. + \frac{1}{|\Delta_1|} \sum_{i=1}^{p} \frac{1}{\Gamma(\alpha-\gamma_i+1)} \left| \int_0^1 s^{\alpha-\gamma_i} dH_i(s) \right| \right]$$

$$= (L_1 M_5 r + M_0) M_7. \tag{6.22}$$

Arguing as before, we find

$$|\mathcal{A}_2(u,v)(t)| \le \frac{1}{\Gamma(\beta)} \int_0^t (t-s)^{\beta-1} (L_2 M_6 r + \widetilde{M}_0) \, ds$$

$$+ \frac{t^{\beta-1}}{|\Delta_2|\Gamma(\beta-\delta_0)} \int_0^1 (1-s)^{\beta-\delta_0-1} (L_2 M_6 r + \widetilde{M}_0) \, ds$$

$$+ \frac{t^{\beta-1}}{|\Delta_2|} \sum_{i=1}^{q} \frac{1}{\Gamma(\beta-\delta_i)}$$

$$\times \left| \int_0^1 \left(\int_0^s (s-\tau)^{\beta-\delta_i-1} (L_2 M_6 r + \widetilde{M}_0) \, d\tau \right) dK_i(s) \right|$$

$$= (L_2 M_6 r + \widetilde{M}_0) \left[\frac{t^{\beta}}{\Gamma(\beta+1)} + \frac{t^{\beta-1}}{|\Delta_2|\Gamma(\beta-\delta_0+1)} \right.$$

$$\left. + \frac{t^{\beta-1}}{|\Delta_2|} \sum_{i=1}^{q} \frac{1}{\Gamma(\beta-\delta_i+1)} \left| \int_0^1 s^{\beta-\delta_i} dK_i(s) \right| \right],$$

$$\forall t \in [0,1].$$

Then we have

$$\|\mathcal{A}_2(u,v)\| \le (L_2 M_6 r + \widetilde{M}_0) \left[\frac{1}{\Gamma(\beta+1)} + \frac{1}{|\Delta_2|\Gamma(\beta-\delta_0+1)} \right.$$

$$\left. + \frac{1}{|\Delta_2|} \sum_{i=1}^{q} \frac{1}{\Gamma(\beta-\delta_i+1)} \left| \int_0^1 s^{\beta-\delta_i} dK_i(s) \right| \right]$$

$$= (L_2 M_6 r + \widetilde{M}_0) M_9. \tag{6.23}$$

By relations (6.22) and (6.23), we deduce

$$\|\mathcal{A}(u,v)\|_Y = \|\mathcal{A}_1(u,v)\| + \|\mathcal{A}_2(u,v)\|$$

$$\le (L_1 M_5 r + M_0) M_7 + (L_2 M_6 r + \widetilde{M}_0) M_9 = r,$$

for all $(u,v) \in \overline{B}_r$, which implies that $\mathcal{A}(\overline{B}_r) \subset \overline{B}_r$.

Next, we prove that operator $\mathcal{A}$ is a contraction. For $(u_i, v_i) \in \overline{B}_r$, $i = 1, 2$, and for each $t \in [0, 1]$, we obtain

$$|\mathcal{A}_1(u_1, v_1)(t) - \mathcal{A}_1(u_2, v_2)(t)|$$

$$\leq \left| -\frac{1}{\Gamma(\alpha)} \int_0^t (t-s)^{\alpha-1} \left[f(s, u_1(s), v_1(s), I_{0+}^{\theta_1} u_1(s), I_{0+}^{\sigma_1} v_1(s)) \right. \right.$$

$$\left. - f(s, u_2(s), v_2(s), I_{0+}^{\theta_1} u_2(s), I_{0+}^{\sigma_1} v_2(s)) \right] ds \bigg|$$

$$+ \frac{t^{\alpha-1}}{|\Delta_1| \Gamma(\alpha - \gamma_0)} \int_0^1 (1-s)^{\alpha-\gamma_0-1}$$

$$\times |f(s, u_1(s), v_1(s), I_{0+}^{\theta_1} u_1(s), I_{0+}^{\sigma_1} v_1(s))$$

$$- f(s, u_2(s), v_2(s), I_{0+}^{\theta_1} u_2(s), I_{0+}^{\sigma_1} v_2(s))| \, ds$$

$$+ \frac{t^{\alpha-1}}{|\Delta_1|} \sum_{i=1}^p \frac{1}{\Gamma(\alpha - \gamma_i)} \left| \int_0^1 \left(\int_0^s (s-\tau)^{\alpha-\gamma_i-1} \right. \right.$$

$$\times |f(\tau, u_1(\tau), v_1(\tau), I_{0+}^{\theta_1} u_1(\tau), I_{0+}^{\sigma_1} v_1(\tau))$$

$$\left. \left. - f(\tau, u_2(\tau), v_2(\tau), I_{0+}^{\theta_1} u_2(\tau), I_{0+}^{\sigma_1} v_2(\tau))| \, d\tau \right) dH_i(s) \right|$$

$$\leq \frac{L_1}{\Gamma(\alpha)} \int_0^t (t-s)^{\alpha-1} \left[|u_1(s) - u_2(s)| + |v_1(s) - v_2(s)| \right.$$

$$\left. + |I_{0+}^{\theta_1} u_1(s) - I_{0+}^{\theta_1} u_2(s)| + |I_{0+}^{\sigma_1} v_1(s) - I_{0+}^{\sigma_1} v_2(s)| \right] ds$$

$$+ \frac{t^{\alpha-1} L_1}{|\Delta_1| \Gamma(\alpha - \gamma_0)} \int_0^1 (1-s)^{\alpha-\gamma_0-1} \left[|u_1(s) - u_2(s)| \right.$$

$$+ |v_1(s) - v_2(s)| + |I_{0+}^{\theta_1} u_1(s) - I_{0+}^{\theta_1} u_2(s)|$$

$$\left. + |I_{0+}^{\sigma_1} v_1(s) - I_{0+}^{\sigma_1} v_2(s)| \right] ds$$

$$+ \frac{t^{\alpha-1} L_1}{|\Delta_1|} \sum_{i=1}^p \frac{1}{\Gamma(\alpha - \gamma_i)}$$

$$\times \left| \int_0^1 \left(\int_0^s (s-\tau)^{\alpha-\gamma_i-1} \left[|u_1(\tau) - u_2(\tau)| + |v_1(\tau) - v_2(\tau)| \right. \right. \right.$$

$$\left. \left. \left. + |I_{0+}^{\theta_1} u_1(\tau) - I_{0+}^{\theta_1} u_2(\tau)| + |I_{0+}^{\sigma_1} v_1(\tau) - I_{0+}^{\sigma_1} v_2(\tau)| \right] d\tau \right) dH_i(s) \right|$$

$$\leq \frac{L_1}{\Gamma(\alpha)} \int_0^t (t-s)^{\alpha-1} \Big[\|u_1 - u_2\| + \|v_1 - v_2\|$$

$$+ \frac{1}{\Gamma(\theta_1 + 1)} \|u_1 - u_2\| + \frac{1}{\Gamma(\sigma_1 + 1)} \|v_1 - v_2\| \Big] ds$$

$$+ \frac{t^{\alpha-1} L_1}{|\Delta_1| \Gamma(\alpha - \gamma_0)} \int_0^1 (1-s)^{\alpha-\gamma_0-1} \Big[\|u_1 - u_2\| + \|v_1 - v_2\|$$

$$+ \frac{1}{\Gamma(\theta_1 + 1)} \|u_1 - u_2\| + \frac{1}{\Gamma(\sigma_1 + 1)} \|v_1 - v_2\| \Big] ds$$

$$+ \frac{t^{\alpha-1} L_1}{|\Delta_1|} \sum_{i=1}^p \frac{1}{\Gamma(\alpha - \gamma_i)}$$

$$\times \left| \int_0^1 \left(\int_0^s (s-\tau)^{\alpha-\gamma_i-1} [\|u_1 - u_2\| + \|v_1 - v_2\| \right. \right.$$

$$+ \frac{1}{\Gamma(\theta_1 + 1)} \|u_1 - u_2\| + \frac{1}{\Gamma(\sigma_1 + 1)} \|v_1 - v_2\| \Big] d\tau \Big) dH_i(s) \Big|$$

$$\leq \frac{L_1 M_5}{\Gamma(\alpha)} \int_0^t (t-s)^{\alpha-1} (\|u_1 - u_2\| + \|v_1 - v_2\|) ds$$

$$+ \frac{t^{\alpha-1} L_1 M_5}{|\Delta_1| \Gamma(\alpha - \gamma_0)} \int_0^1 (1-s)^{\alpha-\gamma_0-1} (\|u_1 - u_2\| + \|v_1 - v_2\|) ds$$

$$+ \frac{t^{\alpha-1} L_1 M_5}{|\Delta_1|} \sum_{i=1}^p \frac{1}{\Gamma(\alpha - \gamma_i)} \left| \int_0^1 \left(\int_0^s (s-\tau)^{\alpha-\gamma_i-1} d\tau \right) dH_i(s) \right|$$

$$\times (\|u_1 - u_2\| + \|v_1 - v_2\|)$$

$$= L_1 M_5 \left(\frac{t^\alpha}{\Gamma(\alpha + 1)} + \frac{t^{\alpha-1}}{|\Delta_1| \Gamma(\alpha - \gamma_0 + 1)} \right.$$

$$+ \frac{t^{\alpha-1}}{|\Delta_1|} \sum_{i=1}^p \frac{1}{\Gamma(\alpha - \gamma_i + 1)} \left| \int_0^1 s^{\alpha-\gamma_i} dH_i(s) \right| \right)$$

$$\times (\|u_1 - u_2\| + \|v_1 - v_2\|), \quad \forall t \in [0,1].$$

Then we conclude

$$\|\mathcal{A}_1(u_1, v_1) - \mathcal{A}_1(u_2, v_2)\| \leq L_1 M_5 M_7 (\|u_1 - u_2\| + \|v_1 - v_2\|).$$

$$(6.24)$$

By similar computation, we also find

$$\|\mathcal{A}_2(u_1, v_1) - \mathcal{A}_2(u_2, v_2)\| \le L_2 M_6 M_9(\|u_1 - u_2\| + \|v_1 - v_2\|).$$

(6.25)

Therefore, by (6.24) and (6.25), we obtain

$$
\begin{aligned}
\|\mathcal{A}(u_1, v_1) &- \mathcal{A}(u_2, v_2)\|_Y \\
&= \|\mathcal{A}_1(u_1, v_1) - \mathcal{A}_1(u_2, v_2)\| + \|\mathcal{A}_2(u_1, v_1) - \mathcal{A}_2(u_2, v_2)\| \\
&\le (L_1 M_5 M_7 + L_2 M_6 M_9)(\|u_1 - u_2\| + \|v_1 - v_2\|) \\
&= \Xi \|(u_1, v_1) - (u_2, v_2)\|_Y.
\end{aligned}
$$

By using the condition $\Xi < 1$, we deduce that operator $\mathcal{A}$ is a contraction. By Theorem 1.2.1, we conclude that operator $\mathcal{A}$ has a unique fixed point $(u, v) \in \overline{B}_r$, which is the unique solution of problem (6.12), (6.13) on $[0, 1]$. □

Theorem 6.2.2. *Assume that* (I1) *and*

(I3) *The functions f, $g : [0, 1] \times \mathbb{R}^4 \to \mathbb{R}$ are continuous and there exist real constants c_i, $d_i \ge 0$, $i = 0, \ldots, 4$, and at least one of c_0 and d_0 is positive, such that*

$$|f(t, x_1, x_2, x_3, x_4)| \le c_0 + \sum_{i=1}^{4} c_i |x_i|,$$

$$|g(t, y_1, y_2, y_3, y_4)| \le d_0 + \sum_{i=1}^{4} d_i |y_i|,$$

for all $t \in [0, 1]$, x_i, $y_i \in \mathbb{R}$, $i = 1, \ldots, 4$,

hold. If $\Xi_1 := \max\{M_{11}, M_{12}\} < 1$, where $M_{11} = (c_1 + \frac{c_3}{\Gamma(\theta_1+1)})M_7 + (d_1 + \frac{d_3}{\Gamma(\theta_2+1)})M_9$ and $M_{12} = (c_2 + \frac{c_4}{\Gamma(\sigma_1+1)})M_7 + (d_2 + \frac{d_4}{\Gamma(\sigma_2+1)})M_9$, then the boundary value problem (6.12), (6.13) has at least one solution $(u(t), v(t))$, $t \in [0, 1]$.

Proof. We prove that operator $\mathcal{A}$ is completely continuous. By the continuity of functions f and g, we obtain that operators $\mathcal{A}_1$ and $\mathcal{A}_2$ are continuous, and then $\mathcal{A}$ is a continuous operator. Next, we prove that $\mathcal{A}$ is a compact operator. Let $\Omega \subset Y$ be a bounded set. Then there exist positive constants

L_3 and L_4 such that

$$|f(t, u(t), v(t), I_{0+}^{\theta_1} u(t), I_{0+}^{\sigma_1} v(t))| \leq L_3,$$

$$|g(t, u(t), v(t), I_{0+}^{\theta_2} u(t), I_{0+}^{\sigma_2} v(t))| \leq L_4,$$

for all $(u, v) \in \Omega$ and $t \in [0, 1]$. Therefore, we deduce as in the proof of Theorem 6.2.1 that

$$|\mathcal{A}_1(u, v)(t)| \leq L_3 M_7, \quad |\mathcal{A}_2(u, v)(t)| \leq L_4 M_9,$$

$$\forall t \in [0, 1], \ (u, v) \in \Omega.$$

So, we obtain

$$\|\mathcal{A}_1(u, v)\| \leq L_3 M_7, \quad \|\mathcal{A}_2(u, v)\| \leq L_4 M_9,$$

$$\|\mathcal{A}(u, v)\|_Y \leq L_3 M_7 + L_4 M_9, \quad \forall (u, v) \in \Omega,$$

and then $\mathcal{A}(\Omega)$ is uniformly bounded.

Next, we will prove that the functions from $\mathcal{A}(\Omega)$ are equicontinuous. Let $(u, v) \in \Omega$ and $t_1, t_2 \in [0, 1]$ with $t_1 < t_2$. Then we have

$$|\mathcal{A}_1(u, v)(t_2) - \mathcal{A}_1(u, v)(t_1)|$$

$$\leq \left| -\frac{1}{\Gamma(\alpha)} \int_0^{t_2} (t_2 - s)^{\alpha-1} f(s, u(s), v(s), I_{0+}^{\theta_1} u(s), I_{0+}^{\sigma_1} v(s)) \, ds \right.$$

$$+ \frac{1}{\Gamma(\alpha)} \int_0^{t_1} (t_1 - s)^{\alpha-1} f(s, u(s), v(s), I_{0+}^{\theta_1} u(s), I_{0+}^{\sigma_1} v(s)) \, ds \bigg|$$

$$+ \frac{t_2^{\alpha-1} - t_1^{\alpha-1}}{|\Delta_1| \Gamma(\alpha - \gamma_0)} \left| \int_0^1 (1 - s)^{\alpha - \gamma_0 - 1} \right.$$

$$\times f(s, u(s), v(s), I_{0+}^{\theta_1} u(s), I_{0+}^{\sigma_1} v(s)) \, ds \bigg| + \frac{t_2^{\alpha-1} - t_1^{\alpha-1}}{|\Delta_1|}$$

$$\times \sum_{i=1}^p \frac{1}{\Gamma(\alpha - \gamma_i)} \left| \int_0^1 \left(\int_0^s (s - \tau)^{\alpha - \gamma_i - 1} \right. \right.$$

$$\times f(\tau, u(\tau), v(\tau), I_{0+}^{\theta_1} u(\tau), I_{0+}^{\sigma_1} v(\tau)) \, d\tau \bigg) dH_i(s) \bigg|$$

$$\leq \frac{L_3}{\Gamma(\alpha)} \int_0^{t_1} [(t_2 - s)^{\alpha-1} - (t_1 - s)^{\alpha-1}] ds + \frac{L_3}{\Gamma(\alpha)} \int_{t_1}^{t_2} (t_2 - s)^{\alpha-1} ds$$

$$+ \frac{L_3(t_2^{\alpha-1} - t_1^{\alpha-1})}{|\Delta_1| \Gamma(\alpha - \gamma_0)} \int_0^1 (1 - s)^{\alpha - \gamma_0 - 1} \, ds + \frac{L_3(t_2^{\alpha-1} - t_1^{\alpha-1})}{|\Delta_1|}$$

$$\times \sum_{i=1}^{p} \frac{1}{\Gamma(\alpha - \gamma_i)} \left| \int_0^1 \left(\int_0^s (s - \tau)^{\alpha - \gamma_i - 1} \, d\tau \right) dH_i(s) \right|$$

$$= \frac{L_3}{\Gamma(\alpha + 1)} (t_2^{\alpha} - t_1^{\alpha}) + L_3 M_8 (t_2^{\alpha - 1} - t_1^{\alpha - 1}).$$

Then

$$|\mathcal{A}_1(u, v)(t_2) - \mathcal{A}_1(u, v)(t_1)| \to 0, \quad \text{as } t_2 \to t_1,$$

$$\text{uniformly with respect to } (u, v) \in \Omega.$$

In a similar manner, we find

$$|\mathcal{A}_2(u, v)(t_2) - \mathcal{A}_2(u, v)(t_1)| \leq \frac{L_4}{\Gamma(\beta + 1)} (t_2^{\beta} - t_1^{\beta}) + L_4 M_{10} (t_2^{\beta - 1} - t_1^{\beta - 1}),$$

and so

$$|\mathcal{A}_2(u, v)(t_2) - \mathcal{A}_2(u, v)(t_1)| \to 0, \quad \text{as } t_2 \to t_1,$$

$$\text{uniformly with respect to } (u, v) \in \Omega.$$

Thus, $\mathcal{A}_1(\Omega)$ and $\mathcal{A}_2(\Omega)$ are equicontinuous, and then $\mathcal{A}(\Omega)$ is also equicontinuous. Hence, by the Arzela-Ascoli theorem, we conclude that $\mathcal{A}(\Omega)$ is relatively compact, and then $\mathcal{A}$ is compact. Therefore, we deduce that $\mathcal{A}$ is completely continuous.

We will show next that the set $V = \{(u, v) \in Y, \ (u, v) = \nu \mathcal{A}(u, v), \ 0 < \nu < 1\}$ is bounded. Let $(u, v) \in V$, that is, $(u, v) = \nu \mathcal{A}(u, v)$ for some $\nu \in (0, 1)$. Then for any $t \in [0, 1]$, we have $u(t) = \nu \mathcal{A}_1(u, v)(t)$, $v(t) = \nu \mathcal{A}_2(u, v)(t)$. Hence, we find $|u(t)| \leq |\mathcal{A}_1(u, v)(t)|$ and $|v(t)| \leq |\mathcal{A}_2(u, v)(t)|$ for all $t \in [0, 1]$.

By (I3), we obtain

$$|u(t)| \leq |\mathcal{A}_1(u, v)(t)|$$

$$\leq \frac{1}{\Gamma(\alpha)} \int_0^t (t - s)^{\alpha - 1} \Big[c_0 + c_1 |u(s)| + c_2 |v(s)|$$

$$+ c_3 |I_{0+}^{\theta_1} u(s)| + c_4 |I_{0+}^{\sigma_1} v(s)| \Big] ds$$

$$+ \frac{t^{\alpha - 1}}{|\Delta_1| \Gamma(\alpha - \gamma_0)} \int_0^1 (1 - s)^{\alpha - \gamma_0 - 1} \Big[c_0 + c_1 |u(s)| + c_2 |v(s)|$$

$$+ c_3 |I_{0+}^{\theta_1} u(s)| + c_4 |I_{0+}^{\sigma_1} v(s)| \Big] ds$$

$$+ \frac{t^{\alpha-1}}{|\Delta_1|} \sum_{i=1}^{p} \frac{1}{\Gamma(\alpha-\gamma_i)} \left| \int_0^1 \left(\int_0^s (s-\tau)^{\alpha-\gamma_i-1} \left[c_0 + c_1 |u(\tau)| \right. \right. \right.$$

$$\left. \left. \left. + c_2 |v(\tau)| + c_3 |I_{0+}^{\theta_1} u(\tau)| + c_4 |I_{0+}^{\sigma_1} v(\tau)| \right] d\tau \right) dH_i(s) \right|$$

$$\leq \left(c_0 + c_1 \|u\| + c_2 \|v\| + \frac{c_3}{\Gamma(\theta_1+1)} \|u\| + \frac{c_4}{\Gamma(\sigma_1+1)} \|v\| \right)$$

$$\times \left[\frac{t^\alpha}{\Gamma(\alpha+1)} + \frac{t^{\alpha-1}}{|\Delta_1|\Gamma(\alpha-\gamma_0+1)} \right.$$

$$\left. + \frac{t^{\alpha-1}}{|\Delta_1|} \sum_{i=1}^{p} \frac{1}{\Gamma(\alpha-\gamma_i+1)} \left| \int_0^1 s^{\alpha-\gamma_i} dH_i(s) \right| \right].$$

Then we deduce

$$\|u\| \leq \left[c_0 + \left(c_1 + \frac{c_3}{\Gamma(\theta_1+1)} \right) \|u\| + \left(c_2 + \frac{c_4}{\Gamma(\sigma_1+1)} \right) \|v\| \right] M_7.$$

In a similar manner, we have

$$\|v\| \leq \left[d_0 + \left(d_1 + \frac{d_3}{\Gamma(\theta_2+1)} \right) \|u\| + \left(d_2 + \frac{d_4}{\Gamma(\sigma_2+1)} \right) \|v\| \right] M_9,$$

and therefore

$$\|(u,v)\|_Y \leq c_0 M_7 + d_0 M_9 + M_{11} \|u\| + M_{12} \|v\|$$

$$\leq c_0 M_7 + d_0 M_9 + \Xi_1 \|(u,v)\|_Y.$$

Because $\Xi_1 < 1$, we obtain

$$\|(u,v)\|_Y \leq (c_0 M_7 + d_0 M_8)(1 - \Xi_1)^{-1}, \quad \forall (u,v) \in V.$$

So, we conclude that the set V is bounded.

By Theorem 1.2.5, we deduce that operator $\mathcal{A}$ has at least one fixed point, which is a solution for our problem (6.12), (6.13). $\qquad \square$

Theorem 6.2.3. *Assume that* (I1), (I2) *and*

(I4) *There exist the functions* $\phi_1, \phi_2 \in C([0,1], [0,\infty))$ *such that*

$$|f(t, x_1, x_2, x_3, x_4)| \leq \phi_1(t), \quad |g(t, x_1, x_2, x_3, x_4)| \leq \phi_2(t),$$

for all $t \in [0,1]$, $x_i \in \mathbb{R}$, $i = 1, \ldots, 4$,

hold. If $\Xi_2 := L_1 M_5 \frac{1}{\Gamma(\alpha+1)} + L_2 M_6 \frac{1}{\Gamma(\beta+1)} < 1$, *then problem* (6.12), (6.13) *has at least one solution on* $[0,1]$.

Proof. We fix $r_1 > 0$ such that $r_1 \geq M_7\|\phi_1\| + M_9\|\phi_2\|$. We consider the set $\overline{B}_{r_1} = \{(u,v) \in Y, \ \|(u,v)\|_Y \leq r_1\}$, and we introduce the operators $D = (D_1, D_2) : \overline{B}_{r_1} \to Y$ and $E = (E_1, E_2) : \overline{B}_{r_1} \to Y$, where $D_1, D_2, E_1, E_2 : \overline{B}_{r_1} \to X$ are defined by

$$D_1(u,v)(t) = -\frac{1}{\Gamma(\alpha)} \int_0^t (t-s)^{\alpha-1} f(s, u(s), v(s), I_{0+}^{\theta_1} u(s), I_{0+}^{\sigma_1} v(s)) \, ds,$$

$$E_1(u,v)(t) = \frac{t^{\alpha-1}}{\Delta_1 \Gamma(\alpha - \gamma_0)} \int_0^1 (1-s)^{\alpha-\gamma_0-1}$$
$$\times f(s, u(s), v(s), I_{0+}^{\theta_1} u(s), I_{0+}^{\sigma_1} v(s)) \, ds$$
$$- \frac{t^{\alpha-1}}{\Delta_1} \sum_{i=1}^p \frac{1}{\Gamma(\alpha - \gamma_i)} \int_0^1 \left(\int_0^s (s-\tau)^{\alpha-\gamma_i-1} \right.$$
$$\left. \times f(\tau, u(\tau), v(\tau), I_{0+}^{\theta_1} u(\tau), I_{0+}^{\sigma_1} v(\tau)) \, d\tau \right) dH_i(s),$$

$$\tag{6.26}$$

$$D_2(u,v)(t) = -\frac{1}{\Gamma(\beta)} \int_0^t (t-s)^{\beta-1} g(s, u(s), v(s), I_{0+}^{\theta_2} u(s), I_{0+}^{\sigma_2} v(s)) \, ds,$$

$$E_2(u,v)(t) = \frac{t^{\beta-1}}{\Delta_2 \Gamma(\beta - \delta_0)}$$
$$\times \int_0^1 (1-s)^{\beta-\delta_0-1} g(s, u(s), v(s), I_{0+}^{\theta_2} u(s), I_{0+}^{\sigma_2} v(s)) \, ds$$
$$- \frac{t^{\beta-1}}{\Delta_2} \sum_{i=1}^q \frac{1}{\Gamma(\beta - \delta_i)} \int_0^1 \left(\int_0^s (s-\tau)^{\beta-\delta_i-1} \right.$$
$$\left. \times g(\tau, u(\tau), v(\tau), I_{0+}^{\theta_2} u(\tau), I_{0+}^{\sigma_2} v(\tau)) \, d\tau \right) dK_i(s),$$

for $t \in [0,1]$ and $(u,v) \in \overline{B}_{r_1}$. So $\mathcal{A}_1 = D_1 + E_1$, $\mathcal{A}_2 = D_2 + E_2$ and $\mathcal{A} = D + E$.

By using (I4), we obtain for all $(u_1, v_1), (u_2, v_2) \in \overline{B}_{r_1}$ that

$$\|D(u_1, v_1) + E(u_2, v_2)\|_Y$$
$$\leq \|D(u_1, v_1)\|_Y + \|E(u_2, v_2)\|_Y = \|D_1(u_1, v_1)\| + \|D_2(u_1, v_1)\|$$
$$+ \|E_1(u_2, v_2)\| + \|E_2(u_2, v_2)\| \leq \frac{1}{\Gamma(\alpha+1)}\|\phi_1\| + \frac{1}{\Gamma(\beta+1)}\|\phi_2\|$$

$$+ \left(\frac{1}{|\Delta_1|\Gamma(\alpha - \gamma_0 + 1)} + \frac{1}{|\Delta_1|} \sum_{i=1}^{p} \frac{1}{\Gamma(\alpha - \gamma_i + 1)} \right.$$

$$\left. \times \left| \int_0^1 s^{\alpha - \gamma_i} \, dH_i(s) \right| \right) \|\phi_1\|$$

$$+ \left(\frac{1}{|\Delta_2|\Gamma(\beta - \delta_0 + 1)} + \frac{1}{|\Delta_2|} \sum_{i=1}^{q} \frac{1}{\Gamma(\beta - \delta_i + 1)} \right.$$

$$\left. \times \left| \int_0^1 s^{\beta - \delta_i} \, dK_i(s) \right| \right) \|\phi_2\|$$

$$= M_7 \|\phi_1\| + M_9 \|\phi_2\| \le r_1.$$

Hence, $D(u_1, v_1) + E(u_2, v_2) \in \overline{B}_{r_1}$ for all $(u_1, v_1), (u_2, v_2) \in \overline{B}_{r_1}$.

The operator D is a contraction, because

$$\|D(u_1, v_1) - D(u_2, v_2)\|_Y$$

$$= \|D_1(u_1, v_1) - D_1(u_2, v_2)\| + \|D_2(u_1, v_1) - D_2(u_2, v_2)\|$$

$$\le \left(L_1 M_5 \frac{1}{\Gamma(\alpha + 1)} + L_2 M_6 \frac{1}{\Gamma(\beta + 1)} \right) (\|u_1 - u_2\| + \|v_1 - v_2\|)$$

$$= \Xi_2 \|(u_1, v_1) - (u_2, v_2)\|_Y,$$

for all $(u_1, v_1), (u_2, v_2) \in \overline{B}_{r_1}$, and $\Xi_2 < 1$.

The continuity of f and g implies that operator E is continuous on $\overline{B}_{r_1}$. We prove in what follows that E is compact. The functions from $E(\overline{B}_{r_1})$ are uniformly bounded, because

$$\|E(u, v)\|_Y = \|E_1(u, v)\| + \|E_2(u, v)\|$$

$$\le \left(\frac{1}{|\Delta_1|\Gamma(\alpha - \gamma_0 + 1)} + \frac{1}{|\Delta_1|} \sum_{i=1}^{p} \frac{1}{\Gamma(\alpha - \gamma_i + 1)} \right.$$

$$\left. \times \left| \int_0^1 s^{\alpha - \gamma_i} \, dH_i(s) \right| \right) \|\phi_1\|$$

$$+ \left(\frac{1}{|\Delta_2|\Gamma(\beta - \delta_0 + 1)} + \frac{1}{|\Delta_2|} \sum_{i=1}^{q} \frac{1}{\Gamma(\beta - \delta_i + 1)} \right.$$

$$\left. \times \left| \int_0^1 s^{\beta - \delta_i} \, dK_i(s) \right| \right) \|\phi_2\|$$

$$= M_8 \|\phi_1\| + M_{10} \|\phi_2\|, \quad \forall (u, v) \in \overline{B}_{r_1}.$$

We prove now that the functions from $E(\overline{B}_{r_1})$ are equicontinuous. We denote by

$$
\begin{aligned}
\Psi_{r_1} = \sup \Big\{ |f(t, u, v, x, y)|, \; t \in [0, 1], \; |u| \le r_1, \; |v| \le r_1, \\
|x| \le \frac{r_1}{\Gamma(\theta_1 + 1)}, \; |y| \le \frac{r_1}{\Gamma(\sigma_1 + 1)} \Big\}, \\
\Theta_{r_1} = \sup \Big\{ |g(t, u, v, x, y)|, \; t \in [0, 1], \; |u| \le r_1, \; |v| \le r_1, \\
|x| \le \frac{r_1}{\Gamma(\theta_2 + 1)}, \; |y| \le \frac{r_1}{\Gamma(\sigma_2 + 1)} \Big\}.
\end{aligned}
\tag{6.27}
$$

Then for $(u, v) \in \overline{B}_{r_1}$, and $t_1, t_2 \in [0, 1]$ with $t_1 < t_2$, we obtain

$$
\begin{aligned}
&|E_1(u, v)(t_2) - E_1(u, v)(t_1)| \\
&\le \frac{t_2^{\alpha - 1} - t_1^{\alpha - 1}}{|\Delta_1| \Gamma(\alpha - \gamma_0)} \int_0^1 (1 - s)^{\alpha - \gamma_0 - 1} \Psi_{r_1} \, ds \\
&\quad + \frac{t_2^{\alpha - 1} - t_1^{\alpha - 1}}{|\Delta_1|} \sum_{i=1}^{p} \frac{1}{\Gamma(\alpha - \gamma_i)} \\
&\qquad \times \left| \int_0^1 \left(\int_0^s (s - \tau)^{\alpha - \gamma_i - 1} \Psi_{r_1} \, d\tau \right) dH_i(s) \right| \\
&\le \Psi_{r_1}(t_2^{\alpha - 1} - t_1^{\alpha - 1}) \left[\frac{1}{|\Delta_1| \Gamma(\alpha - \gamma_0 + 1)} + \frac{1}{|\Delta_1|} \sum_{i=1}^{p} \frac{1}{\Gamma(\alpha - \gamma_i + 1)} \right. \\
&\qquad \left. \times \left| \int_0^1 s^{\alpha - \gamma_i} \, dH_i(s) \right| \right] \\
&= \Psi_{r_1} M_8 (t_2^{\alpha - 1} - t_1^{\alpha - 1}), \\
&|E_2(u, v)(t_2) - E_2(u, v)(t_1)| \\
&\le \frac{t_2^{\beta - 1} - t_1^{\beta - 1}}{|\Delta_2| \Gamma(\beta - \delta_0)} \int_0^1 (1 - s)^{\beta - \delta_0 - 1} \Theta_{r_1} \, ds \\
&\quad + \frac{t_2^{\beta - 1} - t_1^{\beta - 1}}{|\Delta_2|} \sum_{i=1}^{q} \frac{1}{\Gamma(\beta - \delta_i)} \\
&\qquad \times \left| \int_0^1 \left(\int_0^s (s - \tau)^{\beta - \delta_i - 1} \Theta_{r_1} \, d\tau \right) dK_i(s) \right|
\end{aligned}
$$

$$\leq \Theta_{r_1}(t_2^{\beta-1} - t_1^{\beta-1})\left[\frac{1}{|\Delta_2|\Gamma(\beta-\delta_0+1)} + \frac{1}{|\Delta_2|}\sum_{i=1}^{q}\frac{1}{\Gamma(\beta-\delta_i+1)}\right.$$

$$\left. \times \left|\int_0^1 s^{\beta-\delta_i}\,dK_i(s)\right|\right]$$

$$= \Theta_{r_1}M_{10}(t_2^{\beta-1} - t_1^{\beta-1}).$$

Hence, we find

$$|E_1(u,v)(t_2) - E_1(u,v)(t_1)| \to 0, \quad |E_2(u,v)(t_2) - E_2(u,v)(t_1)| \to 0,$$

as $t_2 \to t_1$ uniformly with respect to $(u,v) \in \overline{B}_{r_1}$. So, $E_1(\overline{B}_{r_1})$ and $E_2(\overline{B}_{r_1})$ are equicontinuous, and then $E(\overline{B}_{r_1})$ is also equicontinuous. By using the Arzela-Ascoli theorem, we deduce that the set $E(\overline{B}_{r_1})$ is relatively compact. Therefore E is a compact operator on $\overline{B}_{r_1}$. By Theorem 1.2.3, we conclude that there exists a fixed point of operator $D + E(= \mathcal{A})$, which is a solution of problem (6.12), (6.13). $\qquad\square$

Theorem 6.2.4. *Assume that* (I1), (I2) *and* (I4) *hold. If* $\Xi_3 := L_1 M_5 M_8 + L_2 M_6 M_{10} < 1$, *then problem* (6.12), (6.13) *has at least one solution* (u,v) *on* $[0,1]$.

Proof. We consider again a positive number $r_1 \geq M_7\|\phi_1\| + M_9\|\phi_2\|$, and the operators D and E defined on $\overline{B}_{r_1}$ given by (6.26). As in the proof of Theorem 6.2.3, we obtain that $D(u_1, v_1) + E(u_2, v_2) \in \overline{B}_{r_1}$ for all $(u_1, v_1), (u_2, v_2) \in \overline{B}_{r_1}$.

The operator E is a contraction, because

$$\|E(u_1, v_1) - E(u_2, v_2)\|_Y$$

$$= \|E_1(u_1, v_1) - E_1(u_2, v_2)\| + \|E_2(u_1, v_1) - E_2(u_2, v_2)\|$$

$$\leq L_1 M_5 M_8(\|u_1 - u_2\| + \|v_1 - v_2\|)$$

$$\quad + L_2 M_6 M_{10}(\|u_1 - u_2\| + \|v_1 - v_2\|)$$

$$= (L_1 M_5 M_8 + L_2 M_6 M_{10})\|(u_1, v_1) - (u_2, v_2)\|_Y$$

$$= \Xi_3\|(u_1, v_1) - (u_2, v_2)\|_Y,$$

for all $(u_1, v_1), (u_2, v_2) \in \overline{B}_{r_1}$, with $\Xi_3 < 1$.

Next, the continuity of f and g implies that operator D is continuous on $\overline{B}_{r_1}$. We show now that D is compact. The functions from $D(\overline{B}_{r_1})$ are uniformly bounded, because

$$\|D(u,v)\|_Y = \|D_1(u,v)\| + \|D_2(u,v)\| \leq \frac{1}{\Gamma(\alpha+1)}\|\phi_1\|$$

$$+ \frac{1}{\Gamma(\beta+1)}\|\phi_2\|, \quad \forall\,(u,v) \in \overline{B}_{r_1}.$$

Now, we prove that $D(\overline{B}_{r_1})$ is equicontinuous. By using Ψ_{r_1} and Θ_{r_1} defined in (6.27), we find for $(u,v) \in \overline{B}_{r_1}$ and $t_1,\,t_2 \in [0,1]$ with $t_1 < t_2$ that

$$|D_1(u,v)(t_2) - D_1(u,v)(t_1)| \leq \frac{\Psi_{r_1}}{\Gamma(\alpha+1)}(t_2^\alpha - t_1^\alpha),$$

$$|D_2(u,v)(t_2) - D_2(u,v)(t_1)| \leq \frac{\Theta_{r_1}}{\Gamma(\beta+1)}(t_2^\beta - t_1^\beta).$$

Then we obtain

$$|D_1(u,v)(t_2) - D_1(u,v)(t_1)| \to 0, \quad |D_2(u,v)(t_2) - D_2(u,v)(t_1)| \to 0,$$

as $t_2 \to t_1$ uniformly with respect to $(u,v) \in \overline{B}_{r_1}$. We deduce that $D_1(\overline{B}_{r_1})$ and $D_2(\overline{B}_{r_1})$ are equicontinuous, and so $D(\overline{B}_{r_1})$ is equicontinuous. By using the Arzela-Ascoli theorem, we conclude that the set $D(\overline{B}_{r_1})$ is relatively compact. Then D is a compact operator on $\overline{B}_{r_1}$. By Theorem 1.2.3, we deduce that there exists a fixed point of operator $D + E(= \mathcal{A})$, which is a solution of problem (6.12), (6.13). □

Theorem 6.2.5. *Assume that* (I1) *and*

(I5) *The functions* $f,\,g : [0,1] \times \mathbb{R}^4 \to \mathbb{R}$ *are continuous and there exist the constants* $a_i \geq 0$, $i = 0,\ldots,4$ *with at least one nonzero, the constants* $b_i \geq 0$, $i = 0,\ldots,4$ *with at least one nonzero, and* $l_i,\,m_i \in (0,1)$, $i = 1,\ldots,4$ *such that*

$$|f(t,x_1,x_2,x_3,x_4)| \leq a_0 + \sum_{i=1}^{4} a_i|x_i|^{l_i},$$

$$|g(t,y_1,y_2,y_3,y_4)| \leq b_0 + \sum_{i=1}^{4} b_i|y_i|^{m_i},$$

for all $t \in [0,1]$, $x_i,\,y_i \in \mathbb{R}$, $i = 1,\ldots,4$,

hold. Then problem (6.12), (6.13) *has at least one solution.*

Proof. Let $\overline{B}_R = \{(u,v) \in Y, \ \|(u,v)\|_Y \leq R\}$, where

$$R \geq \max \left\{ 10a_0 M_7, (10a_1 M_7)^{\frac{1}{1-l_1}}, (10a_2 M_7)^{\frac{1}{1-l_2}}, \right.$$

$$\left(\frac{10a_3 M_7}{(\Gamma(\theta_1 + 1))^{l_3}} \right)^{\frac{1}{1-l_3}}, \left(\frac{10a_4 M_7}{(\Gamma(\sigma_1 + 1))^{l_4}} \right)^{\frac{1}{1-l_4}},$$

$$10b_0 M_9, (10b_1 M_9)^{\frac{1}{1-m_1}}, (10b_2 M_9)^{\frac{1}{1-m_2}},$$

$$\left. \left(\frac{10b_3 M_9}{(\Gamma(\theta_2 + 1))^{m_3}} \right)^{\frac{1}{1-m_3}}, \left(\frac{10b_4 M_9}{(\Gamma(\sigma_2 + 1))^{m_4}} \right)^{\frac{1}{1-m_4}} \right\}.$$

We prove that $\mathcal{A} : \overline{B}_R \to \overline{B}_R$. For $(u,v) \in \overline{B}_R$, we deduce

$$|\mathcal{A}_1(u,v)(t)| \leq \left(a_0 + a_1 R^{l_1} + a_2 R^{l_2} + a_3 \frac{R^{l_3}}{(\Gamma(\theta_1 + 1))^{l_3}} \right.$$

$$\left. + a_4 \frac{R^{l_4}}{(\Gamma(\sigma_1 + 1))^{l_4}} \right) M_7 \leq \frac{R}{2},$$

$$|\mathcal{A}_2(u,v)(t)| \leq \left(b_0 + b_1 R^{m_1} + b_2 R^{m_2} + b_3 \frac{R^{m_3}}{(\Gamma(\theta_2 + 1))^{m_3}} \right.$$

$$\left. + b_4 \frac{R^{m_4}}{(\Gamma(\sigma_2 + 1))^{m_4}} \right) M_9 \leq \frac{R}{2},$$

for all $t \in [0,1]$. Then we obtain

$$\|\mathcal{A}(u,v)\|_Y = \|\mathcal{A}_1(u,v)\| + \|\mathcal{A}_2(u,v)\| \leq R, \ \forall (u,v) \in \overline{B}_R,$$

which implies that $\mathcal{A}(\overline{B}_R) \subset \overline{B}_R$.

Because the functions f and g are continuous, we conclude that operator $\mathcal{A}$ is continuous on $\overline{B}_R$. In addition, the functions from $\mathcal{A}(\overline{B}_R)$ are uniformly bounded and equicontinuous. Indeed, by using the notations (6.27) with r_1 replaced by R, we find for any $(u,v) \in \overline{B}_R$ and $t_1, t_2 \in [0,1]$, $t_1 < t_2$ that

$$|\mathcal{A}_1(u,v)(t_2) - \mathcal{A}_1(u,v)(t_1)| \leq \frac{\Psi_R}{\Gamma(\alpha + 1)}(t_2^\alpha - t_1^\alpha)$$

$$+ \Psi_R M_8 (t_2^{\alpha-1} - t_1^{\alpha-1}),$$

$$|\mathcal{A}_2(u,v)(t_2) - \mathcal{A}_2(u,v)(t_1)| \leq \frac{\Theta_R}{\Gamma(\beta + 1)}(t_2^\beta - t_1^\beta)$$

$$+ \Theta_R M_{10} (t_2^{\beta-1} - t_1^{\beta-1}).$$

Therefore,

$$|\mathcal{A}_1(u,v)(t_2) - \mathcal{A}_1(u,v)(t_1)| \to 0,$$

$$|\mathcal{A}_2(u,v)(t_2) - \mathcal{A}_2(u,v)(t_1)| \to 0, \text{ as } t_2 \to t_1,$$

uniformly with respect to $(u,v) \in \overline{B}_R$. By the Arzela-Ascoli theorem, we deduce that $\mathcal{A}(\overline{B}_R)$ is relatively compact, and then $\mathcal{A}$ is a compact operator. By Theorem 1.2.4, we conclude that operator A has at least one fixed point (u,v) in $\overline{B}_R$, which is a solution of our problem (6.12), (6.13). $\qquad \square$

Theorem 6.2.6. *Assume that* (I1) *and*

(I6) *The functions* $f, g : [0,1] \times \mathbb{R}^4 \to \mathbb{R}$ *are continuous and there exist* $p_i \geq 0$, $i = 0, \ldots, 4$ *with at least one nonzero,* $q_i \geq 0$, $i = 0, \ldots, 4$ *with at least one nonzero, and nondecreasing functions* $h_i, k_i \in C([0, \infty), [0, \infty))$ $i = 1, \ldots, 4$ *such that*

$$|f(t, x_1, x_2, x_3, x_4)| \leq p_0 + \sum_{i=1}^{4} p_i h_i(|x_i|),$$

$$|g(t, y_1, y_2, y_3, y_4)| \leq q_0 + \sum_{i=1}^{4} q_i k_i(|y_i|),$$

for all $t \in [0,1]$, $x_i, y_i \in \mathbb{R}$, $i = 1, \ldots, 4$,

hold. If there exists $\Xi_0 > 0$ *such that*

$$\left(p_0 + p_1 h_1(\Xi_0) + p_2 h_2(\Xi_0) + p_3 h_3 \left(\frac{\Xi_0}{\Gamma(\theta_1 + 1)} \right) \right.$$

$$\left. + p_4 h_4 \left(\frac{\Xi_0}{\Gamma(\sigma_1 + 1)} \right) \right) M_7$$

$$+ \left(q_0 + q_1 k_1(\Xi_0) + q_2 k_2(\Xi_0) + q_3 k_3 \left(\frac{\Xi_0}{\Gamma(\theta_2 + 1)} \right) \right.$$

$$\left. + q_4 k_4 \left(\frac{\Xi_0}{\Gamma(\sigma_2 + 1)} \right) \right) M_9 < \Xi_0, \tag{6.28}$$

then problem (6.12), (6.13) *has at least one solution on* $[0,1]$.

Proof. We consider the set $\overline{B}_{\Xi_0} = \{(u,v) \in Y, \ \|(u,v)\|_Y \leq \Xi_0\}$, where Ξ_0 is given in the theorem. We will prove that $\mathcal{A} : \overline{B}_{\Xi_0} \to \overline{B}_{\Xi_0}$. For $(u,v) \in \overline{B}_{\Xi_0}$

and $t \in [0, 1]$ we obtain

$$
|\mathcal{A}_1(u, v)(t)| \le \left(p_0 + p_1 h_1(\Xi_0) + p_2 h_2(\Xi_0) + p_3 h_3 \left(\frac{\Xi_0}{\Gamma(\theta_1 + 1)} \right) \right.
$$
$$
\left. + p_4 h_4 \left(\frac{\Xi_0}{\Gamma(\sigma_1 + 1)} \right) \right) M_7,
$$

$$
|\mathcal{A}_2(u, v)(t)| \le \left(q_0 + q_1 k_1(\Xi_0) + q_2 k_2(\Xi_0) + q_3 k_3 \left(\frac{\Xi_0}{\Gamma(\theta_2 + 1)} \right) \right.
$$
$$
\left. + q_4 k_4 \left(\frac{\Xi_0}{\Gamma(\sigma_2 + 1)} \right) \right) M_9,
$$

and then for all $(u, v) \in \overline{B}_{\Xi_0}$ we have

$$
\|\mathcal{A}(u, v)\|_Y \le \left(p_0 + p_1 h_1(\Xi_0) + p_2 h_2(\Xi_0) + p_3 h_3 \left(\frac{\Xi_0}{\Gamma(\theta_1 + 1)} \right) \right.
$$
$$
\left. + p_4 h_4 \left(\frac{\Xi_0}{\Gamma(\sigma_1 + 1)} \right) \right) M_7
$$
$$
+ \left(q_0 + q_1 k_1(\Xi_0) + q_2 k_2(\Xi_0) + q_3 k_3 \left(\frac{\Xi_0}{\Gamma(\theta_2 + 1)} \right) \right.
$$
$$
\left. + q_4 k_4 \left(\frac{\Xi_0}{\Gamma(\sigma_2 + 1)} \right) \right) M_9 < \Xi_0.
$$

Then $\mathcal{A}(\overline{B}_{\Xi_0}) \subset \overline{B}_{\Xi_0}$. In a similar manner used in the proof of Theorem 6.2.5, we can prove that operator $\mathcal{A}$ is completely continuous.

We suppose now that there exists $(u, v) \in \partial B_{\Xi_0}$ such that $(u, v) = \nu \mathcal{A}(u, v)$ for some $\nu \in (0, 1)$. We obtain as above that $\|(u, v)\|_Y \le \|\mathcal{A}(u, v)\|_Y < \Xi_0$, which is a contradiction, because $(u, v) \in \partial B_{\Xi_0}$. Then by Theorem 1.2.6, we conclude that operator $\mathcal{A}$ has a fixed point $(u, v) \in \overline{B}_{\Xi_0}$, and so problem (6.12), (6.13) has at least one solution. $\qquad \square$

6.2.3 *Examples*

Let $\alpha = \frac{5}{2}$ $(n = 3)$, $\beta = \frac{10}{3}$ $(m = 4)$, $\theta_1 = \frac{1}{3}$, $\sigma_1 = \frac{9}{4}$, $\theta_2 = \frac{16}{5}$, $\sigma_2 = \frac{25}{6}$, $\gamma_0 = \frac{4}{3}$, $\gamma_1 = \frac{1}{2}$, $\gamma_2 = \frac{3}{4}$, $\delta_0 = \frac{11}{5}$, $\delta_1 = \frac{1}{6}$, $\delta_2 = \frac{15}{7}$, $H_1(t) = t^3$, $t \in [0, 1]$, $H_2(t) = \{0, \ t \in [0, \frac{1}{3}); \ 2, \ t \in [\frac{1}{3}, 1]\}$, $K_1(t) = \{0, \ t \in [0, \frac{1}{2}); \ 4, \ t \in [\frac{1}{2}, 1], \}$, $K_2(t) = t^2$, $t \in [0, 1]$.

We consider the system of fractional differential equations

$$\begin{cases} D_{0+}^{5/2} u(t) + f(t, u(t), v(t), I_{0+}^{1/3} u(t), I_{0+}^{9/4} v(t)) = 0, & t \in (0,1), \\ D_{0+}^{10/3} v(t) + g(t, u(t), v(t), I_{0+}^{16/5} u(t), I_{0+}^{25/6} v(t)) = 0, & t \in (0,1), \end{cases}$$

(6.29)

with the boundary conditions

$$\begin{cases} u(0) = u'(0) = 0, \quad D_{0+}^{4/3} u(1) = 3 \int_0^1 t^2 D_{0+}^{1/2} u(t)\, dt + 2 D_{0+}^{3/4} u\left(\frac{1}{3}\right), \\ v(0) = v'(0) = v''(0) = 0, \quad D_{0+}^{11/5} v(1) = 4 D_{0+}^{1/6} v\left(\frac{1}{2}\right) \\ \quad + 2 \int_0^1 t D_{0+}^{15/7} v(t)\, dt. \end{cases}$$

(6.30)

We obtain $\Delta_1 \approx -0.83314732 \neq 0$ and $\Delta_2 \approx -0.85088584 \neq 0$. So, assumption (I1) is satisfied. In addition, we have $M_1 \approx 2.11984652$, $M_2 \approx 1.39227116$, $M_3 \approx 1.12892098$, $M_4 \approx 1.03231866$, $M_5 = M_1$, $M_6 = M_3$, $M_7 \approx 1.98819306$, $M_8 \approx 1.68729195$, $M_9 \approx 1.95523852$, $M_{10} \approx 1.84725332$.

Example 1. We consider the functions

$$f(t, x_1, x_2, x_3, x_4) = \frac{1}{\sqrt{4 + t^2}} + \frac{|x_1|}{7(t+1)^3(1 + |x_1|)}$$
$$- \frac{t}{8} \arctan x_2 + \frac{t^2}{t+9} \cos x_3 - \frac{1}{2(t+10)} \sin^2 x_4,$$

$$g(t, y_1, y_2, y_3, y_4) = \frac{2t}{t^2 + 9} - \frac{1}{10} \sin y_1 + \frac{|y_2|}{4(2 + |y_2|)}$$
$$+ \frac{1}{12} \arctan y_3 - \frac{t}{t+20} \cos^2 y_4,$$

for all $t \in [0,1]$, $x_i, y_i \in \mathbb{R}$, $i = 1, \ldots, 4$. We find the inequalities

$$|f(t, x_1, x_2, x_3, x_4) - f(t, \tilde{x}_1, \tilde{x}_2, \tilde{x}_3, \tilde{x}_4)| \leq \frac{1}{7} \sum_{i=1}^4 |x_i - \tilde{x}_i|,$$

$$|g(t, y_1, y_2, y_3, y_4) - g(t, \tilde{y}_1, \tilde{y}_2, \tilde{y}_3, \tilde{y}_4)| \leq \frac{1}{8} \sum_{i=1}^4 |y_i - \tilde{y}_i|,$$

for all $t \in [0,1]$, $x_i, y_i \in \mathbb{R}$, $i = 1, \ldots, 4$. So we have $L_1 = \frac{1}{7}$, $L_2 = \frac{1}{8}$, and $\Xi \approx 0.878 < 1$. Therefore, assumption (I2) is satisfied, and by Theorem 6.2.1 we deduce that problem (6.29), (6.30) has a unique solution $(u(t), v(t))$, $t \in [0,1]$.

Example 2. We consider the functions

$$f(t, x_1, x_2, x_3, x_4) = \frac{t+1}{t^2+3}\left(3\sin t + \frac{1}{4}\sin x_1\right)$$

$$-\frac{1}{(t+3)^2}x_2 + \frac{t}{4}\arctan x_3 - \cos x_4,$$

$$g(t, y_1, y_2, y_3, y_4) = \frac{e^{-t}}{1+t^2} - \frac{1}{3}\sin y_2 + \cos^2 y_3 + \frac{1}{5}\arctan y_4,$$

for all $t \in [0, 1]$, x_i, $y_i \in \mathbb{R}$, $i = 1, \ldots, 4$. Because we have

$$|f(t, x_1, x_2, x_3, x_4)| \le \frac{5}{2} + \frac{1}{8}|x_1| + \frac{1}{9}|x_2| + \frac{1}{4}|x_3|,$$

$$|g(t, y_1, y_2, y_3, y_4)| \le 2 + \frac{1}{3}|y_2| + \frac{1}{5}|y_4|,$$

for all $t \in [0, 1]$, x_i, $y_i \in \mathbb{R}$, $i = 1, \ldots, 4$, then assumption (I3) is satisfied with $c_0 = \frac{5}{2}$, $c_1 = \frac{1}{8}$, $c_2 = \frac{1}{9}$, $c_3 = \frac{1}{4}$, $c_4 = 0$, $d_0 = 2$, $d_1 = 0$, $d_2 = \frac{1}{3}$, $d_3 = 0$, $d_4 = \frac{1}{5}$. Besides, we obtain $M_{11} \approx 0.805142$, $M_{12} \approx 0.885295$ and $\Xi_1 = M_{12} < 1$. Then by Theorem 6.2.2 we conclude that problem (6.29), (6.30) has at least one solution $(u(t), v(t))$, $t \in [0, 1]$.

Example 3. We consider the functions

$$f(t, x_1, x_2, x_3, x_4) = -\frac{1}{4}x_1^{3/5} + \frac{1}{2(1+t)}\arctan x_4^{2/3},$$

$$g(t, y_1, y_2, y_3, y_4) = \frac{e^{-t}}{1+t^4} - \frac{1}{3}|y_2|^{1/2} + \sin|y_3|^{3/4},$$

for all $t \in [0, 1]$, x_i, $y_i \in \mathbb{R}$, $i = 1, \ldots, 4$. Because we obtain

$$|f(t, x_1, x_2, x_3, x_4)| \le \frac{1}{4}|x_1|^{3/5} + \frac{1}{2}|x_4|^{2/3},$$

$$|g(t, y_1, y_2, y_3, y_4)| \le 1 + \frac{1}{3}|y_2|^{1/2} + |y_3|^{3/4},$$

for all $t \in [0, 1]$, x_i, $y_i \in \mathbb{R}$, $i = 1, \ldots, 4$, then assumption (I5) is satisfied with $a_0 = 0$, $a_1 = \frac{1}{4}$, $a_2 = 0$, $a_3 = 0$, $a_4 = \frac{1}{2}$, $b_0 = 1$, $b_1 = 0$, $b_2 = \frac{1}{3}$, $b_3 = 1$, $b_4 = 0$, $l_1 = \frac{3}{5}$, $l_4 = \frac{2}{3}$, $m_2 = \frac{1}{2}$, $m_3 = \frac{3}{4}$. Therefore, by Theorem 6.2.5, we deduce that problem (6.29), (6.30) has at least one solution $(u(t), v(t))$, $t \in [0, 1]$.

Example 4. We consider the functions

$$f(t, x_1, x_2, x_3, x_4) = \frac{(1-t)^3}{10} + \frac{e^{-t}x_2^3}{20(1+x_1^2)} - \frac{t^2 x_3^{1/3}}{5},$$

$$g(t, y_1, y_2, y_3, y_4) = \frac{t^2}{20} + \frac{1-t^2}{25}y_1^2 - \frac{1}{30}y_4^{1/5},$$

for all $t \in [0,1]$, x_i, $y_i \in \mathbb{R}$, $i = 1, \ldots, 4$. Because we have

$$|f(t, x_1, x_2, x_3, x_4)| \leq \frac{1}{10} + \frac{1}{20}|x_2|^3 + \frac{1}{5}|x_3|^{1/3},$$

$$|g(t, y_1, y_2, y_3, y_4)| \leq \frac{1}{20} + \frac{1}{25}|y_1|^2 + \frac{1}{30}|y_4|^{1/5},$$

for all $t \in [0,1]$, x_i, $y_i \in \mathbb{R}$, $i = 1, \ldots, 4$, then assumption (I6) is satisfied with $p_0 = \frac{1}{10}$, $p_1 = 0$, $p_2 = \frac{1}{20}$, $p_3 = \frac{1}{5}$, $p_4 = 0$, $h_1(x) = 0$, $h_2(x) = x^3$, $h_3(x) = x^{1/3}$, $h_4(x) = 0$, $q_0 = \frac{1}{20}$, $q_1 = \frac{1}{25}$, $q_2 = 0$, $q_3 = 0$, $q_4 = \frac{1}{30}$, $k_1(x) = x^2$, $k_2(x) = 0$, $k_3(x) = 0$, $k_4(x) = x^{1/5}$. For $\Xi_0 = 2$, the condition (6.28) is satisfied, because $(p_0 + p_1 h_1(2) + p_2 h_2(2) + p_3 h_3(\frac{2}{\Gamma(\theta_1+1)}) + p_4 h_4(\frac{2}{\Gamma(\sigma_1+1)}))M_7 + (q_0 + q_1 k_1(2) + q_2 k_2(2) + q_3 k_3(\frac{2}{\Gamma(\theta_2+1)}) + q_4 k_4(\frac{2}{\Gamma(\sigma_2+1)}))M_9 \approx 1.96264 < 2$. Therefore, by Theorem 6.2.6 we conclude that problem (6.29), (6.30) has at least one solution $(u(t), v(t))$, $t \in [0,1]$.

Remark 6.2.1. The results presented in this section were published in [81].

6.3 Systems of Riemann–Liouville Fractional Differential Equations with Coupled Boundary Conditions

We consider the nonlinear system of fractional differential equations

$$\begin{cases} D_{0+}^\alpha u(t) + f(t, u(t), v(t), I_{0+}^{\theta_1} u(t), I_{0+}^{\sigma_1} v(t)) = 0, & t \in (0,1), \\ D_{0+}^\beta v(t) + g(t, u(t), v(t), I_{0+}^{\theta_2} u(t), I_{0+}^{\sigma_2} v(t)) = 0, & t \in (0,1), \end{cases} \quad (6.31)$$

with the coupled nonlocal boundary conditions

$$\begin{cases} u(0) = u'(0) = \cdots = u^{(n-2)}(0) = 0, \\ D_{0+}^{\gamma_0} u(1) = \sum_{i=1}^{p} \int_0^1 D_{0+}^{\gamma_i} v(t)\, dH_i(t), \\ v(0) = v'(0) = \cdots = v^{(m-2)}(0) = 0, \\ D_{0+}^{\delta_0} v(1) = \sum_{i=1}^{q} \int_0^1 D_{0+}^{\delta_i} u(t)\, dK_i(t), \end{cases} \quad (6.32)$$

where α, $\beta \in \mathbb{R}$, $\alpha \in (n-1,n]$, $\beta \in (m-1,m]$, n, $m \in \mathbb{N}$, $n \geq 2$, $m \geq 2$, θ_1, θ_2, σ_1, $\sigma_2 > 0$, p, $q \in \mathbb{N}$, $\gamma_i \in \mathbb{R}$ for all $i = 0,\ldots,p$, $0 \leq \gamma_1 < \gamma_2 < \cdots < \gamma_p < \beta - 1$, $\gamma_0 \in [0,\alpha-1)$, $\delta_i \in \mathbb{R}$ for all $i = 0,\ldots,q$, $0 \leq \delta_1 < \delta_2 < \cdots < \delta_q < \alpha - 1$, $\delta_0 \in [0,\beta-1)$, D_{0+}^k denotes the Riemann–Liouville derivative of order k (for $k = \alpha, \beta, \gamma_0, \gamma_i$, $i = 1,\ldots,p$, δ_0, δ_i, $i = 1,\ldots,q$), I_{0+}^ζ is the Riemann–Liouville integral of order ζ (for $\zeta = \theta_1, \sigma_1, \theta_2, \sigma_2$), f and g are nonlinear functions, and the integrals from the boundary conditions (6.32) are Riemann–Stieltjes integrals with H_i for $i = 1,\ldots,p$ and K_i for $i = 1,\ldots,q$ functions of bounded variation.

By using some theorems from the fixed point theory, we will give conditions for the nonlinearities f and g such that problem (6.31), (6.32) has at least one solution $(u,v) \in (C[0,1])^2$.

6.3.1 *Preliminary results*

We consider the system of fractional differential equations

$$\begin{cases} D_{0+}^\alpha u(t) + h(t) = 0, & t \in (0,1), \\ D_{0+}^\beta v(t) + k(t) = 0, & t \in (0,1), \end{cases} \tag{6.33}$$

with the boundary conditions (6.32), where h, $k \in C(0,1) \cap L^1(0,1)$. We denote by

$$\Delta = \frac{\Gamma(\alpha)\Gamma(\beta)}{\Gamma(\alpha-\gamma_0)\Gamma(\beta-\delta_0)}$$
$$- \left(\sum_{i=1}^p \frac{\Gamma(\beta)}{\Gamma(\beta-\gamma_i)} \int_0^1 s^{\beta-\gamma_i-1}\, dH_i(s) \right)$$
$$\times \left(\sum_{i=1}^q \frac{\Gamma(\alpha)}{\Gamma(\alpha-\delta_i)} \int_0^1 s^{\alpha-\delta_i-1}\, dK_i(s) \right).$$

Lemma 6.3.1. *If $\Delta \neq 0$, then the unique solution $(u,v) \in C[0,1] \times C[0,1]$ of problem (6.33), (6.32) is given by*

$$u(t) = -\frac{1}{\Gamma(\alpha)} \int_0^t (t-s)^{\alpha-1} h(s)\, ds + \frac{t^{\alpha-1}}{\Delta}$$
$$\times \left[\frac{\Gamma(\beta)}{\Gamma(\alpha-\gamma_0)\Gamma(\beta-\delta_0)} \int_0^1 (1-s)^{\alpha-\gamma_0-1} h(s)\, ds \right.$$

$$-\frac{\Gamma(\beta)}{\Gamma(\beta-\delta_0)}\sum_{i=1}^{p}\frac{1}{\Gamma(\beta-\gamma_i)}$$

$$\times\int_0^1\left(\int_0^s(s-\tau)^{\beta-\gamma_i-1}k(\tau)\,d\tau\right)dH_i(s)$$

$$+\left(\sum_{i=1}^{p}\frac{\Gamma(\beta)}{\Gamma(\beta-\gamma_i)}\int_0^1 s^{\beta-\gamma_i-1}\,dH_i(s)\right)$$

$$\times\left(\frac{1}{\Gamma(\beta-\delta_0)}\int_0^1(1-s)^{\beta-\delta_0-1}k(s)\,ds\right)$$

$$-\left(\sum_{i=1}^{p}\frac{\Gamma(\beta)}{\Gamma(\beta-\gamma_i)}\int_0^1 s^{\beta-\gamma_i-1}\,dH_i(s)\right)$$

$$\times\left(\sum_{i=1}^{q}\frac{1}{\Gamma(\alpha-\delta_i)}\int_0^1\left(\int_0^s(s-\tau)^{\alpha-\delta_i-1}h(\tau)\,d\tau\right)dK_i(s)\right)\Bigg],$$

$$\tag{6.34}$$

$$v(t)=-\frac{1}{\Gamma(\beta)}\int_0^t(t-s)^{\beta-1}k(s)\,ds+\frac{t^{\beta-1}}{\Delta}\left[\frac{\Gamma(\alpha)}{\Gamma(\alpha-\gamma_0)\Gamma(\beta-\delta_0)}\right.$$

$$\times\int_0^1(1-s)^{\beta-\delta_0-1}k(s)\,ds-\frac{\Gamma(\alpha)}{\Gamma(\alpha-\gamma_0)}$$

$$\times\sum_{i=1}^{q}\frac{1}{\Gamma(\alpha-\delta_i)}\int_0^1\left(\int_0^s(s-\tau)^{\alpha-\delta_i-1}h(\tau)\,d\tau\right)dK_i(s)$$

$$+\left(\sum_{i=1}^{q}\frac{\Gamma(\alpha)}{\Gamma(\alpha-\delta_i)}\int_0^1 s^{\alpha-\delta_i-1}\,dK_i(s)\right)$$

$$\times\left(\frac{1}{\Gamma(\alpha-\gamma_0)}\int_0^1(1-s)^{\alpha-\gamma_0-1}h(s)\,ds\right)$$

$$-\left(\sum_{i=1}^{q}\frac{\Gamma(\alpha)}{\Gamma(\alpha-\delta_i)}\int_0^1 s^{\alpha-\delta_i-1}\,dK_i(s)\right)$$

$$\times\left(\sum_{i=1}^{p}\frac{1}{\Gamma(\beta-\gamma_i)}\int_0^1\left(\int_0^s(s-\tau)^{\beta-\gamma_i-1}k(\tau)\,d\tau\right)dH_i(s)\right)\Bigg].$$

Proof. The solutions $(u, v) \in (C(0, 1) \cap L^1(0, 1))^2$ of system (6.33) are

$$u(t) = -I_{0+}^{\alpha}h(t) + a_1 t^{\alpha-1} + a_2 t^{\alpha-2} + \cdots + a_n t^{\alpha-n},$$

$$v(t) = -I_{0+}^{\beta}k(t) + b_1 t^{\beta-1} + b_2 t^{\beta-2} + \cdots + b_m t^{\beta-m},$$

with $a_1, a_2, \ldots, a_n, b_1, b_2, \ldots, b_m \in \mathbb{R}$. By using the boundary conditions $u(0) = u'(0) = \cdots = u^{(n-2)}(0) = 0$ and $v(0) = v'(0) = \cdots = v^{(m-2)}(0) = 0$, we obtain $a_2 = \cdots = a_n = 0$ and $b_2 = \cdots = b_m = 0$. So the above solutions become

$$u(t) = -\frac{1}{\Gamma(\alpha)} \int_0^t (t-s)^{\alpha-1}h(s)\,ds + a_1 t^{\alpha-1},$$

$$v(t) = -\frac{1}{\Gamma(\beta)} \int_0^t (t-s)^{\beta-1}k(s)\,ds + b_1 t^{\beta-1}. \tag{6.35}$$

For the obtained functions u and v, we have

$$D_{0+}^{\gamma_i}v(t) = b_1 \frac{\Gamma(\beta)}{\Gamma(\beta-\gamma_i)} t^{\beta-\gamma_i-1} - I_{0+}^{\beta-\gamma_i}k(t), \quad i = 1, \ldots, p,$$

$$D_{0+}^{\gamma_0}u(t) = a_1 \frac{\Gamma(\alpha)}{\Gamma(\alpha-\gamma_0)} t^{\alpha-\gamma_0-1} - I_{0+}^{\alpha-\gamma_0}h(t),$$

$$D_{0+}^{\delta_i}u(t) = a_1 \frac{\Gamma(\alpha)}{\Gamma(\alpha-\delta_i)} t^{\alpha-\delta_i-1} - I_{0+}^{\alpha-\delta_i}h(t), \quad i = 1, \ldots, q,$$

$$D_{0+}^{\delta_0}v(t) = b_1 \frac{\Gamma(\beta)}{\Gamma(\beta-\delta_0)} t^{\beta-\delta_0-1} - I_{0+}^{\beta-\delta_0}k(t),$$

$$D_{0+}^{\gamma_0}u(1) = a_1 \frac{\Gamma(\alpha)}{\Gamma(\alpha-\gamma_0)} - I_{0+}^{\alpha-\gamma_0}h(1) = a_1 \frac{\Gamma(\alpha)}{\Gamma(\alpha-\gamma_0)}$$

$$- \frac{1}{\Gamma(\alpha-\gamma_0)} \int_0^1 (1-s)^{\alpha-\gamma_0-1}h(s)\,ds,$$

$$D_{0+}^{\delta_0}v(1) = b_1 \frac{\Gamma(\beta)}{\Gamma(\beta-\delta_0)} - I_{0+}^{\beta-\delta_0}k(1) = b_1 \frac{\Gamma(\beta)}{\Gamma(\beta-\delta_0)}$$

$$- \frac{1}{\Gamma(\beta-\delta_0)} \int_0^1 (1-s)^{\beta-\delta_0-1}k(s)\,ds.$$

By imposing the boundary conditions $D_{0+}^{\gamma_0}u(1) = \sum_{i=1}^p \int_0^1 D_{0+}^{\gamma_i}v(t)\,dH_i(t)$ and $D_{0+}^{\delta_0}v(1) = \sum_{i=1}^q \int_0^1 D_{0+}^{\delta_i}u(t)\,dK_i(t)$, we deduce the following system in a_1 and b_1

$$
\begin{cases}
a_1 \dfrac{\Gamma(\alpha)}{\Gamma(\alpha - \gamma_0)} - b_1 \sum_{i=1}^{p} \int_0^1 \dfrac{\Gamma(\beta)}{\Gamma(\beta - \gamma_i)} t^{\beta - \gamma_i - 1} dH_i(t) \\
\quad = \dfrac{1}{\Gamma(\alpha - \gamma_0)} \int_0^1 (1 - s)^{\alpha - \gamma_0 - 1} h(s)\, ds \\
\quad\quad - \sum_{i=1}^{p} \int_0^1 \dfrac{1}{\Gamma(\beta - \gamma_i)} \left(\int_0^t (t - s)^{\beta - \gamma_i - 1} k(s)\, ds \right) dH_i(t), \\
b_1 \dfrac{\Gamma(\beta)}{\Gamma(\beta - \delta_0)} - a_1 \sum_{i=1}^{q} \int_0^1 \dfrac{\Gamma(\alpha)}{\Gamma(\alpha - \delta_i)} t^{\alpha - \delta_i - 1} dK_i(t) \\
\quad = \dfrac{1}{\Gamma(\beta - \delta_0)} \int_0^1 (1 - s)^{\beta - \delta_0 - 1} k(s)\, ds \\
\quad\quad - \sum_{i=1}^{q} \int_0^1 \dfrac{1}{\Gamma(\alpha - \delta_i)} \left(\int_0^t (t - s)^{\alpha - \delta_i - 1} h(s)\, ds \right) dK_i(t).
\end{cases}
\tag{6.36}
$$

The above system in the unknowns a_1 and b_1 has the determinant Δ, which, by the assumption of lemma, is nonzero. So, the system (6.36) has the unique solution

$$
a_1 = \frac{\Gamma(\beta)}{\Delta \Gamma(\alpha - \gamma_0)\Gamma(\beta - \delta_0)} \int_0^1 (1 - s)^{\alpha - \gamma_0 - 1} h(s)\, ds
$$

$$
- \frac{\Gamma(\beta)}{\Delta \Gamma(\beta - \delta_0)} \sum_{i=1}^{p} \frac{1}{\Gamma(\beta - \gamma_i)}
$$

$$
\times \int_0^1 \left(\int_0^t (t - s)^{\beta - \gamma_i - 1} k(s)\, ds \right) dH_i(t)
$$

$$
+ \frac{1}{\Delta} \left(\sum_{i=1}^{p} \frac{\Gamma(\beta)}{\Gamma(\beta - \gamma_i)} \int_0^1 t^{\beta - \gamma_i - 1}\, dH_i(t) \right)
$$

$$
\times \left(\frac{1}{\Gamma(\beta - \delta_0)} \int_0^1 (1 - s)^{\beta - \delta_0 - 1} k(s)\, ds \right)
$$

$$
- \frac{1}{\Delta} \left(\sum_{i=1}^{p} \frac{\Gamma(\beta)}{\Gamma(\beta - \gamma_i)} \int_0^1 t^{\beta - \gamma_i - 1}\, dH_i(t) \right)
$$

$$
\times \left(\sum_{i=1}^{q} \frac{1}{\Gamma(\alpha - \delta_i)} \int_0^1 \left(\int_0^t (t - s)^{\alpha - \delta_i - 1} h(s)\, ds \right) dK_i(t) \right),
$$

$$b_1 = \frac{\Gamma(\alpha)}{\Delta\Gamma(\alpha - \gamma_0)\Gamma(\beta - \delta_0)} \int_0^1 (1 - s)^{\beta - \delta_0 - 1} k(s)\, ds \tag{6.37}$$

$$- \frac{\Gamma(\alpha)}{\Delta\Gamma(\alpha - \gamma_0)} \sum_{i=1}^q \frac{1}{\Gamma(\alpha - \delta_i)}$$

$$\times \int_0^1 \left(\int_0^t (t - s)^{\alpha - \delta_i - 1} h(s)\, ds \right) dK_i(t)$$

$$+ \frac{1}{\Delta} \left(\sum_{i=1}^q \frac{\Gamma(\alpha)}{\Gamma(\alpha - \delta_i)} \int_0^1 t^{\alpha - \delta_i - 1}\, dK_i(t) \right)$$

$$\times \left(\frac{1}{\Gamma(\alpha - \gamma_0)} \int_0^1 (1 - s)^{\alpha - \gamma_0 - 1} h(s)\, ds \right)$$

$$- \frac{1}{\Delta} \left(\sum_{i=1}^q \frac{\Gamma(\alpha)}{\Gamma(\alpha - \delta_i)} \int_0^1 t^{\alpha - \delta_i - 1}\, dK_i(t) \right)$$

$$\times \left(\sum_{i=1}^p \frac{1}{\Gamma(\beta - \gamma_i)} \int_0^1 \left(\int_0^t (t - s)^{\beta - \gamma_i - 1} k(s)\, ds \right) dH_i(t) \right).$$

Now, replacing the constants a_1 and b_1 given by (6.37) in (6.35) we find the solution $(u, v) \in C[0, 1] \times C[0, 1]$ of problem (6.33), (6.32) presented in (6.34). Conversely, we can easily prove that the functions u, v given by (6.34) satisfy the problem (6.33), (6.32). $\qquad\square$

We introduce now the assumption (J1) for problem (6.31), (6.32) that will be used in our main results.

(J1) $\alpha, \beta \in \mathbb{R}$, $\alpha \in (n - 1, n]$, $\beta \in (m - 1, m]$, $n, m \in \mathbb{N}$, $n \geq 2$, $m \geq 2$, $\theta_1, \theta_2, \sigma_1, \sigma_2 > 0$, $p, q \in \mathbb{N}$, $\gamma_i \in \mathbb{R}$ for all $i = 0, \ldots, p$, $0 \leq \gamma_1 < \gamma_2 < \cdots < \gamma_p < \beta - 1$, $\gamma_0 \in [0, \beta - 1)$, $\delta_i \in \mathbb{R}$ for all $i = 0, \ldots, q$, $0 \leq \delta_1 < \delta_2 < \cdots < \delta_q < \alpha - 1$, $\delta_0 \in [0, \alpha - 1)$, $H_i : [0, 1] \to \mathbb{R}$, $i = 1, \ldots, p$ and $K_j : [0, 1] \to \mathbb{R}$, $j = 1, \ldots, q$ are functions of bounded variation, and $\Delta \neq 0$.

We introduce the following constants:

$$M_1 = 1 + \frac{1}{\Gamma(\theta_1 + 1)}, \quad M_2 = 1 + \frac{1}{\Gamma(\sigma_1 + 1)},$$

$$M_3 = 1 + \frac{1}{\Gamma(\theta_2 + 1)}, \quad M_4 = 1 + \frac{1}{\Gamma(\sigma_2 + 1)},$$

$$M_5 = \max\{M_1, M_2\}, \quad M_6 = \max\{M_3, M_4\},$$

$$M_7 = \frac{1}{\Gamma(\alpha+1)} + \frac{\Gamma(\beta)}{|\Delta|\Gamma(\alpha-\gamma_0+1)\Gamma(\beta-\delta_0)}$$

$$+ \frac{1}{|\Delta|} \left(\sum_{i=1}^{p} \frac{\Gamma(\beta)}{\Gamma(\beta-\gamma_i)} \left| \int_0^1 s^{\beta-\gamma_i-1} dH_i(s) \right| \right)$$

$$\times \left(\sum_{i=1}^{q} \frac{1}{\Gamma(\alpha-\delta_i+1)} \left| \int_0^1 s^{\alpha-\delta_i} dK_i(s) \right| \right),$$

$$M_{10} = \frac{\Gamma(\beta)}{|\Delta|\Gamma(\beta-\delta_0)} \sum_{i=1}^{p} \frac{1}{\Gamma(\beta-\gamma_i+1)} \left| \int_0^1 s^{\beta-\gamma_i} dH_i(s) \right|$$

$$+ \frac{1}{|\Delta|\Gamma(\beta-\delta_0+1)} \left(\sum_{i=1}^{p} \frac{\Gamma(\beta)}{\Gamma(\beta-\gamma_i)} \left| \int_0^1 s^{\beta-\gamma_i-1} dH_i(s) \right| \right),$$

$$M_9 = \frac{\Gamma(\alpha)}{|\Delta|\Gamma(\alpha-\gamma_0)} \sum_{i=1}^{q} \frac{1}{\Gamma(\alpha-\delta_i+1)} \left| \int_0^1 s^{\alpha-\delta_i} dK_i(s) \right|$$

$$+ \frac{1}{|\Delta|\Gamma(\alpha-\gamma_0+1)} \left(\sum_{i=1}^{q} \frac{\Gamma(\alpha)}{\Gamma(\alpha-\delta_i)} \left| \int_0^1 s^{\alpha-\delta_i-1} dK_i(s) \right| \right),$$

$$M_8 = \frac{1}{\Gamma(\beta+1)} + \frac{\Gamma(\alpha)}{|\Delta|\Gamma(\alpha-\gamma_0)\Gamma(\beta-\delta_0+1)}$$

$$+ \frac{1}{|\Delta|} \left(\sum_{i=1}^{q} \frac{\Gamma(\alpha)}{\Gamma(\alpha-\delta_i)} \left| \int_0^1 s^{\alpha-\delta_i-1} dK_i(s) \right| \right)$$

$$\times \left(\sum_{i=1}^{p} \frac{1}{\Gamma(\beta-\gamma_i+1)} \left| \int_0^1 s^{\beta-\gamma_i} dH_i(s) \right| \right),$$

$$M_{11} = M_7 - \frac{1}{\Gamma(\alpha+1)}, \quad M_{12} = M_8 - \frac{1}{\Gamma(\beta+1)}.$$

$$(6.38)$$

We consider the Banach space $X = C[0,1]$ with supremum norm $\|u\| = \sup_{t\in[0,1]} |u(t)|$, and the Banach space $Y = X \times X$ with the norm $\|(u,v)\|_Y = \|u\| + \|v\|$. We introduce the operator $\mathcal{Q} : Y \to Y$ defined by $\mathcal{Q}(u,v) = (\mathcal{Q}_1(u,v), \mathcal{Q}_2(u,v))$ for $(u,v) \in Y$, where the operators $\mathcal{Q}_1$,

$\mathcal{Q}_2 : Y \to X$ are given by

$$\mathcal{Q}_1(u,v)(t) = -\frac{1}{\Gamma(\alpha)} \int_0^t (t-s)^{\alpha-1} \hat{f}_{uv}(s)\, ds$$

$$+ \frac{t^{\alpha-1}\Gamma(\beta)}{\Delta\Gamma(\alpha-\gamma_0)\Gamma(\beta-\delta_0)} \int_0^1 (1-s)^{\alpha-\gamma_0-1} \hat{f}_{uv}(s)\, ds$$

$$- \frac{t^{\alpha-1}\Gamma(\beta)}{\Delta\Gamma(\beta-\delta_0)} \sum_{i=1}^p \frac{1}{\Gamma(\beta-\gamma_i)}$$

$$\times \int_0^1 \left(\int_0^s (s-\tau)^{\beta-\gamma_i-1} \hat{g}_{uv}(\tau)\, d\tau \right) dH_i(s)$$

$$+ \frac{t^{\alpha-1}}{\Delta} \left(\sum_{i=1}^p \frac{\Gamma(\beta)}{\Gamma(\beta-\gamma_i)} \int_0^1 s^{\beta-\gamma_i-1}\, dH_i(s) \right)$$

$$\times \left(\frac{1}{\Gamma(\beta-\delta_0)} \int_0^1 (1-s)^{\beta-\delta_0-1} \hat{g}_{uv}(s)\, ds \right)$$

$$- \frac{t^{\alpha-1}}{\Delta} \left(\sum_{i=1}^p \frac{\Gamma(\beta)}{\Gamma(\beta-\gamma_i)} \int_0^1 s^{\beta-\gamma_i-1}\, dH_i(s) \right)$$

$$\times \left(\sum_{i=1}^q \frac{1}{\Gamma(\alpha-\delta_i)} \int_0^1 \left(\int_0^s (s-\tau)^{\alpha-\delta_i-1} \hat{f}_{uv}(\tau)\, d\tau \right) dK_i(s) \right),$$

$$(6.39)$$

$$\mathcal{Q}_2(u,v)(t) = -\frac{1}{\Gamma(\beta)} \int_0^t (t-s)^{\beta-1} \hat{g}_{uv}(s)\, ds$$

$$+ \frac{t^{\beta-1}\Gamma(\alpha)}{\Delta\Gamma(\alpha-\gamma_0)\Gamma(\beta-\delta_0)} \int_0^1 (1-s)^{\beta-\delta_0-1} \hat{g}_{uv}(s)\, ds$$

$$- \frac{t^{\beta-1}\Gamma(\alpha)}{\Delta\Gamma(\alpha-\gamma_0)} \sum_{i=1}^q \frac{1}{\Gamma(\alpha-\delta_i)}$$

$$\times \int_0^1 \left(\int_0^s (s-\tau)^{\alpha-\delta_i-1} \hat{f}_{uv}(\tau)\, d\tau \right) dK_i(s)$$

$$+ \frac{t^{\beta-1}}{\Delta} \left(\sum_{i=1}^q \frac{\Gamma(\alpha)}{\Gamma(\alpha-\delta_i)} \int_0^1 s^{\alpha-\delta_i-1}\, dK_i(s) \right)$$

$$\times \left(\frac{1}{\Gamma(\alpha-\gamma_0)} \int_0^1 (1-s)^{\alpha-\gamma_0-1} \hat{f}_{uv}(s)\, ds \right)$$

$$-\frac{t^{\beta-1}}{\Delta}\left(\sum_{i=1}^{q}\frac{\Gamma(\alpha)}{\Gamma(\alpha-\delta_i)}\int_0^1 s^{\alpha-\delta_i-1}\,dK_i(s)\right)$$

$$\times\left(\sum_{i=1}^{p}\frac{1}{\Gamma(\beta-\gamma_i)}\int_0^1\left(\int_0^s(s-\tau)^{\beta-\gamma_i-1}\hat{g}_{uv}(\tau)\,d\tau\right)dH_i(s)\right),$$

for $t \in [0,1]$ and $(u,v) \in Y$, where $\hat{f}_{uv}(s) = f(s,u(s),v(s),$ $I_{0+}^{\theta_1}u(s),I_{0+}^{\sigma_1}v(s))$, $\hat{g}_{uv}(s) = g(s,u(s),v(s),I_{0+}^{\theta_2}u(s),I_{0+}^{\sigma_2}v(s))$ for $s \in [0,1]$.

By using Lemma 6.3.1, we see that (u,v) is a solution of problem (6.31), (6.32) if and only if (u,v) is a fixed point of operator $\mathcal{Q}$.

6.3.2 *Existence of solutions*

In this section, we will give some existence results for the solutions of our problem (6.31), (6.32).

Theorem 6.3.1. *Assume that* (J1) *and*

(J2) *The functions* $f,g : [0,1] \times \mathbb{R}^4 \to \mathbb{R}$ *are continuous and there exist* $L_1, L_2 > 0$ *such that*

$$|f(t,x_1,x_2,x_3,x_4) - f(t,y_1,y_2,y_3,y_4)| \le L_1 \sum_{i=1}^{4}|x_i - y_i|,$$

$$|g(t,x_1,x_2,x_3,x_4) - g(t,y_1,y_2,y_3,y_4)| \le L_2 \sum_{i=1}^{4}|x_i - y_i|,$$

for all $t \in [0,1]$, $x_i, y_i \in \mathbb{R}$, $i = 1,\ldots,4$,

hold. If $\Xi := L_1 M_5(M_7 + M_9) + L_2 M_6(M_8 + M_{10}) < 1$, *then problem* (6.31), (6.32) *has a unique solution* $(u(t),v(t))$, $t \in [0,1]$, *where* $M_5,\ldots,M_{10}$ *are given by* (6.38).

Proof. We consider the positive number r given by

$$r = [M_0(M_7 + M_9) + \widetilde{M}_0(M_8 + M_{10})]$$
$$\times [1 - L_1 M_5(M_7 + M_9) - L_2 M_6(M_8 + M_{10})]^{-1},$$

where $M_0 = \sup_{t\in[0,1]}|f(t,0,0,0,0)|$, $\widetilde{M}_0 = \sup_{t\in[0,1]}|g(t,0,0,0,0)|$. We define the set $\overline{B}_r = \{(u,v) \in Y,\ \|(u,v)\|_Y \le r\}$ and we show firstly that

$\mathcal{Q}(\overline{B}_r) \subset \overline{B}_r$. Let $(u, v) \in \overline{B}_r$. By using (J2) and Lemma 6.1.2, for $\hat{f}_{uv}(t)$, we deduce the following inequalities

$$
\begin{aligned}
|\hat{f}_{uv}(t)| &\leq |f(t, u(t), v(t), I_{0+}^{\theta_1} u(t), I_{0+}^{\sigma_1} v(t)) - f(t, 0, 0, 0, 0)| \\
&\quad + |f(t, 0, 0, 0, 0)| \\
&\leq L_1\Big(|u(t)| + |v(t)| + |I_{0+}^{\theta_1} u(t)| + |I_{0+}^{\sigma_1} v(t)|\Big) + M_0 \\
&\leq L_1\Big(\|u\| + \|v\| + \frac{\|u\|}{\Gamma(\theta_1 + 1)} + \frac{\|v\|}{\Gamma(\sigma_1 + 1)}\Big) + M_0 \\
&= L_1(M_1\|u\| + M_2\|v\|) + M_0 \\
&\leq L_1 M_5\|(u, v)\|_Y + M_0 \leq L_1 M_5 r + M_0, \quad \forall t \in [0, 1].
\end{aligned}
$$

Arguing as before, we find

$$
\begin{aligned}
|\hat{g}_{uv}(t)| &\leq |g(t, u(t), v(t), I_{0+}^{\theta_2} u(t), I_{0+}^{\sigma_2} v(t)) - g(t, 0, 0, 0, 0)| \\
&\quad + |g(t, 0, 0, 0, 0)| \\
&\leq L_2\Big(|u(t)| + |v(t)| + |I_{0+}^{\theta_2} u(t)| + |I_{0+}^{\sigma_2} v(t)|\Big) + \widetilde{M}_0 \\
&\leq L_2\Big(\|u\| + \|v\| + \frac{\|u\|}{\Gamma(\theta_2 + 1)} + \frac{\|v\|}{\Gamma(\sigma_2 + 1)}\Big) + \widetilde{M}_0 \\
&= L_2(M_3\|u\| + M_4\|v\|) + \widetilde{M}_0 \\
&\leq L_2 M_6\|(u, v)\|_Y + \widetilde{M}_0 \leq L_2 M_6 r + \widetilde{M}_0, \quad \forall t \in [0, 1].
\end{aligned}
$$

Then by the definition of operators $\mathcal{Q}_1$ and $\mathcal{Q}_2$, we conclude

$$
\begin{aligned}
|\mathcal{Q}_1(u, v)(t)| \leq{} & \frac{1}{\Gamma(\alpha)} \int_0^t (t - s)^{\alpha-1} (L_1 M_5 r + M_0)\, ds \\
& + \frac{t^{\alpha-1}\Gamma(\beta)}{|\Delta|\Gamma(\alpha - \gamma_0)\Gamma(\beta - \delta_0)} \\
& \times \int_0^1 (1 - s)^{\alpha - \gamma_0 - 1} (L_1 M_5 r + M_0)\, ds \\
& + \frac{t^{\alpha-1}\Gamma(\beta)}{|\Delta|\Gamma(\beta - \delta_0)} \sum_{i=1}^p \frac{1}{\Gamma(\beta - \gamma_i)} \\
& \times \left| \int_0^1 \left(\int_0^s (s - \tau)^{\beta - \gamma_i - 1} (L_2 M_6 r + \widetilde{M}_0)\, d\tau \right) dH_i(s) \right|
\end{aligned}
$$

$$+ \frac{t^{\alpha-1}}{|\Delta|} \left(\sum_{i=1}^{p} \frac{\Gamma(\beta)}{\Gamma(\beta-\gamma_i)} \left| \int_0^1 s^{\beta-\gamma_i-1} dH_i(s) \right| \right)$$

$$\times \left(\frac{1}{\Gamma(\beta-\delta_0)} \int_0^1 (1-s)^{\beta-\delta_0-1} (L_2 M_6 r + \widetilde{M_0}) \, ds \right)$$

$$+ \frac{t^{\alpha-1}}{|\Delta|} \left(\sum_{i=1}^{p} \frac{\Gamma(\beta)}{\Gamma(\beta-\gamma_i)} \left| \int_0^1 s^{\beta-\gamma_i-1} dH_i(s) \right| \right)$$

$$\times \left(\sum_{i=1}^{q} \frac{1}{\Gamma(\alpha-\delta_i)} \right)$$

$$\times \left| \int_0^1 \left(\int_0^s (s-\tau)^{\alpha-\delta_i-1} (L_1 M_5 r + M_0) \, d\tau \right) dK_i(s) \right| \right)$$

$$= (L_1 M_5 r + M_0) \left[\frac{1}{\Gamma(\alpha)} \int_0^t (t-s)^{\alpha-1} \, ds \right.$$

$$+ \frac{t^{\alpha-1}\Gamma(\beta)}{|\Delta|\Gamma(\alpha-\gamma_0)\Gamma(\beta-\delta_0)} \int_0^1 (1-s)^{\alpha-\gamma_0-1} \, ds$$

$$+ \frac{t^{\alpha-1}}{|\Delta|} \left(\sum_{i=1}^{p} \frac{\Gamma(\beta)}{\Gamma(\beta-\gamma_i)} \left| \int_0^1 s^{\beta-\gamma_i-1} dH_i(s) \right| \right)$$

$$\left. \times \left(\sum_{i=1}^{q} \frac{1}{\Gamma(\alpha-\delta_i)} \left| \int_0^1 \left(\int_0^s (s-\tau)^{\alpha-\delta_i-1} \, d\tau \right) dK_i(s) \right| \right) \right]$$

$$+ (L_2 M_6 r + \widetilde{M_0}) \left[\frac{t^{\alpha-1}\Gamma(\beta)}{|\Delta|\Gamma(\beta-\delta_0)} \sum_{i=1}^{p} \frac{1}{\Gamma(\beta-\gamma_i)} \right.$$

$$\times \left| \int_0^1 \left(\int_0^s (s-\tau)^{\beta-\gamma_i-1} \, d\tau \right) dH_i(s) \right|$$

$$+ \frac{t^{\alpha-1}}{|\Delta|} \left(\sum_{i=1}^{p} \frac{\Gamma(\beta)}{\Gamma(\beta-\gamma_i)} \left| \int_0^1 s^{\beta-\gamma_i-1} dH_i(s) \right| \right)$$

$$\left. \times \left(\frac{1}{\Gamma(\beta-\delta_0)} \int_0^1 (1-s)^{\beta-\delta_0-1} \, ds \right) \right]$$

$$= (L_1 M_5 r + M_0) \left[\frac{t^{\alpha}}{\Gamma(\alpha+1)} + \frac{t^{\alpha-1}\Gamma(\beta)}{|\Delta|\Gamma(\alpha-\gamma_0+1)\Gamma(\beta-\delta_0)} \right.$$

$$+ \frac{t^{\alpha-1}}{|\Delta|} \left(\sum_{i=1}^{p} \frac{\Gamma(\beta)}{\Gamma(\beta - \gamma_i)} \left| \int_0^1 s^{\beta - \gamma_i - 1} \, dH_i(s) \right| \right)$$

$$\times \left(\sum_{i=1}^{q} \frac{1}{\Gamma(\alpha - \delta_i + 1)} \left| \int_0^1 s^{\alpha - \delta_i} \, dK_i(s) \right| \right) \Bigg]$$

$$+ (L_2 M_6 r + \widetilde{M_0}) \left[\frac{t^{\alpha-1} \Gamma(\beta)}{|\Delta| \Gamma(\beta - \delta_0)} \sum_{i=1}^{p} \frac{1}{\Gamma(\beta - \gamma_i + 1)} \right.$$

$$\times \left| \int_0^1 s^{\beta - \gamma_i} \, dH_i(s) \right|$$

$$+ \frac{t^{\alpha-1}}{|\Delta| \Gamma(\beta - \delta_0 + 1)} \left(\sum_{i=1}^{p} \frac{\Gamma(\beta)}{\Gamma(\beta - \gamma_i)} \left| \int_0^1 s^{\beta - \gamma_i - 1} \, dH_i(s) \right| \right) \Bigg],$$

$$\forall t \in [0,1].$$

Therefore, we obtain

$$\|\mathcal{Q}_1(u,v)\| \le (L_1 M_5 r + M_0) \left[\frac{1}{\Gamma(\alpha+1)} + \frac{\Gamma(\beta)}{|\Delta| \Gamma(\alpha - \gamma_0 + 1) \Gamma(\beta - \delta_0)} \right.$$

$$+ \frac{1}{|\Delta|} \left(\sum_{i=1}^{p} \frac{\Gamma(\beta)}{\Gamma(\beta - \gamma_i)} \left| \int_0^1 s^{\beta - \gamma_i - 1} \, dH_i(s) \right| \right)$$

$$\times \left(\sum_{i=1}^{q} \frac{1}{\Gamma(\alpha - \delta_i + 1)} \left| \int_0^1 s^{\alpha - \delta_i} \, dK_i(s) \right| \right) \Bigg]$$

$$+ (L_2 M_6 r + \widetilde{M_0}) \left[\frac{\Gamma(\beta)}{|\Delta| \Gamma(\beta - \delta_0)} \sum_{i=1}^{p} \frac{1}{\Gamma(\beta - \gamma_i + 1)} \right.$$

$$\times \left| \int_0^1 s^{\beta - \gamma_i} \, dH_i(s) \right| + \frac{1}{|\Delta| \Gamma(\beta - \delta_0 + 1)}$$

$$\times \left(\sum_{i=1}^{p} \frac{\Gamma(\beta)}{\Gamma(\beta - \gamma_i)} \left| \int_0^1 s^{\beta - \gamma_i - 1} \, dH_i(s) \right| \right) \Bigg]$$

$$= (L_1 M_5 r + M_0) M_7 + (L_2 M_6 r + \widetilde{M_0}) M_{10}. \tag{6.40}$$

In a similar manner, we deduce

$$|Q_2(u,v)(t)| \leq \frac{1}{\Gamma(\beta)} \int_0^t (t-s)^{\beta-1}(L_2 M_6 r + \widetilde{M_0})\, ds$$

$$+ \frac{t^{\beta-1}\Gamma(\alpha)}{|\Delta|\Gamma(\alpha-\gamma_0)\Gamma(\beta-\delta_0)}$$

$$\times \int_0^1 (1-s)^{\beta-\delta_0-1}(L_2 M_6 r + \widetilde{M_0})\, ds$$

$$+ \frac{t^{\beta-1}\Gamma(\alpha)}{|\Delta|\Gamma(\alpha-\gamma_0)} \sum_{i=1}^q \frac{1}{\Gamma(\alpha-\delta_i)}$$

$$\times \left| \int_0^1 \left(\int_0^s (s-\tau)^{\alpha-\delta_i-1}(L_1 M_5 r + M_0)\, d\tau \right) dK_i(s) \right|$$

$$+ \frac{t^{\beta-1}}{|\Delta|} \left(\sum_{i=1}^q \frac{\Gamma(\alpha)}{\Gamma(\alpha-\delta_i)} \left| \int_0^1 s^{\alpha-\delta_i-1} dK_i(s) \right| \right)$$

$$\times \left(\frac{1}{\Gamma(\alpha-\gamma_0)} \int_0^1 (1-s)^{\alpha-\gamma_0-1}(L_1 M_5 r + M_0)\, ds \right)$$

$$+ \frac{t^{\beta-1}}{|\Delta|} \left(\sum_{i=1}^q \frac{\Gamma(\alpha)}{\Gamma(\alpha-\delta_i)} \left| \int_0^1 s^{\alpha-\delta_i-1} dK_i(s) \right| \right)$$

$$\times \left(\sum_{i=1}^p \frac{1}{\Gamma(\beta-\gamma_i)} \left| \int_0^1 \left(\int_0^s (s-\tau)^{\beta-\gamma_i-1}(L_2 M_6 r + \widetilde{M_0})\, d\tau \right) dH_i(s) \right| \right)$$

$$= (L_1 M_5 r + M_0) \left[\frac{t^{\beta-1}\Gamma(\alpha)}{|\Delta|\Gamma(\alpha-\gamma_0)} \sum_{i=1}^q \frac{1}{\Gamma(\alpha-\delta_i)} \right.$$

$$\times \left| \int_0^1 \left(\int_0^s (s-\tau)^{\alpha-\delta_i-1}\, d\tau \right) dK_i(s) \right|$$

$$+ \frac{t^{\beta-1}}{|\Delta|} \left(\sum_{i=1}^q \frac{\Gamma(\alpha)}{\Gamma(\alpha-\delta_i)} \left| \int_0^1 s^{\alpha-\delta_i-1} dK_i(s) \right| \right)$$

$$\left. \times \left(\frac{1}{\Gamma(\alpha-\gamma_0)} \int_0^1 (1-s)^{\alpha-\gamma_0-1}\, ds \right) \right]$$

$$+ (L_2 M_6 r + \widetilde{M_0}) \left[\frac{1}{\Gamma(\beta)} \int_0^t (t-s)^{\beta-1}\, ds + \frac{t^{\beta-1}\Gamma(\alpha)}{|\Delta|\Gamma(\alpha-\gamma_0)\Gamma(\beta-\delta_0)} \right.$$

$$\times \int_0^1 (1-s)^{\beta-\delta_0-1}\, ds$$

$$+ \frac{t^{\beta-1}}{|\Delta|} \left(\sum_{i=1}^q \frac{\Gamma(\alpha)}{\Gamma(\alpha-\delta_i)} \left| \int_0^1 s^{\alpha-\delta_i-1}\, dK_i(s) \right| \right)$$

$$\times \left(\sum_{i=1}^p \frac{1}{\Gamma(\beta-\gamma_i)} \left| \int_0^1 \left(\int_0^s (s-\tau)^{\beta-\gamma_i-1}\, d\tau \right) dH_i(s) \right| \right) \Bigg]$$

$$= (L_1 M_5 r + M_0) \left[\frac{t^{\beta-1}\Gamma(\alpha)}{|\Delta|\Gamma(\alpha-\gamma_0)} \sum_{i=1}^q \frac{1}{\Gamma(\alpha-\delta_i+1)} \left| \int_0^1 s^{\alpha-\delta_i}\, dK_i(s) \right| \right.$$

$$+ \frac{t^{\beta-1}}{|\Delta|\Gamma(\alpha-\gamma_0+1)} \left(\sum_{i=1}^q \frac{\Gamma(\alpha)}{\Gamma(\alpha-\delta_i)} \left| \int_0^1 s^{\alpha-\delta_i-1}\, dK_i(s) \right| \right) \Bigg]$$

$$+ (L_2 M_6 r + \widetilde{M_0}) \left[\frac{t^{\beta}}{\Gamma(\beta+1)} + \frac{t^{\beta-1}\Gamma(\alpha)}{|\Delta|\Gamma(\alpha-\gamma_0)\Gamma(\beta-\delta_0+1)} \right.$$

$$+ \frac{t^{\beta-1}}{|\Delta|} \left(\sum_{i=1}^q \frac{\Gamma(\alpha)}{\Gamma(\alpha-\delta_i)} \left| \int_0^1 s^{\alpha-\delta_i-1}\, dK_i(s) \right| \right)$$

$$\times \left(\sum_{i=1}^p \frac{1}{\Gamma(\beta-\gamma_i+1)} \left| \int_0^1 s^{\beta-\gamma_i}\, dH_i(s) \right| \right) \Bigg],$$

for all $t \in [0,1]$. Then we find

$$\|\mathcal{Q}_2(u,v)\| \le (L_1 M_5 r + M_0) \left[\frac{\Gamma(\alpha)}{|\Delta|\Gamma(\alpha-\gamma_0)} \sum_{i=1}^q \frac{1}{\Gamma(\alpha-\delta_i+1)} \right.$$

$$\times \left| \int_0^1 s^{\alpha-\delta_i}\, dK_i(s) \right|$$

$$+ \frac{1}{|\Delta|\Gamma(\alpha-\gamma_0+1)} \left(\sum_{i=1}^q \frac{\Gamma(\alpha)}{\Gamma(\alpha-\delta_i)} \right.$$

$$\times \left. \left| \int_0^1 s^{\alpha-\delta_i-1}\, dK_i(s) \right| \right) \Bigg]$$

$$\tag{6.41}$$

$$+ (L_2 M_6 r + \widetilde{M_0}) \left[\frac{1}{\Gamma(\beta+1)} + \frac{\Gamma(\alpha)}{|\Delta|\Gamma(\alpha-\gamma_0)\Gamma(\beta-\delta_0+1)} \right.$$

$$+ \frac{1}{|\Delta|} \left(\sum_{i=1}^{q} \frac{\Gamma(\alpha)}{\Gamma(\alpha - \delta_i)} \left| \int_0^1 s^{\alpha - \delta_i - 1} \, dK_i(s) \right| \right)$$

$$\times \left(\sum_{i=1}^{p} \frac{1}{\Gamma(\beta - \gamma_i + 1)} \left| \int_0^1 s^{\beta - \gamma_i} \, dH_i(s) \right| \right) \Bigg]$$

$$= (L_1 M_5 r + M_0) M_9 + (L_2 M_6 r + \widetilde{M}_0) M_8.$$

By relations (6.40) and (6.41), we conclude

$$\|\mathcal{Q}(u,v)\|_Y = \|\mathcal{Q}_1(u,v)\| + \|\mathcal{Q}_2(u,v)\|$$

$$\leq (L_1 M_5 r + M_0)(M_7 + M_9)$$

$$+ (L_2 M_6 r + \widetilde{M}_0)(M_8 + M_{10}) = r,$$

for all $(u,v) \in \overline{B}_r$, which implies that $\mathcal{Q}(\overline{B}_r) \subset \overline{B}_r$.

Next, we prove that operator $\mathcal{Q}$ is a contraction. For $(u_i, v_i) \in \overline{B}_r$, $i = 1, 2$, and for each $t \in [0, 1]$, we have

$$|\mathcal{Q}_1(u_1, v_1)(t) - \mathcal{Q}_1(u_2, v_2)(t)|$$

$$\leq \frac{1}{\Gamma(\alpha)} \int_0^t (t - s)^{\alpha - 1} |\hat{f}_{u_1 v_1}(s) - \hat{f}_{u_2 v_2}(s)| \, ds$$

$$+ \frac{t^{\alpha - 1} \Gamma(\beta)}{|\Delta| \Gamma(\alpha - \gamma_0) \Gamma(\beta - \delta_0)} \int_0^1 (1 - s)^{\alpha - \gamma_0 - 1}$$

$$\times |\hat{f}_{u_1 v_1}(s) - \hat{f}_{u_2 v_2}(s)| \, ds$$

$$+ \frac{t^{\alpha - 1} \Gamma(\beta)}{|\Delta| \Gamma(\beta - \delta_0)} \sum_{i=1}^{p} \frac{1}{\Gamma(\beta - \gamma_i)}$$

$$\times \left| \int_0^1 \left(\int_0^s (s - \tau)^{\beta - \gamma_i - 1} |\hat{g}_{u_1 v_1}(\tau) - \hat{g}_{u_2 v_2}(\tau)| \, d\tau \right) dH_i(s) \right|$$

$$\qquad\qquad\qquad\qquad\qquad\qquad\qquad\qquad\qquad (6.42)$$

$$+ \frac{t^{\alpha - 1}}{|\Delta|} \left(\sum_{i=1}^{p} \frac{\Gamma(\beta)}{\Gamma(\beta - \gamma_i)} \left| \int_0^1 s^{\beta - \gamma_i - 1} \, dH_i(s) \right| \right)$$

$$\times \left(\frac{1}{\Gamma(\beta - \delta_0)} \int_0^1 (1 - s)^{\beta - \delta_0 - 1} |\hat{g}_{u_1 v_1}(s) - \hat{g}_{u_2 v_2}(s)| \, ds \right)$$

$$+ \frac{t^{\alpha - 1}}{|\Delta|} \left(\sum_{i=1}^{p} \frac{\Gamma(\beta)}{\Gamma(\beta - \gamma_i)} \left| \int_0^1 s^{\beta - \gamma_i - 1} \, dH_i(s) \right| \right)$$

$$\times \left(\sum_{i=1}^{q} \frac{1}{\Gamma(\alpha - \delta_i)} \left| \int_0^1 \left(\int_0^s (s-\tau)^{\alpha - \delta_i - 1} |\hat{f}_{u_1 v_1}(\tau) \right. \right. \right.$$

$$\left. \left. \left. - \hat{f}_{u_2 v_2}(\tau)| \, d\tau \right) dK_i(s) \right| \right).$$

Because

$$|\hat{f}_{u_1 v_1}(s) - \hat{f}_{u_2 v_2}(s)| \le L_1 \Big(|u_1(s) - u_2(s)| + |v_1(s) - v_2(s)|$$

$$+ |I_{0+}^{\theta_1} u_1(s) - I_{0+}^{\theta_1} u_2(s)| + |I_{0+}^{\sigma_1} v_1(s) - I_{0+}^{\sigma_1} v_2(s)| \Big)$$

$$\le L_1 \left(\|u_1 - u_2\| + \|v_1 - v_2\| + \frac{1}{\Gamma(\theta_1 + 1)} \|u_1 - u_2\| \right.$$

$$\left. + \frac{1}{\Gamma(\sigma_1 + 1)} \|v_1 - v_2\| \right)$$

$$= L_1(M_1 \|u_1 - u_2\| + M_2 \|v_1 - v_2\|)$$

$$\le L_1 M_5 \|(u_1, v_1) - (u_2, v_2)\|_Y, \quad \forall s \in [0, 1],$$

$$|\hat{g}_{u_1 v_1}(s) - \hat{g}_{u_2 v_2}(s)| \le L_2 \Big(|u_1(s) - u_2(s)| + |v_1(s) - v_2(s)|$$

$$+ |I_{0+}^{\theta_2} u_1(s) - I_{0+}^{\theta_2} u_2(s)| + |I_{0+}^{\sigma_2} v_1(s) - I_{0+}^{\sigma_2} v_2(s)| \Big)$$

$$\le L_2 \left(\|u_1 - u_2\| + \|v_1 - v_2\| + \frac{1}{\Gamma(\theta_2 + 1)} \|u_1 - u_2\| \right.$$

$$\left. + \frac{1}{\Gamma(\sigma_2 + 1)} \|v_1 - v_2\| \right)$$

$$= L_2(M_3 \|u_1 - u_2\| + M_4 \|v_1 - v_2\|)$$

$$\le L_2 M_6 \|(u_1, v_1) - (u_2, v_2)\|_Y, \quad \forall s \in [0, 1],$$

the inequality (6.42) gives us

$$|\mathcal{Q}_1(u_1, v_1)(t) - \mathcal{Q}_1(u_2, v_2)(t)| \le L_1 M_5 \|(u_1, v_1) - (u_2, v_2)\|_Y$$

$$\times \left[\frac{t^\alpha}{\Gamma(\alpha + 1)} + \frac{t^{\alpha - 1} \Gamma(\beta)}{|\Delta| \Gamma(\alpha - \gamma_0 + 1) \Gamma(\beta - \delta_0)} \right.$$

$$\left. + \frac{t^{\alpha - 1}}{|\Delta|} \left(\sum_{i=1}^{p} \frac{\Gamma(\beta)}{\Gamma(\beta - \gamma_i)} \left| \int_0^1 s^{\beta - \gamma_i - 1} \, dH_i(s) \right| \right) \right.$$

$$\times \left(\sum_{i=1}^{q} \frac{1}{\Gamma(\alpha - \delta_i + 1)} \left| \int_0^1 s^{\alpha - \delta_i} \, dK_i(s) \right| \right) \right]$$

$$+ L_2 M_6 \| (u_1, v_1) - (u_2, v_2) \|_Y \left[\frac{t^{\alpha-1} \Gamma(\beta)}{|\Delta| \Gamma(\beta - \delta_0)} \sum_{i=1}^{p} \frac{1}{\Gamma(\beta - \gamma_i + 1)} \right.$$

$$\times \left| \int_0^1 s^{\beta - \gamma_i} \, dH_i(s) \right| + \frac{t^{\alpha-1}}{|\Delta| \Gamma(\beta - \delta_0 + 1)}$$

$$\left. \times \sum_{i=1}^{p} \frac{\Gamma(\beta)}{\Gamma(\beta - \gamma_i)} \left| \int_0^1 s^{\beta - \gamma_i - 1} \, dH_i(s) \right| \right],$$

$$\forall t \in [0, 1].$$

Therefore, we obtain

$$\| \mathcal{Q}_1(u_1, v_1) - \mathcal{Q}_1(u_2, v_2) \|$$

$$\leq \left\{ L_1 M_5 \left[\frac{1}{\Gamma(\alpha + 1)} + \frac{\Gamma(\beta)}{|\Delta| \Gamma(\alpha - \gamma_0 + 1) \Gamma(\beta - \delta_0)} \right. \right.$$

$$+ \frac{1}{|\Delta|} \left(\sum_{i=1}^{p} \frac{\Gamma(\beta)}{\Gamma(\beta - \gamma_i)} \left| \int_0^1 s^{\beta - \gamma_i - 1} \, dH_i(s) \right| \right)$$

$$\left. \times \left(\sum_{i=1}^{q} \frac{1}{\Gamma(\alpha - \delta_i + 1)} \left| \int_0^1 s^{\alpha - \delta_i} \, dK_i(s) \right| \right) \right]$$

$$+ L_2 M_6 \left[\frac{\Gamma(\beta)}{|\Delta| \Gamma(\beta - \delta_0)} \sum_{i=1}^{p} \frac{1}{\Gamma(\beta - \gamma_i + 1)} \left| \int_0^1 s^{\beta - \gamma_i} \, dH_i(s) \right| \right.$$

$$\left. \left. + \frac{1}{|\Delta| \Gamma(\beta - \delta_0 + 1)} \sum_{i=1}^{p} \frac{\Gamma(\beta)}{\Gamma(\beta - \gamma_i)} \left| \int_0^1 s^{\beta - \gamma_i - 1} \, dH_i(s) \right| \right] \right\}$$

$$\times \| (u_1, v_1) - (u_2, v_2) \|_Y$$

$$= (L_1 M_5 M_7 + L_2 M_6 M_{10}) \| (u_1, v_1) - (u_2, v_2) \|_Y.$$

$$(6.43)$$

In a similar manner, we deduce

$$| \mathcal{Q}_2(u_1, v_1)(t) - \mathcal{Q}_2(u_2, v_2)(t) |$$

$$\leq \frac{1}{\Gamma(\beta)} \int_0^t (t - s)^{\beta - 1} |\hat{g}_{u_1 v_1}(s) - \hat{g}_{u_2 v_2}(s)| \, ds$$

$$+ \frac{t^{\beta-1} \Gamma(\alpha)}{|\Delta| \Gamma(\alpha - \gamma_0) \Gamma(\beta - \delta_0)} \int_0^1 (1 - s)^{\beta - \delta_0 - 1} |\hat{g}_{u_1 v_1}(s) - \hat{g}_{u_2 v_2}(s)| \, ds$$

$$+ \frac{t^{\beta-1}\Gamma(\alpha)}{|\Delta|\Gamma(\alpha-\gamma_0)} \sum_{i=1}^{q} \frac{1}{\Gamma(\alpha-\delta_i)}$$

$$\times \left| \int_0^1 \left(\int_0^s (s-\tau)^{\alpha-\delta_i-1} |\hat{f}_{u_1 v_1}(\tau) - \hat{f}_{u_2 v_2}(\tau)| \, d\tau \right) dK_i(s) \right|$$

$$+ \frac{t^{\beta-1}}{|\Delta|} \left(\sum_{i=1}^{q} \frac{\Gamma(\alpha)}{\Gamma(\alpha-\delta_i)} \left| \int_0^1 s^{\alpha-\delta_i-1} \, dK_i(s) \right| \right)$$

$$\times \left(\frac{1}{\Gamma(\alpha-\gamma_0)} \int_0^1 (1-s)^{\alpha-\gamma_0-1} |\hat{f}_{u_1 v_1}(s) - \hat{f}_{u_2 v_2}(s)| \, ds \right)$$

$$+ \frac{t^{\beta-1}}{|\Delta|} \left(\sum_{i=1}^{q} \frac{\Gamma(\alpha)}{\Gamma(\alpha-\delta_i)} \left| \int_0^1 s^{\alpha-\delta_i-1} \, dK_i(s) \right| \right)$$

$$\times \left(\sum_{i=1}^{p} \frac{1}{\Gamma(\beta-\gamma_i)} \left| \int_0^1 \left(\int_0^s (s-\tau)^{\beta-\gamma_i-1} |\hat{g}_{u_1 v_1}(\tau) - \hat{g}_{u_2 v_2}(\tau)| \, d\tau \right) dH_i(s) \right| \right)$$

$$\leq L_2 M_6 \|(u_1,v_1) - (u_2,v_2)\|_Y \left[\frac{t^\beta}{\Gamma(\beta+1)} + \frac{t^{\beta-1}\Gamma(\alpha)}{|\Delta|\Gamma(\alpha-\gamma_0)\Gamma(\beta-\delta_0+1)} \right.$$

$$+ \frac{t^{\beta-1}}{|\Delta|} \left(\sum_{i=1}^{q} \frac{\Gamma(\alpha)}{\Gamma(\alpha-\delta_i)} \left| \int_0^1 s^{\alpha-\delta_i-1} \, dK_i(s) \right| \right)$$

$$\times \left. \left(\sum_{i=1}^{p} \frac{1}{\Gamma(\beta-\gamma_i+1)} \left| \int_0^1 s^{\beta-\gamma_i} \, dH_i(s) \right| \right) \right]$$

$$+ L_1 M_5 \|(u_1,v_1) - (u_2,v_2)\|_Y \left[\frac{t^{\beta-1}\Gamma(\alpha)}{|\Delta|\Gamma(\alpha-\gamma_0)} \sum_{i=1}^{q} \frac{1}{\Gamma(\alpha-\delta_i+1)} \right.$$

$$\times \left| \int_0^1 s^{\alpha-\delta_i} \, dK_i(s) \right| + \frac{t^{\beta-1}}{|\Delta|\Gamma(\alpha-\gamma_0+1)} \sum_{i=1}^{q} \frac{\Gamma(\alpha)}{\Gamma(\alpha-\delta_i)}$$

$$\times \left. \left| \int_0^1 s^{\alpha-\delta_i-1} \, dK_i(s) \right| \right], \quad \forall t \in [0,1].$$

Hence, we find

$$\|\mathcal{Q}_2(u_1,v_1) - \mathcal{Q}_2(u_2,v_2)\|$$

$$\leq \left\{ L_2 M_6 \left[\frac{1}{\Gamma(\beta+1)} + \frac{\Gamma(\alpha)}{|\Delta|\Gamma(\alpha-\gamma_0)\Gamma(\beta-\delta_0+1)} \right. \right.$$

$$+ \frac{1}{|\Delta|} \left(\sum_{i=1}^{q} \frac{\Gamma(\alpha)}{\Gamma(\alpha-\delta_i)} \left| \int_0^1 s^{\alpha-\delta_i-1} \, dK_i(s) \right| \right)$$

$$\times \left(\sum_{i=1}^{p} \frac{1}{\Gamma(\beta - \gamma_i + 1)} \left| \int_0^1 s^{\beta - \gamma_i} \, dH_i(s) \right| \right) \Bigg]$$

$$+ L_1 M_5 \left[\frac{\Gamma(\alpha)}{|\Delta| \Gamma(\alpha - \gamma_0)} \sum_{i=1}^{q} \frac{1}{\Gamma(\alpha - \delta_i + 1)} \left| \int_0^1 s^{\alpha - \delta_i} \, dK_i(s) \right| \right.$$

$$\left. + \frac{1}{|\Delta| \Gamma(\alpha - \gamma_0 + 1)} \sum_{i=1}^{q} \frac{\Gamma(\alpha)}{\Gamma(\alpha - \delta_i)} \left| \int_0^1 s^{\alpha - \delta_i - 1} \, dK_i(s) \right| \right] \Bigg\}$$

$$\times \| (u_1, v_1) - (u_2, v_2) \|_Y$$

$$= (L_1 M_5 M_9 + L_2 M_6 M_8) \| (u_1, v_1) - (u_2, v_2) \|_Y. \tag{6.44}$$

Then by using relations (6.43) and (6.44), we obtain

$$\| \mathcal{Q}(u_1, v_1) - \mathcal{Q}(u_2, v_2) \|_Y = \| \mathcal{Q}_1(u_1, v_1) - \mathcal{Q}_1(u_2, v_2) \|$$

$$+ \| \mathcal{Q}_2(u_1, v_1) - \mathcal{Q}_2(u_2, v_2) \|$$

$$\leq [L_1 M_5 (M_7 + M_9) + L_2 M_6 (M_8 + M_{10})]$$

$$\times \| (u_1, v_1) - (u_2, v_2) \|_Y$$

$$= \Xi \| (u_1, v_1) - (u_2, v_2) \|_Y.$$

By using the condition $\Xi < 1$, we deduce that operator $\mathcal{Q}$ is a contraction. By Theorem 1.2.1, we conclude that operator $\mathcal{Q}$ has a unique fixed point $(u, v) \in \overline{B}_r$, which is the unique solution of problem (6.31), (6.32) on $[0, 1]$.   □

Theorem 6.3.2. *Suppose that* (J1) *and*

(J3) *The functions* $f, g : [0, 1] \times \mathbb{R}^4 \to \mathbb{R}$ *are continuous and there exist real constants* a_i, $b_i \geq 0$, $i = 0, \ldots, 4$, *and at least one of* a_0 *and* b_0 *is positive, such that*

$$|f(t, x_1, x_2, x_3, x_4)| \leq a_0 + \sum_{i=1}^{4} a_i |x_i|,$$

$$|g(t, x_1, x_2, x_3, x_4)| \leq b_0 + \sum_{i=1}^{4} b_i |x_i|,$$

for all $t \in [0, 1]$, $x_i \in \mathbb{R}$, $i = 1, \ldots, 4$,

hold. If $\Xi_1 := \max\{M_{13}, M_{14}\} < 1$, *where* $M_{13} = (a_1 + \frac{a_3}{\Gamma(\theta_1 + 1)})(M_7 + M_9) + (b_1 + \frac{b_3}{\Gamma(\theta_2 + 1)})(M_8 + M_{10})$ *and* $M_{14} = (a_2 + \frac{a_4}{\Gamma(\sigma_1 + 1)})(M_7 + M_9) + (b_2 + \frac{b_4}{\Gamma(\sigma_2 + 1)})(M_8 + M_{10})$, *then the boundary value problem* (6.31), (6.32) *has at least one solution* $(u(t), v(t))$, $t \in [0, 1]$.

Proof. We show that operator $\mathcal{Q}$ is completely continuous. Because the functions f and g are continuous, we deduce that the operators $\mathcal{Q}_1$ and $\mathcal{Q}_2$ are continuous, and then $\mathcal{Q}$ is a continuous operator. We will prove next that $\mathcal{Q}$ is a compact operator, that is, it maps bounded sets into relatively compact sets. Let $\Omega \subset Y$ be a bounded set. Then there exist positive constants L_3 and L_4 such that $|\hat{f}_{uv}(t)| \le L_3$ and $|\hat{g}_{uv}(t)| \le L_4$ for all $t \in [0,1]$ and $(u,v) \in \Omega$. Hence, we obtain as in the proof of Theorem 6.3.1 that

$$|\mathcal{Q}_1(u,v)(t)| \le L_3 M_7 + L_4 M_{10}, \quad |\mathcal{Q}_2(u,v)(t)| \le L_3 M_9 + L_4 M_8,$$

for all $t \in [0,1]$ and $(u,v) \in \Omega$. So, we find

$$\|\mathcal{Q}_1(u,v)\| \le L_3 M_7 + L_4 M_{10}, \quad \|\mathcal{Q}_2(u,v)\| \le L_3 M_9 + L_4 M_8,$$

$$\|\mathcal{Q}(u,v)\|_Y \le L_3(M_7 + M_9) + L_4(M_8 + M_{10}), \quad \forall (u,v) \in \Omega,$$

and then $\mathcal{Q}(\Omega)$ is uniformly bounded.

We show now that $\mathcal{Q}(\Omega)$ are equicontinuous. Let $(u,v) \in \Omega$ and $t_1, t_2 \in [0,1]$ with $t_1 < t_2$. Then we have

$$|\mathcal{Q}_1(u,v)(t_2) - \mathcal{Q}_1(u,v)(t_1)|$$

$$\le \left| -\frac{1}{\Gamma(\alpha)} \int_0^{t_2} (t_2 - s)^{\alpha-1} \hat{f}_{uv}(s)\, ds + \frac{1}{\Gamma(\alpha)} \int_0^{t_1} (t_1 - s)^{\alpha-1} \hat{f}_{uv}(s)\, ds \right|$$

$$+ \frac{(t_2^{\alpha-1} - t_1^{\alpha-1})\Gamma(\beta)}{|\Delta|\Gamma(\alpha-\gamma_0)\Gamma(\beta-\delta_0)} \int_0^1 (1-s)^{\alpha-\gamma_0-1} |\hat{f}_{uv}(s)|\, ds$$

$$+ \frac{(t_2^{\alpha-1} - t_1^{\alpha-1})\Gamma(\beta)}{|\Delta|\Gamma(\beta-\delta_0)} \sum_{i=1}^{p} \frac{1}{\Gamma(\beta-\gamma_i)}$$

$$\times \left| \int_0^1 \left(\int_0^s (s-\tau)^{\beta-\gamma_i-1} |\hat{g}_{uv}(\tau)|\, d\tau \right) dH_i(s) \right|$$

$$+ \frac{t_2^{\alpha-1} - t_1^{\alpha-1}}{|\Delta|} \left(\sum_{i=1}^{p} \frac{\Gamma(\beta)}{\Gamma(\beta-\gamma_i)} \left| \int_0^1 s^{\beta-\gamma_i-1}\, dH_i(s) \right| \right)$$

$$\times \left(\frac{1}{\Gamma(\beta-\delta_0)} \int_0^1 (1-s)^{\beta-\delta_0-1} |\hat{g}_{uv}(s)|\, ds \right)$$

$$+ \frac{t_2^{\alpha-1} - t_1^{\alpha-1}}{|\Delta|} \left(\sum_{i=1}^{p} \frac{\Gamma(\beta)}{\Gamma(\beta-\gamma_i)} \left| \int_0^1 s^{\beta-\gamma_i-1}\, dH_i(s) \right| \right)$$

$$\times \left(\sum_{i=1}^{q} \frac{1}{\Gamma(\alpha - \delta_i)} \left| \int_0^1 \left(\int_0^s (s-\tau)^{\alpha - \delta_i - 1} |\hat{f}_{uv}(\tau)| \, d\tau \right) dK_i(s) \right| \right)$$

$$\leq \frac{L_3}{\Gamma(\alpha)} \int_0^{t_1} [(t_2 - s)^{\alpha - 1} - (t_1 - s)^{\alpha - 1}] \, ds + \frac{L_3}{\Gamma(\alpha)} \int_{t_1}^{t_2} (t_2 - s)^{\alpha - 1} \, ds$$

$$+ \frac{L_3(t_2^{\alpha - 1} - t_1^{\alpha - 1})\Gamma(\beta)}{|\Delta|\Gamma(\alpha - \gamma_0)\Gamma(\beta - \delta_0)} \int_0^1 (1-s)^{\alpha - \gamma_0 - 1} \, ds$$

$$+ \frac{L_4(t_2^{\alpha - 1} - t_1^{\alpha - 1})\Gamma(\beta)}{|\Delta|\Gamma(\beta - \delta_0)} \sum_{i=1}^{p} \frac{1}{\Gamma(\beta - \gamma_i)}$$

$$\times \left| \int_0^1 \left(\int_0^s (s-\tau)^{\beta - \gamma_i - 1} \, d\tau \right) dH_i(s) \right|$$

$$+ \frac{L_4(t_2^{\alpha - 1} - t_1^{\alpha - 1})}{|\Delta|} \left(\sum_{i=1}^{p} \frac{\Gamma(\beta)}{\Gamma(\beta - \gamma_i)} \left| \int_0^1 s^{\beta - \gamma_i - 1} dH_i(s) \right| \right)$$

$$\times \left(\frac{1}{\Gamma(\beta - \delta_0)} \int_0^1 (1-s)^{\beta - \delta_0 - 1} \, ds \right)$$

$$+ \frac{L_3(t_2^{\alpha - 1} - t_1^{\alpha - 1})}{|\Delta|} \left(\sum_{i=1}^{p} \frac{\Gamma(\beta)}{\Gamma(\beta - \gamma_i)} \left| \int_0^1 s^{\beta - \gamma_i - 1} \, dH_i(s) \right| \right)$$

$$\times \left(\sum_{i=1}^{q} \frac{1}{\Gamma(\alpha - \delta_i)} \left| \int_0^1 \left(\int_0^s (s-\tau)^{\alpha - \delta_i - 1} \, d\tau \right) dK_i(s) \right| \right)$$

$$= \frac{L_3}{\Gamma(\alpha + 1)} (t_2^{\alpha} - t_1^{\alpha}) + L_3(t_2^{\alpha - 1} - t_1^{\alpha - 1}) \left[\frac{\Gamma(\beta)}{|\Delta|\Gamma(\alpha - \gamma_0 + 1)\Gamma(\beta - \delta_0)} \right.$$

$$+ \frac{1}{|\Delta|} \left(\sum_{i=1}^{p} \frac{\Gamma(\beta)}{\Gamma(\beta - \gamma_i)} \left| \int_0^1 s^{\beta - \gamma_i - 1} \, dH_i(s) \right| \right)$$

$$\left. \times \left(\sum_{i=1}^{q} \frac{1}{\Gamma(\alpha - \delta_i + 1)} \left| \int_0^1 s^{\alpha - \delta_i} \, dK_i(s) \right| \right) \right]$$

$$+ L_4(t_2^{\alpha - 1} - t_1^{\alpha - 1}) \left[\frac{\Gamma(\beta)}{|\Delta|\Gamma(\beta - \delta_0)} \sum_{i=1}^{p} \frac{1}{\Gamma(\beta - \gamma_i + 1)} \left| \int_0^1 s^{\beta - \gamma_i} \, dH_i(s) \right| \right.$$

$$\left. + \frac{\Gamma(\beta)}{|\Delta|\Gamma(\beta - \delta_0 + 1)} \sum_{i=1}^{p} \frac{1}{\Gamma(\beta - \gamma_i)} \left| \int_0^1 s^{\beta - \gamma_i - 1} \, dH_i(s) \right| \right]$$

$$= \frac{L_3}{\Gamma(\alpha + 1)} (t_2^{\alpha} - t_1^{\alpha}) + (L_3 M_{11} + L_4 M_{10})(t_2^{\alpha - 1} - t_1^{\alpha - 1}).$$

Hence, we infer

$$|\mathcal{Q}_1(u,v)(t_2) - \mathcal{Q}_1(u,v)(t_1)| \to 0, \quad \text{as } t_2 \to t_1, \quad \text{uniformly with respect to}$$
$(u,v) \in \Omega.$

In a similar manner, for $(u,v) \in \Omega$ and $t_1, t_2 \in [0,1]$ with $t_1 < t_2$, we obtain

$$|\mathcal{Q}_2(u,v)(t_2) - \mathcal{Q}_2(u,v)(t_1)|$$

$$\leq \frac{L_4}{\Gamma(\beta+1)}(t_2^\beta - t_1^\beta) + L_4(t_2^{\beta-1} - t_1^{\beta-1}) \left[\frac{\Gamma(\alpha)}{|\Delta|\Gamma(\alpha-\gamma_0)\Gamma(\beta-\delta_0+1)} \right.$$

$$+ \frac{1}{|\Delta|} \left(\sum_{i=1}^{q} \frac{\Gamma(\alpha)}{\Gamma(\alpha-\delta_i)} \left| \int_0^1 s^{\alpha-\delta_i-1} \, dK_i(s) \right| \right)$$

$$\left. \times \left(\sum_{i=1}^{p} \frac{1}{\Gamma(\beta-\gamma_i+1)} \left| \int_0^1 s^{\beta-\gamma_i} \, dH_i(s) \right| \right) \right]$$

$$+ L_3(t_2^{\beta-1} - t_1^{\beta-1}) \left[\frac{\Gamma(\alpha)}{|\Delta|\Gamma(\alpha-\gamma_0)} \sum_{i=1}^{q} \frac{1}{\Gamma(\alpha-\delta_i+1)} \left| \int_0^1 s^{\alpha-\delta_i} \, dK_i(s) \right| \right.$$

$$\left. + \frac{\Gamma(\alpha)}{|\Delta|\Gamma(\alpha-\gamma_0+1)} \sum_{i=1}^{q} \frac{1}{\Gamma(\alpha-\delta_i)} \left| \int_0^1 s^{\alpha-\delta_i-1} \, dK_i(s) \right| \right]$$

$$= \frac{L_4}{\Gamma(\beta+1)}(t_2^\beta - t_1^\beta) + (L_4 M_{12} + L_3 M_9)(t_2^{\beta-1} - t_1^{\beta-1}).$$

So, we deduce

$$|\mathcal{Q}_2(u,v)(t_2) - \mathcal{Q}_2(u,v)(t_1)| \to 0, \quad \text{as } t_2 \to t_1, \quad \text{uniformly with respect to}$$
$(u,v) \in \Omega.$

Then $\mathcal{Q}_1(\Omega)$ and $\mathcal{Q}_2(\Omega)$ is equicontinuous, and so $\mathcal{Q}(\Omega)$ is also equicontinuous. Therefore, by the Arzela-Ascoli theorem, we conclude that $\mathcal{Q}(\Omega)$ is relatively compact, and then $\mathcal{Q}$ is compact. We infer that operator $\mathcal{Q}$ is completely continuous.

Next, we will show that the set $U = \{(u,v) \in Y, \ (u,v) = \nu\mathcal{Q}(u,v), \ 0 < \nu < 1\}$ is bounded. Let $(u,v) \in U$, that is $(u,v) = \nu\mathcal{Q}(u,v)$. Then for any $t \in [0,1]$ we get $u(t) = \nu\mathcal{Q}_1(u,v)(t)$, $v(t) = \nu\mathcal{Q}_2(u,v)(t)$. We denote the following functions

$$F_{uv}(s) = a_0 + a_1|u(s)| + a_2|v(s)| + a_3|I_{0+}^{\theta_1}u(s)| + a_4|I_{0+}^{\sigma_1}v(s)|, \quad s \in [0,1],$$
$$G_{uv}(s) = b_0 + b_1|u(s)| + b_2|v(s)| + b_3|I_{0+}^{\theta_2}u(s)| + b_4|I_{0+}^{\sigma_2}v(s)|, \quad s \in [0,1].$$

By (J3), we find

$$|u(t)| \le |\mathcal{Q}_1(u,v)(t)| \le \frac{1}{\Gamma(\alpha)} \int_0^t (t-s)^{\alpha-1} F_{uv}(s)\, ds$$

$$+ \frac{t^{\alpha-1}\Gamma(\beta)}{|\Delta|\Gamma(\alpha-\gamma_0)\Gamma(\beta-\delta_0)} \int_0^1 (1-s)^{\alpha-\gamma_0-1} F_{uv}(s)\, ds$$

$$+ \frac{t^{\alpha-1}\Gamma(\beta)}{|\Delta|\Gamma(\beta-\delta_0)} \sum_{i=1}^p \frac{1}{\Gamma(\beta-\gamma_i)}$$

$$\times \left| \int_0^1 \left(\int_0^s (s-\tau)^{\beta-\gamma_i-1} G_{uv}(\tau)\, d\tau \right) dH_i(s) \right|$$

$$+ \frac{t^{\alpha-1}}{|\Delta|} \left(\sum_{i=1}^p \frac{\Gamma(\beta)}{\Gamma(\beta-\gamma_i)} \left| \int_0^1 s^{\beta-\gamma_i-1} dH_i(s) \right| \right)$$

$$\times \left(\frac{1}{\Gamma(\beta-\delta_0)} \int_0^1 (1-s)^{\beta-\delta_0-1} G_{uv}(s)\, ds \right)$$

$$+ \frac{t^{\alpha-1}}{|\Delta|} \left(\sum_{i=1}^p \frac{\Gamma(\beta)}{\Gamma(\beta-\gamma_i)} \left| \int_0^1 s^{\beta-\gamma_i-1} dH_i(s) \right| \right)$$

$$\times \left(\sum_{i=1}^q \frac{1}{\Gamma(\alpha-\delta_i)} \left| \int_0^1 \left(\int_0^s (s-\tau)^{\alpha-\delta_i-1} F_{uv}(\tau)\, d\tau \right) dK_i(s) \right| \right)$$

$$\le \left(a_0 + a_1\|u\| + a_2\|v\| + \frac{a_3}{\Gamma(\theta_1+1)}\|u\| + \frac{a_4}{\Gamma(\sigma_1+1)}\|v\| \right)$$

$$\times \left[\frac{t^\alpha}{\Gamma(\alpha+1)} + \frac{t^{\alpha-1}\Gamma(\beta)}{|\Delta|\Gamma(\alpha-\gamma_0+1)\Gamma(\beta-\delta_0)} \right.$$

$$+ \frac{t^{\alpha-1}}{|\Delta|} \left(\sum_{i=1}^p \frac{\Gamma(\beta)}{\Gamma(\beta-\gamma_i)} \left| \int_0^1 s^{\beta-\gamma_i-1} dH_i(s) \right| \right)$$

$$\left. \times \left(\sum_{i=1}^q \frac{1}{\Gamma(\alpha-\delta_i+1)} \left| \int_0^1 s^{\alpha-\delta_i} dK_i(s) \right| \right) \right]$$

$$+ \left(b_0 + b_1 \|u\| + b_2 \|v\| + \frac{b_3}{\Gamma(\theta_2 + 1)} \|u\| + \frac{b_4}{\Gamma(\sigma_2 + 1)} \|v\| \right)$$

$$\times \left[\frac{t^{\alpha-1} \Gamma(\beta)}{|\Delta| \Gamma(\beta - \delta_0)} \sum_{i=1}^{p} \frac{1}{\Gamma(\beta - \gamma_i + 1)} \left| \int_0^1 s^{\beta - \gamma_i} \, dH_i(s) \right| \right.$$

$$\left. + \frac{t^{\alpha-1}}{|\Delta| \Gamma(\beta - \delta_0 + 1)} \sum_{i=1}^{p} \frac{\Gamma(\beta)}{\Gamma(\beta - \gamma_i)} \left| \int_0^1 s^{\beta - \gamma_i - 1} \, dH_i(s) \right| \right],$$

$$\forall t \in [0, 1].$$

Therefore, we deduce

$$\|u\| \le \left(a_0 + a_1 \|u\| + a_2 \|v\| + \frac{a_3}{\Gamma(\theta_1 + 1)} \|u\| + \frac{a_4}{\Gamma(\sigma_1 + 1)} \|v\| \right) M_7$$

$$+ \left(b_0 + b_1 \|u\| + b_2 \|v\| + \frac{b_3}{\Gamma(\theta_2 + 1)} \|u\| + \frac{b_4}{\Gamma(\sigma_2 + 1)} \|v\| \right) M_{10}.$$

$$(6.45)$$

In a similar manner, we obtain

$$|v(t)| \le |\mathcal{Q}_2(u, v)(t)| \le \frac{1}{\Gamma(\beta)} \int_0^t (t - s)^{\beta - 1} G_{uv}(s) \, ds$$

$$+ \frac{t^{\beta-1} \Gamma(\alpha)}{|\Delta| \Gamma(\alpha - \gamma_0) \Gamma(\beta - \delta_0)} \int_0^1 (1 - s)^{\beta - \delta_0 - 1} G_{uv}(s) \, ds$$

$$+ \frac{t^{\beta-1} \Gamma(\alpha)}{|\Delta| \Gamma(\alpha - \gamma_0)} \sum_{i=1}^{q} \frac{1}{\Gamma(\alpha - \delta_i)}$$

$$\times \left| \int_0^1 \left(\int_0^s (s - \tau)^{\alpha - \delta_i - 1} F_{uv}(\tau) \, d\tau \right) dK_i(s) \right|$$

$$+ \frac{t^{\beta-1}}{|\Delta|} \left(\sum_{i=1}^{q} \frac{\Gamma(\alpha)}{\Gamma(\alpha - \delta_i)} \left| \int_0^1 s^{\alpha - \delta_i - 1} \, dK_i(s) \right| \right)$$

$$\times \left(\frac{1}{\Gamma(\alpha - \gamma_0)} \int_0^1 (1 - s)^{\alpha - \gamma_0 - 1} F_{uv}(s) \, ds \right)$$

$$+ \frac{t^{\beta-1}}{|\Delta|} \left(\sum_{i=1}^{q} \frac{\Gamma(\alpha)}{\Gamma(\alpha - \delta_i)} \left| \int_0^1 s^{\alpha - \delta_i - 1} \, dK_i(s) \right| \right)$$

$$\times \left(\sum_{i=1}^{p} \frac{1}{\Gamma(\beta - \gamma_i)} \left| \int_0^1 \left(\int_0^s (s - \tau)^{\beta - \gamma_i - 1} G_{uv}(\tau) \, d\tau \right) dH_i(s) \right| \right)$$

$$\leq \left(a_0 + a_1\|u\| + a_2\|v\| + \frac{a_3}{\Gamma(\theta_1+1)}\|u\| + \frac{a_4}{\Gamma(\sigma_1+1)}\|v\| \right)$$

$$\times \left[\frac{t^{\beta-1}\Gamma(\alpha)}{|\Delta|\Gamma(\alpha-\gamma_0)} \sum_{i=1}^{q} \frac{1}{\Gamma(\alpha-\delta_i+1)} \left| \int_0^1 s^{\alpha-\delta_i}\,dK_i(s) \right| \right.$$

$$+ \frac{t^{\beta-1}}{|\Delta|\Gamma(\alpha-\gamma_0+1)} \sum_{i=1}^{q} \frac{\Gamma(\alpha)}{\Gamma(\alpha-\delta_i)} \left| \int_0^1 s^{\alpha-\delta_i-1}\,dK_i(s) \right| \right]$$

$$+ \left(b_0 + b_1\|u\| + b_2\|v\| + \frac{b_3}{\Gamma(\theta_2+1)}\|u\| + \frac{b_4}{\Gamma(\sigma_2+1)}\|v\| \right)$$

$$\times \left[\frac{t^\beta}{\Gamma(\beta+1)} + \frac{t^{\beta-1}\Gamma(\alpha)}{|\Delta|\Gamma(\alpha-\gamma_0)\Gamma(\beta-\delta_0+1)} \right.$$

$$+ \frac{t^{\beta-1}}{|\Delta|} \left(\sum_{i=1}^{q} \frac{\Gamma(\alpha)}{\Gamma(\alpha-\delta_i)} \left| \int_0^1 s^{\alpha-\delta_i-1}\,dK_i(s) \right| \right)$$

$$\times \left. \left(\sum_{i=1}^{p} \frac{1}{\Gamma(\beta-\gamma_i+1)} \left| \int_0^1 s^{\beta-\gamma_i}\,dH_i(s) \right| \right) \right], \quad \forall t \in [0,1].$$

Then we have

$$\|v\| \leq \left(a_0 + a_1\|u\| + a_2\|v\| + \frac{a_3}{\Gamma(\theta_1+1)}\|u\| + \frac{a_4}{\Gamma(\sigma_1+1)}\|v\| \right) M_9$$

$$+ \left(b_0 + b_1\|u\| + b_2\|v\| + \frac{b_3}{\Gamma(\theta_2+1)}\|u\| + \frac{b_4}{\Gamma(\sigma_2+1)}\|v\| \right) M_8.$$

$$(6.46)$$

By (6.45) and (6.46), we infer

$$\|(u,v)\|_Y = \|u\| + \|v\| \leq a_0(M_7 + M_9) + b_0(M_8 + M_{10})$$

$$+ \left[a_1(M_7 + M_9) + \frac{a_3}{\Gamma(\theta_1+1)}(M_7 + M_9) + b_1(M_8 + M_{10}) \right.$$

$$+ \left. \frac{b_3}{\Gamma(\theta_2+1)}(M_8 + M_{10}) \right] \|u\|$$

$$+ \left[a_2(M_7 + M_9) + \frac{a_4}{\Gamma(\sigma_1+1)}(M_7 + M_9) + b_2(M_8 + M_{10}) \right.$$

$$+ \left. \frac{b_4}{\Gamma(\theta_2+1)}(M_8 + M_{10}) \right] \|v\|$$

$$= a_0(M_7 + M_9) + b_0(M_8 + M_{10}) + M_{13}\|u\| + M_{14}\|v\|$$

$$\leq a_0(M_7 + M_9) + b_0(M_8 + M_{10}) + \Xi_1\|(u,v)\|_Y.$$

Because $\Xi_1 < 1$, we find

$$\|(u,v)\|_Y \leq [a_0(M_7 + M_9) + b_0(M_8 + M_{10})](1 - \Xi_1)^{-1}, \quad \forall\, (u,v) \in U.$$

So, we deduce that the set U is bounded.

By using Theorem 1.2.5, we conclude that operator $\mathcal{Q}$ has at least one fixed point, which is a solution for problem (6.31), (6.32). $\qquad\square$

Theorem 6.3.3. *Assume that* (J1), (J2) *and*

(J4) *There exist the functions* ψ_1, $\psi_2 \in C([0,1],[0,\infty))$ *such that*

$$|f(t,x_1,x_2,x_3,x_4)| \leq \psi_1(t), \quad |g(t,x_1,x_2,x_3,x_4)| \leq \psi_2(t),$$

for all $t \in [0,1]$, $x_i \in \mathbb{R}$, $i = 1,\ldots,4$,

hold. If $\Xi_2 := L_1 M_5 \frac{1}{\Gamma(\alpha+1)} + L_2 M_6 \frac{1}{\Gamma(\beta+1)} < 1$, *then problem* (6.31), (6.32) *has at least one solution on* $[0,1]$.

Proof. We fix $r_1 > 0$ such that $r_1 \geq (M_7 + M_9)\|\psi_1\| + (M_8 + M_{10})\|\psi_2\|$. We consider the set $\overline{B}_{r_1} = \{(x,y) \in Y,\ \|(x,y)\|_Y \leq r_1\}$, and we introduce the operators $D = (D_1, D_2) : \overline{B}_{r_1} \to Y$ and $E = (E_1, E_2) : \overline{B}_{r_1} \to Y$, where $D_1, D_2, E_1, E_2 : \overline{B}_{r_1} \to X$ are defined by

$$D_1(u,v)(t) = \frac{1}{\Gamma(\alpha)} \int_0^t (t-s)^{\alpha-1} \hat{f}_{uv}(s)\, ds,$$

$$E_1(u,v)(t) = \frac{t^{\alpha-1}\Gamma(\beta)}{\Delta\Gamma(\alpha-\gamma_0)\Gamma(\beta-\delta_0)} \int_0^1 (1-s)^{\alpha-\gamma_0-1} \hat{f}_{uv}(s)\, ds$$

$$- \frac{t^{\alpha-1}\Gamma(\beta)}{\Delta\Gamma(\beta-\delta_0)} \sum_{i=1}^p \frac{1}{\Gamma(\beta-\gamma_i)}$$

$$\times \int_0^1 \left(\int_0^s (s-\tau)^{\beta-\gamma_i-1} \hat{g}_{uv}(\tau)\, d\tau \right) dH_i(s)$$

$$+ \frac{t^{\alpha-1}}{\Delta} \left(\sum_{i=1}^p \frac{\Gamma(\beta)}{\Gamma(\beta-\gamma_i)} \int_0^1 s^{\beta-\gamma_i-1}\, dH_i(s) \right)$$

$$\times \left(\frac{1}{\Gamma(\beta-\delta_0)} \int_0^1 (1-s)^{\beta-\delta_0-1} \hat{g}_{uv}(s)\, ds \right)$$

$$- \frac{t^{\alpha-1}}{\Delta} \left(\sum_{i=1}^p \frac{\Gamma(\beta)}{\Gamma(\beta-\gamma_i)} \int_0^1 s^{\beta-\gamma_i-1}\, dH_i(s) \right)$$

$$\times \left(\sum_{i=1}^{q} \frac{1}{\Gamma(\alpha - \delta_i)} \int_0^1 \left(\int_0^s (s-\tau)^{\alpha-\delta_i-1} \hat{f}_{uv}(\tau) \, d\tau \right) dK_i(s) \right),$$

$$D_2(u,v)(t) = -\frac{1}{\Gamma(\beta)} \int_0^t (t-s)^{\beta-1} \hat{g}_{uv}(s) \, ds$$

$$E_2(u,v)(t) = \frac{t^{\beta-1}\Gamma(\alpha)}{\Delta\Gamma(\alpha-\gamma_0)\Gamma(\beta-\delta_0)} \int_0^1 (1-s)^{\beta-\delta_0-1} \hat{g}_{uv}(s) \, ds$$

$$-\frac{t^{\beta-1}\Gamma(\alpha)}{\Delta\Gamma(\alpha-\gamma_0)} \sum_{i=1}^{q} \frac{1}{\Gamma(\alpha-\delta_i)}$$

$$\times \int_0^1 \left(\int_0^s (s-\tau)^{\alpha-\delta_i-1} \hat{f}_{uv}(\tau) \, d\tau \right) dK_i(s)$$

$$+\frac{t^{\beta-1}}{\Delta} \left(\sum_{i=1}^{q} \frac{\Gamma(\alpha)}{\Gamma(\alpha-\delta_i)} \int_0^1 s^{\alpha-\delta_i-1} \, dK_i(s) \right)$$

$$\times \left(\frac{1}{\Gamma(\alpha-\gamma_0)} \int_0^1 (1-s)^{\alpha-\gamma_0-1} \hat{f}_{uv}(s) \, ds \right)$$

$$-\frac{t^{\beta-1}}{\Delta} \left(\sum_{i=1}^{q} \frac{\Gamma(\alpha)}{\Gamma(\alpha-\delta_i)} \int_0^1 s^{\alpha-\delta_i-1} \, dK_i(s) \right)$$

$$\times \left(\sum_{i=1}^{p} \frac{1}{\Gamma(\beta-\gamma_i)} \int_0^1 \left(\int_0^s (s-\tau)^{\beta-\gamma_i-1} \hat{g}_{uv}(\tau) \, d\tau \right) dH_i(s) \right),$$

$$(6.47)$$

for all $t \in [0,1]$ and $(u,v) \in \overline{B}_{r_1}$. So $\mathcal{Q}_1 = D_1 + E_1$, $\mathcal{Q}_2 = D_2 + E_2$ and $\mathcal{Q} = D + E$.

By using (J4), we find for all $(u_1, v_1), (u_2, v_2) \in \overline{B}_{r_1}$ as in the proof of Theorem 6.3.1 that

$$\|D(u_1, v_1) + E(u_2, v_2)\|_Y \le \|D(u_1, v_1)\|_Y + \|E(u_2, v_2)\|_Y$$

$$= \|D_1(u_1, v_1)\| + \|D_2(u_1, v_1)\| + \|E_1(u_2, v_2)\| + \|E_2(u_2, v_2)\|$$

$$\le \frac{1}{\Gamma(\alpha+1)} \|\psi_1\| + \frac{1}{\Gamma(\beta+1)} \|\psi_2\| + (M_{11}\|\psi_1\| + M_{10}\|\psi_2\|)$$

$$+ (M_9\|\psi_1\| + M_{12}\|\psi_2\|)$$

$$= (M_7 + M_9)\|\psi_1\| + (M_8 + M_{10})\|\psi_2\| \le r_1.$$

So, $D(u_1, v_1) + E(u_2, v_2) \in \overline{B}_{r_1}$ for all $(u_1, v_1), (u_2, v_2) \in \overline{B}_{r_1}$.

The operator D is a contraction, because

$$\|D(u_1, v_1) - D(u_2, v_2)\|_Y = \|D_1(u_1, v_1) - D_1(u_2, v_2)\|$$
$$+ \|D_2(u_1, v_1) - D_2(u_2, v_2)\|$$
$$\leq \left(L_1 M_5 \frac{1}{\Gamma(\alpha + 1)} + L_2 M_6 \frac{1}{\Gamma(\beta + 1)} \right)$$
$$\times (\|u_1 - u_2\| + \|v_1 - v_2\|) = \Xi_2 \|(u_1, v_1) - (u_2, v_2)\|_Y,$$

for all (u_1, v_1), $(u_2, v_2) \in \overline{B}_{r_1}$, and $\Xi_2 < 1$.

Because the functions f and g are continuous, we obtain that operator E is continuous on $\overline{B}_{r_1}$. We show next that E is compact. The functions from E are uniformly bounded on $\overline{B}_{r_1}$, because

$$\|E(u, v)\|_Y = \|E_1(u, v)\| + \|E_2(u, v)\| \leq (M_{11} + M_9)\|\psi_1\|$$
$$+ (M_{10} + M_{12})\|\psi_2\|, \quad \forall (u, v) \in \overline{B}_{r_1}.$$

We prove next that the functions from $E(\overline{B}_{r_1})$ are equicontinuous. We denote by

$$\Psi_{r_1} = \sup \left\{ |f(t, u, v, x, y)|, \ t \in [0, 1], \ |u| \leq r_1, \ |v| \leq r_1, \right.$$

$$\left. |x| \leq \frac{r_1}{\Gamma(\theta_1 + 1)}, \ |y| \leq \frac{r_1}{\Gamma(\sigma_1 + 1)} \right\},$$

$$\Theta_{r_1} = \sup \left\{ |g(t, u, v, x, y)|, \ t \in [0, 1], \ |u| \leq r_1, \ |v| \leq r_1, \right.$$

$$\left. |x| \leq \frac{r_1}{\Gamma(\theta_2 + 1)}, \ |y| \leq \frac{r_1}{\Gamma(\sigma_2 + 1)} \right\}. \tag{6.48}$$

Then for $(u, v) \in \overline{B}_{r_1}$ and $t_1, t_2 \in [0, 1]$ with $t_1 < t_2$, we deduce

$$|E_1(u, v)(t_2) - E_1(u, v)(t_1)|$$

$$\leq \frac{(t_2^{\alpha-1} - t_1^{\alpha-1})\Gamma(\beta)}{|\Delta|\Gamma(\alpha - \gamma_0)\Gamma(\beta - \delta_0)} \int_0^1 (1 - s)^{\alpha - \gamma_0 - 1} \Psi_{r_1} \, ds$$

$$+ \frac{(t_2^{\alpha-1} - t_1^{\alpha-1})\Gamma(\beta)}{|\Delta|\Gamma(\beta - \delta_0)} \sum_{i=1}^{p} \frac{1}{\Gamma(\beta - \gamma_i)}$$

$$\times \left| \int_0^1 \left(\int_0^s (s-\tau)^{\beta-\gamma_i-1} \Theta_{r_1} \, d\tau \right) dH_i(s) \right|$$

$$+ \frac{t_2^{\alpha-1} - t_1^{\alpha-1}}{|\Delta|} \left(\sum_{i=1}^p \frac{\Gamma(\beta)}{\Gamma(\beta-\gamma_i)} \left| \int_0^1 s^{\beta-\gamma_i-1} \, dH_i(s) \right| \right)$$

$$\times \left(\frac{1}{\Gamma(\beta-\delta_0)} \int_0^1 (1-s)^{\beta-\delta_0-1} \Theta_{r_1} \, ds \right)$$

$$+ \frac{t_2^{\alpha-1} - t_1^{\alpha-1}}{|\Delta|} \left(\sum_{i=1}^p \frac{\Gamma(\beta)}{\Gamma(\beta-\gamma_i)} \left| \int_0^1 s^{\beta-\gamma_i-1} \, dH_i(s) \right| \right)$$

$$\times \left(\sum_{i=1}^q \frac{1}{\Gamma(\alpha-\delta_i)} \left| \int_0^1 \left(\int_0^s (s-\tau)^{\alpha-\delta_i-1} \Psi_{r_1} \, d\tau \right) dK_i(s) \right| \right)$$

$$= \Psi_{r_1} \left(t_2^{\alpha-1} - t_1^{\alpha-1} \right) \left[\frac{\Gamma(\beta)}{|\Delta|\Gamma(\alpha-\gamma_0+1)\Gamma(\beta-\delta_0)} \right.$$

$$+ \frac{1}{|\Delta|} \left(\sum_{i=1}^p \frac{\Gamma(\beta)}{\Gamma(\beta-\gamma_i)} \left| \int_0^1 s^{\beta-\gamma_i-1} \, dH_i(s) \right| \right)$$

$$\times \left. \left(\sum_{i=1}^p \frac{1}{\Gamma(\alpha-\delta_i+1)} \left| \int_0^1 s^{\alpha-\delta_i} \, dK_i(s) \right| \right) \right]$$

$$+ \Theta_{r_1} \left(t_2^{\alpha-1} - t_1^{\alpha-1} \right) \left[\frac{\Gamma(\beta)}{|\Delta|\Gamma(\beta-\delta_0)} \sum_{i=1}^p \frac{1}{\Gamma(\beta-\gamma_i+1)} \right.$$

$$\times \left| \int_0^1 s^{\beta-\gamma_i} \, dH_i(s) \right| + \frac{1}{|\Delta|\Gamma(\beta-\delta_0+1)} \sum_{i=1}^p \frac{\Gamma(\beta)}{\Gamma(\beta-\gamma_i)}$$

$$\times \left. \left| \int_0^1 s^{\beta-\gamma_i-1} \, dH_i(s) \right| \right]$$

$$= M_{11} \Psi_{r_1} \left(t_2^{\alpha-1} - t_1^{\alpha-1} \right) + M_{10} \Theta_{r_1} \left(t_2^{\alpha-1} - t_1^{\alpha-1} \right),$$

$$|E_2(u,v)(t_2) - E_2(u,v)(t_1)| \leq \frac{\left(t_2^{\beta-1} - t_1^{\beta-1} \right) \Gamma(\alpha)}{|\Delta|\Gamma(\alpha-\gamma_0)\Gamma(\beta-\delta_0)}$$

$$\times \int_0^1 (1-s)^{\beta-\delta_0-1} \Theta_{r_1} \, ds + \frac{\left(t_2^{\beta-1} - t_1^{\beta-1} \right) \Gamma(\alpha)}{|\Delta|\Gamma(\alpha-\gamma_0)}$$

$$\times \sum_{i=1}^{q} \frac{1}{\Gamma(\alpha - \delta_i)} \left| \int_0^1 \left(\int_0^s (s-\tau)^{\alpha-\delta_i-1} \Psi_{r_1} \, d\tau \right) dK_i(s) \right|$$

$$+ \frac{t_2^{\beta-1} - t_1^{\beta-1}}{|\Delta|} \left(\sum_{i=1}^{q} \frac{\Gamma(\alpha)}{\Gamma(\alpha - \delta_i)} \left| \int_0^1 s^{\alpha-\delta_i-1} dK_i(s) \right| \right)$$

$$\times \left(\frac{1}{\Gamma(\alpha - \gamma_0)} \int_0^1 (1-s)^{\alpha-\gamma_0-1} \Psi_{r_1} \, ds \right)$$

$$+ \frac{t_2^{\beta-1} - t_1^{\beta-1}}{|\Delta|} \left(\sum_{i=1}^{q} \frac{\Gamma(\alpha)}{\Gamma(\alpha - \delta_i)} \left| \int_0^1 s^{\alpha-\delta_i-1} dK_i(s) \right| \right)$$

$$\times \left(\sum_{i=1}^{p} \frac{1}{\Gamma(\beta - \gamma_i)} \left| \int_0^1 \left(\int_0^s (s-\tau)^{\beta-\gamma_i-1} \Theta_{r_1} \, d\tau \right) dH_i(s) \right| \right)$$

$$= \Psi_{r_1} \left(t_2^{\beta-1} - t_1^{\beta-1} \right) \left[\frac{\Gamma(\alpha)}{|\Delta|\Gamma(\alpha - \gamma_0)} \sum_{i=1}^{q} \frac{1}{\Gamma(\alpha - \delta_i + 1)} \left| \int_0^1 s^{\alpha-\delta_i} dK_i(s) \right| \right.$$

$$\left. + \frac{1}{|\Delta|\Gamma(\alpha - \gamma_0 + 1)} \sum_{i=1}^{q} \frac{\Gamma(\alpha)}{\Gamma(\alpha - \delta_i)} \left| \int_0^1 s^{\alpha-\delta_i-1} dK_i(s) \right| \right]$$

$$+ \Theta_{r_1} \left(t_2^{\beta-1} - t_1^{\beta-1} \right) \left[\frac{\Gamma(\alpha)}{|\Delta|\Gamma(\alpha - \gamma_0)\Gamma(\beta - \delta_0 + 1)} \right.$$

$$+ \frac{1}{|\Delta|} \left(\sum_{i=1}^{q} \frac{\Gamma(\alpha)}{\Gamma(\alpha - \delta_i)} \left| \int_0^1 s^{\alpha-\delta_i-1} dK_i(s) \right| \right)$$

$$\left. \times \left(\sum_{i=1}^{p} \frac{1}{\Gamma(\beta - \gamma_i + 1)} \left| \int_0^1 s^{\beta-\gamma_i} dH_i(s) \right| \right) \right]$$

$$= M_9 \Psi_{r_1} \left(t_2^{\beta-1} - t_1^{\beta-1} \right) + M_{12} \Theta_{r_1} \left(t_2^{\beta-1} - t_1^{\beta-1} \right).$$

Therefore, we infer

$$|E_1(u,v)(t_2) - E_1(u,v)(t_1)| \to 0, \quad |E_2(u,v)(t_2) - E_2(u,v)(t_1)| \to 0,$$

as $t_2 \to t_1$ uniformly with respect to $(u,v) \in \overline{B}_{r_1}$. Then $E_1(\overline{B}_{r_1})$ and $E_2(\overline{B}_{r_1})$ are equicontinuous, and so $E(\overline{B}_{r_1})$ is also equicontinuous. By applying the Arzela-Ascoli theorem, we conclude that the set $E(\overline{B}_{r_1})$ is relatively compact. Hence, E is a compact operator on $\overline{B}_{r_1}$. By using

Theorem 1.2.3, we deduce that there exists a fixed point of operator $D + E(= Q)$, which is a solution of problem (6.31), (6.32).   □

Theorem 6.3.4. *Suppose that* (J1), (J2) *and* (J4) *hold. If* $\Xi_3 :=$ $L_1 M_5 (M_9 + M_{11}) + L_2 M_6 (M_{10} + M_{12}) < 1$, *then problem* (6.31), (6.32) *has at least one solution* (u, v) *on* $[0, 1]$.

Proof. We consider again a positive number $r_1 \geq (M_7 + M_9)\|\psi_1\| + (M_8 + M_{10})\|\psi_2\|$ and the operators D and E defined on $\overline{B}_{r_1}$ given by (6.47). As in the proof of Theorem 6.3.3, we have $D(u_1, v_1) + E(u_2, v_2) \in \overline{B}_{r_1}$ for all (u_1, v_1), $(u_2, v_2) \in \overline{B}_{r_1}$.

The operator E is a contraction, because

$$\|E(u_1, v_1) - E(u_2, v_2)\|_Y = \|E_1(u_1, v_1) - E_1(u_2, v_2)\|$$
$$+ \|E_2(u_1, v_1) - E_2(u_2, v_2)\|$$
$$\leq (L_1 M_5 M_{11} + L_2 M_6 M_{10})\|(u_1, v_1) - (u_2, v_2)\|_Y$$
$$+ (L_1 M_5 M_9 + L_2 M_6 M_{12})\|(u_1, v_1) - (u_2, v_2)\|_Y$$
$$= (L_1 M_5 (M_9 + M_{11}) + L_2 M_6 (M_{10} + M_{12}))\|(u_1, v_1) - (u_2, v_2)\|_Y$$
$$= \Xi_3 \|(u_1, v_1) - (u_2, v_2)\|_Y,$$

for all (u_1, v_1), $(u_2, v_2) \in \overline{B}_{r_1}$, with $\Xi_3 < 1$.

In what follows, the continuity of functions f and g implies that operator D is continuous on $\overline{B}_{r_1}$. We prove now that D is a compact operator. The functions from $D(\overline{B}_{r_1})$ are uniformly bounded, because

$$\|D(u, v)\|_Y = \|D_1(u, v)\| + \|D_2(u, v)\| \leq \frac{1}{\Gamma(\alpha + 1)}\|\psi_1\|$$
$$+ \frac{1}{\Gamma(\beta + 1)}\|\psi_2\|, \quad \forall (u, v) \in \overline{B}_{r_1}.$$

Now, we show that the functions from $D(\overline{B}_{r_1})$ are equicontinuous. By using Ψ_{r_1} and Θ_{r_1} defined by (6.48), we deduce that for $(u, v) \in \overline{B}_{r_1}$ and $t_1, t_2 \in [0, 1]$ with $t_1 < t_2$ that

$$|D_1(u, v)(t_2) - D_1(u, v)(t_1)| \leq \frac{\Psi_{r_1}}{\Gamma(\alpha + 1)}(t_2^\alpha - t_1^\alpha),$$

$$|D_2(u, v)(t_2) - D_2(u, v)(t_1)| \leq \frac{\Theta_{r_1}}{\Gamma(\beta + 1)}(t_2^\beta - t_1^\beta).$$

Therefore, we conclude

$$|D_1(u, v)(t_2) - D_1(u, v)(t_1)| \to 0, \quad |D_2(u, v)(t_2) - D_2(u, v)(t_1)| \to 0,$$

as $t_2 \to t_1$ uniformly with respect to $(u,v) \in \overline{B}_{r_1}$. We infer that $D_1(\overline{B}_{r_1})$ and $D_2(\overline{B}_{r_1})$ are equicontinuous, and so $D(\overline{B}_{r_1})$ is equicontinuous. By using the Arzela-Ascoli theorem, we deduce that the set $D(\overline{B}_{r_1})$ is relatively compact. Then D is a compact operator on $\overline{B}_{r_1}$. By using Theorem 1.2.3, we conclude that there exists a fixed point of operator $D + E (= \mathcal{Q})$, which is a solution of problem (6.31), (6.32). $\qquad\square$

Theorem 6.3.5. *Assume that* (J1) *and*

(J5) *The functions f, $g : [0,1] \times \mathbb{R}^4 \to \mathbb{R}$ are continuous and there exist the constants $c_i \geq 0$, $i = 0, \ldots, 4$ with at least one nonzero, the constants $d_i \geq 0$, $i = 0, \ldots, 4$ with at least one nonzero, and $l_i, m_i \in (0,1)$, $i = 1, \ldots, 4$ such that*

$$|f(t, x_1, x_2, x_3, x_4)| \leq c_0 + \sum_{i=1}^{4} c_i |x_i|^{l_i},$$

$$|g(t, x_1, x_2, x_3, x_4)| \leq d_0 + \sum_{i=1}^{4} d_i |x_i|^{m_i},$$

for all $t \in [0,1]$, $x_i \in \mathbb{R}$, $i = 1, \ldots, 4$,

hold. Then problem (6.31), (6.32) *has at least one solution.*

Proof. Let $\overline{B}_R = \{(x,y) \in Y, \ \|(x,y)\|_Y \leq R\}$, where

$$R \geq \max \left\{ 20c_0 M_7, \ (20c_1 M_7)^{\frac{1}{1-l_1}}, \ (20c_2 M_7)^{\frac{1}{1-l_2}}, \right.$$

$$\left(\frac{20c_3 M_7}{(\Gamma(\theta_1 + 1))^{l_3}} \right)^{\frac{1}{1-l_3}}, \ \left(\frac{20c_4 M_7}{(\Gamma(\sigma_1 + 1))^{l_4}} \right)^{\frac{1}{1-l_4}},$$

$$20d_0 M_{10}, \ (20d_1 M_{10})^{\frac{1}{1-m_1}}, \ (20d_2 M_{10})^{\frac{1}{1-m_2}},$$

$$\left(\frac{20d_3 M_{10}}{(\Gamma(\theta_2 + 1))^{m_3}} \right)^{\frac{1}{1-m_3}}, \ \left(\frac{20d_4 M_{10}}{(\Gamma(\sigma_2 + 1))^{m_4}} \right)^{\frac{1}{1-m_4}},$$

$$20c_0 M_9, \ (20c_1 M_9)^{\frac{1}{1-l_1}}, \ (20c_2 M_9)^{\frac{1}{1-l_2}},$$

$$\left(\frac{20c_3 M_9}{(\Gamma(\theta_1 + 1))^{l_3}} \right)^{\frac{1}{1-l_3}}, \ \left(\frac{20c_4 M_9}{(\Gamma(\sigma_1 + 1))^{l_4}} \right)^{\frac{1}{1-l_4}},$$

$$20d_0 M_8, \ (20d_1 M_8)^{\frac{1}{1-m_1}}, \ (20d_2 M_8)^{\frac{1}{1-m_2}},$$

$$\left(\frac{20d_3 M_8}{(\Gamma(\theta_2+1))^{m_3}}\right)^{\frac{1}{1-m_3}}, \quad \left(\frac{20d_4 M_8}{(\Gamma(\sigma_2+1))^{m_4}}\right)^{\frac{1}{1-m_4}}\Bigg\}.$$

We prove that $\mathcal{Q} : \overline{B}_R \to \overline{B}_R$. For $(u, v) \in \overline{B}_R$, we have

$$|\mathcal{Q}_1(u,v)(t)| \le \left(c_0 + c_1 R^{l_1} + c_2 R^{l_2} + c_3 \frac{R^{l_3}}{(\Gamma(\theta_1+1))^{l_3}} + c_4 \frac{R^{l_4}}{(\Gamma(\sigma_1+1))^{l_4}}\right) M_7$$

$$+ \left(d_0 + d_1 R^{m_1} + d_2 R^{m_2} + d_3 \frac{R^{m_3}}{(\Gamma(\theta_2+1))^{m_3}} + d_4 \frac{R^{m_4}}{(\Gamma(\sigma_2+1))^{m_4}}\right) M_{10} \le \frac{R}{2},$$

$$|\mathcal{Q}_2(u,v)(t)| \le \left(c_0 + c_1 R^{l_1} + c_2 R^{l_2} + c_3 \frac{R^{l_3}}{(\Gamma(\theta_1+1))^{l_3}} + c_4 \frac{R^{l_4}}{(\Gamma(\sigma_1+1))^{l_4}}\right) M_9$$

$$+ \left(d_0 + d_1 R^{m_1} + d_2 R^{m_2} + d_3 \frac{R^{m_3}}{(\Gamma(\theta_2+1))^{m_3}} + d_4 \frac{R^{m_4}}{(\Gamma(\sigma_2+1))^{m_4}}\right) M_8 \le \frac{R}{2},$$

for all $t \in [0, 1]$. Then we obtain

$$\|\mathcal{Q}(u,v))\|_Y = \|\mathcal{Q}_1(u,v)\| + \|\mathcal{Q}_2(u,v)\| \le R, \quad \forall (u,v) \in \overline{B}_R,$$

which implies that $\mathcal{Q}(\overline{B}_R) \subset \overline{B}_R$.

By using the fact that the functions f and g are continuous, we deduce that operator $\mathcal{Q}$ is continuous on $\overline{B}_R$. Besides, the functions from $\mathcal{Q}(\overline{B}_R)$ are uniformly bounded and equicontinuous. Indeed, by using the notations (6.48) with r_1 replaced by R, we find for any $(u, v) \in \overline{B}_R$ and $t_1, t_2 \in [0, 1]$, $t_1 < t_2$ that

$$|\mathcal{Q}_1(u,v)(t_2) - \mathcal{Q}_1(u,v)(t_1)| \le \frac{\Psi_R}{\Gamma(\alpha+1)}(t_2^\alpha - t_1^\alpha)$$

$$+ (\Psi_R M_{11} + \Theta_R M_{10})(t_2^{\alpha-1} - t_1^{\alpha-1}),$$

$$|\mathcal{Q}_2(u,v)(t_2) - \mathcal{Q}_2(u,v)(t_1)| \le \frac{\Theta_R}{\Gamma(\beta+1)}(t_2^\beta - t_1^\beta)$$

$$+ (\Psi_R M_9 + \Theta_R M_{12})(t_2^{\beta-1} - t_1^{\beta-1}).$$

Therefore, we obtain

$$|\mathcal{Q}_1(u,v)(t_2) - \mathcal{Q}_1(u,v)(t_1)| \to 0,$$

$$|\mathcal{Q}_2(u,v)(t_2) - \mathcal{Q}_2(u,v)(t_1)| \to 0, \quad \text{as } t_2 \to t_1,$$

uniformly with respect to $(u, v) \in \overline{B}_R$. By the Arzela-Ascoli theorem, we conclude that $\mathcal{Q}(\overline{B}_R)$ is relatively compact, and then $\mathcal{Q}$ is a compact operator. By using Theorem 1.2.4, we infer that operator

$\mathcal{Q}$ has at least one fixed point (u, v) in $\overline{B}_R$, which is a solution of our problem (6.31), (6.32). $\qquad\qquad\square$

Theorem 6.3.6. *Suppose that* (J1) *and*

(J6) *The functions* $f, g : [0, 1] \times \mathbb{R}^4 \to \mathbb{R}$ *are continuous and there exist* $p_i \geq 0$, $i = 0, \ldots, 4$ *with at least one nonzero,* $q_i \geq 0$, $i = 0, \ldots, 4$ *with at least one nonzero, and nondecreasing functions* $\xi_i, \eta_i \in C([0, \infty), [0, \infty))$ $i = 1, \ldots, 4$ *such that*

$$|f(t, x_1, x_2, x_3, x_4)| \leq p_0 + \sum_{i=1}^{4} p_i \xi_i(|x_i|),$$

$$|g(t, x_1, x_2, x_3, x_4)| \leq q_0 + \sum_{i=1}^{4} q_i \eta_i(|x_i|),$$

for all $t \in [0, 1]$, $x_i \in \mathbb{R}$, $i = 1, \ldots, 4$,

hold. If there exists $\Xi_0 > 0$ *such that*

$$\left(p_0 + p_1 \xi_1(\Xi_0) + p_2 \xi_2(\Xi_0) \right.$$
$$\left. + p_3 \xi_3 \left(\frac{\Xi_0}{\Gamma(\theta_1 + 1)} \right) + p_4 \xi_4 \left(\frac{\Xi_0}{\Gamma(\sigma_1 + 1)} \right) \right) (M_7 + M_9)$$
$$\tag{6.49}$$
$$+ \left(q_0 + q_1 \eta_1(\Xi_0) + q_2 \eta_2(\Xi_0) \right.$$
$$\left. + q_3 \eta_3 \left(\frac{\Xi_0}{\Gamma(\theta_2 + 1)} \right) + q_4 \eta_4 \left(\frac{\Xi_0}{\Gamma(\sigma_2 + 1)} \right) \right) (M_8 + M_{10}) < \Xi_0,$$

then problem (6.31), (6.32) *has at least one solution on* $[0, 1]$.

Proof. We consider the set $\overline{B}_{\Xi_0} = \{(x, y) \in Y, \ \|(x, y)\|_Y \leq \Xi_0\}$, where Ξ_0 is given in the theorem. We will show that $\mathcal{Q} : \overline{B}_{\Xi_0} \to \overline{B}_{\Xi_0}$. For $(u, v) \in \overline{B}_{\Xi_0}$ and $t \in [0, 1]$ we obtain

$$|\mathcal{Q}_1(u, v)(t)| \leq \left(p_0 + p_1 \xi_1(\Xi_0) + p_2 \xi_2(\Xi_0) \right.$$
$$\left. + p_3 \xi_3 \left(\frac{\Xi_0}{\Gamma(\theta_1 + 1)} \right) + p_4 \xi_4 \left(\frac{\Xi_0}{\Gamma(\sigma_1 + 1)} \right) \right) M_7$$
$$+ \left(q_0 + q_1 \eta_1(\Xi_0) + q_2 \eta_2(\Xi_0) + q_3 \eta_3 \left(\frac{\Xi_0}{\Gamma(\theta_2 + 1)} \right) + q_4 \eta_4 \left(\frac{\Xi_0}{\Gamma(\sigma_2 + 1)} \right) \right) M_{10},$$

$$|\mathcal{Q}_2(u,v)(t)| \leq \left(p_0 + p_1\xi_1(\Xi_0) + p_2\xi_2(\Xi_0) \right.$$

$$+ p_3\xi_3\left(\frac{\Xi_0}{\Gamma(\theta_1+1)}\right) + p_4\xi_4\left(\frac{\Xi_0}{\Gamma(\sigma_1+1)}\right) \right) M_9$$

$$+ \left(q_0 + q_1\eta_1(\Xi_0) + q_2\eta_2(\Xi_0) + q_3\eta_3\left(\frac{\Xi_0}{\Gamma(\theta_2+1)}\right) + q_4\eta_4\left(\frac{\Xi_0}{\Gamma(\sigma_2+1)}\right) \right) M_8,$$

and then for all $(u,v) \in \overline{B}_{\Xi_0}$ we find

$$\|\mathcal{Q}(u,v)\|_Y \leq \left(p_0 + p_1\xi_1(\Xi_0) + p_2\xi_2(\Xi_0) + p_3\xi_3\left(\frac{\Xi_0}{\Gamma(\theta_1+1)}\right) \right.$$

$$+ p_4\xi_4\left(\frac{\Xi_0}{\Gamma(\sigma_1+1)}\right) \right) (M_7 + M_9)$$

$$+ \left(q_0 + q_1\eta_1(\Xi_0) + q_2\eta_2(\Xi_0) + q_3\eta_3\left(\frac{\Xi_0}{\Gamma(\theta_2+1)}\right) \right.$$

$$+ q_4\eta_4\left(\frac{\Xi_0}{\Gamma(\sigma_2+1)}\right) \right) (M_8 + M_{10}) < \Xi_0.$$

Hence, $\mathcal{Q}(\overline{B}_{\Xi_0}) \subset \overline{B}_{\Xi_0}$. Using a similar approach as in the proof of Theorem 6.3.5, we can show that operator $\mathcal{Q}$ is completely continuous.

We suppose now that there exists $(u,v) \in \partial B_{\Xi_0}$ such that $(u,v) = \nu\mathcal{Q}(u,v)$ for some $\nu \in (0,1)$. Arguing as above we deduce $\|(u,v)\|_Y \leq \|\mathcal{Q}(u,v)\|_Y < \Xi_0$, which is a contradiction, because $(u,v) \in \partial B_{\Xi_0}$. Then by using Theorem 1.2.6, we conclude that operator $\mathcal{Q}$ has a fixed point $(u,v) \in \overline{B}_{\Xi_0}$, and so problem (6.31), (6.32) has at least one solution. $\square$

6.3.3 *Examples*

Let $\alpha = \frac{3}{2}$ ($n = 2$), $\beta = \frac{7}{3}$ ($m = 3$), $\theta_1 = \frac{1}{4}$, $\sigma_1 = \frac{6}{5}$, $\theta_2 = \frac{17}{4}$, $\sigma_2 = \frac{1}{3}$, $p = 1$, $q = 2$, $\gamma_0 = \frac{1}{6}$, $\gamma_1 = \frac{3}{4}$, $\delta_0 = \frac{8}{7}$, $\delta_1 = \frac{1}{5}$, $\delta_2 = \frac{1}{3}$, $H_1(t) = \{0, \ t \in [0, \frac{1}{2}); \ 3, \ t \in [\frac{1}{2}, 1]\}$, $K_1(t) = -t^2$, $t \in [0,1]$, $K_2(t) = \{0, \ t \in [0, \frac{1}{3}); \ 4, \ t \in [\frac{1}{3}, 1]\}$.

We consider the system of fractional differential equations

$$\begin{cases} D_{0+}^{3/2}u(t) + f(t, u(t), v(t), I_{0+}^{1/4}u(t), I_{0+}^{6/5}v(t)) = 0, & t \in (0,1), \\ D_{0+}^{7/3}v(t) + g(t, u(t), v(t), I_{0+}^{17/4}u(t), I_{0+}^{1/3}v(t)) = 0, & t \in (0,1), \end{cases} \tag{6.50}$$

with the boundary conditions

$$
\begin{cases}
u(0) = 0, \quad D_{0+}^{1/6} u(1) = 3 D_{0+}^{3/4} v \left(\dfrac{1}{2} \right), \\[2mm]
v(0) = v'(0) = 0, \quad D_{0+}^{8/7} v(1) = -2 \displaystyle\int_0^1 t D_{0+}^{1/5} u(t)\, dt + 4 D_{0+}^{1/3} u \left(\dfrac{1}{3} \right).
\end{cases}
$$
$$(6.51)$$

We obtain $\Delta \approx -4.92715202 \neq 0$. So assumption (J1) is satisfied. In addition, we have $M_1 \approx 2.10326265$, $M_2 \approx 1.90760368$, $M_3 \approx 1.02839972$, $M_4 \approx 2.11984652$, $M_5 = M_1$, $M_6 = M_4$, $M_7 \approx 1.81109405$, $M_{10} \approx 0.68108088$, $M_9 \approx 0.9999811$, $M_8 \approx 1.12515265$, $M_{11} \approx 1.05884127$, $M_{12} \approx 0.76520198$.

Example 1. We consider the functions

$$
f(t, x_1, x_2, x_3, x_4) = \frac{1}{\sqrt{9 + t^3}} - \frac{t}{10} \arctan x_1 + \frac{|x_2|}{(t+2)^4 (1 + |x_2|)}
$$
$$
+ \frac{1}{3(t+8)} \sin^2 x_3 - \frac{t^2}{t+12} \cos x_4,
$$

$$
g(t, x_1, x_2, x_3, x_4) = \frac{3t}{t^2 + 4} - \frac{|x_1|}{6(2 + |x_1|)} + \frac{1}{15} \sin x_2
$$
$$
+ \frac{t}{t+24} \cos^2 x_3 - \frac{1}{12} \arctan x_4,
$$

for all $t \in [0,1]$, $x_i \in \mathbb{R}$, $i = 1, \ldots, 4$. We find the inequalities

$$
|f(t, x_1, x_2, x_3, x_4) - f(t, y_1, y_2, y_3, y_4)|
$$
$$
\leq \frac{1}{10} |x_1 - y_1| + \frac{1}{16} |x_2 - y_2| + \frac{1}{12} |x_3 - y_3|
$$
$$
+ \frac{1}{13} |x_4 - y_4| \leq \frac{1}{10} \sum_{i=1}^4 |x_i - y_i|,
$$

$$
|g(t, x_1, x_2, x_3, x_4) - g(t, y_1, y_2, y_3, y_4)|
$$
$$
\leq \frac{1}{12} |x_1 - y_1| + \frac{1}{15} |x_2 - y_2| + \frac{2}{25} |x_3 - y_3|
$$
$$
+ \frac{1}{12} |x_4 - y_4| \leq \frac{1}{12} \sum_{i=1}^4 |x_i - y_i|,
$$

for all $t \in [0,1]$, x_i, $y_i \in \mathbb{R}$, $i = 1, \ldots, 4$. So we have $L_1 = \frac{1}{10}$, $L_2 = \frac{1}{12}$ and $\Xi = L_1 M_5 (M_7 + M_9) + L_2 M_6 (M_8 + M_{10}) \approx 0.91032 < 1$. Therefore, assumption (J2) is satisfied, and by Theorem 6.3.1, we deduce that problem (6.50), (6.51) has a unique solution $(u(t), v(t))$, $t \in [0, 1]$.

Example 2. We consider the functions

$$f(t, x_1, x_2, x_3, x_4) = \frac{t+2}{t^2+5} \left(2 \sin t + \frac{1}{5} \cos x_1 \right)$$

$$- \frac{1}{(t+5)^2} x_2 - \frac{t}{6} \arctan x_3 + \frac{1}{7} \sin x_4,$$

$$g(t, x_1, x_2, x_3, x_4) = \frac{e^{-t}}{2+t^3} + \frac{1}{4} \cos^2 x_2 - \frac{1}{5} \sin x_3 + \frac{1}{9} \arctan x_4,$$

for all $t \in [0,1]$, $x_i \in \mathbb{R}$, $i = 1, \ldots, 4$. Because we have

$$|f(t, x_1, x_2, x_3, x_4)| \leq \frac{11}{10} + \frac{1}{25}|x_2| + \frac{1}{6}|x_3| + \frac{1}{7}|x_4|,$$

$$|g(t, x_1, x_2, x_3, x_4)| \leq \frac{3}{4} + \frac{1}{5}|x_3| + \frac{1}{9}|x_4|,$$

for all $t \in [0,1]$, $x_i \in \mathbb{R}$, $i = 1, \ldots, 4$, the assumption (J3) is satisfied with $a_0 = \frac{11}{10}$, $a_1 = 0$, $a_2 = \frac{1}{25}$, $a_3 = \frac{1}{6}$, $a_4 = \frac{1}{7}$, $b_0 = \frac{3}{4}$, $b_1 = b_2 = 0$, $b_3 = \frac{1}{5}$, $b_4 = \frac{1}{9}$. In addition, we obtain $M_{13} \approx 0.52715168$, $M_{14} \approx 0.70166538$ and $\Xi_1 = \max\{M_{13}, M_{14}\} = M_{14} < 1$. Then by Theorem 6.3.2, we conclude that problem (6.50), (6.51) has at least one solution $(u(t), v(t))$, $t \in [0, 1]$.

Example 3. We consider the functions

$$f(t, x_1, x_2, x_3, x_4) = -\frac{1}{5}|x_2|^{3/4} + \frac{1}{3(1+t^2)} \arctan |x_3|^{1/2},$$

$$g(t, x_1, x_2, x_3, x_4) = \frac{e^{-t}}{1+t^3} - \frac{1}{3} x_1^{4/5} + \sin x_4^{2/3},$$

for all $t \in [0,1]$, $x_i \in \mathbb{R}$, $i = 1, \ldots, 4$. Because we obtain

$$|f(t, x_1, x_2, x_3, x_4)| \leq \frac{1}{5}|x_2|^{3/4} + \frac{1}{3}|x_3|^{1/2},$$

$$|g(t, x_1, x_2, x_3, x_4)| \leq 1 + \frac{1}{3}|x_1|^{4/5} + |x_4|^{2/3},$$

for all $t \in [0,1]$, $x_i \in \mathbb{R}$, $i = 1, \ldots, 4$, then assumption (J5) is satisfied with $c_0 = c_1 = 0$, $c_2 = \frac{1}{5}$, $c_2 = \frac{1}{3}$, $c_4 = 0$, $d_0 = 1$, $d_1 = \frac{1}{3}$, $d_2 = d_3 = 0$, $d_4 = 1$, $l_2 = \frac{3}{4}$, $l_3 = \frac{1}{2}$, $m_1 = \frac{4}{5}$, $m_4 = \frac{2}{3}$. Therefore, by Theorem 6.3.5, we deduce that problem (6.50), (6.51) has at least one solution $(u(t), v(t))$, $t \in [0, 1]$.

Example 4. We consider the functions

$$f(t, x_1, x_2, x_3, x_4) = \frac{t^3}{25} + \frac{e^{-t}x_1^4}{20(1 + x_2^2)} - \frac{t^2 x_4^{1/3}}{10},$$

$$g(t, x_1, x_2, x_3, x_4) = \frac{(1 - t)^4}{20} - \frac{1 - t^2}{15}x_2^2 - \frac{1}{25}x_3^{2/5},$$

for all $t \in [0, 1]$, $x_i \in \mathbb{R}$, $i = 1, \ldots, 4$. Because we have

$$|f(t, x_1, x_2, x_3, x_4)| \leq \frac{1}{25} + \frac{1}{20}|x_1|^4 + \frac{1}{10}|x_4|^{1/3},$$

$$|g(t, x_1, x_2, x_3, x_4)| \leq \frac{1}{20} + \frac{1}{15}|x_2|^2 + \frac{1}{25}|x_3|^{2/5},$$

for all $t \in [0, 1]$, $x_i \in \mathbb{R}$, $i = 1, \ldots, 4$, then assumption (J6) is satisfied with $p_0 = \frac{1}{25}$, $p_1 = \frac{1}{20}$, $p_2 = p_3 = 0$, $p_4 = \frac{1}{10}$, $q_0 = \frac{1}{20}$, $q_1 = 0$, $q_2 = \frac{1}{15}$, $q_3 = \frac{1}{25}$, $q_4 = 0$, $\xi_1(x) = x^4$, $\xi_4(x) = x^{1/3}$, $\eta_2(x) = x^2$, $\eta_3(x) = x^{2/5}$ for $x \geq 0$. For $\Xi_0 = 1$, the condition (6.49) is satisfied, because $(\frac{1}{25} + \frac{1}{20} + \frac{1}{10}(\frac{1}{\Gamma(11/5)})^{1/3})(M_7 + M_9) + (\frac{1}{20} + \frac{1}{15} + \frac{1}{25}(\frac{1}{\Gamma(21/4)})^{2/5})(M_8 + M_{10}) \approx 0.75328 < 1$. Then by Theorem 6.3.6, we conclude that problem (6.50), (6.51) has at least one solution $(u(t), v(t))$, $t \in [0, 1]$.

Remark 6.3.1. The results presented in this section will be published in [84].

Chapter 7

Existence of Solutions for Caputo Fractional Boundary Value Problems

In this chapter, we study the existence of solutions for some Caputo fractional differential equations and inclusions, and systems of Caputo fractional differential equations subject to nonlocal boundary conditions which contain Riemann-Stieltjes integrals.

7.1 Sequential Caputo Fractional Differential Equations and Inclusions with Nonlocal Boundary Conditions

In this section, we consider the nonlinear Caputo type sequential fractional integro-differential equation and inclusion:

$$({}^cD_{0+}^\alpha + \lambda {}^cD_{0+}^{\alpha-1})u(t) = f(t, u(t), {}^cD_{0+}^p u(t), I_{0+}^q u(t)), \quad t \in (0,1), \quad (7.1)$$

$$({}^cD_{0+}^\alpha + \lambda {}^cD_{0+}^{\alpha-1})u(t) \in F(t, u(t), I_{0+}^q u(t)), \quad t \in (0,1), \quad (7.2)$$

subject to the nonlocal boundary conditions

$$u(0) = h(u), \quad u'(0) = u''(0) = 0, \quad a I_{0+}^\beta u(\xi) = \int_0^1 u(s)\, dH(s), \quad (7.3)$$

where $\alpha \in (3,4]$, $p \in (0,1)$, $q > 0$, $\lambda > 0$, $\xi \in (0,1]$, $a \in \mathbb{R}$, $\beta > 0$, ${}^cD_{0+}^\alpha$, ${}^cD_{0+}^{\alpha-1}$ and ${}^cD_{0+}^p$ are Caputo fractional derivatives of orders $\alpha, \alpha - 1$ and p, respectively, I_{0+}^q and I_{0+}^β are Riemann-Liouville fractional integrals of order

q and β, respectively, H is a bounded variation function, h is a nonlinear function defined on $C[0,1]$, f is a nonlinear function and F is a nonlinear multivalued function which satisfy some assumptions.

We give conditions for the nonlinearities f and F, and for the functions h and H, such that the problems (7.1), (7.3) and (7.2), (7.3) have at least one solution. For the proofs of our main theorems, we use the contraction mapping principle (Theorem 1.2.1) and the Krasnosel'skii fixed point theorem for the sum of two operators (Theorem 1.2.3) in the case of fractional equations, and the nonlinear alternative of Leray-Schauder type for Kakutani maps (Theorem 1.2.14) and the Covitz–Nadler fixed point theorem (Theorem 1.2.15) in the case of fractional inclusions.

In the last few years, nonlocal boundary value problems for sequential fractional differential equations, integro-differential equations and inclusions, and systems of such equations have been studied by many researchers. In [10], by using the fixed point theory, the authors investigated the existence of solutions for the sequential fractional integro-differential equation

$$({}^{c}D_{0+}^{\alpha} + k\,{}^{c}D_{0+}^{\alpha-1})u(t) = pf(t, u(t)) + qI_{0+}^{\beta}g(t, u(t)), \quad 0 < t < 1,$$

with the boundary conditions $u(0) = 0$, $u(1) = 0$, or $u'(0) + ku(0) = a$, $u(1) = b$, $a, b \in \mathbb{R}$, or $u(0) = a$, $u'(0) = u'(1)$, $a \in \mathbb{R}$, where ${}^{c}D_{0+}^{\alpha}$ denotes the Caputo fractional derivative of order $\alpha \in (1, 2]$, I_{0+}^{β} denotes the Riemann-Liouville fractional integral of order $\beta \in (0, 1)$, f, g are given continuous functions, $k \neq 0$ and p, q are real constants. The word "sequential" is used in the sense that the operator ${}^{c}D_{0+}^{\alpha} + k\,{}^{c}D_{0+}^{\alpha-1}$ can be written as the composition of operators ${}^{c}D_{0+}^{\alpha-1}$ and $D + k$. In [16], the authors studied the existence of solutions for the sequential fractional differential equation

$$({}^{c}D_{0+}^{\alpha} + k\,{}^{c}D_{0+}^{\alpha-1})x(t) = f(t, x(t)), \quad t \in [0, 1],$$

with the boundary conditions

$$x(0) = 0, \quad x'(0) = 0, \quad x(\zeta) = aI_{0+}^{\beta}x(\eta), \tag{7.4}$$

where $\alpha \in (2, 3]$, $\beta > 0$, $0 < \eta < \zeta < 1$, f is a given continuous function, and k, a are appropriate positive real constants. They use the Banach contraction mapping principle, the Krasnosel'skii fixed point theorem and the nonlinear alternative of Leray-Schauder type. In [12], by using some fixed point theorems, the authors investigated the existence of solutions for the Caputo type sequential fractional differential inclusion

$$({}^{c}D_{0+}^{\alpha} + k\,{}^{c}D_{0+}^{\alpha-1})x(t) \in F(t, x(t)), \quad t \in [0, 1],$$

with the nonlocal Riemann-Liouville fractional integral boundary conditions (7.4), where $\alpha \in (2,3]$, $F : [0,1] \times \mathbb{R} \to \mathcal{P}(\mathbb{R})$ is a multivalued map, and $\mathcal{P}(\mathbb{R})$ is the family of all nonempty subsets of $\mathbb{R}$. In [14], the authors studied the existence of solutions for the sequential fractional differential equations and inclusions

$$({}^cD_{0+}^q + k{}^cD_{0+}^{q-1})x(t) = f(t, x(t), {}^cD_{0+}^\delta x(t), I_{0+}^\gamma x(t)), \quad t \in [0,1],$$

$$({}^cD_{0+}^q + k{}^cD_{0+}^{q-1})x(t) \in F(t, x(t), {}^cD_{0+}^\delta x(t), I_{0+}^\gamma x(t)), \quad t \in [0,1],$$

supplemented with semi-periodic and nonlocal integro-multi-point boundary conditions involving Riemann-Liouville integral given by

$$x(0) = x(1), \quad x'(0) = 0, \quad \sum_{i=1}^{m} a_i x(\zeta_i) = \lambda I_{0+}^\beta x(\eta),$$

where $q \in (2,3]$, $\delta, \gamma \in (0,1)$, $k > 0$, $\beta > 0$, $0 < \eta < \zeta_1 < \cdots < \zeta_m < 1$, $f : [0,1] \times \mathbb{R}^3 \to \mathbb{R}$ is a given continuous function, $F : [0,1] \times \mathbb{R}^3 \to \mathcal{P}(\mathbb{R})$ is a multivalued map, and λ, a_i, $i = 1, \ldots, m$ are real constants. Some standard fixed point theorems for single-valued and multivalued maps are applied in [14].

7.1.1 *Auxiliary results*

We consider the linear sequential fractional differential equation

$$({}^cD_{0+}^\alpha + \lambda{}^cD_{0+}^{\alpha-1})u(t) = x(t), \quad t \in (0,1), \tag{7.5}$$

supplemented with the nonlocal integral boundary conditions

$$u(0) = u_0, \quad u'(0) = u''(0) = 0, \quad aI_{0+}^\beta u(\xi) = \int_0^1 u(s)\,dH(s), \tag{7.6}$$

where $x \in C[0,1]$, $\alpha \in (3,4]$, $\xi \in (0,1]$, $\lambda > 0$, $a \in \mathbb{R}$, $u_0 \in \mathbb{R}$, and H is a bounded variation function.

In the sequel, we denote by

$$\Delta = \frac{a}{\Gamma(\beta)} \int_0^\xi (\xi - s)^{\beta-1}(\lambda^2 s^2 - 2\lambda s + 2 - 2e^{-\lambda s})\,ds$$

$$- \int_0^1 (\lambda^2 s^2 - 2\lambda s + 2 - 2e^{-\lambda s})\,dH(s), \tag{7.7}$$

$$\gamma(t) = \frac{\lambda^2 t^2 - 2\lambda t + 2 - 2e^{-\lambda t}}{\Delta}, \quad t \in [0,1], \quad \text{for} \quad \Delta \neq 0.$$

Lemma 7.1.1. *If $x \in C[0,1]$ and $\Delta \neq 0$, then the unique solution $u \in C^4[0,1]$ of the boundary value problem (7.5), (7.6) is given by*

$$u(t) = u_0 \left\{ e^{-\lambda t} + \gamma(t) \left[\int_0^1 e^{-\lambda s} \, dH(s) - \frac{a}{\Gamma(\beta)} \int_0^\xi (\xi - s)^{\beta - 1} e^{-\lambda s} \, ds \right] \right\}$$

$$+ \gamma(t) \left[\int_0^1 \left(\int_0^s e^{-\lambda(s-\tau)} I_{0+}^{\alpha-1} x(\tau) \, d\tau \right) dH(s) \right.$$

$$\left. - \frac{a}{\Gamma(\beta)} \int_0^\xi (\xi - s)^{\beta - 1} \left(\int_0^s e^{-\lambda(s-\tau)} I_{0+}^{\alpha-1} x(\tau) \, d\tau \right) ds \right]$$

$$+ \int_0^t e^{-\lambda(t-s)} I_{0+}^{\alpha-1} x(s) \, ds, \tag{7.8}$$

where Δ and γ are given by (7.7).

Proof. Equation (7.5) can equivalently be written as

$$^{c}D_{0+}^{\alpha}(u(t) + \lambda D^{-1} u(t)) = x(t),$$

where $D^{-1} u(t) = I_{0+}^1 u(t) = \int_0^t u(s) \, ds$. Then, by Lemma 1.1.4, the general solution of problem (7.5), (7.6) is

$$u(t) = -\lambda \int_0^t u(s) \, ds + \frac{1}{\Gamma(\alpha)} \int_0^t (t - s)^{\alpha - 1} x(s) \, ds$$

$$+ a_0 + a_1 t + a_2 t^2 + a_3 t^3, \quad t \in [0, 1], \tag{7.9}$$

where $a_i \in \mathbb{R}$ $(i = 0, 1, 2, 3)$ are unknown arbitrary constants.

Differentiating (7.9), we obtain

$$u'(t) = -\lambda u(t) + \frac{1}{\Gamma(\alpha - 1)} \int_0^t (t - s)^{\alpha - 2} x(s) \, ds + a_1 + 2a_2 t + 3a_3 t^2,$$

which can alternatively be written as

$$(e^{\lambda t} u(t))' = \frac{e^{\lambda t}}{\Gamma(\alpha - 1)} \int_0^t (t - s)^{\alpha - 2} x(s) \, ds + a_1 e^{\lambda t} + 2a_2 t e^{\lambda t} + 3a_3 t^2 e^{\lambda t}. \tag{7.10}$$

Integrating (7.10) from 0 to t, we get

$$u(t) = e^{-\lambda t}c_0 + \frac{c_1}{\lambda}(1 - e^{-\lambda t}) + \frac{c_2}{\lambda^2}(\lambda t - 1 + e^{-\lambda t})$$

$$+ \frac{c_3}{\lambda^3}(\lambda^2 t^2 - 2\lambda t + 2 - 2e^{-\lambda t})$$

$$+ \int_0^t e^{-\lambda(t-s)} I_{0+}^{\alpha-1} x(s)\, ds, \quad t \in [0,1], \tag{7.11}$$

where $c_0 = u(0)$, $c_1 = a_1$, $c_2 = 2a_2$, $c_3 = 3a_3$.

Using the boundary conditions $u(0) = u_0$, $u'(0) = u''(0) = 0$ in (7.11), we find that $c_0 = u_0$, $c_1 = 0$ and $c_2 = 0$ and consequently (7.11) becomes

$$u(t) = u_0 e^{-\lambda t} + \frac{c_3}{\lambda^3}(\lambda^2 t^2 - 2\lambda t + 2 - 2e^{-\lambda t})$$

$$+ \int_0^t e^{-\lambda(t-s)} I_{0+}^{\alpha-1} x(s)\, ds, \quad t \in [0,1]. \tag{7.12}$$

Now using the boundary condition $aI_{0+}^\beta u(\xi) = \int_0^1 u(s)\, dH(s)$, we obtain

$$\frac{c_3}{\lambda^3}\left[\frac{a}{\Gamma(\beta)} \int_0^\xi (\xi - s)^{\beta-1}(\lambda^2 s^2 - 2\lambda s + 2 - 2e^{-\lambda s})\, ds \right.$$

$$\left. - \int_0^1 (\lambda^2 s^2 - 2\lambda s + 2 - 2e^{-\lambda s})\, dH(s) \right]$$

$$= u_0 \left[\int_0^1 e^{-\lambda s}\, dH(s) - \frac{a}{\Gamma(\beta)} \int_0^\xi (\xi - s)^{\beta-1} e^{-\lambda s}\, ds \right]$$

$$+ \int_0^1 \left(\int_0^s e^{-\lambda(s-\tau)} I_{0+}^{\alpha-1} x(\tau)\, d\tau \right) dH(s)$$

$$- \frac{a}{\Gamma(\beta)} \int_0^\xi (\xi - s)^{\beta-1} \left(\int_0^s e^{-\lambda(s-\tau)} I_{0+}^{\alpha-1} x(\tau)\, d\tau \right) ds,$$

which, in view of the notation $\Delta \neq 0$ given by (7.7), takes the form

$$\frac{c_3}{\lambda^3} = \frac{u_0}{\Delta}\left[\int_0^1 e^{-\lambda s}\, dH(s) - \frac{a}{\Gamma(\beta)} \int_0^\xi (\xi - s)^{\beta-1} e^{-\lambda s}\, ds \right]$$

$$+ \frac{1}{\Delta} \int_0^1 \left(\int_0^s e^{-\lambda(s-\tau)} I_{0+}^{\alpha-1} x(\tau)\, d\tau \right) dH(s)$$

$$- \frac{a}{\Delta\Gamma(\beta)} \int_0^\xi (\xi - s)^{\beta-1} \left(\int_0^s e^{-\lambda(s-\tau)} I_{0+}^{\alpha-1} x(\tau)\, d\tau \right) ds.$$

Substituting the above value of c_3/λ^3 in (7.12) and using the expression for $\gamma(t)$ given by (7.7), we obtain

$$u(t) = u_0 e^{-\lambda t} + \gamma(t) \left\{ u_0 \left[\int_0^1 e^{-\lambda s}\, dH(s) - \frac{a}{\Gamma(\beta)} \int_0^\xi (\xi - s)^{\beta-1} e^{-\lambda s}\, ds \right] \right.$$

$$+ \int_0^1 \left(\int_0^s e^{-\lambda(s-\tau)} I_{0+}^{\alpha-1} x(\tau)\, d\tau \right) dH(s)$$

$$\left. - \frac{a}{\Gamma(\beta)} \int_0^\xi (\xi - s)^{\beta-1} \left(\int_0^s e^{-\lambda(s-\tau)} I_{0+}^{\alpha-1} x(\tau)\, d\tau \right) ds \right\}$$

$$+ \int_0^t e^{-\lambda(t-s)} I_{0+}^{\alpha-1} x(s)\, ds.$$

The converse of the lemma follows by direct computation. This completes the proof. □

7.1.2   *Existence of solutions for problem (7.1), (7.3)*

In this section, we investigate the existence and uniqueness of solutions for the fractional differential equation (7.1) with the boundary conditions (7.3).

We consider the space $X = \{u \in C[0,1], {}^cD_{0+}^p u \in C[0,1]\}$ equipped with $\|u\|_X = \|u\| + \|{}^cD_{0+}^p u\|$, where $\|w\| = \sup_{t \in [0,1]} |w(t)|$ for $w \in C[0,1]$. Obviously $(X, \|\cdot\|_X)$ is a Banach space.

By Lemma 7.1.1, we introduce the operator $A : X \to X$ defined by

$$(Au)(t) = h(u) \left\{ e^{-\lambda t} + \gamma(t) \left[\int_0^1 e^{-\lambda s}\, dH(s) \right. \right.$$

$$\left. \left. - \frac{a}{\Gamma(\beta)} \int_0^\xi (\xi - s)^{\beta-1} e^{-\lambda s}\, ds \right] \right\}$$

$$+ \gamma(t) \left[\int_0^1 \left(\int_0^s e^{-\lambda(s-\tau)} I_{0+}^{\alpha-1} \hat{f}_u(\tau)\, d\tau \right) dH(s) \right.$$

$$\left. - \frac{a}{\Gamma(\beta)} \int_0^\xi (\xi - s)^{\beta-1} \left(\int_0^s e^{-\lambda(s-\tau)} I_{0+}^{\alpha-1} \hat{f}_u(\tau)\, d\tau \right) ds \right]$$

$$+ \int_0^t e^{-\lambda(t-s)} I_{0+}^{\alpha-1} \hat{f}_u(s)\, ds,$$

for $t \in [0,1]$ and $u \in X$, where $\hat{f}_u(s) = f(s, u(s), {}^cD_{0+}^p u(s), I_{0+}^q u(s))$, $s \in [0,1]$.

Note that the function u is a solution of problem (7.1), (7.3) if and only if u is a fixed point of operator A.

Now, we enlist the assumptions that we need in the sequel.

(H1) $\alpha \in (3,4]$, $p \in (0,1)$, $q > 0$, $\lambda > 0$, $\xi \in (0,1]$, $a \in \mathbb{R}$, $\beta > 0$, $\Delta \neq 0$
(given by (7.7)), and $H : [0,1] \to \mathbb{R}$ is a function of bounded variation.

(H2) The function $f : [0,1] \times \mathbb{R}^3 \to \mathbb{R}$ is continuous and there exists $L_1 > 0$
such that

$$|f(t,x,y,z) - f(t,x_1,y_1,z_1)| \leq L_1(|x - x_1| + |y - y_1| + |z - z_1|),$$

for all $t \in [0,1]$, $x,y,z,x_1,y_1,z_1 \in \mathbb{R}$.

(H3) The function $f : [0,1] \times \mathbb{R}^3 \to \mathbb{R}$ is continuous and there exists a
function $g \in C([0,1], \mathbb{R}_+)$ such that

$$|f(t,x,y,z)| \leq g(t), \quad \forall (t,x,y,z) \in [0,1] \times \mathbb{R}^3.$$

(H4) For the function $h : C[0,1] \to \mathbb{R}$, there exists $L_3 > 0$ such that

$$|h(x) - h(y)| \leq L_3\|x - y\|, \quad \forall x,y \in C[0,1].$$

(H5) The function $h : C[0,1] \to \mathbb{R}$ is continuous and there exists $L_2 > 0$
such that

$$|h(x)| \leq L_2\|x\|, \quad \forall x \in C[0,1].$$

Furthermore, we set the notations:

$$l = \sup_{t \in [0,1]} |\gamma(t)| = \frac{\lambda^2 - 2\lambda + 2 - 2e^{-\lambda}}{|\Delta|} > 0,$$

$$m = \sup_{t \in [0,1]} |\gamma'(t)| = \frac{2\lambda(\lambda - 1 + e^{-\lambda})}{|\Delta|} > 0,$$

$$M_1 = 1 + l\left[\left|\int_0^1 e^{-\lambda s}\, dH(s)\right| + \frac{|a|}{\Gamma(\beta)}\int_0^\xi (\xi - s)^{\beta-1}e^{-\lambda s}\, ds\right],$$

$$M_2 = \frac{1}{\lambda}\left|\int_0^1 (1 - e^{-\lambda s})\, dH(s)\right| + \frac{|a|}{\lambda\Gamma(\beta)}\int_0^\xi (\xi - s)^{\beta-1}(1 - e^{-\lambda s})\, ds,$$

$$M_1^* = \lambda + m\left[\left|\int_0^1 e^{-\lambda s}\, dH(s)\right| + \frac{|a|}{\Gamma(\beta)}\int_0^\xi (\xi - s)^{\beta-1}e^{-\lambda s}\, ds\right],$$

$$L_0 = 1 + \frac{1}{\Gamma(q+1)}. \tag{7.13}$$

Theorem 7.1.1. *Assume that* (H1), (H2), (H4) *and* (H5) *hold. If*

$$\widetilde{\Lambda} := \max\{L_2, L_3\} \left(M_1 + \frac{1}{\Gamma(2-p)} M_1^* \right) + \frac{L_1 L_0}{\Gamma(\alpha)} \left(l M_2 + \frac{1}{\lambda}(1 - e^{-\lambda}) \right)$$

$$+ \frac{L_1 L_0}{\Gamma(2-p)\Gamma(\alpha)} \left(m M_2 + 2 - e^{-\lambda} \right) < 1, \tag{7.14}$$

then problem (7.1), (7.3) *has a unique solution on* $[0, 1]$*, where* $l, m, L_0, M_1, M_2, M_1^*$ *are given by* (7.13).

Proof. Let us fix $r > 0$ such that

$$r \geq \frac{M_0}{\Gamma(\alpha)} \left[l M_2 + \frac{1}{\lambda}\left(1 - e^{-\lambda}\right) + \frac{1}{\Gamma(2-p)} \left(m M_2 + 2 - e^{-\lambda} \right) \right]$$

$$\times \left\{ 1 - \left[L_2 \left(M_1 + \frac{1}{\Gamma(2-p)} M_1^* \right) + \frac{L_1 L_0}{\Gamma(\alpha)} \left(l M_2 + \frac{1}{\lambda}\left(1 - e^{-\lambda}\right) \right) \right. \right.$$

$$\left. \left. + \frac{L_1 L_0}{\Gamma(2-p)\Gamma(\alpha)} \left(m M_2 + 2 - e^{-\lambda} \right) \right] \right\}^{-1}, \tag{7.15}$$

where $M_0 = \sup_{t \in [0,1]} |f(t, 0, 0, 0)|$.

We first consider the set $\overline{B}_r = \{u \in X, \|u\|_X \leq r\}$ and show that $A(\overline{B}_r) \subset \overline{B}_r$. Let $u \in \overline{B}_r$. For $\hat{f}_u$, we obtain

$$|\hat{f}_u(t)| = |f(t, u(t), {}^c D_{0+}^p u(t), I_{0+}^q u(t)|$$

$$\leq |f(t, u(t), {}^c D_{0+}^p u(t), I_{0+}^q u(t)) - f(t, 0, 0, 0)| + |f(t, 0, 0, 0)|$$

$$\leq L_1(|u(t)| + |{}^c D_{0+}^p u(t)| + |I_{0+}^q u(t)|) + M_0$$

$$\leq L_1 \left(\|u\| + \|{}^c D_{0+}^p u\| + \frac{1}{\Gamma(q+1)} \|u\| \right) + M_0$$

$$\leq L_1 \left(1 + \frac{1}{\Gamma(q+1)} \right) \|u\|_X + M_0 \leq L_1 L_0 r + M_0, \quad \forall t \in [0, 1].$$

Then we have

$$|(Au)(t)| \leq |h(u)| \left\{ e^{-\lambda t} + |\gamma(t)| \left[\left| \int_0^1 e^{-\lambda s} \, dH(s) \right| \right. \right.$$

$$\left. \left. + \frac{|a|}{\Gamma(\beta)} \int_0^\xi (\xi - s)^{\beta-1} e^{-\lambda s} \, ds \right] \right\}$$

$$+ |\gamma(t)| \left[\left| \int_0^1 \left(\int_0^s e^{-\lambda(s-\tau)} \left| I_{0+}^{\alpha-1} \hat{f}_u(\tau) \right| \, d\tau \right) dH(s) \right| \right.$$

$$+ \frac{|a|}{\Gamma(\beta)} \int_0^\xi (\xi - s)^{\beta-1} \left(\int_0^s e^{-\lambda(s-\tau)} |I_{0+}^{\alpha-1} \hat{f}_u(\tau)| \, d\tau \right) ds \Bigg]$$

$$+ \int_0^t e^{-\lambda(t-s)} |I_{0+}^{\alpha-1} \hat{f}_u(s)| \, ds$$

$$\leq L_2 \|u\| \left\{ 1 + l \left[\left| \int_0^1 e^{-\lambda s} \, dH(s) \right| \right. \right.$$

$$\left. \left. + \frac{|a|}{\Gamma(\beta)} \int_0^\xi (\xi - s)^{\beta-1} e^{-\lambda s} \, ds \right] \right\}$$

$$+ \frac{l(L_1 L_0 r + M_0)}{\Gamma(\alpha)} \left[\left| \int_0^1 \left(\int_0^s e^{-\lambda(s-\tau)} \, d\tau \right) dH(s) \right| \right.$$

$$\left. + \frac{|a|}{\Gamma(\beta)} \int_0^\xi (\xi - s)^{\beta-1} \left(\int_0^s e^{-\lambda(s-\tau)} \, d\tau \right) ds \right]$$

$$+ \frac{L_1 L_0 r + M_0}{\Gamma(\alpha)} \int_0^t e^{-\lambda(t-s)} \, ds$$

$$\leq L_2 r \left\{ 1 + l \left[\left| \int_0^1 e^{-\lambda s} \, dH(s) \right| + \frac{|a|}{\Gamma(\beta)} \int_0^\xi (\xi - s)^{\beta-1} e^{-\lambda s} \, ds \right] \right\}$$

$$+ \frac{l(L_1 L_0 r + M_0)}{\Gamma(\alpha)} \left[\left| \int_0^1 \frac{1}{\lambda} \left(1 - e^{-\lambda s} \right) dH(s) \right| \right.$$

$$\left. + \frac{|a|}{\Gamma(\beta)} \int_0^\xi (\xi - s)^{\beta-1} \frac{1}{\lambda} (1 - e^{-\lambda s}) \, ds \right]$$

$$+ \frac{L_1 L_0 r + M_0}{\lambda \Gamma(\alpha)} (1 - e^{-\lambda t}), \quad \forall t \in [0,1],$$

which, on taking the norm for $t \in [0,1]$, yields

$$\|Au\| \leq L_2 r \left\{ 1 + l \left[\left| \int_0^1 e^{-\lambda s} \, dH(s) \right| + \frac{|a|}{\Gamma(\beta)} \int_0^\xi (\xi - s)^{\beta-1} e^{-\lambda s} \, ds \right] \right\}$$

$$+ \frac{l(L_1 L_0 r + M_0)}{\Gamma(\alpha)} \left[\frac{1}{\lambda} \left| \int_0^1 (1 - e^{-\lambda s}) \, dH(s) \right| \right.$$

$$\left. + \frac{|a|}{\lambda \Gamma(\beta)} \int_0^\xi (\xi - s)^{\beta-1} (1 - e^{-\lambda s}) \, ds \right]$$

$$+ \frac{L_1 L_0 r + M_0}{\lambda \Gamma(\alpha)} (1 - e^{-\lambda}) = r L_2 M_1$$

$$+ \frac{L_1 L_0 r + M_0}{\Gamma(\alpha)} \left(l M_2 + \frac{1 - e^{-\lambda}}{\lambda} \right),$$

where l, L_0, M_1 and M_2 are given by (7.13).

Furthermore, one can find that

$$(Au)'(t) = h(u) \left\{ -\lambda e^{-\lambda t} + \gamma'(t) \left[\int_0^1 e^{-\lambda s} dH(s) \right. \right.$$

$$\left. - \frac{a}{\Gamma(\beta)} \int_0^\xi (\xi - s)^{\beta - 1} e^{-\lambda s} ds \right] \right\}$$

$$+ \gamma'(t) \left[\int_0^1 \left(\int_0^s e^{-\lambda(s - \tau)} I_{0+}^{\alpha - 1} \hat{f}_u(\tau) d\tau \right) dH(s) \right.$$

$$\left. - \frac{a}{\Gamma(\beta)} \int_0^\xi (\xi - s)^{\beta - 1} \left(\int_0^s e^{-\lambda(s - \tau)} I_{0+}^{\alpha - 1} \hat{f}_u(\tau) d\tau \right) ds \right]$$

$$+ I_{0+}^{\alpha - 1} \hat{f}_u(t) - \lambda \int_0^t e^{-\lambda(t - s)} I_{0+}^{\alpha - 1} \hat{f}_u(s) ds, \quad \forall t \in (0, 1),$$

where $\gamma'(t) = \frac{2\lambda(\lambda t - 1 + e^{-\lambda t})}{\Delta}$, with $|\gamma'(t)| = \frac{2\lambda(\lambda t - 1 + e^{-\lambda t})}{|\Delta|}$ for all $t \in [0, 1]$, and $\sup_{t \in [0,1]} |\gamma'(t)| = m$ (m is given by (7.13)).

Then, we obtain

$$|(Au)'(t)| \le |h(u)| \left\{ \lambda e^{-\lambda t} + |\gamma'(t)| \left[\left| \int_0^1 e^{-\lambda s} dH(s) \right| \right. \right.$$

$$\left. + \frac{|a|}{\Gamma(\beta)} \int_0^\xi (\xi - s)^{\beta - 1} e^{-\lambda s} ds \right] \right\}$$

$$+ |\gamma'(t)| \left[\left| \int_0^1 \left(\int_0^s e^{-\lambda(s - \tau)} |I_{0+}^{\alpha - 1} \hat{f}_u(\tau)| d\tau \right) dH(s) \right| \right.$$

$$\left. + \frac{|a|}{\Gamma(\beta)} \int_0^\xi (\xi - s)^{\beta - 1} \left(\int_0^s e^{-\lambda(s - \tau)} |I_{0+}^{\alpha - 1} \hat{f}_u(\tau)| d\tau \right) ds \right]$$

$$+ |I_{0+}^{\alpha - 1} \hat{f}_u(t)| + \lambda \int_0^t e^{-\lambda(t - s)} |I_{0+}^{\alpha - 1} \hat{f}_u(s)| ds$$

$$\leq L_2 \|u\| \left\{ \lambda + m \left[\left| \int_0^1 e^{-\lambda s} \, dH(s) \right| \right.\right.$$

$$\left.\left. + \frac{|a|}{\Gamma(\beta)} \int_0^\xi (\xi - s)^{\beta-1} e^{-\lambda s} \, ds \right] \right\}$$

$$+ \frac{m(L_1 L_0 r + M_0)}{\Gamma(\alpha)} \left[\left| \int_0^1 \left(\int_0^s e^{-\lambda(s-\tau)} \, d\tau \right) dH(s) \right| \right.$$

$$\left. + \frac{|a|}{\Gamma(\beta)} \int_0^\xi (\xi - s)^{\beta-1} \left(\int_0^s e^{-\lambda(s-\tau)} \, d\tau \right) ds \right] + \frac{L_1 L_0 r + M_0}{\Gamma(\alpha)}$$

$$+ \frac{L_1 L_0 r + M_0}{\Gamma(\alpha)} \lambda \int_0^t e^{-\lambda(t-s)} \, ds$$

$$\leq L_2 r \left\{ \lambda + m \left[\left| \int_0^1 e^{-\lambda s} \, dH(s) \right| + \frac{|a|}{\Gamma(\beta)} \int_0^\xi (\xi - s)^{\beta-1} e^{-\lambda s} \, ds \right] \right\}$$

$$+ \frac{m(L_1 L_0 r + M_0)}{\Gamma(\alpha)} \left[\frac{1}{\lambda} \left| \int_0^1 (1 - e^{-\lambda s}) \, dH(s) \right| \right.$$

$$\left. + \frac{|a|}{\lambda \Gamma(\beta)} \int_0^\xi (\xi - s)^{\beta-1} (1 - e^{-\lambda s}) \, ds \right]$$

$$+ \frac{L_1 L_0 r + M_0}{\Gamma(\alpha)} (2 - e^{-\lambda t}), \quad \forall t \in (0,1).$$

Hence, we conclude

$$\|(Au)'\| \leq r L_2 M_1^* + \frac{L_1 L_0 r + M_0}{\Gamma(\alpha)} (m M_2 + 2 - e^{-\lambda}),$$

where M_1^* is given by (7.13).

By the definition of Caputo fractional derivative with $p \in (0,1)$, we deduce

$$|{}^c D_{0+}^p (Au)(t)| \leq \int_0^t \frac{(t-s)^{-p}}{\Gamma(1-p)} |(Au)'(s)| \, ds$$

$$\leq \left[r L_2 M_1^* + \frac{L_1 L_0 r + M_0}{\Gamma(\alpha)} (m M_2 + 2 - e^{-\lambda}) \right] \int_0^t \frac{(t-s)^{-p}}{\Gamma(1-p)} \, ds$$

$$= \frac{1}{\Gamma(2-p)} \left[r L_2 M_1^* + \frac{L_1 L_0 r + M_0}{\Gamma(\alpha)} (m M_2 + 2 - e^{-\lambda}) \right], \quad \forall t \in [0,1].$$

Therefore, by using the definition of r from (7.15), we obtain

$$\|Au\|_X = \|Au\| + \|{}^cD_{0+}^p(Au)\| \leq rL_2M_1 + \frac{L_1L_0r + M_0}{\Gamma(\alpha)}\left(lM_2 + \frac{1-e^{-\lambda}}{\lambda}\right)$$

$$+\frac{1}{\Gamma(2-p)}\left[rL_2M_1^* + \frac{L_1L_0r + M_0}{\Gamma(\alpha)}(mM_2 + 2 - e^{-\lambda})\right] \leq r.$$

So, we conclude that A maps $\overline{B}_r$ into itself.

Now, for $u, v \in \overline{B}_r$ we have

$$|(Au)(t) - (Av)(t)| \leq |h(u) - h(v)|$$

$$\times \left\{e^{-\lambda t} + |\gamma(t)|\left[\left|\int_0^1 e^{-\lambda s}\,dH(s)\right| + \frac{|a|}{\Gamma(\beta)}\int_0^\xi (\xi - s)^{\beta-1}e^{-\lambda s}\,ds\right]\right\}$$

$$+ |\gamma(t)|\left[\left|\int_0^1\left(\int_0^s e^{-\lambda(s-\tau)}|I_{0+}^{\alpha-1}\hat{f}_u(\tau) - I_{0+}^{\alpha-1}\hat{f}_v(\tau)|\,d\tau\right)dH(s)\right|\right.$$

$$\left.+ \frac{|a|}{\Gamma(\beta)}\int_0^\xi (\xi - s)^{\beta-1}\left(\int_0^s e^{-\lambda(s-\tau)}|I_{0+}^{\alpha-1}\hat{f}_u(\tau) - I_{0+}^{\alpha-1}\hat{f}_v(\tau)|\,d\tau\right)ds\right]$$

$$+ \int_0^t e^{-\lambda(t-s)}|I_{0+}^{\alpha-1}\hat{f}_u(s) - I_{0+}^{\alpha-1}\hat{f}_v(s)|\,ds$$

$$\leq L_3\|u - v\|\left\{1 + l\left[\left|\int_0^1 e^{-\lambda s}\,dH(s)\right| + \frac{|a|}{\Gamma(\beta)}\int_0^\xi (\xi - s)^{\beta-1}e^{-\lambda s}\,ds\right]\right\}$$

$$+ \frac{l}{\Gamma(\alpha)}\|\hat{f}_u - \hat{f}_v\|\left[\left|\int_0^1\left(\int_0^s e^{-\lambda(s-\tau)}d\tau\right)dH(s)\right|\right.$$

$$\left.+ \frac{|a|}{\Gamma(\beta)}\int_0^\xi (\xi - s)^{\beta-1}\left(\int_0^s e^{-\lambda(s-\tau)}d\tau\right)ds\right]$$

$$+ \frac{1}{\Gamma(\alpha)}\|\hat{f}_u - \hat{f}_v\|\int_0^t e^{-\lambda(t-s)}ds, \quad \forall t \in [0,1].$$

Observe that

$$|\hat{f}_u(s) - \hat{f}_v(s)| \leq L_1(|u(s) - v(s)| + |{}^cD_{0+}^p u(s)$$

$$- {}^cD_{0+}^p v(s)| + |I_{0+}^q u(s) - I_{0+}^q v(s)|)$$

$$\leq L_1\left(\|u - v\| + \|{}^cD_{0+}^p u - {}^cD_{0+}^p v\| + \|I_{0+}^q u - I_{0+}^q v\|\right)$$

$$\leq L_1 \left(\|u - v\| + \|{}^c D_{0+}^p u - {}^c D_{0+}^p v\| + \frac{1}{\Gamma(q+1)} \|u - v\| \right)$$

$$\leq L_1 \left(1 + \frac{1}{\Gamma(q+1)} \right) \|u - v\|_X = L_1 L_0 \|u - v\|_X,$$

$$\forall s \in [0,1].$$

Thus, we obtain

$$|(Au)(t) - (Av)(t)|$$

$$\leq L_3 M_1 \|u - v\| + \frac{l L_1 L_0}{\Gamma(\alpha)} \|u - v\|_X \left[\frac{1}{\lambda} \left| \int_0^1 (1 - e^{-\lambda s}) \, dH(s) \right| \right.$$

$$+ \frac{|a|}{\lambda \Gamma(\beta)} \int_0^\xi (\xi - s)^{\beta-1} (1 - e^{-\lambda s}) \, ds \Bigg] + \frac{L_1 L_0 \|u - v\|_X}{\lambda \Gamma(\alpha)} (1 - e^{-\lambda t})$$

$$\leq L_3 M_1 \|u - v\| + \frac{l L_1 L_0 M_2}{\Gamma(\alpha)} \|u - v\|_X + \frac{L_1 L_0 \|u - v\|_X}{\lambda \Gamma(\alpha)} (1 - e^{-\lambda}),$$

$$\forall t \in [0,1],$$

which implies that

$$\|Au - Av\| \leq \left[L_3 M_1 + \frac{l L_1 L_0 M_2}{\Gamma(\alpha)} + \frac{L_1 L_0}{\lambda \Gamma(\alpha)} (1 - e^{-\lambda}) \right] \|u - v\|_X.$$

Also, for all $t \in (0,1)$, we have

$$|(Au)'(t) - (Av)'(t)|$$

$$\leq |h(u) - h(v)| \left\{ \lambda + m \left[\left| \int_0^1 e^{-\lambda s} \, dH(s) \right| \right. \right.$$

$$\left. \left. + \frac{|a|}{\Gamma(\beta)} \int_0^\xi (\xi - s)^{\beta-1} e^{-\lambda s} \, ds \right] \right\}$$

$$+ \frac{m}{\Gamma(\alpha)} \|\hat{f}_u - \hat{f}_v\| \left[\left| \int_0^1 \left(\int_0^s e^{-\lambda(s-\tau)} \, d\tau \right) dH(s) \right| \right.$$

$$\left. + \frac{|a|}{\Gamma(\beta)} \int_0^\xi (\xi - s)^{\beta-1} \left(\int_0^s e^{-\lambda(s-\tau)} \, d\tau \right) ds \right]$$

$$+ \frac{1}{\Gamma(\alpha)} \|\hat{f}_u - \hat{f}_v\| \left(1 + \lambda \int_0^t e^{-\lambda(t-s)} \, ds \right)$$

$$\leq L_3 M_1^* \|u - v\|_X + \frac{m L_1 L_0 \|u - v\|_X}{\Gamma(\alpha)} \left[\frac{1}{\lambda} \left| \int_0^1 (1 - e^{-\lambda s}) \, dH(s) \right| \right.$$

$$\left. + \frac{|a|}{\lambda \Gamma(\beta)} \int_0^\xi (\xi - s)^{\beta - 1} (1 - e^{-\lambda s}) \, ds \right]$$

$$+ \frac{L_1 L_0 \|u - v\|_X}{\Gamma(\alpha)} + \frac{L_1 L_0 (1 - e^{-\lambda t})}{\Gamma(\alpha)} \|u - v\|_X.$$

Thus, we have

$$\|(Au)' - (Av)'\| \leq \left(L_3 M_1^* + \frac{m L_1 L_0 M_2}{\Gamma(\alpha)} + \frac{L_1 L_0 (2 - e^{-\lambda})}{\Gamma(\alpha)} \right) \|u - v\|_X,$$

which implies that

$$|{}^c D_{0+}^p (Au)(t) - {}^c D_{0+}^p (Av)(t)|$$

$$\leq \int_0^t \frac{(t - s)^{-p}}{\Gamma(1 - p)} |(Au)'(s) - (Av)'(s)| \, ds$$

$$\leq \left(L_3 M_1^* + \frac{m L_1 L_0 M_2}{\Gamma(\alpha)} + \frac{L_1 L_0 (2 - e^{-\lambda})}{\Gamma(\alpha)} \right) \frac{1}{\Gamma(2 - p)} \|u - v\|_X,$$

$$\forall t \in [0, 1],$$

and

$$\|{}^c D_{0+}^p (Au) - {}^c D_{0+}^p (Av)\|$$

$$\leq \left(L_3 M_1^* + \frac{m L_1 L_0 M_2}{\Gamma(\alpha)} + \frac{L_1 L_0 (2 - e^{-\lambda})}{\Gamma(\alpha)} \right) \frac{1}{\Gamma(2 - p)} \|u - v\|_X.$$

From the above inequalities, we obtain

$$\|Au - Av\|_X = \|Au - Av\| + \|{}^c D_{0+}^p (Au) - {}^c D_{0+}^p (Av)\|$$

$$\leq \left[L_3 \left(M_1 + \frac{M_1^*}{\Gamma(2 - p)} \right) + \frac{L_1 L_0}{\Gamma(\alpha)} \left(l M_2 + \frac{1}{\lambda} (1 - e^{-\lambda}) \right) \right.$$

$$\left. + \frac{L_1 L_0}{\Gamma(2 - p) \Gamma(\alpha)} (m M_2 + 2 - e^{-\lambda}) \right] \|u - v\|_X = \widetilde{\Lambda} \|u - v\|_X.$$

In view of the condition (7.14), that is, $\widetilde{\Lambda} < 1$, we deduce that A is a contraction. Then we conclude by the contraction mapping principle (Theorem 1.2.1) that the operator A has a unique fixed point $u \in \overline{B}_r$, which corresponds to the unique solution u of problem (7.1), (7.3) on $[0, 1]$. This completes the proof. $\square$

Theorem 7.1.2. *Assume that* (H1), (H2), (H3) *and* (H5) *hold. If*

$$\tilde{\Lambda}_1 := L_2 M_1 + \frac{1}{\Gamma(2-p)} L_2 M_1^* < 1,$$

and

$$\tilde{\Lambda}_2 := \frac{L_1 L_0}{\Gamma(\alpha)} \left(l M_2 + \frac{1}{\lambda}(1 - e^{-\lambda}) \right) + \frac{L_1 L_0}{\Gamma(2-p)\Gamma(\alpha)} (m M_2 + 2 - e^{-\lambda}) < 1,$$

then problem (7.1), (7.3) *has at least one solution on* $[0, 1]$.

Proof. Let us fix $r_1 > 0$ such that

$$r_1 \geq \left\{ \frac{1}{\Gamma(\alpha)} \left(l M_2 + \frac{1 - e^{-\lambda}}{\lambda} \right) \|g\| + \frac{m M_2 + 2 - e^{-\lambda}}{\Gamma(2-p)\Gamma(\alpha)} \|g\| \right\}$$

$$\times \left\{ 1 - \left(L_2 M_1 + \frac{1}{\Gamma(2-p)} L_2 M_1^* \right) \right\}^{-1}, \tag{7.16}$$

where g is the function given in the assumption (H3).

We consider the set $\overline{B}_{r_1} = \{u \in X, \|u\|_X \leq r_1\}$ and define the operators $A_1, A_2 : \overline{B}_{r_1} \to X$ as

$$(A_1 u)(t) = h(u) \left\{ e^{-\lambda t} + \gamma(t) \left[\int_0^1 e^{-\lambda s} \, dH(s) \right.\right.$$

$$\left.\left. - \frac{a}{\Gamma(\beta)} \int_0^\xi (\xi - s)^{\beta - 1} e^{-\lambda s} \, ds \right] \right\},$$

$$(A_2 u)(t) = \gamma(t) \left[\int_0^1 \left(\int_0^s e^{-\lambda(s-\tau)} I_{0+}^{\alpha - 1} \hat{f}_u(\tau) \, d\tau \right) dH(s) \right. \tag{7.17}$$

$$\left. - \frac{a}{\Gamma(\beta)} \int_0^\xi (\xi - s)^{\beta - 1} \left(\int_0^s e^{-\lambda(s-\tau)} I_{0+}^{\alpha - 1} \hat{f}_u(\tau) \, d\tau \right) ds \right]$$

$$+ \int_0^t e^{-\lambda(t-s)} I_{0+}^{\alpha - 1} \hat{f}_u(s) \, ds,$$

for all $t \in [0, 1]$ and $u \in \overline{B}_{r_1}$, where $\hat{f}_u(s) = f(s, u(s), {}^c D_{0+}^p u(s), I_{0+}^q u(s))$, $s \in [0, 1]$.

Using the assumptions (H3), (H5) and applying the arguments employed in the proof of Theorem 7.1.1, for all $u, v \in \overline{B}_{r_1}$, we obtain

$$\|A_1u + A_2v\| \leq \|A_1u\| + \|A_2v\| \leq r_1L_2M_1 + \frac{lM_2\|g\|}{\Gamma(\alpha)}$$

$$+ \frac{(1 - e^{-\lambda})\|g\|}{\lambda\Gamma(\alpha)},$$

$$\|(A_1u)' + (A_2v)'\| \leq r_1L_2M_1^* + \frac{\|g\|}{\Gamma(\alpha)}(mM_2 + 2 - e^{-\lambda}),$$

$$\|{}^cD_{0+}^p(A_1u + A_2v)\| \leq \frac{1}{\Gamma(2-p)}\left[r_1L_2M_1^* + \frac{\|g\|}{\Gamma(\alpha)}(mM_2 + 2 - e^{-\lambda})\right].$$

From the above inequalities and the definition of r_1, we find that

$$\|A_1u + A_2v\|_X = \|A_1u + A_2v\| + \|{}^cD_{0+}^p(A_1u + A_2v)\|$$

$$\leq r_1L_2M_1 + \frac{1}{\Gamma(\alpha)}\left(lM_2 + \frac{1 - e^{-\lambda}}{\lambda}\right)\|g\|$$

$$+ \frac{1}{\Gamma(2-p)}\left[r_1L_2M_1^* + \frac{mM_2 + 2 - e^{-\lambda}}{\Gamma(\alpha)}\|g\|\right] \leq r_1,$$

$$\forall u, v \in \overline{B}_{r_1}.$$

Hence, $A_1u + A_2v \in \overline{B}_{r_1}$ for all $u, v \in \overline{B}_{r_1}$.

Next, we show that the operator A_2 is a contraction. As in the previous theorem, we can obtain

$$\|A_2u - A_2v\| \leq \left[\frac{lL_1L_0M_2}{\Gamma(\alpha)} + \frac{L_1L_0}{\lambda\Gamma(\alpha)}(1 - e^{-\lambda})\right]\|u - v\|_X,$$

$$\|(A_2u)' - (A_2v)'\| \leq \left(\frac{mL_1L_0M_2}{\Gamma(\alpha)} + \frac{L_1L_0(2 - e^{-\lambda})}{\Gamma(\alpha)}\right)\|u - v\|_X,$$

$$\|{}^cD_{0+}^p(A_2u) - {}^cD_{0+}^p(A_2v)\| \leq \frac{1}{\Gamma(2-p)}\left(\frac{mL_1L_0M_2}{\Gamma(\alpha)} + \frac{L_1L_0(2 - e^{-\lambda})}{\Gamma(\alpha)}\right)$$

$$\times \|u - v\|_X,$$

which imply that

$$\|A_2u - A_2v\|_X \leq \left(\frac{lL_1L_0M_2}{\Gamma(\alpha)} + \frac{L_1L_0}{\lambda\Gamma(\alpha)}(1 - e^{-\lambda})\right.$$

$$+ \frac{mL_1L_0M_2}{\Gamma(2-p)\Gamma(\alpha)} + \left.\frac{L_1L_0(2 - e^{-\lambda})}{\Gamma(2-p)\Gamma(\alpha)}\right)$$

$$\times \|u - v\|_X = \widetilde{\Lambda}_2\|u - v\|_X,$$

for all $u, v \in \overline{B}_{r_1}$, with $\widetilde{\Lambda}_2 < 1$. This shows that A_2 is a contraction.

Continuity of h implies that the operator A_1 is continuous on $\overline{B}_{r_1}$. Next, we prove that A_1 is compact. The operator A_1 is uniformly bounded on $\overline{B}_{r_1}$ as

$$\|A_1u\| \leq r_1L_2M_1, \quad \|(A_1u)'\| \leq r_1L_2M_1^*,$$

$$\|{}^cD_{0+}^p(A_1u)\| \leq \frac{1}{\Gamma(2-p)}r_1L_2M_1^*,$$

In consequence, we get

$$\|A_1u\|_X \leq r_1L_2M_1 + \frac{1}{\Gamma(2-p)}r_1L_2M_1^*, \quad \forall u \in \overline{B}_{r_1}.$$

Next, for $u \in \overline{B}_{r_1}$, $t_1, t_2 \in [0,1]$, $t_1 < t_2$, we obtain

$$|(A_1u)(t_2) - (A_1u)(t_1)|$$

$$\leq L_2r_1\left\{|e^{-\lambda t_2} - e^{-\lambda t_1}| + |\gamma(t_2) - \gamma(t_1)|\left[\left|\int_0^1 e^{-\lambda s}dH(s)\right|\right.\right.$$

$$+ \left.\left.\frac{|a|}{\Gamma(\beta)}\int_0^\xi (\xi - s)^{\beta-1}e^{-\lambda s}\,ds\right]\right\},$$

$$|{}^cD_{0+}^p(A_1u)(t_2) - {}^cD_{0+}^p(A_1u)(t_1)|$$

$$\leq \frac{L_2r_1M_1^*}{\Gamma(1-p)}\left|\int_0^{t_1} [(t_2 - s)^{-p} - (t_1 - s)^{-p}]\,ds\right|$$

$$+ \frac{L_2r_1M_1^*}{\Gamma(1-p)}\left|\int_{t_1}^{t_2} (t_2 - s)^{-p}\,ds\right|$$

$$\leq \frac{L_2r_1M_1^*}{\Gamma(2-p)}[2(t_2 - t_1)^{1-p} + t_2^{1-p} - t_1^{1-p}].$$

Clearly, $|(A_1u)(t_2)-(A_1u)(t_1)| \to 0$ and $|{}^cD_{0+}^p(A_1u)(t_2)-{}^cD_{0+}^p(A_1u)(t_1)| \to$ 0, as $t_2 \to t_1$ uniformly with respect to $u \in \overline{B}_{r_1}$. This shows that A_1 is equicontinuous on $\overline{B}_{r_1}$, and so, by using Arzela-Ascoli theorem, the set $A_1(\overline{B}_{r_1})$ is relatively compact. Thus, the operator A_1 is compact on $\overline{B}_{r_1}$. As all the assumptions of Theorem 1.2.3 are satisfied, the conclusion of Theorem 1.2.3 implies that there exists a fixed point of operator $A_1 + A_2$, which is a solution of problem (7.1), (7.3) on $[0,1]$. This completes the proof. □

Theorem 7.1.3. *Assume that* (H1), (H3), (H4) *and* (H5) *hold. If*

$$\widetilde{\Lambda}_3 := \max\{L_2, L_3\}\left(M_1 + \frac{1}{\Gamma(2-p)}M_1^*\right) < 1$$

then problem (7.1), (7.3) *has at least one solution on* $[0,1]$.

Proof. We consider the number r_1 given by (7.16), and the operators A_1, A_2 defined on $\overline{B}_{r_1}$ and given by (7.17). As in the proof of Theorem 7.1.2, we can obtain that $A_1u + A_2v \in \overline{B}_{r_1}$ for all $u, v \in \overline{B}_{r_1}$.

In order to establish that the operator A_1 is a contraction, one can find that

$$\|A_1u - A_1v\| \le L_3M_1\|u - v\|_X, \quad \|(A_1u)' - (A_1v)'\| \le L_3M_1^*\|u - v\|_X,$$

$$\|{}^cD_{0+}^p(A_1u) - {}^cD_{0+}^p(A_1v)\| \le \frac{1}{\Gamma(2-p)}L_3M_1^*\|u - v\|_X.$$

In consequence, we have

$$\|A_1u - A_1v\|_X \le \left(L_3M_1 + \frac{1}{\Gamma(2-p)}L_3M_1^*\right)\|u - v\|_X \le \widetilde{\Lambda}_3\|u - v\|_X,$$

for all $u, v \in \overline{B}_{r_1}$, with $\widetilde{\Lambda}_3 < 1$. This implies that A_1 is a contraction.

It follows from the continuity of f that the operator A_2 is continuous on $\overline{B}_{r_1}$. Next, we prove that A_2 is compact. Note that the operator A_2 is uniformly bounded on $\overline{B}_{r_1}$ as

$$\|A_2u\| \le \frac{lM_2\|g\|}{\Gamma(\alpha)} + \frac{(1 - e^{-\lambda})\|g\|}{\lambda\Gamma(\alpha)},$$

$$\|(A_2u)'\| \le \frac{\|g\|}{\Gamma(\alpha)}(mM_2 + 2 - e^{-\lambda}),$$

$$\|{}^cD_{0+}^p(A_2u)\| \le \frac{\|g\|}{\Gamma(2-p)\Gamma(\alpha)}(mM_2 + 2 - e^{-\lambda}).$$

Consequently,

$$\|A_2 u\|_X \leq \frac{lM_2\|g\|}{\Gamma(\alpha)} + \frac{(1-e^{-\lambda})\|g\|}{\lambda\Gamma(\alpha)} + \frac{\|g\|}{\Gamma(2-p)\Gamma(\alpha)}(mM_2 + 2 - e^{-\lambda}),$$

$$\forall u \in \overline{B}_{r_1}.$$

Now, we prove that A_2 is equicontinuous on $\overline{B}_{r_1}$. We denote by $\Lambda_{r_1} = \sup\{|f(t,x,y,z)|, \ t \in [0,1], \ |x|, |y|, |z| \leq r_1\}$. Then, for $u \in \overline{B}_{r_1}$ and $t_1, t_2 \in [0,1]$ with $t_1 < t_2$, we obtain

$$|(A_2 u)(t_2) - (A_2 u)(t_1)|$$

$$\leq \Lambda_{r_1}|\gamma(t_2) - \gamma(t_1)|$$

$$\times \left[\left| \int_0^1 \left(\int_0^s e^{-\lambda(s-\tau)} \left(\int_0^\tau \frac{(\tau-\zeta)^{\alpha-2}}{\Gamma(\alpha-1)} \, d\zeta \right) d\tau \right) dH(s) \right| \right.$$

$$\left. + \frac{|a|}{\Gamma(\beta)} \int_0^\xi (\xi-s)^{\beta-1} \left(\int_0^s e^{-\lambda(s-\tau)} \left(\int_0^\tau \frac{(\tau-\zeta)^{\alpha-2}}{\Gamma(\alpha-1)} \, d\zeta \right) d\tau \right) ds \right]$$

$$+ \left| \int_0^{t_1} \left(e^{-\lambda(t_2-s)} - e^{-\lambda(t_1-s)} \right) I_{0+}^{\alpha-1} \hat{f}_u(s) \, ds \right|$$

$$+ \left| \int_{t_1}^{t_2} e^{-\lambda(t_2-s)} I_{0+}^{\alpha-1} \hat{f}_u(s) \, ds \right|$$

$$\leq \frac{\Lambda_{r_1}}{\Gamma(\alpha)} \left\{ |\gamma(t_2) - \gamma(t_1)| \left[\left| \int_0^1 \left(\int_0^s e^{-\lambda(s-\tau)} \tau^{\alpha-1} \, d\tau \right) dH(s) \right| \right. \right.$$

$$\left. + \frac{|a|}{\Gamma(\beta)} \int_0^\xi (\xi-s)^{\beta-1} \left(\int_0^s e^{-\lambda(s-\tau)} \tau^{\alpha-1} \, d\tau \right) ds \right]$$

$$\left. + \int_0^{t_1} \left(e^{-\lambda(t_1-s)} - e^{-\lambda(t_2-s)} \right) s^{\alpha-1} \, ds + \int_{t_1}^{t_2} e^{-\lambda(t_2-s)} s^{\alpha-1} \, ds \right\}$$

$$\leq \frac{\Lambda_{r_1}}{\Gamma(\alpha)} \left\{ |\gamma(t_2) - \gamma(t_1)| \left[\frac{1}{\alpha} \left| \int_0^1 s^\alpha \, dH(s) \right| \right. \right.$$

$$\left. + \frac{|a|}{\alpha\Gamma(\beta)} \int_0^\xi (\xi-s)^{\beta-1} s^\alpha \, ds \right]$$

$$\left. + \frac{t_1^{\alpha-1}}{\lambda} \left[1 - e^{-\lambda t_1} - e^{-\lambda(t_2-t_1)} + e^{-\lambda t_2} \right] + \frac{t_2^{\alpha-1}}{\lambda} \left(1 - e^{-\lambda(t_2-t_1)} \right) \right\}.$$

In addition, we have

$$|(A_2 u)'(t)| \leq \frac{m \Lambda_{r_1}}{\Gamma(\alpha)} \left[\frac{1}{\lambda} \left| \int_0^1 (1 - e^{-\lambda s}) \, dH(s) \right| \right.$$

$$+ \frac{|a|}{\lambda \Gamma(\beta)} \int_0^\xi (\xi - s)^{\beta - 1} (1 - e^{-\lambda s}) \, ds \left. \right]$$

$$+ \frac{\Lambda_{r_1}}{\Gamma(\alpha)} (2 - e^{-\lambda}) = \frac{m \Lambda_{r_1} M_2}{\Gamma(\alpha)} + \frac{\Lambda_{r_1}}{\Gamma(\alpha)} (2 - e^{-\lambda}) =: \Lambda_0,$$

and

$$|{}^c D_{0+}^p (A_2 u)(t_2) - {}^c D_{0+}^p (A_2 u)(t_1)|$$

$$= \left| \int_0^{t_2} \frac{(t_2 - s)^{-p}}{\Gamma(1 - p)} (A_2 u)'(s) \, ds - \int_0^{t_1} \frac{(t_1 - s)^{-p}}{\Gamma(1 - p)} (A_2 u)'(s) \, ds \right|$$

$$\leq \frac{\Lambda_0}{\Gamma(1 - p)} \left[\left| \int_0^{t_1} [(t_2 - s)^{-p} - (t_1 - s)^{-p}] \, ds \right| + \left| \int_{t_1}^{t_2} (t_2 - s)^{-p} \, ds \right| \right]$$

$$\leq \frac{\Lambda_0}{\Gamma(2 - p)} [2(t_2 - t_1)^{1-p} + t_2^{1-p} - t_1^{1-p}].$$

Clearly,

$$|(A_2 u)(t_2) - (A_2 u)(t_1)| \to 0, \quad \text{and} \quad |{}^c D_{0+}^p (A_2 u)(t_2) - {}^c D_{0+}^p (A_2 u)(t_1)| \to 0,$$

as $t_2 \to t_1$ uniformly with respect to $u \in \overline{B}_{r_1}$.

From the foregoing arguments, we deduce that A_2 is equicontinuous on $\overline{B}_{r_1}$, and so, by using Arzela-Ascoli theorem, the set $A_2(\overline{B}_{r_1})$ is relatively compact. In consequence, operator A_2 is compact on $\overline{B}_{r_1}$. Thus, all the assumptions of Theorem 1.2.3 are satisfied (with the roles of A_1 and A_2 interchanged). Hence, by Theorem 1.2.3, we deduce that there exists a fixed point of operator $A_1 + A_2$, which is a solution of the boundary value problem (7.1), (7.3) on $[0, 1]$. This completes the proof. $\qquad \square$

7.1.3   *Existence of solutions for problem (7.2), (7.3)*

In this section, we investigate the existence of solutions for the fractional inclusion (7.2) with the boundary conditions (7.3).

Definition 7.1.1. A function $u \in C^4[0,1]$ is a solution of the boundary value problem (7.2), (7.3) if $u(0) = h(u)$, $u'(0) = u''(0) = 0$, $aI_{0+}^\beta u(\xi) = \int_0^1 u(s)\, dH(s)$, and there exists a function $v \in S_{F,u}$ such that

$$u(t) = h(u)\left\{e^{-\lambda t} + \gamma(t)\left[\int_0^1 e^{-\lambda s}\, dH(s) - \frac{a}{\Gamma(\beta)}\int_0^\xi (\xi - s)^{\beta-1}e^{-\lambda s}\, ds\right]\right\}$$

$$+ \gamma(t)\left[\int_0^1\left(\int_0^s e^{-\lambda(s-\tau)}I_{0+}^{\alpha-1}v(\tau)\, d\tau\right)dH(s)\right.$$

$$\left. - \frac{a}{\Gamma(\beta)}\int_0^\xi (\xi - s)^{\beta-1}\left(\int_0^s e^{-\lambda(s-\tau)}I_{0+}^{\alpha-1}v(\tau)\, d\tau\right)ds\right]$$

$$+ \int_0^t e^{-\lambda(t-s)}I_{0+}^{\alpha-1}v(s)\, ds, \quad \forall t \in [0,1].$$

In the above definition, the set of selections $S_{F,u}$ of F is defined by
$S_{F,u} = \{v \in L^1(0,1),\ v(t) \in F(t, u(t), I_{0+}^q u(t)) \text{ for a.a. } t \in [0,1]\}$.

7.1.3.1 *The upper semicontinuous case*

Here, we prove the existence of solutions for problem (7.2), (7.3) when the multivalued map F has convex values by means of the nonlinear alternative of Leray-Schauder type for Kakutani maps (Theorem 1.2.14).

In the next result, we need the following assumptions.

(A1) $F : [0,1] \times \mathbb{R}^2 \to \mathcal{P}(\mathbb{R})$ is L^1-Carathéodory and has nonempty compact and convex values.

(A2) There exist a function $\phi \in C([0,1], \mathbb{R}_+)$ and a nondecreasing, subhomogeneous function $\Omega : \mathbb{R}_+ \to \mathbb{R}_+$ (that is, $\Omega(\mu x) \le \mu\Omega(x)$ for all $\mu \ge 1$ and $x \in \mathbb{R}_+$) such that

$$\|F(t, x, y)\|_{\mathcal{P}} \overset{\text{def}}{=} \sup\{|w|,\ w \in F(t, x, y)\} \le \phi(t)\Omega(|x| + |y|)$$

for each $(t, x, y) \in [0,1] \times \mathbb{R}^2$.

(A3) There exists a constant $M > 0$ such that

$$M\left[ML_2M_1 + \frac{L_0\|\phi\|\Omega(M)}{\Gamma(\alpha)}\left(lM_2 + \frac{1}{\lambda}(1 - e^{-\lambda})\right)\right]^{-1} > 1,$$

where l, L_0, M_1, M_2 are given by (7.13).

Theorem 7.1.4. *Suppose that the conditions* (H1), (H5) *and* (A1)–(A3) *hold. Then the boundary value problem* (7.2), (7.3) *has at least one solution on* $[0, 1]$.

Proof. We define the operator $\Omega_F : C[0, 1] \to \mathcal{P}(C[0, 1])$ by

$$\Omega_F(u) = \{w \in C[0, 1], \ w(t) \in N(u)(t), \ \forall t \in [0, 1]\}, \quad u \in C[0, 1], \quad (7.18)$$

where

$$
N(u)(t) = \left\{ h(u) \left\{ e^{-\lambda t} + \gamma(t) \left[\int_0^1 e^{-\lambda s} \, dH(s) \right. \right. \right.
$$
$$
\left. - \frac{a}{\Gamma(\beta)} \int_0^\xi (\xi - s)^{\beta-1} e^{-\lambda s} \, ds \right] \right\}
$$
$$
+ \gamma(t) \left[\int_0^1 \left(\int_0^s e^{-\lambda(s-\tau)} I_{0+}^{\alpha-1} v(\tau) \, d\tau \right) dH(s) \right.
$$
$$
\left. - \frac{a}{\Gamma(\beta)} \int_0^\xi (\xi - s)^{\beta-1} \left(\int_0^s e^{-\lambda(s-\tau)} I_{0+}^{\alpha-1} v(\tau) \, d\tau \right) ds \right]
$$
$$
+ \int_0^t e^{-\lambda(t-s)} I_{0+}^{\alpha-1} v(s) \, ds, \ v \in S_{F,u} \right\}, \quad t \in [0, 1], \quad u \in C[0, 1].
$$

We will show that Ω_F satisfies the hypothesis of Theorem 1.2.14 in several steps. First of all, we observe that $\Omega_F(u)$ is convex for each $u \in C[0, 1]$ as $S_{F,u}$ is convex (F has convex values). In the next step, we show that Ω_F maps bounded sets (balls) into bounded sets in $C[0, 1]$. For a positive number r, let $\overline{B}_r = \{u \in C[0, 1], \ \|u\| \leq r\}$ be a bounded ball in $C[0, 1]$. Then, for $u \in \overline{B}_r$ and for each $w \in \Omega_F(u)$, there exits $v \in S_{F,u}$ such that $w(t) \in N(u)(t)$ for all $t \in [0, 1]$. Thus

$$
w(t) = h(u) \left\{ e^{-\lambda t} + \gamma(t) \left[\int_0^1 e^{-\lambda s} \, dH(s) - \frac{a}{\Gamma(\beta)} \int_0^\xi (\xi - s)^{\beta-1} e^{-\lambda s} \, ds \right] \right\}
$$
$$
+ \gamma(t) \left[\int_0^1 \left(\int_0^s e^{-\lambda(s-\tau)} I_{0+}^{\alpha-1} v(\tau) \, d\tau \right) dH(s) \right.
$$
$$
\left. - \frac{a}{\Gamma(\beta)} \int_0^\xi (\xi - s)^{\beta-1} \left(\int_0^s e^{-\lambda(s-\tau)} I_{0+}^{\alpha-1} v(\tau) \, d\tau \right) ds \right]
$$
$$
+ \int_0^t e^{-\lambda(t-s)} I_{0+}^{\alpha-1} v(s) \, ds, \quad \forall t \in [0, 1].
$$

Then, for $t \in [0, 1]$, we have

$$|w(t)| \leq L_2 \|u\| \left\{ 1 + |\gamma(t)| \left[\left| \int_0^1 e^{-\lambda s} \, dH(s) \right| \right.\right.$$

$$\left.\left. + \frac{|a|}{\Gamma(\beta)} \int_0^{\xi} (\xi - s)^{\beta-1} e^{-\lambda s} \, ds \right] \right\}$$

$$+ |\gamma(t)| \left[\left| \int_0^1 \left(\int_0^s e^{-\lambda(s-\tau)} \frac{\|v\| \tau^{\alpha}}{\Gamma(\alpha)} \, d\tau \right) dH(s) \right| \right.$$

$$\left. + \frac{|a|}{\Gamma(\beta)} \int_0^{\xi} (\xi - s)^{\beta-1} \left(\int_0^s e^{-\lambda(s-\tau)} \frac{\|v\| \tau^{\alpha}}{\Gamma(\alpha)} \, d\tau \right) ds \right]$$

$$+ \int_0^t e^{-\lambda(t-s)} \frac{\|v\| s^{\alpha}}{\Gamma(\alpha)} \, ds$$

$$\leq L_2 r \left\{ 1 + l \left[\left| \int_0^1 e^{-\lambda s} \, dH(s) \right| + \frac{|a|}{\Gamma(\beta)} \int_0^{\xi} (\xi - s)^{\beta-1} e^{-\lambda s} \, ds \right] \right\}$$

$$+ l \left[\left| \int_0^1 \left(\int_0^s e^{-\lambda(s-\tau)} \frac{1}{\Gamma(\alpha)} \, d\tau \right) dH(s) \right| \|v\| \right.$$

$$\left. + \frac{|a|}{\Gamma(\beta)} \int_0^{\xi} (\xi - s)^{\beta-1} \left(\int_0^s e^{-\lambda(s-\tau)} \frac{1}{\Gamma(\alpha)} \, d\tau \right) ds \|v\| \right]$$

$$+ \frac{\|v\|}{\Gamma(\alpha)} \int_0^t e^{-\lambda(t-s)} \, ds$$

$$\leq L_2 r M_1 + \frac{l}{\Gamma(\alpha)} \left[\frac{1}{\lambda} \left| \int_0^1 (1 - e^{-\lambda s}) \, dH(s) \right| \right.$$

$$\left. + \frac{|a|}{\lambda \Gamma(\beta)} \int_0^{\xi} (\xi - s)^{\beta-1} (1 - e^{-\lambda s}) \, ds \right]$$

$$\times L_0 \|\phi\| \Omega(\|u\|_X) + \frac{1}{\lambda \Gamma(\alpha)} (1 - e^{-\lambda}) L_0 \|\phi\| \Omega(\|u\|_X)$$

$$\leq L_2 r M_1 + \frac{l}{\Gamma(\alpha)} M_2 L_0 \|\phi\| \Omega(r) + \frac{1}{\lambda \Gamma(\alpha)} (1 - e^{-\lambda}) L_0 \|\phi\| \Omega(r),$$

which yields

$$\|w\| \leq L_2 r M_1 + \frac{l M_2 L_0}{\Gamma(\alpha)} \|\phi\| \Omega(r) + \frac{1}{\lambda \Gamma(\alpha)} (1 - e^{-\lambda}) L_0 \|\phi\| \Omega(r),$$

for all $w \in \Omega_F(u)$ and all $u \in \overline{B}_r$. So $\Omega_F(\overline{B}_r)$ is bounded.

Now, we show that Ω_F maps bounded sets into equicontinuous sets of $C[0,1]$. Let $\overline{B}_r = \{u \in C[0,1], \|u\| \le r\}$ be a bounded ball in $C[0,1]$, and let $u \in \overline{B}_r$ and $t_1, t_2 \in [0,1]$ with $t_1 < t_2$. For each $w \in \Omega_F(u)$, and $v(t) \in F(t, u(t), I_{0+}^q u(t))$, $t \in [0,1]$, we obtain

$$|w(t_2) - w(t_1)| \le L_2 \|u\| \left| e^{-\lambda t_2} - e^{-\lambda t_1} + (\gamma(t_2) - \gamma(t_1)) \right.$$

$$\times \left. \left[\int_0^1 e^{-\lambda s} dH(s) - \frac{a}{\Gamma(\beta)} \int_0^\xi (\xi - s)^{\beta-1} e^{-\lambda s} ds \right] \right|$$

$$+ |\gamma(t_2) - \gamma(t_1)| \left[\left| \int_0^1 \left(\int_0^s e^{-\lambda(s-\tau)} |I_{0+}^{\alpha-1} v(\tau)| d\tau \right) dH(s) \right| \right.$$

$$+ \frac{|a|}{\Gamma(\beta)} \int_0^\xi (\xi - s)^{\beta-1} \left(\int_0^s e^{-\lambda(s-\tau)} |I_{0+}^{\alpha-1} v(\tau)| d\tau \right) ds \right]$$

$$+ \left| \int_0^{t_2} e^{-\lambda(t_2-s)} I_{0+}^{\alpha-1} v(s) ds - \int_0^{t_1} e^{-\lambda(t_1-s)} I_{0+}^{\alpha-1} v(s) ds \right|$$

$$\le L_2 r \left[|e^{-\lambda t_2} - e^{-\lambda t_1}| + \frac{M_1 - 1}{l} |\gamma(t_2) - \gamma(t_1)| \right]$$

$$+ \|v\| \left\{ |\gamma(t_2) - \gamma(t_1)| \left[\left| \int_0^1 \left(\int_0^s e^{-\lambda(s-\tau)} \frac{\tau^{\alpha-1}}{\Gamma(\alpha)} d\tau \right) dH(s) \right| \right. \right.$$

$$+ \frac{|a|}{\Gamma(\beta)} \int_0^\xi (\xi - s)^{\beta-1} \left(\int_0^s e^{-\lambda(s-\tau)} \tau^{\alpha-1} d\tau \right) ds \right]$$

$$+ \frac{1}{\Gamma(\alpha)} \int_0^{t_1} \left(e^{-\lambda(t_1-s)} - e^{-\lambda(t_2-s)} \right) s^{\alpha-1} ds$$

$$+ \frac{1}{\Gamma(\alpha)} \int_{t_1}^{t_2} e^{-\lambda(t_2-s)} s^{\alpha-1} ds \right\}$$

$$\le L_2 r \left[|e^{-\lambda t_2} - e^{-\lambda t_1}| + \frac{M_1 - 1}{l} |\gamma(t_2) - \gamma(t_1)| \right] + \frac{L_0 \|\phi\| \Omega(\|u\|_X)}{\Gamma(\alpha)}$$

$$\times \left\{ |\gamma(t_2) - \gamma(t_1)| \left[\frac{1}{\alpha} \left| \int_0^1 s^\alpha dH(s) \right| + \frac{|a|}{\alpha \Gamma(\beta)} \int_0^\xi (\xi - s)^{\beta-1} s^\alpha ds \right] \right.$$

$$+ \frac{t_1^{\alpha-1}}{\lambda} \left[1 - e^{-\lambda t_1} - e^{-\lambda(t_2-t_1)} + e^{-\lambda t_2} \right] + \frac{t_2^{\alpha-1}}{\lambda} \left(1 - e^{-\lambda(t_2-t_1)} \right) \right\}.$$

As the right-hand side of the last inequality tends to zero as $t_2 \to t_1$, independently of $u \in \overline{B}_r$, so $\Omega_F(u)$ is a equicontinuous set of $C[0,1]$.

In view of the foregoing arguments, it follows by the Arzela-Ascoli theorem that $\Omega_F : C[0,1] \to \mathcal{P}(C[0,1])$ is compact.

In order to establish that Ω_F is upper semicontinuous, it is sufficient to show that Ω_F has a closed graph (by [29, Proposition 1.2]). Let $u_n \to u_*$, $w_n \in \Omega_F(u_n)$ and $w_n \to w_*$. Then we will show that $w_* \in \Omega_F(u_*)$. For $w_n \in \Omega_F(u_n)$, there exists $v_n \in S_{F,u_n}$ such that for each $t \in [0,1]$, we have

$$
w_n(t) = h(u_n) \left\{ e^{-\lambda t} + \gamma(t) \left[\int_0^1 e^{-\lambda s}\, dH(s) \right.\right.
$$

$$
\left.\left. - \frac{a}{\Gamma(\beta)} \int_0^\xi (\xi - s)^{\beta-1} e^{-\lambda s}\, ds \right] \right\}
$$

$$
+ \gamma(t) \left[\int_0^1 \left(\int_0^s e^{-\lambda(s-\tau)} I_{0+}^{\alpha-1} v_n(\tau)\, d\tau \right) dH(s) \right.
$$

$$
\left. - \frac{a}{\Gamma(\beta)} \int_0^\xi (\xi - s)^{\beta-1} \left(\int_0^s e^{-\lambda(s-\tau)} I_{0+}^{\alpha-1} v_n(\tau)\, d\tau \right) ds \right]
$$

$$
+ \int_0^t e^{-\lambda(t-s)} I_{0+}^{\alpha-1} v_n(s)\, ds.
$$

We will show that there exists $v_* \in S_{F,u_*}$ such that for each $t \in [0,1]$,

$$
w_*(t) = h(u_*) \left\{ e^{-\lambda t} + \gamma(t) \left[\int_0^1 e^{-\lambda s}\, dH(s) \right.\right.
$$

$$
\left.\left. - \frac{a}{\Gamma(\beta)} \int_0^\xi (\xi - s)^{\beta-1} e^{-\lambda s}\, ds \right] \right\}
$$

$$
+ \gamma(t) \left[\int_0^1 \left(\int_0^s e^{-\lambda(s-\tau)} I_{0+}^{\alpha-1} v_*(\tau)\, d\tau \right) dH(s) \right.
$$

$$
- \frac{a}{\Gamma(\beta)} \int_0^\xi (\xi - s)^{\beta-1}
$$

$$
\left. \times \left(\int_0^s e^{-\lambda(s-\tau)} I_{0+}^{\alpha-1} v_*(\tau)\, d\tau \right) ds \right]
$$

$$
+ \int_0^t e^{-\lambda(t-s)} I_{0+}^{\alpha-1} v_*(s)\, ds. \tag{7.19}
$$

We consider the linear operator $\Psi : L^1(0,1) \to C[0,1]$ given by

$$\Psi(w)(t) = \gamma(t) \left[\int_0^1 \left(\int_0^s e^{-\lambda(s-\tau)} I_{0+}^{\alpha-1} w(\tau) \, d\tau \right) dH(s) \right.$$

$$\left. - \frac{a}{\Gamma(\beta)} \int_0^\xi (\xi - s)^{\beta-1} \left(\int_0^s e^{-\lambda(s-\tau)} I_{0+}^{\alpha-1} w(\tau) \, d\tau \right) ds \right]$$

$$+ \int_0^t e^{-\lambda(t-s)} I_{0+}^{\alpha-1} w(s) \, ds,$$

for $t \in (0,1)$, $w \in L^1(0,1)$.

Notice that the operator Ψ is continuous. Indeed, for $\widetilde{w}_n, \widetilde{w}_* \in L^1(0,1)$ with $\widetilde{w}_n \to \widetilde{w}_*$ in $L^1(0,1)$, we obtain

$$|\Psi(\widetilde{w}_n)(t) - \Psi(\widetilde{w}_*)(t)|$$

$$\leq l \left[\left| \int_0^1 \left(\int_0^s e^{-\lambda(s-\tau)} |I_{0+}^{\alpha-1}(\widetilde{w}_n(\tau) - \widetilde{w}_*(\tau))| \, d\tau \right) dH(s) \right| \right.$$

$$\left. + \frac{|a|}{\Gamma(\beta)} \int_0^\xi (\xi - s)^{\beta-1} \left(\int_0^s e^{-\lambda(s-\tau)} |I_{0+}^{\alpha-1}(\widetilde{w}_n(\tau) - \widetilde{w}_*(\tau))| \, d\tau \right) ds \right]$$

$$+ \int_0^t e^{-\lambda(t-s)} |I_{0+}^{\alpha-1}(\widetilde{w}_n(s) - \widetilde{w}_*(s))| \, ds, \quad \forall t \in [0,1],$$

which implies that $\Psi(\widetilde{w}_n) \to \Psi(\widetilde{w}_*)$ in $C[0,1]$.

Then, as argued in [71] (see also Lemma 4.3 from [14]), we deduce that $\Psi \circ S_{F,u}$ has a closed graph.

For the above functions u_n and w_n, we have

$$w_n - h(u_n) \left\{ e^{-\lambda t} + \gamma(t) \left[\int_0^1 e^{-\lambda s} \, dH(s) \right. \right.$$

$$\left. \left. - \frac{a}{\Gamma(\beta)} \int_0^\xi (\xi - s)^{\beta-1} e^{-\lambda s} \, ds \right] \right\} \in \Psi(S_{F,u_n}).$$

Since $u_n \to u$, there exists $v_* \in S_{F,u_*}$ such that

$$w_*(t) - h(u_*) \left\{ e^{-\lambda t} + \gamma(t) \left[\int_0^1 e^{-\lambda s} \, dH(s) \right. \right.$$

$$\left. \left. - \frac{a}{\Gamma(\beta)} \int_0^\xi (\xi - s)^{\beta-1} e^{-\lambda s} \, ds \right] \right\}$$

$$= \gamma(t) \left[\int_0^1 \left(\int_0^s e^{-\lambda(s-\tau)} I_{0+}^{\alpha-1} v_*(\tau) \, d\tau \right) dH(s) \right.$$

$$\left. - \frac{a}{\Gamma(\beta)} \int_0^\xi (\xi - s)^{\beta-1} \left(\int_0^s e^{-\lambda(s-\tau)} I_{0+}^{\alpha-1} v_*(\tau) \, d\tau \right) ds \right]$$

$$+ \int_0^t e^{-\lambda(t-s)} I_{0+}^{\alpha-1} v_*(s) \, ds,$$

or equivalently, we deduce the relation (7.19).

Finally, we show that there exists an open set $U \subset C[0,1]$ with $u \notin \theta\Omega_F(u)$ for any $\theta \in (0,1)$ and all $u \in \partial U$. Let $\theta \in (0,1)$ and $u \in \theta\Omega_F(u)$. Then there exists $v \in L^1(0,1)$ with $v \in S_{F,u}$ such that for all $t \in [0,1]$, we obtain

$$\|u\| \leq \|u\| L_2 M_1 + \frac{l M_2 L_0 \|\phi\| \Omega(\|u\|)}{\Gamma(\alpha)} + \frac{1}{\lambda\Gamma(\alpha)} (1 - e^{-\lambda}) L_0 \|\phi\| \Omega(\|u\|),$$

which can be written as

$$\|u\| \left[\|u\| L_2 M_1 + \frac{L_0 \|\phi\| \Omega(\|u\|)}{\Gamma(\alpha)} \left(l M_2 + \frac{1}{\lambda}(1 - e^{-\lambda}) \right) \right]^{-1} \leq 1.$$

In view of assumption (A3), there exists M such that $\|u\| \neq M$. Let us define the set $U = \{u \in C[0,1], \|u\| < M\}$. The operator $\Omega_F : \overline{U} \to \mathcal{P}(C[0,1])$ is upper semicontinuous and compact. From the definition of U, there exists no $u \in \partial U$ such that $u \in \theta\Omega_F(u)$ for some $\theta \in (0,1)$. Therefore, by the nonlinear alternative of Leray-Schauder type (Theorem 1.2.14), we conclude that Ω_F has a fixed point $u \in \overline{U}$ which is a solution of problem (7.2), (7.3). $\qquad\square$

7.1.3.2 *The Lipschitz case*

We discuss here the existence of solutions for problem (7.2), (7.3) with a nonconvex valued right-hand side by applying the fixed point theorem due to Covitz and Nadler (Theorem 1.2.15) for multivalued functions.

Before proceeding for the main result, let us state the assumptions needed for it.

(I1) $F : [0,1] \times \mathbb{R}^2 \to \mathcal{P}_{cp}(\mathbb{R})$ satisfies the condition $F(\cdot, x, y) : [0,1] \to \mathcal{P}_{cp}(\mathbb{R})$ is measurable for each $(x,y) \in \mathbb{R}^2$.

(I2) $H_d(F(t,x,y), F(t,\bar{x},\bar{y})) \leq \varphi(t)[|x - \bar{x}| + |y - \bar{y}|]$, for a.a. $t \in [0,1]$ and for all $x,y,\bar{x},\bar{y} \in \mathbb{R}$ with $\varphi \in C([0,1], \mathbb{R}_+)$, and $d(0, F(t,0,0)) \leq \varphi(t)$ for a.a. $t \in [0,1]$.

In assumption (I1), the set $\mathcal{P}_{cp}(\mathbb{R}) = \{Y \in \mathcal{P}(\mathbb{R}), Y \text{ is compact}\}$, and in assumption (I2), H_d is the Hausdorff metric, where d is the Euclidean metric in $\mathbb{R}$ defined by $d(x,y) = |x - y|$ for $x, y \in \mathbb{R}$.

Theorem 7.1.5. *Assume that the hypotheses* (H1), (H4), (I1) *and* (I2) *hold. If*

$$\theta_0 := L_3 M_1 + \frac{L_0 \|\varphi\|}{\Gamma(\alpha)} \left(l M_2 + \frac{1}{\lambda}(1 - e^{-\lambda}) \right) < 1, \qquad (7.20)$$

where l, L_0, M_1, M_2 *are given by* (7.13), *then the boundary value problem* (7.2), (7.3) *has at least one solution on* $[0,1]$.

Proof. Observe that the set $S_{F,u}$ is nonempty for each $u \in C[0,1]$ by assumption (I1), and thus F has a measurable selection (see [25, Theorem III.6]). We now show that the operator $\Omega_F : C[0,1] \to \mathcal{P}(C[0,1])$ defined by (7.18) satisfies the assumptions of Theorem 1.2.15. To establish that $\Omega_F(u) \in \mathcal{P}_{cl}(C[0,1])$ for each $u \in C[0,1]$, let $\{w_n\}_{n \geq 0} \subset \Omega_F(u)$ be such that $w_n \to w$ as $n \to \infty$ in $C[0,1]$. Then $w \in C[0,1]$ and there exists $v_n \in S_{F,u}$ such that for each $t \in [0,1]$, we have

$$
w_n(t) = h(u) \left\{ e^{-\lambda t} + \gamma(t) \left[\int_0^1 e^{-\lambda s} \, dH(s) - \frac{a}{\Gamma(\beta)} \int_0^\xi (\xi - s)^{\beta - 1} e^{-\lambda s} \, ds \right] \right\}
$$
$$
+ \dot{\gamma}(t) \left[\int_0^1 \left(\int_0^s e^{-\lambda(s-\tau)} I_{0+}^{\alpha-1} v_n(\tau) \, d\tau \right) dH(s) \right.
$$
$$
\left. - \frac{a}{\Gamma(\beta)} \int_0^\xi (\xi - s)^{\beta - 1} \left(\int_0^s e^{-\lambda(s-\tau)} I_{0+}^{\alpha-1} v_n(\tau) \, d\tau \right) ds \right]
$$
$$
+ \int_0^t e^{-\lambda(t-s)} I_{0+}^{\alpha-1} v_n(s) \, ds,
$$

with $v_n(t) \in F(t, u(t), I_{0+}^q u(t))$, $t \in [0,1]$.

Since F has compact values, therefore, we can pass onto a subsequence (denoted in a same way) to obtain that v_n converges to v in $L^1(0,1)$. Thus, $v \in S_{F,u}$ and for each $t \in [0,1]$, we have $w_n(t) \to w(t)$, where

$$
w(t) = h(u) \left\{ e^{-\lambda t} + \gamma(t) \left[\int_0^1 e^{-\lambda s} \, dH(s) - \frac{a}{\Gamma(\beta)} \int_0^\xi (\xi - s)^{\beta - 1} e^{-\lambda s} \, ds \right] \right\}
$$
$$
+ \gamma(t) \left[\int_0^1 \left(\int_0^s e^{-\lambda(s-\tau)} I_{0+}^{\alpha-1} v(\tau) \, d\tau \right) dH(s) \right.
$$

$$- \frac{a}{\Gamma(\beta)} \int_0^\xi (\xi - s)^{\beta-1} \left(\int_0^s e^{-\lambda(s-\tau)} I_{0+}^{\alpha-1} v(\tau) \, d\tau \right) ds \Bigg]$$

$$+ \int_0^t e^{-\lambda(t-s)} I_{0+}^{\alpha-1} v(s) \, ds.$$

Hence, $w \in \Omega_F(u)$.

Next, we show that Ω_F is a contraction, that is,

$$H_{d_1}(\Omega_F(u), \Omega_F(\bar{u})) \leq \theta_0 \|u - \bar{u}\|, \quad \forall u, \, \bar{u} \in C[0,1],$$

where θ_0 is defined in (7.20), and d_1 is the metric induced by the norm $\|\cdot\|$ in $C[0,1]$.

For this, let $u, \bar{u} \in C[0,1]$ and $w_1 \in \Omega_F(u)$. Then there exists $v_1 \in S_{F,u}$ such that for all $t \in [0,1]$, we obtain

$$w_1(t) = h(u) \left\{ e^{-\lambda t} + \gamma(t) \left[\int_0^1 e^{-\lambda s} \, dH(s) - \frac{a}{\Gamma(\beta)} \int_0^\xi (\xi - s)^{\beta-1} e^{-\lambda s} \, ds \right] \right\}$$

$$+ \gamma(t) \left[\int_0^1 \left(\int_0^s e^{-\lambda(s-\tau)} I_{0+}^{\alpha-1} v_1(\tau) \, d\tau \right) dH(s) \right.$$

$$\left. - \frac{a}{\Gamma(\beta)} \int_0^\xi (\xi - s)^{\beta-1} \left(\int_0^s e^{-\lambda(s-\tau)} I_{0+}^{\alpha-1} v_1(\tau) \, d\tau \right) ds \right]$$

$$+ \int_0^t e^{-\lambda(t-s)} I_{0+}^{\alpha-1} v_1(s) \, ds.$$

By (I2), we have

$$H_d(F(t, u(t), I_{0+}^q u(t)), F(t, \bar{u}(t), I_{0+}^q \bar{u}(t)))$$

$$\leq \varphi(t)[|u(t) - \bar{u}(t)| + |I_{0+}^q u(t) - I_{0+}^q \bar{u}(t)|],$$

for a.a. $t \in [0,1]$. Then there exists $\psi \in F(t, \bar{u}(t), I_{0+}^q \bar{u}(t))$ such that

$$|v_1(t) - \psi| \leq \varphi(t)[|u(t) - \bar{u}(t)| + |I_{0+}^q u(t) - I_{0+}^q \bar{u}(t)|], \quad \text{a.a. } t \in [0,1].$$

We define $\widetilde{U} : [0,1] \to \mathcal{P}(\mathbb{R})$ by $\widetilde{U}(t) = \{\psi \in \mathbb{R}, \ |v_1(t) - \psi| \leq \varphi(t)[|u(t) - \bar{u}(t)| + |I_{0+}^q u(t) - I_{0+}^q \bar{u}(t)|]\}$. As the multivalued operator $V(t) = \widetilde{U}(t) \cap F(t, \bar{u}(t), I_{0+}^q \bar{u}(t))$ is measurable (see [25, Proposition III.4]), there exists a function $v_2(t)$ which is a measurable selection for $V(t)$. Hence,

$v_2(t) \in F(t, \bar{u}(t), I_{0+}^q \bar{u}(t))$ for a.a. $t \in [0,1]$, and

$$|v_1(t) - v_2(t)| \leq \varphi(t)[|u(t) - \bar{u}(t)| + |I_{0+}^q u(t) - I_{0+}^q \bar{u}(t)|], \quad \text{a.a. } t \in [0,1].$$

Let us define the function $w_2(t)$, $t \in [0,1]$ by

$$w_2(t) = h(\bar{u}) \left\{ e^{-\lambda t} + \gamma(t) \left[\int_0^1 e^{-\lambda s}\, dH(s) - \frac{a}{\Gamma(\beta)} \int_0^\xi (\xi - s)^{\beta-1} e^{-\lambda s}\, ds \right] \right\}$$

$$+ \gamma(t) \left[\int_0^1 \left(\int_0^s e^{-\lambda(s-\tau)} I_{0+}^{\alpha-1} v_2(\tau)\, d\tau \right) dH(s) \right.$$

$$- \frac{a}{\Gamma(\beta)} \int_0^\xi (\xi - s)^{\beta-1} \left(\int_0^s e^{-\lambda(s-\tau)} I_{0+}^{\alpha-1} v_2(\tau)\, d\tau \right) ds \right]$$

$$+ \int_0^t e^{-\lambda(t-s)} I_{0+}^{\alpha-1} v_2(s)\, ds.$$

Then we conclude that

$$|w_1(t) - w_2(t)|$$

$$\leq |h(u) - h(\bar{u})| M_1$$

$$+ |\gamma(t)| \left[\left| \int_0^1 \left(\int_0^s e^{-\lambda(s-\tau)} |I_{0+}^{\alpha-1} v_1(\tau) - I_{0+}^{\alpha-1} v_2(\tau)|\, d\tau \right) dH(s) \right| \right.$$

$$+ \frac{|a|}{\Gamma(\beta)} \int_0^\xi (\xi - s)^{\beta-1} \left(\int_0^s e^{-\lambda(s-\tau)} |I_{0+}^{\alpha-1} v_1(\tau) - I_{0+}^{\alpha-1} v_2(\tau)|\, d\tau \right) ds \right]$$

$$+ \int_0^t e^{-\lambda(t-s)} |I_{0+}^{\alpha-1} v_1(s) - I_{0+}^{\alpha-1} v_2(s)|\, ds$$

$$\leq L_3 M_1 \|u - \bar{u}\| + \left\{ \frac{\|\varphi\| l}{\Gamma(\alpha)} \left[\left| \int_0^1 \left(\int_0^s e^{-\lambda(s-\tau)} \tau^{\alpha-1}\, d\tau \right) dH(s) \right| \right. \right.$$

$$+ \frac{|a|}{\Gamma(\beta)} \int_0^\xi (\xi - s)^{\beta-1} \left(\int_0^s e^{-\lambda(s-\tau)} \tau^{\alpha-1}\, d\tau \right) ds \right]$$

$$+ \frac{\|\varphi\|}{\Gamma(\alpha)} \int_0^t e^{-\lambda(t-s)} s^{\alpha-1}\, ds \right\} \left(1 + \frac{1}{\Gamma(q+1)} \right) \|u - \bar{u}\|$$

$$\leq \left[L_3 M_1 + \frac{L_0 \|\varphi\|}{\Gamma(\alpha)} \left(l M_2 + \frac{1 - e^{-\lambda}}{\lambda} \right) \right] \|u - \bar{u}\|, \quad \forall t \in [0,1].$$

Therefore,

$$\|w_1 - w_2\| \le \theta_0 \|u - \bar{u}\|.$$

By interchanging the roles of u and $\bar{u}$, we obtain a similar relation, and thus we get

$$H_{d_1}(\Omega_F(u), \Omega_F(\bar{u})) \le \theta_0 \|u - \bar{u}\|.$$

In view of the condition $\theta_0 < 1$ (given by (7.20)), it follows that Ω_F is a contraction, and therefore by Theorem 1.2.15, Ω_F has a fixed point u, which is a solution of problem (7.2), (7.3). This completes the proof. $\square$

7.1.4 Examples

(I) Let us fix $\alpha = \frac{7}{2}$, $p = \frac{3}{4}$, $q = \frac{6}{5}$, $\beta = \frac{5}{4}$, $\xi = \frac{1}{2}$, $a = 1$, $\lambda = 4$ and $H(s) = \left\{ 1, \text{ if } s \in [0, \frac{1}{3}); \ 5, \text{ if } s \in [\frac{1}{3}, \frac{2}{3}); \ 14, \text{ if } s \in [\frac{2}{3}, 1] \right\}$ and consider the following fractional differential equation

$$({}^cD_{0+}^{7/2} + 4\,{}^cD_{0+}^{5/2})u(t) = f(t, u(t), {}^cD_{0+}^{3/4}u(t), I_{0+}^{6/5}u(t)), \quad t \in (0, 1), \quad (7.21)$$

supplemented with the boundary conditions

$$\begin{cases} u(0) = h(u), \quad u'(0) = u''(0) = 0, \\ 4u\left(\frac{1}{3}\right) + 9u\left(\frac{2}{3}\right) = \frac{1}{\Gamma\left(\frac{5}{4}\right)} \int_0^{1/2} \left(\frac{1}{2} - s\right)^{1/4} u(s)\,ds. \end{cases} \quad (7.22)$$

Using the given values, it is found that $\Delta \approx -34.94855476$, $(\Delta \ne 0)$, $l \approx 0.28508672$, $m \approx 0.69091627$, $L_0 \approx 1.90760368$, $M_1 \approx 1.52905173$, $M_2 \approx 2.87883365$, $M_1^* \approx 5.28217286$. Now we set the values of $h(u)$ and $f(t, x, y, z)$ in (7.21) and illustrate the obtained results.

(a) Let $h(u) = \frac{1}{12}\sin u\left(\frac{1}{2}\right)$ and the function f be defined by

$$f(t, x, y, z) = \frac{1}{\sqrt{t + 81}}\left(\frac{|x|}{1 + |x|} - \arctan y\right) + \frac{z}{9} - \frac{5t}{t^2 + 3},$$

for $t \in [0, 1]$, $x, y, z \in \mathbb{R}$. For these functions, we have $L_2 = L_3 = \frac{1}{12}$ and

$$|f(t, x, y, z) - f(t, x_1, y_1, z_1)| \le \frac{1}{9}(|x - x_1| + |y - y_1| + |z - z_1|),$$

for all $t \in [0, 1]$ and $x, y, z, x_1, y_1, z_1 \in \mathbb{R}$. So $L_1 = \frac{1}{9}$. In addition we obtain $\tilde{\Lambda} \approx 0.960446 < 1$. Thus, all the conditions of Theorem 7.1.1 are satisfied. Therefore, by Theorem 7.1.1, there exists a unique solution for problem (7.21), (7.22) on $[0, 1]$.

(b) Consider $h(u) = \frac{|u(1/2)|}{8(1+|u(1/2)|)}$ and the function f defined by

$$f(t,x,y,z) = \frac{1}{\sqrt{4+t^2}}\sin t + \frac{x^{2/3}}{4(2+x^{2/3})} - \frac{y^{4/7}}{2(3+y^{4/7})} - \frac{1}{3(1+t)}\cos z^{2/5},$$

for all $t \in [0,1]$ and $x,y,z \in \mathbb{R}$.

For the above functions, we have $L_2 = L_3 = \frac{1}{8}$ and $|f(t,x,y,z)| \leq g(t)$, for all $t \in [0,1]$ and $x,y,z \in \mathbb{R}$, with $g(t) = \frac{|\sin t|}{\sqrt{4+t^2}} + \frac{3}{4} + \frac{1}{3(1+t)}$, $t \in [0,1]$. In addition we obtain $\widetilde{\Lambda}_3 \approx 0.919584 < 1$. As all the conditions of Theorem 7.1.3 are satisfied, we deduce by the conclusion of Theorem 7.1.3 that problem (7.21), (7.22) has at least one solution on $[0,1]$.

(II) Fixing $\alpha = \frac{7}{2}$, $q = \frac{6}{5}$, $\beta = \frac{5}{4}$, $\xi = \frac{1}{2}$, $a = 1$, $\lambda = 4$ and $H(s) = \{1$, if $s \in [0, \frac{1}{3})$; 5, if $s \in [\frac{1}{3}, \frac{2}{3})$; 14, if $s \in [\frac{2}{3}, 1]\}$, we consider the following fractional differential inclusion

$$({}^cD_{0+}^{7/2} + 4\,{}^cD_{0+}^{5/2})u(t) \in F(t, u(t), I_{0+}^{6/5}u(t)), \quad t \in (0,1), \tag{7.23}$$

equipped with the boundary conditions (7.22).

With the given data, we find that $\Delta \approx -34.94855476$, $(\Delta \neq 0)$, $l \approx 0.28508672$, $L_0 \approx 1.90760368$, $M_1 \approx 1.52905173$, $M_2 \approx 2.87883365$.

(a) In order to apply Theorem 7.1.4, we consider $h(u) = -\frac{1}{2}\cos u\left(\frac{1}{3}\right)$, and the multivalued function

$$F(t,x,y) = \left[\frac{1}{\sqrt{400+t^2}}(\sin x - y + 1), \frac{1}{\sqrt{900+t^2}}\left(x + \sin y + \frac{1}{2}\right)\right],$$

for all $t \in [0,1]$, $x,y \in \mathbb{R}$.

It is easy to find that $L_2 = \frac{1}{2}$, $\|F(t,x,y)\|_{\mathcal{P}} \leq \phi(t)\Omega(|x| + |y|)$ for all $t \in [0,1]$ and $x,y \in \mathbb{R}$, where $\phi(t) = \frac{1}{\sqrt{400+t^2}}$ and $\Omega(u) = u + 1$. So, Ω is a nondecreasing and subhomogeneous function, and $\|\phi\| = \frac{1}{20}$. If we take $M \geq 0.15$, then the condition given by (A3) is satisfied. Thus all the conditions of Theorem 7.1.4 are satisfied, and consequently, there exists at least one solution u for problem (7.23), (7.22) on $[0,1]$.

(b) For the illustration of Theorem 7.1.5, we introduce the function $h(u) = \frac{1}{2}\arctan u(1)$, and the multivalued function

$$F(t,x,y) = \left[0, \frac{1}{10+t^3}\left(\frac{|x|}{4+|x|} + \arctan y\right) + \frac{1}{12+t}\right],$$

for $t \in [0,1]$, $x, y \in \mathbb{R}$. Clearly $L_3 = \frac{1}{2}$, and

$$H_d(F(t,x,y), F(t,\bar{x},\bar{y})) \leq \varphi(t) (|x - \bar{x}| + |y - \bar{y}|),$$

for all $t \in [0,1]$, $x, y, \bar{x}, \bar{y} \in \mathbb{R}$ with $\varphi(t) = \frac{1}{10+t^3}$, $t \in [0,1]$. Furthermore, $\|\varphi\| = \frac{1}{10}$, $d(0, F(t,0,0)) \leq \varphi(t)$ for $t \in [0,1]$, and $\theta_0 \approx 0.825722 < 1$. Hence all the assumptions of Theorem 7.1.5 are satisfied, and consequently problem (7.23), (7.22) has at least one solution on $[0,1]$.

Remark 7.1.1. The results presented in this section were published in [9].

7.2 Sequential Caputo Fractional Integro-Differential Systems with Coupled Integral Boundary Conditions

In this section, we investigate the system of nonlinear Caputo type sequential fractional integro-differential equations

$$\begin{cases} ({}^cD_{0+}^{\alpha} + \lambda {}^cD_{0+}^{\alpha-1})u(t) = f(t, u(t), v(t), {}^cD_{0+}^{p_1}v(t), I_{0+}^{q_1}v(t)), & t \in (0,1), \\ ({}^cD_{0+}^{\beta} + \mu {}^cD_{0+}^{\beta-1})v(t) = g(t, u(t), {}^cD_{0+}^{p_2}u(t), I_{0+}^{q_2}u(t), v(t)), & t \in (0,1), \end{cases}$$
$$(7.24)$$

with the coupled boundary conditions

$$\begin{cases} u(0) = u'(0) = u''(0) = 0, \quad u(1) = \int_0^1 u(s)\, dH_1(s) + \int_0^1 v(s)\, dH_2(s), \\ v(0) = v'(0) = v''(0) = 0, \quad v(1) = \int_0^1 u(s)\, dK_1(s) + \int_0^1 v(s)\, dK_2(s), \end{cases}$$
$$(7.25)$$

where $\alpha, \beta \in (3,4]$, $p_1, p_2 \in (0,1)$, $q_1, q_2 > 0$, $\lambda, \mu > 0$, ${}^cD_{0+}^k$ denotes the Caputo derivative of fractional order k (for $k = \alpha, \alpha - 1, \beta, \beta - 1, p_1, p_2$), I_{0+}^j represents the Riemann-Liouville integral of fractional order j (for $j = q_1, q_2$), the integrals from (7.25) are Riemann-Stieltjes integrals, and H_1, H_2, K_1, K_2 are functions of bounded variation.

Under some assumptions on the functions f and g, we prove the existence, and the existence and uniqueness of solutions for problem (7.24), (7.25), by applying the Leray-Schauder alternative (Theorem 1.2.5) and the Banach contraction principle (Theorem 1.2.1).

The system (7.24) with $\lambda = \mu$, $\alpha, \beta \in (2,3]$, $p_1, p_2 \in (0,1)$, $q_1, q_2 \in (0,1)$, with the nonlocal six-point coupled Riemann-Liouville type integral boundary conditions

$$\begin{cases} u(0) = u'(0) = 0, & a_1 u(1) + a_2 u(\zeta) = a I_{0+}^p v(\eta), \\ v(0) = v'(0) = 0, & b_1 v(1) + b_2 v(z) = b I_{0+}^q u(\theta), \end{cases}$$

where $p > 0$, $\zeta, \eta \in (0,1)$, $q > 0$, $z, \theta \in (0,1)$, a, b, a_i, b_i, $i = 1, 2$, are real constants, was investigated in [5]. The system (7.24) with $\alpha, \beta \in (2,3]$ and the functions f and g dependent only on t, u and v, subject to the boundary conditions $u(0) = u'(0) = 0$, $u(1) = av(\xi)$, $v(0) = v'(0) = 0$, $v(1) = bu(\eta)$, $\xi, \eta \in (0,1)$ has been studied in [61]. The system (7.24) with $\lambda = \mu$, $\alpha, \beta \in (2,3]$ and the functions f and g dependent only on t, u and v, subject to the boundary conditions

$$\begin{cases} u(0) = 0, & u'(0) = 0, & u(\zeta) = a I_{0+}^{\beta_1} u(\eta), \\ v(0) = 0, & v'(0) = 0, & v(z) = b I_{0+}^{\gamma_1} v(\theta), \end{cases}$$

where $\beta_1 > 0$, $0 < \eta < \zeta < 1$, $\gamma_1 > 0$, $0 < \theta < z < 1$, and a, b are real constants, was investigated in [11].

7.2.1   *Preliminary results*

We consider the fractional differential system

$$\begin{cases} ({}^cD_{0+}^\alpha + \lambda \, {}^cD_{0+}^{\alpha-1})u(t) = x(t), & t \in (0,1), \\ ({}^cD_{0+}^\beta + \mu \, {}^cD_{0+}^{\beta-1})v(t) = y(t), & t \in (0,1), \end{cases} \tag{7.26}$$

with the coupled boundary conditions (7.25), where $x, y \in C[0,1]$.

We denote by

$$a = \frac{1}{\lambda^3}(\lambda^2 - 2\lambda + 2 - 2e^{-\lambda}) - \frac{1}{\lambda^3}\int_0^1 (\lambda^2 s^2 - 2\lambda s + 2 - 2e^{-\lambda s}) \, dH_1(s),$$

$$b = \frac{1}{\mu^3}\int_0^1 (\mu^2 s^2 - 2\mu s + 2 - 2e^{-\mu s}) \, dH_2(s),$$

$$c = \frac{1}{\lambda^3}\int_0^1 (\lambda^2 s^2 - 2\lambda s + 2 - 2e^{-\lambda s}) \, dK_1(s),$$

$$d = \frac{1}{\mu^3}(\mu^2 - 2\mu + 2 - 2e^{-\mu})$$

$$- \frac{1}{\mu^3}\int_0^1 (\mu^2 s^2 - 2\mu s + 2 - 2e^{-\mu s}) \, dK_2(s),$$

$$\Delta_0 = ad - bc, \quad \gamma_1(t) = \frac{\lambda^2 t^2 - 2\lambda t + 2 - 2e^{-\lambda t}}{\lambda^3 \Delta_0},$$

$$t \in [0,1], \quad (\text{for } \Delta_0 \neq 0),$$

$$\gamma_2(t) = \frac{\mu^2 t^2 - 2\mu t + 2 - 2e^{-\mu t}}{\mu^3 \Delta_0}, \quad t \in [0,1], \quad (\text{for } \Delta_0 \neq 0).$$

$$(7.27)$$

Lemma 7.2.1. *Let* $x, y \in C[0,1]$ *and* $\Delta_0 \neq 0$. *Then the solution* $(u,v) \in (C^4[0,1])^2$ *of the boundary value problem* (7.26), (7.25) *is given by*

$$u(t) = \gamma_1(t) \left\{ d \left[-\int_0^1 e^{-\lambda(1-s)} I_{0+}^{\alpha-1} x(s)\, ds \right. \right.$$

$$+ \int_0^1 \left(\int_0^s e^{-\lambda(s-\tau)} I_{0+}^{\alpha-1} x(\tau)\, d\tau \right) dH_1(s)$$

$$\left. + \int_0^1 \left(\int_0^s e^{-\mu(s-\tau)} I_{0+}^{\beta-1} y(\tau)\, d\tau \right) dH_2(s) \right]$$

$$+ b \left[-\int_0^1 e^{-\mu(1-s)} I_{0+}^{\beta-1} y(s)\, ds \right.$$

$$+ \int_0^1 \left(\int_0^s e^{-\lambda(s-\tau)} I_{0+}^{\alpha-1} x(\tau)\, d\tau \right) dK_1(s)$$

$$\left. \left. + \int_0^1 \left(\int_0^s e^{-\mu(s-\tau)} I_{0+}^{\beta-1} y(\tau)\, d\tau \right) dK_2(s) \right] \right\}$$

$$+ \int_0^t e^{-\lambda(t-s)} I_{0+}^{\alpha-1} x(s)\, ds, \quad t \in [0,1],$$

$$v(t) = \gamma_2(t) \left\{ a \left[-\int_0^1 e^{-\mu(1-s)} I_{0+}^{\beta-1} y(s)\, ds \right. \right.$$

$$+ \int_0^1 \left(\int_0^s e^{-\lambda(s-\tau)} I_{0+}^{\alpha-1} x(\tau)\, d\tau \right) dK_1(s)$$

$$\left. + \int_0^1 \left(\int_0^s e^{-\mu(s-\tau)} I_{0+}^{\beta-1} y(\tau)\, d\tau \right) dK_2(s) \right]$$

$$+ c \left[-\int_0^1 e^{-\lambda(1-s)} I_{0+}^{\alpha-1} x(s)\, ds \right.$$

$$+ \int_0^1 \left(\int_0^s e^{-\lambda(s-\tau)} I_{0+}^{\alpha-1} x(\tau)\, d\tau \right) dH_1(s)$$

$$+ \int_0^1 \left(\int_0^s e^{-\mu(s-\tau)} I_{0+}^{\beta-1} y(\tau) \, d\tau \right) dH_2(s) \bigg] \bigg\}$$

$$+ \int_0^t e^{-\mu(t-s)} I_{0+}^{\beta-1} y(s) \, ds, \quad t \in [0,1], \tag{7.28}$$

where a, b, c, d, Δ_0 and the functions γ_1, γ_2 are given by (7.27).

Proof. The system (7.26) is equivalent to the following system

$$\begin{cases} {}^c D_{0+}^{\alpha}(u(t) + \lambda D^{-1} u(t)) = x(t), & t \in (0,1), \\ {}^c D_{0+}^{\beta}(v(t) + \mu D^{-1} v(t)) = y(t), & t \in (0,1). \end{cases} \tag{7.29}$$

The general solution of problem (7.29), (7.25) can be written as

$$u(t) + \lambda D^{-1} u(t) = \frac{1}{\Gamma(\alpha)} \int_0^t (t-s)^{\alpha-1} x(s) \, ds + \tilde{a}_0 + \tilde{a}_1 t + \tilde{a}_2 t^2 + \tilde{a}_3 t^3,$$

$$t \in [0,1],$$

$$v(t) + \mu D^{-1} v(t) = \frac{1}{\Gamma(\beta)} \int_0^t (t-s)^{\beta-1} y(s) \, ds + \tilde{b}_0 + \tilde{b}_1 t + \tilde{b}_2 t^2 + \tilde{b}_3 t^3,$$

$$t \in [0,1],$$

or equivalently

$$u(t) = -\lambda \int_0^t u(s) \, ds + \frac{1}{\Gamma(\alpha)} \int_0^t (t-s)^{\alpha-1} x(s) \, ds + \tilde{a}_0 + \tilde{a}_1 t + \tilde{a}_2 t^2 + \tilde{a}_3 t^3,$$

$$v(t) = -\mu \int_0^t v(s) \, ds + \frac{1}{\Gamma(\beta)} \int_0^t (t-s)^{\beta-1} y(s) \, ds + \tilde{b}_0 + \tilde{b}_1 t + \tilde{b}_2 t^2 + \tilde{b}_3 t^3,$$

with $\tilde{a}_i, \tilde{b}_i \in \mathbb{R}, \ i = 0, \dots, 3$.

By differentiating the above relations, we obtain

$$u'(t) = -\lambda u(t) + \frac{1}{\Gamma(\alpha-1)} \int_0^t (t-s)^{\alpha-2} x(s) \, ds + \tilde{a}_1 + 2\tilde{a}_2 t + 3\tilde{a}_3 t^2,$$

$$v'(t) = -\mu v(t) + \frac{1}{\Gamma(\beta-1)} \int_0^t (t-s)^{\beta-2} y(s) \, ds + \tilde{b}_1 + 2\tilde{b}_2 t + 3\tilde{b}_3 t^2,$$

and then

$$(e^{\lambda t} u(t))' = \frac{e^{\lambda t}}{\Gamma(\alpha-1)} \int_0^t (t-s)^{\alpha-2} x(s) \, ds + \tilde{a}_1 e^{\lambda t} + 2\tilde{a}_2 t e^{\lambda t} + 3\tilde{a}_3 t^2 e^{\lambda t},$$

$$(e^{\mu t} v(t))' = \frac{e^{\mu t}}{\Gamma(\beta-1)} \int_0^t (t-s)^{\beta-2} y(s) \, ds + \tilde{b}_1 e^{\mu t} + 2\tilde{b}_2 t e^{\mu t} + 3\tilde{b}_3 t^2 e^{\mu t}.$$

Integrating from 0 to t, we have

$$u(t) = \tilde{c}_0 e^{-\lambda t} + \frac{\tilde{c}_1}{\lambda}(1 - e^{-\lambda t}) + \frac{\tilde{c}_2}{\lambda^2}(\lambda t - 1 + e^{-\lambda t})$$

$$+ \frac{\tilde{c}_3}{\lambda^3}(\lambda^2 t^2 - 2\lambda t + 2 - 2e^{-\lambda t})$$

$$+ \int_0^t e^{-\lambda(t-s)} I_{0+}^{\alpha-1} x(s)\, ds,$$

$$v(t) = \tilde{d}_0 e^{-\mu t} + \frac{\tilde{d}_1}{\mu}(1 - e^{-\mu t}) + \frac{\tilde{d}_2}{\mu^2}(\mu t - 1 + e^{-\mu t})$$

$$+ \frac{\tilde{d}_3}{\mu^3}(\mu^2 t^2 - 2\mu t + 2 - 2e^{-\mu t})$$

$$+ \int_0^t e^{-\mu(t-s)} I_{0+}^{\beta-1} y(s)\, ds,$$

where $\tilde{c}_0 = u(0)$, $\tilde{d}_0 = v(0)$, $\tilde{c}_1 = \tilde{a}_1$, $\tilde{c}_2 = 2\tilde{a}_2$, $\tilde{c}_3 = 3\tilde{a}_3$, $\tilde{d}_1 = \tilde{b}_1$, $\tilde{d}_2 = 2\tilde{b}_2$, $\tilde{d}_3 = 3\tilde{b}_3$.

Using the boundary conditions $u(0) = u'(0) = u''(0) = 0$ and $v(0) = v'(0) = v''(0) = 0$ we deduce that $\tilde{c}_0 = \tilde{c}_1 = \tilde{c}_2 = 0$ and $\tilde{d}_0 = \tilde{d}_1 = \tilde{d}_2 = 0$. Then the above solution becomes

$$u(t) = \frac{\tilde{c}_3}{\lambda^3}(\lambda^2 t^2 - 2\lambda t + 2 - 2e^{-\lambda t}) + \int_0^t e^{-\lambda(t-s)} I_{0+}^{\alpha-1} x(s)\, ds, \quad t \in [0,1],$$

$$v(t) = \frac{\tilde{d}_3}{\mu^3}(\mu^2 t^2 - 2\mu t + 2 - 2e^{-\mu t}) + \int_0^t e^{-\mu(t-s)} I_{0+}^{\beta-1} y(s)\, ds, \quad t \in [0,1].$$

$$(7.30)$$

Now by using the boundary conditions $u(1) = \int_0^1 u(s)\, dH_1(s) + \int_0^1 v(s)\, dH_2(s)$, $v(1) = \int_0^1 u(s)\, dK_1(s) + \int_0^1 v(s)\, dK_2(s)$, we obtain the following system in the unknowns $\tilde{c}_3$ and $\tilde{d}_3$

$$\frac{\tilde{c}_3}{\lambda^3}(\lambda^2 - 2\lambda + 2 - 2e^{-\lambda}) + \int_0^1 e^{-\lambda(1-s)} I_{0+}^{\alpha-1} x(s)\, ds$$

$$= \frac{\tilde{c}_3}{\lambda^3}\int_0^1 (\lambda^2 s^2 - 2\lambda s + 2 - 2e^{-\lambda s})\, dH_1(s)$$

$$+ \int_0^1 \left(\int_0^s e^{-\lambda(s-\tau)} I_{0+}^{\alpha-1} x(\tau)\, d\tau \right) dH_1(s)$$

$$+ \frac{\widetilde{d}_3}{\mu^3} \int_0^1 (\mu^2 s^2 - 2\mu s + 2 - 2e^{-\mu s}) \, dH_2(s)$$

$$+ \int_0^1 \left(\int_0^s e^{-\mu(s-\tau)} I_{0+}^{\beta-1} y(\tau) \, d\tau \right) dH_2(s),$$

$$\frac{\widetilde{d}_3}{\mu^3} (\mu^2 - 2\mu + 2 - 2e^{-\mu}) + \int_0^1 e^{-\mu(1-s)} I_{0+}^{\beta-1} y(s) \, ds$$

$$= \frac{\widetilde{c}_3}{\lambda^3} \int_0^1 (\lambda^2 s^2 - 2\lambda s + 2 - 2e^{-\lambda s}) \, dK_1(s)$$

$$+ \int_0^1 \left(\int_0^s e^{-\lambda(s-\tau)} I_{0+}^{\alpha-1} x(\tau) \, d\tau \right) dK_1(s)$$

$$+ \frac{\widetilde{d}_3}{\mu^3} \int_0^1 (\mu^2 s^2 - 2\mu s + 2 - 2e^{-\mu s}) \, dK_2(s)$$

$$+ \int_0^1 \left(\int_0^s e^{-\mu(s-\tau)} I_{0+}^{\beta-1} y(\tau) \, d\tau \right) dK_2(s),$$

or

$$\begin{cases} a\widetilde{c}_3 - b\widetilde{d}_3 = - \int_0^1 e^{-\lambda(1-s)} I_{0+}^{\alpha-1} x(s) \, ds \\[2mm] \qquad + \int_0^1 \left(\int_0^s e^{-\lambda(s-\tau)} I_{0+}^{\alpha-1} x(\tau) \, d\tau \right) dH_1(s) \\[2mm] \qquad + \int_0^1 \left(\int_0^s e^{-\mu(s-\tau)} I_{0+}^{\beta-1} y(\tau) \, d\tau \right) dH_2(s), \\[2mm] -c\widetilde{c}_3 + d\widetilde{d}_3 = - \int_0^1 e^{-\mu(1-s)} I_{0+}^{\beta-1} y(s) \, ds \\[2mm] \qquad + \int_0^1 \left(\int_0^s e^{-\lambda(s-\tau)} I_{0+}^{\alpha-1} x(\tau) \, d\tau \right) dK_1(s) \\[2mm] \qquad + \int_0^1 \left(\int_0^s e^{-\mu(s-\tau)} I_{0+}^{\beta-1} y(\tau) \, d\tau \right) dK_2(s), \end{cases} \tag{7.31}$$

where a, b, c and d are given by (7.27).

The system (7.31) has the determinant $\Delta_0 = ad - bc$ which, by the assumption of this lemma is different from zero. Then the unique solution

$(\widetilde{c}_3, \widetilde{d}_3)$ of (7.31) is

$$
\begin{aligned}
\widetilde{c}_3 = \frac{1}{\Delta_0} \Bigg\{ & d \Bigg[- \int_0^1 e^{-\lambda(1-s)} I_{0+}^{\alpha-1} x(s) \, ds \\
& + \int_0^1 \left(\int_0^s e^{-\lambda(s-\tau)} I_{0+}^{\alpha-1} x(\tau) \, d\tau \right) dH_1(s) \\
& + \int_0^1 \left(\int_0^s e^{-\mu(s-\tau)} I_{0+}^{\beta-1} y(\tau) \, d\tau \right) dH_2(s) \Bigg] \\
& + b \Bigg[- \int_0^1 e^{-\mu(1-s)} I_{0+}^{\beta-1} y(s) \, ds \\
& + \int_0^1 \left(\int_0^s e^{-\lambda(s-\tau)} I_{0+}^{\alpha-1} x(\tau) \, d\tau \right) dK_1(s) \\
& + \int_0^1 \left(\int_0^s e^{-\mu(s-\tau)} I_{0+}^{\beta-1} y(\tau) \, d\tau \right) dK_2(s) \Bigg] \Bigg\}, \\
\widetilde{d}_3 = \frac{1}{\Delta_0} \Bigg\{ & a \Bigg[- \int_0^1 e^{-\mu(1-s)} I_{0+}^{\beta-1} y(s) \, ds \\
& + \int_0^1 \left(\int_0^s e^{-\lambda(s-\tau)} I_{0+}^{\alpha-1} x(\tau) \, d\tau \right) dK_1(s) \\
& + \int_0^1 \left(\int_0^s e^{-\mu(s-\tau)} I_{0+}^{\beta-1} y(\tau) \, d\tau \right) dK_2(s) \Bigg] \\
& + c \Bigg[- \int_0^1 e^{-\lambda(1-s)} I_{0+}^{\alpha-1} x(s) \, ds \\
& + \int_0^1 \left(\int_0^s e^{-\lambda(s-\tau)} I_{0+}^{\alpha-1} x(\tau) \, d\tau \right) dH_1(s) \\
& + \int_0^1 \left(\int_0^s e^{-\mu(s-\tau)} I_{0+}^{\beta-1} y(\tau) \, d\tau \right) dH_2(s) \Bigg] \Bigg\}.
\end{aligned}
$$

Replacing the above constants $\widetilde{c}_3$ and $\widetilde{d}_3$ in (7.30), we obtain the solution $(u(t), v(t))$, $t \in [0, 1]$ of problem (7.26), (7.25) given by (7.28). $\qquad \square$

Lemma 7.2.2. *For $x, y \in C[0,1]$ with $\|x\| = \sup_{t \in [0,1]} |x(t)|$ and $\|y\| = \sup_{t \in [0,1]} |y(t)|$, we have the following inequalities:*

(a) $\left| \int_0^1 e^{-\lambda(1-s)} I_{0+}^{\alpha-1} x(s) \, ds \right| \leq \dfrac{1}{\lambda \Gamma(\alpha)} (1 - e^{-\lambda}) \|x\|$;

(b) $\left| \int_0^1 e^{-\mu(1-s)} I_{0+}^{\beta-1} y(s)\, ds \right| \le \dfrac{1}{\mu\Gamma(\beta)}(1 - e^{-\mu})\|y\|$;

(c) $\left| \int_0^t e^{-\lambda(t-s)} I_{0+}^{\alpha-1} x(s)\, ds \right| \le \dfrac{1}{\lambda\Gamma(\alpha)}(1 - e^{-\lambda})\|x\|, \quad \forall t \in [0,1]$;

(d) $\left| \int_0^t e^{-\mu(t-s)} I_{0+}^{\beta-1} y(s)\, ds \right| \le \dfrac{1}{\mu\Gamma(\beta)}(1 - e^{-\mu})\|y\|, \quad \forall t \in [0,1]$;

(e) $\left| \int_0^1 \left(\int_0^s e^{-\lambda(s-\tau)} I_{0+}^{\alpha-1} x(\tau)\, d\tau \right) dH_1(s) \right|$

$\le \dfrac{\|x\|}{\lambda\Gamma(\alpha)} \left| \int_0^1 s^{\alpha-1}(1 - e^{-\lambda s})dH_1(s) \right|$;

(f) $\left| \int_0^1 \left(\int_0^s e^{-\lambda(s-\tau)} I_{0+}^{\alpha-1} x(\tau)\, d\tau \right) dK_1(s) \right|$

$\le \dfrac{\|x\|}{\lambda\Gamma(\alpha)} \left| \int_0^1 s^{\alpha-1}(1 - e^{-\lambda s})dK_1(s) \right|$;

(g) $\left| \int_0^1 \left(\int_0^s e^{-\mu(s-\tau)} I_{0+}^{\beta-1} y(\tau)\, d\tau \right) dH_2(s) \right|$

$\le \dfrac{\|y\|}{\mu\Gamma(\beta)} \left| \int_0^1 s^{\beta-1}(1 - e^{-\mu s})dH_2(s) \right|$;

(h) $\left| \int_0^1 \left(\int_0^s e^{-\mu(s-\tau)} I_{0+}^{\beta-1} y(\tau)\, d\tau \right) dK_2(s) \right|$

$\le \dfrac{\|y\|}{\mu\Gamma(\beta)} \left| \int_0^1 s^{\beta-1}(1 - e^{-\mu s})dK_2(s) \right|$.

Proof. For the inequality from (a), we have

$$\left| \int_0^1 e^{-\lambda(1-s)} I_{0+}^{\alpha-1} x(s)\, ds \right|$$

$$= \left| \int_0^1 e^{-\lambda(1-s)} \left(\int_0^s \frac{1}{\Gamma(\alpha-1)}(s-\tau)^{\alpha-2} x(\tau)\, d\tau \right) ds \right|$$

$$\le \|x\| \int_0^1 e^{-\lambda(1-s)} \left(\int_0^s \frac{1}{\Gamma(\alpha-1)}(s-\tau)^{\alpha-2}\, d\tau \right) ds$$

$$= \frac{\|x\|}{\Gamma(\alpha)} \int_0^1 e^{-\lambda(1-s)} s^{\alpha-1}\, ds \le \frac{1}{\lambda\Gamma(\alpha)}(1 - e^{-\lambda})\|x\|.$$

For the inequality from (c), we get

$$\left| \int_0^t e^{-\lambda(t-s)} I_{0+}^{\alpha-1} x(s)\, ds \right|$$

$$= \left| \int_0^t e^{-\lambda(t-s)} \left(\int_0^s \frac{1}{\Gamma(\alpha-1)}(s-\tau)^{\alpha-2} x(\tau)\, d\tau \right) ds \right|$$

$$\leq \|x\| \int_0^t e^{-\lambda(t-s)} \left(\int_0^s \frac{1}{\Gamma(\alpha-1)}(s-\tau)^{\alpha-2}\, d\tau \right) ds$$

$$= \frac{\|x\|}{\Gamma(\alpha)} \int_0^t e^{-\lambda(t-s)} s^{\alpha-1}\, ds$$

$$\leq \frac{1}{\lambda\Gamma(\alpha)}(1-e^{-\lambda t})\|x\| \leq \frac{1}{\lambda\Gamma(\alpha)}(1-e^{-\lambda})\|x\|, \quad \forall t \in [0,1].$$

For the inequality from (e), we have

$$\left| \int_0^1 \left(\int_0^s e^{-\lambda(s-\tau)} I_{0+}^{\alpha-1} x(\tau)\, d\tau \right) dH_1(s) \right|$$

$$\leq \frac{\|x\|}{\Gamma(\alpha)} \left| \int_0^1 \left(\int_0^s e^{-\lambda(s-\tau)} \tau^{\alpha-1}\, d\tau \right) dH_1(s) \right|$$

$$\leq \frac{\|x\|}{\Gamma(\alpha)} \left| \int_0^1 \left(\int_0^s e^{-\lambda(s-\tau)} s^{\alpha-1}\, d\tau \right) dH_1(s) \right|$$

$$\leq \frac{\|x\|}{\lambda\Gamma(\alpha)} \left| \int_0^1 s^{\alpha-1}(1-e^{-\lambda s})\, dH_1(s) \right|.$$

In a similar manner, we also deduce the inequalities (b), (d), (f), (g) and (h). The inequalities (a)–(d) were also used in [5]. $\qquad\square$

7.2.2 *Existence of solutions*

We consider the spaces $\mathcal{X} = \{u \in C[0,1],\ {}^cD_{0+}^{p_2} u \in C[0,1]\}$ and $\mathcal{Y} = \{v \in C[0,1],\ {}^cD_{0+}^{p_1} v \in C[0,1]\}$ equipped respectively with the norms $\|u\|_{\mathcal{X}} = \|u\| + \|{}^cD_{0+}^{p_2} u\|$ and $\|v\|_{\mathcal{Y}} = \|v\| + \|{}^cD_{0+}^{p_1} v\|$, where $\|\cdot\|$ is the supremum norm, that is $\|w\| = \sup_{t\in[0,1]} |w(t)|$ for $w \in C[0,1]$. The spaces $(\mathcal{X}, \|\cdot\|_{\mathcal{X}})$ and $(\mathcal{Y}, \|\cdot\|_{\mathcal{Y}})$ are Banach spaces, and the product space $\mathcal{X} \times \mathcal{Y}$ endowed with the norm $\|(u,v)\|_{\mathcal{X}\times\mathcal{Y}} = \|u\|_{\mathcal{X}} + \|v\|_{\mathcal{Y}}$ is also a Banach space.

By using Lemma 7.2.1, we introduce the operator $\mathcal{T} : \mathcal{X} \times \mathcal{Y} \to \mathcal{X} \times \mathcal{Y}$ defined by $\mathcal{T}(u,v) = (\mathcal{T}_1(u,v), \mathcal{T}_2(u,v))$ for $(u,v) \in \mathcal{X} \times \mathcal{Y}$, where the operators $\mathcal{T}_1 : \mathcal{X} \times \mathcal{Y} \to \mathcal{X}$ and $\mathcal{T}_2 : \mathcal{X} \times \mathcal{Y} \to \mathcal{Y}$ are given by

$$
\mathcal{T}_1(u,v)(t) = \gamma_1(t) \left\{ d \left[- \int_0^1 e^{-\lambda(1-s)} \right. \right.
$$

$$
\times I_{0+}^{\alpha-1} f(s, u(s), v(s), {}^cD_{0+}^{p_1} v(s), I_{0+}^{q_1} v(s)) \, ds
$$

$$
+ \int_0^1 \left(\int_0^s e^{-\lambda(s-\tau)} I_{0+}^{\alpha-1} f(\tau, u(\tau), v(\tau), \right.
$$

$$
\left. {}^cD_{0+}^{p_1} v(\tau), I_{0+}^{q_1} v(\tau)) \, d\tau \right) dH_1(s)
$$

$$
+ \int_0^1 \left(\int_0^s e^{-\mu(s-\tau)} I_{0+}^{\beta-1} g(\tau, u(\tau), \right.
$$

$$
\left. \left. {}^cD_{0+}^{p_2} u(\tau), I_{0+}^{q_2} u(\tau), v(\tau)) \, d\tau \right) dH_2(s) \right]
$$

$$
+ b \left[- \int_0^1 e^{-\mu(1-s)} I_{0+}^{\beta-1} g(s, u(s), \right.
$$

$$
{}^cD_{0+}^{p_2} u(s), I_{0+}^{q_2} u(s), v(s)) \, ds
$$

$$
+ \int_0^1 \left(\int_0^s e^{-\lambda(s-\tau)} I_{0+}^{\alpha-1} f(\tau, u(\tau), v(\tau), \right.
$$

$$
\left. {}^cD_{0+}^{p_1} v(\tau), I_{0+}^{q_1} v(\tau)) \, d\tau \right) dK_1(s)
$$

$$
+ \int_0^1 \left(\int_0^s e^{-\mu(s-\tau)} I_{0+}^{\beta-1} g(\tau, u(\tau), \right.
$$

$$
\left. \left. \left. {}^cD_{0+}^{p_2} u(\tau), I_{0+}^{q_2} u(\tau), v(\tau)) \, d\tau \right) dK_2(s) \right] \right\}
$$

$$
+ \int_0^t e^{-\lambda(t-s)} I_{0+}^{\alpha-1} f(s, u(s), v(s),
$$

$$
{}^cD_{0+}^{p_1} v(s), I_{0+}^{q_1} v(s)) \, ds, \quad t \in [0,1], \ (u,v) \in \mathcal{X} \times \mathcal{Y},
$$

$$
\mathcal{T}_2(u,v)(t) = \gamma_2(t) \left\{ a \left[- \int_0^1 e^{-\mu(1-s)} I_{0+}^{\beta-1} g(s, u(s), \right. \right.
$$

$$
{}^cD_{0+}^{p_2} u(s), I_{0+}^{q_2} u(s), v(s)) \, ds
$$

$$
+ \int_0^1 \left(\int_0^s e^{-\lambda(s-\tau)} I_{0+}^{\alpha-1} f(\tau, u(\tau), v(\tau), \right.
$$

$$
{}^cD_{0+}^{p_1}v(\tau), I_{0+}^{q_1}v(\tau))\,d\tau\Biggr)\,dK_1(s)
$$

$$
+\int_0^1\Biggl(\int_0^s e^{-\mu(s-\tau)}I_{0+}^{\beta-1}g(\tau,u(\tau),
$$

$$
{}^cD_{0+}^{p_2}u(\tau), I_{0+}^{q_2}u(\tau),v(\tau))\,d\tau\Biggr)\,dK_2(s)\Biggr]
$$

$$
+c\Biggl[-\int_0^1 e^{-\lambda(1-s)}I_{0+}^{\alpha-1}f(s,u(s),v(s),
$$

$$
{}^cD_{0+}^{p_1}v(s), I_{0+}^{q_1}v(s))\,ds
$$

$$
+\int_0^1\Biggl(\int_0^s e^{-\lambda(s-\tau)}I_{0+}^{\alpha-1}f(\tau,u(\tau),v(\tau),
$$

$$
{}^cD_{0+}^{p_1}v(\tau), I_{0+}^{q_1}v(\tau))\,d\tau\Biggr)\,dH_1(s)
$$

$$
+\int_0^1\Biggl(\int_0^s e^{-\mu(s-\tau)}I_{0+}^{\beta-1}g(\tau,u(\tau),
$$

$$
{}^cD_{0+}^{p_2}u(\tau), I_{0+}^{q_2}u(\tau),v(\tau))\,d\tau\Biggr)\,dH_2(s)\Biggr]\Biggr\}
$$

$$
+\int_0^t e^{-\mu(t-s)}I_{0+}^{\beta-1}g(s,u(s),
$$

$$
{}^cD_{0+}^{p_2}u(s), I_{0+}^{q_2}u(s),v(s))\,ds,\quad t\in[0,1],\ (u,v)\in\mathcal{X}\times\mathcal{Y}.
$$

The pair (u,v) is a solution of problem (7.24), (7.25) if and only if (u,v) is a fixed point of operator $\mathcal{T}$.

We present now the assumptions that we use in this section.

(J1) $\alpha,\beta\in\mathbb{R}$, $3<\alpha\le 4$, $3<\beta\le 4$, $p_1,p_2\in(0,1)$, $q_1,q_2>0$, $\Delta_0\neq 0$ (given by (7.27)), $H_1,H_2,K_1,K_2:[0,1]\to\mathbb{R}$ are functions of bounded variation.

(J2) The functions $f,g:[0,1]\times\mathbb{R}^4\to\mathbb{R}$ are continuous and there exist real constants $a_i,b_i\ge 0$, $i=1,\ldots,4$ and $a_0>0$, $b_0>0$ such that

$$
|f(t,x_1,x_2,x_3,x_4)|\le a_0+a_1|x_1|+a_2|x_2|+a_3|x_3|+a_4|x_4|,
$$

$$
|g(t,x_1,x_2,x_3,x_4)|\le b_0+b_1|x_1|+b_2|x_2|+b_3|x_3|+b_4|x_4|,
$$

for all $t\in[0,1]$ and $x_i\in\mathbb{R}$, $i=1,\ldots,4$.

(J3) The functions $f, g : [0,1] \times \mathbb{R}^4 \to \mathbb{R}$ are continuous and there exist positive constants c_0, d_0 such that

$$|f(t, x_1, x_2, x_3, x_4) - f(t, y_1, y_2, y_3, y_4)|$$
$$\leq c_0(|x_1 - y_1| + |x_2 - y_2| + |x_3 - y_3| + |x_4 - y_4|),$$
$$|g(t, x_1, x_2, x_3, x_4) - g(t, y_1, y_2, y_3, y_4)|$$
$$\leq d_0(|x_1 - y_1| + |x_2 - y_2| + |x_3 - y_3| + |x_4 - y_4|),$$

for all $t \in [0,1]$ and $x_i, y_i \in \mathbb{R}$, $i = 1, \ldots, 4$.

We denote by

$$l_1 = \sup_{t \in [0,1]} |\gamma_1(t)| = \frac{\lambda^2 - 2\lambda + 2 - 2e^{-\lambda}}{\lambda^3 |\Delta_0|} > 0,$$

$$l_2 = \sup_{t \in [0,1]} |\gamma_2(t)| = \frac{\mu^2 - 2\mu + 2 - 2e^{-\mu}}{\mu^3 |\Delta_0|} > 0,$$

$$m_1 = \sup_{t \in [0,1]} |\gamma_1'(t)| = \frac{2\lambda^2 - 2\lambda + 2\lambda e^{-\lambda}}{\lambda^3 |\Delta_0|} > 0,$$

$$m_2 = \sup_{t \in [0,1]} |\gamma_2'(t)| = \frac{2\mu^2 - 2\mu + 2\mu e^{-\mu}}{\mu^3 |\Delta_0|} > 0,$$

$$M_1 = \frac{l_1}{\lambda \Gamma(\alpha)} \left\{ |d| \left[(1 - e^{-\lambda}) + \left| \int_0^1 s^{\alpha-1}(1 - e^{-\lambda s}) dH_1(s) \right| \right] \right.$$
$$\left. + |b| \left| \int_0^1 s^{\alpha-1}(1 - e^{-\lambda s}) dK_1(s) \right| \right\} + \frac{1}{\lambda \Gamma(\alpha)}(1 - e^{-\lambda}),$$

$$M_2 = \frac{l_1}{\mu \Gamma(\beta)} \left\{ |d| \left| \int_0^1 s^{\beta-1}(1 - e^{-\mu s}) dH_2(s) \right| \right.$$
$$\left. + |b| \left[(1 - e^{-\mu}) + \left| \int_0^1 s^{\beta-1}(1 - e^{-\mu s}) dK_2(s) \right| \right] \right\},$$

$$M_1^* = \frac{m_1}{\lambda \Gamma(\alpha)} \left\{ |d| \left[(1 - e^{-\lambda}) + \left| \int_0^1 s^{\alpha-1}(1 - e^{-\lambda s}) dH_1(s) \right| \right] \right.$$
$$\left. + |b| \left| \int_0^1 s^{\alpha-1}(1 - e^{-\lambda s}) dK_1(s) \right| \right\} + \frac{1}{\Gamma(\alpha)}(2 - e^{-\lambda}),$$

$$M_2^* = \frac{m_1}{\mu\Gamma(\beta)} \left\{ |d| \left| \int_0^1 s^{\beta-1}(1 - e^{-\mu s})dH_2(s) \right| \right.$$

$$\left. + |b| \left[(1 - e^{-\mu}) + \left| \int_0^1 s^{\beta-1}(1 - e^{-\mu s})dK_2(s) \right| \right] \right\},$$

$$\widetilde{M}_1 = \frac{l_2}{\lambda\Gamma(\alpha)} \left\{ |a| \left| \int_0^1 s^{\alpha-1}(1 - e^{-\lambda s})dK_1(s) \right| \right.$$

$$\left. + |c| \left[(1 - e^{-\lambda}) + \left| \int_0^1 s^{\alpha-1}(1 - e^{-\lambda s})dH_1(s) \right| \right] \right\},$$

$$\widetilde{M}_2 = \frac{l_2}{\mu\Gamma(\beta)} \left\{ |a| \left[(1 - e^{-\mu}) + \left| \int_0^1 s^{\beta-1}(1 - e^{-\mu s})dK_2(s) \right| \right] \right.$$

$$\left. + |c| \left| \int_0^1 s^{\beta-1}(1 - e^{-\mu s})dH_2(s) \right| \right\} + \frac{1}{\mu\Gamma(\beta)}(1 - e^{-\mu}),$$

$$\widetilde{M}_1^* = \frac{m_2}{\lambda\Gamma(\alpha)} \left\{ |a| \left| \int_0^1 s^{\alpha-1}(1 - e^{-\lambda s})dK_1(s) \right| \right.$$

$$\left. + |c| \left[(1 - e^{-\lambda}) + \left| \int_0^1 s^{\alpha-1}(1 - e^{-\lambda s})dH_1(s) \right| \right] \right\},$$

$$\widetilde{M}_2^* = \frac{m_2}{\mu\Gamma(\beta)} \left\{ |a| \left[(1 - e^{-\mu}) + \left| \int_0^1 s^{\beta-1}(1 - e^{-\mu s})dK_2(s) \right| \right] \right.$$

$$\left. + |c| \left| \int_0^1 s^{\beta-1}(1 - e^{-\mu s})dH_2(s) \right| \right\} + \frac{1}{\Gamma(\beta)}(2 - e^{-\mu}),$$

$$N_1 = M_1 + \widetilde{M}_1 + \frac{M_1^*}{\Gamma(2 - p_2)} + \frac{\widetilde{M}_1^*}{\Gamma(2 - p_1)},$$

$$N_2 = M_2 + \widetilde{M}_2 + \frac{M_2^*}{\Gamma(2 - p_2)} + \frac{\widetilde{M}_2^*}{\Gamma(2 - p_1)},$$

$$\Lambda_2 = a_1 N_1 + \left(b_1 + b_2 + \frac{b_3}{\Gamma(q_2 + 1)} \right) N_2,$$

$$\Lambda_3 = \left(a_2 + a_3 + \frac{a_4}{\Gamma(q_1 + 1)} \right) N_1 + b_4 N_2,$$

$$\Lambda_1 = a_0 N_1 + b_0 N_2, \quad \varrho_1 = 1 + \frac{1}{\Gamma(1 + q_1)}, \quad \varrho_2 = 1 + \frac{1}{\Gamma(1 + q_2)}.$$

Theorem 7.2.1. *Assume that (J1) and (J2) hold. If* $\max\{\Lambda_2, \Lambda_3\} < 1$, *then the boundary value problem (7.24), (7.25) has at least one solution on* $[0, 1]$.

Proof. We firstly prove that the operator $\mathcal{T} : \mathcal{X} \times \mathcal{Y} \to \mathcal{X} \times \mathcal{Y}$ is completely continuous. By continuity of the functions f and g , we deduce that the operators $\mathcal{T}_1$ and $\mathcal{T}_2$ are continuous, and then $\mathcal{T}$ is a continuous operator.

Next we show that $\mathcal{T}$ is uniformly bounded. Let $\Omega \subset \mathcal{X} \times \mathcal{Y}$ be a bounded set. Then there exist positive constants L_1 and L_2 such that

$$|f(t, u(t), v(t), {}^cD_{0+}^{p_1}v(t), I_{0+}^{q_1}v(t))| \le L_1,$$

$$|g(t, u(t), {}^cD_{0+}^{p_2}u(t), I_{0+}^{q_2}u(t), v(t))| \le L_2,$$

for all $(u, v) \in \Omega$ and $t \in [0, 1]$.

Then for any $(u, v) \in \Omega$ and $t \in [0, 1]$, we obtain

$$|\mathcal{T}_1(u, v)(t)| \le |\gamma_1(t)| \Bigg\{ |d| \Bigg[\int_0^1 e^{-\lambda(1-s)}(I_{0+}^{\alpha-1}|f(s, u(s), v(s),$$

$${}^cD_{0+}^{p_1}v(s), I_{0+}^{q_1}v(s))|)ds$$

$$+ \Bigg| \int_0^1 \left(\int_0^s e^{-\lambda(s-\tau)}(I_{0+}^{\alpha-1}|f(\tau, u(\tau), v(\tau),$$

$${}^cD_{0+}^{p_1}v(\tau), I_{0+}^{q_1}v(\tau))|) \, d\tau \right) dH_1(s) \Bigg|$$

$$+ \Bigg| \int_0^1 \left(\int_0^s e^{-\mu(s-\tau)} \left(I_{0+}^{\beta-1}|g(\tau, u(\tau),$$

$${}^cD_{0+}^{p_2}u(\tau), I_{0+}^{q_2}u(\tau), v(\tau))| \right) d\tau \right) dH_2(s) \Bigg| \Bigg]$$

$$+ |b| \Bigg[\int_0^1 e^{-\mu(1-s)} \left(I_{0+}^{\beta-1}|g(s, u(s),$$

$${}^cD_{0+}^{p_2}u(s), I_{0+}^{q_2}u(s), v(s))| \right) ds$$

$$+ \Bigg| \int_0^1 \left(\int_0^s e^{-\lambda(s-\tau)}(I_{0+}^{\alpha-1}|f(\tau, u(\tau), v(\tau),$$

$${}^cD_{0+}^{p_1}v(\tau), I_{0+}^{q_1}v(\tau))|) \, d\tau \right) dK_1(s) \Bigg|$$

$$+ \Bigg| \int_0^1 \left(\int_0^s e^{-\mu(s-\tau)} \left(I_{0+}^{\beta-1}|g(\tau, u(\tau),$$

$${}^cD_{0+}^{p_2}u(\tau), I_{0+}^{q_2}u(\tau), v(\tau))| \right) d\tau \right) dK_2(s) \Bigg| \Bigg] \Bigg\}$$

$$+ \int_0^t e^{-\lambda(t-s)}(I_{0+}^{\alpha-1}|f(s, u(s), v(s),$$

$${}^cD_{0+}^{p_1}v(s), I_{0+}^{q_1}v(s))|) \, ds$$

$$\le l_1 \left\{ |d| \left[L_1 \frac{1}{\lambda \Gamma(\alpha)} (1 - e^{-\lambda}) \right. \right.$$

$$+ L_1 \frac{1}{\lambda \Gamma(\alpha)} \left| \int_0^1 s^{\alpha-1}(1 - e^{-\lambda s}) dH_1(s) \right|$$

$$\left. + L_2 \frac{1}{\mu \Gamma(\beta)} \left| \int_0^1 s^{\beta-1}(1 - e^{-\mu s}) dH_2(s) \right| \right]$$

$$+ |b| \left[L_2 \frac{1}{\mu \Gamma(\beta)} (1 - e^{-\mu}) \right.$$

$$+ L_1 \frac{1}{\lambda \Gamma(\alpha)} \left| \int_0^1 s^{\alpha-1}(1 - e^{-\lambda s}) dK_1(s) \right|$$

$$\left. \left. + L_2 \frac{1}{\mu \Gamma(\beta)} \left| \int_0^1 s^{\beta-1}(1 - e^{-\mu s}) dK_2(s) \right| \right] \right\}$$

$$+ \frac{L_1}{\lambda \Gamma(\alpha)} (1 - e^{-\lambda}) = L_1 M_1 + L_2 M_2.$$

Then $\|\mathcal{T}_1(u,v)\| \le L_1 M_1 + L_2 M_2$, for all $(u,v) \in \Omega$.
From the definition of $\mathcal{T}_1(u,v)$, we have

$$|\mathcal{T}_1'(u,v)(t)| \le |\gamma_1'(t)| \left\{ |d| \left[\int_0^1 e^{-\lambda(1-s)} (I_{0+}^{\alpha-1} |f(s, u(s), v(s), \right. \right.$$

$$^c D_{0+}^{p_1} v(s), I_{0+}^{q_1} v(s))|) ds$$

$$+ \left| \int_0^1 \left(\int_0^s e^{-\lambda(s-\tau)} (I_{0+}^{\alpha-1} |f(\tau, u(\tau), v(\tau), \right. \right.$$

$$\left. \left. ^c D_{0+}^{p_1} v(\tau), I_{0+}^{q_1} v(\tau))|) \, d\tau \right) dH_1(s) \right|$$

$$\left. + \left| \int_0^1 \left(\int_0^s e^{-\mu(s-\tau)} \left(I_{0+}^{\beta-1} |g(\tau, u(\tau), \right. \right. \right.$$

$$\left. \left. \left. ^c D_{0+}^{p_2} u(\tau), I_{0+}^{q_2} u(\tau), v(\tau))| \right) d\tau \right) dH_2(s) \right| \right]$$

$$+ |b| \left[\int_0^1 e^{-\mu(1-s)} \left(I_{0+}^{\beta-1} |g(s, u(s), \right. \right.$$

$$^c D_{0+}^{p_2} u(s), I_{0+}^{q_2} u(s), v(s))| \right) ds$$

$$+ \left| \int_0^1 \left(\int_0^s e^{-\lambda(s-\tau)} (I_{0+}^{\alpha-1} |f(\tau, u(\tau), v(\tau), \right. \right.$$

$$\left. {}^{c}D_{0+}^{p_1}v(\tau), I_{0+}^{q_1}v(\tau))|\right)d\tau\right)dK_1(s)\right|$$

$$+\left|\int_0^1\left(\int_0^s e^{-\mu(s-\tau)}\left(I_{0+}^{\beta-1}|g(\tau,u(\tau),\right.\right.\right.$$

$$\left.\left.\left.{}^{c}D_{0+}^{p_2}u(\tau), I_{0+}^{q_2}u(\tau), v(\tau))|\right)d\tau\right)dK_2(s)\right|\right]\right\}$$

$$+\lambda\int_0^t e^{-\lambda(t-s)}(I_{0+}^{\alpha-1}|f(s,u(s),v(s),{}^{c}D_{0+}^{p_1}v(s), I_{0+}^{q_1}v(s))|)ds$$

$$+\int_0^t \frac{(t-s)^{\alpha-2}}{\Gamma(\alpha-1)}|f(s,u(s),v(s),{}^{c}D_{0+}^{p_1}v(s), I_{0+}^{q_1}v(s))|ds$$

$$\leq m_1\left\{|d|\left[L_1\frac{1}{\lambda\Gamma(\alpha)}(1-e^{-\lambda})\right.\right.$$

$$+L_1\frac{1}{\lambda\Gamma(\alpha)}\left|\int_0^1 s^{\alpha-1}(1-e^{-\lambda s})dH_1(s)\right|$$

$$\left.+L_2\frac{1}{\mu\Gamma(\beta)}\left|\int_0^1 s^{\beta-1}(1-e^{-\mu s})dH_2(s)\right|\right]$$

$$+|b|\left[L_2\frac{1}{\mu\Gamma(\beta)}(1-e^{-\mu})\right.$$

$$+L_1\frac{1}{\lambda\Gamma(\alpha)}\left|\int_0^1 s^{\alpha-1}(1-e^{-\lambda s})dK_1(s)\right|$$

$$\left.\left.+L_2\frac{1}{\mu\Gamma(\beta)}\left|\int_0^1 s^{\beta-1}(1-e^{-\mu s})dK_2(s)\right|\right]\right\}$$

$$+\frac{L_1}{\Gamma(\alpha)}(2-e^{-\lambda})=L_1M_1^*+L_2M_2^*.$$

By the definition of Caputo fractional derivative of order $p_2 \in (0,1)$, we deduce

$$|{}^{c}D_{0+}^{p_2}\mathcal{T}_1(u,v)(t)| \leq \int_0^t \frac{(t-s)^{-p_2}}{\Gamma(1-p_2)}|\mathcal{T}_1'(u,v)(s)|ds$$

$$\leq (L_1M_1^*+L_2M_2^*)\int_0^t \frac{(t-s)^{-p_2}}{\Gamma(1-p_2)}ds$$

$$=\frac{L_1M_1^*+L_2M_2^*}{\Gamma(2-p_2)}, \quad \forall t \in [0,1],$$

from where we obtain

$$\|{}^cD_{0+}^{p_2}\mathcal{T}_1(u,v)\| \leq \frac{L_1 M_1^* + L_2 M_2^*}{\Gamma(2-p_2)}.$$

Therefore we conclude

$$\|\mathcal{T}_1(u,v)\|_{\mathcal{X}} = \|\mathcal{T}_1(u,v)\| + \|{}^cD_{0+}^{p_2}\mathcal{T}_1(u,v)\|$$

$$\leq L_1 M_1 + L_2 M_2 + \frac{1}{\Gamma(2-p_2)}(L_1 M_1^* + L_2 M_2^*). \qquad (7.32)$$

In a similar manner, we have

$$|\mathcal{T}_2(u,v)(t)| \leq l_2 \left\{ |a| \left[L_2 \frac{1}{\mu\Gamma(\beta)}(1-e^{-\mu}) \right. \right.$$

$$+ L_1 \frac{1}{\lambda\Gamma(\alpha)} \left| \int_0^1 s^{\alpha-1}(1-e^{-\lambda s})dK_1(s) \right|$$

$$\left. + L_2 \frac{1}{\mu\Gamma(\beta)} \left| \int_0^1 s^{\beta-1}(1-e^{-\mu s})dK_2(s) \right| \right]$$

$$+ |c| \left[L_1 \frac{1}{\lambda\Gamma(\alpha)}(1-e^{-\lambda}) \right.$$

$$+ L_1 \frac{1}{\lambda\Gamma(\alpha)} \left| \int_0^1 s^{\alpha-1}(1-e^{-\lambda s})dH_1(s) \right|$$

$$\left. \left. + L_2 \frac{1}{\mu\Gamma(\beta)} \left| \int_0^1 s^{\beta-1}(1-e^{-\mu s})dH_2(s) \right| \right] \right\}$$

$$+ L_2 \frac{1}{\mu\Gamma(\beta)}(1-e^{-\mu}) = L_1\widetilde{M_1} + L_2\widetilde{M_2},$$

$$|\mathcal{T}_2'(u,v)(t)| \leq m_2 \left\{ |a| \left[L_2 \frac{1}{\mu\Gamma(\beta)}(1-e^{-\mu}) \right. \right.$$

$$+ L_1 \frac{1}{\lambda\Gamma(\alpha)} \left| \int_0^1 s^{\alpha-1}(1-e^{-\lambda s})dK_1(s) \right|$$

$$\left. + L_2 \frac{1}{\mu\Gamma(\beta)} \left| \int_0^1 s^{\beta-1}(1-e^{-\mu s})dK_2(s) \right| \right]$$

$$+ |c| \left[L_1 \frac{1}{\lambda\Gamma(\alpha)}(1-e^{-\lambda}) \right.$$

$$+ L_1 \frac{1}{\lambda\Gamma(\alpha)} \left| \int_0^1 s^{\alpha-1}(1-e^{-\lambda s})dH_1(s) \right|$$

$$\left. \left. + L_2 \frac{1}{\mu\Gamma(\beta)} \left| \int_0^1 s^{\beta-1}(1-e^{-\mu s})dH_2(s) \right| \right] \right\}$$

$$+ \frac{L_2}{\Gamma(\beta)}(2 - e^{-\mu}) = L_1\widetilde{M}_1^* + L_2\widetilde{M}_2^*,$$

$$|^cD_{0+}^{p_1}\mathcal{T}_2(u,v)(t)| \leq \frac{L_1\widetilde{M}_1^* + L_2\widetilde{M}_2^*}{\Gamma(2 - p_1)}, \quad \forall t \in [0,1].$$

Then we deduce

$$\|\mathcal{T}_2(u,v)\|_{\mathcal{Y}} = \|\mathcal{T}_2(u,v)\| + \|^cD_{0+}^{p_1}\mathcal{T}_2(u,v)\|$$

$$\leq L_2\widetilde{M}_1 + L_2\widetilde{M}_2 + \frac{1}{\Gamma(2 - p_1)}(L_1\widetilde{M}_1^* + L_2\widetilde{M}_2^*). \tag{7.33}$$

From the inequalities (7.32) and (7.33), we conclude that $\mathcal{T}_1$ and $\mathcal{T}_2$ are uniformly bounded, which implies that the operator $\mathcal{T}$ is uniformly bounded.

We will show next that $\mathcal{T}$ is equicontinuous. Let $t_1, t_2 \in [0,1]$, with $t_1 < t_2$. Then we have

$$|\mathcal{T}_1(u,v)(t_2) - \mathcal{T}_1(u,v)(t_1)|$$

$$\leq |\gamma_1(t_2) - \gamma_1(t_1)| \left\{ |d| \left[L_1\frac{1}{\lambda\Gamma(\alpha)}(1 - e^{-\lambda}) \right. \right.$$

$$+ L_1\frac{1}{\lambda\Gamma(\alpha)}\left|\int_0^1 s^{\alpha-1}(1 - e^{-\lambda s})dH_1(s)\right|$$

$$+ L_2\frac{1}{\mu\Gamma(\beta)}\left|\int_0^1 s^{\beta-1}(1 - e^{-\mu s})dH_2(s)\right| \right]$$

$$+ |b|\left[L_2\frac{1}{\mu\Gamma(\beta)}(1 - e^{-\mu}) + L_1\frac{1}{\lambda\Gamma(\alpha)}\left|\int_0^1 s^{\alpha-1}(1 - e^{-\lambda s})dK_1(s)\right| \right.$$

$$\left. \left. + L_2\frac{1}{\mu\Gamma(\beta)}\left|\int_0^1 s^{\beta-1}(1 - e^{-\mu s})dK_2(s)\right| \right] \right\}$$

$$+ \left| \int_0^{t_2} e^{-\lambda(t_2-s)}I_{0+}^{\alpha-1}f(s,u(s),v(s),{^cD_{0+}^{p_1}}v(s),I_{0+}^{q_1}v(s))\,ds \right.$$

$$\left. - \int_0^{t_1} e^{-\lambda(t_1-s)}I_{0+}^{\alpha-1}f(s,u(s),v(s),{^cD_{0+}^{p_1}}v(s),I_{0+}^{q_1}v(s))\,ds \right|$$

$$= \frac{|\lambda^2(t_2^2 - t_1^2) - 2\lambda(t_2 - t_1) + 2 - 2e^{-\lambda t_2} - 2 + 2e^{-\lambda t_1}|}{\lambda^3|\Delta_0|}\Lambda_0$$

$$+ \frac{1}{\Gamma(\alpha - 1)}\int_0^{t_1}(e^{-\lambda(t_2-s)} - e^{-\lambda(t_1-s)})\left|\int_0^s (s - \tau)^{\alpha-2}\right.$$

$$\times |f(\tau, u(\tau), v(\tau), {}^cD_{0+}^{p_1}v(\tau), I_{0+}^{q_1}v(\tau))|\, d\tau \Big|\, ds$$

$$+\frac{1}{\Gamma(\alpha-1)} \int_{t_1}^{t_2} e^{-\lambda(t_2-s)} \Big| \int_0^s (s-\tau)^{\alpha-2}$$

$$\times f(\tau, u(\tau), v(\tau), {}^cD_{0+}^{p_1}v(\tau), I_{0+}^{q_1}v(\tau))\, d\tau \Big|\, ds$$

$$\leq \frac{\lambda^2(t_2^2-t_1^2)+2\lambda(t_2-t_1)+2(e^{-\lambda t_1}-e^{-\lambda t_2})}{\lambda^3|\Delta_0|}\Lambda_0$$

$$+\frac{L_1}{\Gamma(\alpha-1)} \int_0^{t_1} \left[e^{-\lambda(t_2-s)} - e^{-\lambda(t_1-s)} \right] \left(\int_0^s (s-\tau)^{\alpha-2}\, d\tau \right) ds$$

$$+\frac{L_1}{\Gamma(\alpha-1)} \int_{t_1}^{t_2} e^{-\lambda(t_2-s)} \left(\int_0^s (s-\tau)^{\alpha-2}\, d\tau \right) ds$$

$$\leq \frac{\lambda^2(t_2^2-t_1^2)+2\lambda(t_2-t_1)+2e^{-\lambda t_1}(1-e^{-\lambda(t_2-t_1)})}{\lambda^3|\Delta_0|}\Lambda_0$$

$$+\frac{L_1 t_1^{\alpha-1}}{\Gamma(\alpha)} \int_0^{t_1} \left[e^{-\lambda(t_2-s)} - e^{-\lambda(t_1-s)} \right] ds + \frac{L_1 t_2^{\alpha-1}}{\Gamma(\alpha)} \int_{t_1}^{t_2} e^{-\lambda(t_2-s)} ds$$

$$= \frac{\lambda^2(t_2^2-t_1^2)+2\lambda(t_2-t_1)+2e^{-\lambda t_1}(1-e^{-\lambda(t_2-t_1)})}{\lambda^3|\Delta_0|}\Lambda_0$$

$$+\frac{L_1 t_1^{\alpha-1}}{\lambda\Gamma(\alpha)} \left(e^{-\lambda(t_2-t_1)} - e^{-\lambda t_2} - 1 + e^{-\lambda t_1} \right)$$

$$+\frac{L_1 t_2^{\alpha-1}}{\lambda\Gamma(\alpha)} \left(1 - e^{-\lambda(t_2-t_1)} \right),$$

where

$$\Lambda_0 = |d| \left[L_1 \frac{1}{\lambda\Gamma(\alpha)}(1-e^{-\lambda}) + L_1 \frac{1}{\lambda\Gamma(\alpha)} \left| \int_0^1 s^{\alpha-1}(1-e^{-\lambda s})dH_1(s) \right| \right.$$

$$+L_2 \frac{1}{\mu\Gamma(\beta)} \left| \int_0^1 s^{\beta-1}(1-e^{-\mu s})dH_2(s) \right| \right] + |b| \left[L_2 \frac{1}{\mu\Gamma(\beta)}(1-e^{-\mu}) \right.$$

$$+L_1 \frac{1}{\lambda\Gamma(\alpha)} \left| \int_0^1 s^{\alpha-1}(1-e^{-\lambda s})dK_1(s) \right|$$

$$+L_2 \frac{1}{\mu\Gamma(\beta)} \left| \int_0^1 s^{\beta-1}(1-e^{-\mu s})dK_2(s) \right| \right].$$

Then $|\mathcal{T}_1(u,v)(t_2) - \mathcal{T}_1(u,v)(t_1)| \to 0$, as $t_2 \to t_1$.

In addition, we obtain

$$|{}^cD_{0+}^{p_2}\mathcal{T}_1(u,v)(t_2) - {}^cD_{0+}^{p_2}\mathcal{T}_1(u,v)(t_1)|$$

$$\leq \frac{1}{\Gamma(1-p_2)}\int_0^{t_1}\frac{|(t_1-s)^{p_2}-(t_2-s)^{p_2}|}{(t_1-s)^{p_2}(t_2-s)^{p_2}}|\mathcal{T}_1'(u,v)(s)|\,ds$$

$$+\frac{1}{\Gamma(1-p_2)}\int_{t_1}^{t_2}(t_2-s)^{-p_2}|\mathcal{T}_1'(u,v)(s)|ds$$

$$\leq \frac{L_1M_1^*+L_2M_2^*}{\Gamma(1-p_2)}\left[\int_0^{t_1}\frac{-(t_1-s)^{p_2}+(t_2-s)^{p_2}}{(t_1-s)^{p_2}(t_2-s)^{p_2}}\,ds+\int_{t_1}^{t_2}(t_2-s)^{-p_2}ds\right]$$

$$=\frac{L_1M_1^*+L_2M_2^*}{\Gamma(2-p_2)}\left(2(t_2-t_1)^{1-p_2}-t_2^{1-p_2}+t_1^{1-p_2}\right)\to 0,\quad\text{as }t_2\to t_1.$$

In a similar manner, we have

$$|\mathcal{T}_2(u,v)(t_2)-\mathcal{T}_2(u,v)(t_1)|\to 0\quad\text{and,}$$

$$|{}^cD_{0+}^{p_1}\mathcal{T}_2(u,v)(t_2)-{}^cD_{0+}^{p_1}\mathcal{T}_2(u,v)(t_1)|\to 0,\quad\text{as }t_2\to t_1.$$

Thus the operators $\mathcal{T}_1$ and $\mathcal{T}_2$ are equicontinuous, and then $\mathcal{T}$ is also equicontinuous. Hence by Arzela-Ascoli theorem, we deduce that $\mathcal{T}$ is compact. Therefore we conclude that $\mathcal{T}$ is completely continuous.

Now we will prove that the set $F = \{(u,v) \in \mathcal{X}\times\mathcal{Y},\ (u,v) = \nu\mathcal{T}(u,v),\ 0 < \nu < 1\}$ is bounded. Let $(u,v) \in F$, that is $(u,v) = \nu\mathcal{T}(u,v)$ for some $\nu \in (0,1)$. Then for any $t \in [0,1]$, we have $u(t) = \nu\mathcal{T}_1(u,v)(t)$, $v(t) = \nu\mathcal{T}_2(u,v)(t)$. From these last relations we deduce $|u(t)| \leq |\mathcal{T}_1(u,v)(t)|$ and $|v(t)| \leq |\mathcal{T}_2(u,v)(t)|$ for all $t \in [0,1]$.

Then by (J2), we obtain

$$|u(t)| \leq |\gamma_1(t)|\left\{|d|\left[\int_0^1 e^{-\lambda(1-s)}\left(\frac{1}{\Gamma(\alpha-1)}\int_0^s(s-\tau)^{\alpha-2}\right.\right.\right.$$

$$\times\,[a_0+a_1|u(\tau)|+a_2|v(\tau)|$$

$$\left.\left.+a_3|{}^cD_{0+}^{p_1}v(\tau)|+a_4|I_{0+}^{q_1}v(\tau)|]\,d\tau\right)ds$$

$$+\left|\int_0^1\left(\int_0^s e^{-\lambda(s-\tau)}\left(\frac{1}{\Gamma(\alpha-1)}\int_0^\tau(\tau-\varsigma)^{\alpha-2}\right.\right.\right.$$

$$\times\,[a_0+a_1|u(\varsigma)|+a_2|v(\varsigma)|+a_3|{}^cD_{0+}^{p_1}v(\varsigma)|$$

$$\left.\left.\left.+a_4|I_{0+}^{q_1}v(\varsigma)|]\,d\varsigma\right)d\tau\right)dH_1(s)\right|$$

$$+ \left| \int_0^1 \left(\int_0^s e^{-\mu(s-\tau)} \left(\frac{1}{\Gamma(\beta-1)} \int_0^\tau (\tau-\zeta)^{\beta-2} \right. \right. \right.$$

$$\times \left[b_0 + b_1|u(\zeta)| + b_2|{}^cD_{0+}^{p_2}u(\zeta)| + b_3|I_{0+}^{q_2}u(\zeta)| \right.$$

$$\left. \left. \left. + b_4|v(\zeta)| \right] d\zeta \right) d\tau \right) dH_2(s) \right| \Big]$$

$$+ |b| \left[\int_0^1 e^{-\mu(1-s)} \left(\frac{1}{\Gamma(\beta-1)} \int_0^s (s-\tau)^{\beta-2} \right. \right.$$

$$\times \left[b_0 + b_1|u(\tau)| + b_2|{}^cD_{0+}^{p_2}u(\tau)| \right.$$

$$\left. \left. + b_3|I_{0+}^{q_2}u(\tau)| + b_4|v(\tau)| \right] d\tau \right) ds$$

$$+ \left| \int_0^1 \left(\int_0^s e^{-\lambda(s-\tau)} \left(\frac{1}{\Gamma(\alpha-1)} \int_0^\tau (\tau-\zeta)^{\alpha-2} \right. \right. \right.$$

$$\times \left[a_0 + a_1|u(\zeta)| + a_2|v(\zeta)| + a_3|{}^cD_{0+}^{p_1}v(\zeta)| \right.$$

$$\left. \left. \left. + a_4|I_{0+}^{q_1}v(\zeta)| \right] d\zeta \right) d\tau \right) dK_1(s) \right|$$

$$+ \left| \int_0^1 \left(\int_0^s e^{-\mu(s-\tau)} \left(\frac{1}{\Gamma(\beta-1)} \int_0^\tau (\tau-\zeta)^{\beta-2} \right. \right. \right.$$

$$\times \left[b_0 + b_1|u(\zeta)| + b_2|{}^cD_{0+}^{p_2}u(\zeta)| + b_3|I_{0+}^{q_2}u(\zeta)| \right.$$

$$\left. \left. \left. + b_4|v(\zeta)| \right] d\zeta \right) d\tau \right) dK_2(s) \right| \right] \Big\}$$

$$+ \int_0^t e^{-\lambda(t-s)} \left(\frac{1}{\Gamma(\alpha-1)} \int_0^s (s-\tau)^{\alpha-2} \right.$$

$$\times \left[a_0 + a_1|u(\tau)| + a_2|v(\tau)| + a_3|{}^cD_{0+}^{p_1}v(\tau)| \right.$$

$$\left. \left. + a_4|I_{0+}^{q_1}v(\tau)| \right] d\tau \right) ds, \quad \forall t \in [0,1].$$

Therefore, we deduce

$$\|u\| \leq l_1 \left\{ |d| \left[\left(\int_0^1 e^{-\lambda(1-s)} \left(\frac{1}{\Gamma(\alpha-1)} \int_0^s (s-\tau)^{\alpha-2} d\tau \right) ds \right) \right. \right.$$

$$\times \left(a_0 + a_1\|u\|_\mathcal{X} + a_2\|v\|_\mathcal{Y} + a_3\|v\|_\mathcal{Y} + \frac{a_4}{\Gamma(q_1+1)}\|v\|_\mathcal{Y} \right)$$

$$+ \left| \int_0^1 \left(\int_0^s e^{-\lambda(s-\tau)} \left(\frac{1}{\Gamma(\alpha-1)} \int_0^\tau (\tau-\zeta)^{\alpha-2} d\zeta \right) d\tau \right) dH_1(s) \right|$$

$$\times \left(a_0 + a_1\|u\|_{\mathcal{X}} + a_2\|v\|_{\mathcal{Y}} + a_3\|v\|_{\mathcal{Y}} + \frac{a_4}{\Gamma(q_1 + 1)}\|v\|_{\mathcal{Y}} \right)$$

$$+ \left| \int_0^1 \left(\int_0^s e^{-\mu(s-\tau)} \left(\frac{1}{\Gamma(\beta - 1)} \int_0^\tau (\tau - \zeta)^{\beta-2} d\zeta \right) d\tau \right) dH_2(s) \right|$$

$$\times \left(b_0 + b_1\|u\|_{\mathcal{X}} + b_2\|u\|_{\mathcal{X}} + \frac{b_3}{\Gamma(q_2 + 1)}\|u\|_{\mathcal{X}} + b_4\|v\|_{\mathcal{Y}} \right) \Bigg]$$

$$+ |b| \left[\left(\int_0^1 e^{-\mu(1-s)} \left(\frac{1}{\Gamma(\beta - 1)} \int_0^s (s - \tau)^{\beta-2} d\tau \right) ds \right) \right.$$

$$\times \left(b_0 + b_1\|u\|_{\mathcal{X}} + b_2\|u\|_{\mathcal{X}} + \frac{b_3}{\Gamma(q_2 + 1)}\|u\|_{\mathcal{X}} + b_4\|v\|_{\mathcal{Y}} \right)$$

$$+ \left| \int_0^1 \left(\int_0^s e^{-\lambda(s-\tau)} \left(\frac{1}{\Gamma(\alpha - 1)} \int_0^\tau (\tau - \zeta)^{\alpha-2} d\zeta \right) d\tau \right) dK_1(s) \right|$$

$$\times \left(a_0 + a_1\|u\|_{\mathcal{X}} + a_2\|v\|_{\mathcal{Y}} + a_3\|v\|_{\mathcal{Y}} + \frac{a_4}{\Gamma(q_1 + 1)}\|v\|_{\mathcal{Y}} \right)$$

$$+ \left| \int_0^1 \left(\int_0^s e^{-\mu(s-\tau)} \left(\frac{1}{\Gamma(\beta - 1)} \int_0^\tau (\tau - \zeta)^{\beta-2} d\zeta \right) d\tau \right) dK_2(s) \right|$$

$$\times \left(b_0 + b_1\|u\|_{\mathcal{X}} + b_2\|u\|_{\mathcal{X}} + \frac{b_3}{\Gamma(q_2 + 1)}\|u\|_{\mathcal{X}} + b_4\|v\|_{\mathcal{Y}} \right) \Bigg] \Bigg\}$$

$$+ \left(\int_0^t e^{-\lambda(t-s)} \left(\frac{1}{\Gamma(\alpha - 1)} \int_0^s (s - \tau)^{\alpha-2} d\tau \right) ds \right)$$

$$\times \left(a_0 + a_1\|u\|_{\mathcal{X}} + a_2\|v\|_{\mathcal{Y}} + a_3\|v\|_{\mathcal{Y}} + \frac{a_4}{\Gamma(q_1 + 1)}\|v\|_{\mathcal{Y}} \right)$$

$$\leq l_1 \left\{ |d| \left[\frac{1}{\lambda\Gamma(\alpha)} (1 - e^{-\lambda}) \left(a_0 + a_1\|u\|_{\mathcal{X}} + a_2\|v\|_{\mathcal{Y}} \right. \right. \right.$$

$$+ a_3\|v\|_{\mathcal{Y}} + \frac{a_4}{\Gamma(q_1 + 1)}\|v\|_{\mathcal{Y}} \Bigg)$$

$$+ \frac{1}{\lambda\Gamma(\alpha)} \left| \int_0^1 s^{\alpha-1}(1 - e^{-\lambda s}) dH_1(s) \right| \left(a_0 + a_1\|u\|_{\mathcal{X}} \right.$$

$$+ a_2\|v\|_{\mathcal{Y}} + a_3\|v\|_{\mathcal{Y}} + \frac{a_4}{\Gamma(q_1 + 1)}\|v\|_{\mathcal{Y}} \Bigg)$$

$$+ \frac{1}{\mu\Gamma(\beta)} \left| \int_0^1 s^{\beta-1}(1 - e^{-\mu s}) dH_2(s) \right| \left(b_0 + b_1\|u\|_{\mathcal{X}} + b_2\|u\|_{\mathcal{X}} \right.$$

$$+ \frac{b_3}{\Gamma(q_2+1)}\|u\|_{\mathcal{X}} + b_4\|v\|_{\mathcal{Y}}\Big)\Big]$$

$$+ |b|\left[\frac{1}{\mu\Gamma(\beta)}(1 - e^{-\mu})\left(b_0 + b_1\|u\|_{\mathcal{X}} + b_2\|u\|_{\mathcal{X}}\right.\right.$$

$$+ \frac{b_3}{\Gamma(q_2+1)}\|u\|_{\mathcal{X}} + b_4\|v\|_{\mathcal{Y}}\Big)$$

$$+ \frac{1}{\lambda\Gamma(\alpha)}\left|\int_0^1 s^{\alpha-1}(1 - e^{-\lambda s})dK_1(s)\right|\left(a_0 + a_1\|u\|_{\mathcal{X}}\right.$$

$$+ a_2\|v\|_{\mathcal{Y}} + a_3\|v\|_{\mathcal{Y}} + \frac{a_4}{\Gamma(q_1+1)}\|v\|_{\mathcal{Y}}\Big)$$

$$+ \frac{1}{\mu\Gamma(\beta)}\left|\int_0^1 s^{\beta-1}(1 - e^{-\mu s})dK_2(s)\right|\left(b_0 + b_1\|u\|_{\mathcal{X}}\right.$$

$$\left.\left.\left. + b_2\|u\|_{\mathcal{X}} + \frac{b_3}{\Gamma(q_2+1)}\|u\|_{\mathcal{X}} + b_4\|v\|_{\mathcal{Y}}\right)\right]\right\}$$

$$+ \frac{1}{\lambda\Gamma(\alpha)}(1 - e^{-\lambda})\left(a_0 + a_1\|u\|_{\mathcal{X}} + a_2\|v\|_{\mathcal{Y}}\right.$$

$$+ a_3\|v\|_{\mathcal{Y}} + \frac{a_4}{\Gamma(q_1+1)}\|v\|_{\mathcal{Y}}\Big)$$

$$= M_1\left[a_0 + a_1\|u\|_{\mathcal{X}} + \left(a_2 + a_3 + \frac{a_4}{\Gamma(q_1+1)}\right)\|v\|_{\mathcal{Y}}\right]$$

$$+ M_2\left[b_0 + \left(b_1 + b_2 + \frac{b_3}{\Gamma(q_2+1)}\right)\|u\|_{\mathcal{X}} + b_4\|v\|_{\mathcal{Y}}\right].$$

In a similar manner, we obtain

$$\|u'\| \le M_1^*\left[a_0 + a_1\|u\|_{\mathcal{X}} + \left(a_2 + a_3 + \frac{a_4}{\Gamma(q_1+1)}\right)\|v\|_{\mathcal{Y}}\right]$$

$$+ M_2^*\left[b_0 + \left(b_1 + b_2 + \frac{b_3}{\Gamma(q_2+1)}\right)\|u\|_{\mathcal{X}} + b_4\|v\|_{\mathcal{Y}}\right],$$

which gives us

$$\|^cD_{0+}^{p_2}u\| \le \frac{1}{\Gamma(2-p_2)}\left\{M_1^*\left[a_0 + a_1\|u\|_{\mathcal{X}} + \left(a_2 + a_3 + \frac{a_4}{\Gamma(q_1+1)}\right)\|v\|_{\mathcal{Y}}\right]\right.$$

$$\left. + M_2^*\left[b_0 + \left(b_1 + b_2 + \frac{b_3}{\Gamma(q_2+1)}\right)\|u\|_{\mathcal{X}} + b_4\|v\|_{\mathcal{Y}}\right]\right\}.$$

Then we conclude

$$\|u\|_{\mathcal{X}} = \|u\| + \|^c D_{0+}^{p_2} u\| \le \left(M_1 + \frac{M_1^*}{\Gamma(2-p_2)} \right) \left[a_0 + a_1 \|u\|_{\mathcal{X}} \right.$$

$$+ \left(a_2 + a_3 + \frac{a_4}{\Gamma(q_1+1)} \right) \|v\|_{\mathcal{Y}} \right] + \left(M_2 + \frac{M_2^*}{\Gamma(2-p_2)} \right)$$

$$\times \left[b_0 + \left(b_1 + b_2 + \frac{b_3}{\Gamma(q_2+1)} \right) \|u\|_{\mathcal{X}} + b_4 \|v\|_{\mathcal{Y}} \right]. \qquad (7.34)$$

In a similar manner, we deduce for the function v

$$\|v\|_{\mathcal{Y}} = \|v\| + \|^c D_{0+}^{p_1} v\| \le \left(\widetilde{M}_1 + \frac{\widetilde{M}_1^*}{\Gamma(2-p_1)} \right) \left[a_0 + a_1 \|u\|_{\mathcal{X}} \right.$$

$$+ \left(a_2 + a_3 + \frac{a_4}{\Gamma(q_1+1)} \right) \|v\|_{\mathcal{Y}} \right] + \left(\widetilde{M}_2 + \frac{\widetilde{M}_2^*}{\Gamma(2-p_1)} \right)$$

$$\times \left[b_0 + \left(b_1 + b_2 + \frac{b_3}{\Gamma(q_2+1)} \right) \|u\|_{\mathcal{X}} + b_4 \|v\|_{\mathcal{Y}} \right]. \qquad (7.35)$$

Therefore, by (7.34) and (7.35), we obtain

$$\|(u,v)\|_{\mathcal{X} \times \mathcal{Y}} = \|u\|_{\mathcal{X}} + \|v\|_{\mathcal{Y}}$$

$$\le \|u\|_{\mathcal{X}} \left[a_1 \left(M_1 + \widetilde{M}_1 + \frac{M_1^*}{\Gamma(2-p_2)} + \frac{\widetilde{M}_1^*}{\Gamma(2-p_1)} \right) \right.$$

$$+ \left(b_1 + b_2 + \frac{b_3}{\Gamma(q_2+1)} \right)$$

$$\times \left(M_2 + \widetilde{M}_2 + \frac{M_2^*}{\Gamma(2-p_2)} + \frac{\widetilde{M}_2^*}{\Gamma(2-p_1)} \right) \right]$$

$$+ \|v\|_{\mathcal{Y}} \left[\left(a_2 + a_3 + \frac{a_4}{\Gamma(q_1+1)} \right) \right.$$

$$\times \left(M_1 + \widetilde{M}_1 + \frac{M_1^*}{\Gamma(2-p_2)} + \frac{\widetilde{M}_1^*}{\Gamma(2-p_1)} \right)$$

$$+ b_4 \left(M_2 + \widetilde{M}_2 + \frac{M_2^*}{\Gamma(2-p_2)} + \frac{\widetilde{M}_2^*}{\Gamma(2-p_1)} \right) \right]$$

$$+ a_0 \left(M_1 + \widetilde{M}_1 + \frac{M_1^*}{\Gamma(2-p_2)} + \frac{\widetilde{M}_1^*}{\Gamma(2-p_1)} \right)$$

$$+ b_0 \left(M_2 + \widetilde{M}_2 + \frac{M_2^*}{\Gamma(2 - p_2)} + \frac{\widetilde{M}_2^*}{\Gamma(2 - p_1)} \right)$$

$$\leq \Lambda_1 + \max\{\Lambda_2, \Lambda_3\} \|(u, v)\|_{\mathcal{X} \times \mathcal{Y}}. \tag{7.36}$$

So by using the assumption $\max\{\Lambda_2, \Lambda_3\} < 1$, we conclude

$$\|(u, v)\|_{\mathcal{X} \times \mathcal{Y}} \leq \frac{\Lambda_1}{1 - \max\{\Lambda_2, \Lambda_3\}}.$$

Hence, we deduce that the set F is bounded.

By using Theorem 1.2.5, we conclude that the operator $\mathcal{T}$ has at least one fixed point, which is a solution for our problem (7.24), (7.25). This completes the proof. $\qquad \square$

Next, we will present an existence and uniqueness result for our problem (7.24), (7.25), based on the Banach contraction mapping principle (Theorem 1.2.1).

We introduce the constants

$$r_1 = \sup_{t \in [0,1]} |f(t, 0, 0, 0, 0)|, \quad r_2 = \sup_{t \in [0,1]} |g(t, 0, 0, 0, 0)|,$$

$$A_1 = c_0 \varrho_1 M_1 + d_0 \varrho_2 M_2, \quad A_1^* = c_0 \varrho_1 M_1^* + d_0 \varrho_2 M_2^*,$$

$$\widetilde{A}_1 = c_0 \varrho_1 \widetilde{M}_1 + d_0 \varrho_2 \widetilde{M}_2, \quad \widetilde{A}_1^* = c_0 \varrho_1 \widetilde{M}_1^* + d_0 \varrho_2 \widetilde{M}_2^*,$$

$$D_1 = r_1 M_1 + r_2 M_2, \quad D_1^* = r_1 M_1^* + r_2 M_2^*,$$

$$\widetilde{D}_1 = r_1 \widetilde{M}_1 + r_2 \widetilde{M}_2, \quad \widetilde{D}_1^* = r_1 \widetilde{M}_1^* + r_2 \widetilde{M}_2^*.$$

Theorem 7.2.2. *Assume that* (J1) *and* (J3) *hold. If*

$$A_1 + \widetilde{A}_1 + \frac{A_1^*}{\Gamma(2 - p_2)} + \frac{\widetilde{A}_1^*}{\Gamma(2 - p_1)} < 1, \tag{7.37}$$

then problem (7.24), (7.25) *has a unique solution.*

Proof. We consider the positive number r given by

$$r = \left(D_1 + \widetilde{D}_1 + \frac{D_1^*}{\Gamma(2 - p_2)} + \frac{\widetilde{D}_1^*}{\Gamma(2 - p_1)} \right)$$

$$\times \left(1 - \left(A_1 + \widetilde{A}_1 + \frac{A_1^*}{\Gamma(2 - p_2)} + \frac{\widetilde{A}_1^*}{\Gamma(2 - p_1)} \right) \right)^{-1}.$$

We will show that $\mathcal{T}(\overline{B}_r) \subset \overline{B}_r$, where $\overline{B}_r = \{(u,v) \in \mathcal{X} \times \mathcal{Y},$ $\|(u,v)\|_{\mathcal{X} \times \mathcal{Y}} \leq r\}$. For $(u,v) \in \overline{B}_r$, we obtain

$$|f(t, u(t), v(t), {}^c D_{0+}^{p_1} v(t), I_{0+}^{q_1} v(t))|$$

$$\leq |f(t, u(t), v(t), {}^c D_{0+}^{p_1} v(t), I_{0+}^{q_1} v(t)) - f(t, 0, 0, 0, 0)|$$

$$+ |f(t, 0, 0, 0, 0)| \leq c_0(|u(t)| + |v(t)| + |{}^c D_{0+}^{p_1} v(t)| + |I_{0+}^{q_1} v(t)|) + r_1$$

$$\leq c_0 \left(\|u\|_{\mathcal{X}} + \|v\|_{\mathcal{Y}} + \frac{1}{\Gamma(q_1+1)} \|v\|_{\mathcal{Y}} \right) + r_1$$

$$= c_0(\|u\|_{\mathcal{X}} + \varrho_1 \|v\|_{\mathcal{Y}}) + r_1 \leq c_0 \varrho_1 \|(u,v)\|_{\mathcal{X} \times \mathcal{Y}} + r_1 \leq c_0 \varrho_1 r + r_1.$$

In a similar manner, we have

$$|g(t, u(t), {}^c D_{0+}^{p_2} u(t), I_{0+}^{q_2} u(t), v(t))| \leq d_0 \varrho_2 r + r_2.$$

Then

$$|\mathcal{T}_1(u,v)(t)| \leq (c_0 \varrho_1 r + r_1) M_1 + (d_0 \varrho_2 r + r_2) M_2$$

$$= (c_0 \varrho_1 M_1 + d_0 \varrho_2 M_2) r + r_1 M_1 + r_2 M_2 = A_1 r + D_1,$$

$$\forall t \in [0,1],$$

and

$$|\mathcal{T}_1'(u,v)(t)| \leq (c_0 \varrho_1 r + r_1) M_1^* + (d_0 \varrho_2 r + r_2) M_2^*$$

$$= (c_0 \varrho_1 M_1^* + d_0 \varrho_2 M_2^*) r + r_1 M_1^* + r_2 M_2^* = A_1^* r + D_1^*,$$

$$\forall t \in [0,1],$$

which gives us

$$|{}^c D_{0+}^{p_2} \mathcal{T}_1(u,v)(t)| \leq \int_0^t \frac{(t-s)^{-p_2}}{\Gamma(1-p_2)} |\mathcal{T}_1'(u,v)(s)| ds$$

$$\leq \frac{1}{\Gamma(2-p_2)} (A_1^* r + D_1^*), \quad \forall t \in [0,1].$$

Therefore, we deduce

$$\|\mathcal{T}_1(u,v)\|_{\mathcal{X}} = \|\mathcal{T}_1(u,v)\| + \|{}^c D_{0+}^{p_2} \mathcal{T}_1(u,v)\|$$

$$\leq \left(A_1 + \frac{A_1^*}{\Gamma(2-p_2)} \right) r + D_1 + \frac{D_1^*}{\Gamma(2-p_2)}. \tag{7.38}$$

In a similar manner, we obtain

$$|T_2(u,v)(t)| \le \tilde{A}_1 r + \tilde{D}_1, \quad |T_2'(u,v)(t)| \le \tilde{A}_1^* r + \tilde{D}_1^*,$$

$$|{}^c D_{0+}^{p_1} T_2(u,v)(t)| \le \int_0^t \frac{(t-s)^{-p_1}}{\Gamma(1-p_1)} |T_2'(u,v)(s)|\, ds$$

$$\le \frac{1}{\Gamma(2-p_1)}(\tilde{A}_1^* r + \tilde{D}_1^*), \quad \forall t \in [0,1].$$

Then, we conclude

$$\|T_2(u,v)\|_Y = \|T_2(u,v)\| + \|{}^c D_{0+}^{p_1} T_2(u,v)\|$$

$$\le \left(\tilde{A}_1 + \frac{\tilde{A}_1^*}{\Gamma(2-p_1)}\right) r + \tilde{D}_1 + \frac{\tilde{D}_1^*}{\Gamma(2-p_1)}. \qquad (7.39)$$

By relations (7.38) and (7.39), we deduce

$$\|T(u,v)\|_{X \times Y} = \|T_1(u,v)\|_X + \|T_2(u,v)\|_Y$$

$$\le \left(A_1 + \tilde{A}_1 + \frac{A_1^*}{\Gamma(2-p_2)} + \frac{\tilde{A}_1^*}{\Gamma(2-p_1)}\right) r$$

$$+ D_1 + \tilde{D}_1 + \frac{D_1^*}{\Gamma(2-p_2)} + \frac{\tilde{D}_1^*}{\Gamma(2-p_1)} = r,$$

which implies $T(\overline{B}_r) \subset \overline{B}_r$.

Next, we prove that the operator T is a contraction. For $(u_i, v_i) \in \overline{B}_r$, $i = 1, 2$, and for each $t \in [0,1]$, we have

$$|T_1(u_1, v_1)(t) - T_1(u_2, v_2)(t)|$$

$$\le |\gamma_1(t)| \left\{ |d| \left[\int_0^1 e^{-\lambda(1-s)} \left(\frac{1}{\Gamma(\alpha-1)} \int_0^s (s-\tau)^{\alpha-2} \right. \right. \right.$$

$$\times |f(\tau, u_1(\tau), v_1(\tau), {}^c D_{0+}^{p_1} v_1(\tau), I_{0+}^{q_1} v_1(\tau))$$

$$\left. - f(\tau, u_2(\tau), v_2(\tau), {}^c D_{0+}^{p_1} v_2(\tau), I_{0+}^{q_1} v_2(\tau))| \, d\tau \right) ds$$

$$+ \left| \int_0^1 \left(\int_0^s e^{-\lambda(s-\tau)} \left(\frac{1}{\Gamma(\alpha-1)} \int_0^\tau (\tau-\zeta)^{\alpha-2} \right. \right. \right.$$

$$\times |f(\zeta, u_1(\zeta), v_1(\zeta), {}^c D_{0+}^{p_1} v_1(\zeta), I_{0+}^{q_1} v_1(\zeta))$$

$$\left. - f(\zeta, u_2(\zeta), v_2(\zeta), {}^c D_{0+}^{p_1} v_2(\zeta), I_{0+}^{q_1} v_2(\zeta))| \, d\zeta \right) d\tau \right) dH_1(s) \right|$$

$$+ \left| \int_0^1 \left(\int_0^s e^{-\mu(s-\tau)} \left(\frac{1}{\Gamma(\beta-1)} \int_0^\tau (\tau - \zeta)^{\beta-2} \right. \right. \right.$$

$$\times \left| g(\zeta, u_1(\zeta), {}^c D_{0+}^{p_2} u_1(\zeta), I_{0+}^{q_2} u_1(\zeta), v_1(\zeta)) \right.$$

$$\left. \left. \left. - g(\zeta, u_2(\zeta), {}^c D_{0+}^{p_2} u_2(\zeta), I_{0+}^{q_2} u_2(\zeta), v_2(\zeta)) \right| d\zeta \right) d\tau \right) dH_2(s) \right| \right]$$

$$+ |b| \left[\int_0^1 e^{-\mu(1-s)} \left(\frac{1}{\Gamma(\beta-1)} \int_0^s (s-\tau)^{\beta-2} \right. \right.$$

$$\times \left| g(\tau, u_1(\tau), {}^c D_{0+}^{p_2} u_1(\tau), I_{0+}^{q_2} u_1(\tau), v_1(\tau)) \right.$$

$$\left. \left. - g(\tau, u_2(\tau), {}^c D_{0+}^{p_2} u_2(\tau), I_{0+}^{q_2} u_2(\tau), v_2(\tau)) \right| d\tau \right) ds$$

$$+ \left| \int_0^1 \left(\int_0^s e^{-\lambda(s-\tau)} \left(\frac{1}{\Gamma(\alpha-1)} \int_0^\tau (\tau - \zeta)^{\alpha-2} \right. \right. \right.$$

$$\times \left| f(\zeta, u_1(\zeta), v_1(\zeta), {}^c D_{0+}^{p_1} v_1(\zeta), I_{0+}^{q_1} v_1(\zeta)) \right.$$

$$\left. \left. \left. - f(\zeta, u_2(\zeta), v_2(\zeta), {}^c D_{0+}^{p_1} v_2(\zeta), I_{0+}^{q_1} v_2(\zeta)) \right| d\zeta \right) d\tau \right) dK_1(s) \right|$$

$$+ \left| \int_0^1 \left(\int_0^s e^{-\mu(s-\tau)} \left(\frac{1}{\Gamma(\beta-1)} \int_0^\tau (\tau - \zeta)^{\beta-2} \right. \right. \right.$$

$$\times \left| g(\zeta, u_1(\zeta), {}^c D_{0+}^{p_2} u_1(\zeta), I_{0+}^{q_2} u_1(\tau), v_1(\zeta)) \right.$$

$$\left. \left. \left. - g(\zeta, u_2(\zeta), {}^c D_{0+}^{p_2} u_2(\zeta), I_{0+}^{q_2} u_2(\zeta), v_2(\zeta)) | d\zeta \right) d\tau \right) dK_2(s) \right| \right] \right\}$$

$$+ \int_0^t e^{-\lambda(t-s)} \left(\frac{1}{\Gamma(\alpha-1)} \int_0^s (s-\tau)^{\alpha-2} | f(\tau, u_1(\tau), v_1(\tau), \right.$$

$${}^c D_{0+}^{p_1} v_1(\tau), I_{0+}^{q_1} v_1(\tau)) - f(\tau, u_2(\tau), v_2(\tau), {}^c D_{0+}^{p_1} v_2(\tau), I_{0+}^{q_1} v_2(\tau)) | d\tau \right) ds$$

$$\leq M_1 c_0 (\|u_1 - u_2\| + \|v_1 - v_2\| + \|{}^c D_{0+}^{p_1} v_1$$

$$- {}^c D_{0+}^{p_1} v_2\| + \|I_{0+}^{q_1} v_1 - I_{0+}^{q_1} v_2\|)$$

$$+ M_2 d_0 (\|u_1 - u_2\| + \|{}^c D_{0+}^{p_2} u_1 - {}^c D_{0+}^{p_2} u_2\|$$

$$+ \|I_{0+}^{q_2} u_1 - I_{0+}^{q_2} u_2\| + \|v_1 - v_2\|)$$

$$\leq M_1 c_0 (\|u_1 - u_2\| + \varrho_1 \|v_1 - v_2\| + \|{}^c D_{0+}^{p_1} v_1 - {}^c D_{0+}^{p_1} v_2\|)$$

$$+ M_2 d_0 (\varrho_2 \|u_1 - u_2\| + \|{}^c D_{0+}^{p_2} u_1 - {}^c D_{0+}^{p_2} u_2\| + \|v_1 - v_2\|)$$

$$\leq A_1 (\|u_1 - u_2\|_X + \|v_1 - v_2\|_Y).$$

Then we obtain

$$|\mathcal{T}_1'(u_1, v_1)(t) - \mathcal{T}_1'(u_2, v_2)(t)| \leq A_1^*(\|u_1 - u_2\|_\mathcal{X} + \|v_1 - v_2\|_\mathcal{Y}),$$

which gives us

$$|^cD_{0+}^{p_2}\mathcal{T}_1(u_1, v_1)(t) - {}^cD_{0+}^{p_2}\mathcal{T}_1(u_2, v_2)(t)|$$

$$\leq \int_0^t \frac{(t-s)^{-p_2}}{\Gamma(1-p_2)}|\mathcal{T}_1'(u_1, v_1)(s) - \mathcal{T}_1'(u_2, v_2)(s)|\, ds$$

$$\leq \frac{A_1^*}{\Gamma(2-p_2)}(\|u_1 - u_2\|_\mathcal{X} + \|v_1 - v_2\|_\mathcal{Y}).$$

From the above inequalities, we conclude

$$\|\mathcal{T}_1(u_1, v_1) - \mathcal{T}_1(u_2, v_2)\|_\mathcal{X}$$

$$= \|\mathcal{T}_1(u_1, v_1) - \mathcal{T}_1(u_2, v_2)\| + \|^cD_{0+}^{p_2}\mathcal{T}_1(u_1, v_1)$$

$$- {}^cD_{0+}^{p_2}\mathcal{T}_1(u_2, v_2)\|$$

$$\leq \left(A_1 + \frac{A_1^*}{\Gamma(2-p_2)}\right)(\|u_1 - u_2\|_\mathcal{X} + \|v_1 - v_2\|_\mathcal{Y}). \tag{7.40}$$

In a similar manner, we deduce

$$\|\mathcal{T}_2(u_1, v_1) - \mathcal{T}_2(u_2, v_2)\|_\mathcal{Y} \leq \left(\widetilde{A}_1 + \frac{\widetilde{A}_1^*}{\Gamma(2-p_1)}\right)(\|u_1 - u_2\|_\mathcal{X} + \|v_1 - v_2\|_\mathcal{Y}).$$

$$\tag{7.41}$$

Therefore, by (7.40) and (7.41) we obtain

$$\|\mathcal{T}(u_1, v_1) - \mathcal{T}(u_2, v_2)\|_{\mathcal{X} \times \mathcal{Y}}$$

$$= \|\mathcal{T}_1(u_1, v_1) - \mathcal{T}_1(u_2, v_2)\|_\mathcal{X} + \|\mathcal{T}_2(u_1, v_1) - \mathcal{T}_2(u_2, v_2)\|_\mathcal{Y}$$

$$\leq \left(A_1 + \widetilde{A}_1 + \frac{A_1^*}{\Gamma(2-p_2)} + \frac{\widetilde{A}_1^*}{\Gamma(2-p_1)}\right)(\|u_1 - u_2\|_\mathcal{X} + \|v_1 - v_2\|_\mathcal{Y}).$$

By using the condition (7.37), we deduce that $\mathcal{T}$ is a contraction. By the Banach fixed point theorem (Theorem 1.2.1), we conclude that the operator $\mathcal{T}$ has a unique fixed point, which is the unique solution of problem (7.24), (7.25). This completes the proof. $\square$

7.2.3   *Examples*

Let $\alpha = 7/2$, $\beta = 10/3$, $p_1 = 1/3$, $p_2 = 1/2$, $q_1 = 3/2$, $q_2 = 8/3$, $\lambda = 1$, $\mu = 2$, $H_1(s) = s^2/2$ for all $s \in [0,1]$,

$$H_2(s) = \begin{cases} 0, & 0 \le s < \dfrac{1}{2}, \\ 3, & \dfrac{1}{2} \le s \le 1, \end{cases} \qquad K_1(s) = \begin{cases} 0, & 0 \le s < \dfrac{1}{3}, \\ 2, & \dfrac{1}{3} \le s \le 1, \end{cases}$$

$$K_2(s) = \begin{cases} 0, & 0 \le s < \dfrac{1}{2}, \\ -\dfrac{s^3}{3}, & \dfrac{1}{2} \le s \le 1. \end{cases}$$

We consider the system of fractional differential equations

$$\begin{cases} (^cD_{0+}^{7/2} + {}^cD_{0+}^{5/2})u(t) = f(t, u(t), v(t), {}^cD_{0+}^{1/3}v(t), I_{0+}^{3/2}v(t)), & t \in (0,1), \\ (^cD_{0+}^{10/3} + 2{}^cD_{0+}^{7/3})v(t) = g(t, u(t), {}^cD_{0+}^{1/2}u(t), I_{0+}^{8/3}u(t), v(t)), & t \in (0,1), \end{cases} \tag{7.42}$$

with the boundary conditions

$$\begin{cases} u(0) = u'(0) = u''(0) = 0, & u(1) = \int_0^1 su(s)\,ds + 3v\left(\dfrac{1}{2}\right), \\ v(0) = v'(0) = v''(0) = 0, & v(1) = 2u\left(\dfrac{1}{3}\right) - \int_{1/2}^1 s^2 v(s)\,ds. \end{cases} \tag{7.43}$$

We have $a = (1 - 2e^{-1}) - \int_0^1 (s^3 - 2s^2 + 2s - 2se^{-s})\,ds \approx 0.20939002$, $b = \frac{3}{8}(1 - 2e^{-1}) \approx 0.09909042$, $c = \frac{2}{9}(13 - 18e^{-1/3}) \approx 0.02276365$, $d = \frac{1}{4}(1 - e^{-2}) + \frac{1}{4}\int_{1/2}^1 (2s^4 - 2s^3 + s^2 - s^2 e^{-2s})\,ds \approx 0.25358146$, $\Delta_0 \approx 0.05084177$ ($\Delta_0 \ne 0$), $\gamma_1(t) = \frac{1}{\Delta_0}(t^2 - 2t + 2 - 2e^{-t})$, $\gamma_2(t) = \frac{1}{4\Delta_0}(2t^2 - 2t + 1 - e^{-2t})$. Besides, we deduce $l_1 = \frac{1}{\Delta_0}(1 - 2e^{-1}) \approx 5.1973236$, $l_2 = \frac{1}{4\Delta_0}(1 - e^{-2}) \approx 4.25174399$, $m_1 = \frac{1}{\Delta_0}2e^{-1} \approx 14.47154415$, $m_2 = \frac{1}{2\Delta_0}(1 + e^{-2}) \approx 11.16537977$, $M_1 \approx 0.49529109$, $M_2 \approx 0.18317499$, $M_1^* \approx 1.34059325$, $M_2^* \approx 0.51003654$, $\widetilde{M}_1 \approx 0.03173319$, $\widetilde{M}_2 \approx 0.32453092$, $\widetilde{M}_1^* \approx 0.08333359$, $\widetilde{M}_2^* \approx 1.11476371$, $N_1 \approx 2.13203307$, $N_2 \approx 2.34109666$.

Example 1. We consider the functions

$$f(t, x_1, x_2, x_3, x_4) = \frac{t}{t^2 + 1}\left(3\cos t + \frac{1}{8}\sin(x_1 + x_2)\right)$$

$$- \frac{1}{8(t+1)^2}x_3 + \frac{1}{10}\arctan x_4,$$

$$g(t, x_1, x_2, x_3, x_4) = \frac{1}{(t+2)^3}\left(5e^{-t} + \frac{x_1}{2} + 2x_2\right)$$

$$-\frac{t}{6}\sin(x_3 + x_4),$$

for all $t \in [0,1]$, $x_1, x_2, x_3, x_4 \in \mathbb{R}$. We obtain the inequalities

$$|f(t, x_1, x_2, x_3, x_4)| \le \frac{3}{2} + \frac{1}{16}|x_1| + \frac{1}{16}|x_2| + \frac{1}{8}|x_3| + \frac{1}{10}|x_4|,$$

$$|g(t, x_1, x_2, x_3, x_4)| \le \frac{5}{8} + \frac{1}{16}|x_1| + \frac{1}{4}|x_2| + \frac{1}{6}|x_3| + \frac{1}{6}|x_4|,$$

for all $t \in [0,1]$, $x_1, x_2, x_3, x_4 \in \mathbb{R}$. So we have $a_0 = \frac{3}{2}$, $a_1 = \frac{1}{16}$, $a_2 = \frac{1}{16}$, $a_3 = \frac{1}{8}$, $a_4 = \frac{1}{10}$, $b_0 = \frac{5}{8}$, $b_1 = \frac{1}{16}$, $b_2 = \frac{1}{4}$, $b_3 = \frac{1}{6}$, $b_4 = \frac{1}{6}$.

Because $\Lambda_2 \approx 0.962094$ and $\Lambda_3 \approx 0.950322$, we deduce that the condition $\max\{\Lambda_2, \Lambda_3\} < 1$ is satisfied, and then by Theorem 7.2.1, we conclude that problem (7.42), (7.43) has at least one solution $(u(t), v(t))$, $t \in [0,1]$.

Example 2. We consider the functions

$$f(t, x_1, x_2, x_3, x_4) = \frac{t}{2} + \frac{1}{8(t+1)^2}\left(\frac{|x_1|}{1+|x_1|} - x_2\right)$$

$$+ \frac{1}{32}\sin^2 x_3 - \frac{t}{9}\arctan x_4,$$

$$g(t, x_1, x_2, x_3, x_4) = \frac{t^2}{t^3+1} - \frac{1}{16}\sin x_1 + \frac{1}{10}x_2$$

$$+ \frac{1}{8\sqrt{4+t^2}}\cos x_3 - \frac{|x_4|}{6(1+|x_4|)},$$

for all $t \in [0,1]$, $x_1, x_2, x_3, x_4 \in \mathbb{R}$.

We obtain the following inequalities

$$|f(t, x_1, x_2, x_3, x_4) - f(t, y_1, y_2, y_3, y_4)|$$

$$\le \frac{1}{8}|x_1 - y_1| + \frac{1}{8}|x_2 - y_2| + \frac{1}{16}|x_3 - y_3| + \frac{1}{9}|x_4 - y_4|$$

$$\le \frac{1}{8}(|x_1 - y_1| + |x_2 - y_2| + |x_3 - y_3| + |x_4 - y_4|),$$

$$|g(t, x_1, x_2, x_3, x_4) - g(t, y_1, y_2, y_3, y_4)|$$

$$\leq \frac{1}{16}|x_1 - y_1| + \frac{1}{10}|x_2 - y_2| + \frac{1}{16}|x_3 - y_3| + \frac{1}{6}|x_4 - y_4|$$

$$\leq \frac{1}{6}(|x_1 - y_1| + |x_2 - y_2| + |x_3 - y_3| + |x_4 - y_4|),$$

for all $t \in [0, 1]$ and $x_1, x_2, x_3, x_4 \in \mathbb{R}$.

Here $c_0 = \frac{1}{8}$ and $d_0 = \frac{1}{6}$. Besides, we deduce $\varrho_1 \approx 1.75225278$, $\varrho_2 \approx 1.24923974$, $A_1 \approx 0.14662264$, $A_1^* \approx 0.39982527$, $\widetilde{A}_1 \approx 0.07452006$, $\widetilde{A}_1^* \approx 0.25035388$, and $A_1 + \widetilde{A}_1 + \frac{A_1^*}{\Gamma(2-p_2)} + \frac{\widetilde{A}_1^*}{\Gamma(2-p_1)} \approx 0.949622 (< 1)$. Thus all the conditions of Theorem 7.2.2 are satisfied. Therefore, by Theorem 7.2.2 we conclude that problem (7.42), (7.43) has a unique solution $(u(t), v(t))$, $t \in [0, 1]$.

Remark 7.2.1. The results presented in this section were published in [7].

7.3   Caputo Fractional Differential Systems with Coupled Nonlocal Boundary Conditions

In this section, we investigate the system of nonlinear fractional differential equations

$$\begin{cases} {}^cD_{0+}^{\alpha} u(t) = f(t, v(t), {}^cD_{0+}^p v(t)), & t \in (0, 1), \\ {}^cD_{0+}^{\beta} v(t) = g(t, u(t), {}^cD_{0+}^q u(t)), & t \in (0, 1), \end{cases} \tag{7.44}$$

supplemented with the coupled nonlocal boundary conditions

$$\begin{cases} u(0) = \varphi(v), & u'(0) = 0, \ldots, u^{(n-2)}(0) = 0, & u(1) = \lambda I_{0+}^{\gamma} v(\xi), \\ v(0) = \psi(u), & v'(0) = 0, \ldots, v^{(m-2)}(0) = 0, & v(1) = \mu I_{0+}^{\delta} u(\eta), \end{cases} \tag{7.45}$$

where $\alpha \in (n-1, n]$, $\beta \in (m-1, m]$, $n, m \in \mathbb{N}$, $n, m \geq 2$, $p, q \in (0, 1)$, $\lambda, \mu \in \mathbb{R}$, $\gamma, \delta > 0$, $\xi, \eta \in [0, 1]$, $f, g : [0, 1] \times \mathbb{R} \times \mathbb{R} \to \mathbb{R}$, $\varphi, \psi : C[0, 1] \to \mathbb{R}$ are continuous functions, ${}^cD_{0+}^{\theta}$ denotes the Caputo derivative of fractional order θ ($\theta = \alpha, \beta, p, q$), and I_{0+}^{γ}, I_{0+}^{δ} are the Riemann-Liouville fractional integrals of orders γ and δ, respectively. If $n = 2$ and $m = 2$, then the boundary conditions (7.45) take the form $u(0) = \varphi(v)$, $u(1) = \lambda I_{0+}^{\gamma} v(\xi)$ and $v(0) = \psi(u)$, $v(1) = \mu I_{0+}^{\delta} u(\eta)$, respectively, that is, the derivatives of u and v at the point 0 do not appear in these conditions. Under some assumptions on the functions f and g, we will establish the existence of solutions for problem (7.44), (7.45) by using the Schauder fixed point theorem (Theorem 1.2.4) and the nonlinear alternative of Leray-Schauder type (Theorem 1.2.6).

The system (7.44) with $\alpha, \beta \in (1, 2)$ subject to the uncoupled boundary conditions

$$\begin{cases} u(0) = \varphi(u), & u(1) = aI_{0+}^{\theta_1}u(\eta), \\ v(0) = \psi(v), & v(1) = bI_{0+}^{\theta_2}v(\xi), \end{cases}$$

with $\xi, \eta \in [0, 1]$, $\theta_1, \theta_2 > 0$, $a, b \in \mathbb{R}$, was studied in [94]. In [13], the authors investigated the system of fractional differential equations equipped with nonlocal coupled boundary conditions

$$\begin{cases} {}^cD_{0+}^{\alpha}x(t) = f(t, x(t), y(t), {}^cD_{0+}^{\gamma}y(t)), & t \in [0, T], \\ {}^cD_{0+}^{\beta}y(t) = g(t, x(t), {}^cD_{0+}^{\delta}x(t), y(t)), & t \in [0, T], \\ x(0) = h(y), & \int_0^T y(s)\, ds = \mu_1 x(\eta), \\ y(0) = \phi(x), & \int_0^T x(s)\, ds = \mu_2 y(\xi), \end{cases} \tag{7.46}$$

where $\alpha, \beta \in (1, 2]$, $\gamma, \delta \in (0, 1)$, $\eta, \xi \in (0, T)$, $f, g : [0, T] \times \mathbb{R} \times \mathbb{R} \times \mathbb{R} \to \mathbb{R}$, $h, \phi : C([0, T], \mathbb{R}) \to \mathbb{R}$ are continuous functions, μ_1, μ_2 are real constants. By using the Banach contraction mapping principle and the Leray-Schauder nonlinear alternative, the authors proved the existence and uniqueness, and the existence of solutions for problem (7.46).

7.3.1 *Auxiliary results*

We consider the fractional differential system

$$\begin{cases} {}^cD_{0+}^{\alpha}u(t) = x(t), & t \in (0, 1), \\ {}^cD_{0+}^{\beta}v(t) = y(t), & t \in (0, 1), \end{cases} \tag{7.47}$$

with the coupled boundary conditions

$$\begin{cases} u(0) = u_0, & u'(0) = 0, \ldots, u^{(n-2)}(0) = 0, & u(1) = \lambda I_{0+}^{\gamma}v(\xi), \\ v(0) = v_0, & v'(0) = 0, \ldots, v^{(m-2)}(0) = 0, & v(1) = \mu I_{0+}^{\delta}u(\eta), \end{cases} \tag{7.48}$$

where $\alpha \in (n-1, n]$, $\beta \in (m-1, m]$, $n, m \in \mathbb{N}$, $n, m \geq 2$, $\lambda, \mu \in \mathbb{R}$, $\gamma, \delta > 0$, $\xi, \eta \in [0, 1]$, $u_0, v_0 \in \mathbb{R}$, and x, y are real functions.

In the sequel we set $\widetilde{\Delta} = 1 - \dfrac{\lambda\mu\xi^{\gamma+m-1}\eta^{\delta+n-1}(m-1)!(n-1)!}{\Gamma(\gamma+m)\Gamma(\delta+n)}.$

Lemma 7.3.1. *If* $x, y \in C[0,1]$, $u_0, v_0 \in \mathbb{R}$ *and* $\widetilde{\Delta} \neq 0$, *then the unique solution* $(u, v) \in C^n[0,1] \times C^m[0,1]$ *of the boundary value problem* (7.47), (7.48) *is given by*

$$u(t) = u_0 + I_{0+}^\alpha x(t) + \frac{t^{n-1}}{\widetilde{\Delta}}\left[-u_0 - I_{0+}^\alpha x(1) + \frac{\lambda v_0\xi^\gamma}{\Gamma(\gamma+1)}\right.$$

$$+ \lambda I_{0+}^{\beta+\gamma} y(\xi) - \frac{\lambda v_0\xi^{\gamma+m-1}(m-1)!}{\Gamma(\gamma+m)}$$

$$- \frac{\lambda\xi^{\gamma+m-1}(m-1)!}{\Gamma(\gamma+m)} I_{0+}^\beta y(1) + \frac{\lambda\mu u_0\xi^{\gamma+m-1}\eta^\delta(m-1)!}{\Gamma(\delta+1)\Gamma(\gamma+m)}$$

$$\left.+ \frac{\lambda\mu\xi^{\gamma+m-1}(m-1)!}{\Gamma(\gamma+m)} I_{0+}^{\alpha+\delta} x(\eta)\right], \quad t \in [0,1],$$

$$v(t) = v_0 + I_{0+}^\beta y(t) + \frac{t^{m-1}}{\widetilde{\Delta}}\left[-v_0 - I_{0+}^\beta y(1) + \frac{\mu u_0\eta^\delta}{\Gamma(\delta+1)}\right.$$

$$+ \mu I_{0+}^{\alpha+\delta} x(\eta) - \frac{\mu u_0\eta^{\delta+n-1}(n-1)!}{\Gamma(\delta+n)}$$

$$- \frac{\mu\eta^{\delta+n-1}(n-1)!}{\Gamma(\delta+n)} I_{0+}^\alpha x(1) + \frac{\lambda\mu v_0\xi^\gamma\eta^{\delta+n-1}(n-1)!}{\Gamma(\gamma+1)\Gamma(\delta+n)}$$

$$\left.+ \frac{\lambda\mu\eta^{\delta+n-1}(n-1)!}{\Gamma(\delta+n)} I_{0+}^{\beta+\gamma} y(\xi)\right], \quad t \in [0,1]. \tag{7.49}$$

Proof. The solution $u \in C^n[0,1]$ of equation ${}^cD_{0+}^\alpha u(t) = x(t)$ is

$$u(t) = \widetilde{c}_1 + \widetilde{c}_2 t + \cdots + \widetilde{c}_n t^{n-1} + I_{0+}^\alpha x(t), \quad t \in [0,1],$$

where $\widetilde{c}_1, \widetilde{c}_2, \ldots, \widetilde{c}_n \in \mathbb{R}$. By using the conditions $u(0) = u_0$, $u'(0) = \cdots = u^{(n-2)}(0) = 0$, we obtain $\widetilde{c}_1 = u_0$, $\widetilde{c}_2 = \cdots = \widetilde{c}_{n-1} = 0$, and so we deduce $u(t) = u_0 + \widetilde{c}_n t^{n-1} + I_{0+}^\alpha x(t)$, $t \in [0,1]$.

In a similar manner the solution $v \in C^m[0,1]$ of equation ${}^cD_{0+}^\beta v(t) = y(t)$ satisfying the conditions $v(0) = v_0$, $v'(0) = \cdots = v^{(m-2)}(0) = 0$ is $v(t) = v_0 + \widetilde{d}_m t^{m-1} + I_{0+}^\beta y(t)$, $t \in [0,1]$, with $\widetilde{d}_m \in \mathbb{R}$.

Imposing the conditions $u(1) = \lambda I_{0+}^\gamma v(\xi)$ and $v(1) = \mu I_{0+}^\delta u(\eta)$ on the above functions u and v, we obtain the following system in the unknown constants $\widetilde{c}_n$ and $\widetilde{d}_m$

$$
\begin{cases}
\widetilde{c}_n - \dfrac{\lambda \widetilde{d}_m \xi^{\gamma+m-1}(m-1)!}{\Gamma(\gamma+m)} = -u_0 - I_{0+}^{\alpha}x(1) + \dfrac{\lambda v_0 \xi^{\gamma}}{\Gamma(\gamma+1)} + \lambda I_{0+}^{\beta+\gamma}y(\xi), \\[4mm]
-\dfrac{\mu \widetilde{c}_n \eta^{\delta+n-1}(n-1)!}{\Gamma(\delta+n)} + \widetilde{d}_m = -v_0 - I_{0+}^{\beta}y(1) + \dfrac{\mu u_0 \eta^{\delta}}{\Gamma(\delta+1)} + \mu I_{0+}^{\alpha+\delta}x(\eta).
\end{cases}
$$

In view of the given assumption $\widetilde{\Delta} \neq 0$, the unique solution $(\widetilde{c}_n, \widetilde{d}_m)$ of the above system is

$$
\begin{aligned}
\widetilde{c}_n = \frac{1}{\widetilde{\Delta}} \Bigg[&-u_0 - I_{0+}^{\alpha}x(1) + \frac{\lambda v_0 \xi^{\gamma}}{\Gamma(\gamma+1)} + \lambda I_{0+}^{\beta+\gamma}y(\xi) \\
&- \frac{\lambda v_0 \xi^{\gamma+m-1}(m-1)!}{\Gamma(\gamma+m)} - \frac{\lambda \xi^{\gamma+m-1}(m-1)!}{\Gamma(\gamma+m)} I_{0+}^{\beta}y(1) \\
&+ \frac{\lambda \mu u_0 \xi^{\gamma+m-1}\eta^{\delta}(m-1)!}{\Gamma(\delta+1)\Gamma(\gamma+m)} + \frac{\lambda \mu \xi^{\gamma+m-1}(m-1)!}{\Gamma(\gamma+m)} I_{0+}^{\alpha+\delta}x(\eta) \Bigg],
\end{aligned}
$$

$$
\begin{aligned}
\widetilde{d}_m = \frac{1}{\widetilde{\Delta}} \Bigg[&-v_0 - I_{0+}^{\beta}y(1) + \frac{\mu u_0 \eta^{\delta}}{\Gamma(\delta+1)} + \mu I_{0+}^{\alpha+\delta}x(\eta) \\
&- \frac{\mu u_0 \eta^{\delta+n-1}(n-1)!}{\Gamma(\delta+n)} - \frac{\mu \eta^{\delta+n-1}(n-1)!}{\Gamma(\delta+n)} I_{0+}^{\alpha}x(1) \\
&+ \frac{\lambda \mu v_0 \xi^{\gamma}\eta^{\delta+n-1}(n-1)!}{\Gamma(\gamma+1)\Gamma(\delta+n)} + \frac{\lambda \mu \eta^{\delta+n-1}(n-1)!}{\Gamma(\delta+n)} I_{0+}^{\beta+\gamma}y(\xi) \Bigg].
\end{aligned}
$$

Substituting the values of $\widetilde{c}_n$ and $\widetilde{d}_m$ in the expressions for u and v, we obtain the solution (u, v) of problem (7.47), (7.48) given by formula (7.49). $\square$

By direct computation, we can obtain the following result.

Lemma 7.3.2. *If $x, y \in C[0, 1]$, then the following inequalities are satisfied*

(a) $|I_{0+}^{\alpha}x(t)| \leq \dfrac{\|x\|}{\Gamma(\alpha+1)};$ (b) $|I_{0+}^{\alpha}x(1)| \leq \dfrac{\|x\|}{\Gamma(\alpha+1)};$

(c) $|I_{0+}^{\alpha+\delta}x(\eta)| \leq \dfrac{\eta^{\alpha+\delta}\|x\|}{\Gamma(\alpha+\delta+1)};$ (d) $|I_{0+}^{\beta}y(t)| \leq \dfrac{\|y\|}{\Gamma(\beta+1)};$

(e) $|I_{0+}^{\beta}y(1)| \leq \dfrac{\|y\|}{\Gamma(\beta+1)};$ (f) $|I_{0+}^{\beta+\gamma}y(\xi)| \leq \dfrac{\xi^{\beta+\gamma}\|y\|}{\Gamma(\beta+\gamma+1)},$

where $\|x\| = \sup_{t\in[0,1]} |x(t)|$ and $\|y\| = \sup_{t\in[0,1]} |y(t)|$.

7.3.2   *Existence of solutions*

We consider the spaces $X = \{u \in C[0,1],\ {}^cD^q_{0+}u \in C[0,1]\}$ and $Y = \{v \in C[0,1],\ {}^cD^p_{0+}v \in C[0,1]\}$ equipped respectively with the norms $\|u\|_X = \|u\| + \|{}^cD^q_{0+}u\|$ and $\|v\|_Y = \|v\| + \|{}^cD^p_{0+}v\|$, where $\|\cdot\|$ is the supremum norm, that is $\|w\| = \sup_{t\in[0,1]}|w(t)|$ for $w \in C[0,1]$. The spaces $(X, \|\cdot\|_X)$ and $(Y, \|\cdot\|_Y)$ are Banach spaces, and the product space $X \times Y$ endowed with the norm $\|(u,v)\|_{X\times Y} = \|u\|_X + \|v\|_Y$ is also a Banach space.

By using Lemma 7.3.1, we introduce the operator $\mathcal{Q} : X \times Y \to X \times Y$ defined by $\mathcal{Q}(u,v) = (\mathcal{Q}_1(u,v), \mathcal{Q}_2(u,v))$ for $(u,v) \in X \times Y$, where the operators $\mathcal{Q}_1 : X \times Y \to X$ and $\mathcal{Q}_2 : X \times Y \to Y$ are given by

$$
\mathcal{Q}_1(u,v)(t) = \varphi(v)\left[1 - \frac{t^{n-1}}{\widetilde{\Delta}} + \frac{t^{n-1}\lambda\mu\xi^{\gamma+m-1}\eta^\delta(m-1)!}{\widetilde{\Delta}\Gamma(\delta+1)\Gamma(\gamma+m)}\right]
$$

$$
+ \psi(u)\frac{t^{n-1}\lambda\xi^\gamma}{\widetilde{\Delta}}\left[\frac{1}{\Gamma(\gamma+1)} - \frac{\xi^{m-1}(m-1)!}{\Gamma(\gamma+m)}\right]
$$

$$
+ I^\alpha_{0+}f(t,v(t),{}^cD^p_{0+}v(t)) + \frac{t^{n-1}}{\widetilde{\Delta}}\left[-I^\alpha_{0+}f(t,v(t),{}^cD^p_{0+}v(t))|_{t=1}\right.
$$

$$
+ \frac{\lambda\mu\xi^{\gamma+m-1}(m-1)!}{\Gamma(\gamma+m)}I^{\alpha+\delta}_{0+}f(t,v(t),{}^cD^p_{0+}v(t))|_{t=\eta}
$$

$$
+ \lambda I^{\beta+\gamma}_{0+}g(t,u(t),{}^cD^q_{0+}u(t))|_{t=\xi}
$$

$$
\left. - \frac{\lambda\xi^{\gamma+m-1}(m-1)!}{\Gamma(\gamma+m)}I^\beta_{0+}g(t,u(t),{}^cD^q_{0+}u(t))|_{t=1}\right],
$$

$$
\mathcal{Q}_2(u,v)(t) = \varphi(v)\frac{t^{m-1}\mu\eta^\delta}{\widetilde{\Delta}}\left[\frac{1}{\Gamma(\delta+1)} - \frac{\eta^{n-1}(n-1)!}{\Gamma(\delta+n)}\right]
$$

$$
+ \psi(u)\left[1 - \frac{t^{m-1}}{\widetilde{\Delta}} + \frac{t^{m-1}\lambda\mu\xi^\gamma\eta^{\delta+n-1}(n-1)!}{\widetilde{\Delta}\Gamma(\gamma+1)\Gamma(\delta+n)}\right]
$$

$$
+ I^\beta_{0+}g(t,u(t),{}^cD^q_{0+}u(t)) + \frac{t^{m-1}}{\widetilde{\Delta}}\left[\mu I^{\alpha+\delta}_{0+}f(t,v(t),{}^cD^p_{0+}v(t))|_{t=\eta}\right.
$$

$$
\left. - \frac{\mu\eta^{\delta+n-1}(n-1)!}{\Gamma(\delta+n)}I^\alpha_{0+}f(t,v(t),{}^cD^p_{0+}v(t))|_{t=1}\right.
$$

$$- I_{0+}^{\beta} g(t, u(t), {}^{c}D_{0+}^{q} u(t))|_{t=1}$$

$$+ \frac{\lambda \mu \eta^{\delta+n-1}(n-1)!}{\Gamma(\delta+n)} I_{0+}^{\beta+\gamma} g(t, u(t), {}^{c}D_{0+}^{q} u(t))|_{t=\xi} \Bigg],$$

for all $(u, v) \in X \times Y$ and $t \in [0, 1]$.

The pair (u, v) is a solution of problem (7.44), (7.45) if and only if (u, v) is a fixed point of operator $\mathcal{Q}$.

Now we enlist the assumptions that we need in this section.

(K1) $\alpha, \beta \in \mathbb{R}$, $\alpha \in (n-1, n]$, $\beta \in (m-1, m]$, $n, m \in \mathbb{N}$, $n, m \geq 2$; $p, q \in (0, 1)$; $\gamma, \delta > 0$; $\lambda, \mu \in \mathbb{R}$; $\xi, \eta \in [0, 1]$; $\widetilde{\Delta} = 1 - \frac{\lambda \mu \xi^{\gamma+m-1} \eta^{\delta+n-1}(m-1)!(n-1)!}{\Gamma(\gamma+m)\Gamma(\delta+n)} \neq 0$.

(K2) $f, g : [0, 1] \times \mathbb{R} \times \mathbb{R} \to \mathbb{R}$ are continuous functions and there exist the constants $a_i, b_i \geq 0$, $i = 0, 1, 2$, and $l_j, m_j \in (0, 1)$, $j = 1, 2$ such that

$$|f(t, u, v)| \leq a_0 + a_1 |u|^{l_1} + a_2 |v|^{l_2}, \quad \forall t \in [0, 1], \ u, v \in \mathbb{R},$$

$$|g(t, u, v)| \leq b_0 + b_1 |u|^{m_1} + b_2 |v|^{m_2}, \quad \forall t \in [0, 1], \ u, v \in \mathbb{R}.$$

(K3) $\varphi, \psi : C[0, 1] \to \mathbb{R}$ are continuous functions, $\varphi(0) = \psi(0) = 0$ and there exist constants $L_1, L_2 > 0$, $\theta_1, \theta_2 \in (0, 1)$ such that

$$|\varphi(x)| \leq L_1 \|x\|^{\theta_1}, \quad |\psi(x)| \leq L_2 \|x\|^{\theta_2}, \quad \forall x \in C[0, 1].$$

(K4) $f, g : [0, 1] \times \mathbb{R} \times \mathbb{R} \to \mathbb{R}$ are continuous functions, and there exist $c_i, d_i \geq 0$, $i = 0, 1, 2$ and nondecreasing functions $h_j, k_j \in C([0, \infty), [0, \infty))$, $j = 1, 2$ such that

$$|f(t, u, v)| \leq c_0 + c_1 h_1(|u|) + c_2 h_2(|v|), \quad \forall t \in [0, 1], \ u, v \in \mathbb{R},$$

$$|g(t, u, v)| \leq d_0 + d_1 k_1(|u|) + d_2 k_2(|v|), \quad \forall t \in [0, 1], \ u, v \in \mathbb{R}.$$

(K5) There exists $L_0 > 0$ such that

$$(M_1 + \widetilde{M_1} + N_1 + \widetilde{N_1}) L_0^{\theta_1} + (M_2 + \widetilde{M_2} + N_2 + \widetilde{N_2}) L_0^{\theta_2}$$

$$+ A_1(M_3 + \widetilde{M_3} + N_3 + \widetilde{N_3}) + A_2(M_4 + \widetilde{M_4} + N_4 + \widetilde{N_4}) < L_0,$$

where $A_1 = c_0 + c_1 h_1(L_0) + c_2 h_2(L_0)$, $A_2 = d_0 + d_1 k_1(L_0) + d_2 k_2(L_0)$, θ_1, θ_2 are given in (K3), c_i, d_i, $i = 0, 1, 2$ and h_j, k_j, $j = 1, 2$ are given in (K4), and the constants $M_i, \widetilde{M_i}, N_i, \widetilde{N_i}$, $i = 1, 2, 3, 4$ are given below.

For computational convenience, we introduce the following notations:

$$M_1 = L_1 \left(1 + \frac{1}{|\widetilde{\Delta}|} + \frac{\lambda\mu\xi^{\gamma+m-1}\eta^\delta(m-1)!}{|\widetilde{\Delta}|\Gamma(\delta+1)\Gamma(\gamma+m)} \right),$$

$$M_2 = \frac{L_2\lambda\xi^\gamma}{|\widetilde{\Delta}|} \left(\frac{1}{\Gamma(\gamma+1)} + \frac{\xi^{m-1}(m-1)!}{\Gamma(\gamma+m)} \right),$$

$$M_3 = \frac{1}{\Gamma(\alpha+1)} + \frac{1}{|\widetilde{\Delta}|\Gamma(\alpha+1)} + \frac{\lambda\mu\xi^{\gamma+m-1}\eta^{\alpha+\delta}(m-1)!}{|\widetilde{\Delta}|\Gamma(\gamma+m)\Gamma(\alpha+\delta+1)},$$

$$M_4 = \frac{\lambda\xi^{\beta+\gamma}}{|\widetilde{\Delta}|\Gamma(\beta+\gamma+1)} + \frac{\lambda\xi^{\gamma+m-1}(m-1)!}{|\widetilde{\Delta}|\Gamma(\gamma+m)\Gamma(\beta+1)},$$

$$\widetilde{M_1} = \frac{L_1(n-1)}{\Gamma(2-q)|\widetilde{\Delta}|} \left(1 + \frac{\lambda\mu\xi^{\gamma+m-1}\eta^\delta(m-1)!}{\Gamma(\delta+1)\Gamma(\gamma+m)} \right),$$

$$\widetilde{M_2} = \frac{L_2(n-1)\lambda\xi^\gamma}{\Gamma(2-q)|\widetilde{\Delta}|} \left(\frac{1}{\Gamma(\gamma+1)} + \frac{\xi^{m-1}(m-1)!}{\Gamma(\gamma+m)} \right),$$

$$\widetilde{M_3} = \frac{1}{\Gamma(2-q)} \left(\frac{1}{\Gamma(\alpha)} + \frac{n-1}{|\widetilde{\Delta}|\Gamma(\alpha+1)} + \frac{(n-1)\lambda\mu\xi^{\gamma+m-1}\eta^{\alpha+\delta}(m-1)!}{|\widetilde{\Delta}|\Gamma(\gamma+m)\Gamma(\alpha+\delta+1)} \right),$$

$$\widetilde{M_4} = \frac{n-1}{\Gamma(2-q)|\widetilde{\Delta}|} \left(\frac{\lambda\xi^{\beta+\gamma}}{\Gamma(\beta+\gamma+1)} + \frac{\lambda\xi^{\gamma+m-1}(m-1)!}{\Gamma(\gamma+m)\Gamma(\beta+1)} \right),$$

$$N_1 = \frac{L_1\mu\eta^\delta}{|\widetilde{\Delta}|} \left(\frac{1}{\Gamma(\delta+1)} + \frac{\eta^{n-1}(n-1)!}{\Gamma(\delta+n)} \right),$$

$$N_2 = L_2 \left(1 + \frac{1}{|\widetilde{\Delta}|} + \frac{\lambda\mu\xi^\gamma\eta^{\delta+n-1}(n-1)!}{|\widetilde{\Delta}|\Gamma(\gamma+1)\Gamma(\delta+n)} \right),$$

$$N_3 = \frac{\mu\eta^{\alpha+\delta}}{|\widetilde{\Delta}|\Gamma(\alpha+\delta+1)} + \frac{\mu\eta^{\delta+n-1}(n-1)!}{|\widetilde{\Delta}|\Gamma(\delta+n)\Gamma(\alpha+1)},$$

$$N_4 = \frac{1}{\Gamma(\beta+1)} + \frac{1}{|\widetilde{\Delta}|\Gamma(\beta+1)} + \frac{\lambda\mu\xi^{\beta+\gamma}\eta^{\delta+n-1}(n-1)!}{|\widetilde{\Delta}|\Gamma(\delta+n)\Gamma(\beta+\gamma+1)},$$

$$\widetilde{N_1} = \frac{L_1(m-1)\mu\eta^\delta}{\Gamma(2-p)|\widetilde{\Delta}|} \left(\frac{1}{\Gamma(\delta+1)} + \frac{\eta^{n-1}(n-1)!}{\Gamma(\delta+n)} \right),$$

$$\widetilde{N_2} = \frac{L_2(m-1)}{\Gamma(2-p)|\widetilde{\Delta}|} \left(1 + \frac{\lambda\mu\xi^\gamma\eta^{\delta+n-1}(n-1)!}{\Gamma(\gamma+1)\Gamma(\delta+n)} \right),$$

$$\widetilde{N}_3 = \frac{m-1}{\Gamma(2-p)|\widetilde{\Delta}|} \left(\frac{\mu\eta^{\alpha+\delta}}{\Gamma(\alpha+\delta+1)} + \frac{\mu\eta^{\delta+n-1}(n-1)!}{\Gamma(\delta+n)\Gamma(\alpha+1)} \right),$$

$$\widetilde{N}_4 = \frac{1}{\Gamma(2-p)} \left(\frac{1}{\Gamma(\beta)} + \frac{m-1}{|\widetilde{\Delta}|\Gamma(\beta+1)} + \frac{(m-1)\lambda\mu\eta^{\delta+n-1}\xi^{\beta+\gamma}(n-1)!}{|\widetilde{\Delta}|\Gamma(\delta+n)\Gamma(\beta+\gamma+1)} \right).$$

Theorem 7.3.1. *Assume that* (K1)–(K3) *hold. Then problem* (7.44), (7.45) *has at least one solution on* $[0,1]$.

Proof. Let $\overline{B}_R = \{(u,v) \in X \times Y, \|(u,v)\|_{X \times Y} \leq R\}$, where

$$R \geq \max \left\{ [8(M_1 + \widetilde{M}_1)]^{\frac{1}{1-\theta_1}}, [8(M_2 + \widetilde{M}_2)]^{\frac{1}{1-\theta_2}}, \right.$$

$$24a_0(M_3 + \widetilde{M}_3), [24a_1(M_3 + \widetilde{M}_3)]^{\frac{1}{1-l_1}},$$

$$[24a_2(M_3 + \widetilde{M}_3)]^{\frac{1}{1-l_2}}, 24b_0(M_4 + \widetilde{M}_4),$$

$$[24b_1(M_4 + \widetilde{M}_4)]^{\frac{1}{1-m_1}}, [24b_2(M_4 + \widetilde{M}_4)]^{\frac{1}{1-m_2}},$$

$$[8(N_1 + \widetilde{N}_1)]^{\frac{1}{1-\theta_1}}, [8(N_2 + \widetilde{N}_2)]^{\frac{1}{1-\theta_2}},$$

$$24a_0(N_3 + \widetilde{N}_3), [24a_1(N_3 + \widetilde{N}_3)]^{\frac{1}{1-l_1}},$$

$$[24a_2(N_3 + \widetilde{N}_3)]^{\frac{1}{1-l_2}}, 24b_0(N_4 + \widetilde{N}_4),$$

$$\left. [24b_1(N_4 + \widetilde{N}_4)]^{\frac{1}{1-m_1}}, [24b_2(N_4 + \widetilde{N}_4)]^{\frac{1}{1-m_2}} \right\}.$$

Let us first show that $\mathcal{Q} : \overline{B}_R \to \overline{B}_R$. For $(u,v) \in \overline{B}_R$, it follows by Lemma 7.3.2 that

$$|\mathcal{Q}_1(u,v)(t)| \leq |\varphi(v)| \left(1 + \frac{1}{|\widetilde{\Delta}|} + \frac{\lambda\mu\xi^{\gamma+m-1}\eta^\delta(m-1)!}{|\widetilde{\Delta}|\Gamma(\delta+1)\Gamma(\gamma+m)} \right)$$

$$+ \frac{|\psi(u)|\lambda\xi^\gamma}{|\widetilde{\Delta}|} \left(\frac{1}{\Gamma(\gamma+1)} + \frac{\xi^{m-1}(m-1)!}{\Gamma(\gamma+m)} \right)$$

$$+ |I_{0+}^\alpha f(t,v(t),{}^cD_{0+}^p v(t))| + \frac{1}{|\widetilde{\Delta}|} \left[|I_{0+}^\alpha f(t,v(t),{}^cD_{0+}^p v(t))|_{t=1}| \right.$$

$$+ \frac{\lambda\mu\xi^{\gamma+m-1}(m-1)!}{\Gamma(\gamma+m)} |I_{0+}^{\alpha+\delta} f(t,v(t),{}^cD_{0+}^p v(t))|_{t=\eta}|$$

$$+ \lambda|I_{0+}^{\beta+\gamma} g(t,u(t),{}^cD_{0+}^q u(t))|_{t=\xi}|$$

$$+ \left. \frac{\lambda\xi^{\gamma+m-1}(m-1)!}{\Gamma(\gamma+m)} |I_{0+}^\beta g(t,u(t),{}^cD_{0+}^q u(t))|_{t=1}| \right]$$

$$\leq L_1 \|v\|^{\theta_1} \left(1 + \frac{1}{|\widetilde{\Delta}|} + \frac{\lambda\mu\xi^{\gamma+m-1}\eta^{\delta}(m-1)!}{|\widetilde{\Delta}|\Gamma(\delta+1)\Gamma(\gamma+m)} \right)$$

$$+ \frac{L_2}{|\widetilde{\Delta}|} \|u\|^{\theta_2} \lambda\xi^{\gamma} \left(\frac{1}{\Gamma(\gamma+1)} + \frac{\xi^{m-1}(m-1)!}{\Gamma(\gamma+m)} \right)$$

$$+ \frac{1}{\Gamma(\alpha+1)} \left(a_0 + a_1\|v\|^{l_1} + a_2\|{}^cD_{0+}^p v\|^{l_2} \right)$$

$$+ \frac{1}{|\widetilde{\Delta}|} \left[\frac{1}{\Gamma(\alpha+1)} \left(a_0 + a_1\|v\|^{l_1} + a_2\|{}^cD_{0+}^p v\|^{l_2} \right) \right.$$

$$+ \frac{\lambda\mu\xi^{\gamma+m-1}\eta^{\alpha+\delta}(m-1)!}{\Gamma(\gamma+m)\Gamma(\alpha+\delta+1)} \left(a_0 + a_1\|v\|^{l_1} + a_2\|{}^cD_{0+}^p v\|^{l_2} \right)$$

$$+ \frac{\lambda\xi^{\beta+\gamma}}{\Gamma(\beta+\gamma+1)} \left(b_0 + b_1\|u\|^{m_1} + b_2\|{}^cD_{0+}^q u\|^{m_2} \right)$$

$$\left. + \frac{\lambda\xi^{\gamma+m-1}(m-1)!}{\Gamma(\gamma+m)\Gamma(\beta+1)} \left(b_0 + b_1\|u\|^{m_1} + b_2\|{}^cD_{0+}^q u\|^{m_2} \right) \right]$$

$$\leq L_1 R^{\theta_1} \left(1 + \frac{1}{|\widetilde{\Delta}|} + \frac{\lambda\mu\xi^{\gamma+m-1}\eta^{\delta}(m-1)!}{|\widetilde{\Delta}|\Gamma(\delta+1)\Gamma(\gamma+m)} \right)$$

$$+ \frac{L_2 R^{\theta_2} \lambda\xi^{\gamma}}{|\widetilde{\Delta}|} \left(\frac{1}{\Gamma(\gamma+1)} + \frac{\xi^{m-1}(m-1)!}{\Gamma(\gamma+m)} \right)$$

$$+ (a_0 + a_1 R^{l_1} + a_2 R^{l_2}) \left(\frac{1}{\Gamma(\alpha+1)} + \frac{1}{|\widetilde{\Delta}|\Gamma(\alpha+1)} \right.$$

$$\left. + \frac{\lambda\mu\xi^{\gamma+m-1}\eta^{\alpha+\delta}(m-1)!}{|\widetilde{\Delta}|\Gamma(\gamma+m)\Gamma(\alpha+\delta+1)} \right)$$

$$+ (b_0 + b_1 R^{m_1} + b_2 R^{m_2}) \left(\frac{\lambda\xi^{\beta+\gamma}}{|\widetilde{\Delta}|\Gamma(\beta+\gamma+1)} \right.$$

$$\left. + \frac{\lambda\xi^{\gamma+m-1}(m-1)!}{|\widetilde{\Delta}|\Gamma(\gamma+m)\Gamma(\beta+1)} \right)$$

$$= M_1 R^{\theta_1} + M_2 R^{\theta_2} + M_3(a_0 + a_1 R^{l_1} + a_2 R^{l_2})$$

$$+ M_4(b_0 + b_1 R^{m_1} + b_2 R^{m_2}), \quad \forall t \in [0,1].$$

On the other hand, we have

$$^cD_{0+}^q \mathcal{Q}_1(u,v)(t) = \frac{1}{\Gamma(1-q)} \int_0^t \frac{(\mathcal{Q}_1(x,y))'(s)}{(t-s)^q}\, ds,$$

where

$$(\mathcal{Q}_1(u,v))'(t) = \varphi(v)\left(-\frac{(n-1)t^{n-2}}{\widetilde{\Delta}} + \frac{(n-1)t^{n-2}\lambda\mu\xi^{\gamma+m-1}\eta^\delta(m-1)!}{\widetilde{\Delta}\Gamma(\delta+1)\Gamma(\gamma+m)}\right)$$

$$+ \psi(u)\frac{(n-1)t^{n-2}\lambda\xi^\gamma}{\widetilde{\Delta}}\left(\frac{1}{\Gamma(\gamma+1)} - \frac{\xi^{m-1}(m-1)!}{\Gamma(\gamma+m)}\right)$$

$$+ I_{0+}^{\alpha-1}f(t,v(t),{}^cD_{0+}^p v(t)) + \frac{(n-1)t^{n-2}}{\widetilde{\Delta}}$$

$$\times\left[-I_{0+}^\alpha f(t,v(t),{}^cD_{0+}^p v(t))|_{t=1}\right.$$

$$+ \frac{\lambda\mu\xi^{\gamma+m-1}(m-1)!}{\Gamma(\gamma+m)}I_{0+}^{\alpha+\delta}f(t,v(t),{}^cD_{0+}^p v(t))|_{t=\eta}$$

$$+ \lambda I_{0+}^{\beta+\gamma}g(t,u(t),{}^cD_{0+}^q u(t))|_{t=\xi}$$

$$\left.- \frac{\lambda\xi^{\gamma+m-1}(m-1)!}{\Gamma(\gamma+m)}I_{0+}^\beta g(t,u(t),{}^cD_{0+}^q u(t))|_{t=1}\right],$$

$$\forall t \in (0,1).$$

Then, by Lemma 7.3.2, we obtain

$$|(\mathcal{Q}_1(u,v))'(t)| \le |\varphi(v)|\left(\frac{n-1}{|\widetilde{\Delta}|} + \frac{(n-1)\lambda\mu\xi^{\gamma+m-1}\eta^\delta(m-1)!}{|\widetilde{\Delta}|\Gamma(\delta+1)\Gamma(\gamma+m)}\right)$$

$$+ \frac{|\psi(u)|(n-1)\lambda\xi^\gamma}{|\widetilde{\Delta}|}\left(\frac{1}{\Gamma(\gamma+1)} + \frac{\xi^{m-1}(m-1)!}{\Gamma(\gamma+m)}\right)$$

$$+ |I_{0+}^{\alpha-1}f(t,v(t),{}^cD_{0+}^p v(t))| + \frac{n-1}{|\widetilde{\Delta}|}$$

$$\times\left[|I_{0+}^\alpha f(t,v(t),{}^cD_{0+}^p v(t))|_{t=1}|\right.$$

$$+ \frac{\lambda\mu\xi^{\gamma+m-1}(m-1)!}{\Gamma(\gamma+m)}|I_{0+}^{\alpha+\delta}f(t,v(t),{}^cD_{0+}^p v(t))|_{t=\eta}|$$

$$+ \lambda|I_{0+}^{\beta+\gamma}g(t,u(t),{}^cD_{0+}^q u(t))|_{t=\xi}|$$

$$\left.+ \frac{\lambda\xi^{\gamma+m-1}(m-1)!}{\Gamma(\gamma+m)}|I_{0+}^\beta g(t,u(t),{}^cD_{0+}^q u(t))|_{t=1}|\right]$$

$$\leq L_1\|v\|^{\theta_1}\left(\frac{n-1}{|\widetilde{\Delta}|}+\frac{(n-1)\lambda\mu\xi^{\gamma+m-1}\eta^\delta(m-1)!}{|\widetilde{\Delta}|\Gamma(\delta+1)\Gamma(\gamma+m)}\right)$$

$$+\frac{L_2\|u\|^{\theta_2}(n-1)\lambda\xi^\gamma}{|\widetilde{\Delta}|}\left(\frac{1}{\Gamma(\gamma+1)}+\frac{\xi^{m-1}(m-1)!}{\Gamma(\gamma+m)}\right)$$

$$+\frac{1}{\Gamma(\alpha)}\left(a_0+a_1\|v\|^{l_1}+a_2\|{}^cD_{0+}^p v\|^{l_2}\right)$$

$$+\frac{n-1}{|\widetilde{\Delta}|}\left[\frac{1}{\Gamma(\alpha+1)}\left(a_0+a_1\|v\|^{l_1}+a_2\|{}^cD_{0+}^p v\|^{l_2}\right)\right.$$

$$+\frac{\lambda\mu\xi^{\gamma+m-1}(m-1)!\eta^{\alpha+\delta}}{\Gamma(\gamma+m)\Gamma(\alpha+\delta+1)}\left(a_0+a_1\|v\|^{l_1}+a_2\|{}^cD_{0+}^p v\|^{l_2}\right)$$

$$+\frac{\lambda\xi^{\beta+\gamma}}{\Gamma(\beta+\gamma+1)}\left(b_0+b_1\|u\|^{m_1}+b_2\|{}^cD_{0+}^q u\|^{m_2}\right)$$

$$+\left.\frac{\lambda\xi^{\gamma+m-1}(m-1)!}{\Gamma(\gamma+m)\Gamma(\beta+1)}\left(b_0+b_1\|u\|^{m_1}+b_2\|{}^cD_{0+}^q u\|^{m_2}\right)\right]$$

$$\leq L_1R^{\theta_1}\left(\frac{n-1}{|\widetilde{\Delta}|}+\frac{(n-1)\lambda\mu\xi^{\gamma+m-1}\eta^\delta(m-1)!}{|\widetilde{\Delta}|\Gamma(\delta+1)\Gamma(\gamma+m)}\right)$$

$$+\frac{L_2R^{\theta_2}(n-1)\lambda\xi^\gamma}{|\widetilde{\Delta}|}\left(\frac{1}{\Gamma(\gamma+1)}+\frac{\xi^{m-1}(m-1)!}{\Gamma(\gamma+m)}\right)$$

$$+(a_0+a_1R^{l_1}+a_2R^{l_2})\left(\frac{1}{\Gamma(\alpha)}+\frac{n-1}{|\widetilde{\Delta}|\Gamma(\alpha+1)}\right.$$

$$+\left.\frac{(n-1)\lambda\mu\xi^{\gamma+m-1}\eta^{\alpha+\delta}(m-1)!}{|\widetilde{\Delta}|\Gamma(\gamma+m)\Gamma(\alpha+\delta+1)}\right)$$

$$+(b_0+b_1R^{m_1}+b_2R^{m_2})\left(\frac{(n-1)\lambda\xi^{\beta+\gamma}}{|\widetilde{\Delta}|\Gamma(\beta+\gamma+1)}\right.$$

$$+\left.\frac{(n-1)\lambda\xi^{\gamma+m-1}(m-1)!}{|\widetilde{\Delta}|\Gamma(\gamma+m)\Gamma(\beta+1)}\right),\quad\forall t\in(0,1).$$

In consequence, we obtain

$$
\begin{aligned}
|{}^{c}D_{0+}^{q}\mathcal{Q}_1(u,v)(t)| &\leq \frac{1}{\Gamma(1-q)}\int_0^t (t-s)^{-q}\|(\mathcal{Q}_1(u,v))'\|ds \\
&\leq \frac{1}{\Gamma(2-q)}\|(\mathcal{Q}_1(u,v))'\| \\
&\leq \frac{L_1 R^{\theta_1}}{\Gamma(2-q)}\left(\frac{n-1}{|\widetilde{\Delta}|}+\frac{(n-1)\lambda\mu\xi^{\gamma+m-1}\eta^{\delta}(m-1)!}{|\widetilde{\Delta}|\Gamma(\delta+1)\Gamma(\gamma+m)}\right) \\
&\quad +\frac{L_2 R^{\theta_2}(n-1)\lambda\xi^{\gamma}}{|\widetilde{\Delta}|\Gamma(2-q)}\left(\frac{1}{\Gamma(\gamma+1)}+\frac{\xi^{m-1}(m-1)!}{\Gamma(\gamma+m)}\right) \\
&\quad +\frac{1}{\Gamma(2-q)}(a_0+a_1 R^{l_1}+a_2 R^{l_2})\left(\frac{1}{\Gamma(\alpha)}+\frac{n-1}{|\widetilde{\Delta}|\Gamma(\alpha+1)}\right. \\
&\quad \left.+\frac{(n-1)\lambda\mu\xi^{\gamma+m-1}\eta^{\alpha+\delta}(m-1)!}{|\widetilde{\Delta}|\Gamma(\gamma+m)\Gamma(\alpha+\delta+1)}\right) \\
&\quad +\frac{1}{\Gamma(2-q)}(b_0+b_1 R^{m_1}+b_2 R^{m_2})\left(\frac{(n-1)\lambda\xi^{\beta+\gamma}}{|\widetilde{\Delta}|\Gamma(\beta+\gamma+1)}\right. \\
&\quad \left.+\frac{(n-1)\lambda\xi^{\gamma+m-1}(m-1)!}{|\widetilde{\Delta}|\Gamma(\gamma+m)\Gamma(\beta+1)}\right) \\
&= \widetilde{M}_1 R^{\theta_1}+\widetilde{M}_2 R^{\theta_2}+\widetilde{M}_3(a_0+a_1 R^{l_1}+a_2 R^{l_2}) \\
&\quad +\widetilde{M}_4(b_0+b_1 R^{m_1}+b_2 R^{m_2}), \quad \forall t\in[0,1].
\end{aligned}
$$

Thus we have

$$
\begin{aligned}
\|\mathcal{Q}_1(u,v)\|_X &= \|\mathcal{Q}_1(u,v)\|+\|{}^{c}D_{0+}^{q}\mathcal{Q}_1(u,v)\| \\
&\leq (M_1+\widetilde{M}_1)R^{\theta_1}+(M_2+\widetilde{M}_2)R^{\theta_2} \\
&\quad +(M_3+\widetilde{M}_3)(a_0+a_1 R^{l_1}+a_2 R^{l_2}) \\
&\quad +(M_4+\widetilde{M}_4)(b_0+b_1 R^{m_1}+b_2 R^{m_2}) \\
&\leq \frac{R}{8}+\frac{R}{8}+\frac{R}{8}+\frac{R}{8}=\frac{R}{2}.
\end{aligned}
$$

In a similar manner, we can obtain

$$\|Q_2(u,v)\|_Y = \|Q_2(u,v)\| + \|^cD_{0+}^p Q_2(u,v)\|$$
$$\leq (N_1 + \tilde{N}_1)R^{\theta_1} + (N_2 + \tilde{N}_2)R^{\theta_2}$$
$$+ (N_3 + \tilde{N}_3)(a_0 + a_1 R^{l_1} + a_2 R^{l_2})$$
$$+ (N_4 + \tilde{N}_4)(b_0 + b_1 R^{m_1} + b_2 R^{m_2})$$
$$\leq \frac{R}{8} + \frac{R}{8} + \frac{R}{8} + \frac{R}{8} = \frac{R}{2}.$$

By the foregoing arguments, we deduce that

$$\|Q(u,v)\|_{X \times Y} = \|Q_1(u,v)\|_X + \|Q_2(u,v)\|_Y \leq \frac{R}{2} + \frac{R}{2} = R,$$

which implies that $Q : \overline{B}_R \to \overline{B}_R$.

From the continuity of the functions f, g, φ and ψ, we can easily show that the operator Q is continuous.

Next we will show that the operator $Q : \overline{B}_R \to \overline{B}_R$ is equicontinuous. We denote $\Lambda_1 = \max_{t \in [0,1], |u| \leq R, |v| \leq R} |f(t, u, v)|$ and $\Lambda_2 = \max_{t \in [0,1], |u| \leq R, |v| \leq R} |g(t, u, v)|$. For any $(u,v) \in \overline{B}_R$ and $t_1, t_2 \in [0,1]$, $t_1 < t_2$, we have

$$|Q_1(u,v)(t_2) - Q_1(u,v)(t_1)|$$

$$\leq \left| \varphi(v) \left(1 - \frac{t_2^{n-1}}{\tilde{\Delta}} + \frac{t_2^{n-1} \lambda \mu \xi^{\gamma+m-1} \eta^\delta (m-1)!}{\tilde{\Delta} \Gamma(\delta+1) \Gamma(\gamma+m)} \right. \right.$$

$$\left. \left. - 1 + \frac{t_1^{n-1}}{\tilde{\Delta}} - \frac{t_1^{n-1} \lambda \mu \xi^{\gamma+m-1} \eta^\delta (m-1)!}{\tilde{\Delta} \Gamma(\delta+1) \Gamma(\gamma+m)} \right) \right|$$

$$+ \left| \psi(u) \lambda \xi^\gamma \frac{t_2^{n-1} - t_1^{n-1}}{\tilde{\Delta}} \left(\frac{1}{\Gamma(\gamma+1)} - \frac{\xi^{m-1}(m-1)!}{\Gamma(\gamma+m)} \right) \right|$$

$$+ \left| I_{0+}^\alpha f(t, v(t), {}^cD_{0+}^p v(t))|_{t=t_2} - I_{0+}^\alpha f(t, v(t), {}^cD_{0+}^p v(t))|_{t=t_1} \right|$$

$$+ \frac{t_2^{n-1} - t_1^{n-1}}{|\tilde{\Delta}|} \left[|I_{0+}^\alpha f(t, v(t), {}^cD_{0+}^p v(t))|_{t=1}| \right.$$

$$+ \frac{\lambda \mu \xi^{\gamma+m-1}(m-1)!}{\Gamma(\gamma+m)} |I_{0+}^{\alpha+\delta} f(t, v(t), {}^cD_{0+}^p v(t))|_{t=\eta}|$$

$$+ \lambda |I_{0+}^{\beta+\gamma} g(t, u(t), {}^cD_{0+}^q u(t))|_{t=\xi}|$$

$$+ \frac{\lambda \xi^{\gamma+m-1}(m-1)!}{\Gamma(\gamma+m)} |I_{0+}^\beta g(t, u(t), {}^cD_{0+}^q u(t))|_{t=1}| \right]$$

$$\leq L_1\|v\|^{\theta_1}\frac{t_2^{n-1}-t_1^{n-1}}{|\widetilde{\Delta}|}\left(1+\frac{\lambda\mu\xi^{\gamma+m-1}\eta^\delta(m-1)!}{\Gamma(\delta+1)\Gamma(\gamma+m)}\right)$$

$$+L_2\|u\|^{\theta_2}\frac{t_2^{n-1}-t_1^{n-1}}{|\widetilde{\Delta}|}\lambda\xi^\gamma\left(\frac{1}{\Gamma(\gamma+1)}+\frac{\xi^{m-1}(m-1)!}{\Gamma(\gamma+m)}\right)$$

$$+\frac{1}{\Gamma(\alpha)}\left|\int_0^{t_2}(t_2-s)^{\alpha-1}f(s,v(s),{}^cD_{0+}^pv(s))\,ds\right.$$

$$\left.-\int_0^{t_1}(t_1-s)^{\alpha-1}f(s,v(s),{}^cD_{0+}^pv(s))\,ds\right|$$

$$+\frac{t_2^{n-1}-t_1^{n-1}}{|\widetilde{\Delta}|}\left[\frac{\Lambda_1}{\Gamma(\alpha+1)}+\frac{\lambda\mu\xi^{\gamma+m-1}\eta^{\alpha+\delta}(m-1)!\Lambda_1}{\Gamma(\gamma+m)\Gamma(\alpha+\delta+1)}\right.$$

$$+\frac{\lambda\xi^{\beta+\gamma}\Lambda_2}{\Gamma(\beta+\gamma+1)}+\frac{\lambda\xi^{\gamma+m-1}(m-1)!\Lambda_2}{\Gamma(\gamma+m)\Gamma(\beta+1)}\Bigg]$$

$$\leq L_1R^{\theta_1}\frac{t_2^{n-1}-t_1^{n-1}}{|\widetilde{\Delta}|}\left(1+\frac{\lambda\mu\xi^{\gamma+m-1}\eta^\delta(m-1)!}{\Gamma(\delta+1)\Gamma(\gamma+m)}\right)$$

$$+L_2R^{\theta_2}\frac{t_2^{n-1}-t_1^{n-1}}{|\widetilde{\Delta}|}\lambda\xi^\gamma\left(\frac{1}{\Gamma(\gamma+1)}+\frac{\xi^{m-1}(m-1)!}{\Gamma(\gamma+m)}\right)$$

$$+\frac{1}{\Gamma(\alpha)}\left|\int_0^{t_1}[(t_2-s)^{\alpha-1}-(t_1-s)^{\alpha-1}]f(s,v(s),{}^cD_{0+}^pv(s))\,ds\right|$$

$$+\frac{1}{\Gamma(\alpha)}\left|\int_{t_1}^{t_2}(t_2-s)^{\alpha-1}f(s,v(s),{}^cD_{0+}^pv(s))\,ds\right|$$

$$+\frac{t_2^{n-1}-t_1^{n-1}}{|\widetilde{\Delta}|}\left[\frac{\Lambda_1}{\Gamma(\alpha+1)}+\frac{\lambda\mu\xi^{\gamma+m-1}\eta^{\alpha+\delta}(m-1)!\Lambda_1}{\Gamma(\gamma+m)\Gamma(\alpha+\delta+1)}\right.$$

$$+\frac{\lambda\xi^{\beta+\gamma}\Lambda_2}{\Gamma(\beta+\gamma+1)}+\frac{\lambda\xi^{\gamma+m-1}(m-1)!\Lambda_2}{\Gamma(\gamma+m)\Gamma(\beta+1)}\Bigg],$$

which further implies that

$$|\mathcal{Q}_1(u,v)(t_2)-\mathcal{Q}_1(u,v)(t_1)|$$

$$\leq L_1R^{\theta_1}\frac{t_2^{n-1}-t_1^{n-1}}{|\widetilde{\Delta}|}\left(1+\frac{\lambda\mu\xi^{\gamma+m-1}\eta^\delta(m-1)!}{\Gamma(\delta+1)\Gamma(\gamma+m)}\right)$$

$$+L_2R^{\theta_2}\lambda\xi^\gamma\frac{t_2^{n-1}-t_1^{n-1}}{|\widetilde{\Delta}|}\left(\frac{1}{\Gamma(\gamma+1)}+\frac{\xi^{m-1}(m-1)!}{\Gamma(\gamma+m)}\right)$$

$$+ \frac{\Lambda_1}{\Gamma(\alpha)} \left| \int_0^{t_1} [(t_2 - s)^{\alpha-1} - (t_1 - s)^{\alpha-1}] \, ds \right|$$

$$+ \frac{\Lambda_1}{\Gamma(\alpha)} \left| \int_{t_1}^{t_2} (t_2 - s)^{\alpha-1} \, ds \right|$$

$$+ \frac{t_2^{n-1} - t_1^{n-1}}{|\widetilde{\Delta}|} \left[\frac{\Lambda_1}{\Gamma(\alpha+1)} + \frac{\lambda\mu\xi^{\gamma+m-1}\eta^{\alpha+\delta}(m-1)!\Lambda_1}{\Gamma(\gamma+m)\Gamma(\alpha+\delta+1)} \right.$$

$$\left. + \frac{\lambda\xi^{\beta+\gamma}\Lambda_2}{\Gamma(\beta+\gamma+1)} + \frac{\lambda\xi^{\gamma+m-1}(m-1)!\Lambda_2}{\Gamma(\gamma+m)\Gamma(\beta+1)} \right]$$

$$= L_1 R^{\theta_1} \frac{t_2^{n-1} - t_1^{n-1}}{|\widetilde{\Delta}|} \left(1 + \frac{\lambda\mu\xi^{\gamma+m-1}\eta^{\delta}(m-1)!}{\Gamma(\delta+1)\Gamma(\gamma+m)} \right)$$

$$+ L_2 R^{\theta_2} \lambda\xi^{\gamma} \frac{t_2^{n-1} - t_1^{n-1}}{|\widetilde{\Delta}|} \left(\frac{1}{\Gamma(\gamma+1)} + \frac{\xi^{m-1}(m-1)!}{\Gamma(\gamma+m)} \right)$$

$$+ \frac{\Lambda_1}{\Gamma(\alpha+1)} (t_2^{\alpha} - t_1^{\alpha}) + \frac{t_2^{n-1} - t_1^{n-1}}{|\widetilde{\Delta}|} \left[\frac{\Lambda_1}{\Gamma(\alpha+1)} \right.$$

$$+ \frac{\lambda\mu\xi^{\gamma+m-1}\eta^{\alpha+\delta}(m-1)!\Lambda_1}{\Gamma(\gamma+m)\Gamma(\alpha+\delta+1)}$$

$$\left. + \frac{\lambda\xi^{\beta+\gamma}\Lambda_2}{\Gamma(\beta+\gamma+1)} + \frac{\lambda\xi^{\gamma+m-1}(m-1)!\Lambda_2}{\Gamma(\gamma+m)\Gamma(\beta+1)} \right]. \tag{7.50}$$

On the other hand we obtain

$$|{}^c D_{0+}^q \mathcal{Q}_1(u,v)(t_2) - {}^c D_{0+}^q \mathcal{Q}_1(u,v)(t_1)|$$

$$= \frac{1}{\Gamma(1-q)} \left| \int_0^{t_2} \frac{(\mathcal{Q}_1(u,v))'(s)}{(t_2-s)^q} \, ds - \int_0^{t_1} \frac{(\mathcal{Q}_1(u,v))'(s)}{(t_1-s)^q} \, ds \right|$$

$$\leq \frac{1}{\Gamma(1-q)} \left| \int_0^{t_1} \left[\frac{1}{(t_2-s)^q} - \frac{1}{(t_1-s)^q} \right] (\mathcal{Q}_1(u,v))'(s) \, ds \right|$$

$$+ \frac{1}{\Gamma(1-q)} \left| \int_{t_1}^{t_2} \frac{(\mathcal{Q}_1(u,v))'(s)}{(t_2-s)^q} \, ds \right|.$$

As

$$|(\mathcal{Q}_1(u,v))'(t)| \leq L_1 R^{\theta_1} \left(\frac{n-1}{|\widetilde{\Delta}|} + \frac{(n-1)\lambda\mu\xi^{\gamma+m-1}\eta^{\delta}(m-1)!}{|\widetilde{\Delta}|\Gamma(\delta+1)\Gamma(\gamma+m)} \right)$$

$$+ L_2 R^{\theta_2} \lambda\xi^{\gamma} \frac{n-1}{|\widetilde{\Delta}|} \left(\frac{1}{\Gamma(\gamma+1)} + \frac{\xi^{m-1}(m-1)!}{\Gamma(\gamma+m)} \right)$$

$$+\frac{\Lambda_1}{\Gamma(\alpha)}+\frac{n-1}{|\widetilde{\Delta}|}\left[\frac{\Lambda_1}{\Gamma(\alpha+1)}+\frac{\lambda\mu\xi^{\gamma+m-1}\eta^{\alpha+\delta}(m-1)!\Lambda_1}{\Gamma(\gamma+m)\Gamma(\alpha+\delta+1)}\right.$$

$$\left.+\frac{\lambda\xi^{\beta+\gamma}\Lambda_2}{\Gamma(\beta+\gamma+1)}+\frac{\lambda\xi^{\gamma+m-1}(m-1)!\Lambda_2}{\Gamma(\gamma+m)\Gamma(\beta+1)}\right]=:D_1,$$

$$\forall t\in(0,1),$$

we deduce by the above inequalities that

$$|^cD_{0+}^q\mathcal{Q}_1(u,v)(t_2)-{}^cD_{0+}^q\mathcal{Q}_1(u,v)(t_1)|$$

$$\leq\frac{D_1}{\Gamma(1-q)}\left[\int_0^{t_1}[(t_1-s)^{-q}-(t_2-s)^{-q}]\,ds+\int_{t_1}^{t_2}(t_2-s)^{-q}\,ds\right]$$

$$=\frac{D_1}{\Gamma(2-q)}[2(t_2-t_1)^{1-q}-t_2^{1-q}+t_1^{1-q}].\qquad(7.51)$$

In a similar manner, we find that

$$|\mathcal{Q}_2(u,v)(t_2)-\mathcal{Q}_2(u,v)(t_1)|$$

$$\leq L_1R^{\theta_1}\mu\eta^\delta\frac{t_2^{m-1}-t_1^{m-1}}{|\widetilde{\Delta}|}\left(\frac{1}{\Gamma(\delta+1)}+\frac{\eta^{n-1}(n-1)!}{\Gamma(\delta+n)}\right)$$

$$+L_2R^{\theta_2}\frac{t_2^{m-1}-t_1^{m-1}}{|\widetilde{\Delta}|}\left(1+\frac{\lambda\mu\xi^\gamma\eta^{\delta+n-1}(n-1)!}{\Gamma(\gamma+1)\Gamma(\delta+n)}\right)$$

$$+\frac{\Lambda_2}{\Gamma(\beta)}(t_2^\beta-t_1^\beta)+\frac{t_2^{m-1}-t_1^{m-1}}{|\widetilde{\Delta}|}\left[\frac{\mu\eta^{\alpha+\delta}\Lambda_1}{\Gamma(\alpha+\delta+1)}+\frac{\mu\eta^{\delta+n-1}(n-1)!\Lambda_1}{\Gamma(\delta+n)\Gamma(\alpha+1)}\right.$$

$$\left.+\frac{\Lambda_2}{\Gamma(\beta+1)}+\frac{\lambda\mu\eta^{\delta+n-1}\xi^{\beta+\gamma}(n-1)!\Lambda_2}{\Gamma(\delta+n)\Gamma(\beta+\gamma+1)}\right].\qquad(7.52)$$

Using the estimate

$$|(\mathcal{Q}_2(u,v))'(t)|\leq L_1R^{\theta_1}\mu\eta^\delta\frac{m-1}{|\widetilde{\Delta}|}\left(\frac{1}{\Gamma(\delta+1)}+\frac{\eta^{n-1}(n-1)!}{\Gamma(\delta+n)}\right)$$

$$+L_2R^{\theta_2}\frac{m-1}{|\widetilde{\Delta}|}\left(1+\frac{\lambda\mu\xi^\gamma\eta^{\delta+n-1}(n-1)!}{\Gamma(\gamma+1)\Gamma(\delta+n)}\right)$$

$$+\frac{\Lambda_2}{\Gamma(\beta)}+\frac{m-1}{|\widetilde{\Delta}|}\left[\frac{\mu\eta^{\alpha+\delta}\Lambda_1}{\Gamma(\alpha+\delta+1)}+\frac{\mu\eta^{\delta+n-1}(n-1)!\Lambda_1}{\Gamma(\delta+n)\Gamma(\alpha+1)}\right.$$

$$\left.+\frac{\Lambda_2}{\Gamma(\beta+1)}+\frac{\lambda\mu\eta^{\delta+n-1}\xi^{\beta+\gamma}(n-1)!\Lambda_2}{\Gamma(\delta+n)\Gamma(\beta+\gamma+1)}\right]=:D_2,$$

$$\forall t\in(0,1),$$

we get

$$|{}^cD_{0+}^p Q_2(u,v)(t_2) - {}^cD_{0+}^p Q_2(u,v)(t_1)|$$

$$\leq \frac{D_2}{\Gamma(2-p)}[2(t_2-t_1)^{1-p} - t_2^{1-p} + t_1^{1-p}]. \qquad (7.53)$$

By the relations (7.50)–(7.53), we deduce that $Q : \overline{B}_R \to \overline{B}_R$ is equicontinuous. Thus the Arzela-Ascoli theorem applies and that the set $Q(\overline{B}_R)$ is relatively compact, and thus Q is a completely continuous operator. Therefore, by the Schauder fixed point theorem (Theorem 1.2.4), we deduce that the operator Q has at least one fixed point (u,v) in $\overline{B}_R$, which is a solution of problem (7.44), (7.45). $\qquad \square$

Theorem 7.3.2. *Assume that* (K1), (K3), (K4) *and* (K5) *hold. Then problem* (7.44), (7.45) *has at least one solution on* $[0,1]$.

Proof. With L_0 given by (K5), we consider a set $\overline{B}_{L_0} = \{(u,v) \in X \times Y, \|(u,v)\|_{X \times Y} \leq L_0\}$ and show that $Q(\overline{B}_{L_0}) \subset \overline{B}_{L_0}$. For $(u,v) \in \overline{B}_{L_0}$ and $t \in [0,1]$, we obtain

$$|Q_1(u,v)(t)| \leq L_1 L_0^{\theta_1}\left(1 + \frac{1}{|\widetilde{\Delta}|} + \frac{\lambda\mu\xi^{\gamma+m-1}\eta^\delta(m-1)!}{|\widetilde{\Delta}|\Gamma(\delta+1)\Gamma(\gamma+m)}\right)$$

$$+ L_2 L_0^{\theta_2}\frac{\lambda\xi^\gamma}{|\widetilde{\Delta}|}\left(\frac{1}{\Gamma(\gamma+1)} + \frac{\xi^{m-1}(m-1)!}{\Gamma(\gamma+m)}\right)$$

$$+ (c_0 + c_1 h_1(L_0) + c_2 h_2(L_0))\left(\frac{1}{\Gamma(\alpha+1)}\right.$$

$$+ \frac{1}{|\widetilde{\Delta}|\Gamma(\alpha+1)} + \frac{\lambda\mu\xi^{\gamma+m-1}\eta^{\alpha+\delta}(m-1)!}{|\widetilde{\Delta}|\Gamma(\gamma+m)\Gamma(\alpha+\delta+1)}\right)$$

$$+ (d_0 + d_1 k_1(L_0) + d_2 k_2(L_0))\left(\frac{\lambda\xi^{\beta+\gamma}}{|\widetilde{\Delta}|\Gamma(\beta+\gamma+1)}\right.$$

$$+ \frac{\lambda\xi^{\gamma+m-1}(m-1)!}{|\widetilde{\Delta}|\Gamma(\gamma+m)\Gamma(\beta+1)}\right)$$

$$= M_1 L_0^{\theta_1} + M_2 L_0^{\theta_2} + M_3(c_0 + c_1 h_1(L_0) + c_2 h_2(L_0))$$

$$+ M_4(d_0 + d_1 k_1(L_0) + d_2 k_2(L_0)),$$

and

$$|^cD_{0+}^q \mathcal{Q}_1(u,v)(t)| \leq \frac{L_1 L_0^{\theta_1}}{\Gamma(2-q)} \left(\frac{n-1}{|\widetilde{\Delta}|} + \frac{(n-1)\lambda\mu\xi^{\gamma+m-1}\eta^\delta(m-1)!}{|\widetilde{\Delta}|\Gamma(\delta+1)\Gamma(\gamma+m)} \right)$$

$$+ \frac{L_2 L_0^{\theta_2}(n-1)\lambda\xi^\gamma}{|\widetilde{\Delta}|\Gamma(2-q)} \left(\frac{1}{\Gamma(\gamma+1)} + \frac{\xi^{m-1}(m-1)!}{\Gamma(\gamma+m)} \right)$$

$$+ \frac{1}{\Gamma(2-q)}(c_0 + c_1 h_1(L_0) + c_2 h_2(L_0)) \left(\frac{1}{\Gamma(\alpha)} \right.$$

$$+ \frac{n-1}{|\widetilde{\Delta}|\Gamma(\alpha+1)} + \frac{(n-1)\lambda\mu\xi^{\gamma+m-1}\eta^{\alpha+\gamma}(m-1)!}{|\widetilde{\Delta}|\Gamma(\gamma+m)\Gamma(\alpha+\delta+1)} \right)$$

$$+ \frac{1}{\Gamma(2-q)}(d_0 + d_1 k_1(L_0) + d_2 k_2(L_0))$$

$$\times \left(\frac{(n-1)\lambda\xi^{\beta+\gamma}}{|\widetilde{\Delta}|\Gamma(\beta+\gamma+1)} + \frac{(n-1)\lambda\xi^{\gamma+m-1}(m-1)!}{|\widetilde{\Delta}|\Gamma(\gamma+m)\Gamma(\beta+1)} \right)$$

$$= \widetilde{M}_1 L_0^{\theta_1} + \widetilde{M}_2 L_0^{\theta_2} + \widetilde{M}_3(c_0 + c_1 h_1(L_0) + c_2 h_2(L_0))$$

$$+ \widetilde{M}_4(d_0 + d_1 k_1(L_0) + d_2 k_2(L_0)).$$

In view of the above estimates, we find that

$$\begin{aligned}
\|\mathcal{Q}_1(u,v)\|_X &= \|\mathcal{Q}_1(u,v)\| + \|^cD_{0+}^q \mathcal{Q}_1(u,v)\| \\
&\leq (M_1 + \widetilde{M}_1)L_0^{\theta_1} + (M_2 + \widetilde{M}_2)L_0^{\theta_2} \\
&\quad + (M_3 + \widetilde{M}_3)(c_0 + c_1 h_1(L_0) + c_2 h_2(L_0)) \\
&\quad + (M_4 + \widetilde{M}_4)(d_0 + d_1 k_1(L_0) + d_2 k_2(L_0)).
\end{aligned} \tag{7.54}$$

In a similar manner, we obtain

$$\begin{aligned}
\|\mathcal{Q}_2(u,v)\|_Y &\leq (N_1 + \widetilde{N}_1)L_0^{\theta_1} + (N_2 + \widetilde{N}_2)L_0^{\theta_2} \\
&\quad + (N_3 + \widetilde{N}_3)(c_0 + c_1 h_1(L_0) + c_2 h_2(L_0)) \\
&\quad + (N_4 + \widetilde{N}_4)(d_0 + d_1 k_1(L_0) + d_2 k_2(L_0)).
\end{aligned} \tag{7.55}$$

Using (7.54) and (7.55), we get

$$\begin{aligned}
\|\mathcal{Q}(u,v)\|_{X\times Y} &\leq (M_1 + \widetilde{M}_1 + N_1 + \widetilde{N}_1)L_0^{\theta_1} \\
&\quad + (M_2 + \widetilde{M}_2 + N_2 + \widetilde{N}_2)L_0^{\theta_2} \\
&\quad + (M_3 + \widetilde{M}_3 + N_3 + \widetilde{N}_3)(c_0 + c_1 h_1(L_0) + c_2 h_2(L_0))
\end{aligned}$$

$$+ (M_4 + \widetilde{M}_4 + N_4 + \widetilde{N}_4)(d_0 + d_1 k_1(L_0) + d_2 k_2(L_0))$$
$$= (M_1 + \widetilde{M}_1 + N_1 + \widetilde{N}_1)L_0^{\theta_1} + (M_2 + \widetilde{M}_2 + N_2 + \widetilde{N}_2)L_0^{\theta_2}$$
$$+ A_1(M_3 + \widetilde{M}_3 + N_3 + \widetilde{N}_3)$$
$$+ A_2(M_4 + \widetilde{M}_4 + N_4 + \widetilde{N}_4) < L_0,$$

which shows that $\mathcal{Q}(\overline{B}_{L_0}) \subset \overline{B}_{L_0}$. As argued in the proof of Theorem 7.3.1, it can be shown that the operator $\mathcal{Q}$ is completely continuous.

Next we suppose that there exists $(u, v) \in \partial B_{L_0}$ such that $(u, v) = \nu\mathcal{Q}(u, v)$ for some $\nu \in (0, 1)$. Then

$$\|(u, v)\|_{X \times Y} \leq \|\mathcal{Q}(u, v)\|_{X \times Y} \leq (M_1 + \widetilde{M}_1 + N_1 + \widetilde{N}_1)L_0^{\theta_1}$$
$$+ (M_2 + \widetilde{M}_2 + N_2 + \widetilde{N}_2)L_0^{\theta_2} + A_1(M_3 + \widetilde{M}_3 + N_3 + \widetilde{N}_3)$$
$$+ A_2(M_4 + \widetilde{M}_4 + N_4 + \widetilde{N}_4) < L_0,$$

which contradicts that $(u, v) \in \partial B_{L_0}$. Thus, by the nonlinear alternative of Leray-Schauder type (Theorem 1.2.6), we deduce that the operator $\mathcal{Q}$ has a fixed point $(u, v) \in \overline{B}_{L_0}$, and so problem (7.44), (7.45) has at least one solution on $[0, 1]$. $\qquad\square$

7.3.3   *Examples*

Example 1. Letting $\alpha = 10/3$ $(n = 3)$, $\beta = 9/2$ $(m = 4)$, $p = 3/4$, $q = 1/2$, $\gamma = 5/3$, $\delta = 11/5$, $\xi = 1/2$, $\eta = 1/3$, $\lambda = 1$ and $\mu = 2$, we consider the following system of Caputo fractional differential equations

$$\begin{cases} {}^cD_{0+}^{10/3}u(t) = \dfrac{1}{\sqrt{4 + t^2}} \sin t + \dfrac{1}{4}(v(t))^{1/3} - \dfrac{1}{3(1 + t)} \left({}^cD_{0+}^{3/4}v(t)\right)^{2/5}, \\ \qquad t \in (0, 1), \\ {}^cD_{0+}^{9/2}v(t) = \dfrac{e^{-t}}{1 + t^2} - \dfrac{1}{2}(u(t))^{2/3} + \dfrac{1}{5(3 + t)} \arctan\left({}^cD_{0+}^{1/2}u(t)\right)^{1/7}, \\ \qquad t \in (0, 1), \end{cases}$$
$$\tag{7.56}$$

supplemented with the boundary conditions

$$\begin{cases} u(0) = 2\left(\displaystyle\int_0^1 v(t)\, ds\right)^{1/5}, \quad u'(0) = 0, \quad u(1) = I_{0+}^{5/3}v(1/2), \\ v(0) = 4\left(\displaystyle\int_0^1 u(t)\, dt\right)^{1/3}, \quad v'(0) = v''(0) = 0, \quad v(1) = 2I_{0+}^{11/5}u(1/3). \end{cases}$$
$$\tag{7.57}$$

With the given data, it is found that $\widetilde{\Delta} \approx 0.9999958$ ($\widetilde{\Delta} \neq 0$) and

$$|f(t, u, v)| = \left| \frac{1}{\sqrt{4 + t^2}} \sin t + \frac{1}{4} u^{1/3} - \frac{1}{3(1 + t)} v^{2/5} \right|$$

$$\leq \frac{1}{2} + \frac{1}{4} |u|^{1/3} + \frac{1}{3} |v|^{2/5},$$

$$|g(t, u, v)| = \left| \frac{e^{-t}}{1 + t^2} - \frac{1}{2} u^{2/3} + \frac{1}{5(3 + t)} \arctan(v^{1/7}) \right|$$

$$\leq 1 + \frac{1}{2} |u|^{2/3} + \frac{1}{15} |v|^{1/7},$$

for all $t \in [0, 1]$, $u, v \in \mathbb{R}$ and $\varphi(0) = \psi(0) = 0$, $|\varphi(v)| \leq 2\|v\|^{1/5}$, $|\psi(u)| \leq 4\|u\|^{1/3}$, where

$$\varphi(v) = 2 \left(\int_0^1 v(t)\, dt \right)^{1/5}, \quad \psi(u) = 4 \left(\int_0^1 u(t)\, dt \right)^{1/3}, \quad \forall u, v \in C[0, 1].$$

Here $l_1 = 1/3$, $l_2 = 2/5$, $m_1 = 2/3$, $m_2 = 1/7$, $\theta_1 = 1/5$, $\theta_2 = 1/3$, $L_1 = 2$ and $L_2 = 4$. Clearly the assumptions (K1)–(K3) are satisfied. Thus, by Theorem 7.3.1, we deduce that the problem (7.56), (7.57) has at least one solution on $[0, 1]$.

Example 2. Let us choose $\alpha = 5/2$ ($n = 2$), $\beta = 11/3$ ($m = 3$), $p = 1/2$, $q = 1/3$, $\gamma = 9/4$, $\delta = 7/2$, $\lambda = 2$, $\mu = 3$, $\xi = 1/5$, $\eta = 1/2$, and consider the system of Caputo fractional differential equations

$$\begin{cases} {}^cD_{0+}^{5/2} u(t) = \dfrac{(1 + t)^3}{200} - \dfrac{v^4(t)}{500(1 + |v(t)|)} + \dfrac{t^2}{400} \left({}^cD_{0+}^{1/2} v(t) \right)^3, \\[4mm] {}^cD_{0+}^{11/3} v(t) = \dfrac{(1 - t)^2}{300} + \dfrac{(1 - t)u^2(t)}{400(1 + u^2(t))} - \dfrac{t^3}{100} \left({}^cD_{0+}^{1/3} u(t) \right)^4, \end{cases} \tag{7.58}$$

with the boundary conditions

$$\begin{cases} u(0) = \dfrac{1}{300} \left(\max_{t \in [0,1]} |v(t)| \right)^{1/2}, \quad u(1) = 2I_{0+}^{9/4} v(1/5), \\[4mm] v(0) = \dfrac{1}{200} \left(\max_{t \in [0,1]} |u(t)| \right)^{1/4}, \quad v'(0) = 0, \quad v(1) = 3I_{0+}^{7/2} u(1/2), \end{cases} \tag{7.59}$$

where

$$f(t, u, v) = \frac{(1+t)^3}{200} - \frac{u^4}{500(1+|u|)} + \frac{t^2 v^3}{400},$$

$$g(t, u, v) = \frac{(1-t)^2}{300} + \frac{(1-t)u^2}{400(1+u^2)} - \frac{t^3 v^4}{100}, \quad \forall t \in [0, 1], \quad u, v \in \mathbb{R},$$

$$\varphi(x) = \frac{1}{300} \left(\max_{t \in [0,1]} |x(t)| \right)^{1/2}, \quad \psi(x) = \frac{1}{200} \left(\max_{t \in [0,1]} |x(t)| \right)^{1/4},$$

$$\forall x \in C[0, 1].$$

Obviously

$$|f(t, u, v)| \leq \frac{1}{25} + \frac{1}{500}|u|^4 + \frac{1}{400}|v|^3,$$

$$|g(t, u, v)| \leq \frac{1}{300} + \frac{1}{400}|u|^2 + \frac{1}{100}|v|^4,$$

$$|\varphi(x)| \leq \frac{1}{300}\|x\|^{1/2}, \quad |\psi(x)| \leq \frac{1}{200}\|x\|^{1/4},$$

for all $t \in [0, 1]$, $u, v \in \mathbb{R}$, $x \in C[0, 1]$ and that $h_1(x) = x^4$, $h_2(x) = x^3$, $k_1(x) = x^2$, $k_2(x) = x^4$. One can notice that $c_0 = \frac{1}{25}$, $c_1 = \frac{1}{500}$, $c_2 = \frac{1}{400}$, $d_0 = \frac{1}{300}$, $d_1 = \frac{1}{400}$, $d_2 = \frac{1}{100}$, $\theta_1 = \frac{1}{2}$, $\theta_2 = \frac{1}{4}$, $L_1 = \frac{1}{300}$, $L_2 = \frac{1}{200}$. Using the given values, we obtain $\widetilde{\Delta} \approx 0.99999969$, $M_1 \approx 0.00666668$, $M_2 \approx 0.00010554$, $M_3 \approx 0.60180232$, $M_4 \approx 8.49974 \times 10^{-6}$, $\widetilde{M_1} \approx 0.00369245$, $\widetilde{M_2} \approx 0.00011691$, $\widetilde{M_3} \approx 1.16661255$, $\widetilde{M_4} \approx 9.41543 \times 10^{-6}$, $N_1 \approx 0.00008443$, $N_2 \approx 0.01000027$, $N_3 \approx 0.00082728$, $N_4 \approx 0.13594897$, $\widetilde{N_1} \approx 0.00019054$, $\widetilde{N_2} \approx 0.01128439$, $\widetilde{N_3} \approx 0.00186696$, and $\widetilde{N_4} \approx 0.43463894$. Taking $L_0 = 3$, we find that $A_1 \approx 0.26949999$ and $A_2 \approx 0.83583333$, and the assumption (K5) holds true as

$$(M_1 + \widetilde{M_1} + N_1 + \widetilde{N_1})L_0^{\theta_1} + (M_2 + \widetilde{M_2} + N_2 + \widetilde{N_2})L_0^{\theta_2}$$

$$+ A_1(M_3 + \widetilde{M_3} + N_3 + \widetilde{N_3}) + A_2(M_4 + \widetilde{M_4} + N_4 + \widetilde{N_4}) \approx 1.001 < 3.$$

Therefore, the conclusion of Theorem 7.3.2 applies and consequently problem (7.58), (7.59) has at least one solution on $[0, 1]$.

Remark 7.3.1. The results presented in this section were published in [8].

Bibliography

[1] R. P. Agarwal, B. Andrade, and C. Cuevas, Weighted pseudo-almost periodic solutions of a class of semilinear fractional differential equations, *Nonlinear Anal. Real World Appl.* **11** (2010) 3532–3554.

[2] R. P. Agarwal and R. Luca, Positive solutions for a semipositone singular Riemann-Liouville fractional differential problem, *Int. J. Nonlinear Sci. Numer. Simul.* **20**(7-8) (2019) 823–832.

[3] R. P. Agarwal, Y. Zhou, and Y. He, Existence of fractional neutral functional differential equations, *Comput. Math. Appl.* **59** (2010) 1095–1100.

[4] A. Aghajani, Y. Jalilian, and J. J. Trujillo, On the existence of solutions of fractional integro-differential equations, *Fract. Calc. Appl. Anal.* **15**(1) (2012) 44–69.

[5] B. Ahmad, A. Alsaedi, S. Aljoudi, and S. K. Ntouyas, A six-point nonlocal boundary value problem of nonlinear coupled sequential fractional integro-differential equations and coupled integral boundary conditions, *J. Appl. Math. Comput.* **56**(1–2) (2018) 367–389.

[6] B. Ahmad, A. Alsaedi, S. K. Ntouyas, and J. Tariboon, *Hadamard-Type Fractional Differential Equations, Inclusions and Inequalities* (Springer, Switzerland, 2017).

[7] B. Ahmad and R. Luca, Existence of solutions for a sequential fractional integro-differential system with coupled integral boundary conditions, *Chaos Solitons Fractals* **104** (2017) 378–388.

[8] B. Ahmad and R. Luca, Existence of solutions for a system of fractional differential equations with coupled nonlocal boundary conditions, *Fract. Calc. Appl. Anal.* **21**(2) (2018) 423–441.

[9] B. Ahmad and R. Luca, Existence of solutions for sequential fractional integro-differential equations and inclusions with nonlocal boundary conditions, *Appl. Math. Comput.* **339** (2018) 516–534.

439

[10] B. Ahmad and J. J. Nieto, Boundary value problems for a class of sequential integrodifferential equations of fractional order, *J. Funct. Space Appl.* **2013**, Article ID: 149659 (2013) 1–8.

[11] B. Ahmad and S. K. Ntouyas, Existence results for a coupled system of Caputo type sequential fractional differential equations with nonlocal integral boundary conditions, *Appl. Math. Comput.* **266** (2015) 615–622.

[12] B. Ahmad and S. K. Ntouyas, Existence results for Caputo type sequential fractional differential inclusions with nonlocal integral boundary conditions, *J. Appl. Math. Comput.* **50**(1-2) (2016) 157–174.

[13] B. Ahmad, S. Ntouyas, and A. Alsaedi, On a coupled system of fractional differential equations with coupled nonlocal and integral boundary conditions, *Chaos Solitons Fractals* **83** (2016) 234–241.

[14] B. Ahmad, S. K. Ntouyas, and A. Alsaedi, Sequential fractional differential equations and inclusions with semi-periodic and nonlocal integro-multipoint boundary conditions, *J. King Saud Univ. Sci.* **31**(2) (2019) 184–193.

[15] S. Aljoudi, B. Ahmad, J. J. Nieto, and A. Alsaedi, A coupled system of Hadamard type sequential fractional differential equations with coupled strip conditions, *Chaos Solitons Fractals* **91** (2016) 39–46.

[16] A. Alsaedi, S. K. Ntouyas, R. P. Agarwal, and B. Ahmad, On Caputo type sequential fractional differential equations with nonlocal integral boundary conditions, *Adv. Difference Equ.* **2015**(33) (2015) 1–12.

[17] H. Amann, Fixed point equations and nonlinear eigenvalue problems in ordered Banach spaces, *SIAM Rev.* **18** (1976) 620–709.

[18] A. A. M. Arafa, S. Z. Rida, and M. Khalil, Fractional modeling dynamics of HIV and $CD4^+$ T-cells during primary infection, *Nonlinear Biomed. Phys.* **6**(1) (2012) 1–7.

[19] Z. Bai, On positive solutions of a nonlocal fractional boundary value problem, *Nonlinear Anal.* **72** (2010) 916–924.

[20] K. Balachandran and J. J. Trujillo, The nonlocal Cauchy problem for nonlinear fractional integrodifferential equations in Banach spaces, *Nonlinear Anal.* **72** (2010) 4587–4593.

[21] D. Baleanu, K. Diethelm, E. Scalas, and J. J. Trujillo, *Fractional Calculus Models and Numerical Methods*, Complexity, Nonlinearity and Chaos (World Scientific, Boston, 2012).

[22] D. Baleanu, O. G. Mustafa, and R. P. Agarwal, An existence result for a superlinear fractional differential equation, *Appl. Math. Lett.* **23** (2010) 1129–1132.

[23] T. A. Burton, A fixed point theorem of Krasnoselskii, *Appl. Math. Lett.* **11**(1) (1998) 85–88.

[24] J. Caballero, I. Cabrera, and K. Sadarangani, Positive solutions of nonlinear fractional differential equations with integral boundary value conditions, *Abstr. Appl. Anal.* **2012** Article ID: 303545 (2012) 1–11.

[25] C. Castaing and M. Valadier, *Convex Analysis and Measurable Multifunctions*, Lecture Notes in Mathematics, Vol. 580 (Springer-Verlag, Berlin, 1977).

[26] K. Cole, Electric conductance of biological systems, In *Proc. Cold Spring Harbor Symp. Quantitative Biology* (Col Springer Harbor Laboratory Press, New York, 1993), pp. 107–116.

[27] H. Covitz and S. B. Nadler Jr., Multivalued contraction mappings in generalized metric spaces, *Israel J. Math.* **8** (1970) 5–11.

[28] S. Das, *Functional Fractional Calculus for System Identification and Controls* (Springer, New York, 2008).

[29] K. Deimling, *Nonlinear Functional Analysis* (Springer, Berlin, 1985).

[30] K. Diethelm, *The Analysis of Fractional Differential Equations: An Application-Oriented Exposition Using Differential Operators of Caputo Type* (Springer, Berlin, 2010).

[31] Y. Ding and H. Ye, A fractional-order differential equation model of HIV infection of CD4$^+$ T-cells, *Math. Comp. Model.* **50** (2009) 386–392.

[32] V. Djordjevic, J. Jaric, B. Fabry, J. Fredberg, and D. Stamenovic, Fractional derivatives embody essential features of cell rheological behavior, *Ann. Biomed. Eng.* **31** (2003) 692–699.

[33] M. El-Shahed and J. J. Nieto, Nontrivial solutions for a nonlinear multipoint boundary value problem of fractional order, *Comput. Math. Appl.* **59** (2010) 3438–3443.

[34] H. A. Fallahgoul, S. M. Focardi, and F. J. Fabozzi, *Fractional Calculus and Fractional Processes with Applications to Financial Economics: Theory and Application* (Elsevier/Academic Press, London, UK, 2017).

[35] Y. Feng, J. R. Graef, L. Kong, and M. Wang, The forward and inverse problems for a fractional boundary value problem, *Appl. Anal.* **97**(14) (2018) 2474–2484.

[36] Z. M. Ge and C. Y. Ou, Chaos synchronization of fractional order modified Duffing systems with parameters excited by a chaotic signal, *Chaos Solitons Fractals* **35** (2008) 705–717.

[37] W. Glockle and T. Nonnenmacher, A fractional calculus approach to self-similar protein dynamics, *Biophys. J.* **68** (1995) 46–53.

[38] S. R. Grace, J. R. Graef, and E. Tunc, On the boundedness of nonoscillatory solutions of certain fractional differential equations with positive and negative terms, *Appl. Math. Lett.* **97** (2019) 114–120.

[39] J. R. Graef, L. Kong, Q. Kong, and M. Wang, Uniqueness of positive solutions of fractional boundary value problems with non-homogeneous integral boundary conditions, *Fract. Calc. Appl. Anal.* **15**(3) (2012) 509–528.

[40] A. Granas and J. Dugundji, *Fixed Point Theory* (Springer-Verkag, New York, 2005).

[41] D. Guo and V. Lakshmikantham, *Nonlinear Problems in Abstract Cones* (Academic Press, New York, 1988).

[42] L. Guo, L. Liu, and Y. Wu, Iterative unique positive solutions for singular p-Laplacian fractional differential equation system with several parameters, *Nonlinear Anal. Model. Control.* **23**(2) (2018) 182–203.

[43] J. Henderson and R. Luca, *Boundary Value Problems for Systems of Differential, Difference and Fractional Equations: Positive solutions* (Elsevier, Amsterdam, 2016).

[44] J. Henderson and R. Luca, Existence of positive solutions for a system of semipositone fractional boundary value problems, *Electr. J. Qualit. Theory Differ. Equ.* **2016**(22) (2016) 1–28.

[45] J. Henderson and R. Luca, On a system of Riemann–Liouville fractional boundary value problems, *Commun. Appl. Nonlinear Anal.* **23**(2) (2016) 1–19.

[46] J. Henderson and R. Luca, Existence of positive solutions for a singular fractional boundary value problem, *Nonlinear Anal. Model. Control* **22**(1) (2017) 99–114.

[47] J. Henderson and R. Luca, Existence of nonnegative solutions for a fractional integro-differential equation, *Results Math.* **72** (2017) 747–763.

[48] J. Henderson and R. Luca, Systems of Riemann–Liouville fractional equations with multi-point boundary conditions, *Appl. Math. Comput.* **309** (2017) 303–323.

[49] J. Henderson and R. Luca, Positive solutions for a system of fractional boundary value problems, in *Proc. Equadiff 2017 Conf.*, eds. Karol Mikula, Daniel Sevcovic, and Jozef Urban, Bratislava July 24–28, 2017 (Slovak University of Technology, SPEKTRUM STU Publishing, 2017), pp. 1–10.

[50] J. Henderson and R. Luca, Positive solutions for a system of coupled fractional boundary value problems, *Lith. Math. J.* **58**(1) (2018) 15–32.

[51] J. Henderson, R. Luca, and A. Tudorache, Existence and nonexistence of positive solutions for coupled Riemann–Liouville fractional boundary value problems, *Discrete Dyn. Nat. Soc.* **2016**, Article ID: 2823971 (2016) 1–12.

[52] J. Henderson, R. Luca, and A. Tudorache, Positive solutions for a system of fractional differential equations with multi-point boundary conditions, *Romai J.* **13**(2) (2017) 85–100.

[53] J. Henderson, R. Luca, and A. Tudorache, Positive solutions for a system of fractional differential equations with parameters and coupled multi-point boundary conditions, *Int. J. Appl. Phys. Math.* **8**(4) (2018) 53–65.

[54] B. Henry and S. Wearne, Existence of Turing instabilities in a two-species fractional reaction-diffusion system, *SIAM J. Appl. Math.* **62** (2002) 870–887.

[55] R. Herrmann, *Fractional Calculus: An Introduction for Physicists* (World Scientific. Singapore, 2011).

[56] N. Heymans and J. C. Bauwens, Fractal rheological models and fractional differential equations for viscoelastic behavior, *Rheol. Acta* **33** (1994) 210–219.

[57] R. Hilfer, *Anomalous Transport: Foundations and Applications*, eds. R. Klages, G. Radons, and I. M. Sokolov, (Wiley-VCH, 2008), pp. 17–74.

[58] M. Javidi and B. Ahmad, Dynamic analysis of time fractional order phytoplankton-toxic phytoplankton-zooplankton system, *Ecol. Model.* **318** (2015) 8–18.

[59] M. Jia and X. Liu, Multiplicity of solutions for integral boundary value problems of fractional differential equations with upper and lower solutions, *Appl. Math. Comput.* **232** (2014) 313–323.

[60] Y. Jia and X. Zhang, Positive solutions for a class of fractional differential equation multi-point boundary value problems with changing sign nonlinearity, *J. Appl. Math. Comput.* **47** (2015) 15–31.

[61] J. Jiang and L. Liu, Existence of solutions for a sequential fractional differential system with coupled boundary conditions, *Bound. Value Probl.* **2016**(159) (2016) 1–15.

[62] J. Jiang, L. Liu, and Y. Wu, Symmetric positive solutions to singular system with multi-point coupled boundary conditions, *Appl. Math. Comp.* **220**(1) (2013) 536–548.

[63] J. Jiang, L. Liu, and Y. Wu, Positive solutions to singular fractional differential system with coupled boundary conditions, *Commun. Nonlinear Sci. Numer. Simul.* **18**(11) (2013) 3061–3074.

[64] D. Jiang and C. Yuan, The positive properties of the Green function for Dirichlet-type boundary value problems of nonlinear fractional differential equations and its application, *Nonlinear Anal.* **72** (2010) 710–719.

[65] A. A. Kilbas, H. M. Srivastava, and J. J. Trujillo, Theory and Applications of Fractional Differential Equations, North-Holland Mathematics Studies, Vol. 204 (Elsevier Science B.V., Amsterdam, 2006).

[66] J. Klafter, S. C. Lim, and R. Metzler (eds.), *Fractional Dynamics in Physics* (World Scientific, Singapore, 2011).

[67] M. A. Krasnosel'skii, Some problems of nonlinear analysis, *Amer. Math. Soc. Transl.* **10**(2) (1958) 345–409.

[68] M. A. Krasnoselskii, *Positive Solution of Operator Equations* (Noordhoff, Groningen, 1964).

[69] M. A. Krasnosel'skii and P. P. Zabreiko, *Geometrical Methods of Nonlinear Analysis* (Springer, New York, 1984).

[70] V. Lakshmikantham, S. Leela, and J. V. Devi, *Theory of Fractional Dynamic Systems* (Cambridge Academic Publishers, Cambridge, UK, 2009).

[71] A. Lasota and Z. Opial, An application of the Kakutani-Ky Fan theorem in the theory of ordinary differential equations, *Bull. Acad. Polon. Sci. Ser. Sci. Math. Astronom. Phys.* **13** (1965) 781–786.

[72] R. Leggett and L. Williams, Multiple positive fixed points of nonlinear operators on ordered Banach spaces, *Indiana Univ. Math. J.* **28** (1979) 673–688.

[73] S. Liang and J. Zhang, Positive solutions for boundary value problems of nonlinear fractional differential equation, *Nonlinear Anal.* **71** (2009) 5545–5550.

[74] L. Liu, H. Li, C. Liu, and Y. Wu, Existence and uniqueness of positive solutions for singular fractional differential systems with coupled integral boundary value problems, *J. Nonlinear Sci. Appl.* **10** (2017) 243–262.

[75] S. Liu, J. Liu, Q. Dai, and H. Li, Uniqueness results for nonlinear fractional differential equations with infinite-point integral boundary conditions, *J. Nonlinear Sci. Appl.* **10** (2017) 1281–1288.

[76] X. Liu, M. Jia, and W. Ge, The method of lower and upper solutions for mixed fractional four-point boundary value problem with p-Laplacian operator, *Appl. Math. Lett.* **65** (2017) 56–62.

[77] R. Luca, Positive solutions for a system of Riemann–Liouville fractional differential equations with multi-point fractional boundary conditions, *Bound. Value Probl.* **2017**(102) (2017) 1–35.

[78] R. Luca, On a class of nonlinear singular Riemann–Liouville fractional differential equations, *Results Math.* **73**(125) (2018) 1–15.

[79] R. Luca, Positive solutions for a system of fractional differential equations with p-Laplacian operator and multi-point boundary conditions, *Nonlinear Anal. Model. Control.* **23**(5) (2018) 771–801.

[80] R. Luca, On a system of fractional boundary value problems with p-Laplacian operator, *Dyn. Syst. Appl.* **28**(3) (2019) 691–713.

[81] R. Luca, Existence of solutions for a system of fractional boundary value problems, *Math. Commun.* **25** (2020) 87–105.

[82] R. Luca, Existence of solutions for a fractional nonlocal boundary value problem, *Carpathian J. Math.* **36**(3) (2020) 453–462.

[83] R. Luca, Existence and multiplicity of positive solutions for a singular Riemann-Liouville fractional differential problem, accepted in *Filomat*.

[84] R. Luca, On a system of Riemann-Liouville fractional differential equations with coupled nonlocal boundary conditions, submitted.

[85] R. Luca and A. Tudorache, Nonnegative solutions for a Riemann–Liouville fractional boundary value problem, *Open J. Appl. Sci.* **9** (2019) 749–760.

[86] R. L. Magin, *Fractional Calculus in Bioengineering* (Begell House, Chicago, IL, USA, 2006).

[87] F. Mainardi, *Fractional Calculus and Waves in Linear Viscoelasticity* (World Scientific, Singapore, 2010).

[88] T. Matsuzaki and M. Nakagawa, A chaos neuron model with fractional differential equation, *J. Phys. Soc. Jpn.* **72** (2003) 2678–2684.

[89] R. Metzler and J. Klafter, The random walks guide to anomalous diffusion: a fractional dynamics approach, *Phys. Rep.* **339** (2000) 1–77.

[90] M. Ostoja-Starzewski, Towards thermoelasticity of fractal media, *J. Therm. Stress.* **30** (2007) 889–896.

[91] S. Picozzi and B. J. West, Fractional Langevin model of memory in financial markets, *Phys. Rev. E* **66** (2002) 46–118.

[92] I. Podlubny, *Fractional Differential Equations* (Academic Press, San Diego, 1999).

[93] R. Pu, X. Zhang, Y. Cui, P. Li, and W. Wang, Positive solutions for singular semipositone fractional differential equation subject to multi-point boundary conditions, *J. Funct. Spaces* **2017**, Article ID: 5892616 (2017) 1–7.

[94] T. Qi, Y. Liu, and Y. Cui, Existence of solutions for a class of coupled fractional differential systems with nonlocal boundary conditions, *J. Funct. Spaces* **2017**, Article ID: 6703860 (2017) 1–9.

[95] J. Sabatier, O. P. Agrawal, and J. A. T. Machado (eds.), *Advances in Fractional Calculus: Theoretical Developments and Applications in Physics and Engineering* (Springer, Dordrecht, 2007).

[96] S. G. Samko, A. A. Kilbas, and O. I. Marichev, *Fractional Integrals and Derivatives. Theory and Applications* (Gordon and Breach, Yverdon, 1993).

[97] R. Schumer, D. Benson, M. M. Meerschaert, and S. W. Wheatcraft, Eulerian derivative of the fractional advection-dispersion equation, *J. Contam. Hydrol.* **48** (2001) 69–88.

[98] C. Shen, H. Zhou, and L. Yang, Positive solution of a system of integral equations with applications to boundary value problems of differential equations, *Adv. Difference Equ.* **2016**(260) (2016) 1–26.

[99] D. R. Smart, *Fixed Point Theorems* (Cambridge University Press, Cambridge, 1980).

[100] I. M. Sokolov, J. Klafter, and A. Blumen, A fractional kinetics, *Phys. Today* **55** (2002) 48–54.

[101] J. A. Tenreiro Machado and V. Kiryakova, The chronicles of fractional calculus, *Fract. Calc. Appl. Anal.* **20**(2) (2017) 307–336.

[102] A. Tudorache and R. Luca, On a singular Riemann-Liouville fractional boundary value problem with parameters, accepted in *Nonlinear Anal. Model. Control.*

[103] Y. Wang, Existence and multiplicity of positive solutions for a class of singular fractional nonlocal boundary value problems, *Bound. Value Probl.* **2019**(92) (2019) 1–18.

[104] Y. Wang and L. Liu, Uniqueness and existence of positive solutions for the fractional integro-differential equation, *Bound. Value Probl.* **2017**(12) (2017) 1–17.

[105] Y. Wang, L. Liu, and Y. Wu, Positive solutions for a nonlocal fractional differential equation, *Nonlinear Anal.* **74** (2011) 3599–3605.

[106] Y. Wang, L. Liu, and Y. Wu, Positive solutions for a class of higher-order singular semipositone fractional differential systems with coupled integral boundary conditions and parameters, *Adv. Difference Equ.* **2014**(268) (2014) 1–24.

[107] J. Webb and K. Lan, Eigenvalue criteria for existence of multiple positive solutions of nonlinear boundary value problems of local and nonlocal type, *Topol. Methods Nonlinear Anal.* **27** (2006) 91–116.

[108] S. Xie and Y. Xie, Positive solutions of higher-order nonlinear fractional differential systems with nonlocal boundary conditions, *J. Appl. Anal. Comput.* **6**(4) (2016) 1211–1227.

[109] J. Xu and Z. Wei, Positive solutions for a class of fractional boundary value problems, *Nonlinear Anal. Model. Control.* **21** (2016) 1–17.

[110] C. Yuan, Multiple positive solutions for $(n-1,1)$-type semipositone conjugate boundary value problems of nonlinear fractional differential equations, *Electron. J. Qual. Theory Differ. Equ.* **36** (2010) 1–12.

[111] C. Yuan, Two positive solutions for $(n-1,1)$-type semipositone integral boundary value problems for coupled systems of nonlinear fractional

differential equations, *Commun. Nonlinear Sci. Numer. Simul.* **17**(2) (2012) 930–942.

[112] C. Yuan, D. Jiang, D. O'Regan, and R. P. Agarwal, Multiple positive solutions to systems of nonlinear semipositone fractional differential equations with coupled boundary conditions, *Electron. J. Qual. Theory Differ. Equ.* **2012**(13) (2012) 1–17.

[113] G. M. Zaslavsky, *Hamiltonian Chaos and Fractional Dynamics* (Oxford University Press, New York, USA, 2008).

[114] K. Zhang, Nontrivial solutions of fourth-order singular boundary value problems with sign-changing nonlinear terms, *Topol. Meth. Nonlinear Anal.* **40** (2012) 53–70.

[115] L. Zhang, B. Ahmad, and G. Wang, Existence and approximation of positive solutions for nonlinear fractional integro-differential boundary value problems on an unbounded domain, *Appl. Comput. Math.* **15** (2016) 149–158.

[116] L. Zhang, Z. Sun, and X. Hao, Positive solutions for a singular fractional nonlocal boundary value problem, *Adv. Difference Equ.* **2018**(381) (2018) 1–8.

[117] X. Zhang, Positive solutions for a class of singular fractional differential equation with infinite-point boundary conditions, *Appl. Math. Lett.* **39** (2015) 22–27.

[118] X. Zhang and Q. Zhong, Triple positive solutions for nonlocal fractional differential equations with singularities both on time and space variables, *Appl. Math. Lett.* **80** (2018) 12–19.

[119] Y. Zhou, *Basic Theory of Fractional Differential Equations* (World Scientific, Singapore, 2014).

[120] Y. Zhou and Y. Xu, Positive solutions of three-point boundary value problems for systems of nonlinear second order ordinary differential equations, *J. Math. Anal. Appl.* **320** (2006) 578–590.

Index

Printed in the United States
by Baker & Taylor Publisher Services

Printed in the United States
by Baker & Taylor Publisher Services